Anatomic Atlas of Aquatic and Wetland Plant Stems

Fritz H. Schweingruber • Andrea Kučerová
Lubomír Adamec • Jiří Doležal

Anatomic Atlas of Aquatic and Wetland Plant Stems

Springer

Fritz H. Schweingruber
Institute of Forest, Snow
and Landscape Research – WSL
Birmensdorf, Switzerland

Lubomír Adamec
Czech Academy of Sciences
Institute of Botany
Třeboň, Czech Republic

Andrea Kučerová
Czech Academy of Sciences
Institute of Botany
Třeboň, Czech Republic

Jiří Doležal
Czech Academy of Sciences
Institute of Botany, Třeboň
Czech Republic; Faculty of Science
University of South Bohemia
České Budějovice, Czech Republic

ISBN 978-3-030-33419-2 ISBN 978-3-030-33420-8 (eBook)
https://doi.org/10.1007/978-3-030-33420-8

This Springer imprint is published by the registered company Springer Nature Switzerland AG
The registered company address is: Gewerbestrasse 11, 6330 Cham, Switzerland

Contents

List of abbreviations

ae	aerenchyma
ca	cambium
ccy	central cylinder
ch	chlorenchyma
co	cortex
cr/cry	crystal
cu	cuticle
di	(ray) dilatation
dss	dark-stained substances
duct	duct
en	endodermis
ep	epidermis
ew	early wood
f	fiber
ft	fiber tracheid
ge	gelatinous fiber
gi	girder
gr	growth ring
he	helical thickenings
in	intercellulars, intercellular spaces
ivp	intervessel pit
la	laticifers

med phl	medullary phloem
mu	mucilage
nu	nucleus
p	perforation
pa	parenchyma
peri	pericycle
ph	phloem
phe	phellem
pit	pit
pith	pith
r	ray
rd	resin duct
sc	sclereid
scl	sclerenchyma
sf	septate fiber
si	sieve tube, sieve element
ta	tannins
tr	tracheid
v	vessel
vab	vascular bundle
vrp	vessel-ray pits
xy	xylem

Introduction

General introduction

The transition of plants from aquatic to terrestrial habitats occurred approximately 400 million years ago in the Silurian geological period. The first tracheophytes, which were very small plants with conducting tissues, broke the barrier between water and air. This represented a transition from rather constant environmental conditions to unpredictable and changeable ones. It is therefore not surprising that the stems of the first terrestrial plants already had almost all of the characteristics associated with recently living vascular plants: a protective peripheral layer, an epidermis, a parenchymatic cortex and a central cylinder, consisting of tracheids which conduct water and sieve elements which transport assimilates. Upright plant growth was possible by enforcing the cell walls with lignin. All of these features have survived an evolutionary crisis and are the basis of terrestrial Pteridophytes, Gymnosperms and Angiosperms. Most of the today's plant species grow in dry habitats and have done for the past 300 million years. However, several Pteridophytes and many Angiosperms ventured back into their ancestral wet or aquatic habitats (Taylor et al. 2009). While most types of air-conducting tissues developed in the Cretaceous period, approximately 100 million years ago, a small spectrum of anatomical adaptations spanning the entire 400 million year evolution process is presented here.

Why an atlas of aquatic and wetland plants?

The anatomy of wood, meaning the xylem of adult tree stems, has been investigated by many authors and presented in atlases, ranging from tropical commercial trees to the temperate woody taxa, since the 19th century. Since archeologists, historians and palaeobotanists recognize wood as a historical information source, the microscopic wood structures of shrubs and dwarf shrubs from all over the world have been described for reconstructing human behavior and environmental conditions e.g. by Greguss (1945), Schweingruber (1990), Neumann et al. (2001), Benková et al. (2004) and Signorini et al. (2014). The analysis of the non-lignified plant stems in herbs has been of recent interest to ecologists (Schweingruber et al. 2011; Schweingruber et al. 2013; Crivellaro and Schweingruber 2015; Doležal et al. 2018; Schweingruber and Berger 2019) because many of these plants also develop annual rings as they age. However, all of these studies concentrate mainly on plants in terrestrial, dry habitats.

The stem/shoot anatomy of aquatic and wetland plants has been largely overlooked, although millions of km^2 of the Earth consists of sea shores, rivers, lakes, bogs and swamps and are inhabited by aquatic (hydrophytes) and wetland plants (helophytes; Sculthorpe 1967; Pfandenhauer and Klötzli 2014). Schenk (1886) made the first synthesis of the anatomy of aquatic plants and later initiated many detailed studies. Anatomically close to aquatic plants are many halophytes (Grigore et al. 2014). However, the anatomical structure and the differentiation between aquatic and wetland plants is, for many taxa, unclear. Detailed black-and-white anatomical drawings of the cross-sections of shoots for a large spectrum of aquatic plants (67 species) from the New York region of North America were presented by Ogden (1974). The anatomy of selected monocotyledon species of aquatic and wetland plants with an emphasis on leaf, petiole and root anatomy was presented in five extensive volumes (Vol. IV, V, VII, VIII and IX) within the series Anatomy of the Monocotyledons issued by the Clarendon Press, Oxford, U.K., in 1969-2002 (Cutler 1969; Metcalfe 1971; Tomlinson 1982; Rudall 1995; Keating 2002). Schweingruber and Berger (2019) characterized the anatomy of the whole spectrum of monocotyledonous plants in Central Europe.

Over the last 150 years, many authors have described the aerenchyma in aquatic and wetland plants and its formation (Evans 2003; Seago et al. 2005; Jongduk et al. 2008); however, an overview of the anatomical variability of the total shoot structure in plants in different environments is still missing. Therefore, the present atlas with 397 species fills this large plant-anatomical gap.

The purpose of the book

- The major goal of the atlas is to present and combine macroscopic characteristics, plant distribution and microscopic structures in transverse and longitudinal sections of shoots and root collars in several hundred aquatic and wetland vascular plants by comparable photographic images.
- The anatomical variation within the shoots of a large number of plant species within a wide taxonomic and ecological spectrum of aquatic and wet environments of the Central European temperate zone will be demonstrated. Many of the species presented have worldwide distribution.
- The present atlas clearly shows that the entire anatomical spectrum across all aquatic and wet environments can only be evaluated and grasped by using the analysis of numerous species belonging to many different taxonomic and ecological units.
- Transverse and longitudinal sections, stained with cellulose- and lignin-specific dyes, demonstrate that modern preparation techniques are an additional useful approach for the interpretation of ecophysiological results or traits.

F. H. Schweingruber et al., *Anatomic Atlas of Aquatic and Wetland Plant Stems*, https://doi.org/10.1007/978-3-030-33420-8_1

Hydrobotanical terminology of aquatic and wetland vascular plants and wetlands: definitions, synonyms, specifications, classifications

Before we start dealing with the anatomy of *aquatic* and *wetland plants* it is important to define and specify what is meant by these terms throughout this book. Looking at various hydrobotanical literature sources, one can be rather lost and confused as both terms are very broad, imprecise and ambiguous (cf. e.g., Hejný 1960; Sculthorpe 1967; Hutchinson 1975; Denny 1987; Cook 1990; Gopal et al. 1990; Hejný et al. 1998; Pokorný and Květ 2004). It seems logical that *wetland plants* grow in *wetlands* but the definitions and concepts of wetlands are also rather ambiguous. On the one hand, wetlands have been defined in a very broad sense by the IUCN at the Ramsar Convention in Iran in 1971 as:

"Areas of marsh, fen, peatland or water, whether natural or artificial, permanent or temporary, with water that is static or flowing, fresh, brackish or salt, including areas of marine water the depth of which at low tide does not exceed 6 m." (see Hejný et al. 1998).

Thus, this broad concept also includes all freshwater streams and water bodies. On the other hand, newer, more practical and narrower definitions which exclude aquatic habitats from the definition of wetlands are becoming accepted. For example, for the IBP Wetland synthesis, the following definition was therefore adopted:

"A wetland is an area dominated by herbaceous macrophytes, which photosynthesise predominantly in the aerial environment and root in a soil which, generally speaking, is entirely saturated with water throughout the greater part of the growing season." (Hejný et al. 1998).

In this book, in line with this definition and the generally prevailing understanding of the term 'wetland' and 'wetland plants' in the world literature (e.g., Hutchinson 1975; Cook 1990; Hejný et al. 1998; Pokorný and Květ 2004), we shall strictly use the term 'wetland' *in the narrower sense, only for wet or temporary waterlogged (semi)terrestrial habitats* as opposed to freshwater 'aquatic habitats' comprising both running and standing waters. Nevertheless, there is not a strict boundary between these two terms because gradual transitions between 'wetland' and 'aquatic' habitats commonly occur.

Wetlands can be defined and specified according to the characteristics of different wet (semi)terrestrial habitats and mainly wetland plants are concentrated in these wetlands. An alternative approach can thus also be taken: wetlands are such wet or temporarily waterlogged (semi)terrestrial habitats, which host ecologically specific plant species and communities which are characteristically adapted (morphologically, anatomically, ecophysiologically) to the conditions of wet soils or temporary waterlogging (*sensu* Hejný et al. 1998). The basis of the biological characteristics of wetlands is the wetland vegetation. This is because the plants are primary producers, they indicate environmental conditions and they largely determine ecosystem structure, including the species composition of all the other biotic communities.

Both wetland and aquatic vegetation usually exhibit distinct spatial patterns and/or zonation, which are structured along the water-table gradient and its more or less regular seasonal oscillation (Hejný et al. 1998). These spatial patterns are termed 'ecophases'. In the hydrobotanical literature, four ecophases leading from the wet land to deeper water are distinguished: terrestrial, limosal, littoral and hydrophase. The limosal ecophase is marked by the maximum and minimum levels of fluctuating water. Time sequences of different ecophases are called 'ecoperiods' or 'hydroperiods'. The regularly repeating time sequences of the same ecoperiods may be called 'ecocycles' (Hejný 1960).

The classification of aquatic and wetland vascular plants into characteristic ecological groups is as ambiguous and vague as the classification of aquatic and wetland habitats is (cf. e.g., Hejný 1960; Sculthorpe 1967; Hutchinson 1975; Denny 1987; Cook 1990; Hejný et al. 1998; Pokorný and Květ 2004). All aquatic and wetland vascular plants dealt with in this book are 'macrophytes' as they are visible to the naked eye (Vollenweider et al. 1969; Westlake 1975; Hejný et al. 1998; Pokorný and Květ 2004). Most of the attempts have classified aquatic and wetland vascular plants according to their life-form, which is dependent on the relation of the plant to the position and fluctuations of the water level, and/or to the atmosphere, and to the substrate (rooting; Sculthorpe 1967; Hutchinson 1975; Denny 1987; Cook 1990; Hejný et al. 1998; Pokorný and Květ 2004). According to the combination of life-form and growth form, which includes structural patterns and survival of ecologically unfavourable situations, Hejný (1960) divided all Central European aquatic and wetland plants into 10 categories given Greek names. Although these categories are relatively precise and well reflect the ecological traits of the plants (also including their developmental aspects during the life cycle), they have not gained widespread acceptance. Otherwise, the terms 'hydrophytes' or 'water plants' commonly used in the hydrobotanical or hydrobiological literature (e.g., Sculthorpe 1967; Hutchinson 1975; Cook 1990; Hejný et al. 1998; Pokorný and Květ 2004) can be considered synonyms to 'aquatic plants' though they are less precise. For the detailed specification of growth forms of aquatic plants, see Hutchinson (1975) and Hejný et al. (1998).

Instead, a relatively simple classification of aquatic plants (accepting the narrower definition of wetlands) based on life form (relation of the plant to water level) and rooting has been introduced by Denny (1987) and has been accepted, with minor modification, e.g., by Cook (1990), Pokorný and Květ (2004) and many others. This classification of aquatic plants into four main categories has shown to be very simple, practical and thus widely accepted by hydrobotanists. It is, though, often imprecise and ambiguous in many borderline cases and it does not reflect marked developmental or seasonal changes of ecologically plastic species, either:

1. Submerged plants: ('submersed' plants in American usage) have all their photosynthetic parts under water and only flowers are commonly emerged. According to the ability to root in the bottom, two asymmetric subgroups are distinguished:
i. *rooted submerged plants*: include the dominant majority of submerged plants;
ii. *non-rooted submerged plants*: ('free-swimming' after Cook 1990) include only either some or all species of a few genera (*Utricularia, Aldrovanda, Ceratophyllum, Lemna*) which float below the water surface or can be loosely anchored to the bottom by their shoots.

2. Rooted floating-leaved plants: ('floating' after Cook 1990; 'nymphaeids' after Hejný et al. 1998) the majority of leaves are floating, the minority are submerged but the plants are well rooted in the bottom.

3. Surface-floating plants: ('free-floating' after Cook 1990; 'pleuston' after Hutchinson 1975; Hejný et al. 1998) are floating on the surface and their foliage is in full contact with air; they mostly have roots but usually are not attached to the bottom.

4. Emergent plants: ('emersed' plants in American usage, 'helophytes' after Hutchinson 1975; Hejný et al. 1998) most or all photosynthetic parts of leaves and/or stems are emerged and in contact with air, they are bottom-rooted and usually possess thick rhizomes. The water depth usually does not exceed 1.5-2 m.

Finally, there is a very abundant and ecologically diverse group of plants which strictly grow – permanently or temporarily – in the terrestrial ecophase in wet soils. Unlike emergent plants (helophytes) these are only rarely waterlogged and are sometimes termed as 'hygrophytes' or 'uliginophytes' (Hejný 1960). Many of them grow in bogs, fens, wetland forests or wet meadows (e.g., *Drosera rotundifolia, Andromeda polifolia, Carex canescens, Vaccinium uliginosum, Empetrum nigrum, Leucojum verum, Carex brizoides, C. remota, Cardamine amara, Lysimachia vulgaris*) and in the epilittoral zone of lakes and fishponds (e.g., *Juncus articulatus*), commonly also in wet sandy or clayish soils (e.g., *Lycopodiella inundata, Centunculus minimus*). In line with the literature (Hejný et al. 1998), we call them **terrestrial wetland plants**. Throughout the whole book, we shall consistently use the above classification system for aquatic plants consisting of four main groups, and for the terrestrial species, the term 'terrestrial wetland plant' is employed.

Submerged and rooted floating-leaved plants often exhibit flexible traits merging into each other that make these distinctions rather formal and arbitrary and, thus, the two categories are referred to as 'euhydrophytes' (Denny 1987; Pokorný and Květ 2004). Typically, *Potamogeton natans* bears both submerged and floating foliage and could be thus classified within both categories. Moreover, "every theory is grey" and there are several aquatic species in the European flora which cannot be classified reliably within any of the four categories. Perhaps the best example for such a species is *Stratiotes aloides*, the adult specimens of which float closely below the water surface (as submerged plants) at 0.5-3 m depth. But the apical parts of their leaves grow permanently above water with different ecophysiology (as emergent plants). These plants could also resemble surface-floating plants but they are well bottom-rooted.

Another great problem with the ecological classification of the majority of aquatic vascular plants is associated with the enormous ecological plasticity and adaptability of their life forms as the water level changes (see e.g., Sculthorpe 1967; Hutchinson 1975; Hejný et al. 1998). For example, adult species of *Nuphar* and *Nymphaea* (water lilies) normally grow in ca. 1-2 m deep water as typical rooted, floating-leaved plants. However, when the water level sinks to only ca. 0.1-0.2 m, a part of their foliage emerges above the surface, reminiscent of typical emergent plants. They can even grow satisfactorily as terrestrial wetland plants over the whole season when the water level drops to the bottom. Another excellent example of this plasticity is *Phragmites australis* (common reed), which grows typically in shallow water (0.1-0.5 m) as an emergent plant. However, it can also grow permanently in the terrestrial ecophase of wetlands as well as fully submerged in 2 m deep water in oligotrophic lakes.

Similarly, most aquatic plants exhibit markedly broad ecological plasticity during their normal development (ontogeny) within the seasonal life cycle and this regular developmental plasticity is commonly associated with distinctly changing life forms, often from the submerged or hydric ecophase to the emergent or terrestrial one (Hejný 1960; Sculthorpe 1967; Hutchinson 1975). This amphibious life style is very common among aquatic plants so that Hutchinson (1975) and Hejný et al. (1998) have adopted a special aquatic plant group termed 'amphiphytes' growing at the borderline between water and land and adapted to both environments. A classic example of such developmental plasticity is the biennial plant *Oenanthe aquatica*: its seeds germinate in spring under deeper water, giving rise to completely submerged juvenile plantlets with filamentous leaves resembling those of strictly submerged species (like *Myriophyllum* spp.). Later, taller plants reach the water surface and can be considered as rooted floating-leaved plants. Afterwards, tall subadult plants can grow at declining water level as typical emergent plants with aerial foliage. They frequently complete their life cycle and set seeds as terrestrial wetland plants. Obviously, this great developmental plasticity associated with ecophysiological and anatomical/morphological adaptations makes the classification of discrete amphibious plants according to the accepted classification system very difficult. Therefore, the above five main categories of wetland and aquatic plants can better characterize a discrete stand of an aquatic species in a given phenological phase rather than that species in itself. Yet, it is possible to underline and mark the dominant life form during any plant's life cycle associated with plant maturity, flowering and fruiting.

Throughout the Atlas, we consistently use the botanical nomenclature according to Danihelka et al. 2012, which incorporated numerous taxonomic and nomenclatural novelties. However, where possible and for better clarity, we also state commonly used synonyms of species or generic names (e.g., *Ranunculus /*syn. *Batrachium/*).

Criteria for selection of plant species included and their countries

In this Atlas, 397 species of aquatic and wetland vascular plants growing in Central Europe are included. Clearly, the first criterion for inclusion in the atlas was the availability of plant material for anatomical preparations. Secondly, we have included all species that have been growing in Central Europe for so long that they have become a component of natural plant communities (e.g., *Elodea canadensis, Acorus calamus*). On the list of species included, we have not listed terrestrial species (regardless of their availability) with weak or ambiguous wetland character. We have listed only such species which are either aquatic or terrestrial wetland ones, i.e., which *regularly* grow in wet soils at least for a part of the growing season and, thus, are anatomically and ecophysiologically adapted to these wetland conditions. While admitting that the difference between wetland and non-wetland plants is gradual and rather arbitrary, we have therefore not listed species such as *Carex hirta,*

Vaccinium vitis-idaea, Calluna vulgaris, Arnica montana or Pyrolaceae. They usually do not grow in wetland soils but rather in heathlands, forests, meadows or mountains. These species can also occasionally grow for short periods in wet soils after rain, but obviously they are not adapted to wetland conditions. Similarly, the mountain meadow species usually meet wet soil conditions during the spring season when snow melts before the start of the proper growing season.

In this atlas, the site area of aquatic and wetland vascular plants to be dealt with has been selected as seven countries attributable to Central Europe (CE). We have confined our interests to and focused on Central Europe for several reasons. Perhaps the most important reason was the availability of the selected aquatic and wetland vascular plants for the Atlas. As our author team is based in the Czech Republic and Switzerland, it would be difficult to provide many dozens of typical aquatic and wetland plant species from other parts of Europe. Therefore, the majority of plant material for this Atlas has been conveniently collected from the Collection of aquatic and wetland plants, which is kept at the Institute of Botany of the Czech Academy of Sciences at Třeboň in South Bohemia, Czech Republic. This is the largest European collection and was founded in 1976, currently holding ca. 400 species of aquatic and wetland vascular plants originating mainly from the Czech Republic or adjacent countries. The term Central Europe is geographically confined to the part of Europe along the 50th parallel between the river Rhine bed in the West and the northern Carpathian range in the East. We have thus decided to arbitrarily select seven countries which cover the whole Central European area. They are: Germany (DE), Switzerland (CH), Austria (AT), Czech Republic (CZ), Slovakia (SK), Hungary (HU) and Poland (PL), which will be referred to using the common abbreviations. Moreover, high-level hydrobotanical research traditionally exists in all of these countries. We provided the morphological and ecological descriptions of each species according to the following sources: Botany.cz (online); Database of the Czech flora and vegetation (online); Floraweb (online); Fitter and Fitter (1984); Hejný and Slavík (1988, 1990–2008); Haeupler and Muer (2000); Kubát et al. (2002). In addition, we compiled the distribution of the species in Central European countries (Dostál 1989; Mirek et al. 2002, 2006; Aeschimann et al. 2004; Fischer et al. 2008; Király 2009; Tokarska-Guzik et al. 2012; Kaźmierczakowa et al. 2014, 2016; Bartha et al. 2015; Floraweb (online); Info Flora (online)), but added also the world distribution according to Dostál (1989); Meusel and Jäger (online); Online Atlas of the British and Irish Flora (online).

The monograph concentrates on shoots, root collars or rhizomes of 397 vascular plants. Included are 11 Pteridophytes of 7 families, 210 dicotyledonous species of 43 dicotyledonous families belonging to 15 plant orders, and 176 monocotyledonous species of 18 families belonging to 4 plant orders.

Aquatic and wetland vegetation – basic types and description

Oligotrophic shallow lakes
Low, well-rooted, perennial amphibious aquatic plants (e.g. *Isoëtes* spp., *Eleocharis acicularis*, *Littorella uniflora*) occur at the bottom of oligotrophic and mesotrophic, highly transparent water bodies or in the littoral zone. Most species form low productive monospecific stands and they are most common in the oceanic parts of Europe.

Here stands of *Littorella uniflora* (Plantaginaceae) along the shores of an oligotrophic water reservoir (Láz, Czech Republic). This species is well adapted to regular water draw-down.

Ponds and fishponds
Usually mesotrophic to eutrophic shallow water bodies where different ecological groups of aquatic plants (submerged, free-floating, natant and emergent species) occur along the gradient of water depth. The spatial and temporal variability of aquatic vegetation may be high, depending on water level, water chemistry, transparency and fish stock.

Here stands of *Nymphoides peltata* (Menyanthaceae) in a small fishpond in the Czech Republic, with reeds in the littoral zone.

Alluvial pools and oxbows
Aquatic plants well-adapted to high water level fluctuation are typical of shallow, often periodic small water bodies in riverine floodplains. The dominant species (*Batrachium* spp., *Callitriche* spp., *Hottonia palustris*, *Nuphar lutea*) can have both aquatic and terrestrial growth forms depending on water depth. These water bodies are commonly lined by stands dominated by *Salix* spp. and reeds (*Phalaris arundinacea*).

Here *Nuphar lutea* (Nymphaeaceae) in an alluvial oxbow, in the Lužnice river floodplain, Czech Republic.

Reedbeds

Monodominant stands of tall grasses (e.g., *Phragmites australis*, *Glyceria maxima*) and other grass-like plants (e.g., *Carex* spp., *Schoenoplectus* spp., *Typha* spp.) usually occur in littoral zones of standing or slowly flowing waters. Flexible stems (culms) are resistant to disturbance by wave action and strong winds. Most of the dominant species have an extensive rhizome system through which they can rapidly spread.

Here *Phragmites australis* (Poaceae) forming monodominant stands in a littoral zone of the Rožmberk fishpond, the largest fishpond in the Czech Republic.

Alder and willow carrs

Alder-dominated wetland forests and willow carrs develop at sites with a permanently high water table. *Alnus glutinosa*, *Salix aurita* or *S. cinerea* dominate, with the herb layer consisting of wetland plants such as herbs (*Cardamine amara*, *Caltha palustris*, *Iris pseudacorus*, *Lysimachia thyrsiflora*, *L. vulgaris*), tall sedges (*Carex acutiformis*, *C. elongata*, *C. paniculata*) and ferns (*Dryopteris carthusiana*, *Thelypteris palustris*).

Here an *Alnus glutinosa* forest in a terminal phase of development with many gaps dominated by wetland plants (e.g, *Iris pseudacorus*), Lužnice river floodplain, Czech Republic.

Stream vegetation

Aquatic plants adapted to fast running waters have roots firmly anchored in the bottom, long, elastic and firm stems, and submerged leaves either divided into long narrow segments (e.g., *Batrachium fluitans, Myriophyllum alterniflorum*) or thin and membraneous (e.g., *Potamogeton crispus, P. perfoliatus*). Some species can also develop natant leaves floating on the water surface (*Callitriche hamulata, Potamogeton nodosus*). All species can spread through vegetative fragments.

Here, *Callitriche hamulata* (Plantaginaceae) stands in a small stream in the National Park Labské pískovce, Czech Republic.

Springs

The well oxygenated, flowing water and the rather stable, low water temperature throughout the year determine the plant species composition both in calcium-poor and calcium-rich springs. Bryophytes are more common in stands with low nutrient content but suppressed by high tree litter. Low-productive, sometimes evergreen herbs (*Montia* spp., *Epilobium nutans*) are typical for unshaded springs in the submontane to montane areas.

Here a small spring area with *Montia fontana* (Montiaceae) in an open meadow spring, Nature Reserve Boží Dar, Czech Republic.

Tall-herb vegetation

Tall-forbs (*Aconitum napellus*, *Cicerbita alpina*, *Delphinium elatum*), grasses (*Calamagrostis villosa*, *Poa chaixii*) and tall-ferns (*Athyrium distentifolium*) occur in naturally treeless habitats from the mountainous to the subalpine belt, mainly along mountain streams, in glacial cirques, on avalanche tracks in leeward slopes or in canopy openings in subalpine forests. Deep snow cover serves as frost protection for plants in winter and as a source of moisture during snow melt, soils are therefore well supplied with water and nutrients.

Here *Cicerbita alpina* (Asteraceae) in a subalpine forest in the Central Alps.

Fens and peaty meadows

Rich sedge-moss vegetation with an admixture of dicot herbs, cottongrasses and horsetails occur in shallow peat deposits, usually in terrain depressions or spring areas. Vascular plants are usually of the low-productive, heliophytic and low competitive type. Peaty soils are permanently saturated with water and are poor in nitrogen and phosphorus. Species composition varies according to the water pH and mineral content. Traditional management was mowing for hay production, which helped to eliminate shrubs and trees. Aquatic carnivorous plants (*Utricularia* spp.), some Cyperaceae and submerged bryophytes may occur in oligotrophic dystrophic waters of shallow fen pools.

Here *Dactylorhiza majalis* (Orchidaceae) and *Eriophorum angustifolium* (Cyperaceae) in a peaty meadow, Czech Republic.

Bogs and wooded peatbogs
Bogs are dominated by bryophytes and vascular plants which tolerate permanently waterlogged soils, low pH and low concentrations of nutrients. A few species from Cyperaceae (*Carex* spp., *Eriophorum vaginatum*, *Trichophorum cespitosum*), dwarf evergreen shrubs (*Andromeda polifolia*, *Vaccinium oxycoccos*, *V. uliginosum*) and carnivorous plants (*Drosera anglica*, *D. rotundifolia*) are present. Here a peaty pool (with submerged *Sphagnum* spp. and *Carex limosa*) surrounded by *Trichophorum cespitosum* lawns, Šumava National Park, Czech Republic.

Alpine snow beds
Usually species-poor communities develop in late-melting depressions or valleys with short snow-free periods (2–3 months), constantly saturated with water. These damp and cool habitats, with small fluctuations of temperature, consist of hygrophilic, frost-sensitive species, mainly of oceanic distribution. Different species occur on silicate (*Salix herbacea*, *Gnaphalium supinum*, *Sibbaldia procumbens*) or calcareous bedrock (*Alchemilla fissa*, *Salix retusa*, *Salix reticulata*).

Here snow bed vegetation in 3000 m a.s.l. in the Central Alps.

Drained water body bottoms
Annual wetland herbs or graminoids are well-adapted to irregularly occurring wet sediments when the water level drops down to the bottom. Their seeds are able to survive for many years in the seed bank. Low-growing plants (e.g. *Carex bohemica*, *Coleanthus subtilis*, *Tillaea aquatica*) occur under nutrient-poor conditions or at an initial stage of succession after water drop-down. Here stands of the tiny grass *Coleanthus subtilis* (Poaceae) on a drained fishpond bottom, Třeboň Basin Biosphere Reserve, Czech Republic.

Ecophysiological characteristics of aquatic and wetland vascular plants

Ecophysiological characteristics of aquatic and wetland vascular plants will be reviewed according to the ecological groups of these plants, which possess specific ecological adaptations, different to those in terrestrial plants. In the different plant groups, typical ecophysiological adaptations of photosynthesis, mineral nutrition, gas exchange, tolerance of wet soils and waterlogging, and overwintering strategies will be briefly reviewed.

The aquatic environment is characterised by a series of conspicuous physico-chemical properties which are crucially different from those in the air (Sand-Jensen 1989; Madsen and Sand-Jensen 1991; Pedersen et al. 2013). Firstly, in still water, there is very slow diffusion of solutes (diffusion coefficients of solutes in water are by about 4 orders of magnitude lower that those in air). As a result, a stagnant boundary layer is created on the surface of submerged plants and all solutes have to pass through this by slow diffusion. This boundary layer can be only 10 μm thick in flowing water, but is around 500 μm thick in standing water. The concentration of biologically important gases (O_2, CO_2, ethylene) dissolved in the water is generally very different from that in the air, which contains a relatively constant concentration of both O_2 (21%) and CO_2 (ca. 0.039% or 390 ppm). Moreover, the solubility of gases dissolved in water depends strongly on temperature and, for CO_2, also on pH. On the other hand, unlike air, many important organic and mineral substances are dissolved in the water (e.g., humic acids, sugars, mineral ions, HCO_3^-) and can thus be taken up by plant shoots directly. Due to its properties, the aquatic environment is rather isolated from the atmosphere and, therefore, the physiological processes of aquatic organisms can crucially and rapidly change the water chemistry and influence species diversity by a feedback mechanism. Other important differences between the two environments are of a physical nature: the specific heat capacity of water (per volume) is more than two orders of magnitude higher than that of air, which causes a much higher thermal ballasting effect in water bodies. Finally, light transparency of waters inhabited by scattered organisms (algae, water blooms) or turbid due to suspended particles or coloured by substances such as humic acids in dystrophic (humic) waters is greatly reduced. This results in submerged plants being commonly limited by light at their sites, especially in deeper waters.

1. Submerged aquatic plants

Submerged vascular plants - though taxonomically very diverse - form an ecological group with very marked and distinct anatomical and functional traits (Sculthorpe 1967; Hutchinson 1975). However, the ancestors of recent submerged vascular plants had been growing as terrestrial plants for millions of years and as late as in the Mesozoic and Tertiary periods, they secondarily re-entered the aquatic environment and became adapted to it. Yet, they have retained some anatomical features e.g. thin cuticles or morphological features inherent in their terrestrial ancestors. An example of this is that the majority of submerged species flower above the water surface as aerial pollination is evolutionally rather conservative. Only a minority of genera (e.g., *Ceratophyllum, Najas, Zannichellia, Groenlandia, Callitriche* spp.) do flower under water and they have very reduced or achlamydeic flowers, and pollen transfer is ensured by hydrozoogamy or hydrogamy. The main adaptive anatomical traits of submerged plants represent a relatively low proportion of root biomass (ca. 10-30% of total plant biomass) and some genera (*Ceratophyllum, Utricularia, Aldrovanda*) are rootless. In all organs, a great reduction of mechanical tissues and vascular bundles (mainly of xylem which can be lacking in some genera) occurred. Leaves of submerged plants usually do not bear stomata. In many species, they are only 2-3 cell layers thick and chloroplasts are present in the epidermis. All organs are connected by gas spaces of different sizes. In many species, functional heterophylly takes place: submerged leaves are usually filamentous, bar-shaped or narrow ribbon-shaped (to increase their specific area), while natant leaves are wide or elliptical and leathery and bear stomata on their upper sides.

2. Rooted floating-leaved plants

Rooted floating-leaved plants form an intermediary ecological group, with some typical anatomical and functional traits of both submerged and emergent plants. Properties of submerged plants include thin cuticles, an absence of stomata on their submerged organs or ventral leaf sides and numerouos chloroplasts in the lower leaf side epidermis (casual stomata on the lower side of their floating leaves are remnants from ancestral terrestrial leaves). Properties of emergent plants include abundant roots and thick rhizomes, leathery floating leaves with a thick cuticle, numerous stomata and few chloroplasts on their dorsal side and flowering strictly above the water surface (Sculthorpe 1967; Hutchinson 1975). The thick, waxy cuticle helps the leaves to shed water drops from their surface, rendering the stomata functional. Thus, a good deal of their gas exchange (O_2, CO_2) occurs through stomata as in terrestrial leaves. A great number of species (e.g., *Nuphar pumila, Potamogeton natans*) form also thin, membraneous or needle-shaped submerged leaves which persist also in adult individuals. On the other hand, the leaves become emergent at a great density of floating leaves or in shallow water. A great majority of these species can transiently grow well in the terrestrial ecophase. *Trapa natans* forms markedly inflated leaf petioles, creating a float for the rosette-shaped shoot. As opposed to submerged plants, rooted floating-leaved plants take up the majority of their mineral nutrients by roots from the sediment (Sculthorpe 1967). As most rooted floating-leaved species root in deeper, more or less organic sediments with permanent hypoxia/anoxia, the main adaptive traits within this plant group are associated with tolerance of these conditions - they ensure continuous O_2 diffusion or mass-flow to roots and rhizomes from floating leaves through the continuum of air spaces (Sculthorpe 1967). Very young roots and apices of rhizomes consist mainly of compact tissue of low porosity and so have a relatively limited internal atmosphere.

3. Surface-floating plants

Floating leaves of this plant group resemble those of rooted floating-leaved species. Leaves are distinctly bifacial and conspicuously dorsiventral: the upper waxy epidermis has a rather thick cuticle and few chloroplasts and bears numerous stomata, while the lower epidermis in contact with the water is thin-walled and often rich in chloroplasts (Sculthorpe 1967; Hutchinson 1975). The palisade tissue is developed only in the upper half of the leaf. In the lower half of the leaf, a polygonal meshwork of parenchymatic cells is developed, forming a lacunar tissue with a high proportion of air spaces (aerenchyma). All plant organs contain a minimum of mechanical lignified tissues. Generally, the anatomy and morphology of floating leaves contribute to buoyancy of the plants. Except for rootless *Wolffia arrhiza*, all Central European species of this group form roots (or analogous rhizophylls in *Salvinia*). The roots grow in the free water column and in some species (e.g., *Hydrocharis morsus-ranae*), they contain chloroplasts in their epidermis and outer cortex and are greenish. Although the roots are not anchored in the bottom they considerably aid the stability of the plants on the water surface and prevent them from turning upside down. Many species can grow on wet sediments in the terrestrial ecophase for long periods (e.g., Lemnaceae).

All surface-floating plants absorb all of their mineral nutrients from the water only and, therefore, most species are limited to nutrient- and base-rich waters (Sculthorpe 1967). In spite of the non-rooting in the bottom sediments, they exhibit very rapid vegetative propagation, which is a record high among higher plants (e.g., Rejmánková et al. 1990).

4. Emergent plants

These represent a morphologically very diverse group of plants which usually grow in shallow waters with fluctuating water level. The aerial leaves and stems of emergent plants resemble those of terrestrial plants in their form and anatomy (Sculthorpe 1967; Hutchinson 1975). Due to the resistance to mechanical strains, the erect stems and vertical leaves are reinforced by the development of collenchyma, sclerenchyma and lignified vascular bundles. It is common for most species of this ecological group that their young leaves and stems in particular are submerged during their early development and even their adult organs are often flooded for limited time periods. Thus, unlike terrestrial plants, emergent plants are usually tolerant of flooding. New leaves and stems produced under water (mainly of dicots) often differ in structure from the typical aerial organs, but the differences are of degree rather than kind (Sculthorpe 1967). The thickness and hairiness of the leaf, frequency of stomata and extent of the cuticle are often reduced, the shape of the leaf is altered, and the volume of spongy mesophyll increased (e.g., *Alisma gramineum, Lysimachia vulgaris, Mentha aquatica, Oenanthe aquatica*). The greatest amounts of mineral nutrients are taken up by roots from sediments; the contribution of submerged shoots is marginal.

The majority of emergent species root in deeper, partly organic, waterlogged sediments with permanent hypoxia/anoxia. Therefore, the main adaptive traits within this plant group are associated with tolerance of these conditions – the plants ensure continuous O_2 transport to their roots and rhizomes from emergent stems and leaves through the continuum of air spaces in these organs (so called diffusive ventilation; Sculthorpe 1967; Armstrong et al. 1991). At the level of roots, well developed aerenchyma and a barrier to radial oxygen loss along most of their length are the main anatomical adaptations of most wetland plant species growing in waterlogged soils (Končalová 1990; Armstrong et al. 1991).

5. Adaptations of photosynthesis in submerged plants

Generally, the net photosynthetic rate (P_N) of submerged plants and their stands in their typical freshwater habitats is markedly and directly limited by several environmental factors. The main factors usually considered are: dissolved inorganic carbon (DIC) and light availability, temperature and water streaming (e.g., Boston et al. 1989; Sand-Jensen 1989; Madsen and Sand-Jensen 1991; Pokorný and Květ 2004; Pedersen et al. 2013). Submerged plants are mostly adapted to low irradiances and typically possess thin or filamentous leaves with chloroplasts in their epidermis. Similar to shade-adapted plants, their light compensation point of photosynthesis is usually only between 5-12 $\mu mol.m^{-2}.s^{-1}$ (i.e., ca. 1-3 $W.m^{-2}$ of PAR) and their photosynthesis is saturated at only 200-400 $\mu mol.m^{-2}.s^{-1}$, whereas photosynthesis remains unsaturated in their aerial leaves up to full irradiance (Pokorný and Květ 2004; Pedersen et al. 2013).

Concentrations of DIC ([free CO_2] + [HCO_3^-] + [CO_3^{2-}]) in waters inhabited by submerged plants are very variable within two orders of magnitude, from about 0.05 mM in oligotrophic, softwaters to more than 5 mM in hardwaters, but HCO_3^- is the dominant DIC form in most standing and streaming waters and its concentration usually exceeds that of free CO_2 by 5-100 times. It is generally accepted that only CO_2 is the universal DIC form which is used for photosynthesis by *all* submerged (and aquatic) plants including algae (Maberly and Spence 1983; Madsen and Sand-Jensen 1991; Pedersen et al. 2013). However, at a given total alkalinity (TA: in $meq.l^{-1}$; TA = [HCO_3^-] + 2 [CO_3^{2-}] + [OH^-] - [H^+]), CO_2 concentration is very variable and depends strongly on the pH of the water; both [CO_2] and pH depend crucially on biological processes in the water and bottom sediments. Common values of [CO_2] in standing waters with submerged plants at pH 7.5-8.5 reach only about 8-100 µM. In medium-hard waters at pH about 9, [CO_2] is only 2-4 µM, which is comparable with the CO_2 compensation point (concentration) of photosynthesis in many submerged plant species, or even lower (Maberly and Spence 1983).

Generally, under favourable laboratory or field conditions, due to all physico-chemical limitations, a maximum P_N ranges only from 250 to 700 $\mu mol\ O_2.g^{-1}_{DW}.h^{-1} \approx$ 25-70 $\mu mol\ O_2.g^{-1}_{FW}.h^{-1}$ (quartiles; mean ca. 43 $\mu mol.g^{-1}_{DW}.h^{-1}$) in a number of submerged plants (Boston et al. 1989; Pedersen et al. 2013), whereas these values are ca. 5-10 times greater in leaves of terrestrial plants (Sand-Jensen and Krause-Jensen 1997). However, when the terrestrial leaves are submerged in a solution their mean P_N is about 5 times lower than that of submerged plants (Pedersen et al. 2013); an important positive effect also exhibited is a gas film on the surface of submerged terrestrial leaves, improving greatly their gas exchange (Colmer and Pedersen 2008). These comparisons document clearly the extent of adaptations of submerged plants to aquatic photosynthesis.

During their evolution, submerged aquatic plants have evolved various adaptation mechanisms and ecological strategies at the physiological, biochemical, anatomical, morphological or ecological levels, which substantially increase their photosynthetic inorganic carbon uptake from the water. Owing to much higher [HCO_3^-] than [CO_2] in most waters, the main physiological adaptation of photosynthesis in submerged vascular plants worldwide is HCO_3^- use as a DIC source (Maberly and Spence 1983; Boston et al. 1989; Madsen and Sand-Jensen 1991; Maberly and Madsen 2002; Pedersen et al. 2013). In the European flora, nearly two-thirds of vascular submerged aquatic species (mainly of the genera *Potamogeton, Ranunculus/Batrachium, Ceratophyllum, Najas, Elodea*) use HCO_3^- with a greater or lesser affinity as DIC source for photosynthesis (Maberly and Madsen 2002). In summary, the dominant majority of submerged plants regardless of their HCO_3^- use have their CO_2 compensation point (at 20 to 25 °C and sufficient irradiance) within 3-8 μM, while the HCO_3^- use affinity is one to two orders of magnitude lower: the HCO_3^- compensation point usually ranges from 50 to 800 μM (Maberly and Spence 1983). Thus, the photosynthetic uptake of DIC in strict CO_2 users in waters at a medium total alkalinity is commonly ceased at relatively low pH values of around 9, while in HCO_3^- users, at high pH values between ca. 10.0-10.8. Similarly, much greater photosynthetic affinity of submerged plants to CO_2 than HCO_3^- is also clearly characterised by comparing the values of half-saturation constants for CO_2 (ca. 0.1-0.3 mM) and HCO_3^- (ca. 1.3-20 mM; Madsen and Sand-Jensen 1991).

The most common type of HCO_3^- use is "polar leaf" photosynthesis, which occurs in at least half of European HCO_3^- users (Prins and Elzenga 1989; Madsen and Sand-Jensen 1991; Maberly and Madsen 2002; Pedersen et al. 2013). It occurs in submerged species growing mainly in medium to hard standing waters and possessing thin leaves with only 2-3 cell layers, but wider than ca. 1 mm (e.g., *Potamogeton lucens, Elodea canadensis*). Polar leaves are physiologically very different on both sides. In light, they extensively and actively secrete H^+ ions by proton pumps on the lower (abaxial) leaf side, thus acidifying the unstirred boundary layer close to the leaf surface to a pH of 4-5. As a result, HCO_3^- ions diffusing from the bulk medium are permanently titrated to free CO_2 (with the contribution of carboanhydrase present in cell walls) and a high [CO_2], which exceeds many times that in the bulk medium, permanently occurs on the lower leaf surface. Along this concentration gradient, CO_2 molecules diffuse into the leaf up to chloroplasts where they are permanently fixed in the Calvin cycle, but a smaller proportion of CO_2 is released to the bulk medium and lost. However, the same amount of H^+ ions secreted on the lower leaf side must be simultaneously counterbalanced by OH^- release from the leaf cells. Otherwise the protoplasts would be strongly and quickly alkalised by OH^- ions, which are permanently formed in cells of the lower leaf side, due to H^+ secretion. The OH^- ions diffuse to the upper leaf side cells through numerous highly-conductive plasmodesmata and are permanently released on the upper leaf side into the adjacent water. The pH is 10-11 here. The polar leaves thus maintain a pH gradient of ca. 6 pH units on opposite leaf sides. On the upper leaf surface, as a result of very high pH, $CaCO_3$ is precipitated and the upper leaf surface can be covered by a whitish lime encrustation. It is assumed that OH^- ions are released on the upper leaf side passively along the gradient of the electrochemical potential between leaf cells and ambient water. The conspicuous pH polarity across the leaf is also accompanied by an electrical potential difference of up to 10 mV across the leaf (the lower side is positive with respect to the upper side) and by an apoplastic transport of cations from the lower to the upper side (Prins and Elzenga 1989). Summarily, the use of one HCO_3^- ion from ambient water is followed by a release of one OH^- ion – it represents the total uptake of one CO_2 molecule and cannot be stoichiometricaly distinguished from free CO_2 uptake. Moreover, although the water pH can rise over 10 due to HCO_3^- use, the total alkalinity of the water remains constant.

Polar leaf photosynthesis is the most studied adaptation of HCO_3^- use in submerged plants but many other submerged plants with another leaf structure can also use HCO_3^- very efficiently (Prins and Elzenga 1989; Maberly and Madsen 2002). These plants have either thin filamentous, circular in section (*Myriophyllum, Ranunculus/Batrachium, Ceratophyllum*) or thick triangular or ribbon-shaped leaves (*Potamogeton pectinatus, Najas, Stratiotes, Sagittaria, Luronium, Vallisneria*; see Sculthorpe 1967). The geometry of the leaves or their structure renders the operation of the polar leaf photosynthesis impossible. As there are no pH gradients on the surface of these plants, an active uptake of HCO_3^- ions accompanied by OH^- efflux occurs on the whole leaf surface (Prins and Elzenga 1989). However, HCO_3^- use need not occur in all leaves of the plant. For example, in *Stratiotes aloides*, aerial emergent parts of the leaves are adapted to strict CO_2 use, while submerged basal parts of the same leaves exhibit HCO_3^- use (Prins and de Guia 1986). Similarly, in submerged species which also form floating leaves (*Potamogeton alpinus, Luronium natans*), HCO_3^- use occurs only in their submerged leaves, while their floating leaves with stomata are strict CO_2 users (Hyldgaard and Brix 2011).

Polar leaf photosynthesis and direct HCO_3^- uptake are the most distributed adaptive processes representing carbon concentrating mechanisms in vascular submerged plants. The others are C_4-like photosynthesis and CAM carbon fixation (Madsen and Sand-Jensen 1991; Klavsen et al. 2011). In *Hydrilla verticillata* and *Egeria densa*, physiological and biochemical traits of typical terrestrial C_4 photosynthesis were found: low rate of photorespiration, low O_2 sensitivity of photosynthesis, low CO_2 compensation point and an accumulation of C_4 organic acids. However, due to thin leaves consisting of two cell layers, these species do not have the C_4 anatomy (bundle sheath cells) typical of C_4 terrestrials. Instead, they have the C_4 primary carboxylation processes located in the cytosol of leaf cells, while the C_3 secondary carboxylation runs in chloroplasts of the same cells. Thus, the effective operation of this C_4 pathway is based on intracellular separation of primary and secondary carboxylation. In addition to the C_4-like photosynthesis, both species can also efficiently use HCO_3^- by means of polar leaf photosynthesis (Maberly and Madsen 2002).

Crassulacean acid metabolism (CAM) is another biochemical adaptation of a number of submerged or amphibious plants termed isoetids (*Crassula helmsii, Isoëtes* spp., *Littorella uniflora, Lobelia dortmanna, Subularia aquatica*) on the basis of their anatomical and morphological traits (Boston et al. 1989; Madsen and Sand-Jensen 1991; Klavsen et al. 2011). The common features of these plants are rooted rosettes of short and subulate leaves, a great proportion of root biomass (usually ca. 50% of the total plant biomass) and extensive air spaces in the whole plant. At the ecological level, isoetids usually grow in oligotrophic, very soft waters with low [CO_2] on nutrient-barren sandy sediments. In submerged isoetids, CAM shows characteristics comparable with terrestrial succulent CAM plants: the processes of primary C_4 CO_2 fixation and secondary C_3 fixation are not spatially, but temporally separated. The dark C_4 fixation results in an extensive accumulation of malate in the leaf cell vacuoles. In light, as a result of malate decarboxylation, CO_2 is released and is consequently fixed by the C_3 pathway in chloroplasts. As opposed to terrestrial CAM plants, submerged isoetids also assimilate external CO_2 in light in addition to the internal CO_2 pool created by malate decarboxylation (Madsen and Sand-Jensen 1991; Klavsen et al. 2011). Approximately 30-90%

of the total carbon is fixed through the CAM pathway in isoetids.

The CAM of submerged isoetids is also closely associated with their spectacular anatomical adaptation which supplies the leaves with CO_2 from sediments. Their root systems and leaves are interconnected by voluminous internal gas lacunae. They represent >50% in cross section of both organs (Sculthorpe 1967; Winkel and Borum 2009). CO_2 diffuses from the interstitial water in the bottom sediments along the concentration gradient to the gas lacunae through a very thin primary root cortex, which is more permeable to CO_2 than leaf surfaces. Once trapped within the internal lacunal system in the roots, CO_2 originating from the sediment is transported from the roots by fast diffusion in the lacunal system to the leaves and is partly fixed by the CAM pathway. The root uptake from the sediment provides ca. 40-90% of the total fixed carbon for the plants (Boston et al. 1989). However, the CAM pathway operates only in plants adapted to submerged life. In amphibious plants growing in the terrestrial ecophase (e.g., *Littorella uniflora*), the CAM pathway is not manifested, the leaves are provided with stomata and perform C_3 photosynthesis. A similar root uptake of CO_2 – though usually to a lesser extent – is widespread in aquatic plants growing in their submerged form: commonly in seedlings of emergent plants (in the European flora, e.g, *Juncus bulbosus, Eleocharis acicularis, Elatine* spp., *Limosella aquatica,* seedlings of Alismataceae, Sparganiaceae and Typhaceae; Winkel and Borum 2009). In most submerged species, however, the root uptake of CO_2 is <2% of the total.

Many submerged plants partly avoid the limited DIC availability in waters and the high boundary-layer resistance for DIC uptake by making use of a morphological adaptation: they create a fraction of surface-floating or emergent leaves provided with stomata, which can directly fix aerial CO_2 (e.g., the genera *Potamogeton, Ranunculus/Batrachium, Callitriche,* and the family Alismataceae; Sculthorpe 1967; Madsen and Sand-Jensen 1991; Maberly and Madsen 2002; Hyldgaard and Brix 2011). In many submerged species, some ecological strategies also exist, which enable the plants to increase the limited CO_2 availability – these plants grow in microhabitats with a relatively higher $[CO_2]$ (Boston et al. 1989; Madsen and Sand-Jensen 1991; Nielsen and Borum 2008). Several submerged (or amphibious) species (e.g., *Callitriche* spp., *Elatine* spp.) are adapted to growing close to the bottom sediment where the local $[CO_2]$ is higher than in the upper layers. These and other species also commonly grow in very shallow standing waters which are permanently saturated with CO_2 from the air. Similarly, many submerged species grow in streaming waters in which the limitation of photosynthesis is considerably reduced in comparison with standing waters. Finally, it should be added that most of the photosynthetic traits – especially photosynthetic CO_2 and HCO_3^- affinities or C_4-like photosynthesis – are very plastic and variable in submerged plants depending on many ambient conditions, mainly irradiance and DIC availability, during their seasonal development (see Boston et al. 1989; Madsen and Sand-Jensen 1991). At a sufficient irradiance, low DIC (esp. CO_2) availability in the medium usually markedly increases the plants' affinity to both CO_2 and HCO_3^- within a few weeks.

6. Particulars of gas exchange in submerged plants

The internal gas atmosphere in submerged plants plays an important function in photosynthesis, respiration and gas exchange. Unlike floating-leaved and emergent species, this is completely isolated from the aerial atmosphere. The high diffusion resistance of the unstirred boundary layer for gases on the whole plant surface, especially in standing waters, results in a markedly different composition of the internal gas atmosphere (as compared to the ambient water or air) and, also, in a number of ecophysiological functions (e.g., Sorrell and Dromgoole 1987, 1988; Sorrell 1991; Schuette and Klug 1995). In the leaves of most submerged plants (except for isoetids and ribbon-shaped leaves), the proportion of gas spaces is relatively low at only 10-15%, while it ranges from 20-40% in their stems and roots (Sculthorpe 1967). It is accepted that gas spaces in submerged plants, which usually lack mechanical tissues, help to stabilise the vertical position of shoots and contribute to the photosynthetic refixation (recycling) of CO_2 from (photo)respiration and also to O_2 transport to roots. The proportion of re-fixed CO_2 to the total P_N greatly depends on leaf anatomy: it was ca. 15% in *Elodea canadensis* with leaves with only fine intercellulars, 30-40% in *Najas flexilis*, 50% in *Juncus bulbosus*, but 50-70% in submerged isoetids (Wetzel et al. 1984). Gas spaces in the leaves of submerged plants thus represent a conspicuous photosynthetic adaptation which reduces the losses of CO_2 originating from some physiological processes also including root respiration.

Thanks to the boundary layer on the submerged leaves, processes of photosynthesis or dark respiration and low solubility of O_2 in ambient water, the internal gas atmosphere in submerged plants is subjected to marked diurnal changes in its composition. These changes are also accompanied by pressure changes (Schuette and Klug 1995; Adamec, unpubl.). In standing water in the morning, the pressure in the gas lacunae is exactly equal to the atmospheric pressure and O_2 concentration in the gas decreases below 21 %. During the whole morning, the O_2 concentration is increasing much faster in the internal gas than in the ambient water. In late afternoon, both O_2 concentrations are maximal and comparable. In fully submerged *Myriophyllum heterophyllum,* this state was associated with a measurable increase of the pressure by ca. 4 kPa in the internal atmosphere, but no overpressure occurred in plants with an emergent flower spike (Schuette and Klug 1995). Generally, in unstirred water in light, intact submerged plants can form an overpressure of 3 to 25 kPa in the internal atmosphere, whereas a negative pressure of -5 to -10 kPa can be formed in darkness (Sorrell and Dromgoole 1987, 1988; Sorrell 1991). The highest O_2 concentrations found in the internal atmosphere of submerged plants (*Elodea canadensis, Ceratophyllum demersum*) in light reached 40-46% (Adamec, unpubl.). If the shoots are cut off, O_2-enriched air bubbles are released from them and zero overpressure is measured. During the night, O_2 concentration in the internal atmosphere in shoots as well as in roots of submerged plants can drop to only 2-3% and is lower than that in the ambient water. One of the main functions of the internal gas atmosphere in submerged plants is to supply the roots permanently with O_2 (Sorrell and Dromgoole 1987, 1988; Sorrell 1991). Gas spaces in roots do not serve a long-term storage of O_2 as roots are able to consume all O_2 present during ca. 1 h. Therefore, O_2 needed for night respiration of roots must permanently diffuse into them from submerged shoots. But this night O_2 supply to roots is fully dependent on $[O_2]$ in the ambient water around shoots.

7. Particulars of mineral nutrition in submerged plants

Generally, the main adaptation of submerged vascular plants concerning their mineral nutrition is their leafy shoots' ability to take up considerable amounts of mineral nutrients directly from the ambient water, although the availability of mineral nutrients (per unit volume) is usually by 1-3 orders of magnitude greater in bottom sediments than in the water (Sculthorpe 1967; Agami and Waisel 1986; Barko et al. 1991). Submerged plants possess structural adaptations for mineral nutrition. Their roots often constitute only a small proportion of the total biomass (10-30%), or may even be lacking, root hairs are often not developed, and a significant reduction of vascular tissue is typical. Moreover, the contact between leaf cells and the surrounding water is enhanced as the cuticle on the leaves of most submerged plants is very thin or absent (Jeschke 1976). Therefore, the basic question may be raised about the relative importance of roots and leafy shoots of submerged plants for the uptake of mineral nutrients from the sediments and free water (Agami and Waisel 1986; Barko et al. 1991).

Great importance of roots was found experimentally in many submerged species when their growth was 2-7 times more rapid if rooted in minerally-rich as compared to minerally-poor substrates, or if grown in free water or with their roots removed (Agami and Waisel 1986; Barko et al. 1991). Further evidence comes from ecological studies in natural aquatic habitats in which accumulation of a dense submerged plant biomass occurred in water free of phosphorus. Roots of many submerged plants thus possess a great capacity for mineral nutrient uptake from sediments, but a very low capacity when grown in free water. It is now generally accepted (e.g., Barko et al. 1991) that strictly submerged plant species take up the majority of the required N, P, Fe, Mn and other microelements by roots from sediments, whereas the majority of Ca, Mg, K, Na, sulphate and chloride is taken up by leafy shoots from free water. In addition, submerged plant roots alone take up various mineral nutrients from diluted nutrient solutions per unit biomass at approximately the same rate as leafy shoots do. Even though the relative proportion of root biomass is low (10-30%), the importance of roots for the total uptake of N and P is crucial due to a much greater nutrient availability in sediments.

The contribution of roots to the total plant N and P uptake distinctly reflects the ecological type of submerged plants (Agami and Waisel 1986; Schuurkes et al. 1986; Barko et al. 1991). Submerged species which are potentially amphibious, and also isoetids, take up almost all (>85%) of their required N and P from sediments by roots, while the proportion is somewhat smaller in strictly submerged non-isoetid species (e.g., *Egeria densa*) or much lower in acid-tolerant species (*Juncus bulbosus, Agrostis canina*). Moreover, various types of natural waters differ markedly in their ratio of NH_4^+ and NO_3^- concentrations: NO_3^- is the dominant form in well oxygenated waters while the contrary is true for acidic and dystrophic waters. Submerged plants have adapted to these ratios: the species growing in acidic or strongly dystrophic waters (e.g., *Juncus bulbosus, Agrostis canina, Aldrovanda vesiculosa*) greatly (>6 times) prefer NH_4^+ to NO_3^- from their equimolar concentrations, while isoetids and softwater species from oligo- to mesotrophic waters (e.g., *Luronium natans, Baldellia /Echinodorus/ ranunculoides*) prefer by a factor of 2-3 times NO_3^- to NH_4^+ (Schuurkes et al. 1986; Adamec 2000). Rootless submerged species (*Aldrovanda vesiculosa, Utricularia* spp., *Ceratophyllum* spp.) rely only on the uptake of mineral nutrients from the free water and/or on carnivory (Adamec 1997a). Some rootless species can grow in very oligotrophic waters also at zero NO_3^- and at only 0-3 µg.l^{-1} NH_4^+-N and 0-2 µg.l^{-1} PO_4-P concentrations (e.g., *Utricularia australis*; Adamec 2009), which indicates that their NH_4^+ and PO_4 uptake affinities are extremely high.

The uptake and reutilization of K^+ ions in submerged plants exhibit some peculiarities. In rooted submerged species, the dominant proportion of K^+ is taken up directly from the water by leafy shoots. It was even proven in *Hydrilla verticillata* that K^+ ions having been taken up from the water by shoots, are released *via* their roots into a K^+-poor sediment (Barko et al. 1991). K^+ uptake by shoots of HCO_3^- users from the water is dependent on photosynthesis and light and a smaller part of absorbed K^+ is released from shoots during night, while in a strict CO_2 user *Callitriche hamulata*, K^+ uptake occurs both in light and night (Jeschke 1976; Adamec 1997b). Finally, crucial differences exist between the traits of N, P and K reutilization (recycling) from senescent tissues (shoots) in submerged plants. A few rootless submerged species (*Aldrovanda vesiculosa, Utricularia* spp., *Ceratophyllum submersum*) were found to efficiently reutilize N and P from their senescent shoots, but zero or very low reutilization of K was found, even at very low ambient [K^+] in the water (Adamec 2014). The absence of K^+ reutilization in submerged plants is greatly different from the very efficient K^+ reutilization in many terrestrial wetland plants (Adamec 2014).

8. Adaptations of emergent and rooted floating-leaved plants to soil anoxia and waterlogging

Many species of emergent and rooted floating-leaved plants usually grow permanently in aquatic soils which are fertile and highly organic so that the belowground organs of these plants (rhizomes, roots) are subjected to permanent soil anoxia and related phenomena such as low redox potential and phytotoxin accumulation. Such harmful ecological conditions also commonly occur in temporarily waterlogged wetland soils inhabited by terrestrial wetland species (e.g., wetland graminoids and cyperoids; Crawford 1989; Voesenek et al. 2006). Thus, all these aquatic and wetland plants living in waterlogged soils face two major ecophysiological problems: to obtain an adequate oxygen supply for their roots and rhizomes and to protect themselves against soluble phytotoxins (Končalová 1990). These plants possess several adaptations or strategies allowing them to grow successfully in their habitats (e.g., Dacey 1987; Crawford 1989; Končalová 1990; Smits et al. 1990; Armstrong et al. 1991, 1996; Grosse 1996a,b; Voesenek et al. 2006). These adaptations and strategies can be morphological, anatomical, biochemical and physical and they can combine with one another in a single species. Some are constitutive, while others are at least partly inducible.

As a morphological adaptation to avoid soil anoxia, many aquatic or wetland graminoid and cyperoid species (e.g., *Phragmites australis, Typha* spp., *Carex* spp.) growing in shallow water exhibit root dimorphism – they form long soil roots, which are usually thick and poorly branched, and aquatic (superficial) roots with numerous finely branched laterals; aerenchyma is well developed in the soil roots, but not in the aquatic roots, which function mainly in nutrient uptake (Sculthorpe 1967; Končalová 1990; Armstrong et al. 1991).

Enhanced development of internal gas spaces is considered a basic and universal anatomical adaptation in aquatic and terrestrial wetland plant roots and rhizomes. It represents an avoidance strategy allowing the roots to penetrate much deeper aquatic soil layers (Sculthorpe 1967; Crawford 1989; Končalová 1990; Smits et al. 1990; Armstrong et al. 1991). Internal gas spaces provide a conduit for O_2 transport from the shoot system. If there is a gas continuum from shoots to roots *via* rhizomes, even a small proportion of gas spaces (i.e., tissue porosity) can provide at least some O_2 transport in submerged roots. In roots, the transport is mainly diffusive as convective gas flows (i.e., mass flows; such as in emergent and floating-leaved plants) cannot operate to any significant degree in roots. Moreover, concurrently with this longitudinal diffusive O_2 transport in roots, there is a potential for radial O_2 loss from roots to aquatic soil, thus representing an additional factor in root survival: the oxidative removal of soil-borne phytotoxins (Armstrong et al. 1991).

The major anatomical adaptation important for survival in aquatic or temporarily waterlogged soils is the enhanced development of gas spaces in the primary (and/or secondary) cortex of the roots. A majority of emergent and terrestrial wetland plants of diverse taxa have root porosities exceeding 10% and in most cases, extensive aerenchyma develops (Sculthorpe 1967; Armstrong et al. 1991). Aerenchyma development in submerged roots serves two purposes: it can greatly enhance diffusive conductivity of longitudinal transport of gases from and to the shoot and it also leads to a reduction in O_2 consumption per unit root volume. This adaptation enhances further O_2 supply to the apical parts of the root. In some species, aerenchyma is formed only in response to soil flooding, while in others it is an almost constant feature of the root system although it can become more extensive due to flooding (Armstrong et al. 1991). Moreover, in some emergent or terrestrial wetland species, root porosity depends also on the presence of organic matter (detritus) and nitrogen in the soil: elevated N supply reduces the porosity (Končalová 1990). The O_2 transported through the gas spaces is depleted in submerged organs by two processes: respiration of rhizomes and roots and radial O_2 loss across their surface tissues to the anoxic soil. But it thus depletes O_2 in the apical zone of roots where it is most needed by actively growing apical meristems and the subapical zone of roots, involved in nutrient and water uptake (Končalová 1990). Many species tolerant of soil waterlogging efficiently restrict radial O_2 loss along differentiated parts of their roots and rhizomes. They form the exodermis: a subsurface layer of compactly packed cells with thickened and lignified cell walls (Končalová 1990; Armstrong et al. 1991). The lysigenous aerenchyma formation in roots of terrestrial wetland species is triggerred off by the phytohormone ethylene, the biosynthesis of which from its precursor is stimulated after soil waterlogging (Voesenek et al. 2006).

The biochemical (metabolic) adaptation of roots and rhizomes of emergent, floating-leaved and terrestrial wetland plants to permanently anoxic soils, total plant submergence or temporary soil waterlogging, may be considered a very common adaptation in wetland-tolerant plant species, which supplements the main anatomical adaptation – aerenchyma formation (e.g., ap Rees et al. 1987; Crawford 1989; Smits et al. 1990; Braendle and Crawford 1999). Operation of the biochemical adaptation is especially important during floods when the plant system of aerenchyma in stems and leaves is separated by low-oxygen flooding water from the air and cannot supply roots and rhizomes with O_2. The operation of a biochemical adaptation to anoxia was demonstrated when isolated rhizomes of many emergent species (e.g., *Phragmites australis, Typha* spp., *Scirpus* spp.) were able to regenerate the aerial shoots after long periods of anaerobic incubation of 7 to 90 days, whereas other flooding-intolerant species were able to withstand such conditions only for several hours to a few days (Crawford 1989; Braendle and Crawford 1999). Great tolerance of long periods of anoxia was even found in leaves of some emergent species (*Acorus calamus, Iris pseudacorus*).

In general, anoxia in flooded soils is probably the most important stress factor limiting the growth and survival of vascular plants in such habitats. The major biochemical effect of low O_2 concentration in plant tissues in flooded soils is the inhibition of aerobic respiratory chain and, consequently, the loss of sufficient ATP production needed for maintenance of cellular functions (Crawford 1989; Braendle and Crawford 1999). The majority of emergent, floating-leaved and terrestrial wetland plants are able to switch their energy metabolism in roots and rhizomes from aerobic respiration to anaerobic fermentation (glycolysis) as a way of ATP production (ap Rees et al., 1987; Crawford 1989; Smits et al. 1990; Braendle and Crawford 1999). However, two ecophysiological problems are associated with this adaptation in organs living in flooded soils. First, fermentation yields ca. 19 times less ATP per hexose equivalent: much more substrate has to be consumed to produce a given amount of ATP in anoxia than in air ('Pasteur effect') and, thus, the organs can rapidly exhaust their energy resources. Second, reduced end-products of fermentation are accumulated in tissues, can be toxic to these organs and, together with a great release of CO_2, can lead to cell acidosis. In roots of many flood-tolerant species, ethanol, lactate and malate are produced and accumulated as the end-products of anaerobic fermentation (Rees et al. 1987; Smits et al. 1990; Braendle and Crawford 1999). Similarly, the rate of anaerobic fermentation (CO_2 release) in roots of aquatic species tolerant of soil anoxia (nymphaeids) is comparable with that of aerobic respiration (O_2 consumption), while in non-tolerant isoetids, anaerobic fermentation rate reaches only 10-25% of the aerobic respiration (Smits et al. 1990). Thus, the increased capacity of roots and rhizomes to provide sufficient ATP by anaerobic fermentation for prolonged periods is the main biochemical adaptation in anoxia-tolerant species unlike anoxia-intolerant ones. However, plant organs surviving a period of anoxia can also be strongly damaged after their return to air (post-anoxic injury; e.g., Crawford 1989; Braendle and Crawford 1999). The reasons for this post-anoxic injury, which occurs mainly in flood-intolerant plants, is the formation of reactive oxygen species, peroxidative damage to cell membranes, and the oxidation of accumulated ethanol to toxic acetaldehyde by means of catalase and hydrogen peroxide (Crawford 1989; Braendle and Crawford 1999). Flood- and anoxia-tolerant plants are adapted to the post-anoxic injury. They avoid excessive accumulation of ethanol in their anoxic tissues by diffusive losses, and also possess two defence mechanisms (Braendle and Crawford 1999): they produce high concentrations of anti-oxidant substances – glutathion, ascorbate and phenolics – as well as detoxification enzymes (ascorbate peroxidase and superoxide dismutase).

As physical adaptations to soil anoxia and waterlogging, two physical mechanisms participate in pressurised ventilation in emergent and rooted floating-leaved plants: thermal osmosis and humidity-induced convection (Grosse et al. 1991; Armstrong et al. 1996; Grosse 1996a,b). Both mechanisms are based on the separation of a gas (air) by a porous membrane (partition) with a pore diameter ≤0.1 μm when the compartments on the two sides of the membrane differ either in temperature (thermal osmosis, thermal transpiration, or Knudsen diffusion) or humidity (humidity-induced diffusion and convection or hygrometric diffusion and pressurization). Simply, gradients of either temperature or air humidity developed on both sides of the porous membrane create thermodynamic driving forces for diffusion of gases through the membrane (osmosis) until the gas concentrations become equal on its both sides. In closed chambers, the diffusive flows of gases can lead to pressure differences, which are manifested as convective (mass) flows (Armstrong et al. 1996; Grosse 1996a). In young floating leaves of nymphaeids (*Nuphar, Nymphaea, Nelumbo*), tissue layers with minute pores separate the aerenchyma from the ambient atmosphere, while in mature leaves these pores are dilated so that free gas convection through them is possible. Therefore, in light, the air enters the young leaves (influx leaves) along humidity and temperature gradients, flows through the continuous gas-space system in petioles and rhizomes, and passes through the mature leaves (efflux leaves) back to the atmosphere. In *Nelumbo nucifera*, this flow-through can also occur at the level of a single leaf: the air enters lateral parts of the leaf lamina and exits back to the atmosphere at the point of petiole insertion (Dacey 1987). The pressurised ventilation occurs also in emergent plants (e.g., *Phragmites australis, Typha, Eleocharis*) but, presumably, their main mechanism is based on the humidity-induced air convection (Armstrong et al. 1996). In *Phragmites australis*, the ambient air enters the young culms in leaf sheaths, is driven down through the rhizome system and exits through the dead previous year's culms. In the absence of old culms, flows can occur between living culms. In many rooted floating-leaved species *in-situ* in bright light, the inflow rates of the air within 50-7410 $ml.h^{-1}$ were measured per cut leaf or 0.2-8.1 $ml.cm^{-2}.h^{-1}$ per unit leaf area (Grosse et al. 1991; Grosse 1996b). In emergent *Phragmites australis*, the inflow rate within 900-1200 $ml.h^{-1}$ per living shoot and the flow velocity over 14 $mm.s^{-1}$ were measured (Armstrong et al. 1996). A Venturi-induced convection (caused by wind blowing across dead culms) can also take part in the flow-through. The pressurization values in floating leaves of various floating-leaved species measured in the laboratory ranged between ca. 0.1-1.6 kPa (1-16 mbar; Dacey 1987; Grosse et al. 1991). It is considered that the air flow through rhizomes has both a great physiological impact on aerobic respiration of roots and rhizomes of emergent and floating-leaved plants and, as a result of radial oxygen loss from rhizomes and roots, a positive ecological impact on the oxidative regime within the otherwise anoxic rhizosphere (Grosse 1996b). The above mechanisms of pressurised ventilation in aquatic plants also considerably speeds up the release of methane - an important greenhouse gas - from the rhizosphere to the atmosphere (Armstrong et al. 1996).

As an avoidance growth strategy, many floating-leaved and emergent species have developed a physiological mechanism which regulates reaching of the water surface or foliar emergence when the plant is flooded and its leaves stay under water; the aerobic ventilation through the aerenchyma is then inefficient (for reviews see Sculthorpe 1967; Ridge 1987; Jackson 1990; Voesenek et al. 2006). This adaptive growth response of aquatic plants is mediated by gaseous ethylene, which can inhibit root elongation and induce lysigenous aerenchyma formation therein on the one hand, and, on the other hand, stimulate petiole or internode elongation in the flooded plants, thereby enabling their shoots to reach the water surface. Ethylene is moderately soluble and its diffusion in water is very slow. Therefore, in flooded floating-leaved and emergent plants with no contact with air, its internal concentration within the aerenchyma system can increase substantially and quickly (Jackson 1990). Although the ethylene biosynthesis is induced by shortage of O_2 in plant tissues the last step of the biosynthesis, the conversion of 1-aminocyclopropane-1-carboxylic acid to ethylene, requires oxygen and can occur in more oxygenated tissues. The character of the elongation response of petioles or shoot internodes is greatly species-dependent and, moreover, it depends much on the age (ontogeny) of the organs. As soon as the elongating organ attains the water surface the elongation rate is slowed down, but the elongation still continues. The minimum time required for petioles of several floating-leaved species to attain the water surface ranges from 4 to 8 days, while the time needed for the maximum elongation can be as much as 12-28 days (Ridge 1987). Both cell elongation and cell division can participate in the flood-induced organ elongation, but their proportion is highly species-dependent.

9. Ecophysiology of aquatic plants' turions

Turions (winter buds) are dormant vegetative organs formed by perennial aquatic plants as a response to unfavourable ecological conditions - mainly decreased temperature or daylength or nutrient availability (Sculthorpe 1967; Bartley and Spence 1987). Turions as modified shoot apices are spherical (*Utricularia*), rhomboid (*Aldrovanda*) or torpedo-shaped (*Potamogeton* spp., *Caldesia*) tough and sturdy organs, formed at the end of the growing season by extreme condensation of short modified leaves in the shoot apex. In *Potamogeton crispus*, they can be formed exceptionally in early summer as a mode of aestivation. Turions are partly frost resistant and protect fragile plant shoots from freezing and decaying. Dormant, non-hardened turions of 9 aquatic species exhibited extracellular freezing within a narrow temperature range of -7.0 to -10.2 °C, while *Hydrocharis* froze at -3.6 °C (Adamec and Kučerová 2013). The mean freezing temperature of outdoor-hardened turions of 6 species (*Aldrovanda, Utricularia* spp., *Caldesia*) was within a very narrow range of -2.8 to -3.3 °C and was thus higher by 4 to 7 °C than that for the non-hardened turions. Therefore, turions of aquatic species can be hardened by weak frosts and their frost hardiness is based on the shift from frost avoidance in non-hardened turions to frost tolerance.

Two dormancy states and their hormonal patterns were described in detail for turions of Canadian *Utricularia macrorhiza* using bioassays (Winston and Gorham 1979a,b). Turions entered a state of innate dormancy at the end of summer, when their growth was blocked by endogenous factors in the turions. At the end of October, the turions entered a state of imposed dormancy when turion germination depended only on higher temperature. The next stage in turion germination leads to the sprouting of new shoots, which occurs under sufficient temperature and light conditions. These two dormancy states were also described in *Aldrovanda vesiculosa* turions (Adamec 2003).

Turions are also storage organs and in autumn, they accumulate starch (9-70% dry weight), free sugars (in total 7-14% dry weight), reserve proteins and lipids (Adamec 2010; Płachno et al. 2014). In *Hydrocharis morsus-ranae* turions, a great deal of the total organic N was found in the form of arginine (Villanueva et al. 1985). Turions also represent storage organs for mineral nutrients (N, P) though their storage function is presumably less distinct than that for carbohydrates (Adamec 2000, 2010). Turions of all aquatic plant species usually overwinter (hibernate) and break their innate dormancy at the bottom of an aquatic habitat in darkness, under hypoxic or anoxic conditions, being usually lightly covered by organic sediments. However, concerning the overwintering (hibernation) of turions, two distinct ecological strategies may be distinguished (Bartley and Spence 1987; Adamec 2008). Turions of surface-floating or non-rooted aquatic genera (e.g., *Hydrocharis, Utricularia, Aldrovanda, Spirodela*) ripen at the water surface, overwinter at the bottom, but usually germinate and sprout at the water surface in warmer water, while those of aquatic rooted genera (e.g., *Potamogeton, Caldesia, Myriophyllum*) both ripen, germinate, sprout, and root at the bottom in colder water and shade (Sculthorpe 1967; Adamec 2008). Moreover, turions of surface-floating or non-rooted aquatic species differ in the mechanisms of their autumnal sinking and spring floating (rising; Adamec 2003, 2008): Turions of *Aldrovanda vesiculosa, Hydrocharis morsus-ranae,* and *Spirodela polyrhiza* have developed an active mechanism of sinking and rising. It has been suggested that sinking and rising of *Aldrovanda* turions is caused by variable gas volume in the gas spaces of turion leaves (Adamec 2003). On the contrary, *Utricularia* turions are less dense than water and are dragged to the bottom by their decaying mother shoots. By early spring, the turions separate and float up to the water surface (Adamec 2008, 2010).

As in typical storage organs, the dark respiration rate of turions of 6 aquatic species such as *Aldrovanda* and *Utricularia* spp. was rather low compared to dark respiration rates of shoots (leaves) of aquatic plants of the same or other species at a standard temperature (Adamec 2008) - around 30-75% lower on a fresh weight basis and 80-85% lower per unit dry weight. A great proportion of cyanide-resistant respiration (22-90%) was found in these turions. Some subtropical or temperate *Utricularia* species and *Aldrovanda vesiculosa* populations produce non-dormant, quiescent shoot apices ('winter apices'). However, contrary to true dormant turions, respiration rates in non-dormant winter apices of these two genera were comparable with those in growing aquatic plant shoots (Adamec 2008). Net photosynthetic rate of turions of 7 aquatic species (*Aldrovanda, Utricularia* spp., *Caldesia, Hydrocharis*) at the stage of imposed dormancy was negative or only slightly positive under standard light, temperature and CO_2 conditions. But both old and newly formed shoot segments in sprouting turions of rootless carnivorous *Aldrovanda* and *Utricularia* spp. exhibited a very high photosynthetic rate, which was comparable with that in adult plants (Adamec 2011). Thus, very high photosynthetic rate of sprouting turions of rootless submerged plants, together with the storage functions of turions for N, P, S, and Mg, contributes to rapid growth of sprouting turions and rapid production of standing biomass at the beginning of the growing season. *Potamogeton distinctus* turions can start their sprouting in anoxia and this phase is associated with extensive acidification of the apoplast which can regulate the cytoplastic pH (Koizumi et al. 2011).

Preparation methods

Anatomy, age and growth was measured on more than three individual specimens collected in 2013–2015. The material collected in the field has been stored in 40% ethanol to avoid drying artefacts. Each species was carefully labeled. The material has been sectioned without embedding and double stained with Safranin/Astrablue with the principal goal to determine the maximum plant tissue age and the stem anatomy just above ground. Therefore, all the sections are anatomically, taxonomically and ecologically comparable. Information on each species is given.

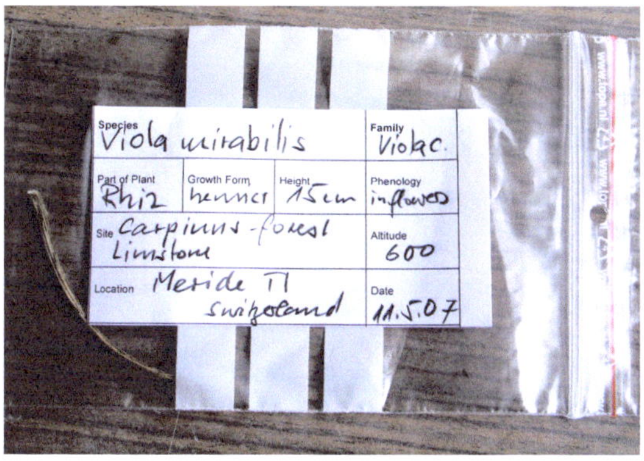

Plastic bag containing 40% ethanol with a stem of a herb sample. The label contains information about the taxon and its morphology, ecology and the sampling date.

A new laboratory microtome designed by Gaertner et al. 2015.

Chemical equipment for staining and dehydration of microsections.

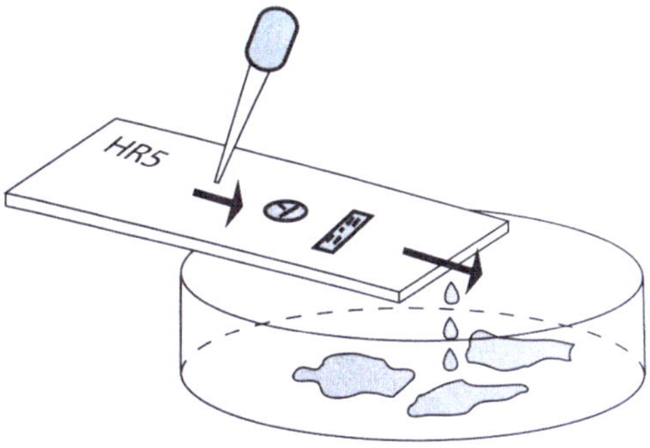

Principle of staining and dehydration on the glass.

Flattening microscopic sections with magnets on an iron plate.

Storing of labeled microscopic slides in boxes.

F. H. Schweingruber et al., *Anatomic Atlas of Aquatic and Wetland Plant Stems*, https://doi.org/10.1007/978-3-030-33420-8_2

Definition of anatomical features

Definition of petiole features of Pteridophytes

Outline of petioles and arrangement of vascular bundles (stele)

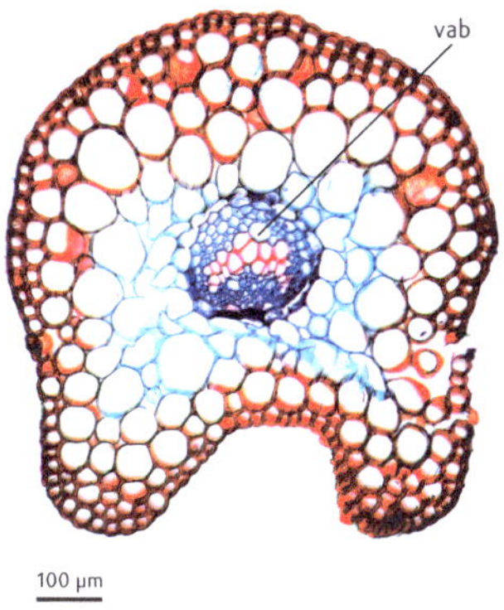

One vascular bundle in the center of the petiole (Protostele). Outline crenate. *Anogramma leptophylla*

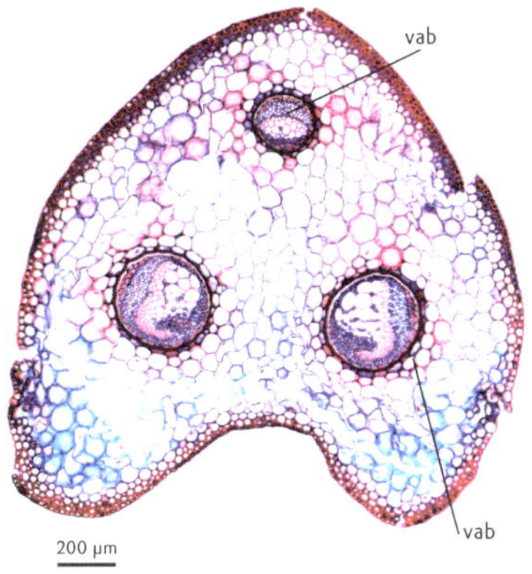

Several vascular bundles in the petiole (Polystele). Outline crenate. *Dryopteris cristata*

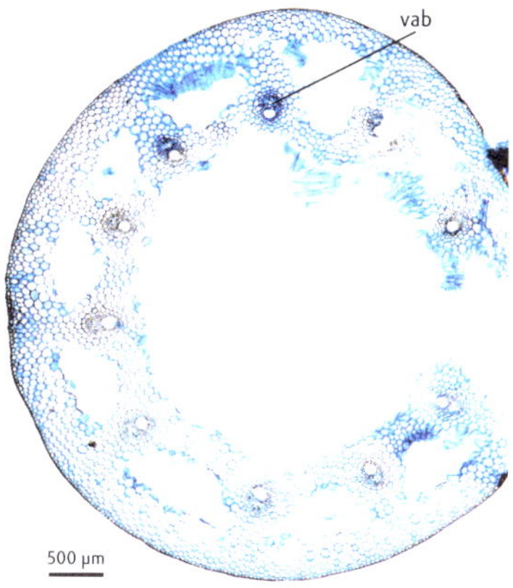

Circular arrangement of vascular bundles (Siphonostele). Outline crenate. *Equisetum fluviatile*

Construction of vascular bundles

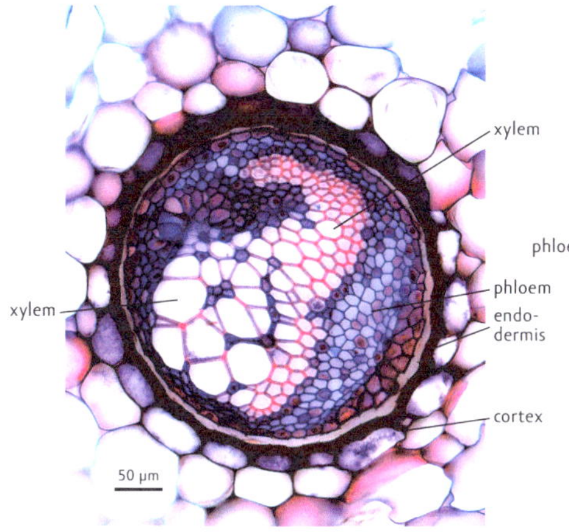

Round concentric vascular bundle with the xylem in the center. *Dryopteris cristata*

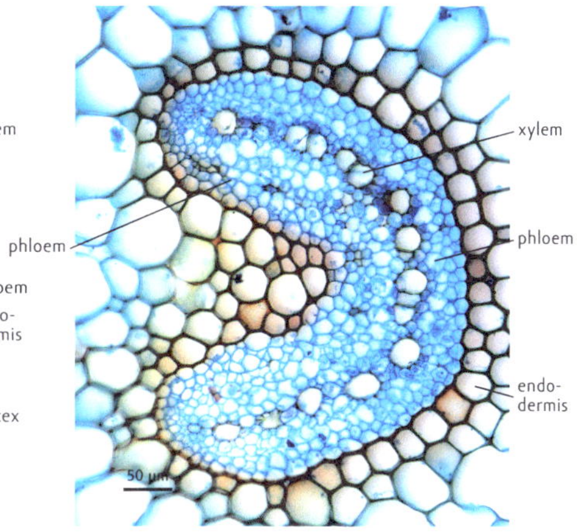

Arc-like concentric vascular bundle with the xylem in the center. *Salvinia natans*

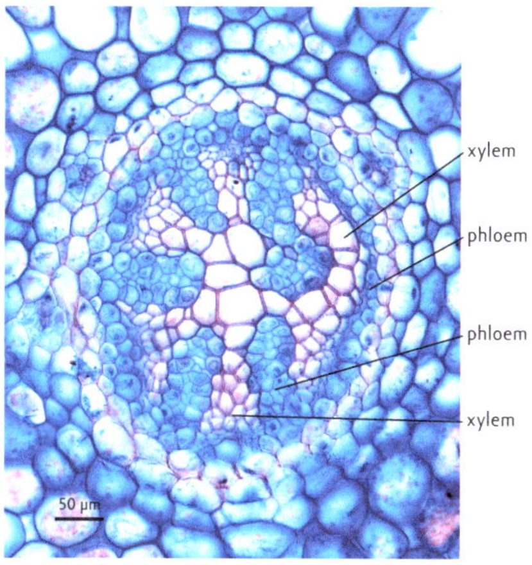

Concentric vascular bundle with star-like arrangement of the xylem and the phloem (Plectostele). *Lycopodium inundatum*

F. H. Schweingruber et al., *Anatomic Atlas of Aquatic and Wetland Plant Stems*, https://doi.org/10.1007/978-3-030-33420-8_3

Boundary between the cortex and vascular bundles – endodermis

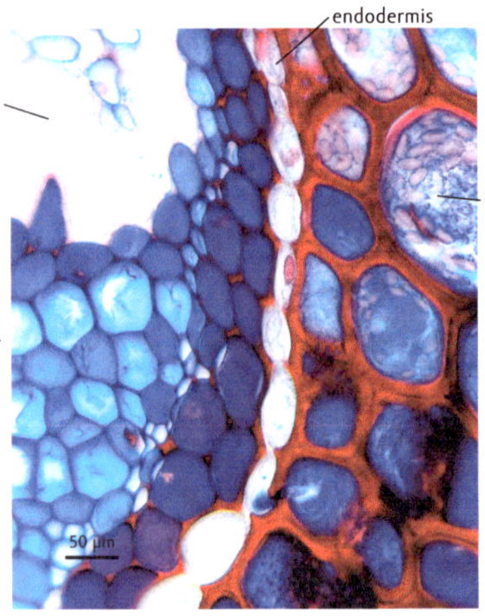

Uniseriate layer with thick-walled centripetal and centrifugal and very thin-walled, unlignified lateral walls. *Thelypteris palustris*

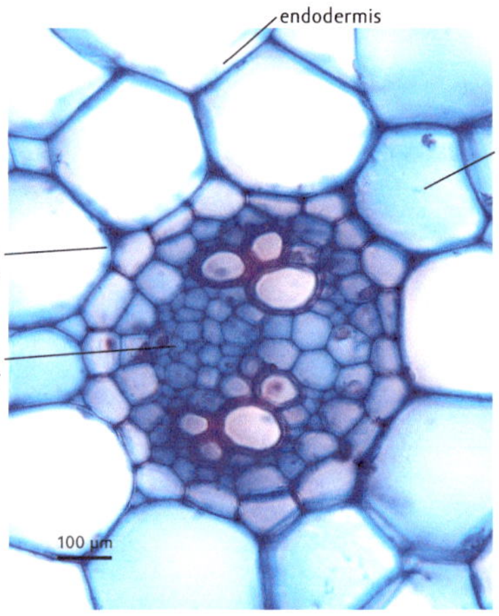

Uniseriate unlignified thin-walled layer. *Pilularia globulifera*

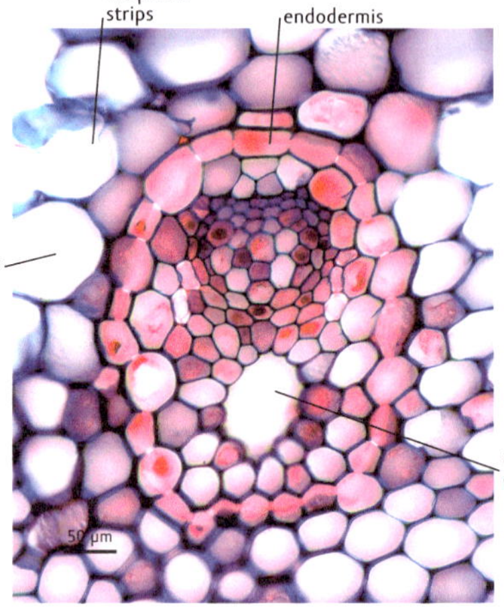

Uniseriate thin-walled layer with Casparian strips. *Equisetum hyemale*

Intercellulars in the cortex, aerenchyma

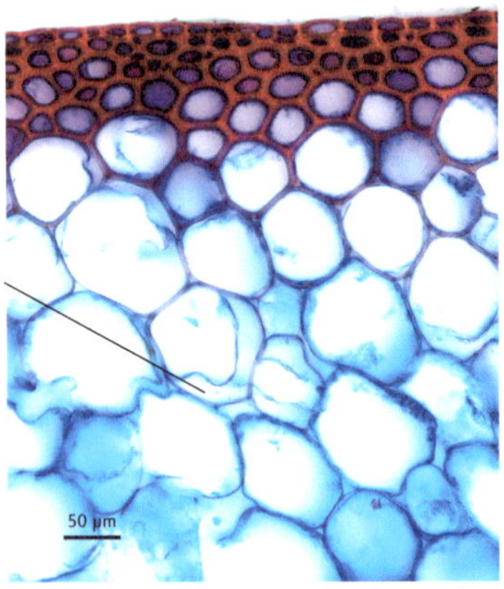

Intercellulars absent or small, triangular between single parenchyma cells. *Thelypteris palustris*

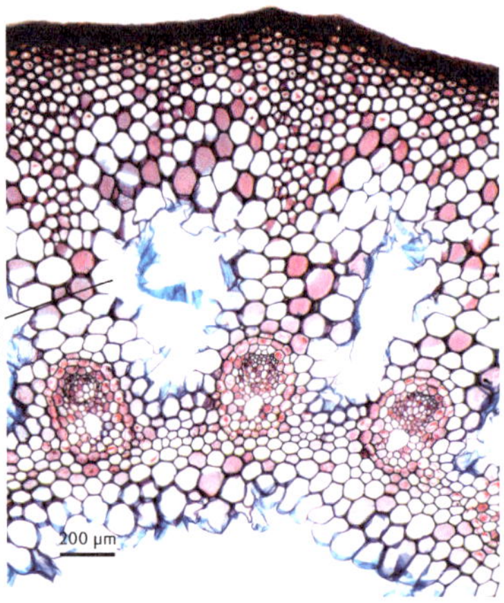

Intercellulars irregularly formed, radially expanded. Developed by cell death (schizogenous). *Equisetum hyemale*

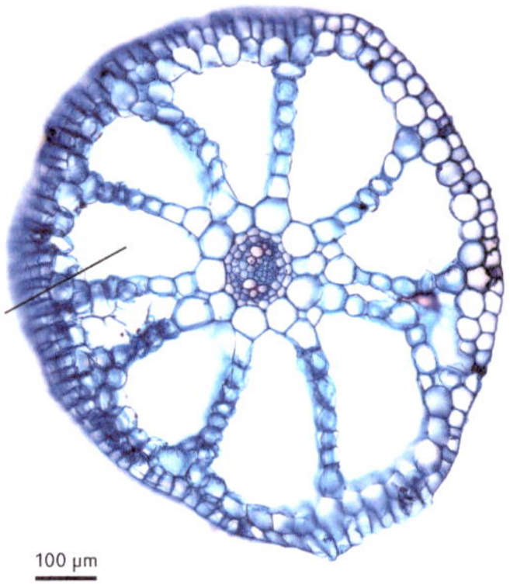

Intercellulars regularly formed, radially arranged. Developed by cell separation (lysigenous). *Pilularia globulifera*

Definition of shoot- and rhizome features of Dicotyledons

Presented are mainly features, which are relevant for aquatic and wetland plants. All occurring features are described by Crivellaro and Schweingruber (2015) and Schweingruber et al. (2011). Here we concentrate on features, which are relevant for wetland plants.

Large cortex, small central cylinders

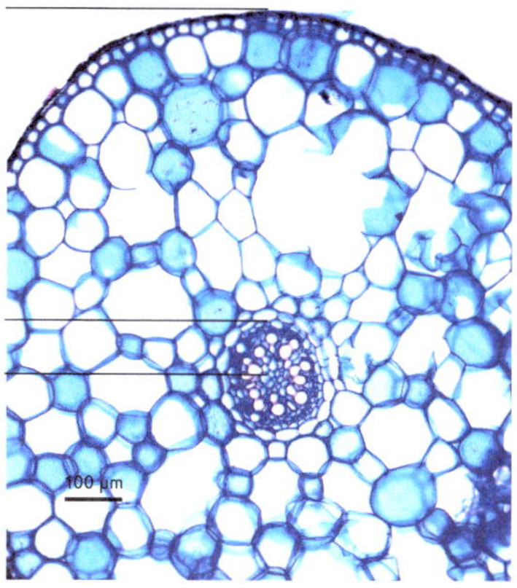

Central cylinder/cortex proportion 1:4. 4 cm tall wetland plant. *Crassula helmsii*

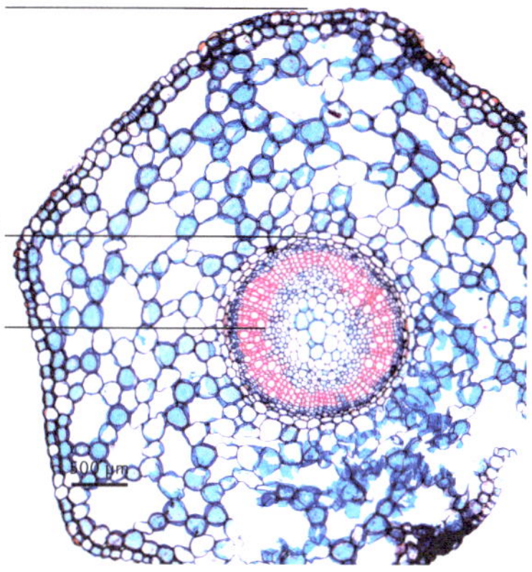

Central cylinder/cortex proportion 1:3. 40 cm tall wetland plant. *Veronica anagallis-aquatica*

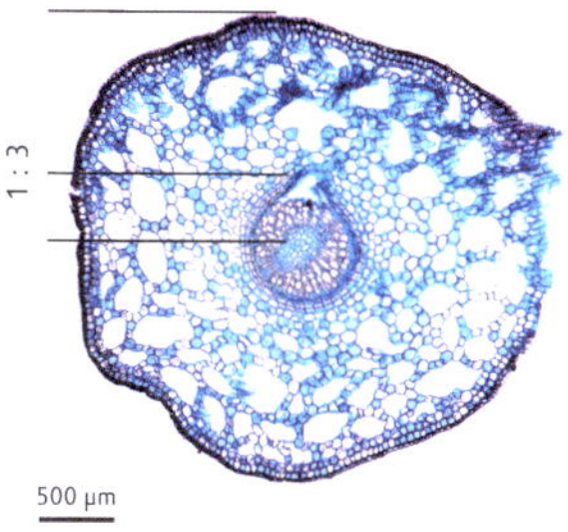

Central cylinder/cortex proportion 1:3. 40 cm tall wetland plant. *Veronica scutellata*

Endodermis-like layer of cells between the cortex and the central cylinder

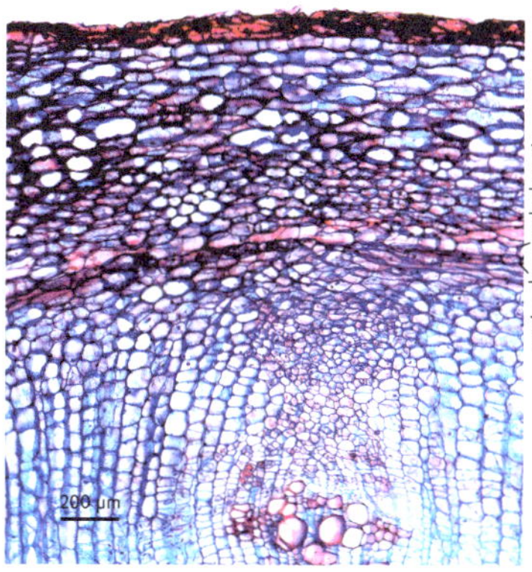

A layer of small cells (pericycle) and a layer of large cells (endodermis) separate the cortex and the central cylinder. 80 cm tall wetland plant. *Cicerbita alpina*

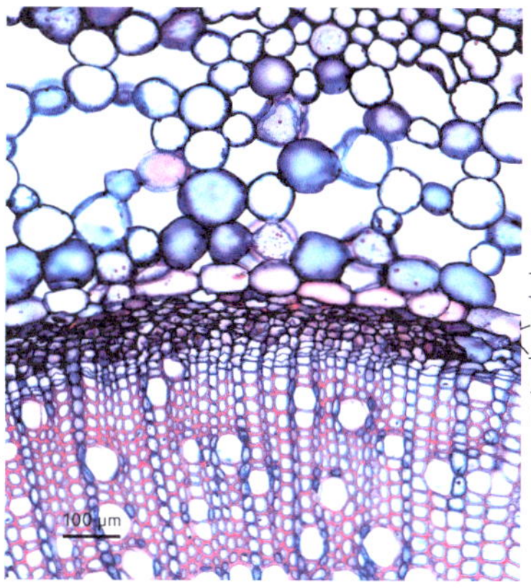

A layer of small cells (pericycle) and a layer of large cells (endodermis). 50 cm tall wetland plant. *Mentha aquatica*

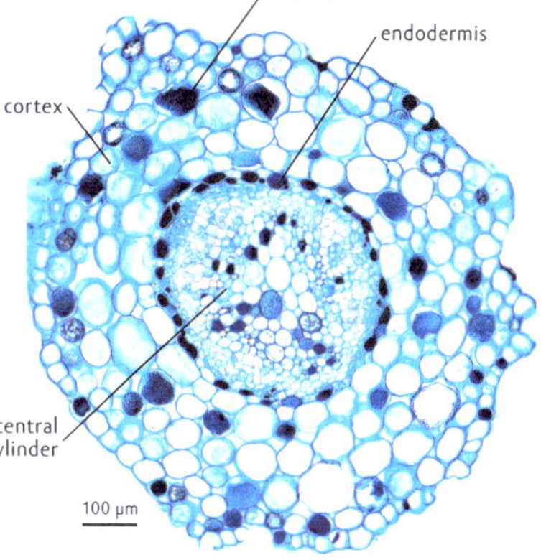

A uniseriate layer with dark staining substances (endodermis). 4 cm tall wetland plant. *Centunculus minimus*

Types of vascular bundles

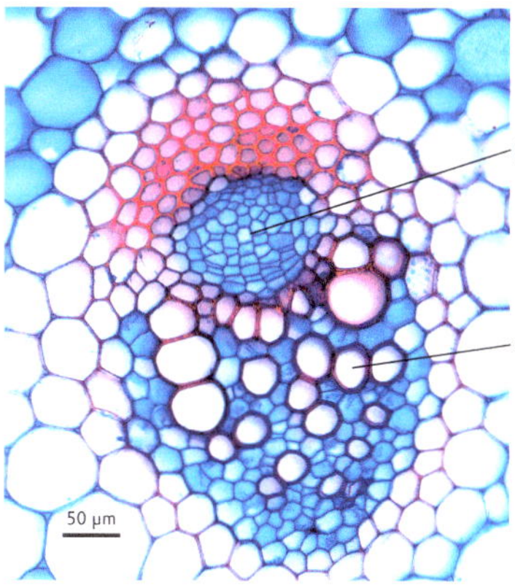

Collateral closed vascular bundle from a dicotyledonous plant. 40 cm tall wetland plant. *Ranunculus sceleratus*, annual shoot

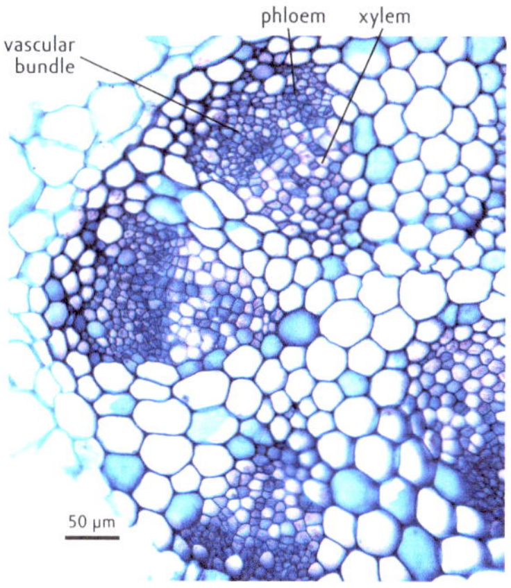

Collateral open vascular bundle from a dicotyledonous plant. The cambium was active for a very short time. 10 cm tall wetland plant. *Chrysosplenium alternifolium*

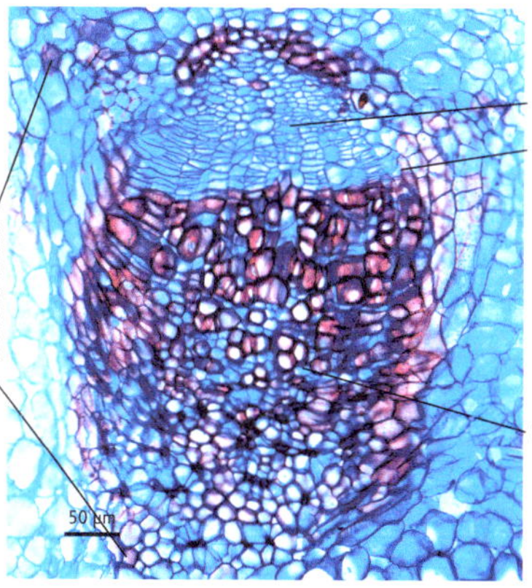

Collateral open vascular bundle of a dicotyledonous plant. *Ranunculus sceleratus*, rhizome

Fibers absent in the xylem

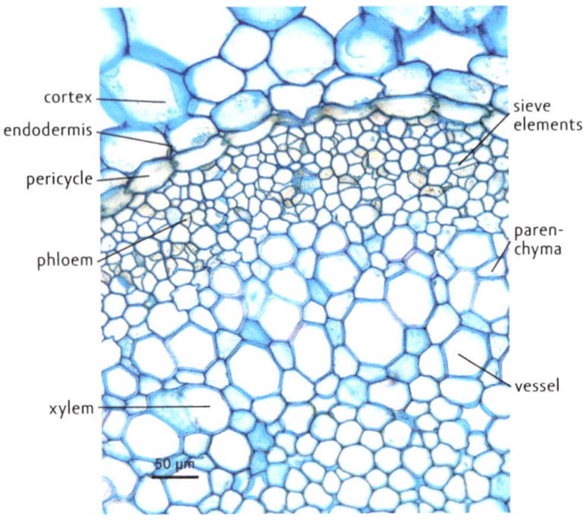

Shoot of a 50 cm tall rooting aquatic plant with emergent parts. *Hippuris vulgaris*

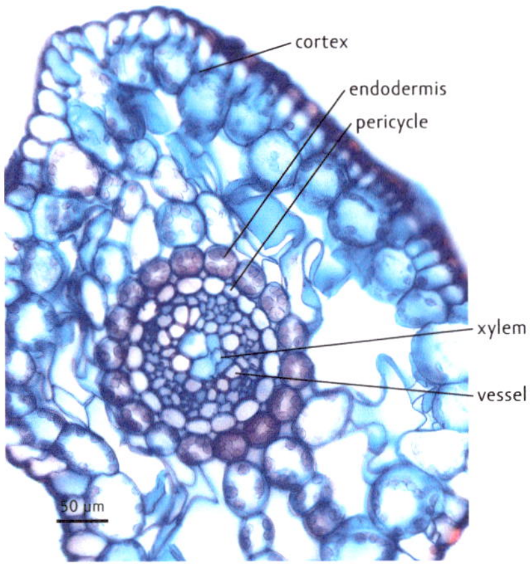

Shoot of a 5 cm tall rooting aquatic plant with emergent parts. *Callitriche palustris*

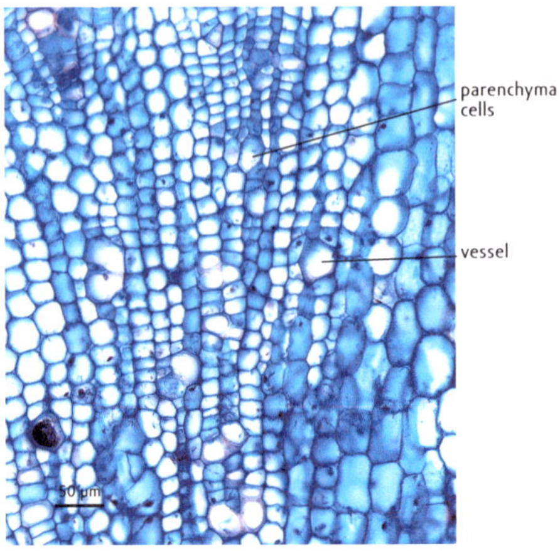

Rhizome of an 80 cm tall herb in a wet meadow. *Cirsium brachycephalum*

Vessels absent in the xylem

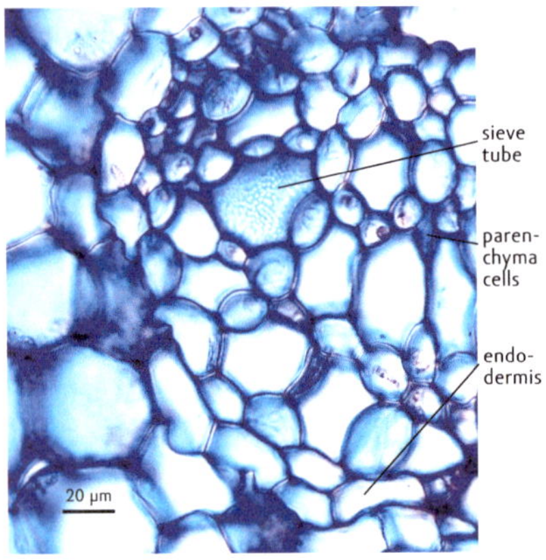

Rhizome of a submerged, rooting aquatic plant with floating leaves. *Nuphar lutea*

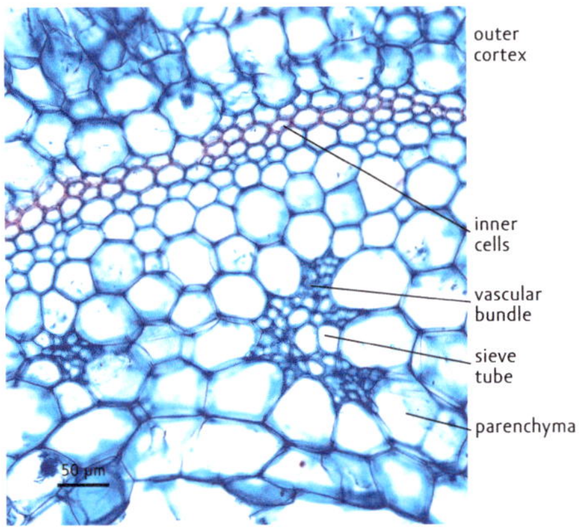

Emergent flower stalk of an unrooted aquatic plant. *Utricularia australis*

Vessel-walls not lignified

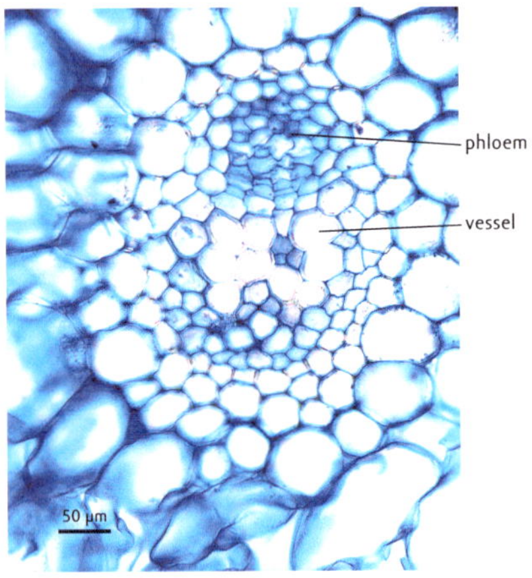

Xylem of a solitary vascular bundle. Shoot of a rooted 20 cm tall aquatic plant. *Batrachium circinatum*

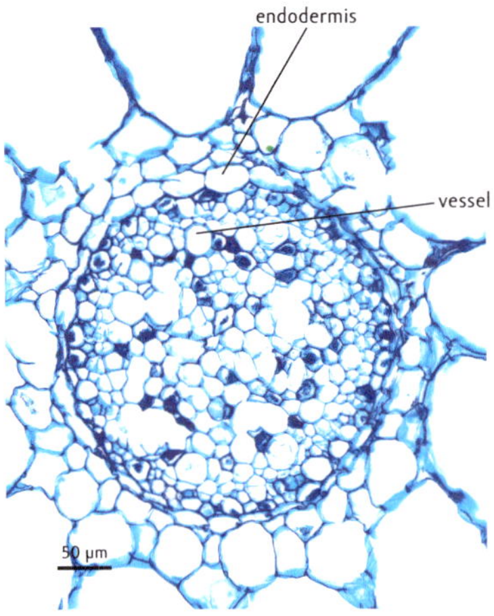

Xylem of a central cylinder of an annual shoot of a 20 cm tall emergent aquatic plant. *Elatine alsinastrum*

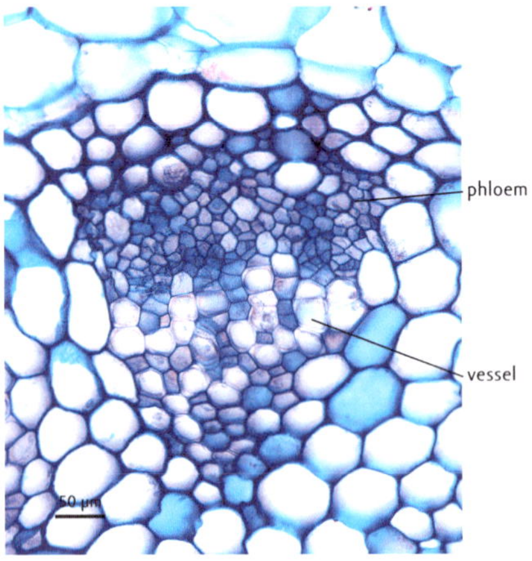

Flower stalk of a small wetland plant. *Chrysosplenium alternifolium*

Cortex consisting of large parenchyma cells

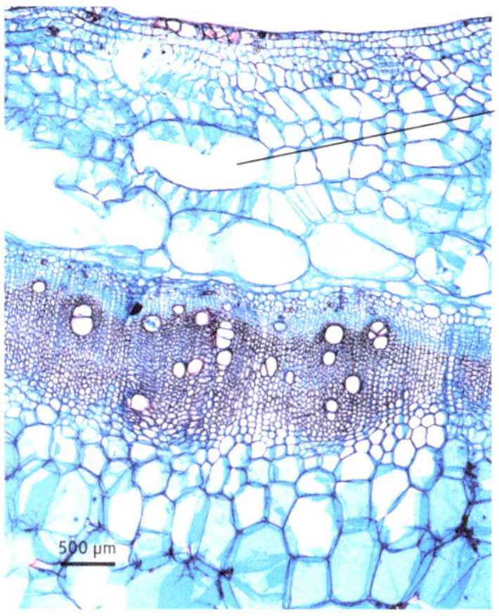

Extremely large parenchyma cells in a 1 m tall wetland plant. *Impatiens glandulifera*

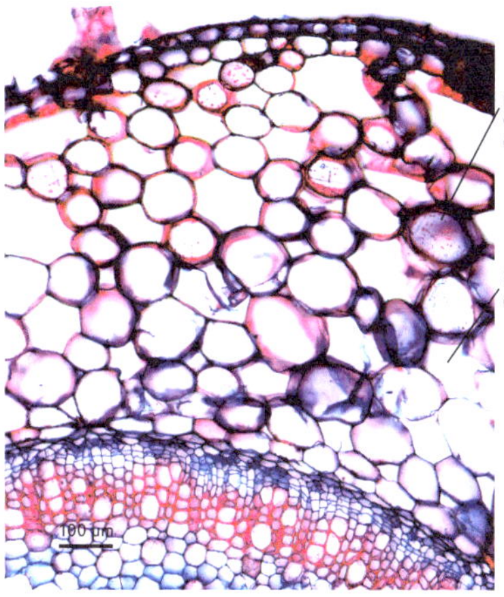

Large parenchyma cells in a 40 cm tall wetland plant. *Veronica anagallis-aquatica*

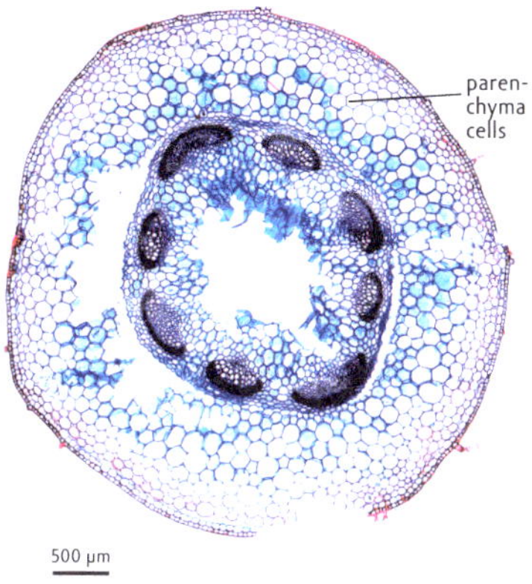

Large parenchyma cells in a 20 cm tall wetland plant. *Stellaria nemorum*

Small to large irregularly formed intercellulars

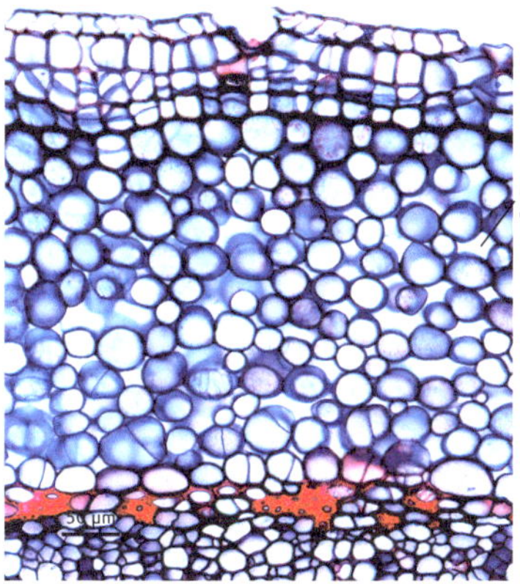

Small intercellulars between unlignified parenchyma cells of a 60 cm tall wetland plant. *Pseudolysimachion maritimum*

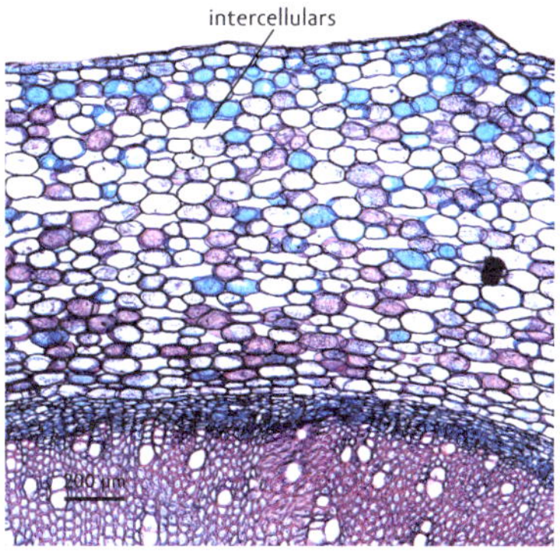

Small to large intercellulars between unlignified parenchyma cells of a 50 cm tall wetland plant. *Lysimachia vulgaris*

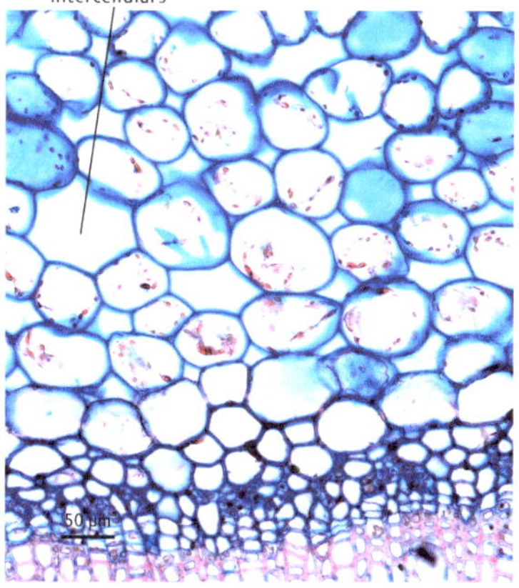

Small to large intercellulars between unlignified parenchyma cells of a 40 cm tall wetland plant. *Pedicularis palustris*

Intercellulars (aerenchyma) circular and honeycomb

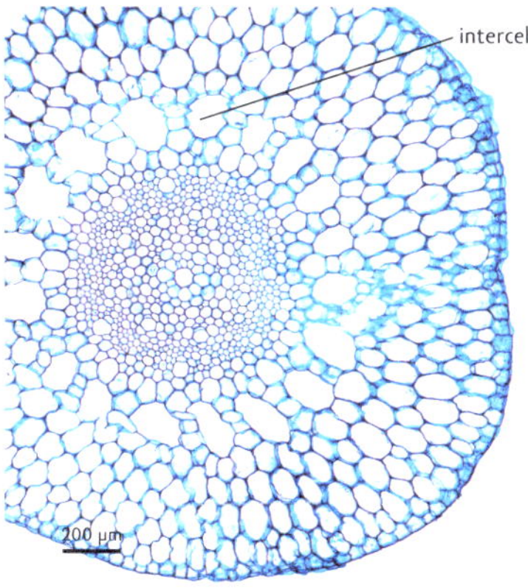

Circularly arranged large intercellulars in a submerged aquatic plant. *Ceratophyllum demersum*

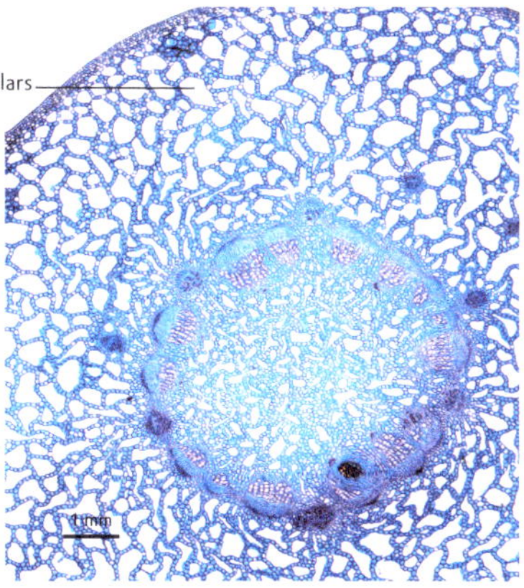

Honey-comb-shaped intercellulars in a wetland plant with emergent parts. *Menyanthes trifoliata*

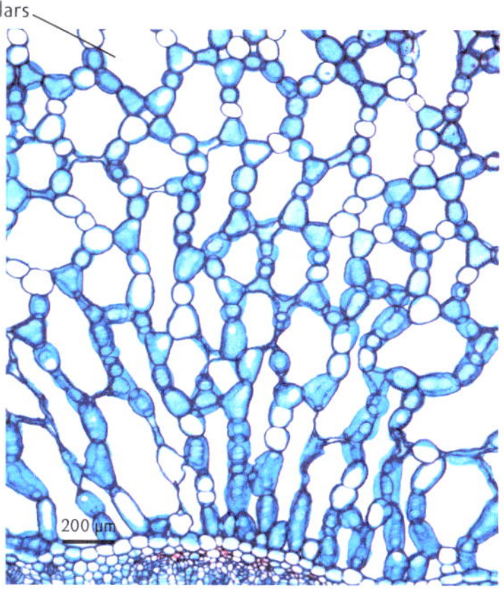

Honey-comb-shaped intercellulars in a wetland plant with emergent parts. *Gratiola officinalis*

Intercellulars (aerenchyma) large, radially extended

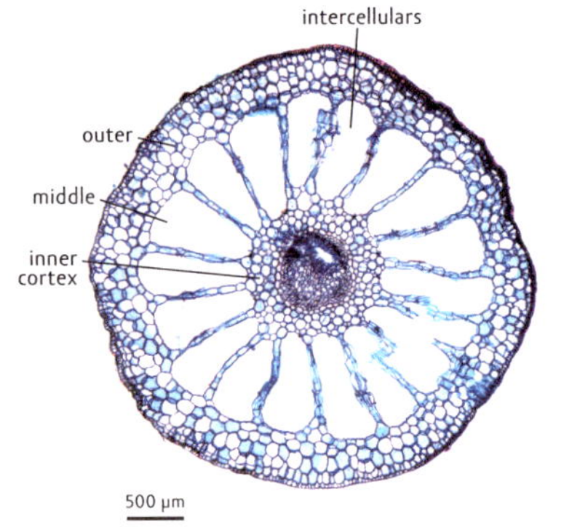

Wheel-like formed large intercellulars in the middle of the cortex of a submerged aquatic plant. *Myriophyllum alterniflorum*

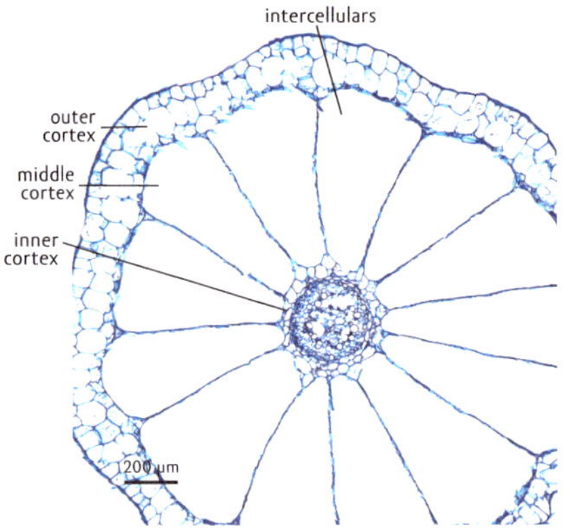

Wheel-like formed large intercellulars in the middle of the cortex of a 20 cm tall emergent aquatic plant. *Elatine alsinastrum*

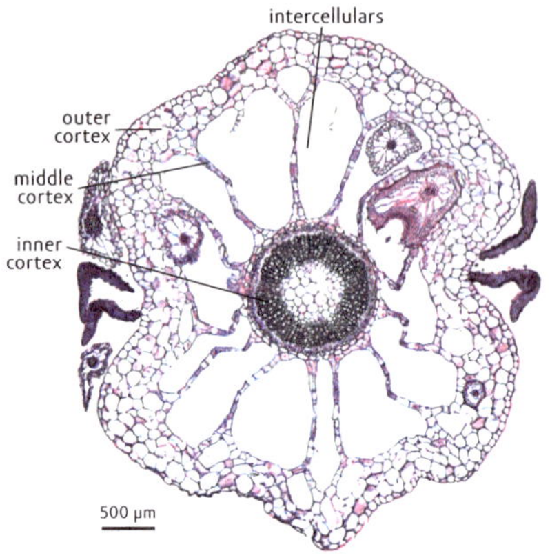

Wheel-like formed large intercellulars in the inner part of the cortex of a 10 cm tall emergent aquatic plant. *Lindernia procumbens*

Air conducting canals

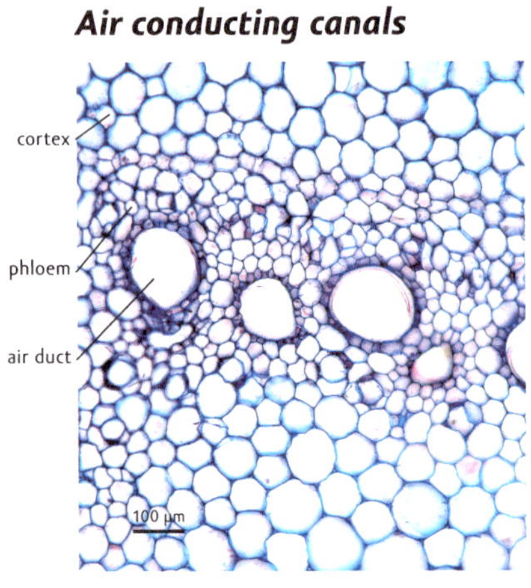

Air ducts in the xylem zone of a rooted aquatic plant with floating leaves. *Trapa natans*

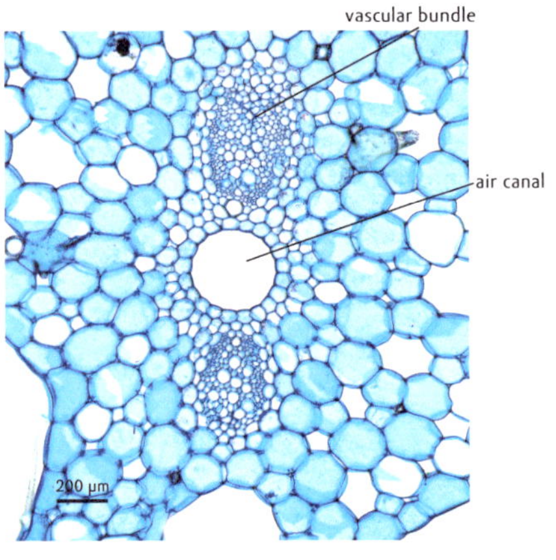

200 cm tall flower stalk of a rooted aquatic plant with floating leaves. *Nymphaea alba*

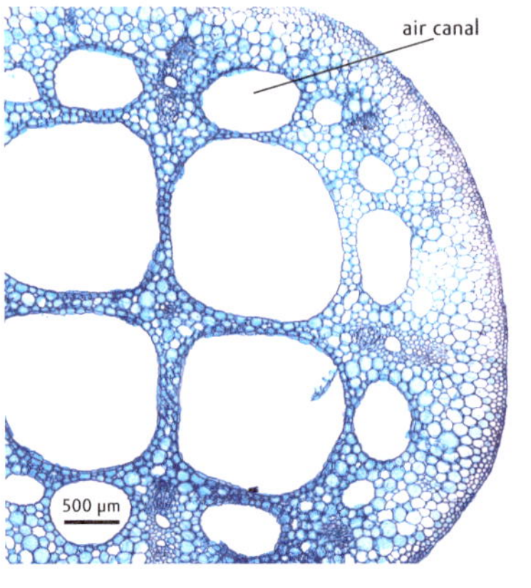

100 cm tall flower stalk of a rooted aquatic plant with floating leaves. *Nymphaea candida*

Crystals

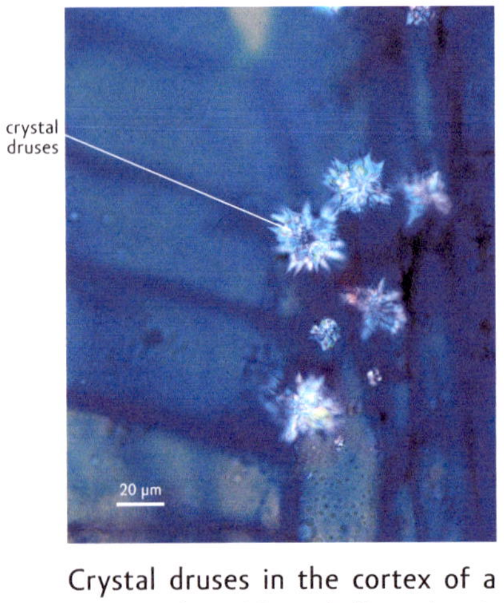

Crystal druses in the cortex of a aquatic plant. *Myriophyllum alterniflorum*

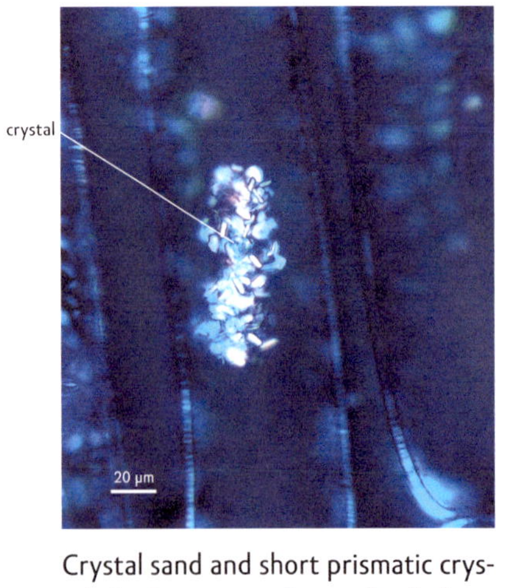

Crystal sand and short prismatic crystals in the cortex of a wetland plant. *Petasites hybridus*

Spicular hairs

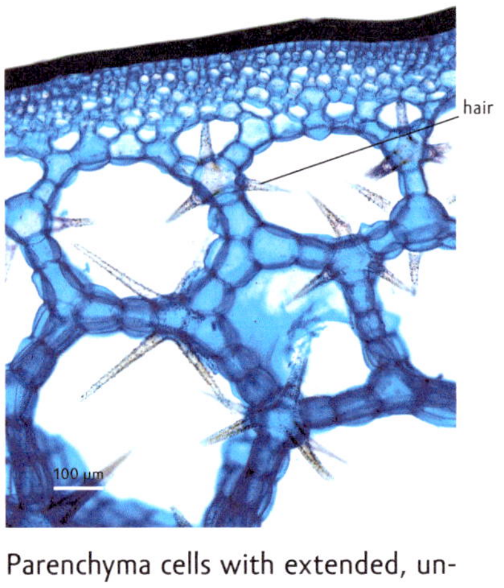

Parenchyma cells with extended, unlignified walls, occupied by small prismatic crystals. 80 cm tall flower stalk of a rooted aquatic plant with floating leaves. *Nuphar pumila*

Definition of culm and rhizome features of Monocotyledons

Presented are only frequently occurring features

Outline

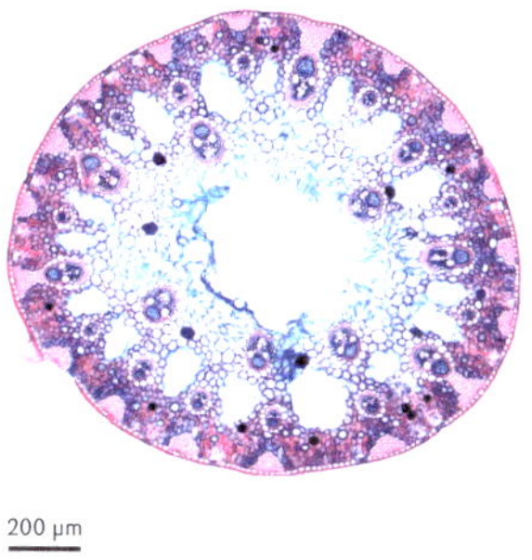

Circular smooth. *Blysmus compressus*

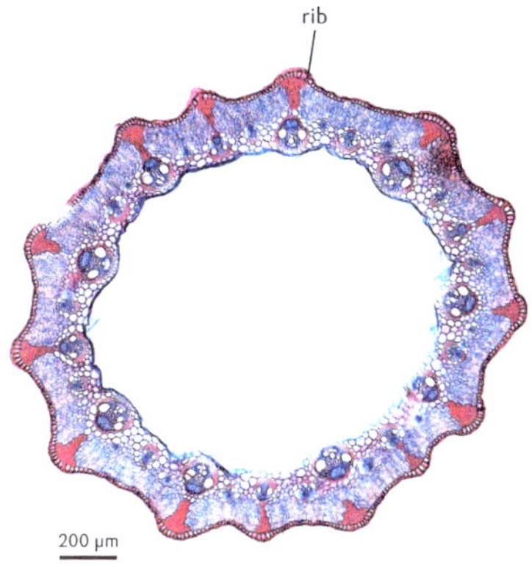

Circular ripped. (Ripping occurs on any other outline form). *Juncus effusus*

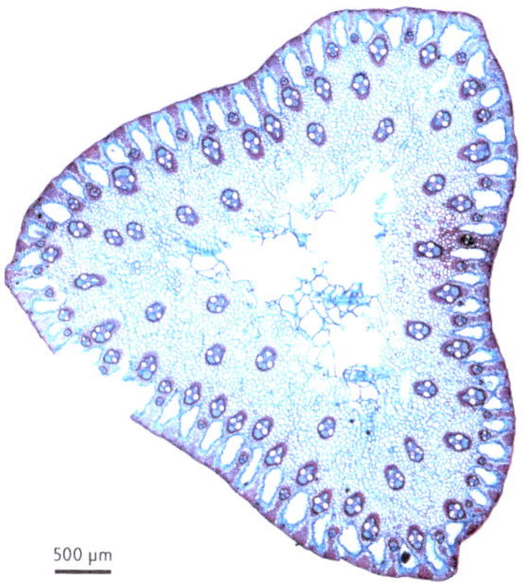

Triangular obtuse. (The edges are often acute). *Cladium mariscus*

Center and central cylinder

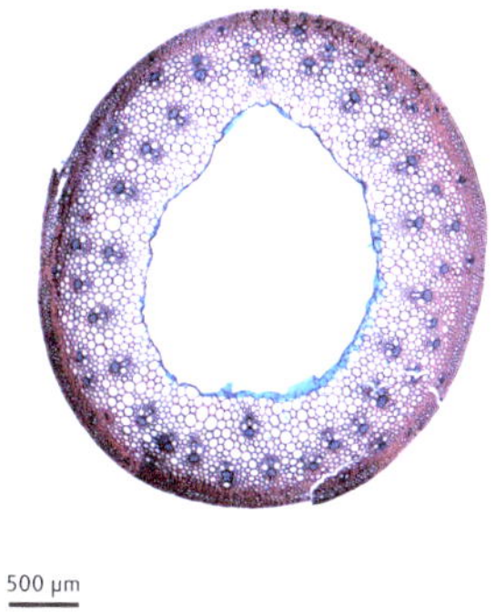

Hollow. *Elymus arenarius*

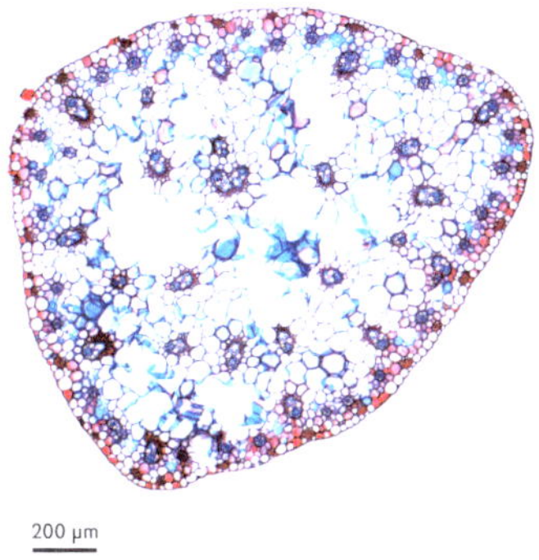

Full, containing parenchyma cells and often vascular bundles. *Cyperus michelianus*

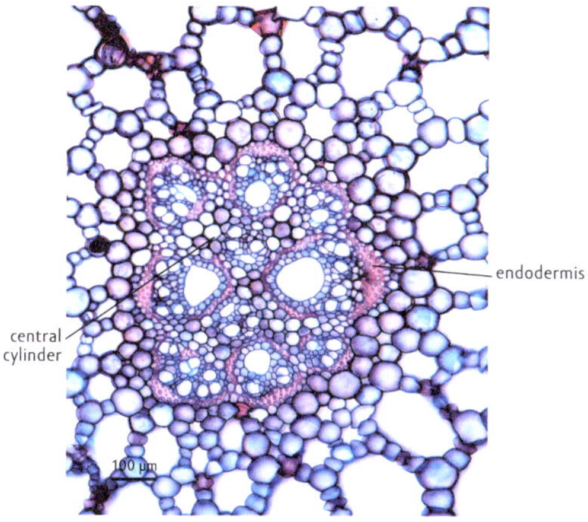

Center consists of a central cylinder, which is separated by an endodermis from the cortex. *Potamogeton polygonifolius*

Culm and rhizome type

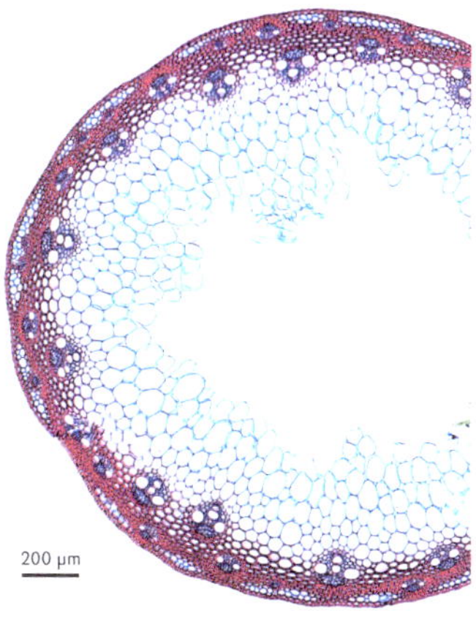

Uniform. Endodermis is absent between lignified and unlignifed parts. *Poa palustris*

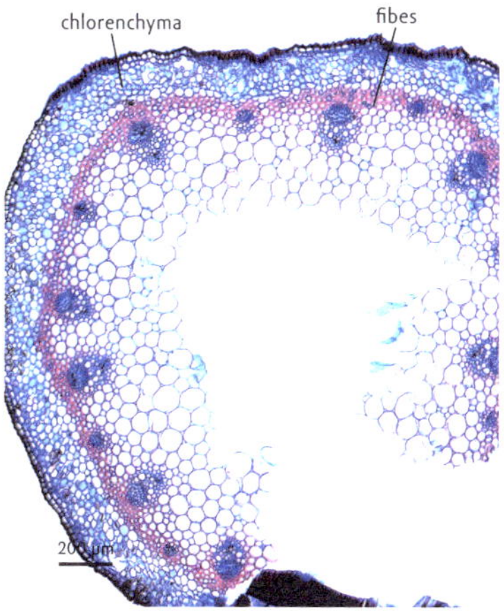

Uniform. The culm is divided in an unlignified and lignified belt. The endodermis is absent. *Triglochin maritima*

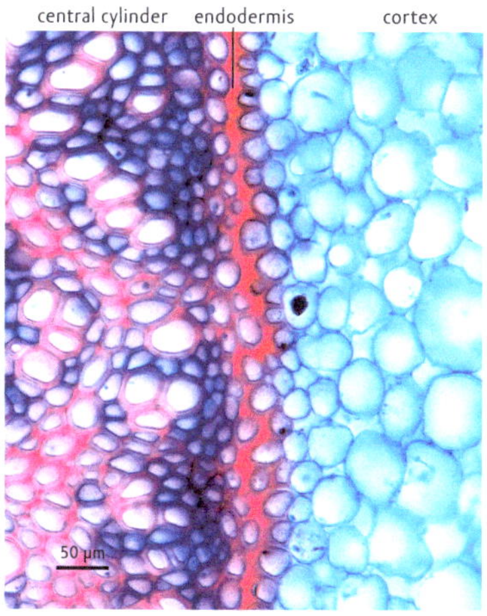

Rhizomes or shoots with a cortex and a central cylinder. The parts are separated by an endodermis. *Carex leporina*

Distribution of vascular bundles

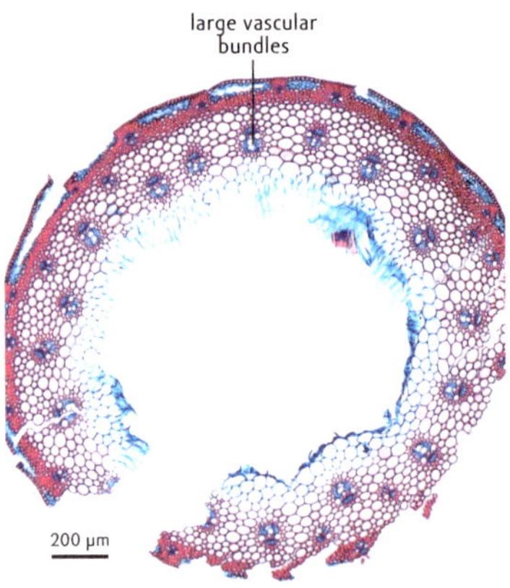

Large vascular bundles are arranged in one circular row. *Phleum alpinum*

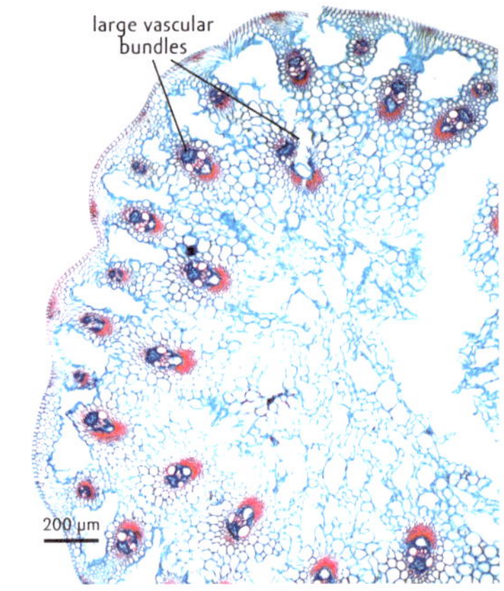

Large vascular bundles are arranged in several circular rows. *Eriophorum vaginatum*

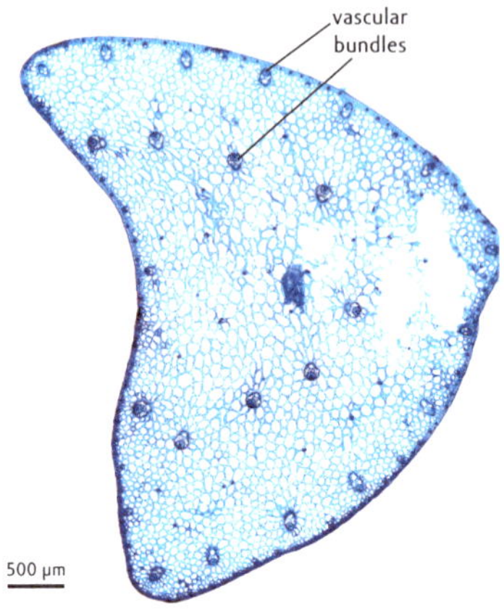

Large vascular bundles are randomly distributed in the whole culm or rhizome. *Acorus calamus*

Types of vascular bundles

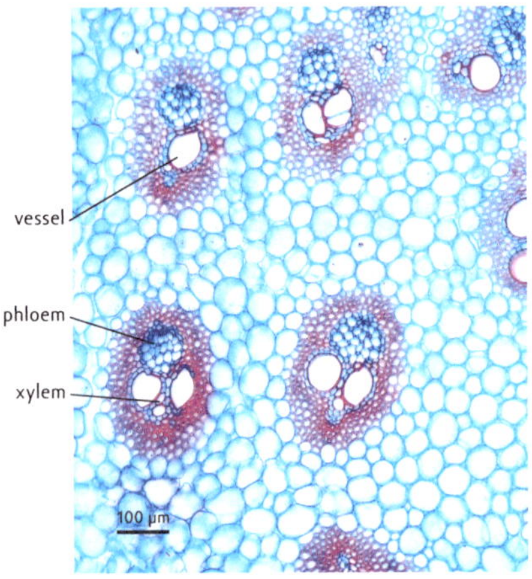

Collateral, closed vascular bundles. The lignified xylem is separated from an unlignified phloem. *Carex riparia*

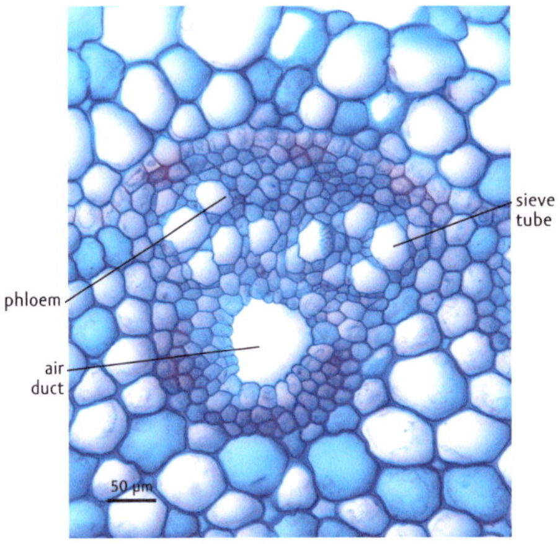

Collateral, closed vascular bundles. The unlignified xylem is separated from an unlignified phloem. *Potamogeton natans*

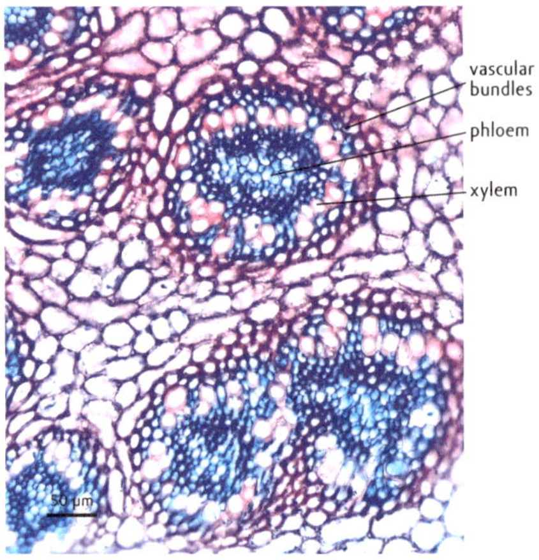

Concentric vascular bundle with an unlignified phloem in the center. *Carex otrubae*

Sclerenchyma sheath around vascular bundles

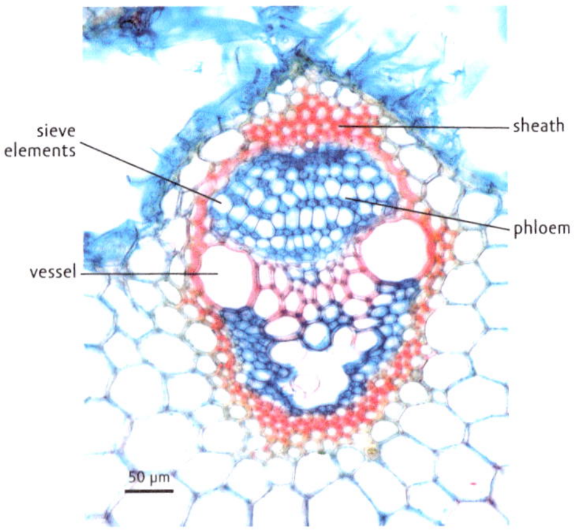

One to two seriate lignified sheath around the vascular bundle. *Carex acutiformis*

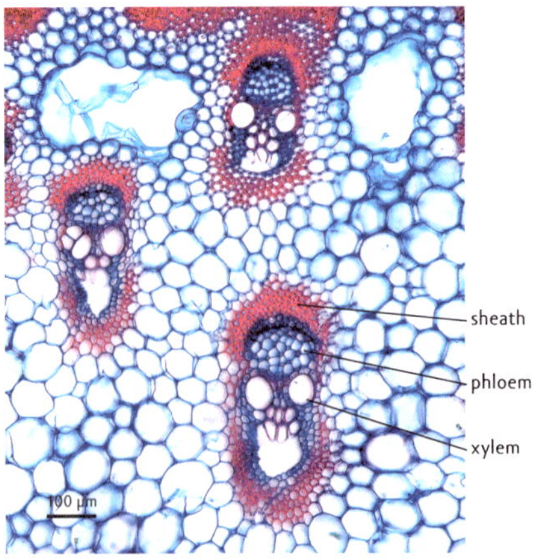

Large multiseriate (>2) lignified sheath around the vascular bundle. *Carex pendula*

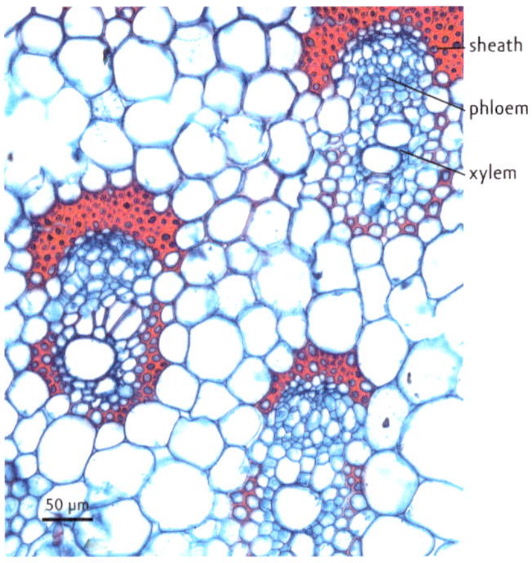

Eccentric lignified sheath. Large part at the peripheral side of vascular bundles. *Typha minima*

Epidermis

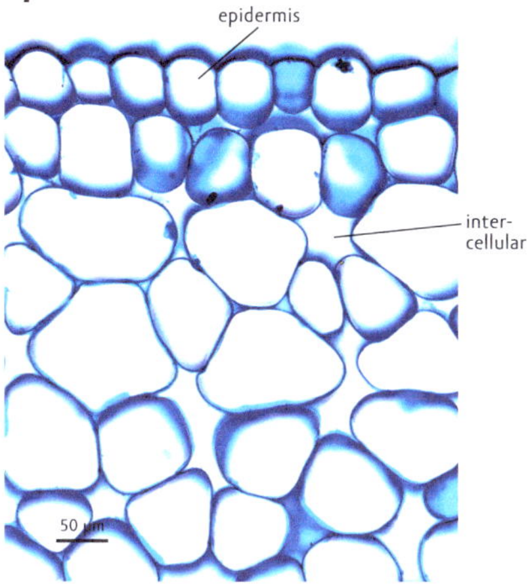

Thin-walled all around. *Epipactis palustris*

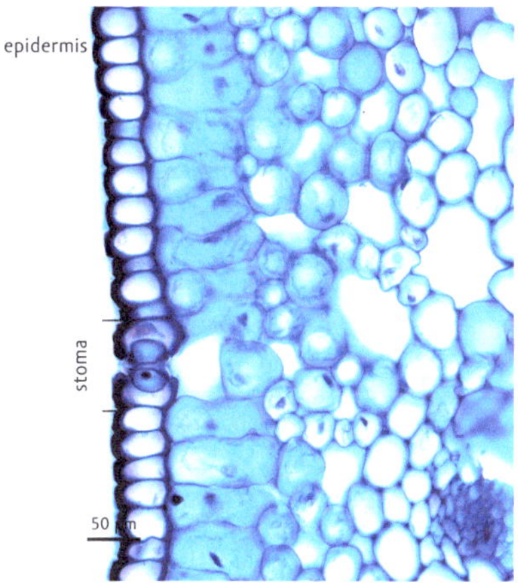

Thick walled at the external side. The walls are mostly lignified and interrupted by stomata. *Leucojum vernum*

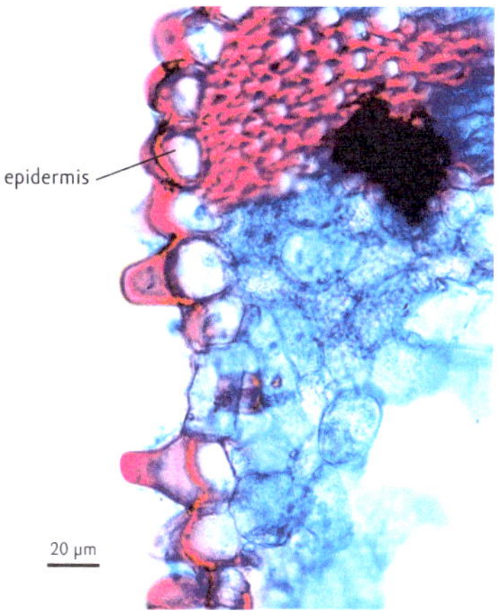

Papillous (bulbiform). *Carex tomentosa*

Endodermis

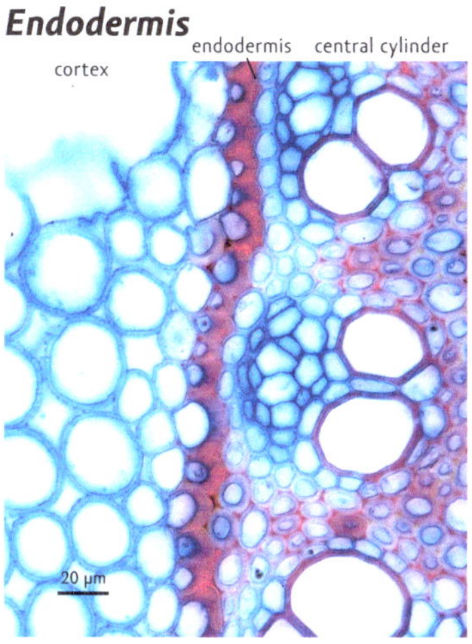

Thick-walled, U-shaped cells. *Carex elata*

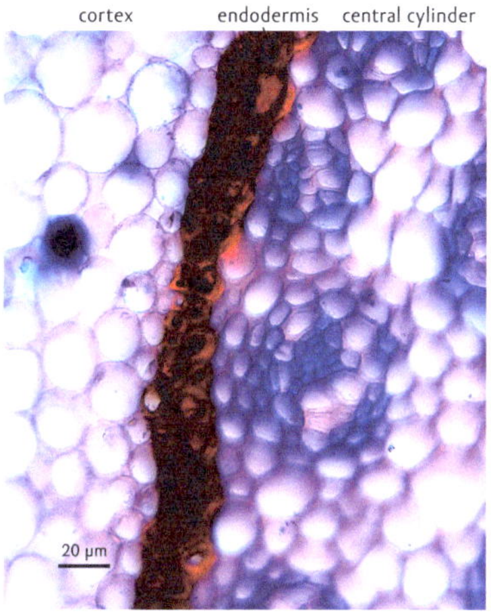

Thick-walled. Two seriate. *Isolepis cernua*

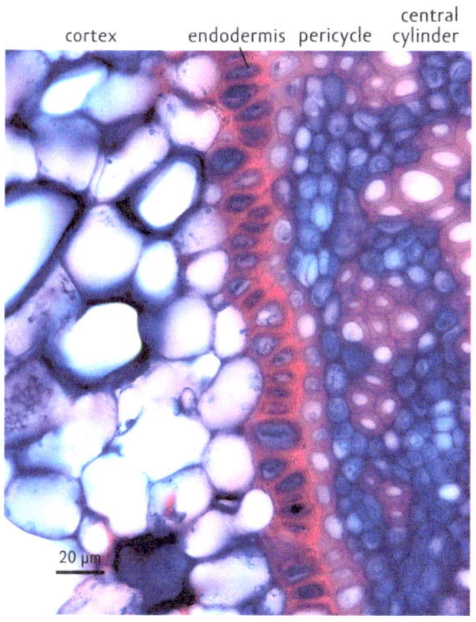

Endodermis and pericycle separate the cortex and the central cylinder. *Carex melanostachya*

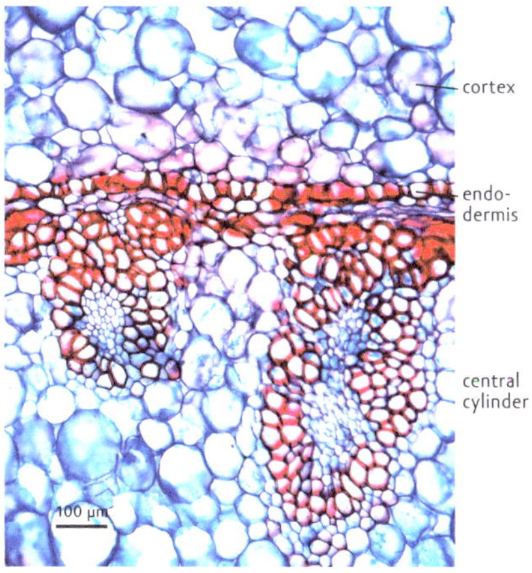

Endodermis irregularly formed. Endodermis absent in the culm. *Iris sibirica*

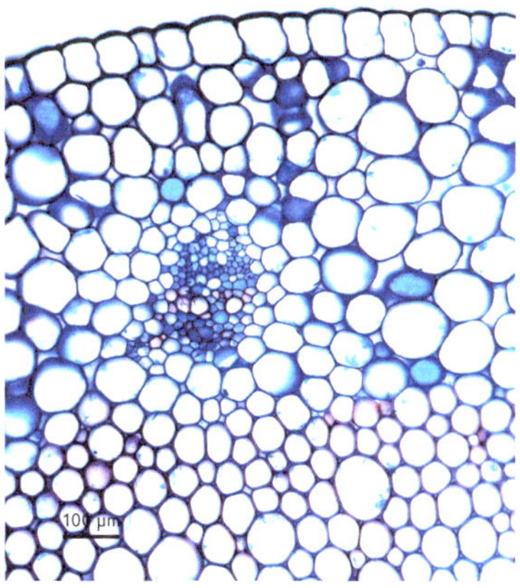

Endodermis absent around the culm. *Dactylorhiza incarnata*

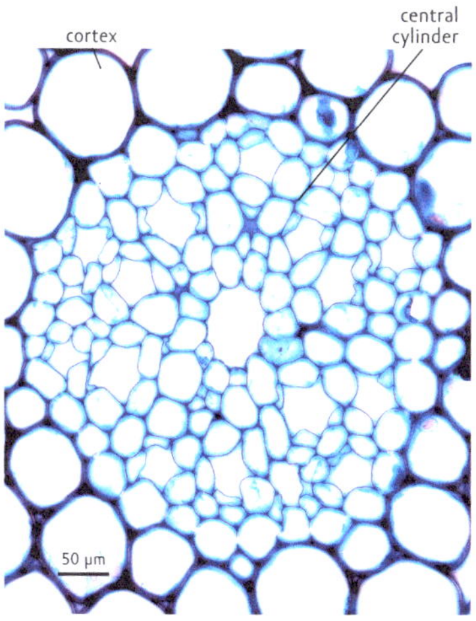

Endodermis absent around the central cylinder. *Elodea nuttallii*

Chlorenchyma (assimilating unlignified tissue)

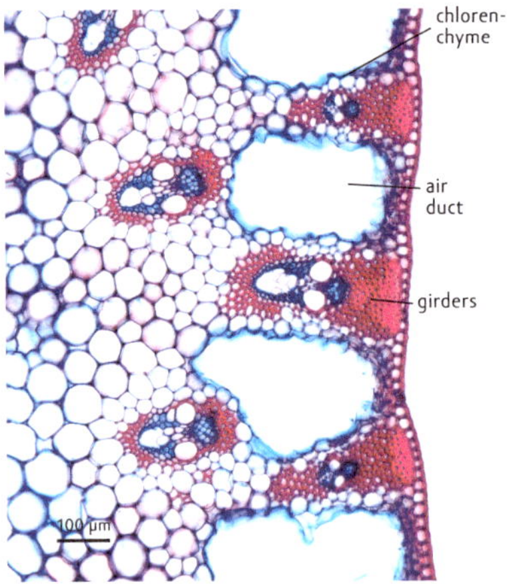

Square or radial rectangular. *Carex pseudocyperus*

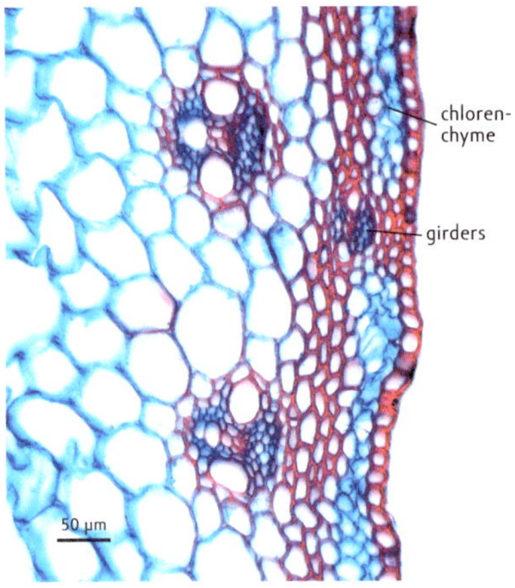

Tangentially enlarged. *Agrostis gigantea*

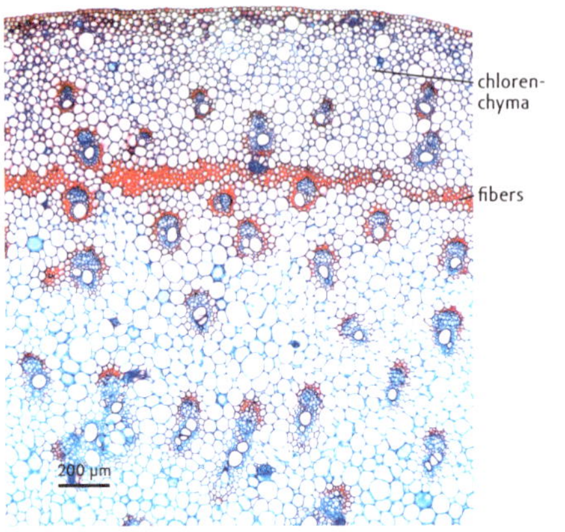

Forming a continuous belt. *Sparganium erectum*

Sclerenchymatic belt at the periphery of the culm

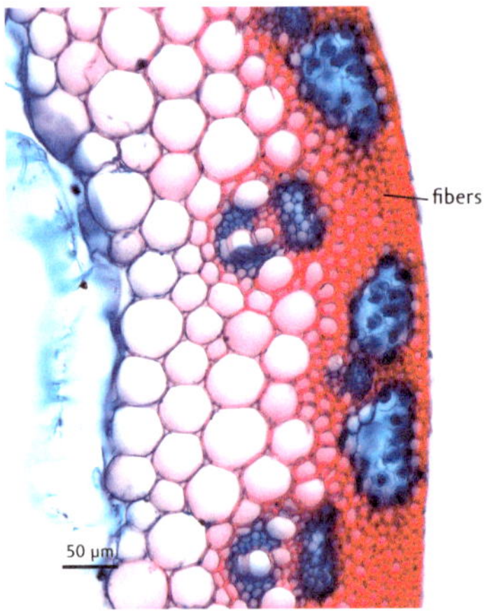

Large and thick-walled. *Deschampsia cespitosa*

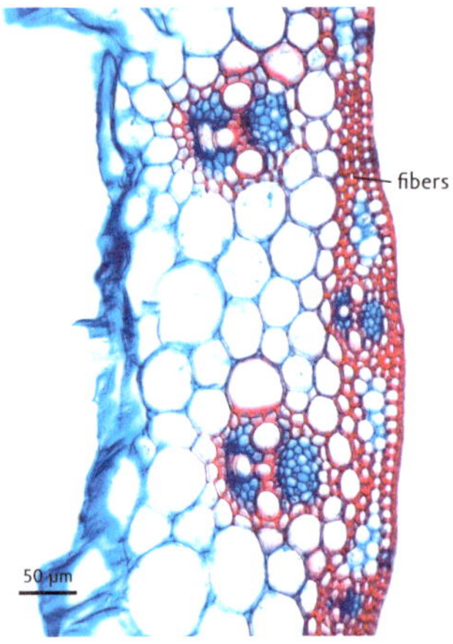

Small and cells rather thin-walled. *Alopecurus geniculatus*

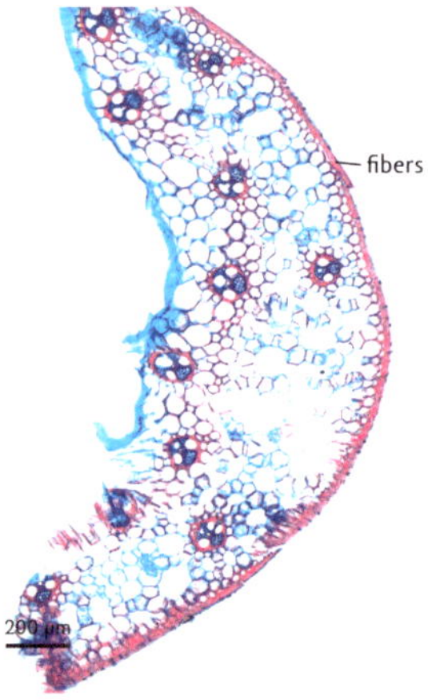

Sclerenchymatic belt almost absent. *Glyceria fluitans*

Girders (axial oriented strings consisting of thick-walled fibers)

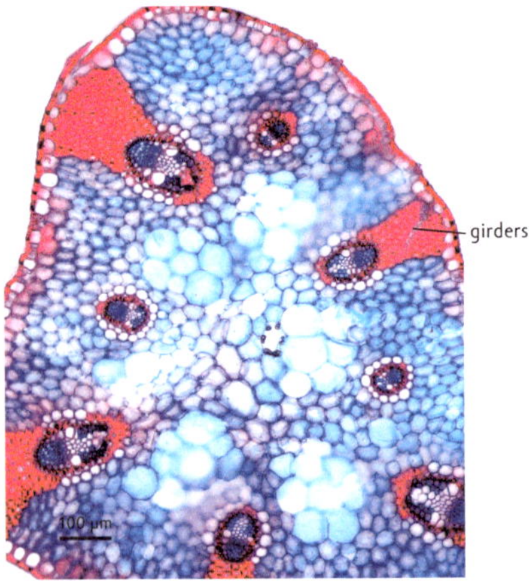

Rectangular to square. *Carex pauciflora*

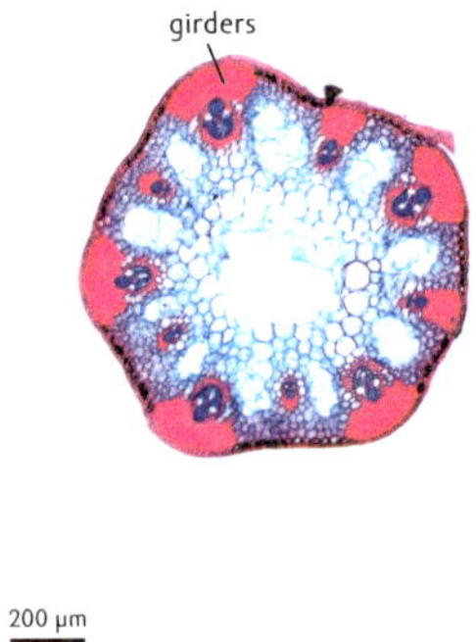

Triangular. *Carex pulicaris*

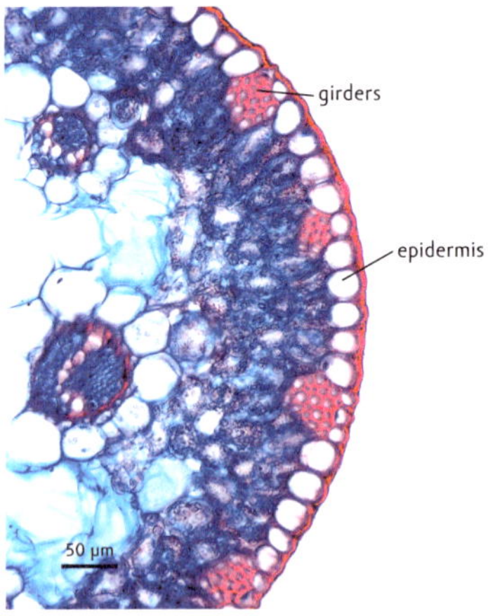

Round, small at the periphery. *Isolepis cernua*

Aerenchyma in the cortex

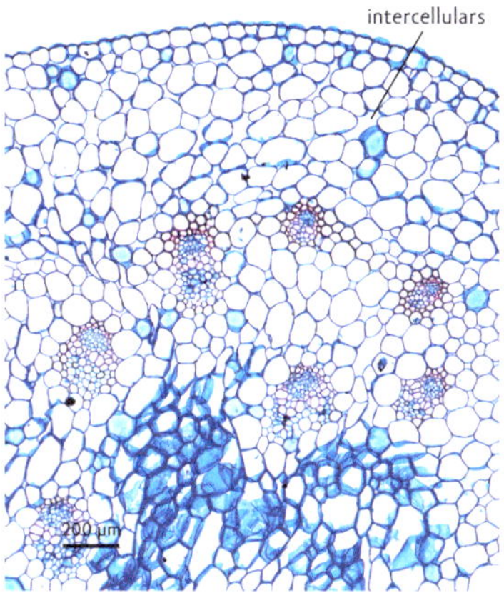

Small, irregularly formed intercellulars between parenchyma cells. *Epipactis palustris*

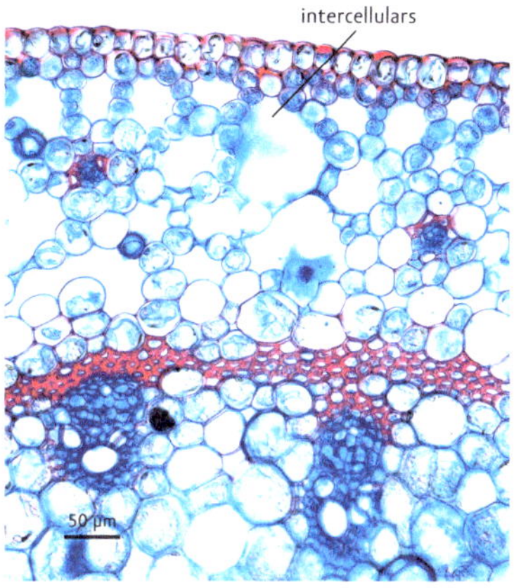

Large, irregularly formed intercellulars between parenchyma cells. *Sparganium natans*

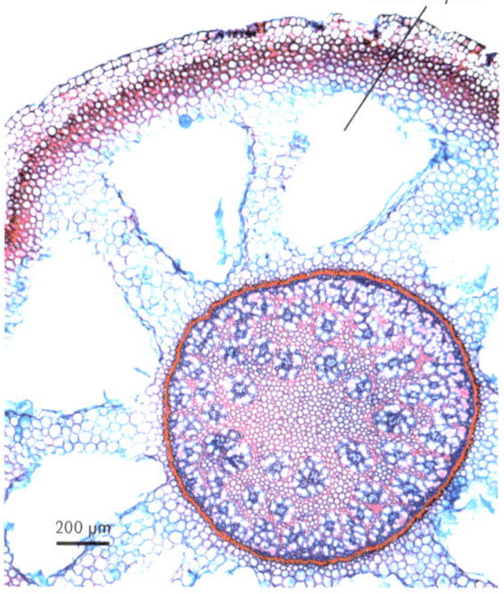

Very large, radially expanded, wheel-like aerenchyma (schizogenous). *Juncus gerardii*

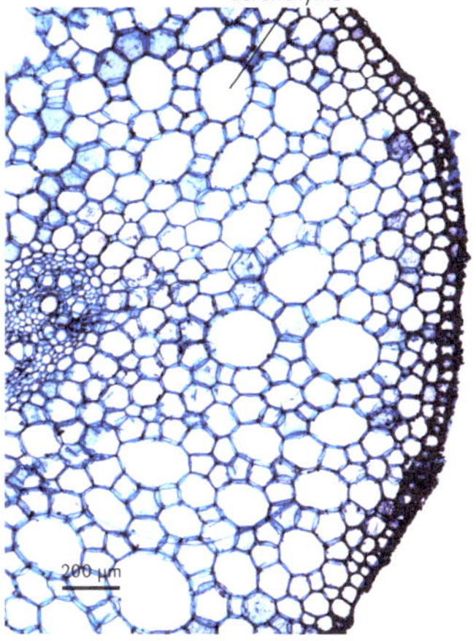

Large, irregularly distributed round to oval cavities (lysigenous aerenchyma). *Groenlandia densa*

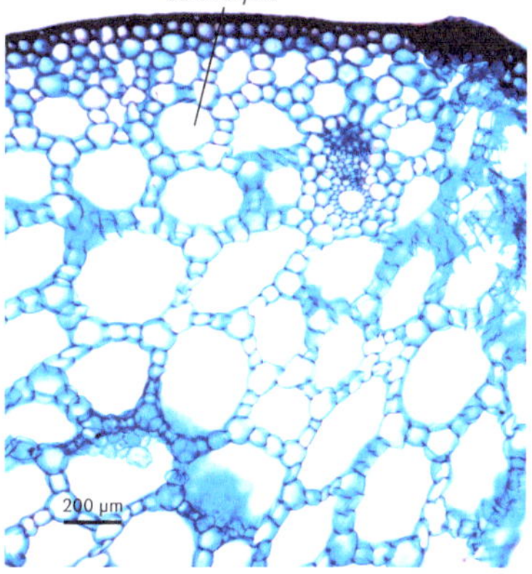

Net-like (honeycomb) aerenchyma. *Orontium aquaticum*

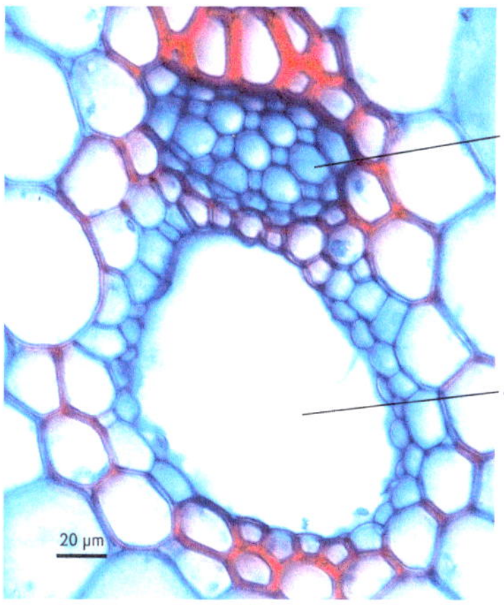

Air ducting canal in the vascular bundle. *Echinochloa crus-galli*

Cell contents

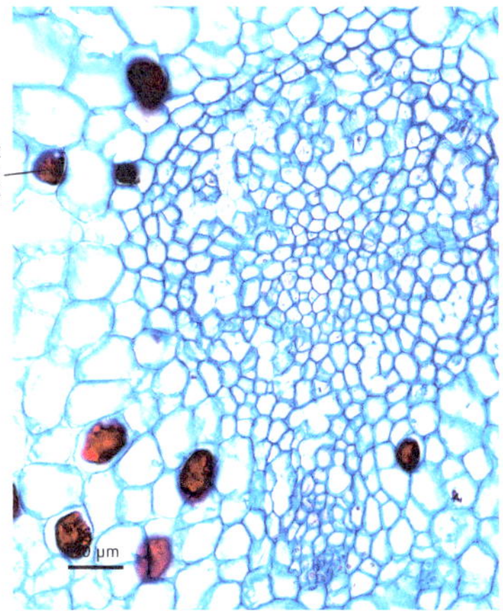

Dark staining substances in single parenchyma cells. *Gladiolus imbricatus*

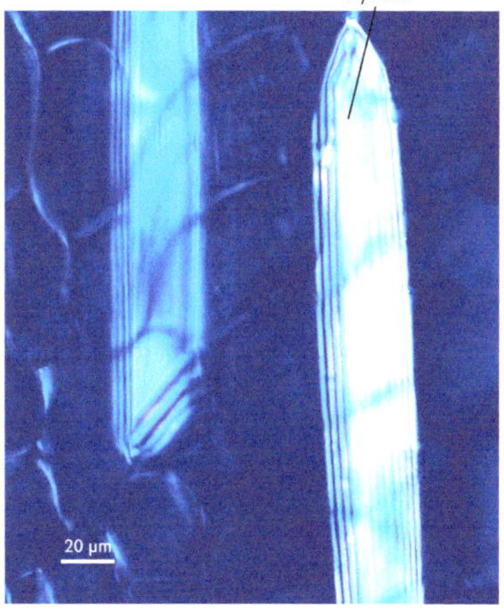

Extremely long crystals. *Iris sibirica*

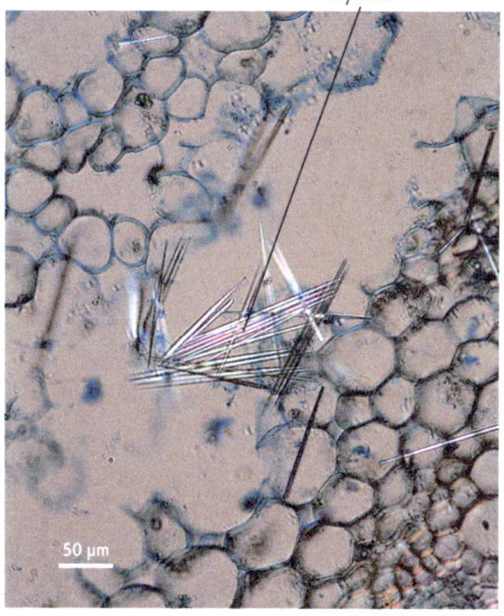

Needle-like crystals (rhaphids). *Vallisneria spiralis*

Anatomical descriptions of Pteridophytes

Dryopteris cristata (L.) A. Gray

Crested Buckler-fern — Dryopteridaceae

Perennial wetland fern forming small tufts, deciduous. Leaves 1–2 pinnate, narrow, pale green, 30–90 cm long. Inner fertile leaves almost erect. Stalks with pale brown scales. Spore cases roundish.

Distribution and ecology

Euro-Siberian temperate element; also in N. America. Native in all CE countries. Occurs in sligthly acidic fens in *Sphagnum* lawns. It is able to tolerate partial shading of invading reeds, birch and willow shrubs.

Anatomy of the petiole

The lower part of the petiole has a diameter of 1.5–2 mm (1, 2). Secondary growth is absent.

Three round vascular bundles are surrounded by a thin-walled parenchymatic tissue (3, 4). Vascular bundles are composed of a xylem with angularly lignified tracheids with a diameter up to 50 µm and a few unlignified, small parenchyma cells (2). Tracheids have scalariform pits (5). The surrounding phloem consists of angular, unlignified sieve elements and parenchyma cells. Bundles are surrounded by a layer of living parenchyma cells (round nuclei) and a living endodermis (elongated nuclei) with an extremely thick-walled peripheral wall and extremely thin-walled lateral walls (6). The external endodermis wall is perforated by bordered pits (7).

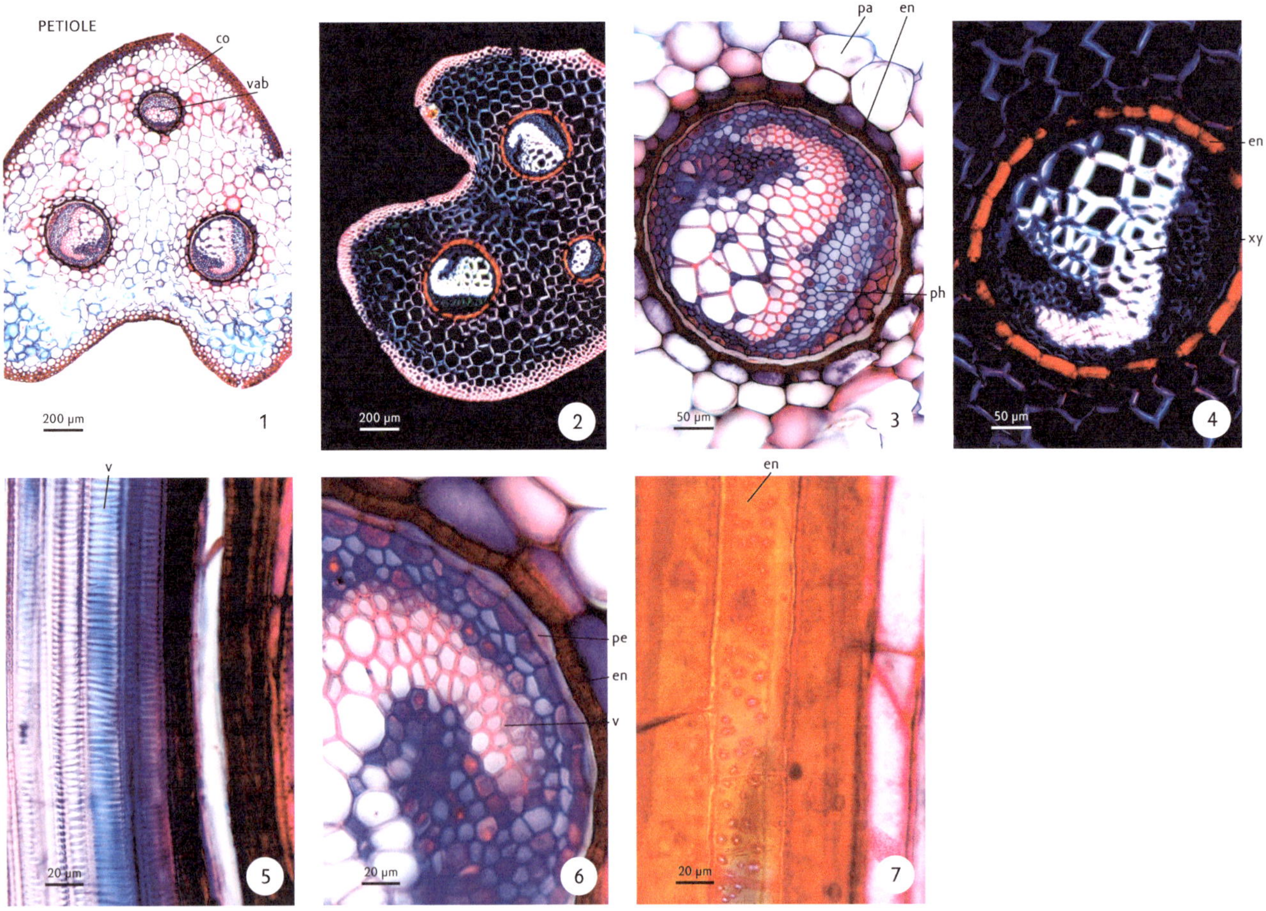

F. H. Schweingruber et al., *Anatomic Atlas of Aquatic and Wetland Plant Stems*, https://doi.org/10.1007/978-3-030-33420-8_4

Equisetum fluviatile L.

Water Horsetail

Amphibious perennial wetland horsetail, deciduous. Stems usually little branched or unbranched, smooth, yellow-green, up to 150 cm high. Sheaths tight, green, with black short teeth. Blunt cones on green stems.

Distribution and ecology

Circumpolar boreo-temperate element. Native in all CE countries. Occurs in ponds, lakes or ditches. Often a pioneer species in freshwater successions, tolerating a wide range of water depth, pH, nutrient concentrations and substrate types. Iron resistant plant.

Anatomy of the rhizome

The rhizome has a diameter of 2 mm (1). Secondary growth is absent.

The hollow pith is surrounded by a circle of vascular bundles, a large cortex and an epidermis (1, 2). A very few dispersed lignified vessels with ring-like thickenings (3) and a complex of parenchyma cells and sieve tubes are located laterally and peripherally of the large canal (Carinal canal) (2). Vascular bundles are surrounded by an endodermis (2). The inner part of the cortex has large schizogenous spaces (valecular canals) (1). The outer part of the cortex and the epidermis consist of thick-walled cells (4).

Roots have a central vascular bundle, consisting of few lignified vessels and two groups of phloem. The cortex and the central cylinder are separated by an endodermis (4).

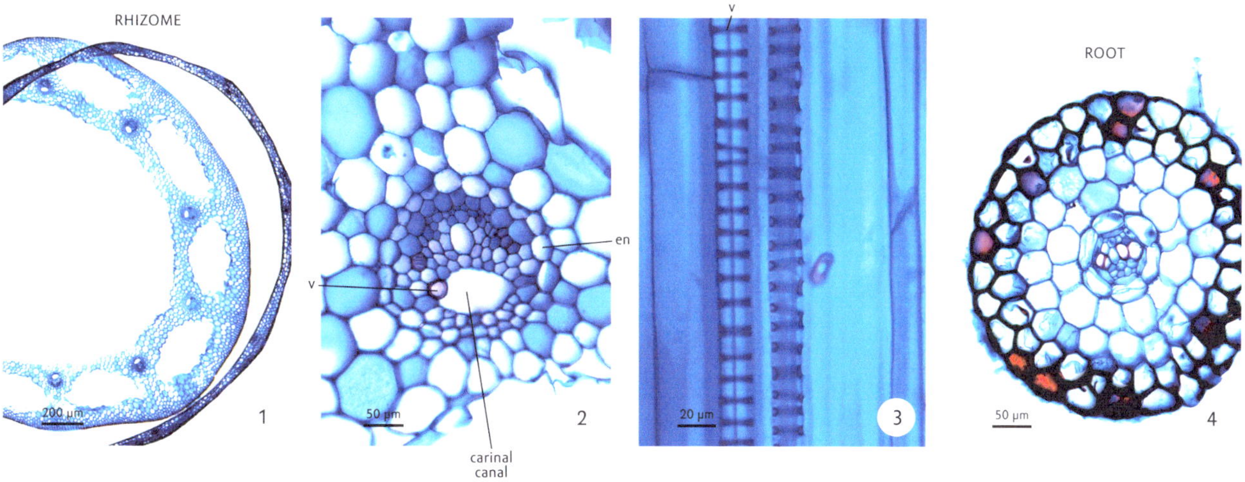

F. H. Schweingruber et al., *Anatomic Atlas of Aquatic and Wetland Plant Stems*, https://doi.org/10.1007/978-3-030-33420-8_5

Equisetum hyemale L.

Rough Horsetail

Equisetaceae

Perennial wetland horsetail, wintergreen, often forming colonies. Stems unbranched, stiff, dark green, up to 100 cm high. Sheaths black with a green central band, untoothed. Pointed cones on green stems.

Distribution and ecology

Circumpolar boreo-temperate element. Native in all CE countries from lowlands to subalpine zone, usually in open woodlands along streams and rivers. Prefers heavy soils with a high mineral and silica content.

Anatomy of a rhizome

The horizontal rhizome has a diameter of 3–4 mm (1). Secondary growth is absent.

The hollow pith is surrounded by a circle of vascular bundles, a large cortex and an epidermis (1, 2).

The xylem of the vascular bundles consists of a large canal (Carinal canal), a very few lignified vessels and a round complex of sieve tubes and parenchyma cells (3). The vascular bundles are surrounded by a distinct endodermis containing Casparian strips (3). The inner part of the cortex has large schizogenous spaces (valecular canals) and the outer part consists of round, thick-walled cells (4). Epidermis cells are very thick-walled (4).

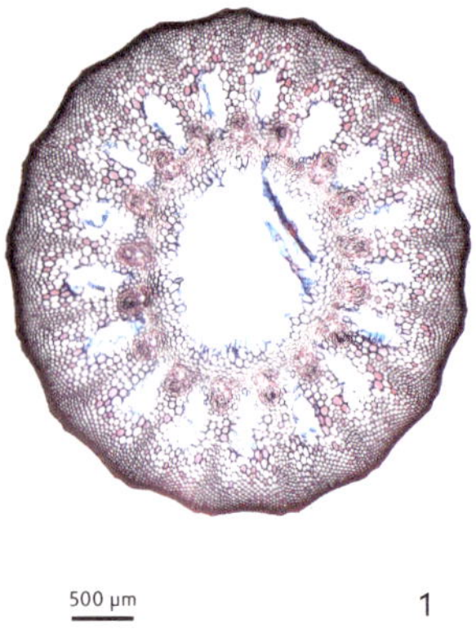

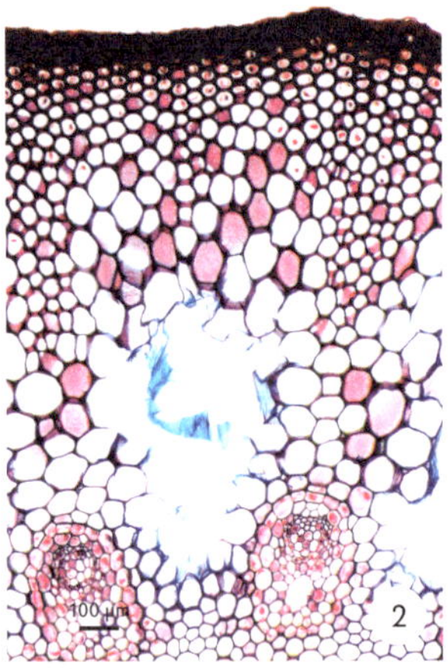

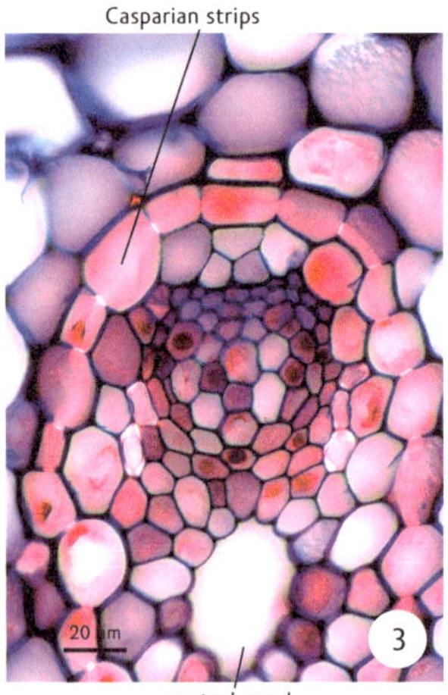

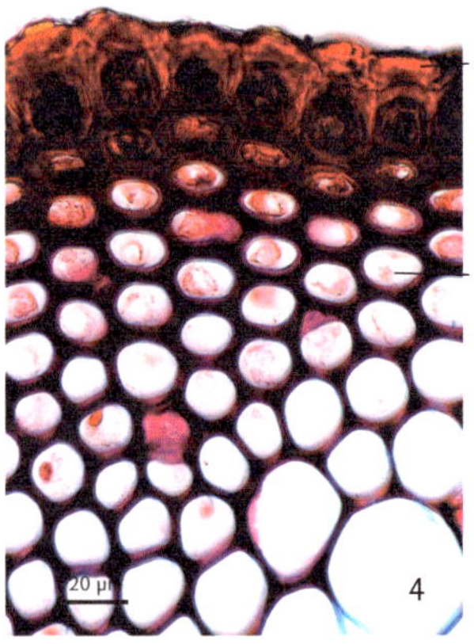

Equisetum palustre L.

Marsh Horsetail

Equisetaceae

Perennial wetland horsetail, deciduous. Stems with whorls of branches, to 50 cm high, the central hollow very small. Sheaths loose, the teeth black with broad white edges. Pointed cones on green stems.

Distribution and ecology

Circumpolar boreo-temperate element. Native in all CE countries from lowlands to subalpine zone, usually in marshes, ditches, wet meadows and damp pastures. Contains alkaloids toxic for livestock.

Anatomy of the rhizome

The horizontal rhizome has a diameter of 3 mm (1). Secondary growth is absent.

The pith is surrounded by a circle of poorly differentiated vascular bundles, a large cortex and an epidermis (1–4). A very few dispersed unlignified vessels and a complex of parenchyma cells and sieve tubes are located laterally and peripherally of the large canal (Carinal canal) (3, 4). The central part (stele) is surrounded by an endodermis. Vessels are located laterally of the carinal canals (5, 6). The inner part of the cortex has large schizogenous spaces (valecular canals) (1, 2). The outer part of the cortex is anatomically variable. It consists of thick-walled (1) or thin-walled cells (2). Epidermis cells are thick-walled (1, 2).

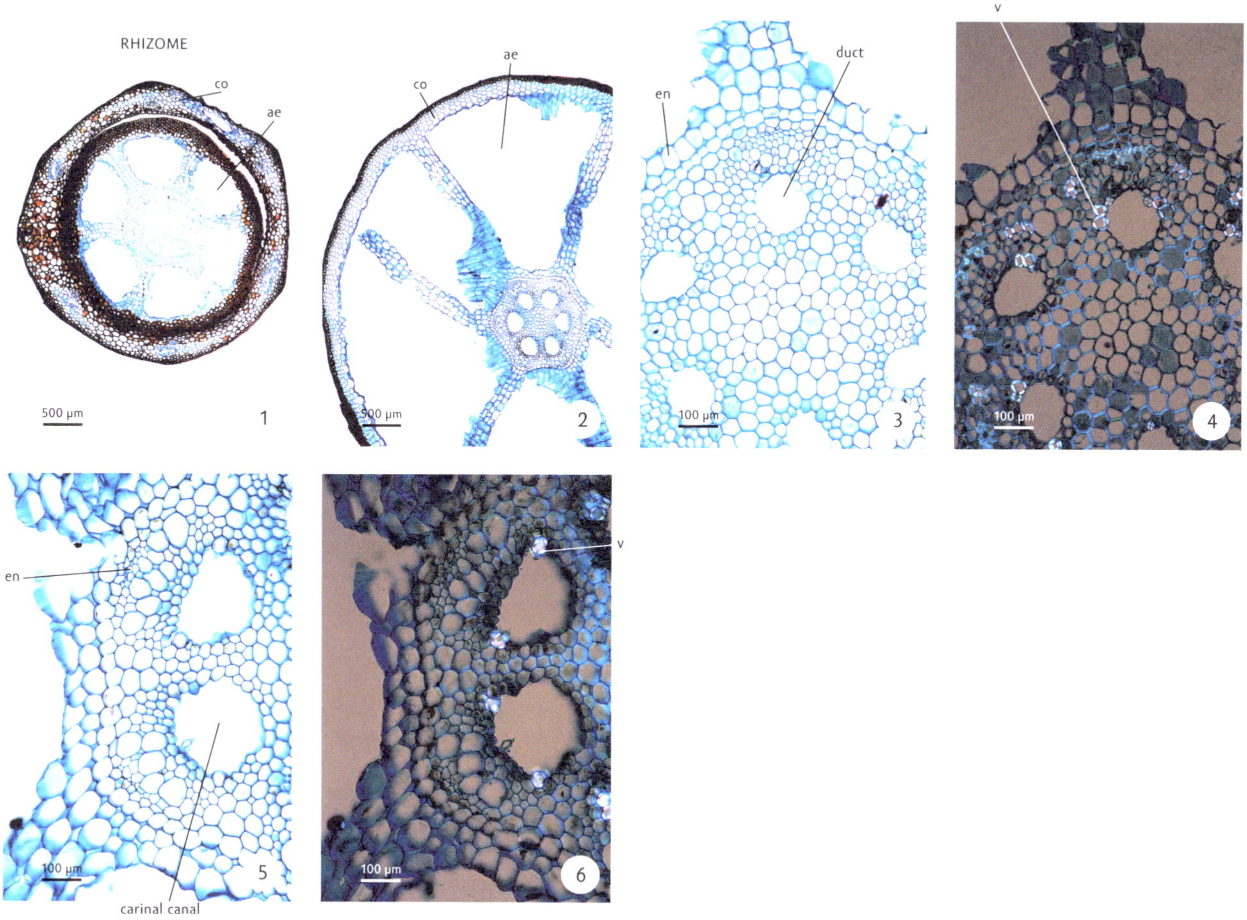

Equisetum variegatum Schleich. **Variegated Horsetail** Equisetaceae

Perennial wetland horsetail, wintergreen, tufted. Stems more or less prostrate, often curly, branched only at base, hollow, dark green, 10–30 cm high. Sheaths black above, green below, the teeth whitish with a black midrib. Pointed cones on green stems.

Distribution and ecology

Circumpolar boreo-arctic, montane element. Native in all CE countries from montane to subalpine zone. Occurs on base-rich fens, dune-slacks, upland flushes, also on banks of gravel pits. Prefers open sites with low competition and calcareous soils. Facultatively mycotrophic.

Anatomy of a rhizome

The rhizome has a diameter of 2 mm (1). Secondary growth is absent.

The pith is surrounded by a circle of vascular bundles, a large cortex and an epidermis (1). A very few dispersed, poorly lignified vessels with ring-like thickenings (3, 5) and a complex of parenchyma cells and sieve tubes are located laterally and peripherally of the large canal (Carinal canal) (2–4). Vascular bundles are surrounded by an endodermis with distinct Casparian strips (4). The inner part of the cortex has large schizogenous spaces (valecular canals) (1, 2). The outer part of the cortex is anatomically variable. It consists of thick-walled (1) or thin-walled cells (6). Epidermis cells are thick-walled (1, 6).

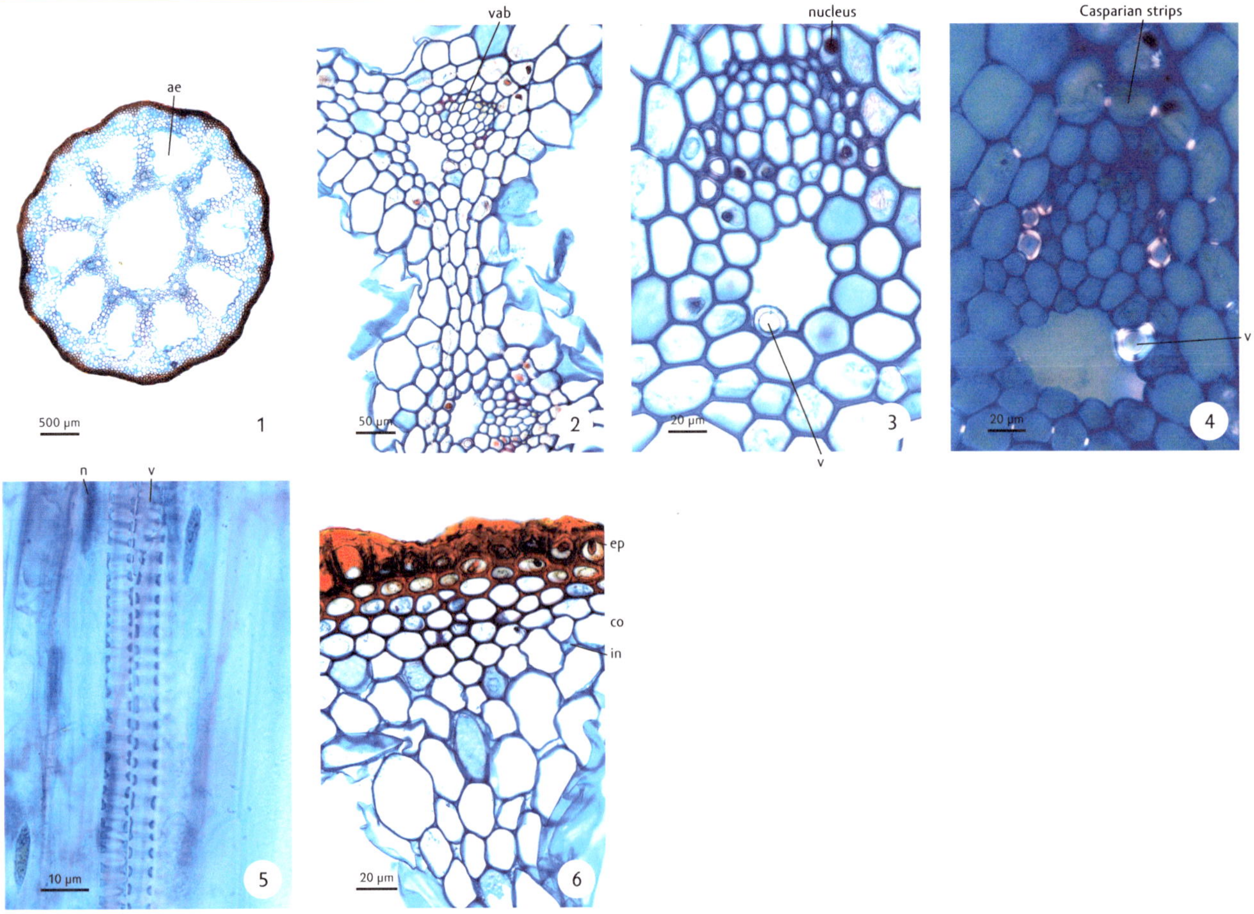

Perennial submerged plant, stems short and corm-like, hollow roots at base, erect rosette of subulate stiff leaves 6–15 cm long with minute ligule on upperside near base. Heterosporous sporangia embedded in leaf-base below the ligule in older leaves. Leaves contain 4 air cavities connected with a distinct central air cavity in roots.

Distribution and ecology

Euro-Siberian boreo-montane distribution and native also in N. America. Very rarely in AT, CH, CZ, DE and PL. Oligotrophic mountainous lakes (>1000 m a.s.l.), mostly stony substrates, down to 6 m depth. Psychrophilous, germination of megaspores lasts for two seasons, sensitive to overgrowing of the substrate by filamentous algae. CAM and diffusive CO_2 uptake by roots facilitate photosynthetic CO_2 fixation.

Anatomy of the basal part of a leaf

The leaf has a diameter of 1 mm (1). The leaf is composed of a closed concentric vascular bundle, a large cortex and an epidermis (1). Xylem and phloem are difficult to distinguish on cross-sections (2). Radial sections show that vessel-walls have annual thickenings (3). The center of the stele contains a large intercellular. Intercellulars of the cortex are laterally separated by radial strips of parenchyma cells (1). Aerenchymatic chambers in the cortex are horizontally separated by a thick-walled stellate cellular tissue with extraordinary filaments around pits (4). The leaf is externally surrounded by an epidermis and a hypodermis (1).

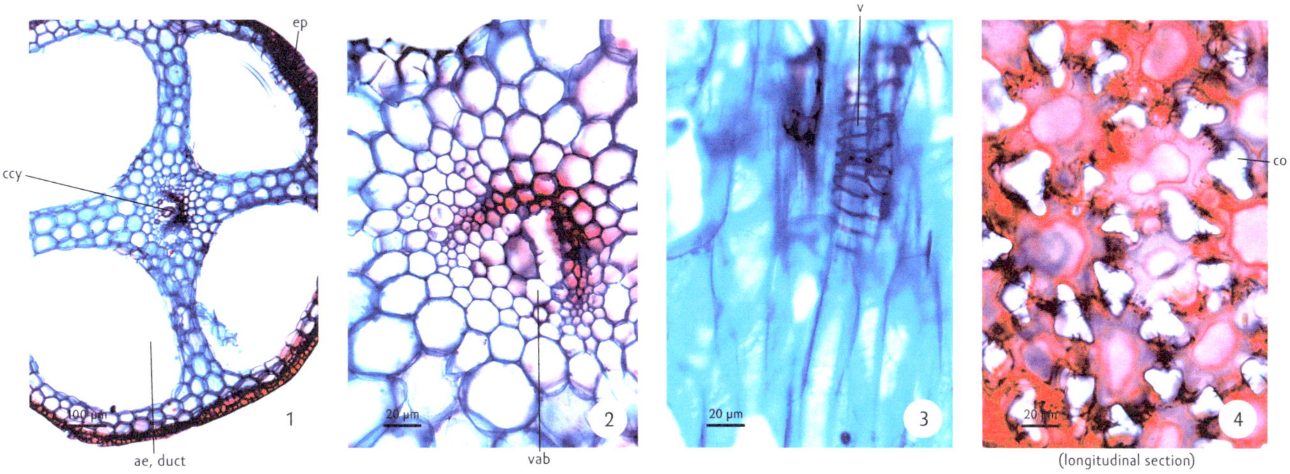

F. H. Schweingruber et al., *Anatomic Atlas of Aquatic and Wetland Plant Stems*, https://doi.org/10.1007/978-3-030-33420-8_6

Lycopodiella inundata (L.) Holub

(syn. *Lycopodium inundatum*)

Marsh clubmoss

Lycopodiaceae

Perennial wetland homosporous clubmoss. Procumbent, dichotomically branched, non-flattened sterile stems up to 20 cm long. Erect, fertile stems (8–10 cm long) with sporangium-bearing leaves weakly differentiated into apical cones, leaves spirally arranged. Procumbent stems dying back quickly behind. Non-branched roots bear numerous root hairs.

Distribution and ecology

Mainly suboceanic Europe and N. America. Scattered rarely in all CE countries, from lowlands to uplands. Wet barren sandy or peaty soils, old shallow sand-pits, margins of peat bogs and fens, very often transiently as a pioneering species after sand or fen extraction on opened, disturbed areas, sometimes also slightly submerged. A strict heliophyte, competitively very weak, quickly retreats due to overgrowing by other herbs. Prefers slightly acid substrates. Ecological requirements mostly unknown.

Anatomy of the stem

The stem has a diameter of 1.5–2 mm (1, 2). Secondary growth is absent.

The stem is composed of a central cylinder, a large cortex and an epidermis (1).

The central cylinder is a plectostele, consisting of irregularly radial arranged tracheids and radial extended phloem groups (1, 2). Tracheids of the xylem are angular (2). Their walls are laterally perforated by scalariform intervessel-pits (3).

Phloem groups contain unlignified sieve elements and parenchyma cells. The plectostele is surrounded by a layer of living unlignified parenchyma cells (round nuclei) and a layer of thin-walled, lignified cells (2). The very large cortex is composed of large unlignified parenchyma cells, large intercellulars (4) and large mucilage conducting ducts (1).

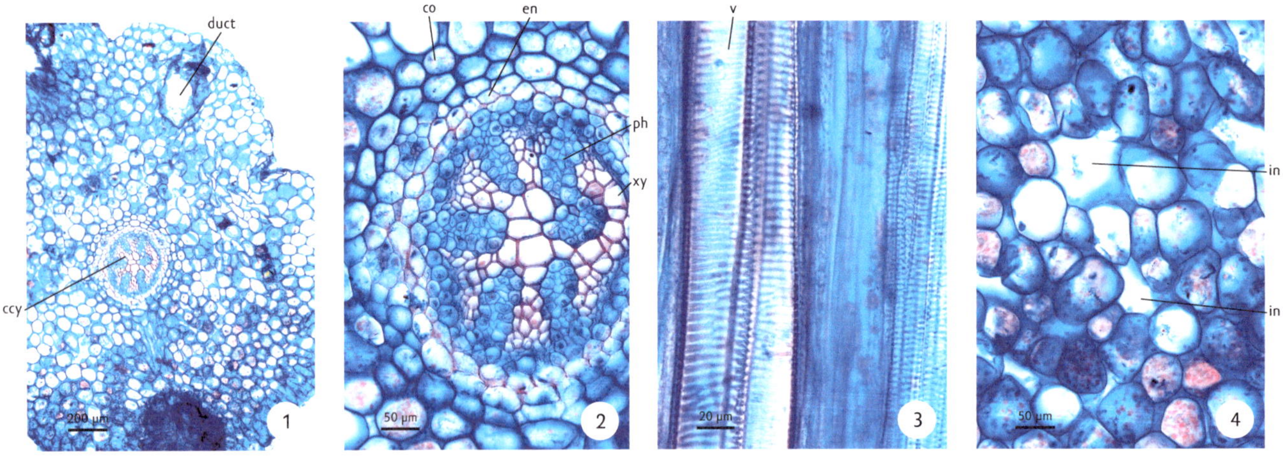

F. H. Schweingruber et al., *Anatomic Atlas of Aquatic and Wetland Plant Stems*, https://doi.org/10.1007/978-3-030-33420-8_7

Marsilea quadrifolia L.

Four leaf clover

Perennial amphibious heterosporous fern. Long belowground rhizome with four-parted leaves resembling clover on a 10–25 cm long tubular petiole, young leaves are spirally curled, roots with long root hairs. Sporocarps ellipsoid, 4–5 mm long on 1–2 cm long stalks, attached to base of petioles.

Distribution and ecology

Cosmopolitan, mainly Mediterranean distribution in Europe. Very rarely in AT, CH, DE, HU and SK, extinct in PL. Shallow banks of standing mesotrophic waters in lakes, fishponds, floodplain pools, on base-rich, neutral or alkaline silty or sandy soils mainly in lowlands, submerged or terrestrial form. Competitively very weak, retreats from eutrophic waters. Ecological requirements mostly unknown.

Anatomy of the petiole

The upright submerged petiole has a diameter of <1 mm (1). Secondary growth is absent.

The petiole is composed of a closed concentric vascular bundle, a large aerenchymatic cortex and an epidermis (1). The xylem consists of two radial groups of lignified vessels, which are embedded in parenchyma and phloem cells. Sieve tubes, companion cells and parenchyma cells are hardly to distinguish in transverse sections. The central cylinder and the cortex are separated by a thin-walled endodermis and a thicker-walled pericyle (2, 3). Large intercellulars are separated laterally by radial strips of parenchyma cells (1). The petiole is externally surrounded by an epidermis and a hypodermis.

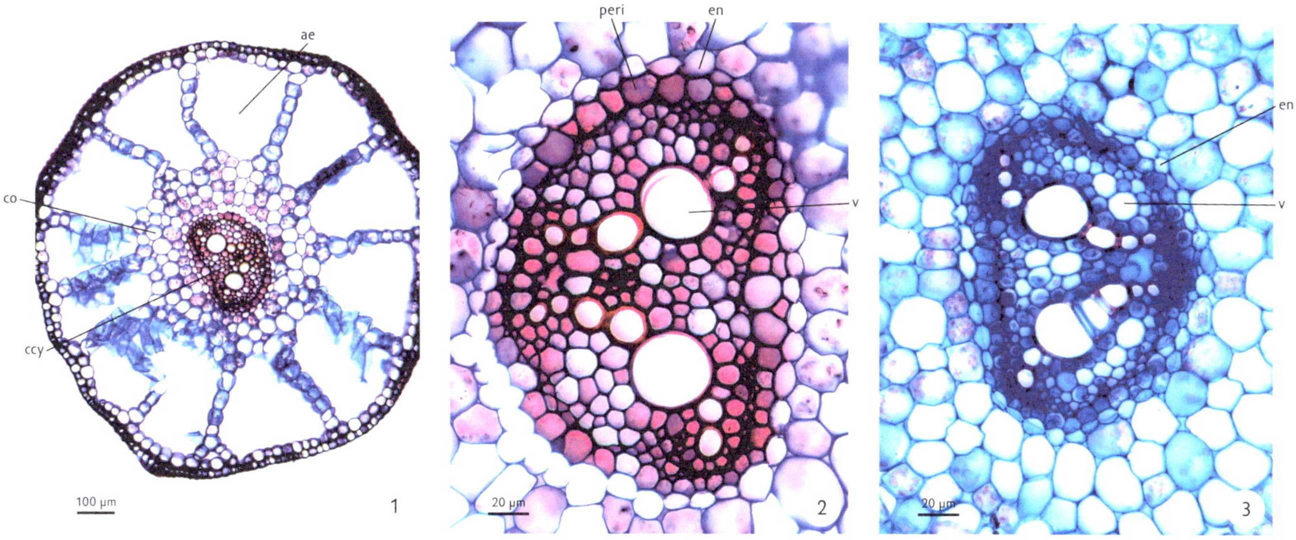

F. H. Schweingruber et al., *Anatomic Atlas of Aquatic and Wetland Plant Stems*, https://doi.org/10.1007/978-3-030-33420-8_8

Pilularia globulifera L.

Pillwort

Marsileaceae

Perennial amphibious plant. Rhizome long, thin, non-branched roots with root hairs, needle-shaped erect leaves 4–20 cm long borne singly, spirally coiled when young. Special short leaves bear heterosporous, globular, 3–5 mm large sporocarps. It overwinters in rhizomes with shortened leaves.

Distribution and ecology

Subatlantic distribution. Very rarely in CZ, PL, DE and CH. Oligo- to mesotrophic, also slightly dystrophic standing waters in fishponds, lakes, reservoirs, usually on acidic peaty or silty substrates near the water surface, terrestrial or slightly submerged. Competitively very weak, retreats from eutrophic waters, ecological requirements mostly unknown.

Anatomy of the basal part of a leaf

The leaf has a diameter of <1 mm (1). Secondary growth is absent.

The leaf is composed of a closed concentric vascular bundle, a large cortex and an epidermis. (1). The xylem consists of two groups of lignified vessels, which are separated by a few parenchyma cells. The phloem is composed of a round group of sieve tubes and companion cells. The central cylinder and the cortex are separated by an endodermis (2). Large intercellulars are separated laterally by radial strips of parenchyma cells (1). The leaf is externally surrounded by an epidermis and a hypodermis (1).

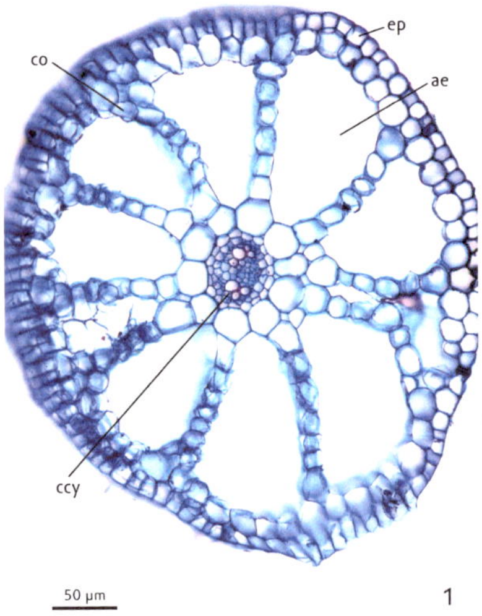

1

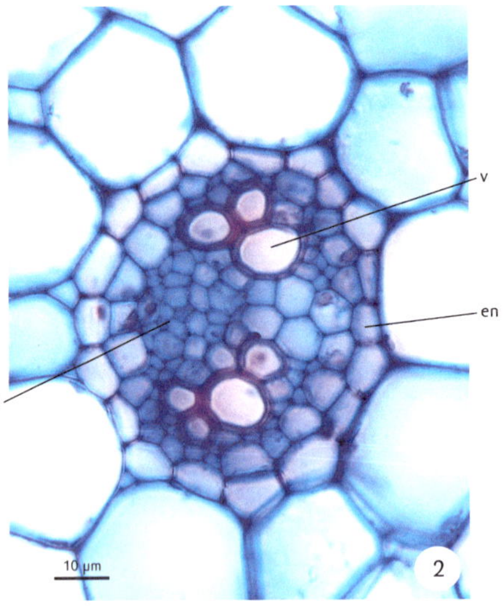

2

Salvinia natans (L.) All.

Floating fern

Annual, free-floating rootless heterosporous fern, overwinters only in the form of spores in sporocarps forming under senescing leaves. Shoots 3–5 cm long bear whorls of three leaves: a pair of floating leaves 12–20 mm long and a submerged, thin and filamentous leaf 3–6 cm long with numerous hairs reminding a root (rhizophyll), floating leaves densely covered by 1–1.5 mm long hairs repelling water. Three-mm large sporocarps contain either small dark brown microspores or larger whitish macrospores, membraneous prothallium ca. 2 mm large, terrestrial form known.

Distribution and ecology

Euro-Siberian and South Asian distribution. Scattered in all 7 CE countries, mainly in lowlands. Standing waters in fishponds, lakes, dam reservoirs, backwater pools, oxbows, fen lakes, and canals, at water depth of 0.05–2 m. Tolerates slightly acidic to alkaline, hard water (pH 6.5–8, total alkalinity 1.5–4 meq/l), clear or slightly dystrophic waters with organic or clayish bottom, prefers mesotrophic to slightly eutrophic waters on neutral or alkaline base-rich soils, thermo- and photophilous.

Anatomy of the stem

The stem has a diameter of 1 mm (1). Secondary growth is absent.

The stem is composed of a closed concentric reniform vascular bundle, a large aerenchymatic cortex and an epidermis (1). The xylem consists of an arc of lignified vessels, which are separated by a few parenchyma cells. The phloem is composed of sieve tubes and small companion cells. The central cylinder and the cortex are separated by a lignified endodermis (2). Large intercellulars are separated laterally by radial strips of parenchyma cells (1). The stem is externally surrounded by an epidermis and a hypodermis (1).

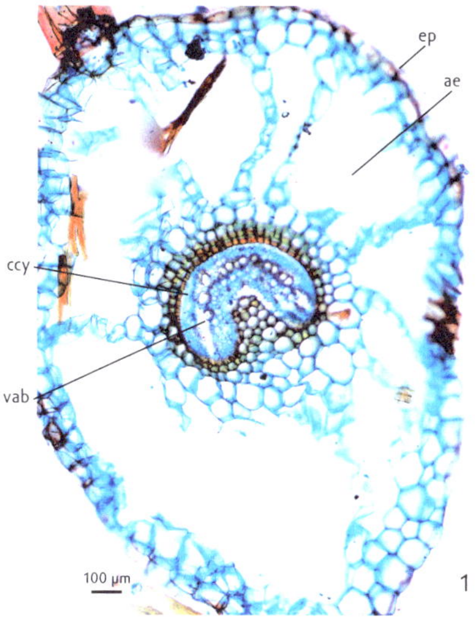

1

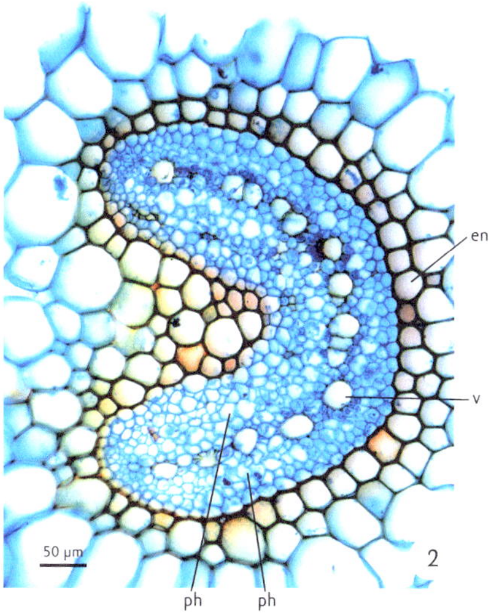

2

F. H. Schweingruber et al., *Anatomic Atlas of Aquatic and Wetland Plant Stems*, https://doi.org/10.1007/978-3-030-33420-8_9

Thelypteris palustris Schott

Marsh Fern Thelypteridaceae

Perennial wetland fern, deciduous, underground with creeping rootstock. Soft, pale green 2-pinnate leaves arise singly, they are 30–100 cm high, stalks almost hairless. Spore cases rounded, close together.

Distribution and ecology

Euro-Siberian boreo-temperate element; widely naturalised outside its native range. Native in all CE countries. Typical of marshes, fens, mires, wet grasslands, pastures, wet woodlands, springs and tall-herb vegetation in mountains. Persistent spore bank.

Anatomy of the petiole

The lower part of the petiole has a diameter of 3 mm (1, 2). Secondary growth is absent.

Two to four oval vascular bundles are surrounded by a thin-walled parenchymatic tissue (3, 4). Vascular bundles are composed of a xylem with angularly lignified thin-walled tracheids with a diameter up to 100 µm and an unlignified, small parenchyma cells (2, 3). Tracheids have scalariform pits. The surrounding phloem consists of angular, unlignified sieve elements and companion cells. Vascular bundles are surrounded by a layer of living, unlignified, thin-walled parenchyma cells and an endodermis (4). Endodermis cell walls are brown stained and fairly thick-walled at its outer and inner part and extremely thin-walled laterally (4). Therefore, most vascular bundles do not resist the sectioning process (1). Cells of the peripheral part of the petiole are brown stained and thick-walled (5).

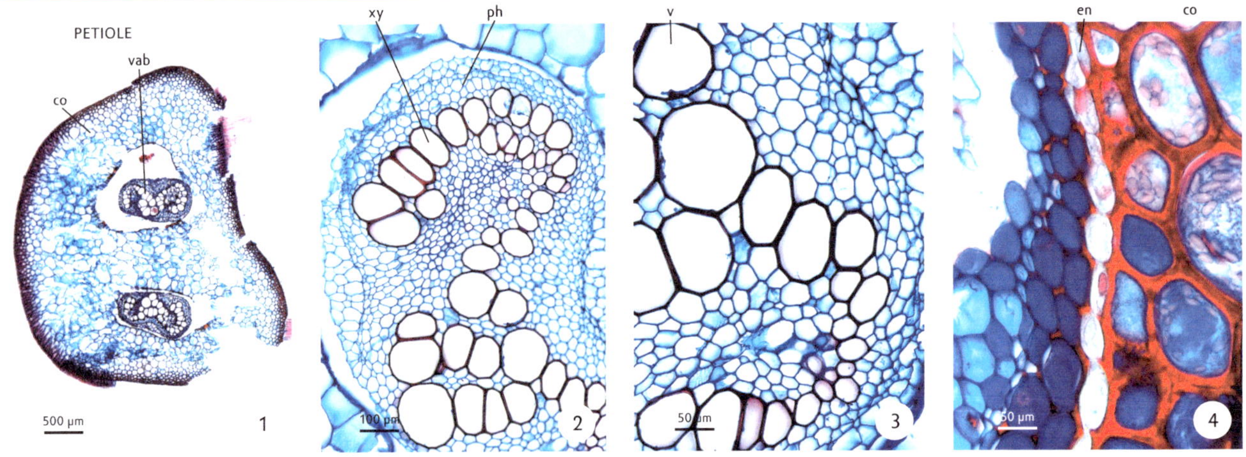

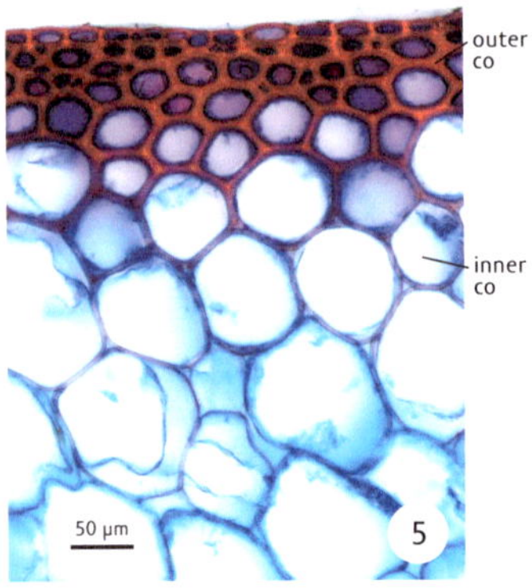

F. H. Schweingruber et al., *Anatomic Atlas of Aquatic and Wetland Plant Stems*, https://doi.org/10.1007/978-3-030-33420-8_10

Anatomical descriptions of Dicotyledons

Beta vulgaris subsp. *maritima* (L.) Arcang.

(syn. *Beta maritima*)

Amaranthaceae
Sea Beet

Biennial or perennial wetland herb. Stem grows during the second vegetation period, procumbent to erect, the upper part branched, 30–80 cm high. Basal leaves long-petiolate, blade ovate, up to 12 cm long and 8 cm wide. Small pentamerous flowers curled up in a ball, greenish to pinkish. Fruit is a hard achene, often enclosed by swollen perianth.

Distribution and ecology

Mediterranean-Atlantic distribution, also W. Asia. Native to DE and PL. Occurs along the coast, on coastal rocks and seabird cliffs, saltmarshes, on sand beaches, in sandy waste places near sea, also on salty soils inland. Prefers nutrient-enriched sites. Halophyte. Gynodioecious.

Anatomy of the root collar

Root collar with a diameter of 4–5 mm. Maximum age unknown. Annual rings are absent.

Presented is a specimen from a sea shore.

The root collar contains concentric-arranged successive cambia, a few xylem and phloem belts and a cortex (1). The xylem consists of solitary vessels, few fibers and much parenchyma (2). Vessels have a diameter of 20–30 µm (2) and a length of 70–100 µm (3). Intervessel pits are large, oval to scalariform and perforation plates are simple (3). Parenchyma is pervasive (1, 2). Fibers occur in groups but are rare (2). Confluent rays are discontinuous (1). Axially elongated groups of crystal sand are tangentially arranged, parallel to vessel groups (5).

The phloem consists of groups of sieve tubes and parenchyma cells (2). Intercellulars in the cortex are small. Crystal sand is frequent (5). The phellem consists of irregularly formed, thin-walled cells (4).

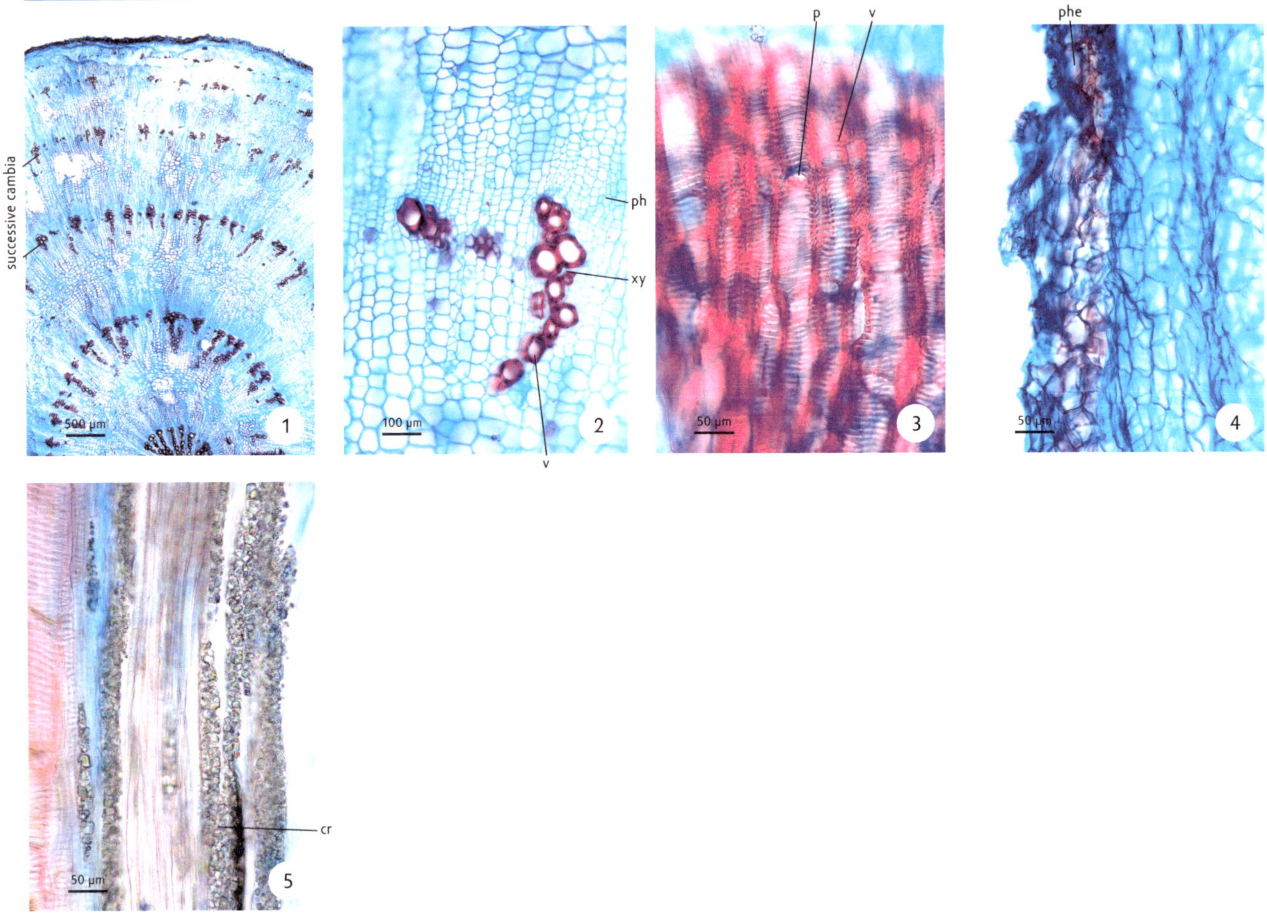

F. H. Schweingruber et al., *Anatomic Atlas of Aquatic and Wetland Plant Stems*, https://doi.org/10.1007/978-3-030-33420-8_11

Chenopodium chenopodioides (L.) Aellen

(syn. *Chenopodium botryodes*)

Amaranthaceae
Saltmarsh Goosefoot

Annual wetland herb. Stem procumbent to ascending, richly branched, 15–50 cm high. Leaves fleshy, triangular, green to reddish on both surfaces, young leaves farinose. Inflorescence is a small, dense cluster of tiny flowers. Flowers both bisexual and female, perianth markedly thin to membraneous. Fruit is a circular to ovoid achene.

Distribution and ecology

W. and S. Europe, Middle and SW Asia. Native to all CE countries except CH, very rare in AT, CZ and SK. Occurs along the Atlantic and Mediterranean coast, inland on salty marshes, waste places and sandy margins of fishponds. Prefers wet soils with high concentration of salts. Obligate halophyte.

Anatomy of the root collar

Polar root and the annual shoot with a diameter of 4–5 mm. Maximum age 1 year.

Presented is a plant from a saline place.

The plant contains an intensively lignified xylem with successive cambia, a cortex and a phellem (1). Successive cambia are arranged in concentric continuous bands in adult and in concentric vascular bundles in juvenile plants. The xylem of the laterally weakly limited vascular bundles consists of vessels in short radial rows and parenchyma (2). Vessels have a diameter of 30–40 µm (2) and a length of 50–100 µm (3). Intervessel pits are round and perforation plates are simple (3). Parenchyma is paratracheal (2). Fibers are thin- to thick-walled. Rays are probably absent (2). Unlignified elongated cells might be axial parenchyma cells or rudimentary rays (4).

The phloem of the vascular bundles is fairly uniform. Sieve tubes and parenchyma cells cannot be properly differentiated (5). The cortex consists of large parenchyma cells and intercellulars are absent (5). The phellem consists of several layers of polygonal cells (1). Crystal sand occurs in the phellem. Crystal sand is frequent in all parts of the shoots (6, 7).

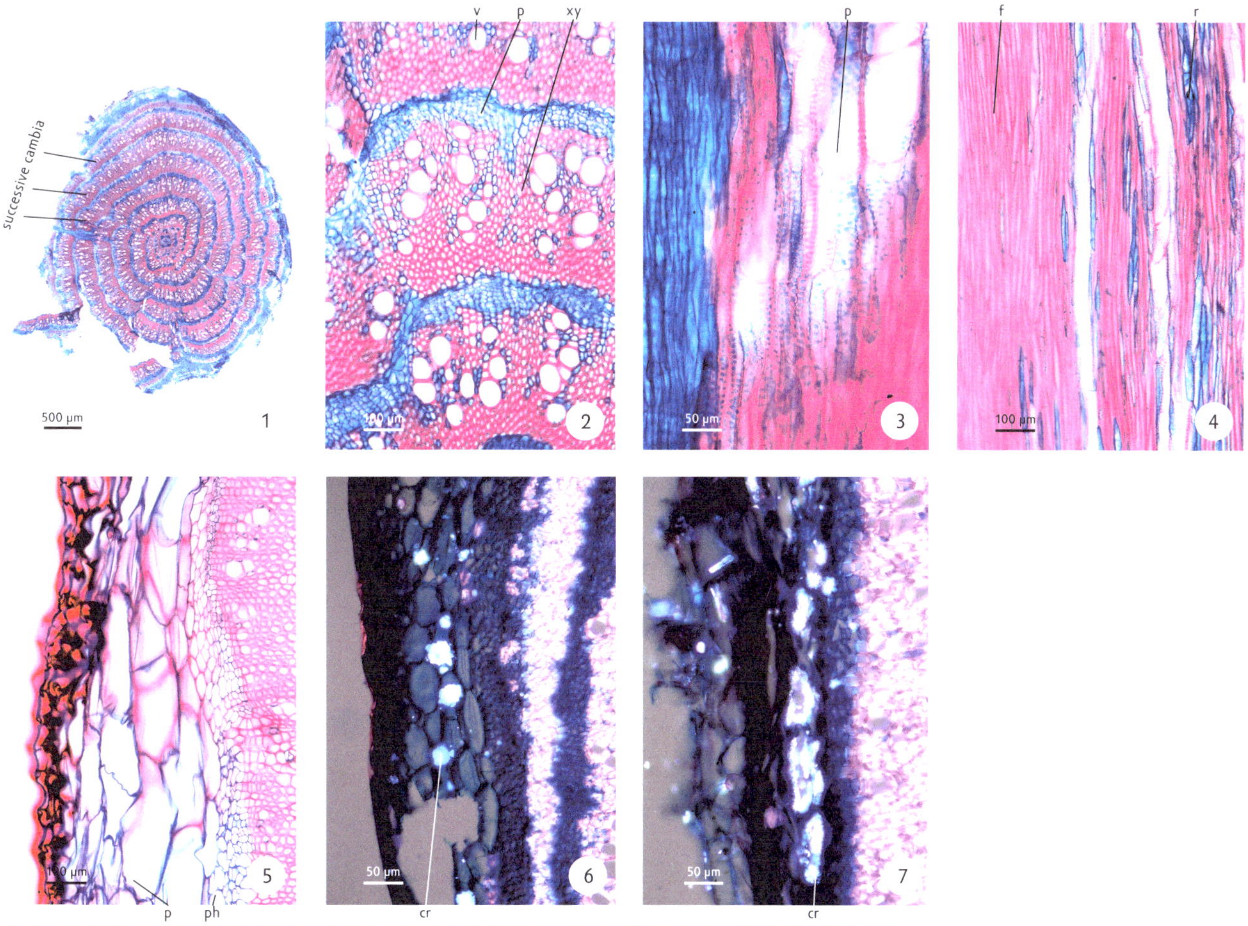

Salicornia europaea agg. L. **Common Glasswort** Amaranthaceae

Annual, rarely biennial wetland herb. Stem succulent, erect or procumbent, glaucous or grass-green, reddish above, richly branched, 10–35 cm high. Dark green when young, becoming yellow-green and ultimately flushed pink or red. Leaves much reduced, constitute only a rim at each stem node. Inflorescence is a terminal spike with central flower distinctly larger than the lateral two. Seeds elliptic with hairy brownish-green surface. Taxonomically complicated complex.

Distribution and ecology

Circumpolar wide-temperate element. Native to DE and PL around the coastline, and rarely in inland salines in AT, DE and HU, extinct in CZ. Typical of saline, brackish or alkaline substrates on sandy or muddy saltmarshes, tidally inundated duneslacks or inland salines. Tolerates hypersaline soil conditions, regular flooding and prolonged submergence. A-V mycorrhiza.

Anatomy of the root collar

Root collar with a diameter of 4–5 mm. Maximum age 1 year.

Presented is a plant from a saline place.

The root collar contains an intensively lignified xylem with successive cambia arranged in single vascular bundles (diffuse foraminate), a cortex and a phellem (1). The xylem of the vascular bundles consists of vessels in short radial rows and parenchyma (2). Vessels have a diameter of 15–20 µm (2) and a length of 70–100 µm (3). Intervessel pits are round and perforation plates are simple (3). Parenchyma is paratracheal (2). Fibers are thin- to thick-walled. Rays are absent (2, 4).

The phloem of the vascular bundles is fairly uniform. Sieve tubes and parenchyma cells cannot be properly differentiated (2). Intercellulars in the cortex are frequent (2, 5). The phellem consists of several layers of polygonal cells (5). Crystals are absent.

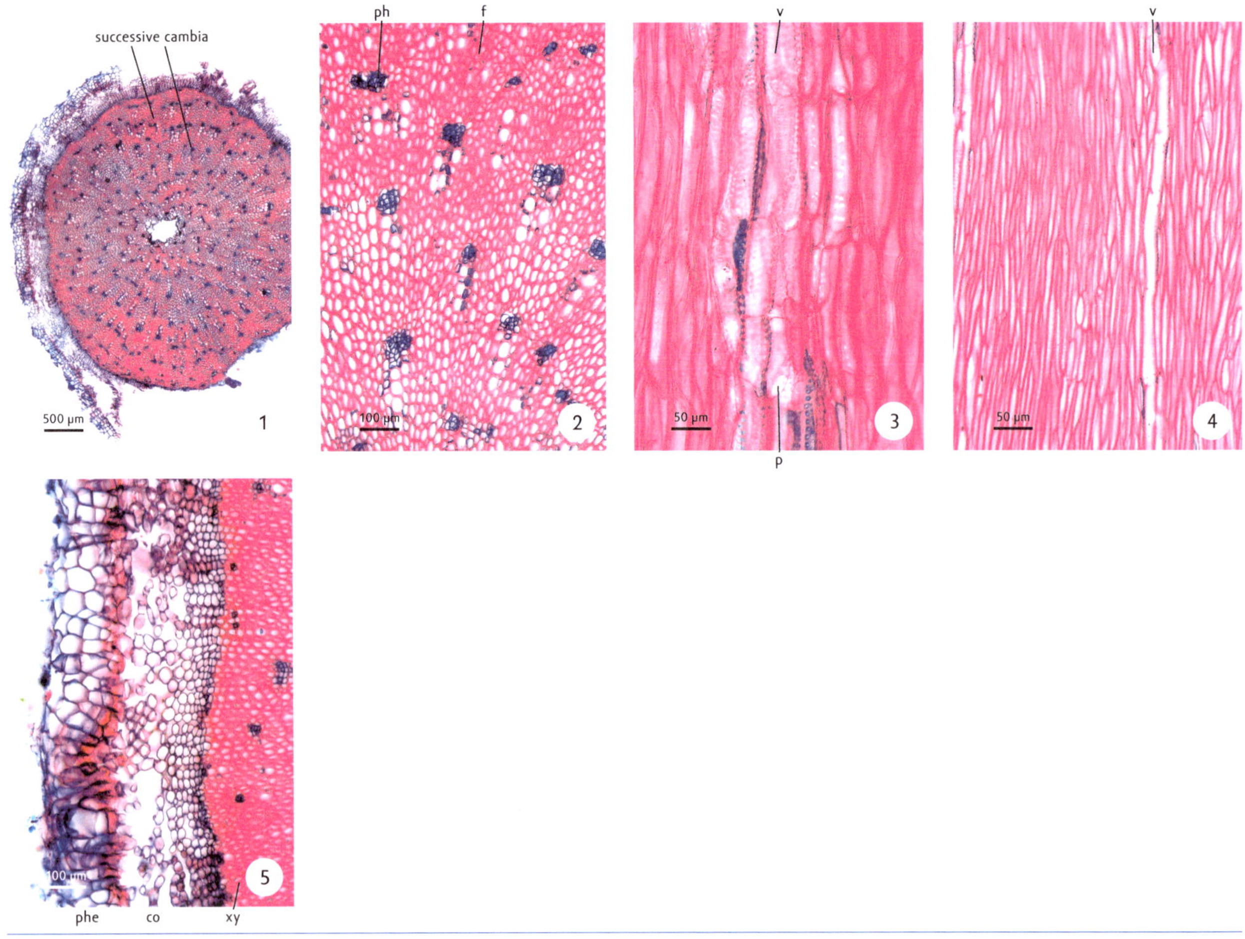

Apium nodiflorum (L.) Lag.

Fool's-water-cress

Apiaceae

Perennial emergent herb. Stem hollow, suberect to procumbent, rooting at lower nodes, up to 1 m high. Leaves unipinnate. Inflorescence is a compound umbel, opposite to leaf, with white flowers, bracts and sepals absent. Fruit has 5 prominent and thick ridges.

Distribution and ecology

Euro-Siberian southern-temperate element; widely naturalised outside its native range. Native in CH and DE, very rare in PL, absent in AT, CZ, HU and SK. Nutrient-enriched sites in lowlands. In shallow water in streams, ditches and marshes, and on seasonally exposed mud at the edges of fishponds, lakes, rivers and canals. Able to saturate the substrate with oxygen.

Anatomy of the basal part of the annual shoot

Round stem with a diameter of 10 mm. Maximal age 1 year.

A circle of vascular bundles and a cortex surround a large pith (1). Vascular bundles are separated laterally by vessel free zones (2, 3). The xylem in rhizomes consists of vessels with a diameter of 50–80 µm and a length of 100–140 µm. Vessels are solitary and in short radial rows (3). Intervessel pits are horizontally enlarged and perforation plates are simple (4). Fibers are absent (3). Axial parenchyma is pervasive (3). Rays are very large between the vascular bundles and absent within them (3).

The small phloem is fairly homogeneous (2, 5). The cortex consists of thin-walled unlignified parenchyma cells (1). Crystals are absent.

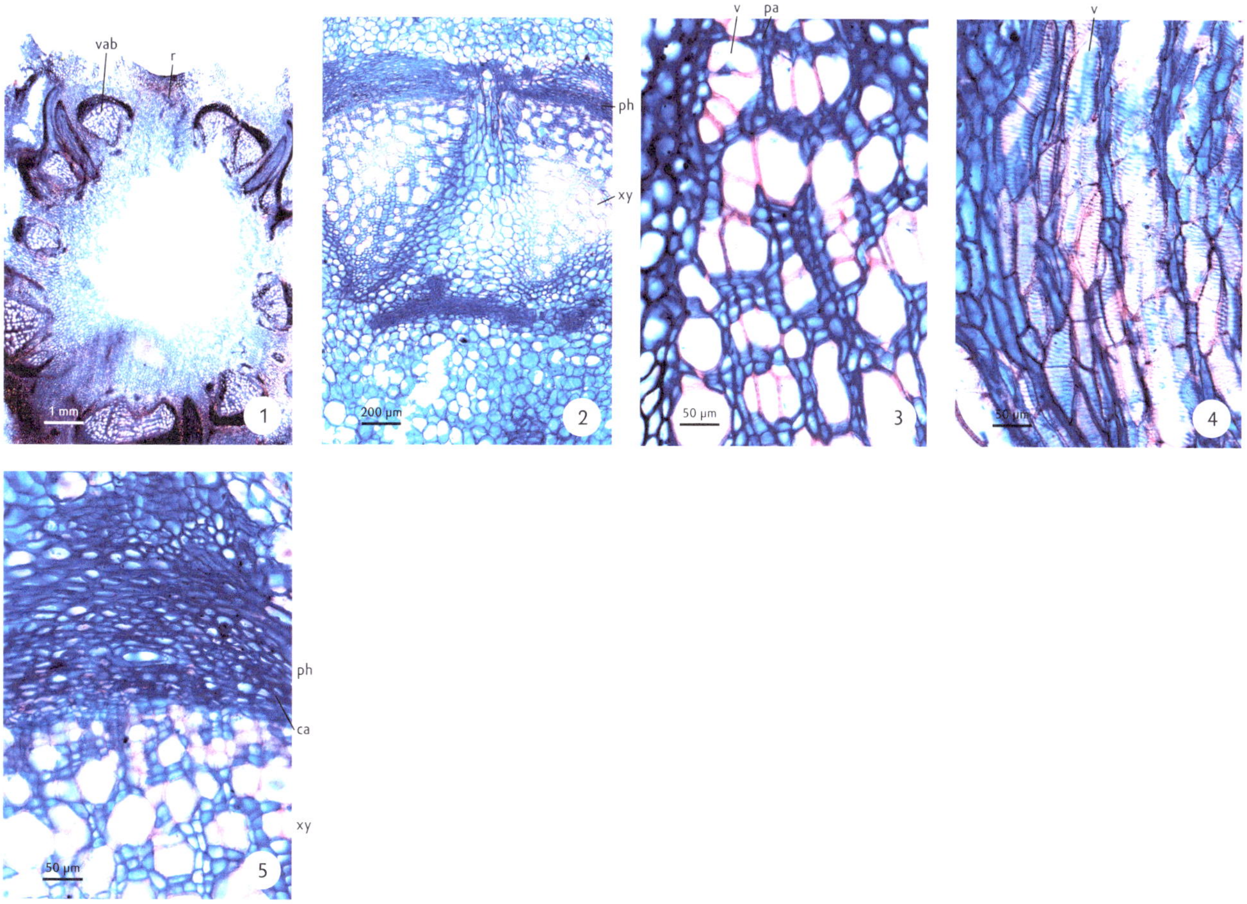

F. H. Schweingruber et al., *Anatomic Atlas of Aquatic and Wetland Plant Stems*, https://doi.org/10.1007/978-3-030-33420-8_12

Berula erecta (Huds.) Coville

Lesser Water-parsnip Apiaceae

Perennial submerged or emergent herb. Stem stoloniferous, hollow, decumbent to erect, up to 1 m high. Lower leaves pinnate with 5–10 pairs of ovate and serrate leaflets, terminal leaflet often trifid. Inflorescence is a compound umbel, terminal and in leaf axils, short-stalked with white flowers, several bracts and bracteoles, sepals not persistent. Fruit is slightly laterally compressed, glabrous.

Distribution and ecology

European temperate element; also in C. Asia and N. America. In all CE countries mainly in lowlands. Prefers slowly flowing mesotrophic and eutrophic waters. Occurs as a submerged aquatic plant in rivers and streams, and as an emergent species at the edges of lakes, rivers, ditches and canals. Vegetative spread by short-lived stolons or rhizomes.

Anatomy of the basal part of the rhizome and the annual shoot

Round stem with a diameter of 10 mm. Maximal age 2 years.

A circle of vascular bundles and a cortex surround a very large pith. The xylem/phloem belt is continuous in the rhizome (2) and is discontinuous in the annual shoot (1). The central cylinder is surrounded by a poorly differentiated pericycle/endodermis in the annual shoot (2), but absent in the rhizome (3). Vascular bundles with different sizes are separated laterally by vessel-free zones in annual shoots (2) and by large rays in the rhizome (3). The xylem in rhizomes consists of vessels with a diameter of 40–60 µm and a length of 120–200 µm. They are arranged in short radial rows (4). Intervessel pits are round to slightly horizontally enlarged and plates are simple (5, 6). Fibers are thin-to thick-walled (4). Pits in fibers are extremely rare. Axial parenchyma is probably absent (4). Rays are very large between the vascular bundles and indistinct within them (4). Large oil ducts occur in the position of the primary xylem (7) and small ducts in the cortex and the pith (2).

The phloem is fairly homogeneous. Sieve tubes collapse in the second year (4).

Aerenchymatic tissues occur in the pith and the cortex. The very thin-walled, unlignified pith cells in the pith burst by the expansion of the stem (2, 3). Large aerenchyma-like intercellulars occur in the cortex (2, 3). The peripheral part of the annual shoot consists of a thin-walled epidermis, a hypodermis and groups of collenchyma (8). Crystal sand and small crystal druses occur in parenchyma cells of the cortex.

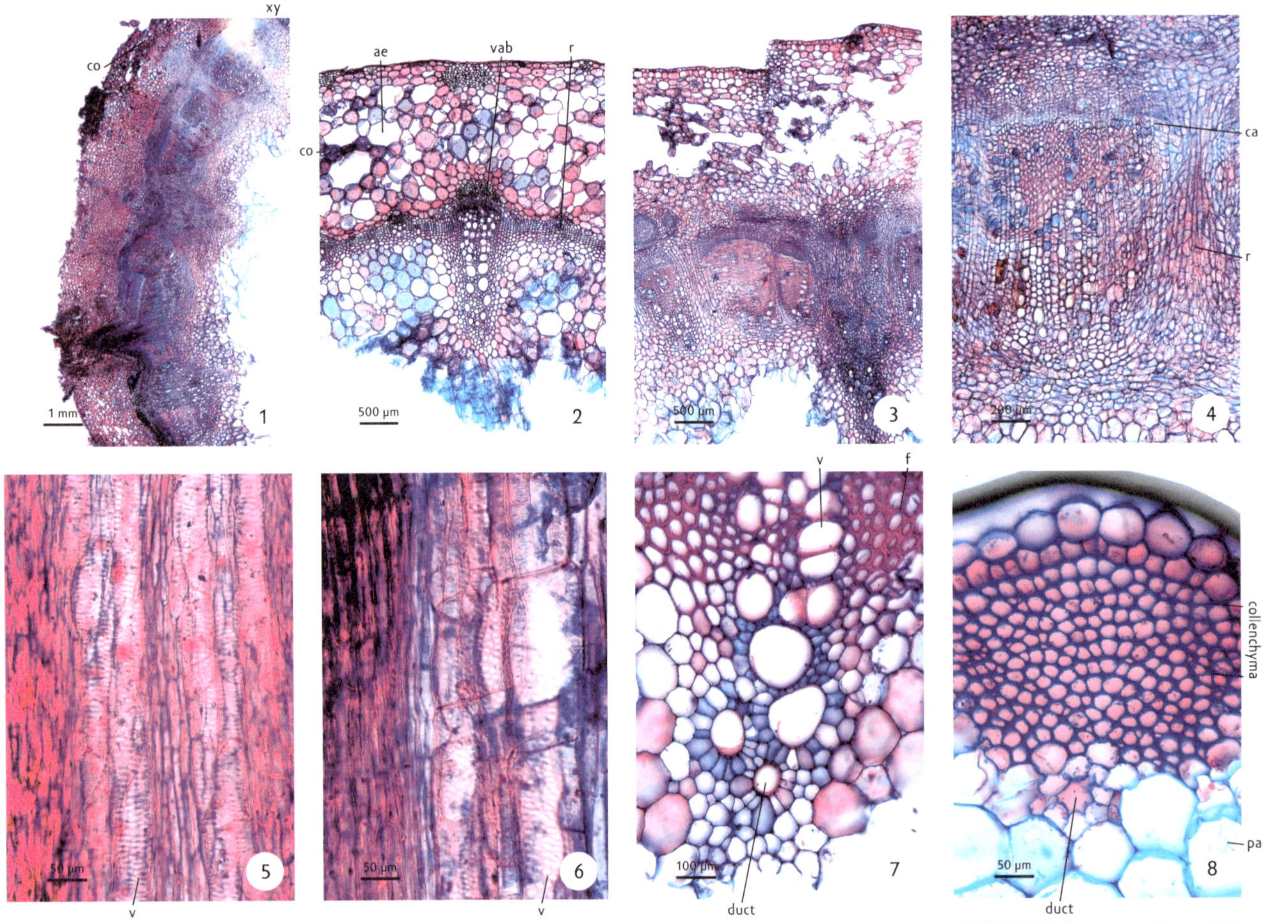

Cicuta virosa L.

Cowbane

Perennial emergent herb. Stem hollow, ridged, erect, up to 1.5 m high. Leaves 2–3 times pinnate, with narrowly elliptic to linear-lanceolate segments and long hollow stalks. Inflorescence is a terminal compound umbel, with white flowers, bracts usually absent, bracteoles numerous, sepals persistent. Fruit is globular with blunt ridges.

Distribution and ecology

Eurasian boreo-temperate element. In all CE countries mainly in lowlands. In shallow water on the margins of standing or slowly flowing water (lakes, fishponds, rivers, streams, ditches) or in deeper water on floating mats of vegetation. Also in tall-herb fens and alder-carrs. Short-term persistent seed bank.

Anatomy of the rhizome and the annual shoot

Round stem with a diameter of 5–10 mm. Maximum age 2 years.

Annual shoot: A circle of vascular bundles and a cortex surround a very large, hollow central zone (1).

The rhizome consists of a central unlignified parenchymatic tissue with a few vessels. A continuous xylem belt forms the outer part of the rhizome (2). Vascular bundles with different sizes are separated laterally by vessel-free fibrous zones in annual shoots (1) and by large rays in the rhizome (2). The xylem belt is separated from the cortex by a cambium (1) or by a phloem (2). The xylem is divided in a parenchymatic central (2) and a fibrous peripheral zone (2). The xylem in rhizomes consists of vessels with a diameter of 50–70 µm (3) and a length of 120–200 µm (4). Vessels are arranged solitary and in short radial rows (3). Intervessel pits are round to slightly horizontally enlarged in the external xylem belt (4) and scalariform in vessels of the central part. Perforation plates are simple (4). Fibers with small, slit-like pits are thin-to thick-walled (4). Axial parenchyma is paratracheal in the external part (4) and pervasive in the center (2). Rays in rhizomes are 4–6-seriate (5), partially confluent and consist of upright cells (6). Oil ducts occur centripetally in the primary xylem (7) and in the cortex (8). The phloem is fairly homogeneous (8). Aerenchymatic tissues occur in the cortex of annual shoots where they form irregular tubes (1). The aerenchyma in the phloem is represented by irregular splits (intercellulars) between bent radial strips of parenchyma cells (8). Crystal sand occurs in parenchyma cells of the cortex and the pith.

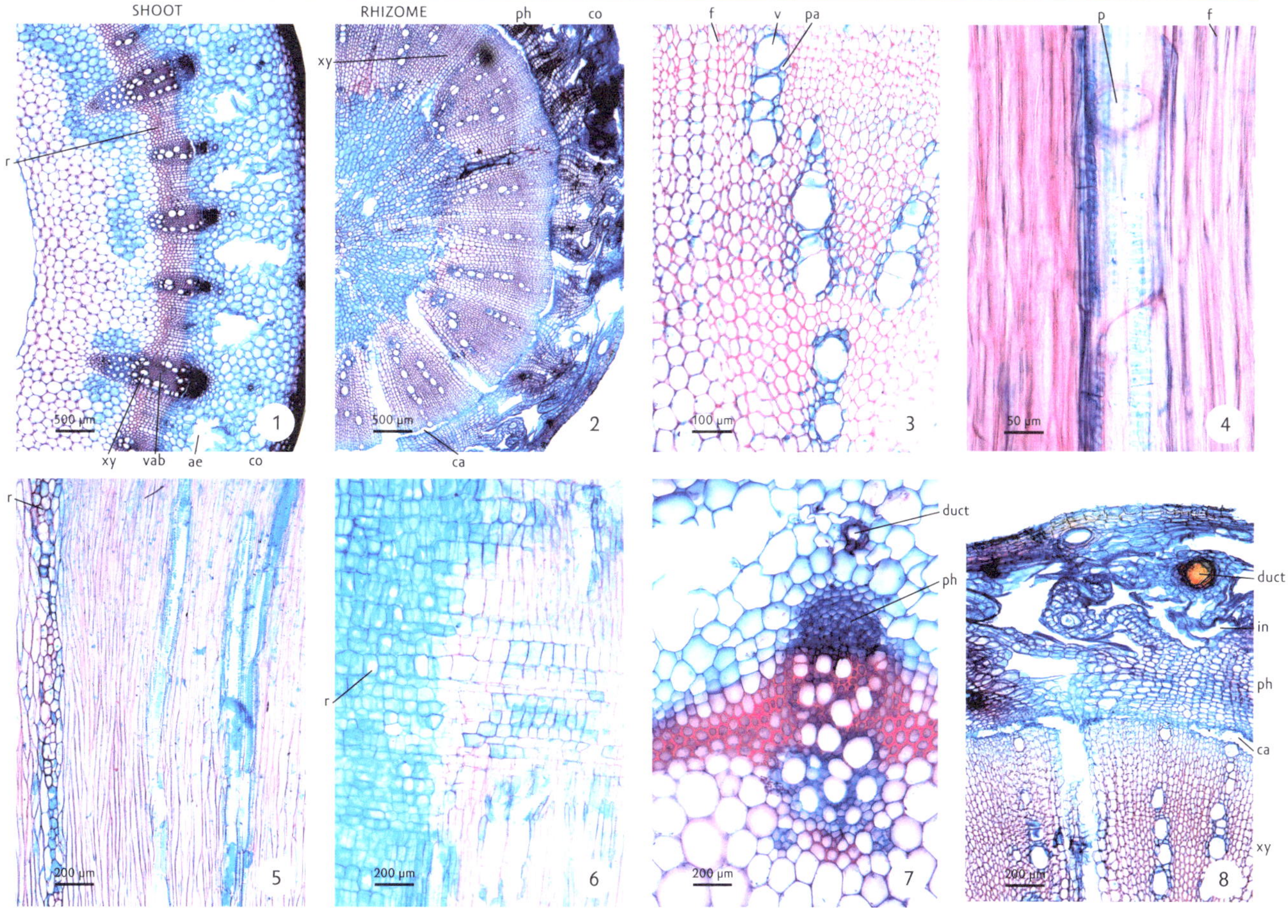

Cnidium dubium (Schkuhr) Thell.

Snowparsley Apiaceae

Biennial to perennial wetland herb. Stem hollow, 30–100 cm high, rounded below, angularly grooved above, little branched. Leaves 2–3 pinnate, with narrowly elliptic segments. Inflorescence is a terminal compound umbel with white flowers, bracts usually absent, bracteoles numerous. Fruit has 5 prominent winged ridges.

Distribution and ecology

Eurasian distribution. In all CE countries except CH, in lowlands. Typical of continental flooded meadows, wet and peaty meadows and edges of lowland forests. River corridor species. Germination stimulated by high irradiances and by daily temperature fluctuations.

Anatomy of the basal part of the annual shoot and the transition between the root and the annual shoot

Round root collar with a diameter of 4–6 mm. Maximum age 2 years.

Shoots and rhizomes are surrounded by a circle of vascular bundles and a large phloem and a cortex (1).

The shoot consists of a central unlignified parenchymatic tissue which is surrounded by radial vascular bundles (1, 2). Vascular bundles consist of vessels, pervasive and paratracheal parenchyma; fibers are absent (2). Vessels have a diameter of 40–60 μm (3) and a length of 100–150 μm (3) and are arranged primarily solitary (3). Intervessel pits are horizontally enlarged and scalariform. Perforation plates are simple. Axial parenchyma is pervasive (2). Vascular bundles are laterally separated by large, unlignified confluent rays (1). Rays are >10-seriate between the bundles (5). Rays are absent within the bundles.

The phloem consists of tangentially and radially arranged irregular groups of sieve elements within the parenchyma (4–6). Irregularly formed intercellulars occur in the cortex (1). Oil ducts occur in the pith, in the phloem and cortex (5). The phellem consists of some layers of rectangular cells.

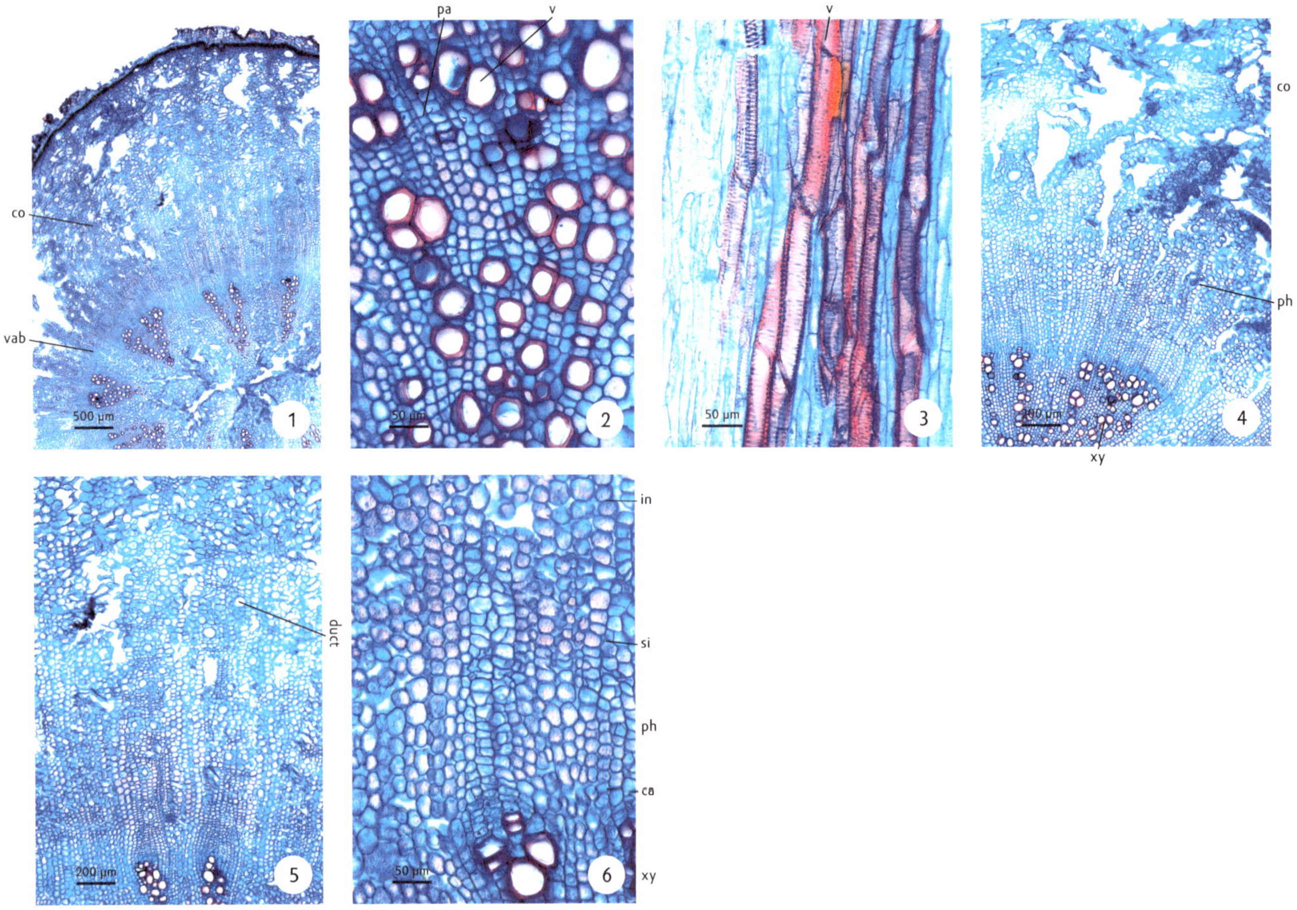

Laserpitium prutenicum L.

Broad-leaved Sermountain

Usually biennial, monocarpic wetland herb. Stem solid, erect, deeply grooved, up to 1 m high. Leaves 2–3 pinnate. Inflorescence is a terminal compound umbel with white flowers, bracts and bracteoles numerous, sepals inconspicuous. Fruit has 4 winged ridges.

Distribution and ecology

European temperate element. In all CE countries from lowlands to foothills. Moist fens or peaty meadows, with fluctuating water table on base-rich soils.

Anatomy of the basal part of the annual shoot and the transition between the root and the annual shoot

Round ribbed stem with a diameter of 5–10 mm. Maximum age 1 year.

Annual shoot: A circle of vascular bundles and a cortex surround a very large, central zone (1). Vascular bundles of different sizes are separated laterally by vessel-free fibrous zones (1).

The rhizome consists of a central primarily unlignified parenchymatic tissue with few vessels and groups of fibers (2). Vessels are solitary arranged and have a diameter of 30–45 µm (3) and a length of 60–80 µm (4). Intervessel pits are round to slightly horizontally enlarged in the external xylem (4) and scalariform of the central parenchymatic part (5). Perforation plates are simple (4). Fibers with small, slit-like pits are thin-to thick-walled (3, 4). Axial parenchyma is paratracheal in the external part and pervasive in the center (2, 3). Rays are 1–4-seriate (6).

The phloem consists of parenchyma cells and irregularly distributed small groups of sieve tubes (6). Rays are partially dilated. Oil ducts occur in the phloem and cortex and in the pith (7).

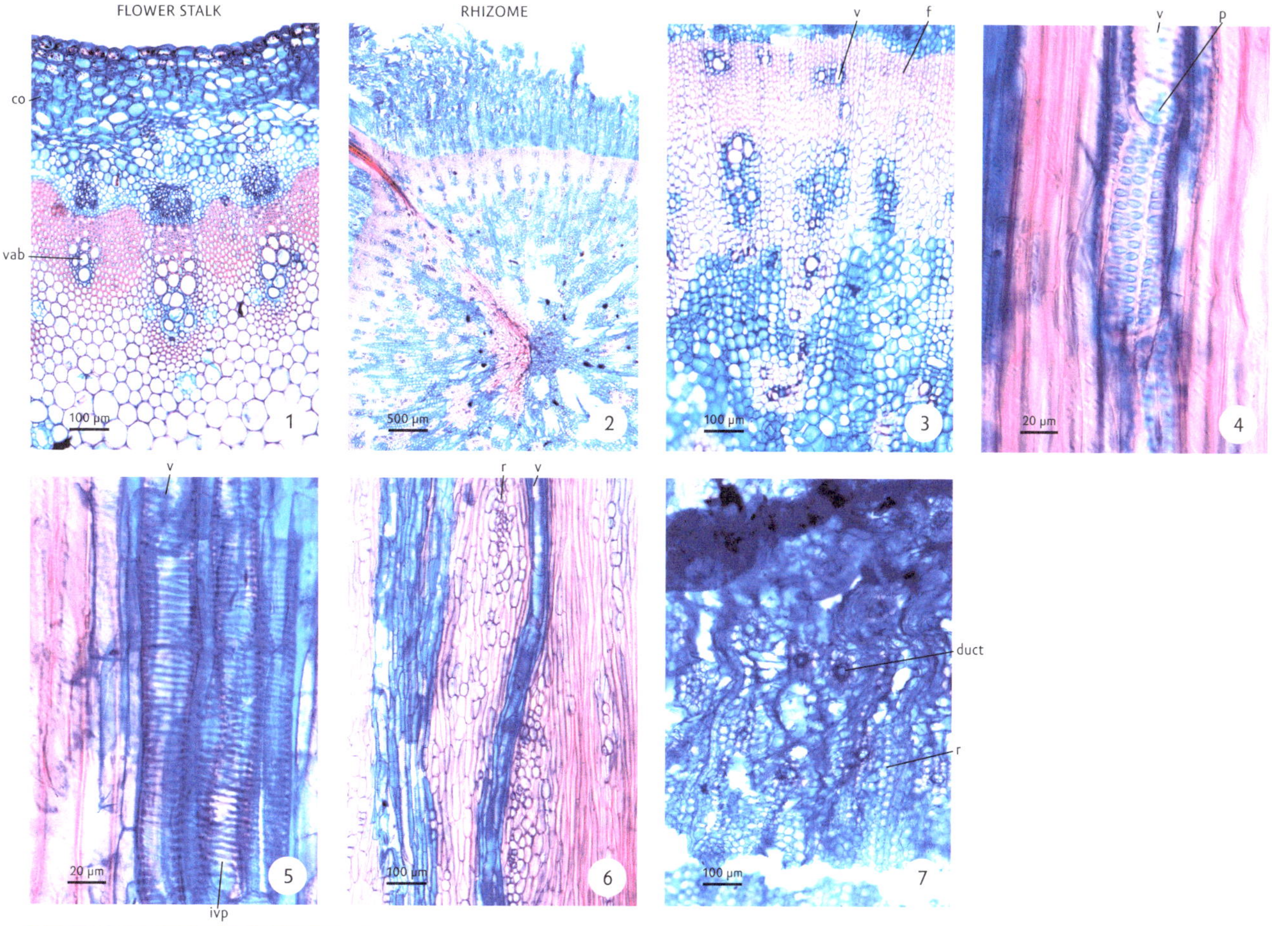

Oenanthe aquatica (L.) Poir.

Fine-leaved Water-dropwort

Annual to biennial emergent herb. Stem hollow, grooved, erect to ascending, up to 1.5 m high. Leaves 2–4 pinnate, with linear to ovate segments. Inflorescence is a compound umbel, terminal and in leaf axils, short-stalked. Flowers white, bracts usually absent, bracteoles numerous, sepals persistent. Fruit has 5 ribs on dorsal side.

Distribution and ecology

Euro-Siberian temperate element; also in E. Asia. In all CE countries from lowlands to foothills. Usually occurs on deep, muddy, often eutrophic substrates in shallow fishponds and ditches, with fluctuating water table. Also in open vegetation by streams and rivers, in marshes and seasonally flooded depressions. The reproduction on denudated bottoms. Non-dormant and light dependent seeds.

Anatomy of the basal part of the annual shoot

Round stem with a diameter of 2–3 mm. Maximum age 1 year (1, 2).

Isolated vascular bundles are concentrically arranged (1, 2).

The xylem of the vascular bundles consists of unlignified parenchyma cells and lignified vessels. Radially arranged, angular vessels have a diameter of 20–30 µm (2). Vessel length not observed. Perforations are simple and intervessel pits scalariform (3). Axial parenchyma is pervasive. Large parenchymatic zones between vascular bundles (rays) (1, 2).

The phloem is homogeneous; parenchyma and sieve tubes cannot be distinguished (4). The cortex contains large intercellular spaces (2).

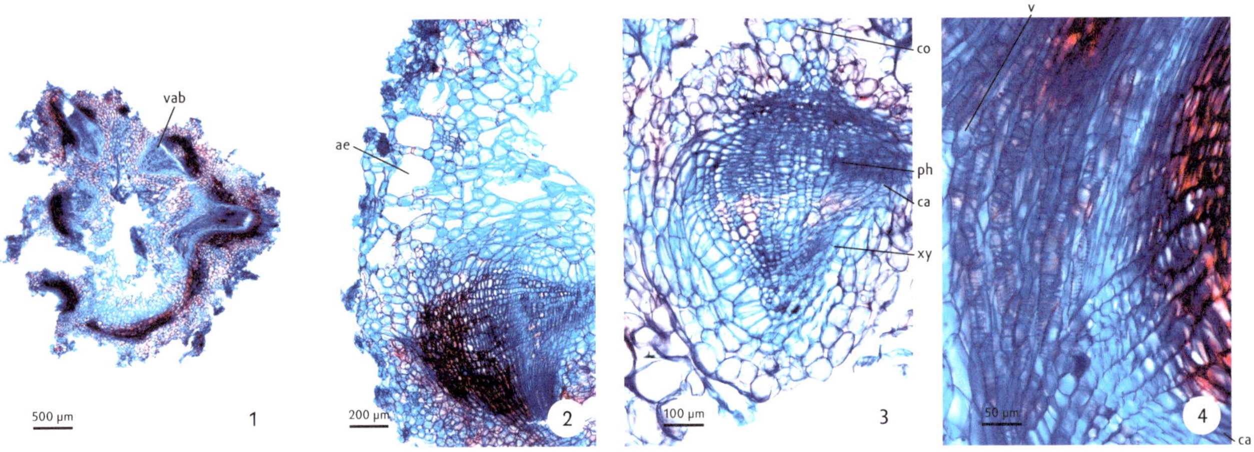

Oenanthe fistulosa L.

Perennial emergent herb, stoloniferous with root-tubers. Stem hollow, erect, 30–60 cm high. Leaves 1–2 pinnate, with linear, untoothed segments. Inflorescence is a terminal compound umbel, few-rayed, flowers white, bracts absent, bracteoles numerous, sepals persistent. Fruit has some spines at truncate apex.

Distribution and ecology

European temperate element. In all CE countries mainly in lowlands, extinct in AT and CZ. Meadows and pastures in the river floodplains, in marshes and fens and in emergent vegetation by rivers, streams, ditches, lakes and fishponds. Damp or wet habitats, usually in areas of winter flooding. It reproduces by seed and spreads by stolons.

Anatomy of the annual shoot and the rhizome

Round stem with a diameter of 7–8 mm. Maximum age 2 years.

Flower stalk: A continuous xylem-bark belt forms the outer part of the stem (1, 2).

Rhizome: A large phloem and cortex surround a pith-less xylem (3). The xylem of the annual shoot consists of an inner part of unlignified parenchyma cells and lignified vessels and an outer part with fibers, vessels and parenchyma (2). Vessels have a diameter of 40–50 µm and a length of 100–150 µm. Intervessel pits are scalariform in the inner, unlignified part and round in opposite position in the outer, lignified part (4). Perforation plates are simple (4). Fibers with small, slit-like pits are thin-to thick-walled (4). Axial parenchyma is paratracheal in the external part (2) and pervasive in the center (4). Rays are >10-seriate between xylem compartments (1) and confluent or indistinct within the xylem (5). Small oil ducts occur in the pith (1, 6) and in the cortex (2).

The phloem is fairly homogeneous or groups of small sieve tubes and larger parenchyma cells occur (7). The cortex is often divided by a small periderm zone in an inner living and an outer dead zone (8) and consists of round thin-walled parenchyma cells and many intercellulars (8).

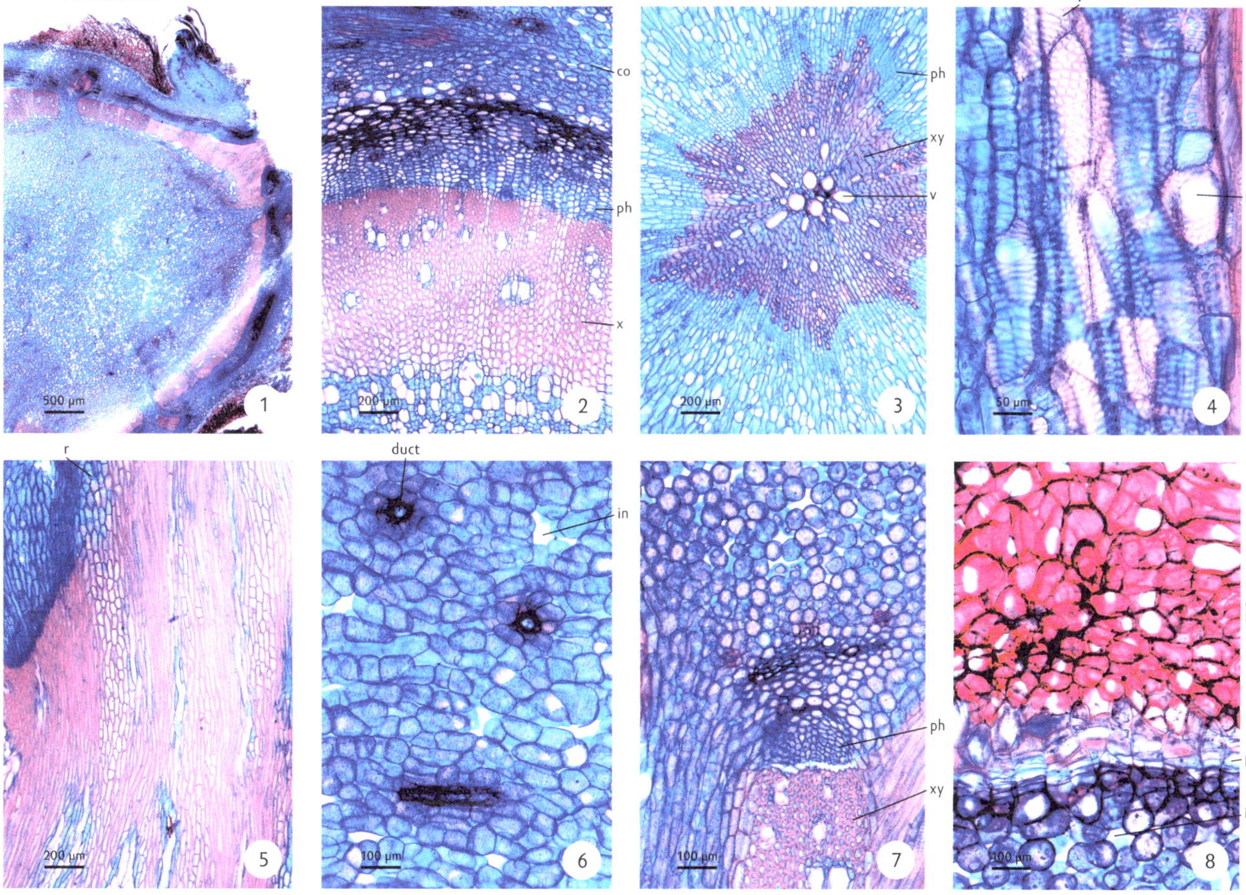

Oenanthe lachenalii C.C.Gmel.

Parsley Water-dropwort Apiaceae

Perennial emergent herb, with root tubers. Stem erect, with thick walls, up to 1 m high. Leaves 1–3 pinnate, with linear segments. Inflorescence is a terminal compound umbel with white flowers. Bracts few, bracteoles numerous and sepals persistent. Fruit is cylindric with 3 spines at apex.

Distribution and ecology

Suboceanic southern-temperate element. Only in CH, DE and PL in lowlands. Mostly near sea in the uppermost parts of salt marshes and the lower reaches of tide. Inland in base-enriched fen-meadows and tall-herb fens. Tolerates brackish water.

Anatomy of the annual shoot

Round stem with a diameter of 8–10 mm. Maximum age 2 years.

An interrupted xylem-bark belt around a large pith forms the outer part of the stem (1).

The xylem of the annual shoot consists of an inner part of unlignified parenchyma cells and lignified vessels and an outer part with fibers, vessels and parenchyma (2). Vessels in the external part are solitary and have a diameter of 40–50 µm and a length of 100–150 µm. Intervessel pits are scalariform in the inner, unlignified part, scalariform and round in opposite position in the outer, lignified part (3). Perforation plates are simple (3). Fibers with small, slit-like pits are thick-walled (2). Axial parenchyma is paratracheal in the external part and pervasive in the center (2). Rays are >10-seriate between xylem compartments and confluent or indistinct within the xylem (2). Ray cells are square (4).

The phloem consists of groups of small sieve tubes and larger parenchyma cells (5). The cortex is often divided by a small phellem zone in an inner and outer zone (6) and consists of round thin-walled parenchyma cells and many intercellulars (6). Cortex with ducts (5).

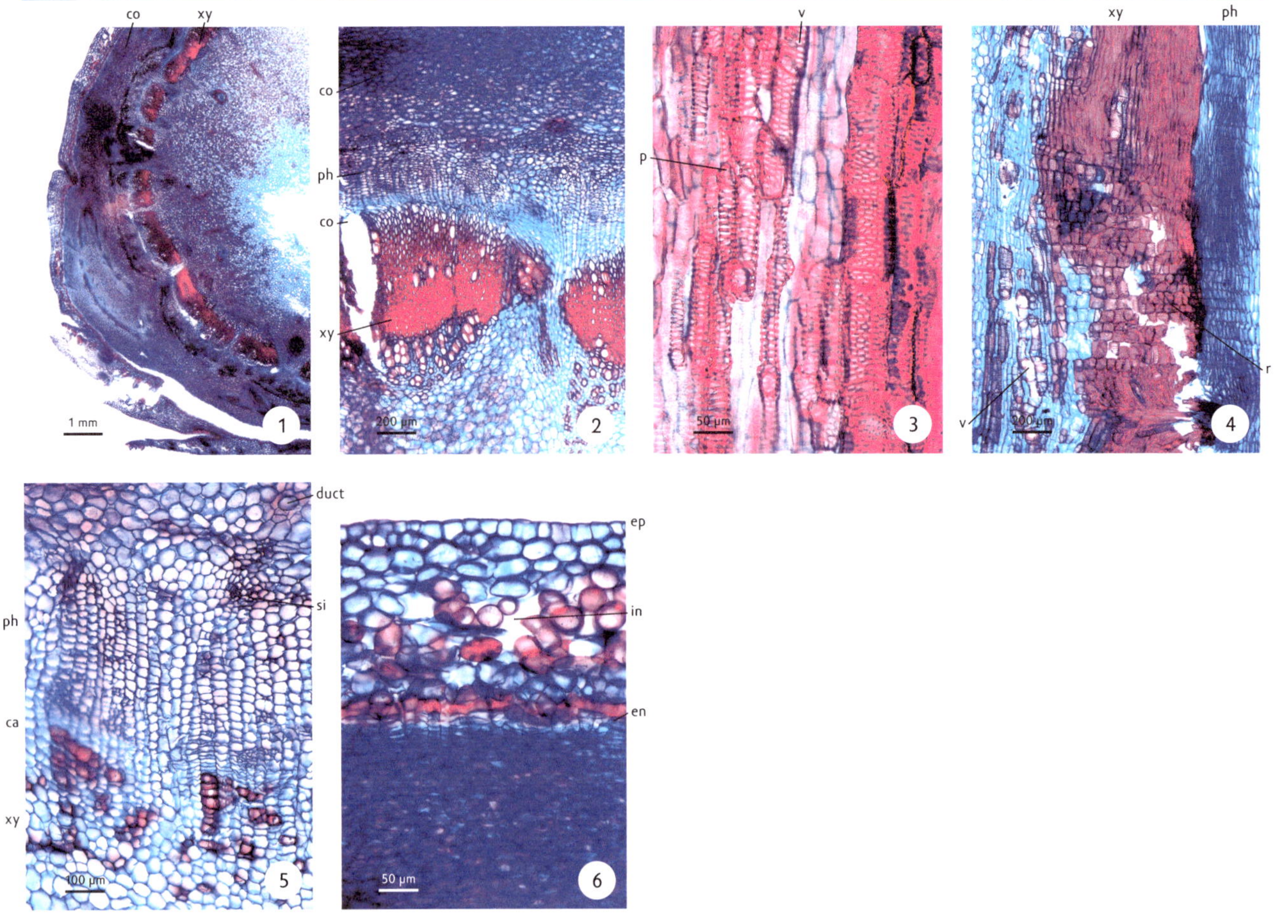

Peucedanum palustre (L.) Moench

Milk-parsley Apiaceae

Biennial or short-lived perennial emergent herb. Stem hollow, ridged, up to 1.5 m high. Leaves triangular in outline, 2–4 times pinnate. Inflorescence is a terminal compound umbel, many-rayed. Flowers greenish-white with several bracts and bracteoles. Fruit is shortly oval with broad wings on edges.

Distribution and ecology

Euro-Siberian, boreo-temperate element. In all CE countries, usually on peaty soils but also on mineral soils rich in organic matter and periodically waterlogged. Typically grows in fens and also in peaty banks of fishponds. Heliophilous, but can survive in alder-carrs. A milky fluid exudes from the broken tissues of young plants. The whole plant contains coumarins. Short-term persistent seed bank.

Anatomy of the annual shoot and the rhizome

Round stem with a diameter of 8–12 mm. Maximum age 2 years (1).

A continuous belt of xylem and a cortex surround a large, central pith (1). Remaining vascular bundles of different sizes are separated laterally by vessel-free parenchymatic zones (1, 2).

The xylem is semi-ring porous and consists of fiber zone with angular vessels and a large marginal parenchyma zone (2). Vessels are arranged solitary and have a diameter of 40–60 µm (2, 3) and a length of 150–200 µm. Intervessel pits are horizontally enlarged to scalariform (5). Perforation plates are simple (5). Fibers with small, slit-like pits are thin-to thick-walled. Some fibers are septate (5). Axial parenchyma is paratracheal and marginal (1, 2). Rays are uni- to biseriate with vascular bundles and 4–8 seriate between them (2, 4).

The phloem consists of parenchyma cells and radially arranged groups of sieve elements (6). Large rays are dilated and the course is undulated (1, 6). Oil ducts occur in the phloem (1, 6). Intercellulars are expressed by irregularly formed radial spaces.

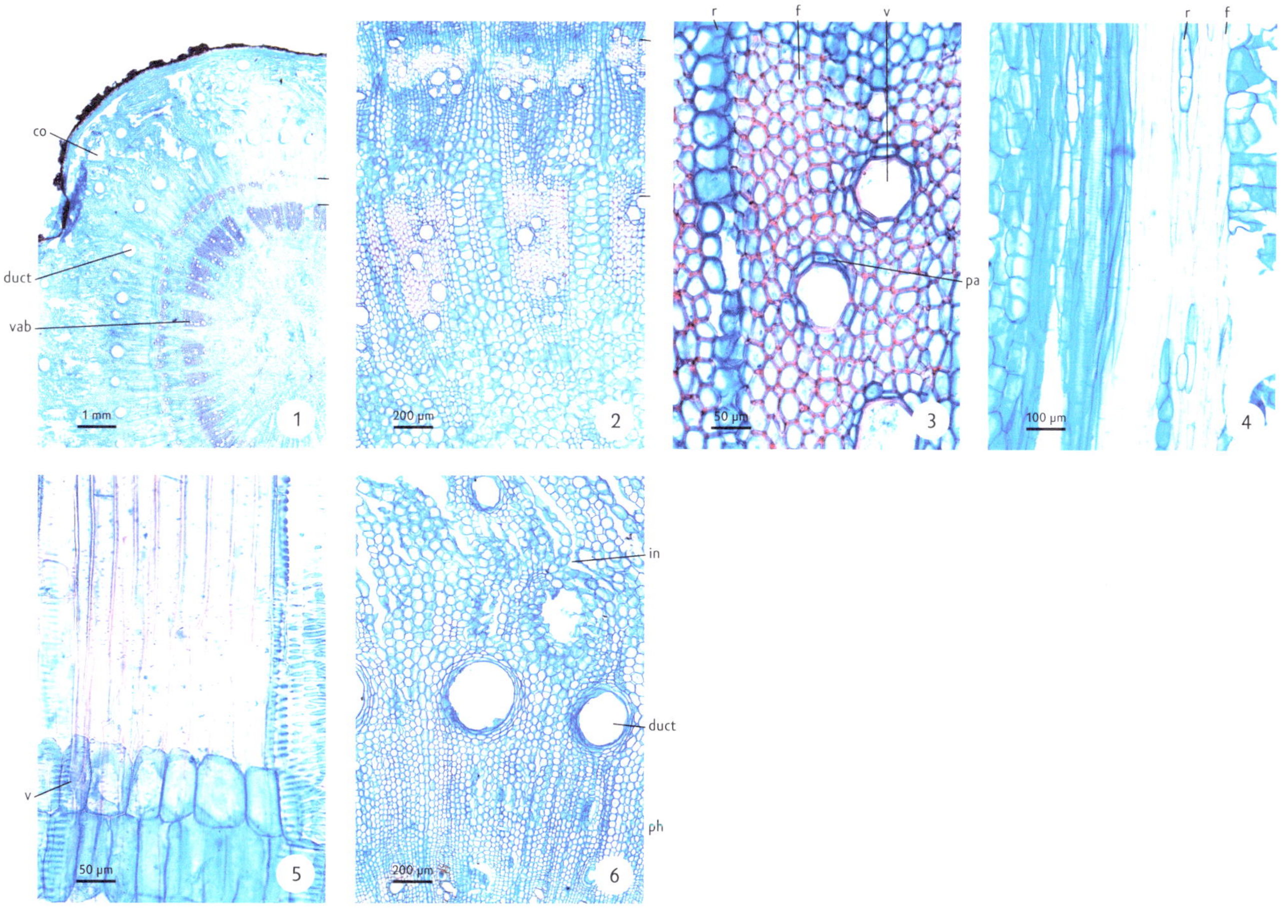

Selinum carvifolia (L.) L.

Cambridge Milk-parsley Apiaceae

Perennial wetland herb. Stem solid, erect, ridged, to 1 m high. Leaves 2–4 times pinnate, leaflets linear-lanceolate with shortly spine-tipped lobes. Inflorescence is a terminal compound umbel of white flowers, without bracts but with many linear bracteoles, sepals absent. Fruit is long-oval, with marked winged ridges.

Distribution and ecology

European temperate element, secondarily in N. America. In all CE countries from lowlands to foothills. In wet habitats like fens, damp meadows and grazed marshy pastures, also on sites with fluctuating water level. Non-deep simple morphophysiological dormancy of seeds.

Anatomy of the annual shoot and the rhizome

Round ribbed rhizome with a diameter of 5–10 mm. Maximum age 2 years (1).

Annual shoot: A circle of vascular bundles and a cortex surround a very large, central zone (1). Vascular bundles of different sizes are separated laterally by vessel-free fibrous zones (1).

The rhizome consists of a central primarily unlignified parenchymatic tissue with few vessels and groups of fibers (2). Vessels are arranged in short radial rows and have a diameter of 40–50 µm (3) and a length of 100–130 µm. Intervessel pits are round to slightly horizontally enlarged in the external xylem and scalariform in the central parenchymatic part (4). Perforation plates are simple. Fibers with small, slit-like pits are thin-to thick-walled (3, 4). Axial parenchyma is paratracheal in the external part and pervasive in the center (3). Rays are 1–4-seriate (5).

The phloem consists of parenchyma cells and irregularly distributed groups of sieve tubes (6). Large rays are dilated. Oil ducts occur in the phloem and cortex (6, 7) and the pith.

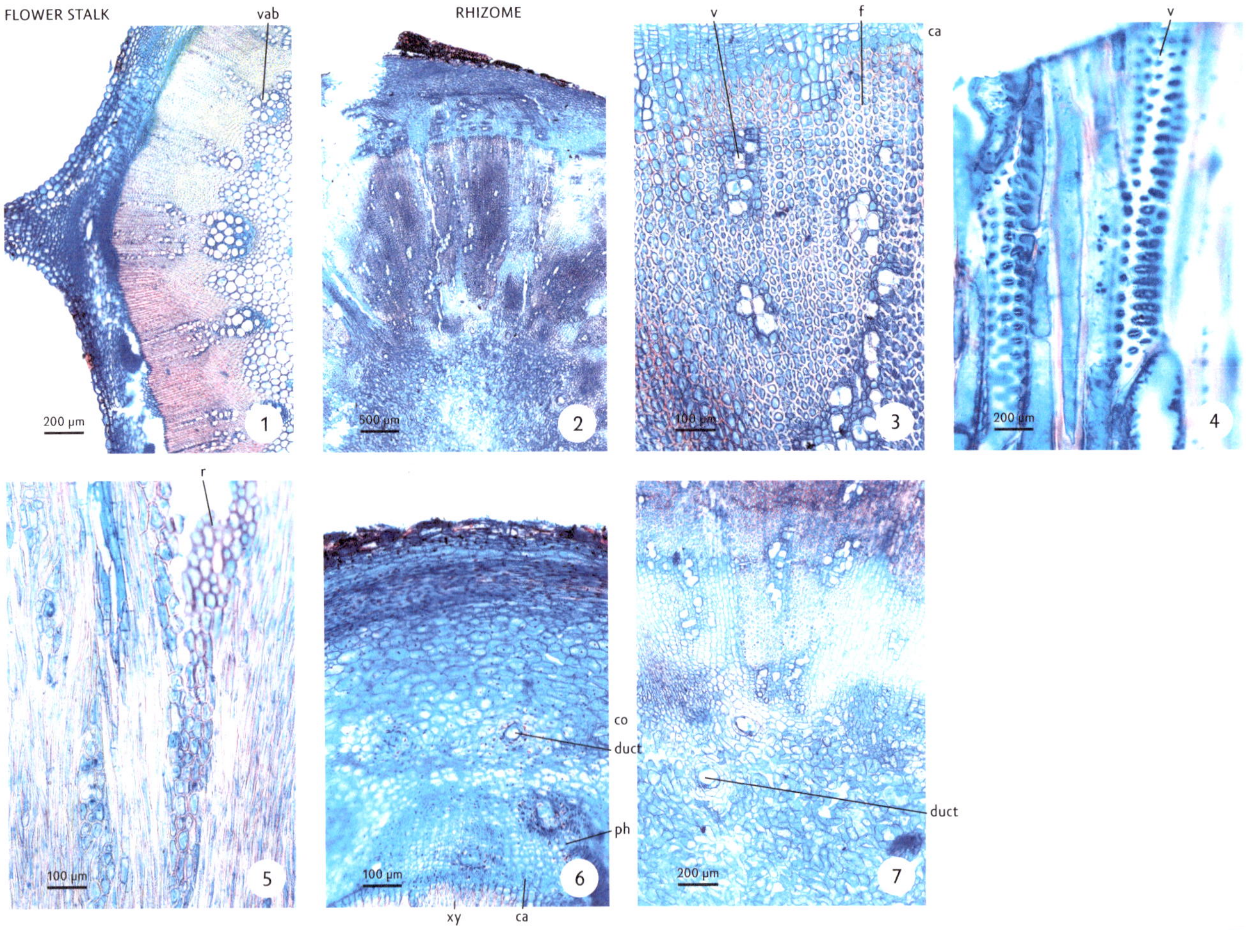

Perennial emergent or wetland herb. Stem hollow, erect, robust, up to 2 m high. Leaves unipinnate, with 4–6 pairs of oval to oblong, toothed leaflets. Inflorescence is a terminal compound umbel of white flowers with several bracts and large bracteoles, sepals persistent. Fruits oval with thick, prominent ridges.

Distribution and ecology

Euro-Siberian temperate element. In all CE countries especially in lowlands. Typical of very wet, tall-herb fens, also floating mats at the margins of still or slowly flowing waters. Common in reed swamps and ditches, growing among other emergent species. It is grazed by stock. Tolerates deep water column.

Anatomy of the annual shoot and the rhizome

Ribbed annual shoot with a diameter of 6–7 mm and round rhizomes with a diameter of 8–15 mm. Maximum age 2 years.

Annual shoots are surrounded by a circle of vascular bundles and a cortex (1).

The rhizome consists of a central unlignified parenchymatic tissue which is surrounded by long radial vascular bundles (2). Vascular bundles become oval in the second year and consist of vessels, pervasive and paratracheal parenchyma and thin-to thick-walled fibers (3). Vessels have a diameter of 30–50 µm (3) and a length of 100–150 µm (4) and are arranged solitary and in short radial rows (3). Intervessel pits are round to slightly horizontally enlarged in the external xylem (4) and scalariform in the central parenchymatic part (4). Perforation plates are simple (4). Fibers with small, slit-like pits are thin-to thick-walled (3, 4). Axial parenchyma is paratracheal in the external part and pervasive in the center (2, 3). Vascular bundles are laterally separated by large, unlignified confluent rays (5). Rays are 2–3-seriate within the bundles (5).

The phloem consists of tangentially and radially arranged irregular groups of sieve tubes within the parenchyma (6). Cortex and phloem are compressed and irregularly formed at the periphery below the phellem (7). Oil ducts occur in the pith and in the phloem and cortex (6). Very small crystal druses occur in the phellem. Collenchyma is located at the periphery of longitudinal ribs of annual shoots (1, 8).

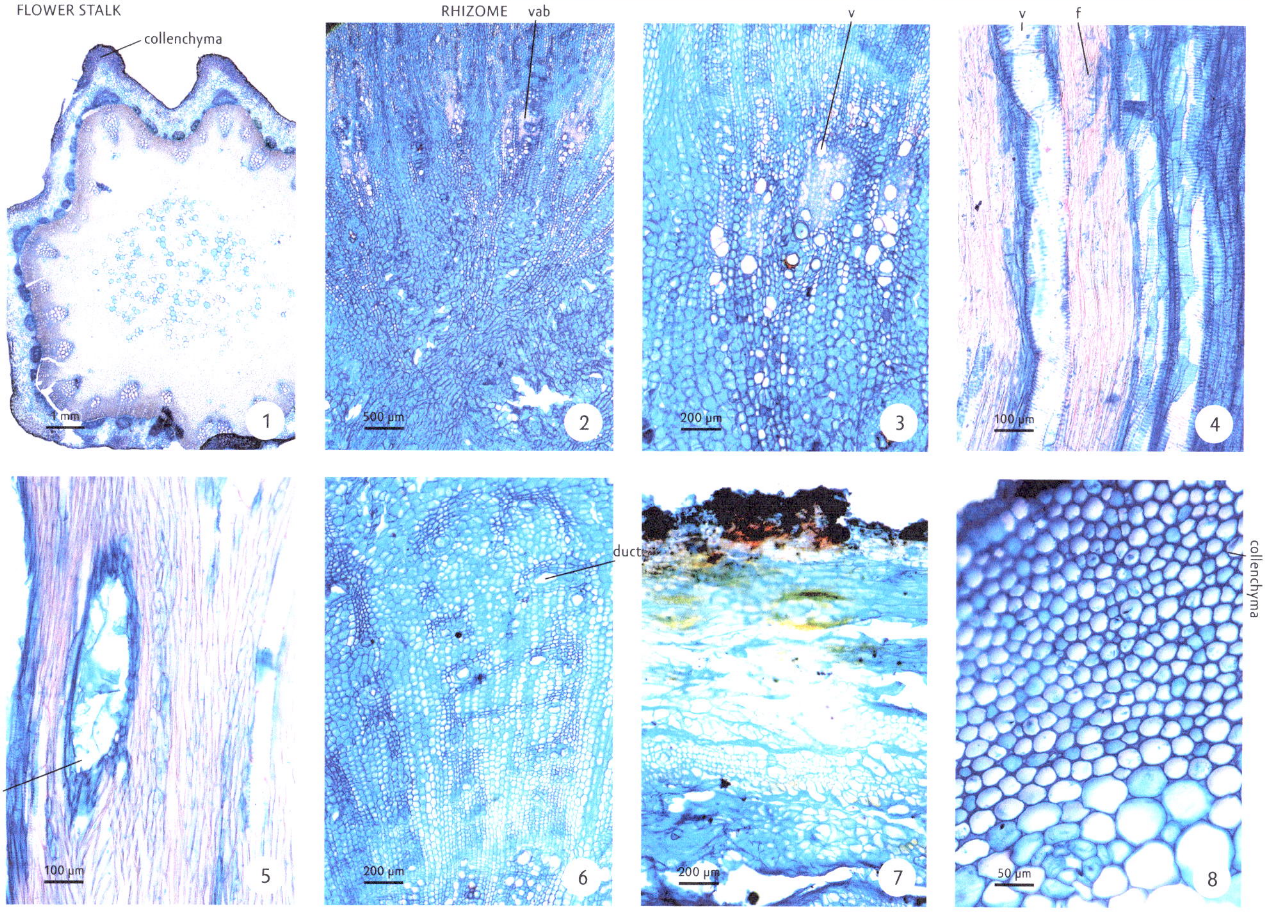

Achillea ptarmica L. **Sneezewort** Asteraceae

Perennial wetland herb. Stem erect, 20–60 cm high, with long underground rhizomes. Leaves unlobed, linear-lanceolate, finely toothed. Flower heads 12–18 mm across, ray-florets oval, white, disc-florets greenish-white. Fruit is an achene.

Distribution and ecology

Eurasian boreo-temperate element. Native in all CE countries, rare in HU. It grows in wet meadows, springs, banks of fish-ponds and slowly flowing streams. Tolerates waterlogged soils and water table fluctuation.

Anatomy of the basal part of the annual shoot

Round stem with a diameter of 3 mm. Maximum age 1 year (1).

The pith is surrounded by a xylem, a small phloem, an aerenchymatic cortex and a small phellem (1). The xylem consists of vessels, thin-to thick-walled fibers and paratracheal parenchyma (2). Vessels have a diameter of 20–30 µm (2) and length of 150–250 µm. Vessels are arranged in short radial rows (2). Intervessel pits are round and perforation plates are simple (4). Fibers are thin-to thick-walled (2) with small slit-like pits and long nuclei (5). Parenchyma is paratracheal (2). Rays are very rare or absent (3).

The phloem consists of parenchyma cells and small groups of sieve elements (6). Round groups of thick-walled fibers are located in the phloem/cortex boundary (1, 7). The cortex consists of unlignified oval cells (6). A few irregularly formed cells represent the phellem (7).

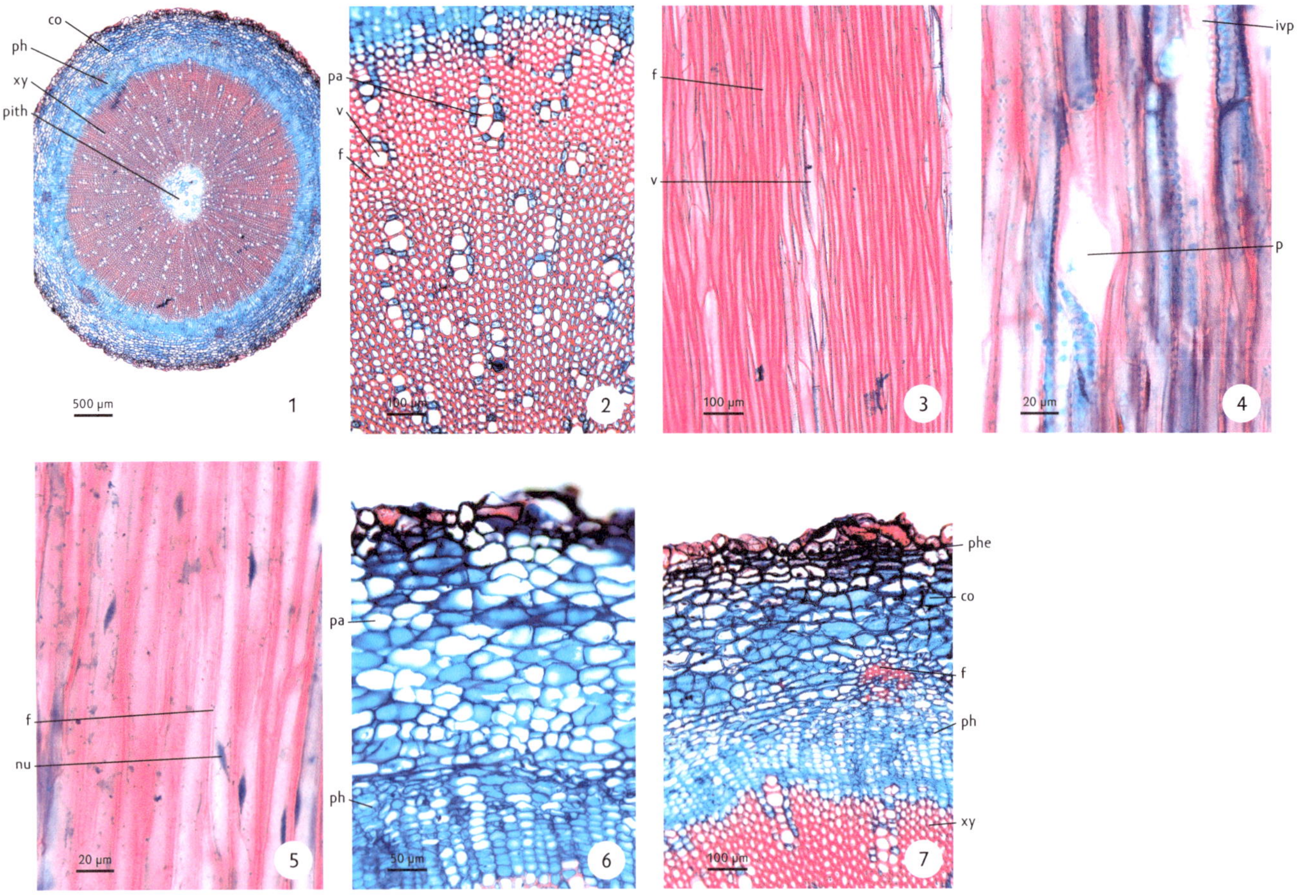

F. H. Schweingruber et al., *Anatomic Atlas of Aquatic and Wetland Plant Stems*, https://doi.org/10.1007/978-3-030-33420-8_13

Adenostyles alliariae (Gouan) A. Kern.

Perennial wetland herb. Stem erect, thick, grooved, 60–150 cm high. Leaves with long petioles, cordate, white-wooled undersides, stem leaf with two basal lobes. Flower heads numerous, in corymbous panicles up to 20 cm across, flowers red-purpur. Fruit is an achene.

Distribution and ecology

European montane distribution. Native in all CE countries, except HU. It grows on mountain springs, tall-herb springs, it can occur also in mountain spruce and beech forests on wet soils. It contains alkaloids.

Anatomy of the rhizome

Round rhizome with a diameter of 9–12 mm. Maximum age 5 years (2).

The large pith of rhizome is surrounded by single vascular bundles, consisting of a xylem, a phloem and a cortex (1). The phellem is absent (1, 6).

The xylem of the vascular bundles consists of vessels, thin- to thick-walled fibers and paratracheal parenchyma (2, 3). Ring boundaries are mainly indicated by semi-ring porosity and marginal parenchyma (2). Vessels are primarily arranged solitary or in small groups (2, 3). Vessels have a diameter of 40–50 µm (2-3). Vessel length varies between 150–200 µm. Intervessel pits are round to slightly horizontally enlarged (4) and perforation plates are simple (4). Rays consist of upright unlignified cells and are 5–10 cells wide with laterally separated sheet cells (5). Large rays separate vascular bundles laterally (1–3).

The phloem consists of parenchyma cells and small groups of sieve elements (6, 7). The cortex is composed of round, unlignified parenchyma and small ducts just outside of the phloem. Intercellulars are absent.

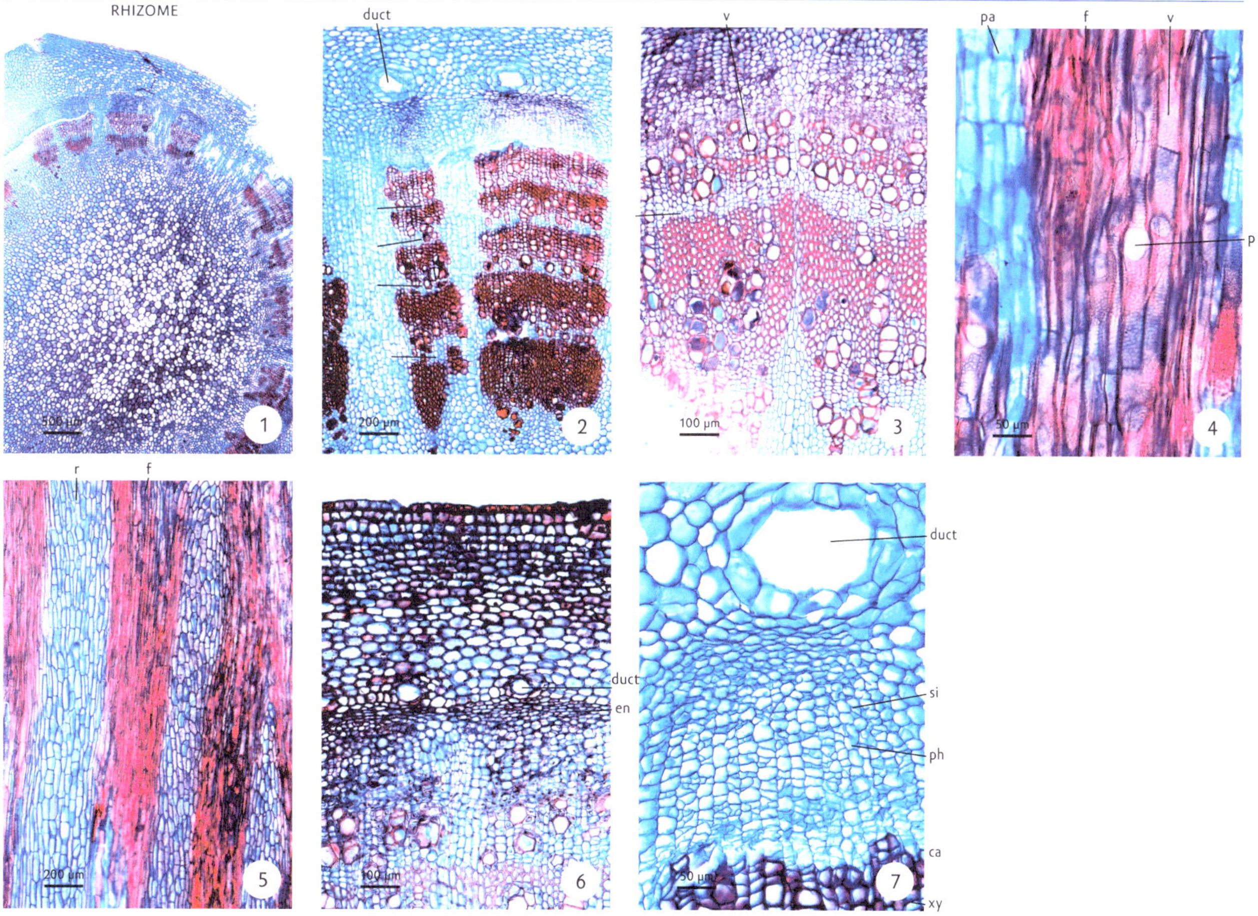

Aster tripolium L. subsp. *pannonicus* (Jacq.) Soó

(syn. *Tripolium pannonicum* subsp. *pannonicum*)

Shortly lived perennial or biennial wetland herb, sometimes annual. Stem erect, branched, 10–70 cm high. Leaves fleshy, with few lateral veins. Flower heads in loose umbel-like corymb, with yellow disc-florets and blue to purple ray-florets. Fruit is a brown achene with simple pappus.

Distribution and ecology

Pannonian endemic plant. In AT, CZ, HU and SK in lowlands, in ungrazed or lightly grazed salt marshes, salt meadows. Prefers soils usually wet in spring, drying during the summer. It flowers in late summer. Obligate inland halophyte.

Anatomy of the basal part of the annual shoot above and below ground

Round stem with a diameter of 2–3 mm. Maximum age 2 years (1).

Annual shoots above ground are surrounded by a circle of vascular bundles and the cortex (1). The basal part consists of dilating vascular bundles and a very large cortex which are laterally separated by large rays (2). The xylem consists of vessels, thin-to thick-walled fibers and paratracheal parenchyma (3, 4). Vessels above ground are solitary and in radial rows (3, 4) they have a diameter of 25–35 µm (3) and those of the basal part 30–50 µm. Vessel length in both parts is 100–150 µm (4). Intervessel pits are round (5) and perforation plates are simple (5). Rays in basal parts are dilating (2) and consist of lignified and unlignified square and upright cells. Vascular bundles in shoots are laterally separated by 5–10 cell wide rays (2). Rays are 3–5 seriate (6) and absent within the bundles (3, 4).

The phloem consists mainly of dilating rays, parenchyma cells, a few sieve tubes (7, 8) and small groups of fibers. The cortex contains irregularly formed air ducts (1, 8). A few thin-walled rectangular cells characterize the phellem.

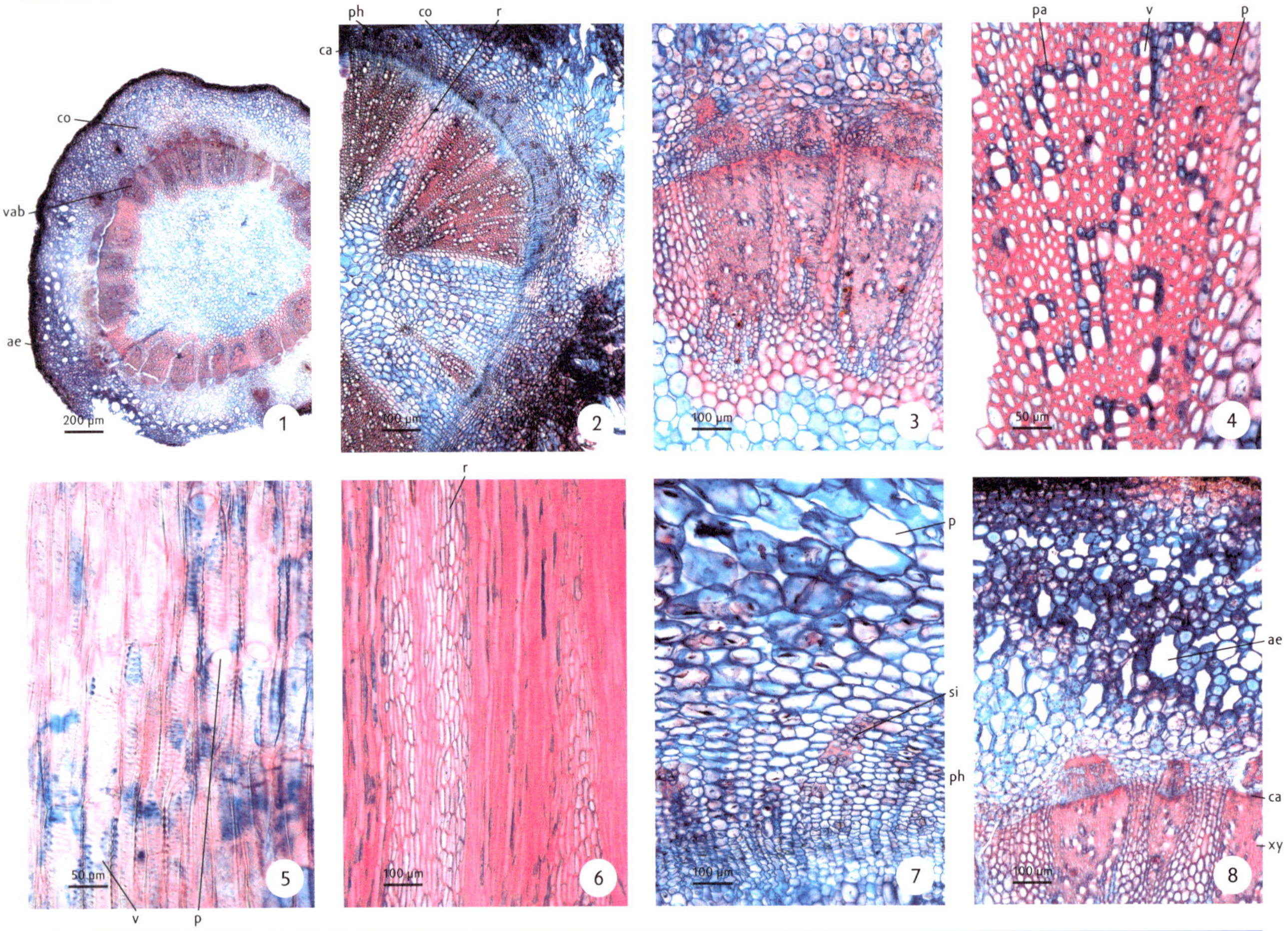

Bidens cernuus L. (syn. *Bidens cernua*) **Nodding Bur-marigold** Asteraceae

Annual wetland herb. Stem erect, 15–60 cm high, usually branched. Leaves opposite, sessile, unlobed, strongly serrate. Flower-heads long-stalked, drooping, 15–25 mm in diameter, with yellow to yellow-brown disc-florets, often with conspicuous golden-yellow ray-florets. Fruit is a four-sided achene, with 3–4 bristles, the achenes and bristles with backward-directed barbs.

Distribution and ecology

Circumpolar temperate element. In all CE countries usually in lowlands. Occurs on wet substrates on the margins of slow-flowing rivers and streams, in drained bottoms of fishponds or water reservoirs, often in places flooded in winter. Short-term seed bank, annual dormancy cycle.

Anatomy of the basal part of the annual shoot and the transition between the root and the annual shoot

Round stem with a diameter of 4–6 mm. Maximum age 1 year (1).

The pith above and below ground is surrounded by a xylem, a small phloem and an aerenchymatic cortex (1, 2). The xylem consists of vessels, thin-to thick-walled fibers and paratracheal parenchyma (3). Vessels have a diameter of 30–50 µm (3) and length of 150–250 µm. Vessels are solitary and arranged in short radial rows (3). Intervessel pits are round and slightly horizontally enlarged. Perforation plates are simple or scalariform with many small bars (aberrant scalariform) (4). Fibers thin-to thick-walled (3) with small slit-like pits (see *Bidens frondosa*) (4). Parenchyma is paratracheal (7). Rays are indistinct in tangential sections (5). Ray cells are upright (6).

The phloem consists of parenchyma cells, a few sieve tubes (7) and dilating rays. There are half-moon like groups of thick-walled fibers in the radial continuation of primary vascular bundles. The stele is surrounded by a poorly differentiated pericycle/endodermis zone (8). The cortex in the above ground shoot is represented by a small-meshed net (8) and the below ground shoot by a large-meshed net of parenchyma cells (2). The phellem is absent.

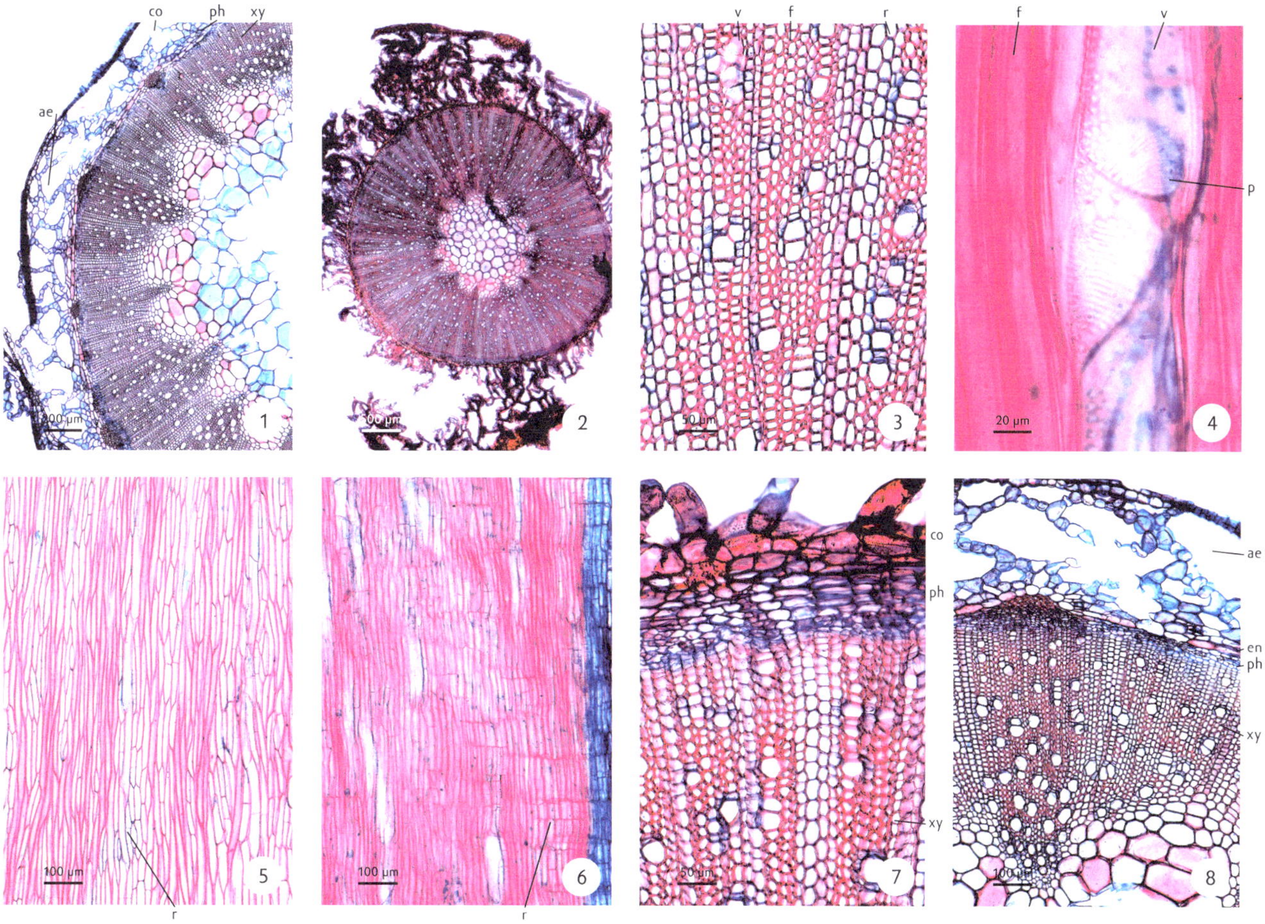

Bidens connatus Willd.

London Bur-marigold

Annual wetland herb. Stem erect, usually branched, 25–150 cm high. Leaves simple, sessile or petiolate, blades elliptic to lanceolate. Flower-heads often born singly. Ray-florets usually absent, disc-florets pale yellow to orange. Fruit is a four-sided achene, blackish to purplish or brown, usually without pappus.

Distribution and ecology

Native of eastern N. America. As a neophyte in CH, CZ, DE and PL, especially in lowlands. It occurs along the rivers, in fishponds and wetlands. It prefers drained, periodically flooded and nitrogen-rich soils.

Anatomy of the transition zone between the root and the annual shoot

Round stem with a diameter of 4–6 mm. Maximum age 1 year1 (1).

The pith above and below ground is surrounded by a xylem, a small phloem and a cortex without intercellular spaces (1). The xylem consists of vessels, thin-to thick-walled fibers and paratracheal parenchyma (2). Vessels have a diameter of 40–70 μm (2) and a length of 150–300 μm. Vessels are solitary and in short radial rows arranged (2). Intervessel pits are round and horizontally enlarged (3). Perforation plates are simple (3) or scalariform with many small bars (aberrant scalariform) (4, 5). Fibers thin-to thick-walled (2) with small slit-like pits. Parenchyma is paratracheal (2). Rays are 2–4-seriate (6). Ray cells are upright.

The phloem consists of parenchyma cells and small groups of sieve tubes (7), a few single fibers, a few oil ducts (8) and dilating rays (7). Intercellular spaces are rare or absent (7). The phellem is absent (1, 7).

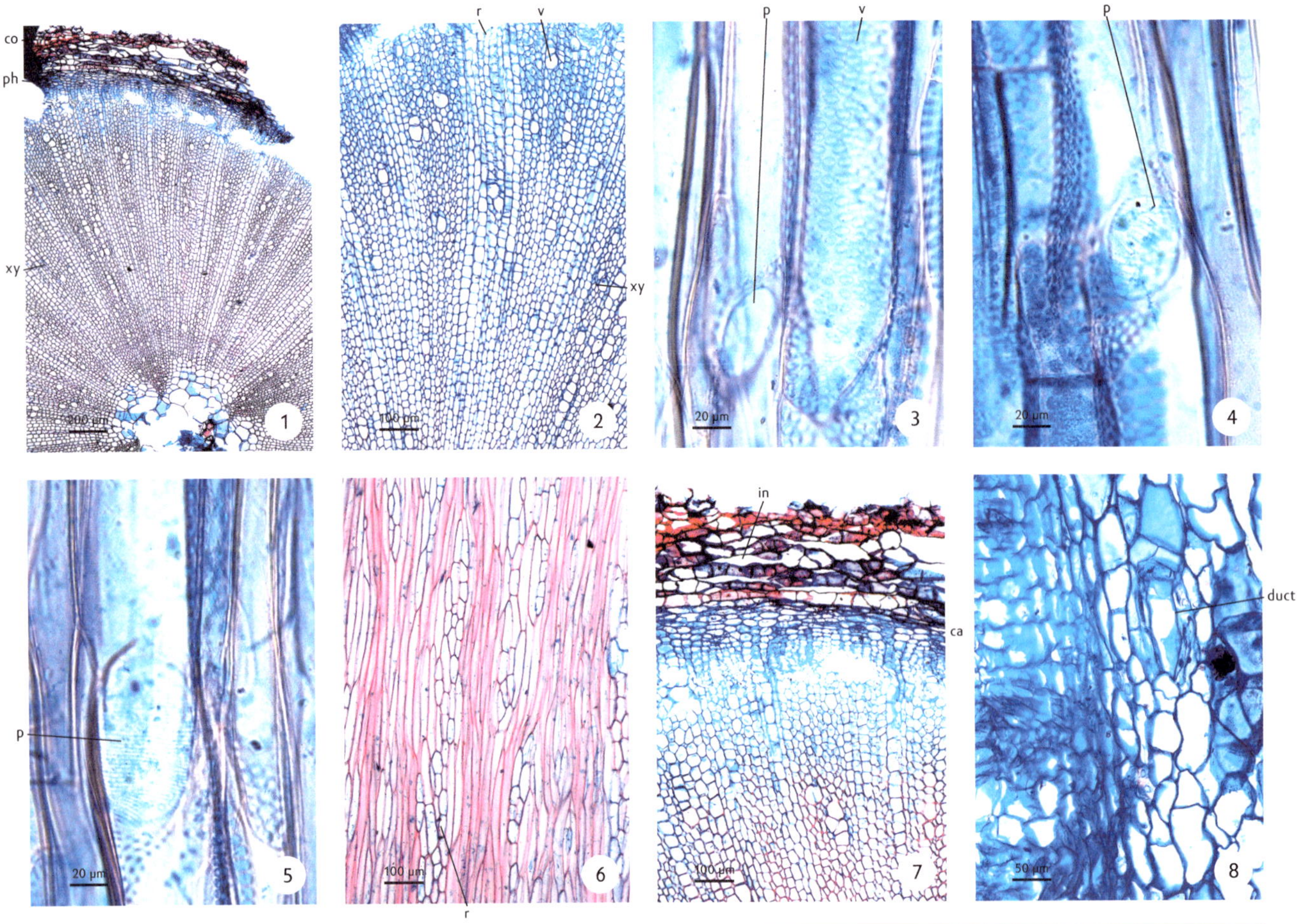

Annual wetland herb. Stem erect, usually branched, up to 1 m high. Leaves pinnate, with 1–2 pairs of lateral leaflets, petioles unwinged. Flower-heads 10–20 mm in diameter, with yellow-brown disc-florets, rarely with yellow ray-florets. Achenes flat, rough, with forward-directed barbs, usually with 2 bristles. Central achenes are brown and elongated, peripheral achenes are black and thicker.

Distribution and ecology

Neophyte, native of N. and S. America; widely naturalised in temperate Europe. Margins of rivers, fishponds, water reservoirs or canals, drained bottoms or wet depressions in fields. Prefers nitrogen-rich soils.

Anatomy of the transition zone between the root and the annual shoot

Round stem with a diameter of 4–6 mm. Maximum age 1 year (1).

The pith above and below ground is surrounded by a xylem, a small phloem and a cortex without intercellular spaces (1). The xylem consists of vessels, thin-to thick-walled fibers and paratracheal parenchyma (2). Vessels have a diameter of 40–70 µm (2) and a length of 150–300 µm. Vessels are solitary and in short radial rows arranged (2). Intervessel pits are round and horizontally enlarged (3, 4). Perforation plates are simple or scalariform with many small bars (aberrant scalariform) (5). Fibers thin-to thick-walled (2) with small slit-like pits (6). Parenchyma is paratracheal (2). Rays are 2–4-seriate (7). Ray cells are upright.

The phloem consists of parenchyma cells, small groups of sieve tubes (8), a few groups of sclereids (8), dilating rays (1) and a few oil ducts. There are half-moon like groups of thick-walled fibers in the radial continuation of primary vascular bundles. Intercellular spaces and phellem are absent (8).

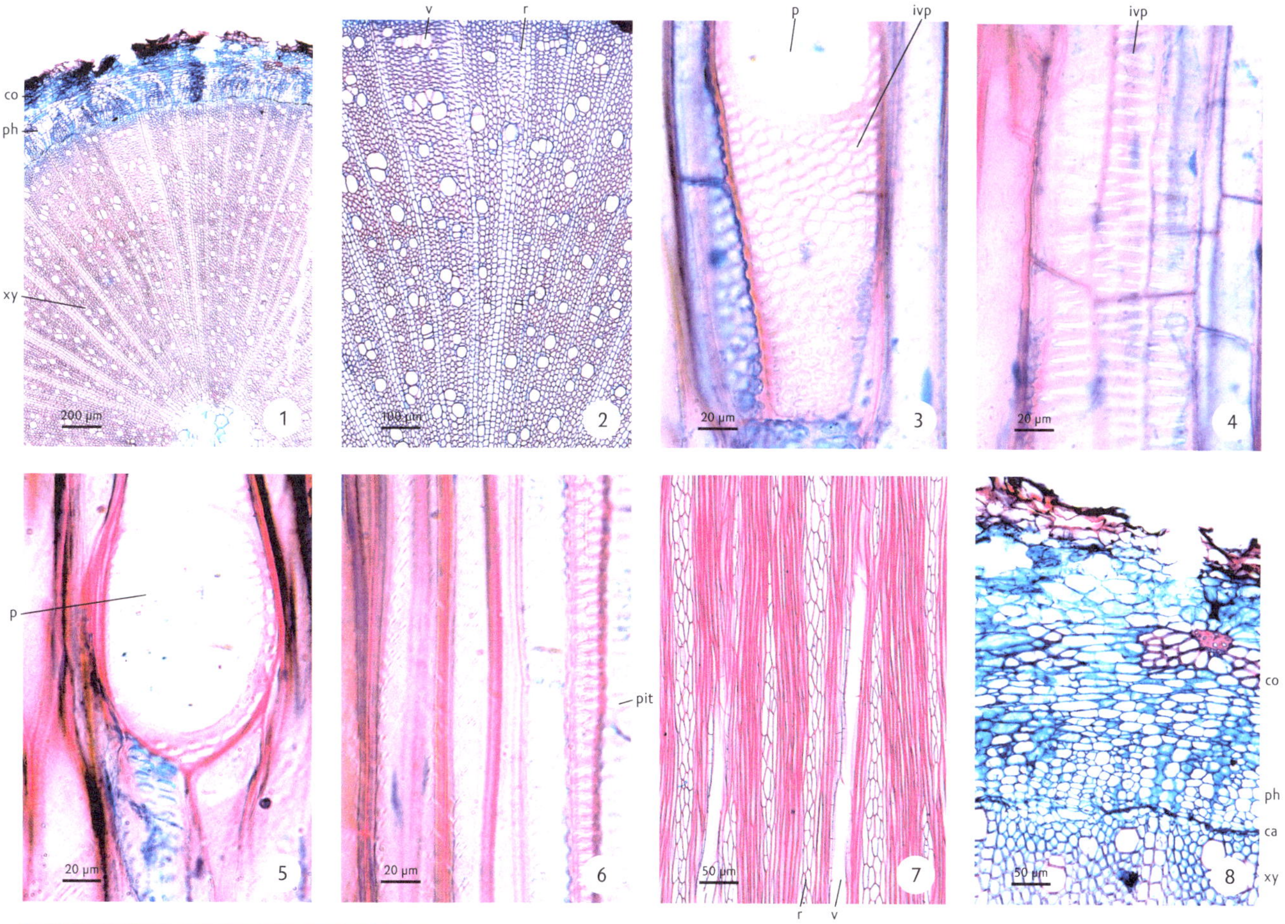

Bidens tripartitus L. (syn. *Bidens tripartita*) **Trifid Bur-marigold** Asteraceae

Annual wetland herb. Stem erect, usually rich branched, up to 120 cm high. Leaves trifoliate, with 1–2 pairs of deeply dissected lateral lobes, with winged petiole. Flower-heads 10–25 mm in diameter, erect, with yellow to yellow-brown disc-florets, very rarely with yellow ray-florets. Achenes flat, obovate, with only 2 barbed spines on top.

Distribution and ecology

Eurasian temperate element. Common in all CE countries at margins of rivers, drained bottoms of fishponds, water reservoirs, ditches, wet depressions in fields. Prefers nutrient-rich mud or gravel, often in areas wet in winter but exposed in summer. It prefers less acidic and drier substrates than *B. cernua*. Persistent seed bank, annual dormancy cycle.

Anatomy of the transition zone between the root and the annual shoot

Round stem with a diameter of 2.5–4 mm. Maximum age 1 year (1).

The pith above and below ground is surrounded by a xylem, a small phloem and a cortex without intercellular spaces (1). The xylem consists of vessels, thin-to thick-walled fibers and paratracheal parenchyma (2). Vessels have a diameter of 30–40 µm (2) and a length of 100–180 µm. Vessels are solitary and in short radial rows arranged (2). Intervessel pits are round and horizontally enlarged. All perforation plates are scalariform with 1–10 small bars (aberrant scalariform) (3). Fibers thin-to thick-walled (2) with small slit-like pits. Parenchyma is paratracheal (2). Rays are 1–4-seriate (4). Ray cells are upright (5).

The phloem consists of parenchyma cells, small groups of sieve tubes (6), a few oil ducts (7) and dilating rays (6). Intercellular spaces are absent (6). The phellem is absent (1).

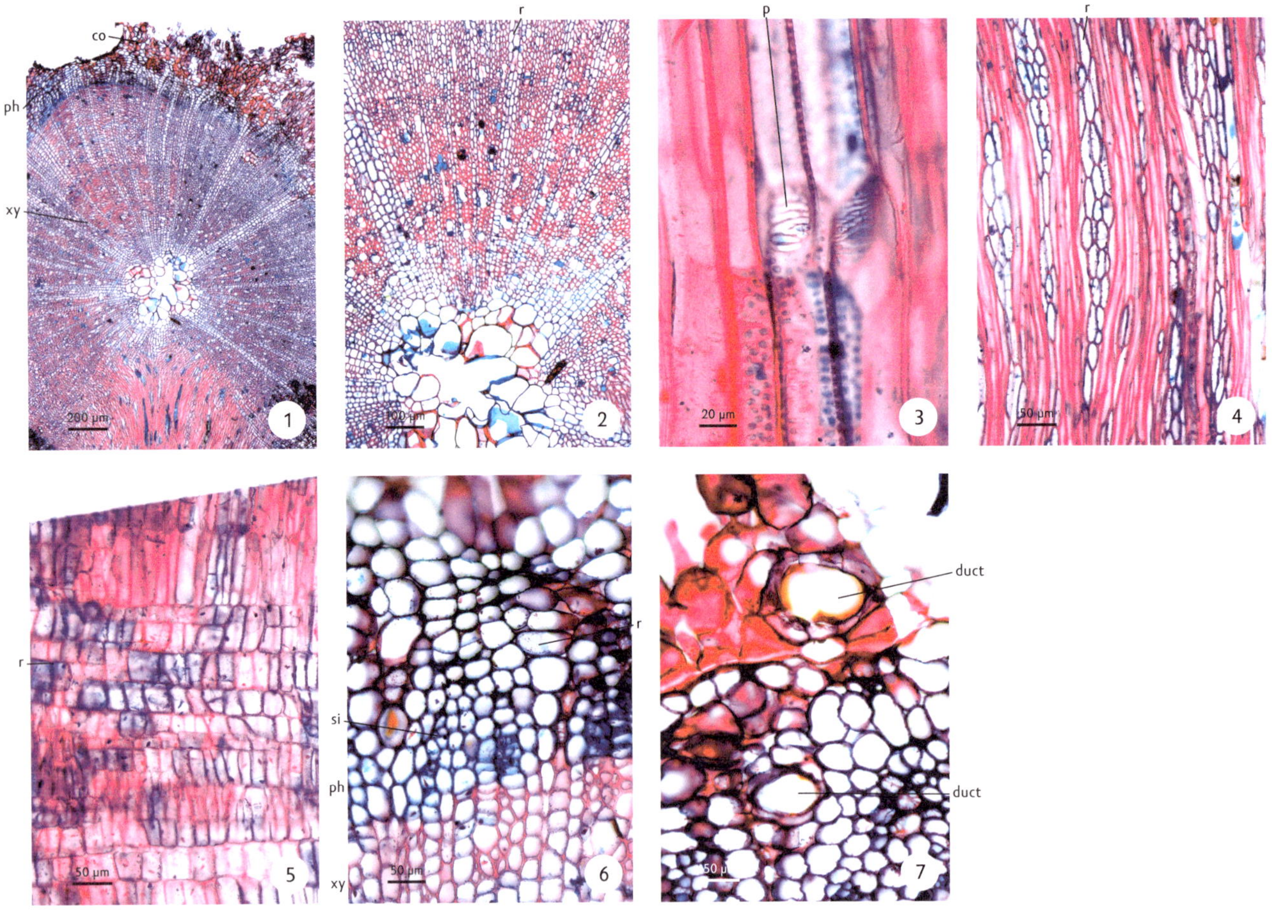

Carduus personata (L.) Jacq.

Perennial wetland herb. Stem erect, usually branched, with spiny wings, glabrous to hairy, 60–200 cm high. Leaves alternate, bottom leaves dissected, stem leaves entire, softly spiny. Flower-heads sessile, 4–6 in clusters at the end of twigs. Only dark red-purple disc florets. Fruit is an achene.

Distribution and ecology

European endemic species. Native in all CE countries except HU, mainly in the uplands and mountains. Prefers wet sandy or stony soils rich in nutrients. Occurs along the rivers and streams, also in wet forests.

Anatomy of the rhizome

Round rhizome with a diameter of 8 mm. Maximum age 2 years (1).

Basal shoots (root collar) are surrounded by a completely lignified xylem belt, a phloem and a cortex with some sclerenchyma (1). The phellem is absent (1, 7). The xylem consists of vessels, thin-to thick-walled fibers and paratracheal parenchyma (2). The primary vascular bundles around the pith are unlignified (1). Vessels are arranged in short radial rows or in small groups (2). Vessels have a diameter of 40–60 µm (2). Vessel length varies between 100–180 µm. Intervessel pits are round in the adult xylem and perforation plates are generally simple and occasionally scalariform (3). Rays consist of square and upright lignified cells (4), they are 5–10 cells wide with laterally separated sheet cells (5). Large rays also separate vascular bundles laterally (1).

The phloem consists of parenchyma cells, a few sieve tubes (7) and few small groups of fibers. Rays are dilated. The cortex consists of oval, unlignified parenchyma cells. Air cavities and ducts are absent (6).

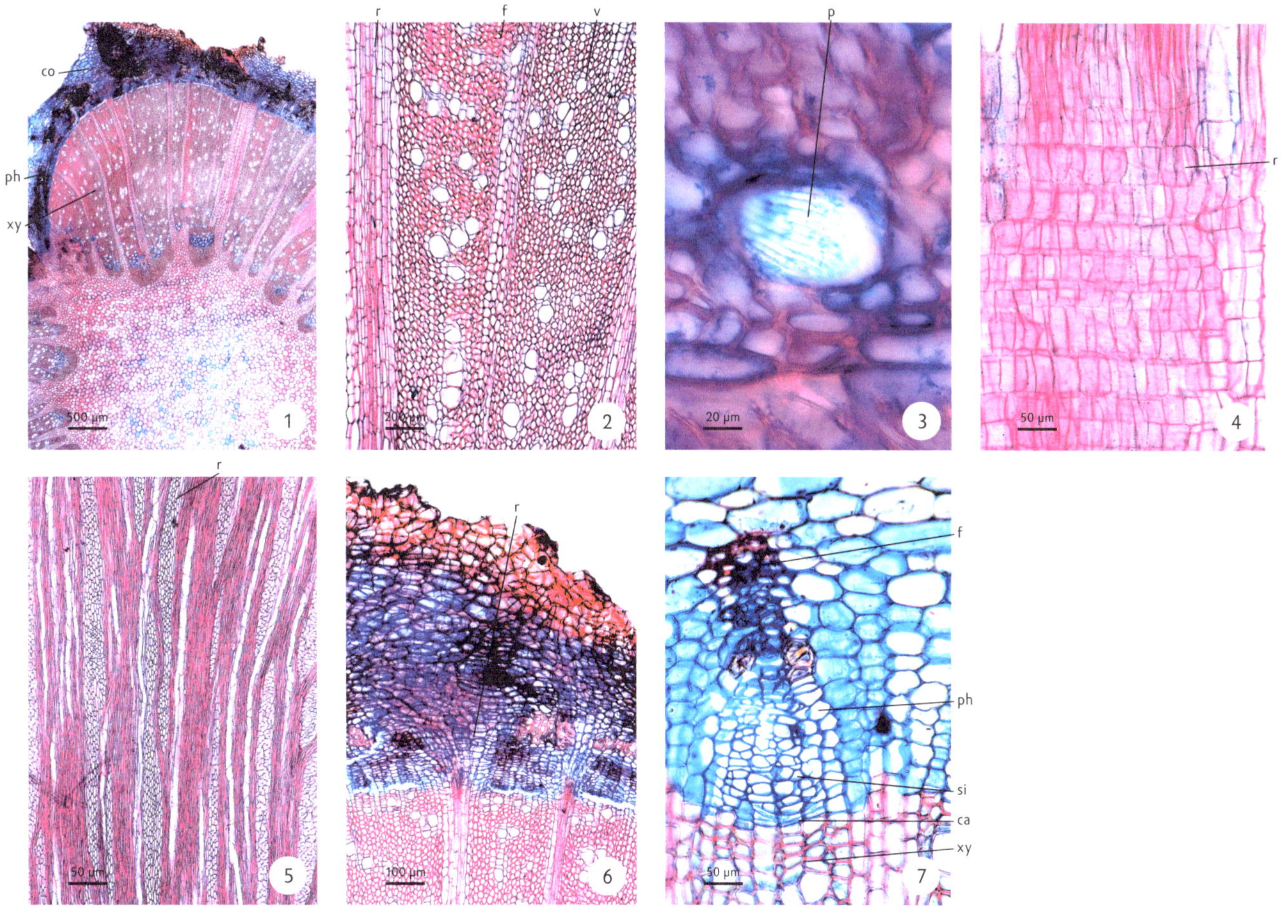

Cicerbita alpina (L.) Wallr.

Alpine Blue-sow-thistle Asteraceae

Perennial wetland rhizomatous herb. Stem erect, unbranched, hollow, 70–130 cm high, glandular-pubescent near the top. Leaves with a sharply triangular terminal lobe and a few smaller laterals. Blue-purple flower-heads in a panicle, only ray florets. Fruit is an achene.

Distribution and ecology

European arcto-alpine element. In all CE countries except HU, from the uplands to subalpine altitudes. A part of tall herb vegetation, also in moist forests, along streams, often where the late snow is present.

Anatomy of the flower stalk and the rhizome

Round rhizome with a diameter of 6–14 mm. Maximum age 2 years.

The large pith of the rhizome is surrounded by single vascular bundles, a cortex and a small phellem (2). The major part of the xylem of the vascular bundles consists of vessels, thick-walled fibers and paratracheal parenchyma (3). The innermost part of the vascular bundles is unlignified (2, 3). Ring boundaries of the diffuse porous wood are mainly indicated by marginal parenchyma (3).

Vessels have a diameter of 30–50 µm (3). Vessel length in both parts is 100–150 µm. Vessels are primarily solitary (3). Intervessel pits are round to scalariform (4). Perforation plates are primarily simple (4) and occasionally scalariform. Rays consist of square and upright unlignified cells, they are >10 cells wide (5) with laterally separated sheet cells (6). Large rays separate vascular bundles laterally (1–3).

The phloem consists of parenchyma cells and groups of sieve tubes (7), small groups of fibers (2) and dilating rays (6). The cortex consists of round and oval parenchyma cells and is surrounded by an endodermis (7). Intercellulars, ducts and sclerenchyma are absent. The phellem consists of few irregularly formed cells (7).

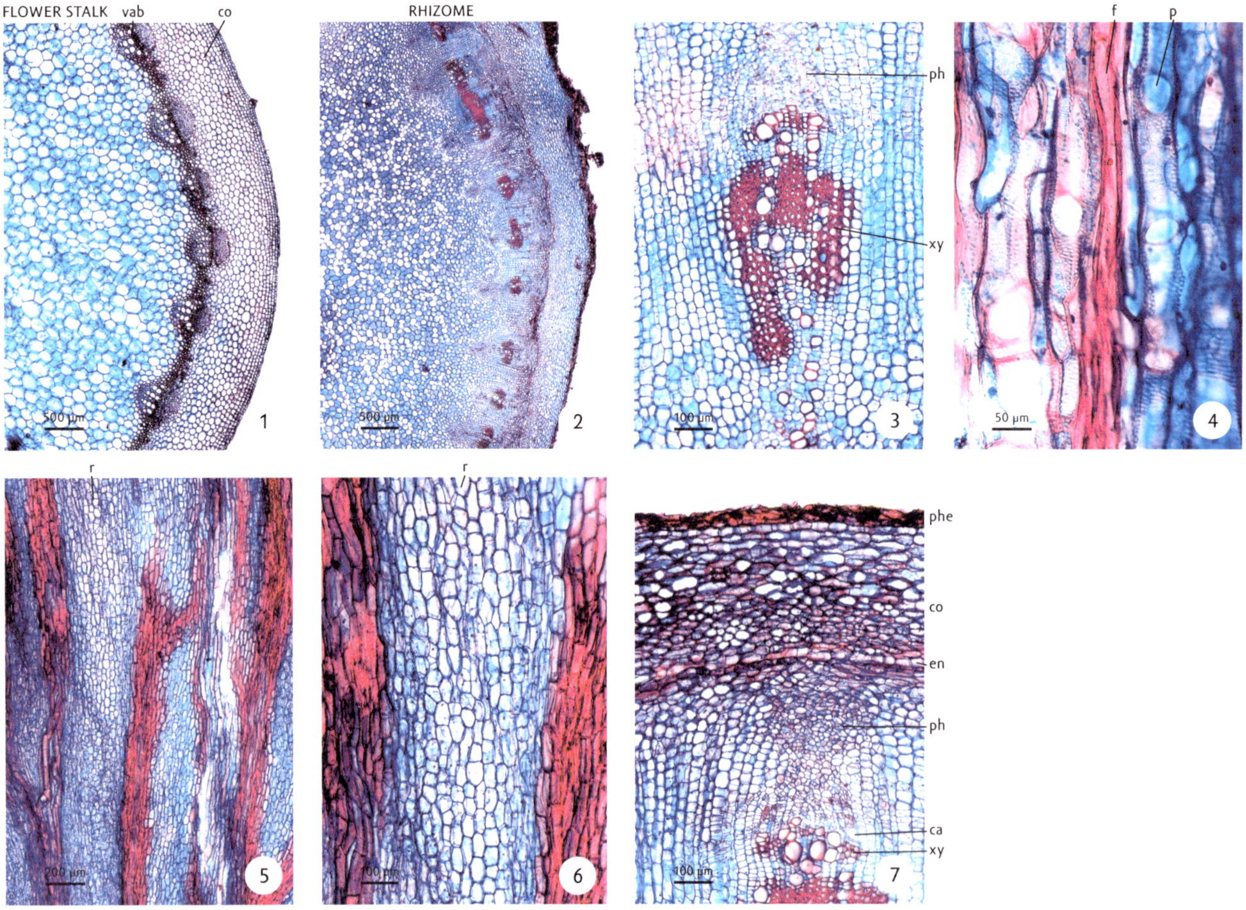

Cirsium brachycephalum Jur.

Mostly biennial wetland herb. Stem erect, simple, grooved and spiny, 50–150 cm high. The lower leaves entire, lanceolate, the middle leaves lanceolate or linear, dissected and glabrous. Flower heads in weak cymose panicles, flower stalks thickly white-felted, florets pink. Fruit is an achene with pappus.

Distribution and ecology

Pannonian endemic species, native in AT, CZ, HU and SK, usually in lowlands. Prefers fen and salty meadows and pastures, usually in wet or marshy soils. Monocarpic plant.

Anatomy of the transition zone between the root and the annual shoot

Round stem with a diameter of 8 mm. Maximum age 2 years (1).

Basal shoots (root collar) are surrounded by a parenchymatic central zone with vessels, a lignified xylem belt, a phloem, a parenchymatic cortex and a phellem (1). The central xylem consists of vessels and paratracheal parenchyma and the peripheral xylem of fibers with vessels (1). Ring boundaries are absent (1, 2).

Vessels are angular and have a diameter of 20–30 µm (4). Vessel length varies between 80–100 µm. Vessels are solitary (3, 4). Intervessel pits are round to scalariform (6) and perforation plates are primarily simple and rarely scalariform (7). Rays consist of square and upright cells, they are 5–10 cells wide with laterally separated sheet cells (5). Large rays separate vascular bundles laterally (2, 3).

The phloem consists of parenchyma cells with a few sieve elements, which are compressed in outer parts (8). The cortex consists of round, unlignified parenchyma. Ducts are located in the cortex, just outside of the phloem.

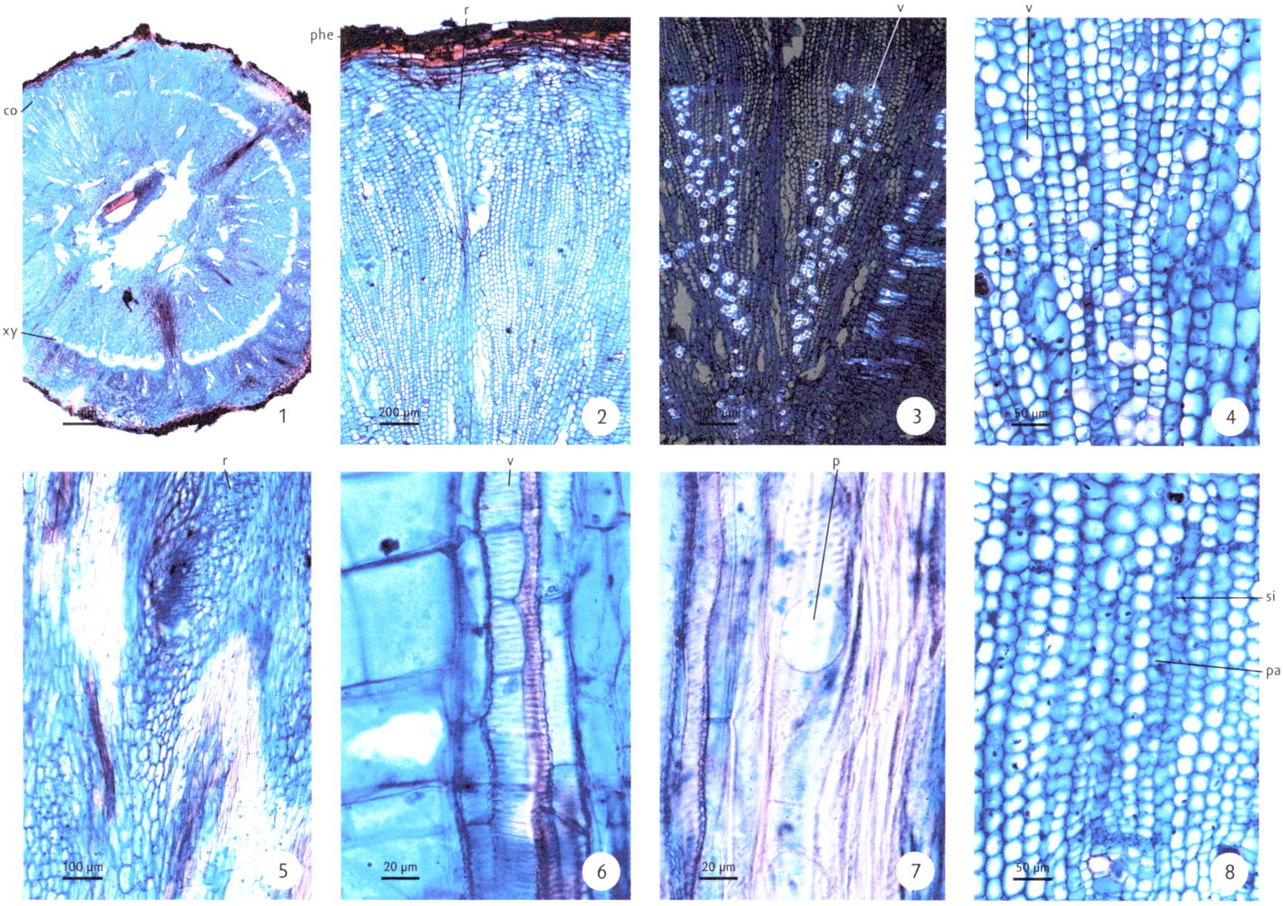

Cirsium canum (L.) All.

Perennial wetland herb. Stem erect, tough, 50–150 cm high, with basal leaf rosette. Leaves softly spiny, unlobed to lobed, mainly in the lower part of stem. Flower heads usually terminal with purple florets. Pale achenes with feathery pappus.

Distribution and ecology

Euro-Siberian temperate element with continental distribution. Native in all CE countries except CH, where it is naturalised. In lowlands and uplands in wet meadows and pastures, in ditches and along water banks.

Anatomy of the flower stalk and the rhizome

Round ribbed annual shoot with a diameter of 5 mm and round rhizome with a diameter of 8 mm. Maximum age 2 years (1).

The annual shoot consists of a circle of collateral vascular bundles, which are surrounded by a belt of unlignified cells in the cortex (1).

Basal shoots (root collar) are surrounded by a dense xylem belt, a phloem and a cortex with some sclerenchyma (2). The phellem is absent.

The xylem consists of vessels, thin-to thick-walled fibers and paratracheal parenchyma (2). The ring boundary diffuse porous xylem is marked by marginal, unlignified parenchyma (2, 3). Vessels have a diameter of 40–60 µm (2, 3). Vessel length in both parts is 100–180 µm. Vessels are arranged solitary or in short radial groups (2, 3). Intervessel pits are round in the lignified xylem and scalariform in the unlignified marginal parenchyma (4). Perforation plates are generally simple (4) and occasionally scalariform. Rays consist of square and upright lignified cells, they are 5–8 cells wide with laterally separated sheet cells (5). Large rays also separate vascular bundles laterally (1). Some rays are dilated (1).

The phloem consists of parenchyma cells, a few sieve tubes, small groups of fibers and dilating rays (1, 6). The cortex consists of irregularly formed, unlignified parenchyma cells. Few ducts are located in the cortex (6). Intercellulars are absent (6).

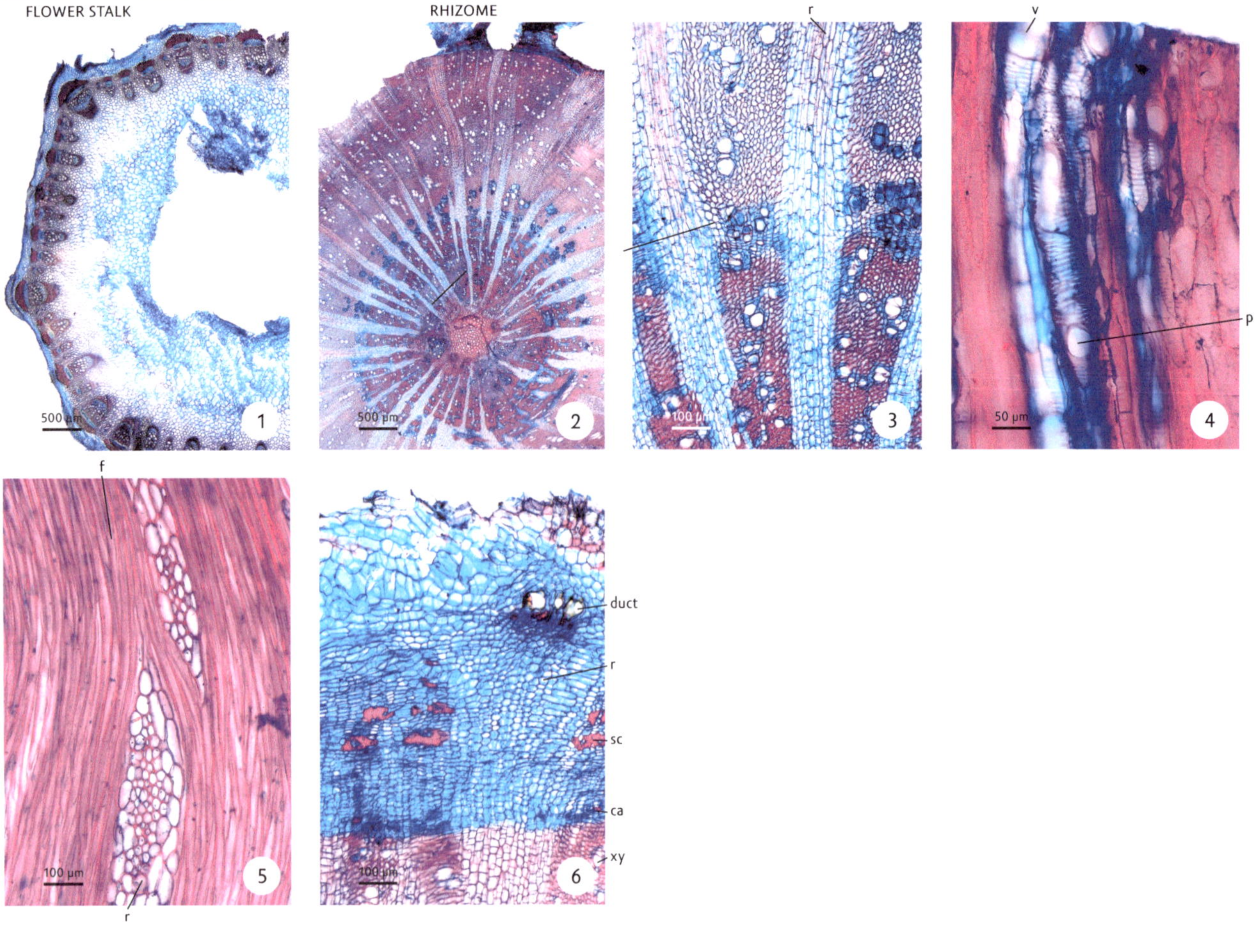

Cirsium heterophyllum (L.) Hill **Melancholy Thistle** Asteraceae

(syn. *Cirsium helenioides*)

Perennial wetland herb. Stem erect, grooved, usually unbranched, spineless, 45–120 cm high. Leaves mainly in the basal part, serrate on the margin. Basal leaves with short petiole, stem leaves sessile. All leaves green, hairless above, thickly white-felted bellow. Flower heads solitary, up to 5 cm in diameter, with red-purple florets, involucre oval. Fruit is an achene with pappus.

Distribution and ecology

Euro-Siberian boreo-montane element. In all CE countries except HU, scattered in the mountains. Humic wet soils in wet meadows, wet ditches and along mountain brooks.

Anatomy of the transition zone between the root and the annual shoot

Round rhizome with a diameter of 8 mm. Maximum age 2 years (1).

Basal shoots (root collar) are surrounded by a completely lignified xylem belt, a phloem and a cortex with some sclerenchyma (1). The phellem is absent (1). The xylem consists of vessels, thin-to thick-walled fibers and paratracheal parenchyma (2). Ring boundaries are indicated by semi-ring porosity and marginal unlignified parenchyma (1, 2).

Vessels are angular and have a diameter of 50–80 µm (2). Vessel length varies between 120–180 µm. Vessels are primarily in tangentially arranged groups (1, 2). Intervessel pits are round to slightly horizontally enlarged (3) and perforation plates primarily simple and occasionally scalariform (3, 4). Rays consist of square and upright cells, they are 5–10 cells wide with laterally separated sheet cells (5). Large rays also separate vascular bundles laterally (1, 2).

The phloem consists of parenchyma cells, a few sieve tubes, small groups of fibers (6) and dilating rays (6) and the cortex of round, unlignified parenchyma with small intercellulars. Large ducts are located in the cortex, just outside of the phloem (6).

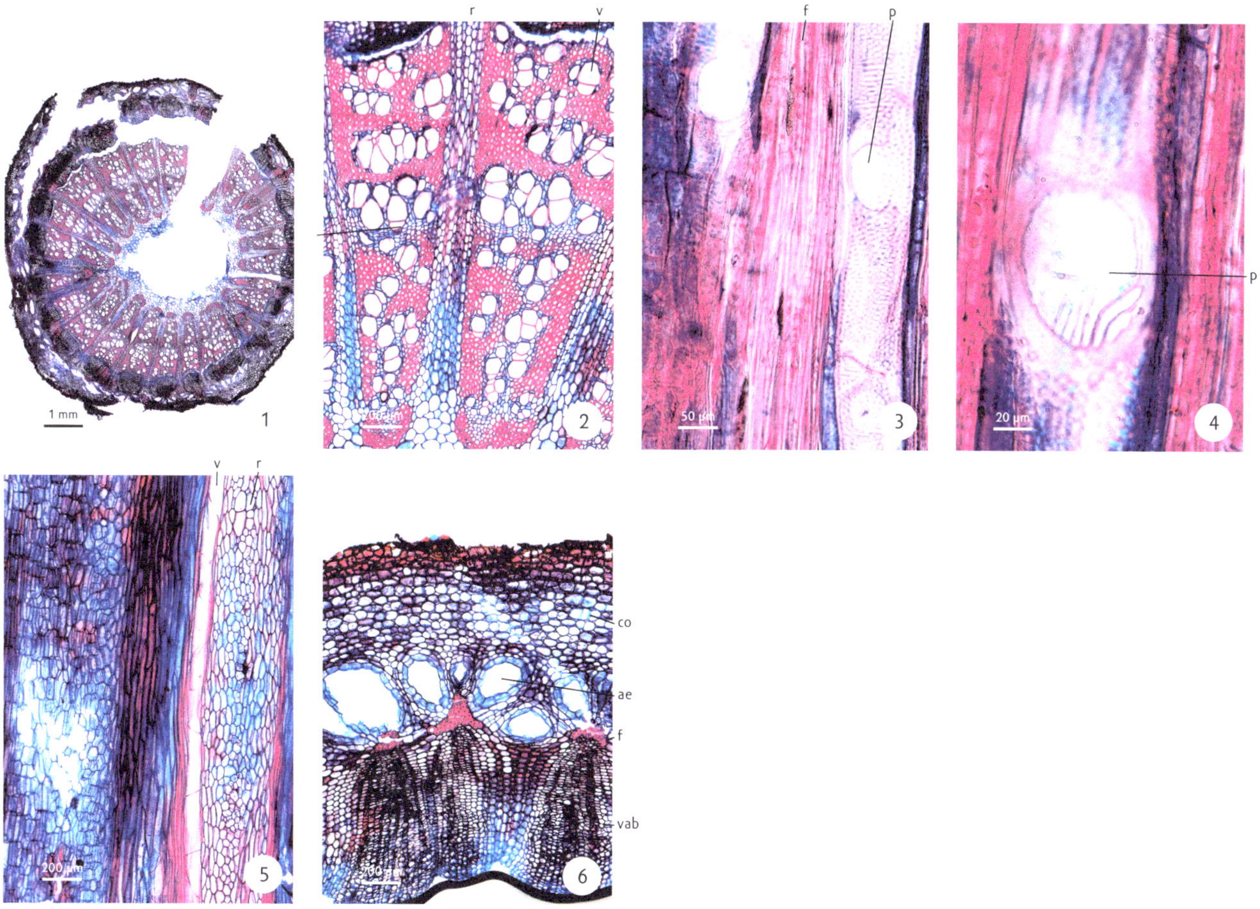

Cirsium oleraceum (L.) Scop.

Cabbage Thistle Asteraceae

Perennial wetland herb. Stem erect, leafy to the top, unwinged, 50–120 cm high. Pale green leaves sligthly to deeply lobed, softly spiny. Flower heads erect, clustered, overtopped by oval yellowish upper leaves. Florets yellow. Fruit is an achene with rusty pappus.

Distribution and ecology

Euro-Siberian temperate distribution. Common in all CE countries. Wet and flooded meadows, ditches, lowland forests, springs, marshes and banks of streams. Prefers nutrient-rich, wet and waterlogged soils.

Anatomy of the rhizome

Round rhizome with a diameter of 7–13 mm. Maximum age 3 years (1).

The large pith of rhizome is surrounded by single vascular bundles, consisting of a xylem, a phloem and a cortex (1) with some sclerenchyma (1, 8). The phellem is absent (1, 8). The xylem of the vascular bundles consists of vessels, thin-to thick-walled fibers and paratracheal parenchyma (2, 3). The innermost part of the vascular bundles is unlignified (3). Ring boundaries are mainly indicated by semi-ring porosity and marginal parenchyma (2). Vessels are primarily arranged solitary or in small groups (2, 3). Vessels have a diameter of 50–80 µm (2–3). Vessel length varies between 120–180 µm. Intervessel pits are round to slightly horizontally enlarged (4) and perforation plates are primarily simple and occasionally (4) scalariform (5).

Rays consist of square and upright unlignified cells and they are 5–10 cells wide with laterally separated sheet cells (6). Large rays also separate vascular bundles laterally (1–3).

The phloem consists of parenchyma cells, a few sieve tubes (7), small groups of fibers, sclereids (8) and dilating rays (6). The cortex is composed of round, unlignified parenchyma, a few small ducts just outside of the phloem. Intercellulars are almost absent.

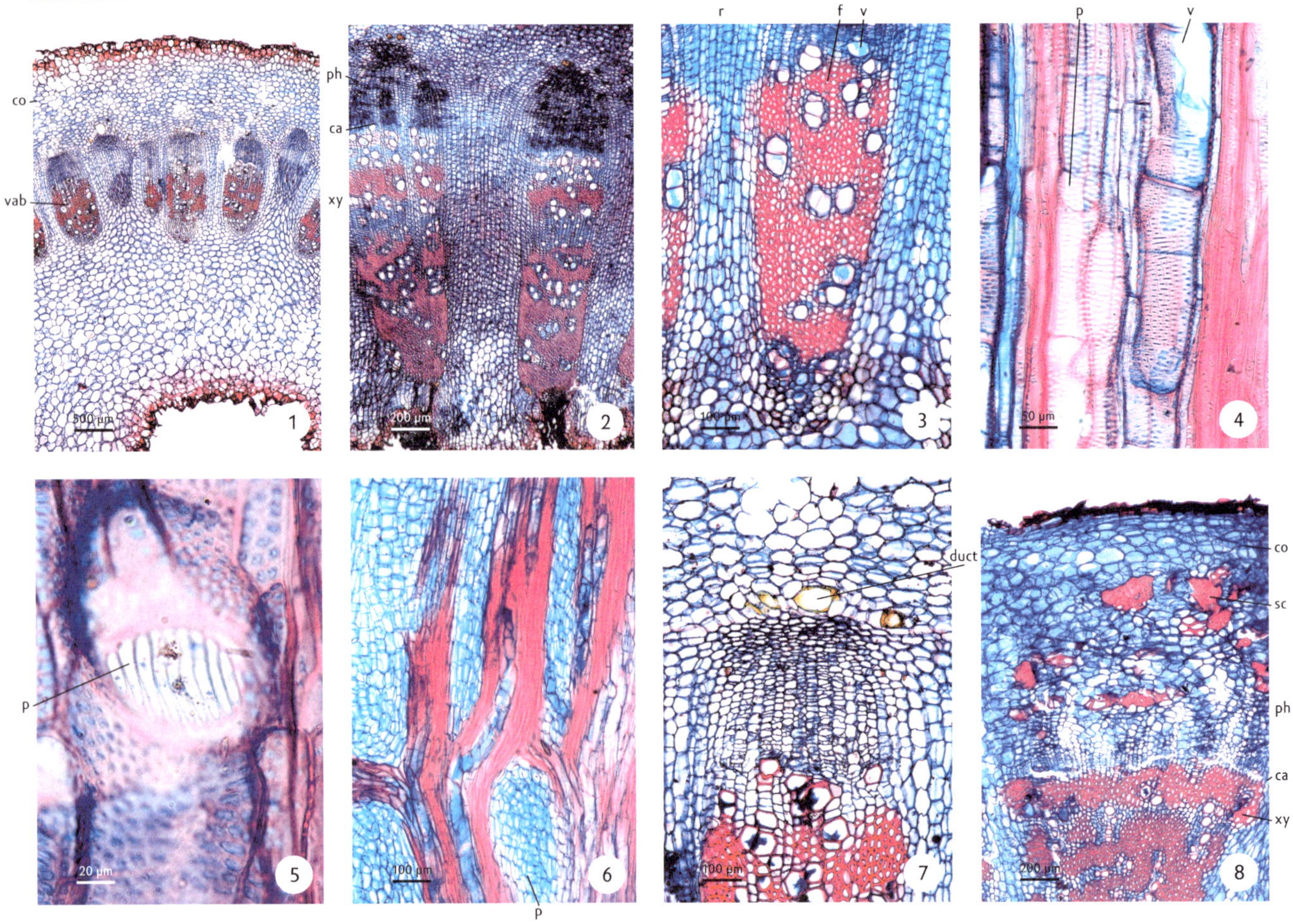

Cirsium palustre (L.) Scop. **Marsh Thistle** Asteraceae

Biennial or monocarpic perennial wetland herb. Stem erect, branched above, continuously spiny-winged, 30–150 cm high. Dark green leaves deeply pinnatified, strongly spiny. Flower heads in crowded clusters, outer bracts purplish, florets dark red-purple or sometimes white. Fruit is an achene with pappus.

Distribution and ecology

Euro-Siberian boreo-temperate element; widely naturalised outside its native range. Common in all CE countries. Typical of marshes, fens, mires, wet grasslands, pastures, wet woodlands, springs and tall-herb vegetation in mountains. Persistent seed bank.

Anatomy of the transition zone between the root and the annual shoot

Round stem with a diameter of 8–12 mm. Maximum age 2 years (1).

Basal shoots (root collar) are surrounded by a circle of radial long vascular bundles, a phloem and a cortex. The phellem is absent.

The xylem consists of vessels, thin-to thick-walled fibers and paratracheal parenchyma (2). Ring boundaries of the diffuse porous xylem are marked by marginal zone with an enhanced amount of unlignified parenchyma (1–3). Vessels have a diameter of 40–50 µm (2, 3) and a length of 100–120 µm. Vessels are arranged in groups (2, 3). Intervessel pits are round to scalariform (4, 5) and perforation plates are simple (4, 5). Rays consist of square and upright, thin-walled lignified cells and are 5–10 cells wide (1–3). Large rays separate vascular bundles laterally (1, 2). Some rays are dilated (1).

Detailed phloem structures could not be observed.

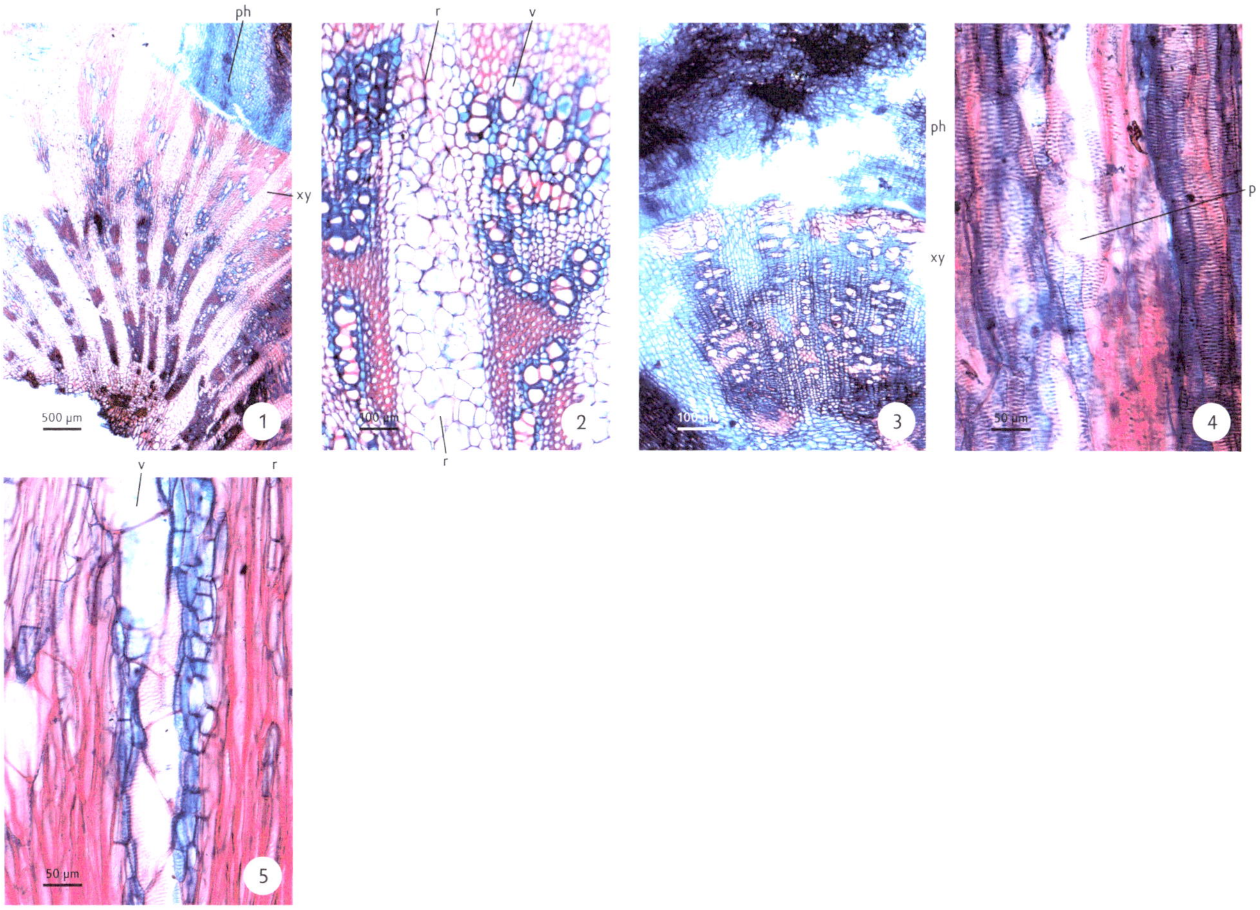

Crepis paludosa L. Moench

Marsh Hawk's-beard

Asteraceae

Perennial wetland herb, with short rhizomes. Stem erect, leafy, glabrous, hollow, branched above only, 30–90 cm high. Lower leaves elliptical-lanceolate, upper leaves with long, back-pointed basal lobes forming arrow-shape. Flower heads few in a loose corymb, erect, florets golden yellow. Fruit is an achene with simple pappus.

Distribution and ecology

European boreo-temperate element. In all CE countries from foothills to the mountains in wet to waterlogged, nutrient-rich soils. Occurs in springs, wet peaty meadows, along brooks and in alder carrs.

Anatomy of the annual shoot and the rhizome

Round stem with a diameter of 3–5 mm. Maximum age 2 years.

Annual shoot: Pentagonal outline. A circle of collateral vascular bundles and a cortex surrounds a very large pith (1). Vascular bundles are laterally connected by a vessel-free belt of fibers. Peripheral, longitudinal unlignified stabilizing strips consist of collenchyma (2).

Root collar: An almost continuous belt of xylem, phloem and cortex surrounds the pith which consists of round unlignified parenchyma cells (3). Ring boundaries are characterized by marginal parenchyma (4). The xylem consists mainly of thin- to thick-walled fibers (4). Vessels have the same dimensions as fibers and are therefore difficult to recognize (4). Intervessel pits are round to scalariform in alternating position and perforation plates primarily simple (5). Vessels have a diameter of 20–25 µm (4) and vessel length is <100 µm. Parenchyma is unlignified and exists only on ring boundaries (4). Rays are absent. The large rays in (3) represent lateral roots.

The phloem consists of parenchyma cells and small groups of sieve tubes (4, 6). The phloem is surrounded by a poorly differentiated pericycle/endodermis zone (6). Large oval, unlignified cells characterize the cortex. Between them, there are small intercellulars. The phellem consists of 2–3 rows of irregularly formed cells.

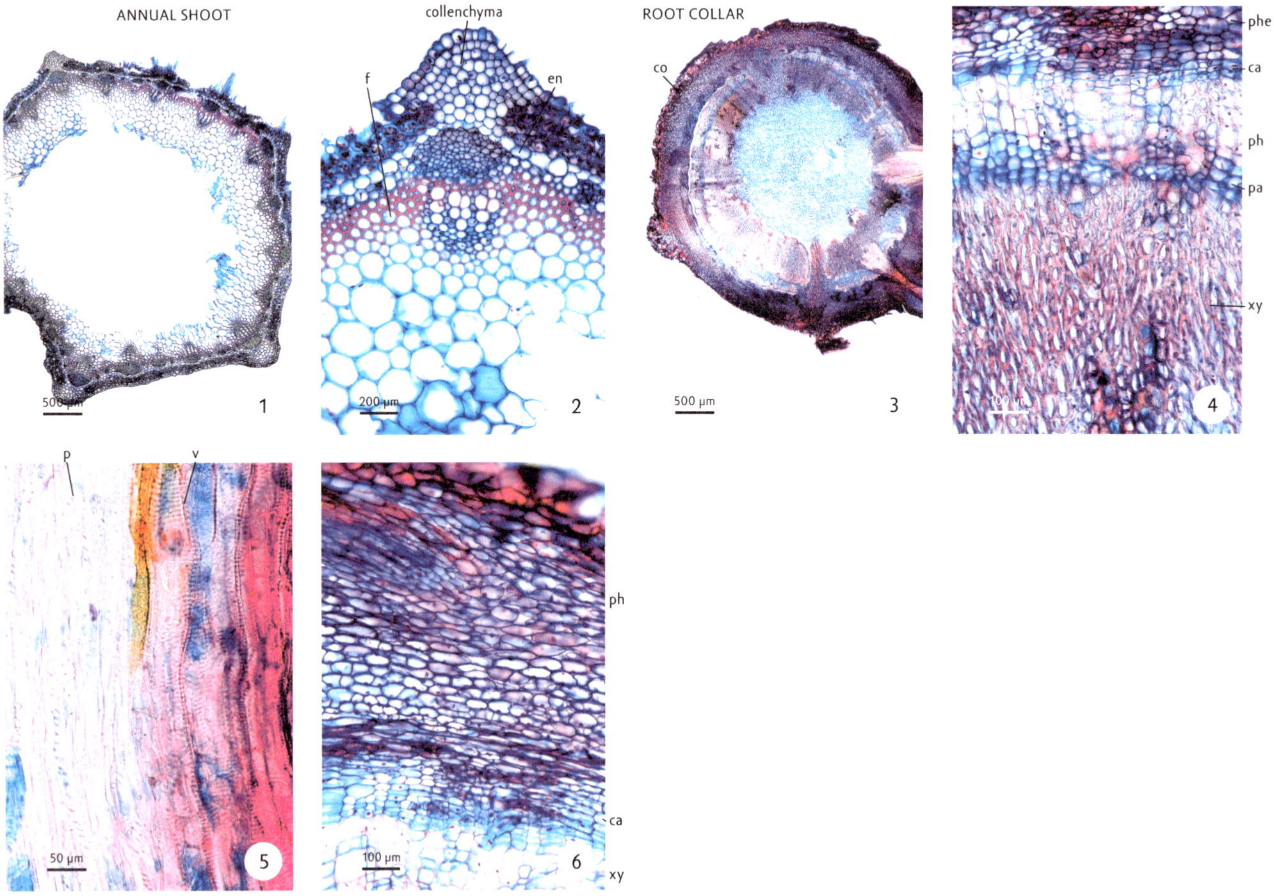

Gnaphalium supinum L. (syn. *Omalotheca supina*) **Dwarf Cudweed** Asteraceae

Perennial wetland herb, dwarf tufted. Stem erect, unbranched, wooly, 2–12 cm high. Leaves sessile, linear to lanceolate, grey-woolled on both sides. Flower heads 1–7 in a compact terminal spike, bracts with broad, brown chaffy margins. Florets whitish or greenish. Fruit is an achene with pappus.

Distribution and ecology

European arctic-alpine distribution; also in C. Asia and N. America. In all CE countries except HU. Occurs especially in snowbeds, also in springs, in alpine meadows and pastures. Prefers wet stony soils rich in nutrients, avoids calcium-rich soils. Heliophilous and psychrophilous species.

Anatomy of the transition between the root and the annual shoot

Round stem with a diameter of 1–2 mm. Indistinct annual rings.

The root collar consists of a pith, a dense, totally lignified xylem belt, a small phloem and a large cortex (1). The xylem consists of vessels, thick and thin- to thick-walled fibers and paratracheal and apotracheal parenchyma (2, 3). Vessels, fibers and parenchyma cells have almost the same diameter and it is often difficult to recognize them (2, 3). The solitary arranged vessels have a diameter of 10–20 μm (2, 3). Vessel length varies from 30–50 μm (4). Intervessel pits are round and scalariform (5) and perforation plates are simple (5). Rays are absent (2).

The phloem consists of parenchyma cells and a few sieve tubes difficult to recognize (2). The stele is surrounded by a poorly differentiated pericycle/endodermis zone (2, 3). The cortex consists of large, round to oval, unlignified parenchyma cells (2). The phellem is probably absent (1).

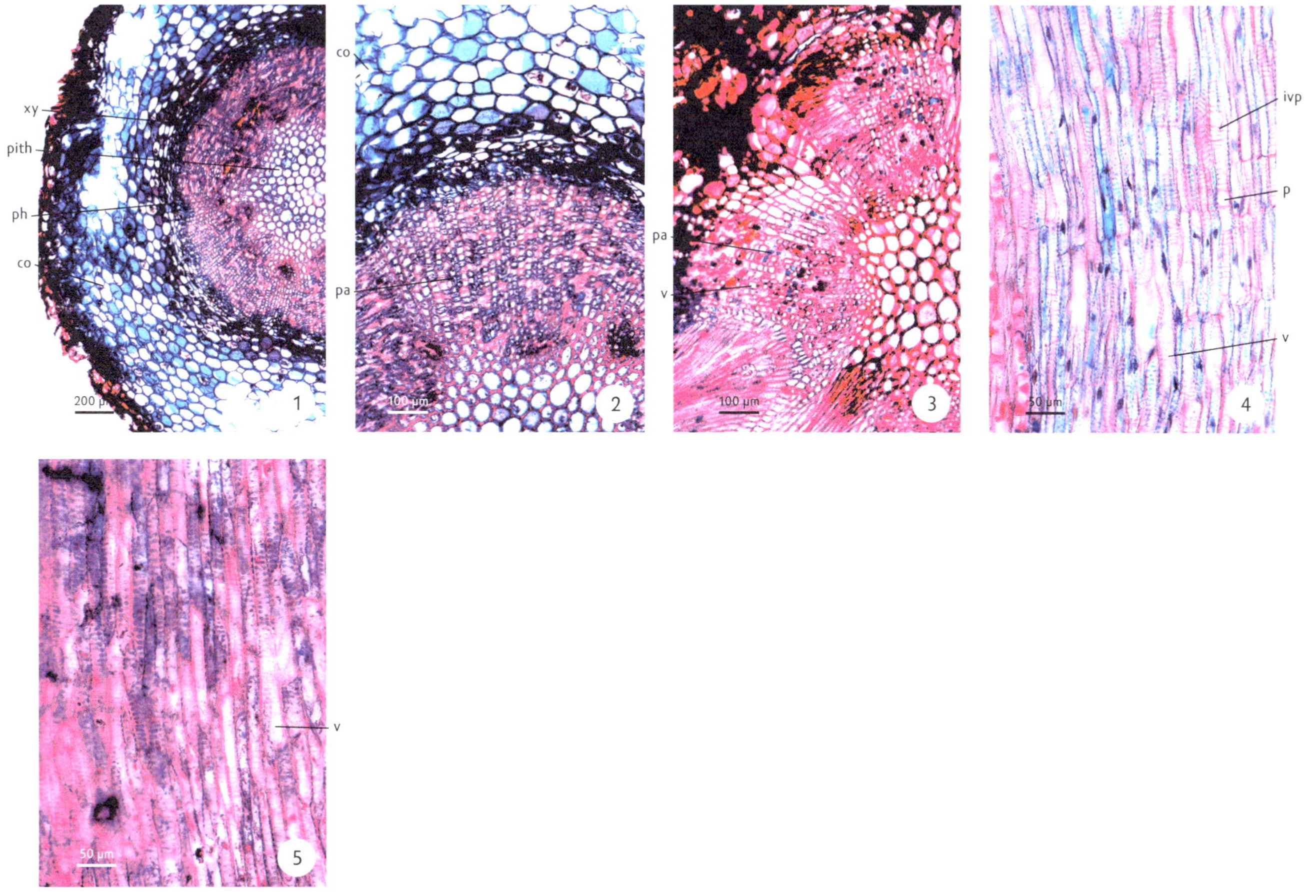

Gnaphalium uliginosum L. (syn. *Filaginella uliginosa*) **Marsh Cudweed** Asteraceae

Annual wetland herb. Stem decumbent to erect, often richly branched from base, 2–30 cm high. Leaves woolly both sides, ovate to lanceolate. Flower heads in dense terminal clusters, surrounded by long leaves, bracts woolly. Florets brownish-yellow. Fruit is an achene with simple pappus.

Distribution and ecology

Eurasian boreo-temperate element, but naturalised in N. America. In all CE countries on open, muddy ground, often subject to waterlogging during the winter. Typical of drained fishpond bottoms, margins of reservoirs, also as a weed in wet depressions in fields. Prefers compacted, mildly to strongly acidic soils.

Anatomy of the transition between the root and the annual shoot

Round stem with a diameter of 1–2 mm. Maximum age 1 year (1).

The root collar consists of a dense totally lignified xylem belt, a small phloem and a large cortex (1, 2). The xylem consists of vessels, thin- to thick-walled fibers and paratracheal parenchyma (3). Vessels are solitary arranged (3) and have a diameter of 20–30 µm (3). Vessel length varies from 80–100 µm (4). Intervessel pits are round (4) and perforation plates are simple (4). Rays are absent (3, 5).

The phloem consists of parenchyma cells and a few sieve tubes (6). Phloem and cortex are separated by an endodermis-like layer of thin-walled cells (7). The cortex consists of large, round to oval, unlignified parenchyma cells, which are separated by small and large intercellulars (7). The phellem is absent (7).

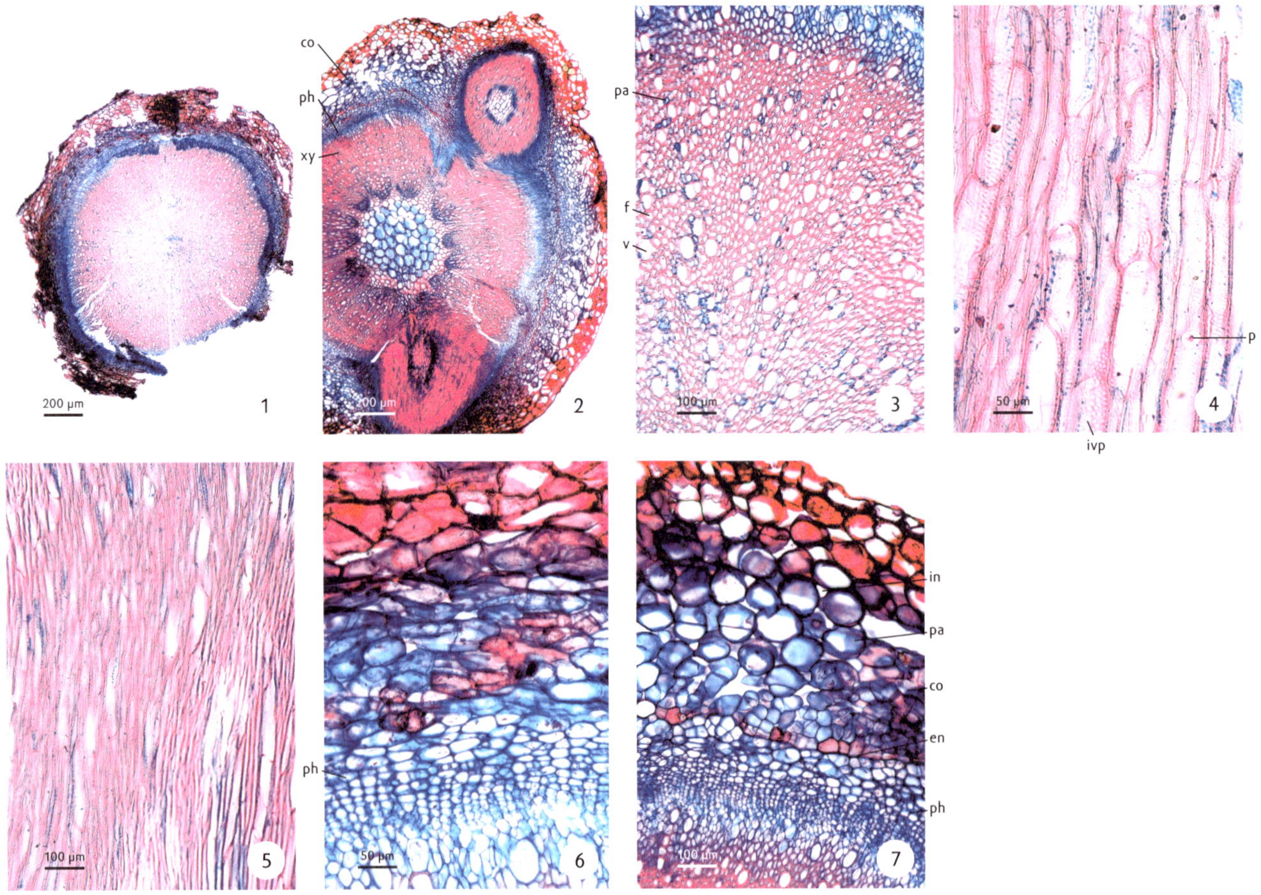

Biennial or perennial wetland herb. Stem erect, branched, hairy, dense leafy, often in polycormones, 20–70 cm high. Stem leaves sessile, linear to elongate, glabrous above, pubescent and glandular beneath. Flower heads 2–5 cm across in subcorymbose panicle, ray- and disc florets yellow. Fruit is a pubescent achene with pappus.

Distribution and ecology

Eurasian temperate element. In all CE countries rather in lowlands. It grows in wet meadows, along water banks, also in ruderal stands. Prefers heavy soils with fluctuating water table. Facultative halophyte.

Anatomy of the transition between the root and the annual shoot

Round stem with a diameter of 4–6 mm. Maximum age 2 years.

Stems above and below ground are surrounded by a xylem, a phloem, a cortex and a small phellem (1). The ring boundary is marked by a marginal parenchyma (2). The xylem consists of vessels, thin- to thick-walled fibers, paratracheal parenchyma and rays (2). Vessels have a diameter of 20–40 µm (3) and vessel length is 150–220 µm. Vessels of the semi-ring porous xylem are arranged in short radial rows (1, 3). Intervessel pits are round and perforation plates simple (3). Fibers are thin- to thick-walled (3) and have small slit-like pits. Rays are uni- and multiseriate (5) and often confluent (4). Ray cells are upright (3).

The phloem consists of parenchyma cells, a few sieve tubes, irregular groups of sclerenchyma and dilating rays (5). The cortex in the above ground shoot is represented by an aerenchymatic tissue, consisting of irregularly formed thin-walled parenchyma cells, single sclereids and very few ducts (1, 5). Sclereids are round in cross-sections (5). Many parenchyma cells are filled with dark staining substances (5) with tangentially enlarged intercellulars (5).

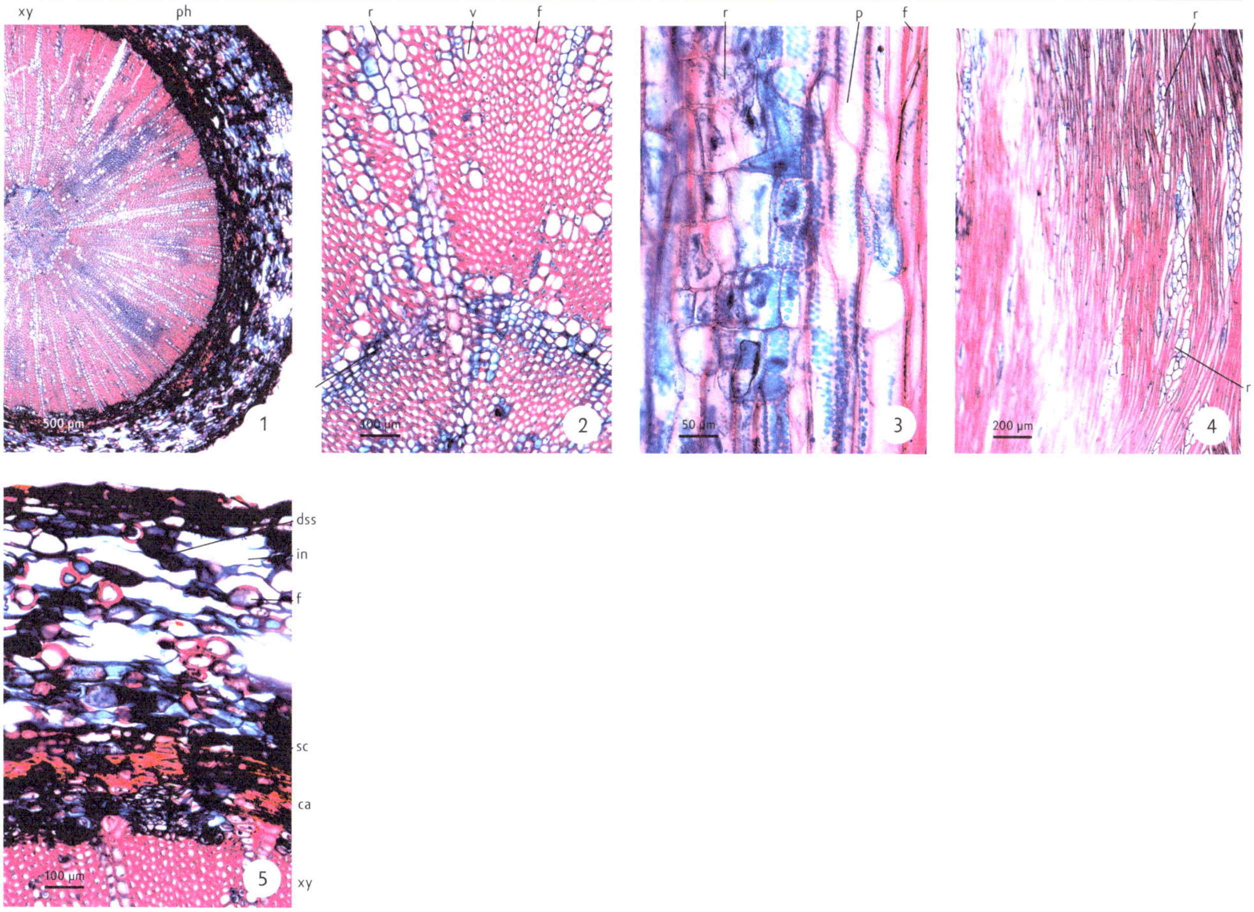

Perennial wetland herb, dioecious and rhizomatous. Stem erect, 30 cm high in flowering time, in fruiting up to 80 cm. Basal leaves large, petiolate, rounded-cordate, serrate on the margin, white-wooly beneath. Stem leaves squamous, glandulous. Flower heads in a raceme with white flowers. Fruit is an achene with pappus.

Distribution and ecology

Native in mountains of C. and SE Europe and SW Asia (the Caucasus). In all CE countries from foothills up to the mountains. It occurs by springs, in ditches, in shady riversides, along forest brooks and forest tracks, occasionally forming large stands.

Anatomy of the rhizome

Round rhizome with a diameter of 9–10 mm. Maximum age 2 years.

The large pith of the rhizome is surrounded by solitary vascular bundles, a cortex and a small phellem (1, 2). The major part of the bundles has a centripetal vessel-free group of thick-walled fibers (3, 4). The vascular bundles consist of vessels, thin- to thick-walled fibers and paratracheal parenchyma (3). Ring boundaries of the diffuse-porous wood are mainly indicated by an abrupt change of thick-walled and thin-walled fibers between the latewood and the earlywood (4). Vessels have a diameter of 30–50 µm (3, 4) and a length of 100–200 µm (5). Vessels are solitary (3, 4). Intervessel pits are scalariform and perforation plates are simple (5). Rays between vascular bundles consist of square and upright unlignified cells (see *P. paradoxus*, 5) and are >10cells wide (6). Rays are absent within the bundles (3, 4).

The phloem consists of parenchyma cells and groups of sieve tubes (7) and occasionally of small groups of fibers (2, 3, 4) and dilating rays (2, 3). Outside of each bundle is a large duct (2, 3, 4, 7). Intercellulars are very small in the cortex (8). The phellem, if present, consists of a few square cells.

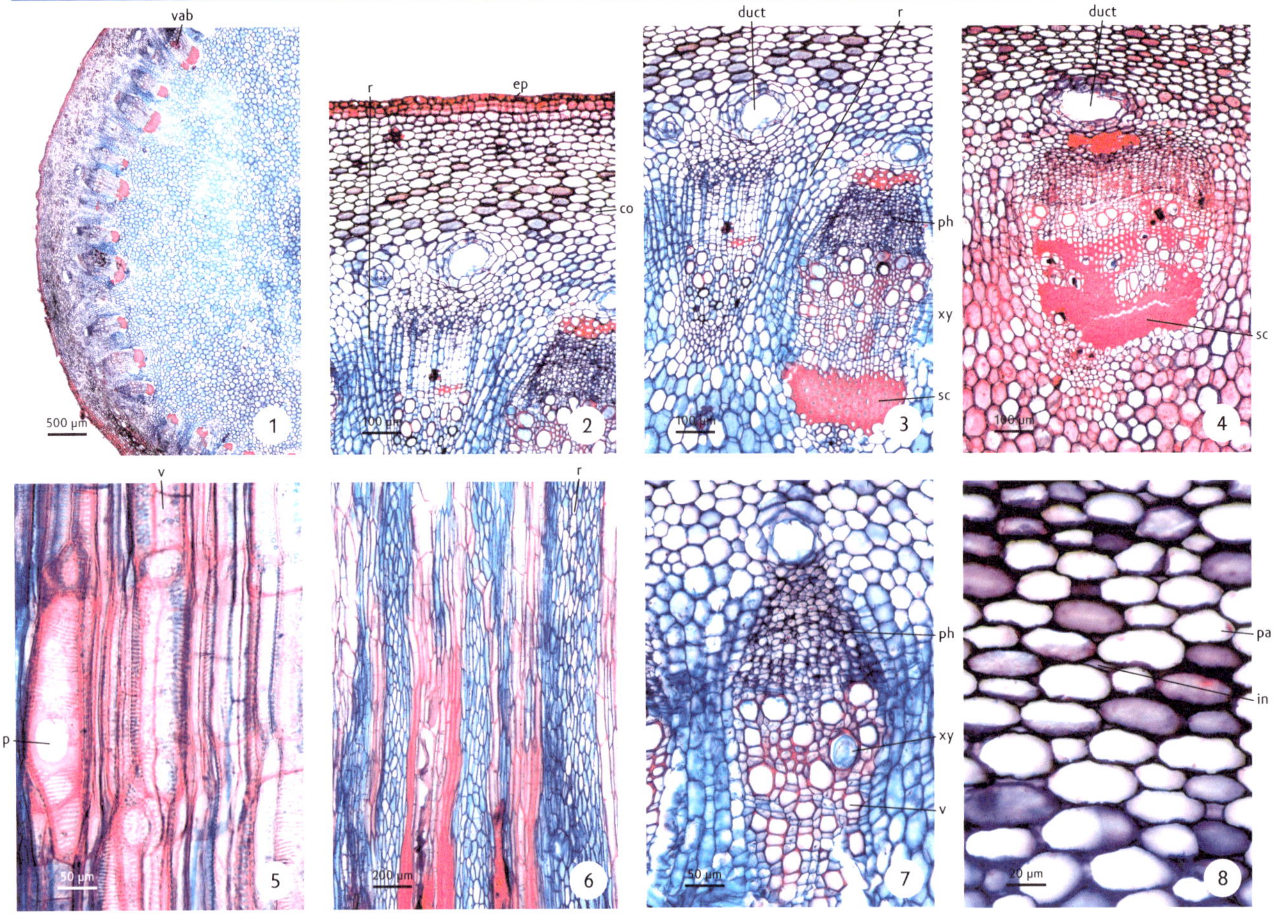

Petasites hybridus (L.) G. Gaertn. et al. **Butterbur** Asteraceae

Perennial wetland herb, dioecious and rhizomatous. Stem erect, with red colour, 10–40 cm high in flowering time, in fruiting 70–100 cm high. Large basal leaves up to 80 cm across, petiolate, serrate on the margin, grey-downy below. Stem-leaves squamous. Flower heads in a raceme, tubular florets reddish-pink. Fruit is an achene with pappus.

Distribution and ecology

European temperate element. In all CE countries from foothills to the mountains, especially in banks of rivers and streams and in wet meadows. Prefers moist, fertile, often alluvial soils. It spreads mostly vegetatively from rhizome fragments. Aromatic and medicinal plant.

Anatomy of the rhizome

Round rhizome with a diameter of 15–40 mm. Maximum age 1–2 years.

The large pith of the rhizome is surrounded by solitary, radially extended vascular bundles, a cortex and a small phellem (1, 2). The major part of the bundles has a centripetal vessel-free group of thick-walled fibers (3). The vascular bundles consist of vessels, thin- to thick-walled fibers, paratracheal and pervasive parenchyma (2). The angular vessels have a diameter of 50–80 µm (2) and a length of 100–200 µm (3). Vessels are in groups (2). Intervessel pits are scalariform and perforation plates are simple (3). Rays between vascular bundles consist of square and upright unlignified cells and are 5–10 cells wide (2). Rays are absent within the bundles (3, 4).

The phloem consists of parenchyma cells, groups of sieve tubes (2), small groups of fibers (2, 3) and dilating rays (2, 3). Outside of each bundle is a duct (1, 2). Intercellulars are small in the cortex (5). The phellem, if present, consists of a few square cells inside of the epidermis (1) with crystal sand and prismatic crystals (4).

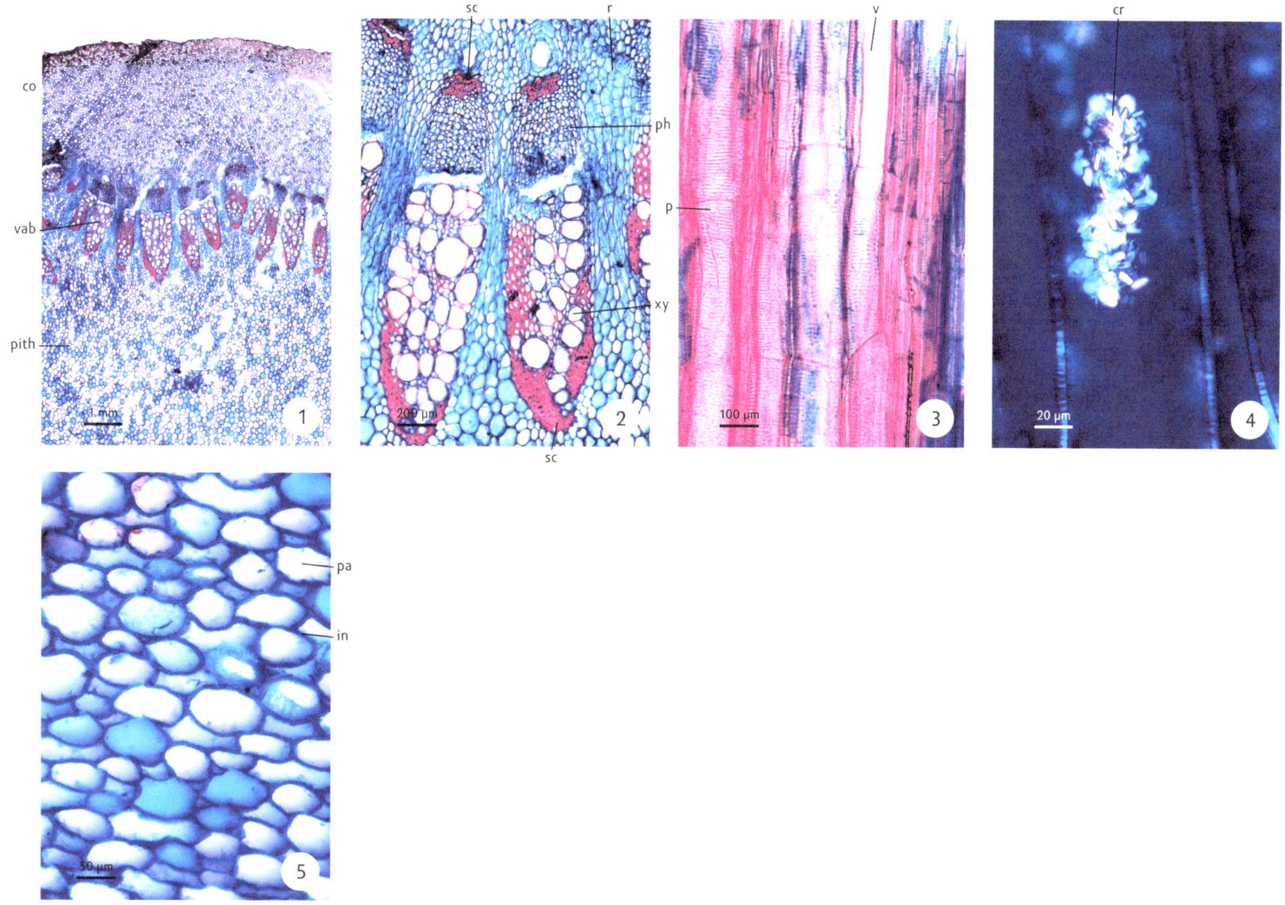

Pseudognaphalium luteoalbum (L.) Hilliard et B. L. Burtt Asteraceae

(syn. *Gnaphalium luteo-album*) **Jersey Cudweed**

Annual or biennial wetland herb. Stem simple or richly branched from the bottom, woolly pubescent, 10–40 cm high. Basal leaves obovate, grey-woolled on both sides. Flower heads in dense terminal leafless clusters of 4–12, florets bright yellow. Fruit is an elongated achene with pappus.

Distribution and ecology

Euro-Siberian, southern-temperate element, widely naturalised outside its native range. In all CE countries scattered in drained fishpond bottoms, margins of ponds and rivers, in wet sandy fields and dune-slacks.

Anatomy of the transition zone between the root and the annual shoot

Round stem with a diameter of 3–4 mm. Maximum age 2 years (1).

The root collar consists of a dense, totally lignified xylem belt, a small phloem, a large cortex and a small phellem (1, 2). The xylem consists of vessels, thin- to thick-walled fibers and paratracheal parenchyma (2). Vessels are arranged in short radial rows (2), they have a diameter of 30–40 µm (2). Vessel length varies from 80–120 µm (3, 4). Intervessel pits are round (3, 4) and perforation plates are simple (4). Rays are absent (5).

The phloem consists of small, unlignified sieve tube groups and lignified parenchyma cells (5). Phloem and cortex are separated by an endodermis-like layer of thin-walled cells (2). The cortex consists of large, round to oval parenchyma cells. Intercellulars are absent (2). The phellem is absent in small and present in large individuals.

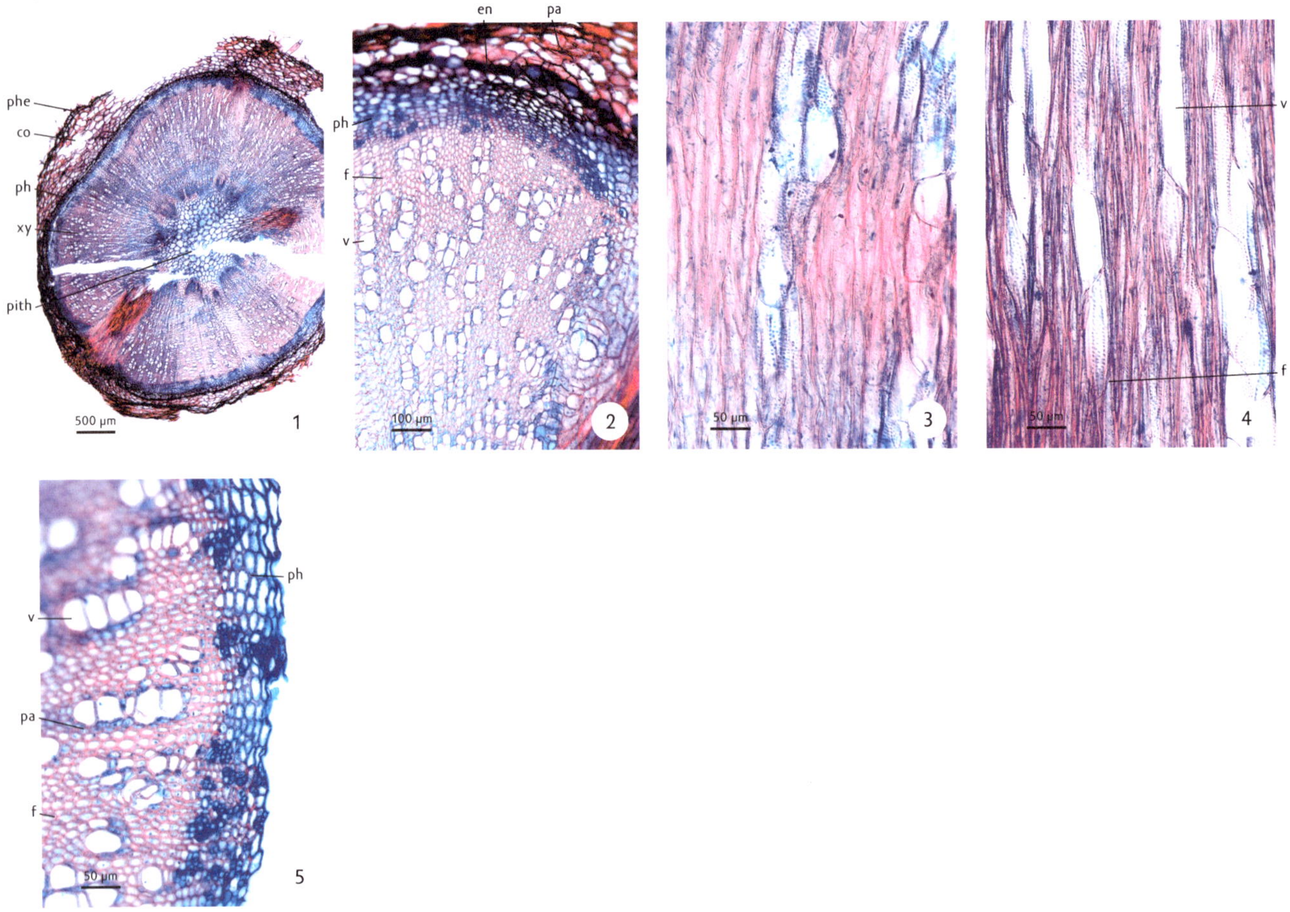

Pulicaria dysenterica (L.) Bernh. **Common Fleabane** Asteraceae

Perennial wetland herb, with runners. Stem erect, branched in the upper part, 30–60 cm high. Leaves alternate, oblong and untoothed, upper leaves cordate-based. Flower heads in a loose corymb. Ray-florets twice as long as disc-florets, all florets golden-yellow. Fruit is a hairy achene with pappus.

Distribution and ecology

Euro-Siberian, southern-temperate element. In all CE countries in wet meadows, tall-herb fens, along rivers, canals and ditches, in floodplain forests and in dune-slacks. It occurs on a wide range of acidic and base-rich soil types, often in slightly saline soils.

Anatomy of the transition between the root and the annual shoot

Round stem with a diameter of 3–5 mm. Maximum age 1 year.

The root collars are surrounded by a continuous belt of xylem, a phloem, an aerenchymatic cortex and a small phellem (1). The ring boundary is marked by 1–2 tangentially flat fibers (2). The xylem consists of vessels, thin- to thick-walled fibers, paratracheal parenchyma and rays (2, 3). Vessels have a diameter of 40–50 µm (2) and vessel length is 120–220 µm. Vessels of the diffuse-porous xylem are solitary or arranged in short radial rows (2, 3).

Intervessel pits are round to scalariform, perforation plates are primarily simple (4) and occasionally scalariform (5). Fibers are thin- to thick-walled (2, 3, 5) and have small slit-like pits (4). Rays are extremely high (>3 mm) and 5–8-seriate (6). Ray cells are upright.

The phloem consists of parenchyma cells and small groups of sieve tubes. The primary phloem is represented by roundish groups of fibers (7). Rays dilate (7, 8). The aerenchymatic cortex is characterized by net-like large intercellulars. Few ducts are located at the boundary between phloem and cortex (7, 8). The phellem consists of 2–3 rows of round cells (8).

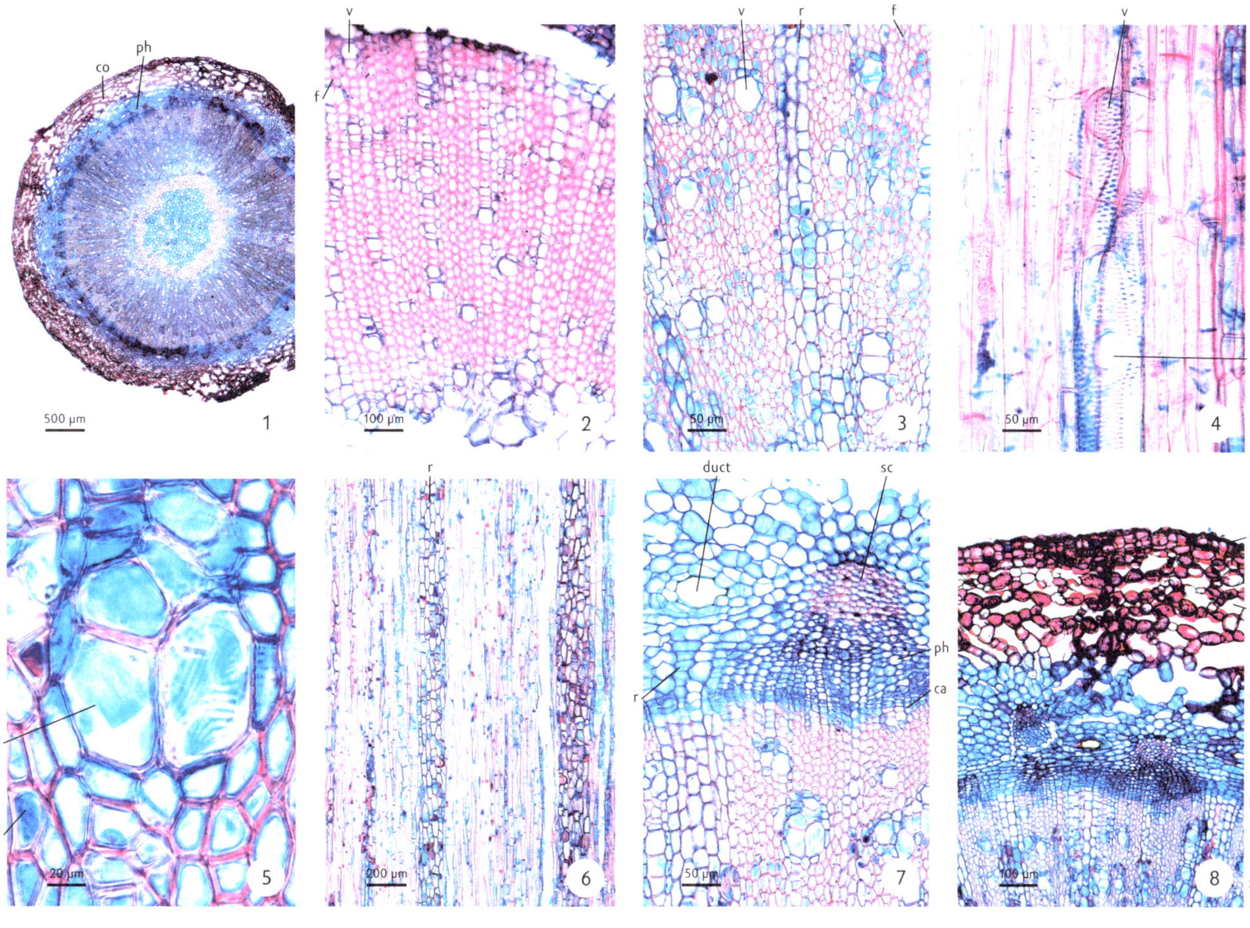

Pulicaria vulgaris Gaertn. **Small Fleabane** Asteraceae

Annual wetland herb. Stem erect, branched, with sticky hairs, 8–40 cm high. Stem leaves alternate, oblong, upper leaves rounded-based. Flower heads many, only 1 cm across. Ray-florets equal to length of disc-florets, all florets pale yellow. Fruit is a hairy achene with pappus.

Distribution and ecology

Euro-Siberian temperate element. Native in all CE countries; especially in lowlands on acidic soils. Occurs in winter-flooded grasslands, in pastures, along banks of ponds and wet tracks. It is declining throughout Europe – largely due to lack of grazing as seedlings need open sites for survival. Germinates both in light and in darkness.

Anatomy of the transition between the root and the annual shoot

Round stem with a diameter of 3–5 mm. Maximum age 1 year.

The root collars are surrounded by a xylem, a phloem, an aerenchymatic cortex and a small phellem (1). The xylem contains single vascular bundles (2) or forms a continuous lignified belt (1, 3). The xylem consists of vessels, thin- to thick-walled fibers, paratracheal parenchyma and rays (1–3). Vessels have a diameter of 15–25 µm (2, 3) and vessel length is 100–150 µm. Vessels of the diffuse-porous xylem are arranged in short (3) or long radial rows (2). Intervessel pits are round and perforation plates are simple (4). Fibers are thin- to thick-walled or thick-walled (2, 3) and have small slit-like pits. Rays are confluent and their width varies between 2 and 6 cells. Ray cells are upright (5).

Phloem and cortex are separated by a distinct thin-walled endodermis (1, 6). The phloem is homogeneous (6, 7). Rays dilate (1, 3). Groups of fibers belong to the cortex (7). The aerenchymatic cortex is characterized by net-like large intercellulars (2). Ducts are absent. The phellem consists of 2–3 rows of round cells (1).

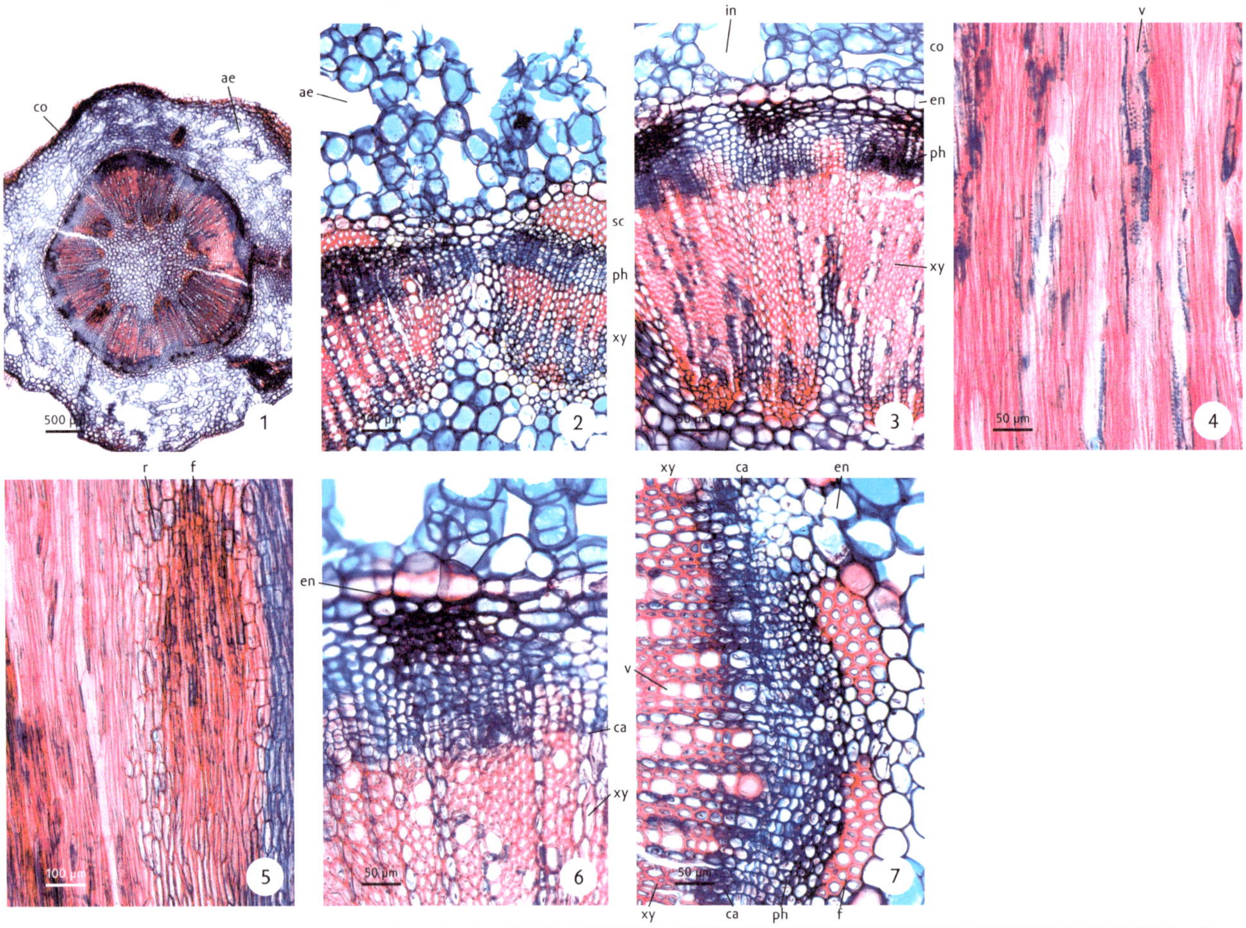

Senecio alpinus (L.) Scop. (syn. *S. cordatus*)

Perennial wetland herb. Stem erect, regularly leafy, 30–100 cm high. Leaves cordate, irregularly serrate, with unwinged petiole. Long stalked flower heads with yellow ray- and disc florets. Fruit is an achene with yellow pappus.

Distribution and ecology

Endemic in the eastern Alps. Native in AT, CH and DE in subalpine and alpine altitudes. Occurs in alpine meadows and pastures, also in banks of streams. Prefers nutrient- and calcium-rich soils. Contains alkaloids toxic for cattle.

Anatomy of the rhizome

Round rhizome with a diameter of 10 mm. Maximum age 4 years.

The large pith of the rhizome is surrounded by solitary dilating vascular bundles, a cortex and a small phellem (1). The initial part of the bundles consists of a few sclereids (2). The vascular bundles consist of vessels, thick-walled fibers and paratracheal parenchyma (3). Ring boundaries of the diffuse-porous wood are indicated by a large zone of marginal parenchyma (4). Vessels have a diameter of 40–50 µm (4). Vessel length is 80–150 µm (5). Vessels are solitary or arranged in small radial multiples (2). Intervessel pits are round to oval and perforation plates are simple (5). Fiber pits are very small. Rays between vascular bundles consist of square and upright unlignified cells and are 6 to >10 cells (1, 4). Rays are absent within the bundles (3). Vascular bundles form remains over several years (1).

The phloem is uniform; parenchyma cells and sieve tubes cannot be distinguished. Fibers are absent (6). Rays dilate (6). Ducts are absent (1, 6). Intercellulars in the cortex are very small (7). The phellem, if present, consists of a few irregularly formed cells.

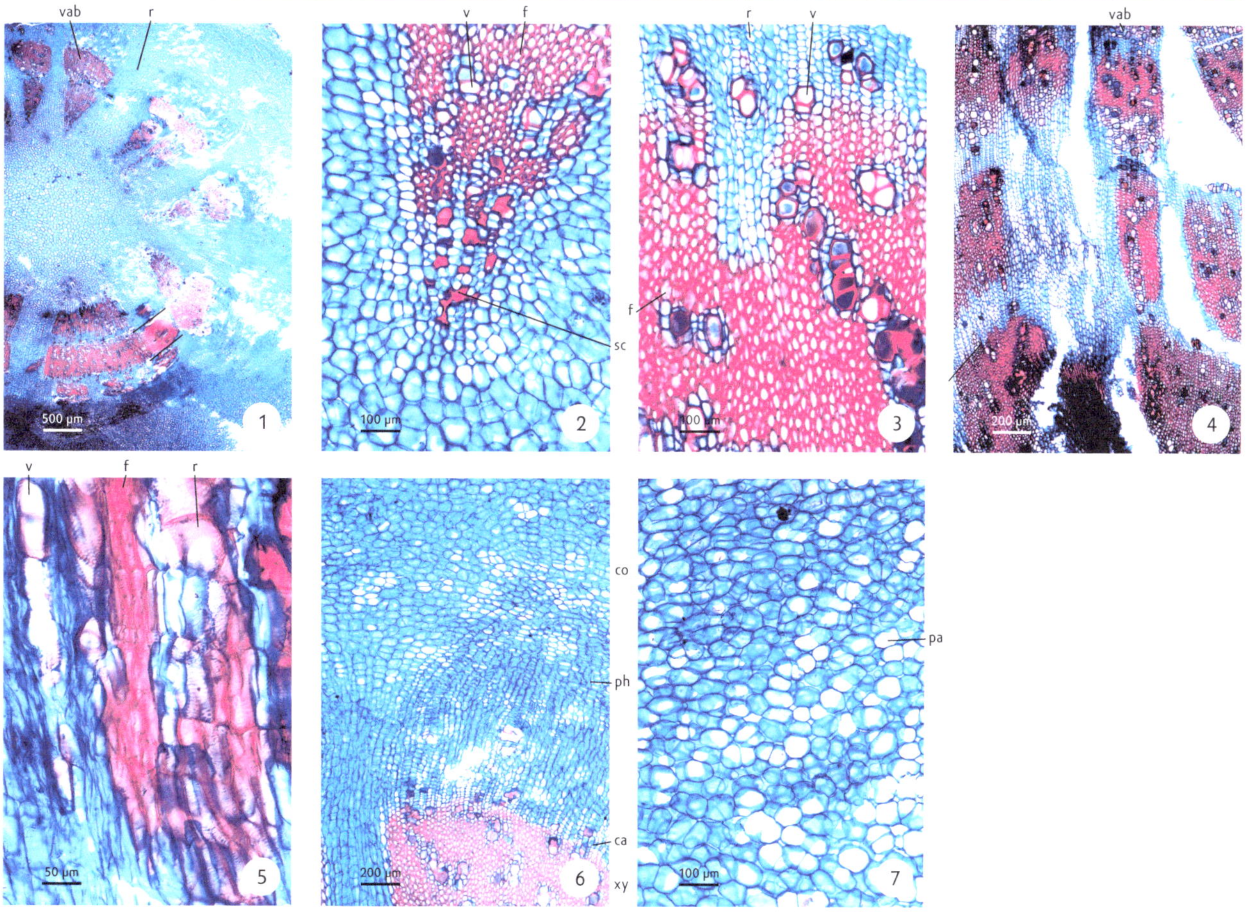

Senecio aquaticus Hill

Marsh Ragwort

Biennial, sometimes perennial wetland herb with short rhizomes. Stem erect, branched in the upper part, 30–70 cm high. Basal leaves elliptical-ovate, undivided, stem-leaves with few pairs of lateral lobes and with a much larger terminal lobe. Loosely branched inflorescence with golden-yellow flower heads. Fruit is an achene with pappus.

Distribution and ecology

European temperate element. In all CE countries except HU and SK, in middle altitudes. Occurs in wet meadows and pastures, in marshes, by streams and fishponds. Prefers fertile, non-calcareous soils. Tolerates fertilization. Contains toxic alkaloids, develops large seedbanks.

Anatomy of the annual shoot and the transition between the root and the annual shoot

Round stem with a diameter of 5–6 mm. Maximum age 1 year.

Annual shoot: Round, ribbed outline. A circle of collateral vascular bundles and a cortex surround a very large pith (1). Vascular bundles are laterally connected by a vessel-free belt of fibers. Peripheral, longitudinal unlignified stabilizing strips consist of collenchyma (1).

Perennial shoot: A continuous belt of xylem, phloem and cortex surrounds the pith, which consists of round unlignified parenchyma cells (2). The xylem is composed primarily of thin- to thick-walled fibers, vessels and rays (3). Vessels have a diameter of 20–30 µm (3) and vessel length is 80–100 µm (4). Vessels of the diffuse-porous xylem are solitary or arranged in short radial rows (2, 3). Intervessel pits are round and perforation plates are simple (4). Parenchyma is scanty paratracheal (3). Ray width is 5–8-seriate (2), mostly confluent and ray cells are upright.

The phloem consists of parenchyma cells and small groups of radially arranged groups of sieve tubes (1, 5). Phloem and cortex are separated by an endodermis (5). Large oval, unlignified cells and small intercellulars characterize the cortex (5). Few ducts are located just outside of the endodermis (1). The phellem is absent.

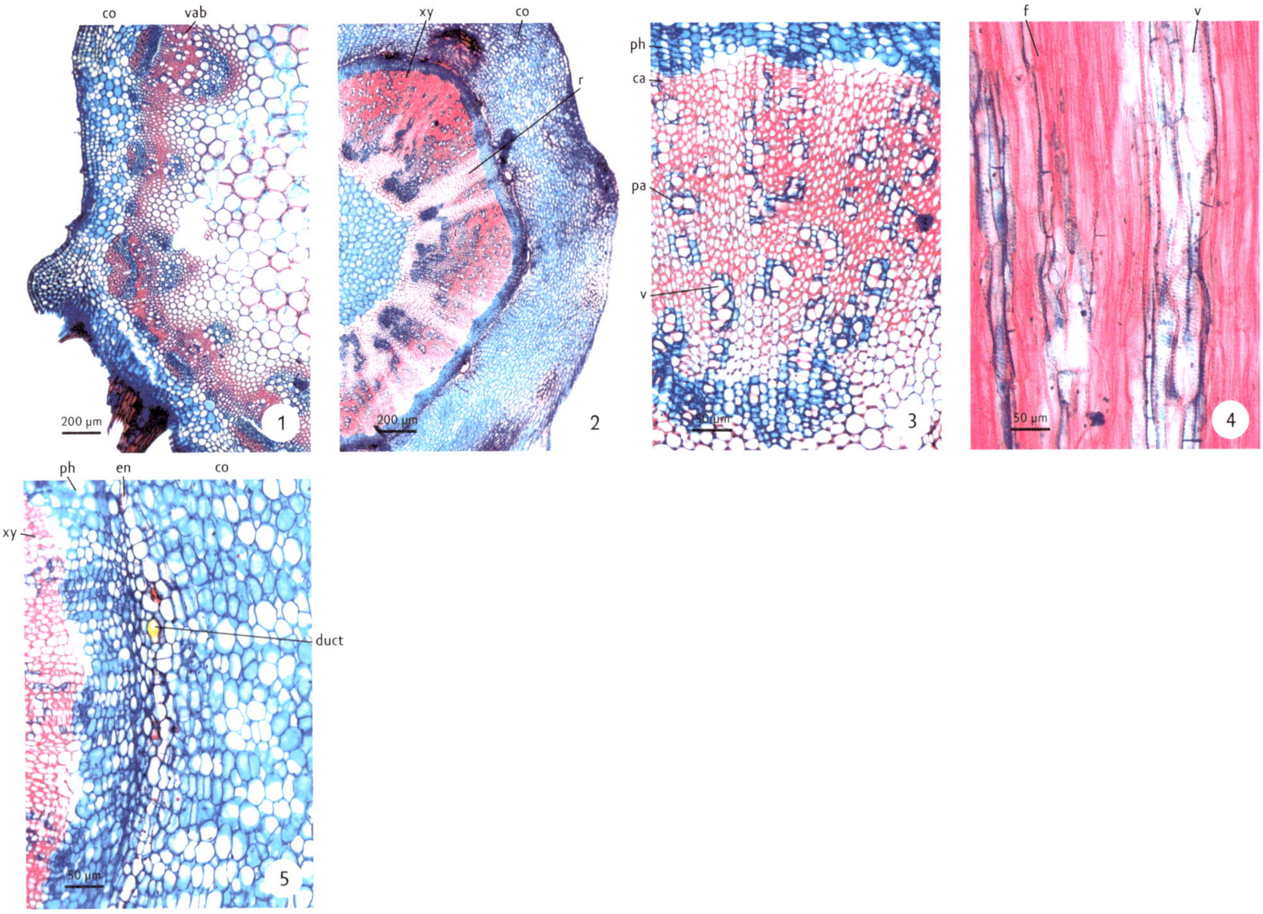

Senecio doria L.

Golden Ragwort

Perennial wetland herb with short rhizomes. Stem erect, glabrous, branched in the upper part 40–150 cm high. Basal leaves oblanceolate to widely lanceolate, distantly dentate to subentire, glabrescent. Flower heads in a rich panicle, ligules golden-yellow. Fruit is an achene with pappus.

Distribution and ecology

Native of the Pannonian lowland and W. Asia. In AU, CZ, HU and SK especially in lowlands. Occurs in inland salt marshes, wet grasslands, in alluvial soils along small streams and along road ditches. Prefers heavy and mineral-rich soils.

Anatomy of the rhizome

Round rhizome with a diameter of 8 mm. Maximum age 4 years.

The large pith of the rhizome is surrounded by solitary vascular bundles, a cortex and a small phellem (1). The vascular bundles consist of vessels, thick-walled fibers and paratracheal parenchyma (2). Ring boundaries of the diffuse-porous wood are indicated by marginal parenchyma (2). Vessels are angular and have a diameter of 30–40 µm (2). Vessel length is 100–150 µm (3). Vessels are solitary, in groups or in small radial multiples (2). Intervessel pits are round to oval and perforation plates are simple (3). Fiber pits are very small. Rays between vascular bundles are >10 cells wide (4), they are laterally bordered by sheet cells and consist of square and upright unlignified cells (5). Vascular bundles form remains over several years (1). Rays are absent within the bundles (2, 4).

The phloem consists of parenchyma cells, groups of sieve tubes (6) and occasionally of groups of fibers (1, 6) and dilating rays (1). Outside of most bundles are one or two large ducts (1, 6). Intercellulars in the cortex are irregularly formed and are as big as the surrounding parenchyma cells (7, 8). The phellem, if present, consists of irregularly formed cells inside of the epidermis (8).

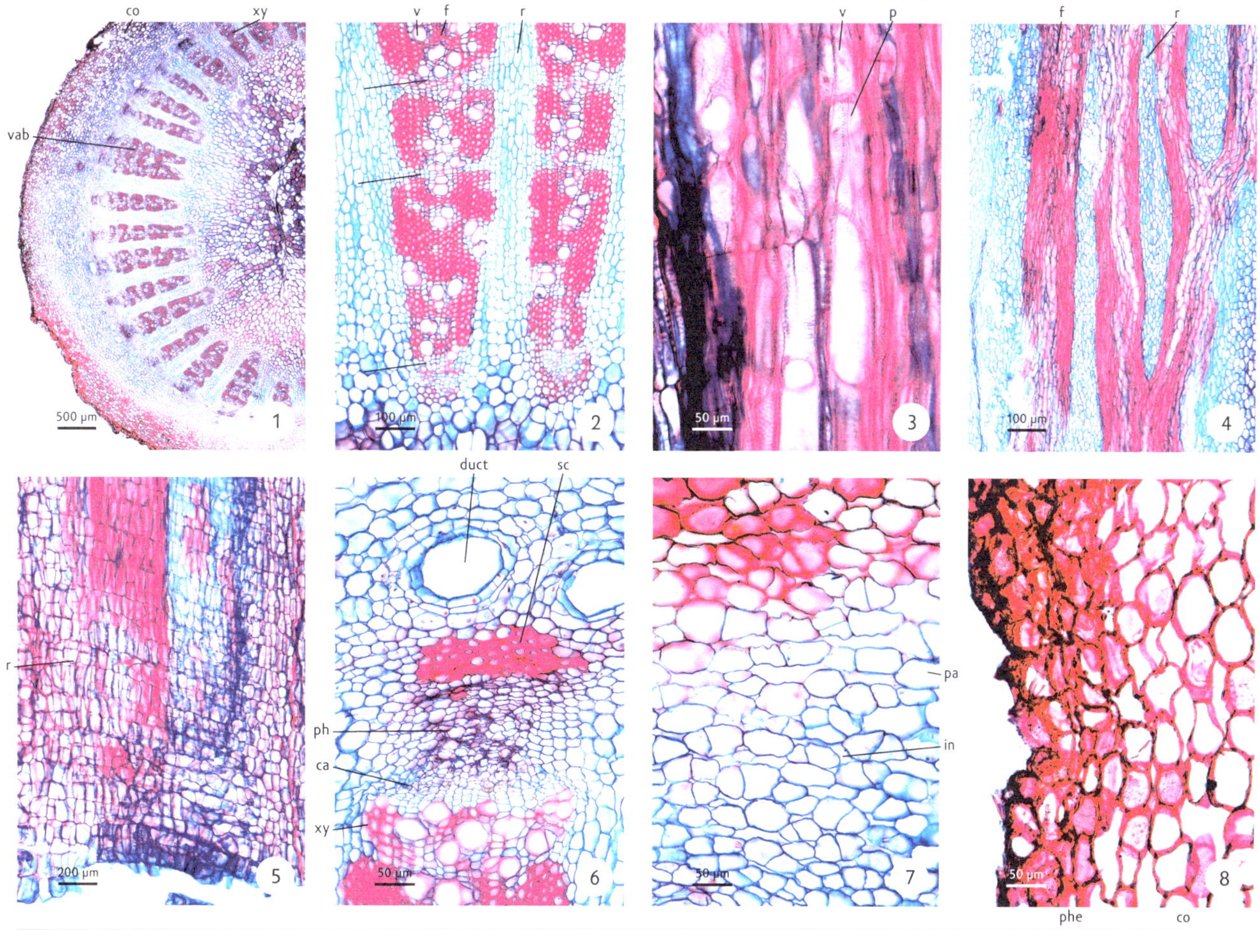

Senecio paludosus L.

Fen Ragwort Asteraceae

Perennial wetland herb. Stem erect, hollow, grooved, 50–220 cm high. Leaves sessile, lanceolate to linear-lanceolate, 9–20 cm long, shortly serrate, clasping stem at base. Both juvenile stem and leaves sligthly wooly pubescent. Flower heads 3–4 cm wide, bright yellow, in a dense erect umbel-like corymb. Fruit is an achene with pappus.

Distribution and ecology

Euro-Siberian temperate element. In all CE countries especially in lowlands. River corridor species. Occurs in sunny or partly shaded stands in wet meadows, tall-herb fens, along banks of slowly flowing waters and in the margin of floodplain forests. Prefers periodically flooded sites, humic or fen soils.

Anatomy of the rhizome

Round rhizome with a diameter of 7 mm. Maximum age 1 year.

The large pith of the rhizome is surrounded by solitary vascular bundles, a cortex and a small phellem (1). The vascular bundles consist of vessels, thick-walled fibers and paratracheal parenchyma (2). Vessels are angular and have a diameter of 30–40 µm (2). Vessel length is 100–150 µm (3). Vessels are solitary, in groups or small radial multiples (2). Intervessel pits are round and perforation plates are simple (3). Fiber pits are very small. Rays between vascular bundles are >10 cells wide (4) and consist of square and upright unlignified cells (5).

The phloem consists of parenchyma cells and groups of sieve tubes (6) and occasionally of groups of fibers (1, 6) and dilating rays (1, 6). Outside of most bundles are one or two large ducts (1, 7). Intercellulars in the cortex are irregularly formed and are as big as the surrounding parenchyma cells (7). The phellem, if present, consists of irregularly formed cells inside of the epidermis (7).

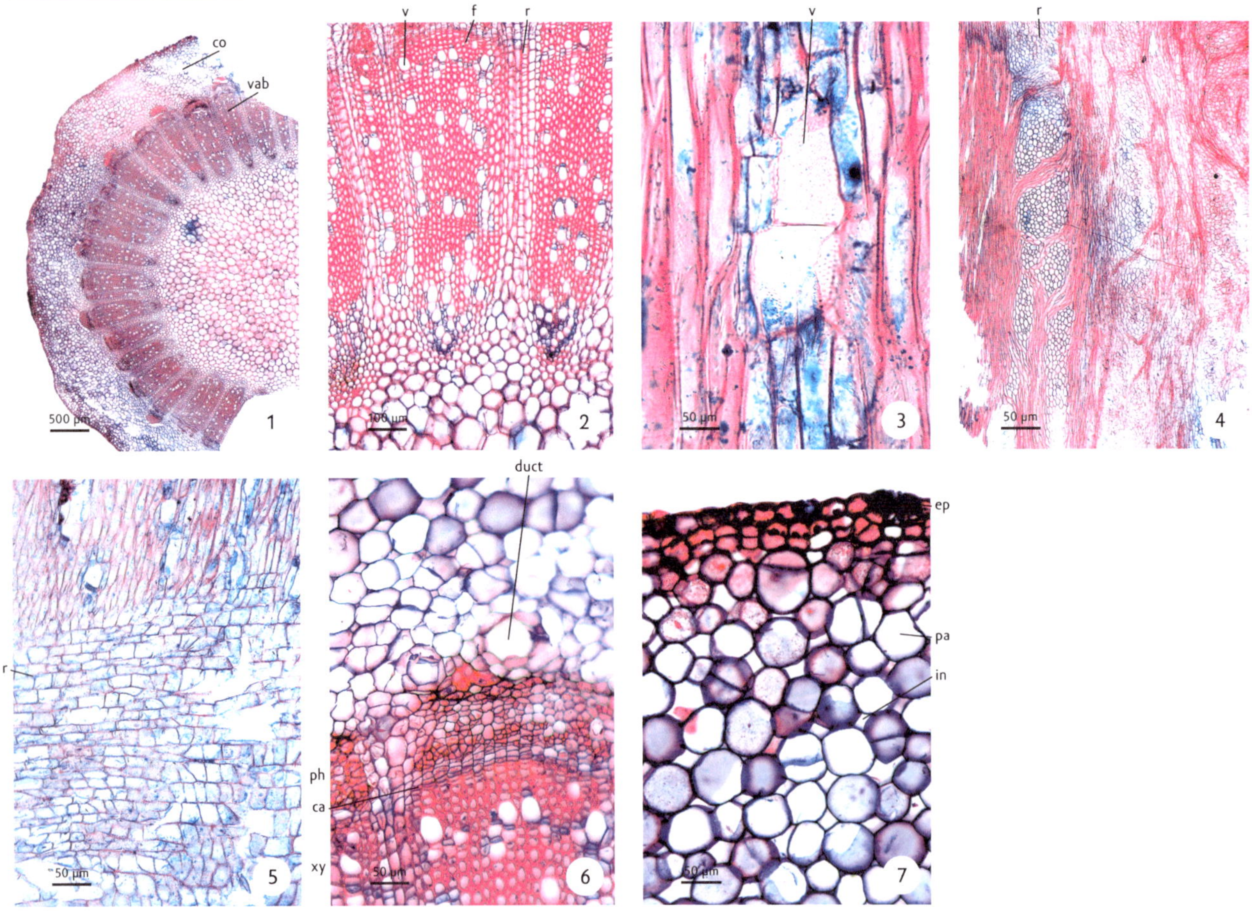

Senecio sarracenicus L. (syn. *Senecio fluviatilis*) **Broad-leaved Ragwort** Asteraceae

Perennial wetland herb with horizontal rhizomes and long runners. Stem erect, full, 60–200 cm high. Leaves lanceolate to oblanceolate, shortly serrate, glabrous, sessile, clasping stem at base. Flower heads in a rich corymb-like panicle, with 6–8 ray-florets only. Brown achenes with pappus.

Distribution and ecology

Euro-Siberian temperate distribution. Native in all CE countries except CH and PL. Occurs in the margin of floodplain forests and reed stands, in wet floodplain meadows. Heliophyte, nitrophilous species. Prefers wet, periodically flooded loamy soils.

Anatomy of the rhizome

Round rhizome with a diameter of 9–11 mm. Maximum age 2 years.

The large pith of the rhizome is surrounded by solitary vascular bundles, a cortex and a small phellem (1). The vascular bundles consist of vessels, thick-walled fibers and paratracheal parenchyma (2). Ring boundaries are indicated by slight semi-ring porosity (2). Vessels are angular and have a diameter of 40–60 µm (2). Vessel length is 100–150 µm (3). Vessels are solitary, in groups or small radial multiples (2). Intervessel pits are round and perforation plates are simple and scalariform (3). Fiber pits are very small. Rays between vascular bundles are >10 cells wide and often laterally limited by sheath cells (5). Rays consist of square and upright unlignified cells (4).

The phloem consists of parenchyma cells and groups of sieve tubes (6) and occasionally of groups of fibers (1, 6) and dilating rays (1, 6). Outside of most bundles are one or two large ducts (1, 6). Intercellulars are absent in the cortex (6). The phellem is absent.

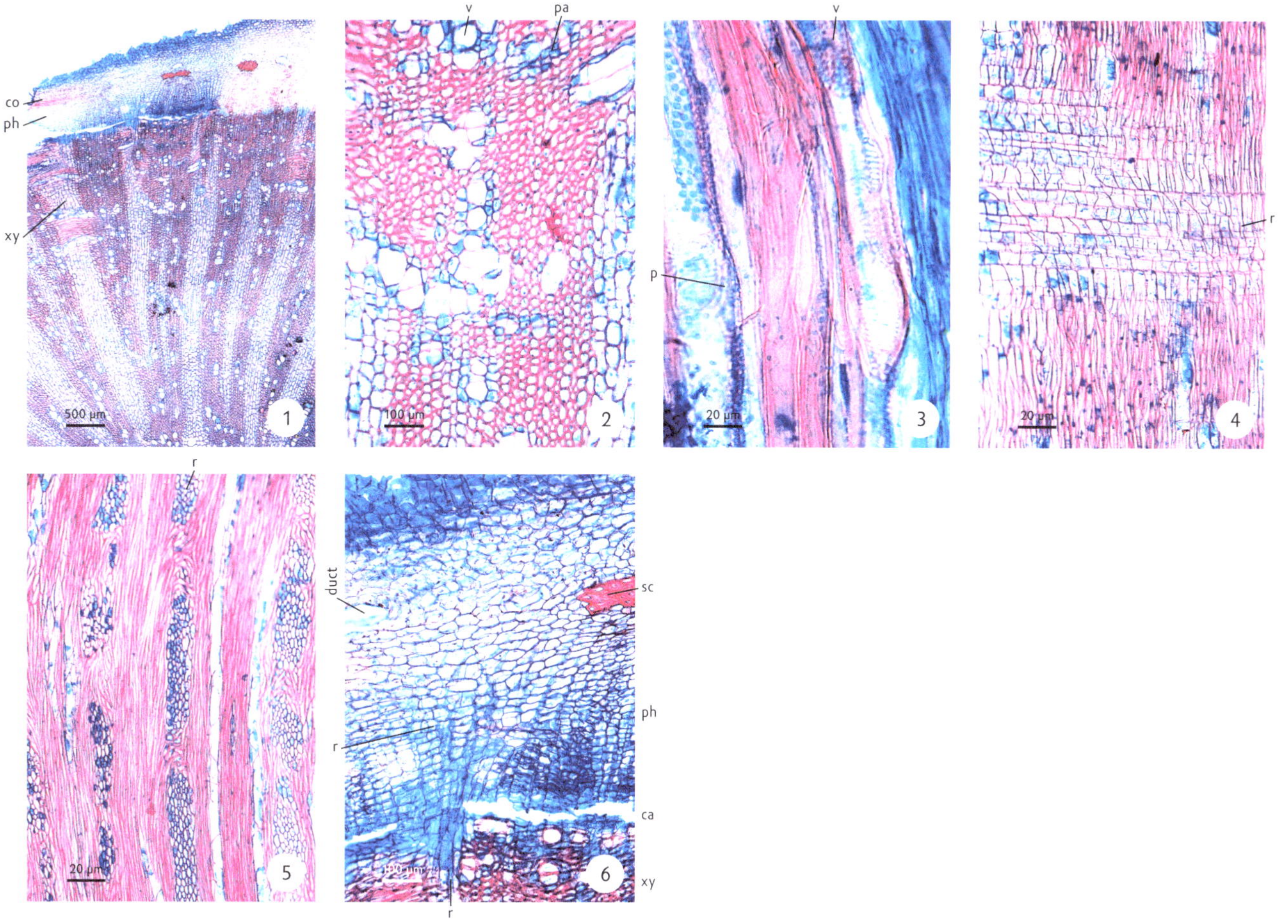

Sonchus palustris L.

Marsh Sow-thistle Asteraceae

Perennial wetland herb with short rhizomes. Stem erect, hollow, sticky-hairy above, robust, up to 250 cm high. Basal leaves deeply pinnately lobed, with arrow-shaped bases. Upper leaves lanceolate, unlobed. Flower heads in large open umbel-like corymbs, florets pale clear yellow, all bracts covered with blackish-green sticky hairs. Fruit is a yellow achene with pappus.

Distribution and ecology

Euro-Siberian temperate element. Native in all CE countries except CH. Occurs in tall vegetation along rivers on wet soils rich in nitrogen, in periodically flooded meadows in lowlands, also at ruderal sites. Moderately tolerant of saline conditions and can grow near tidal river mouths. Tolerates eutrophication.

Anatomy of the rhizome

Round stem with a diameter of 9–10 mm. Maximum age 1 year.

A continuous belt of xylem, phloem and cortex surrounds the pith in parts (l). The pith consists of round unlignified parenchyma cells and includes some groups of sieve tubes (2). The xylem is composed primarily of thin-walled fibers, vessels and rays (3). Vessels have a diameter of 50–70 µm (4) and vessel length is 80–200 µm (5). Vessels of the diffuse-porous xylem are solitary or arranged in short radial rows (1, 3). Intervessel pits are round and perforation plates are primarily simple (5), however a few scalariform plates occur (5). Parenchyma is scanty paratracheal (3). Ray width is 2–5-seriate (6) and ray cells are upright (7).

The phloem consists of parenchyma cells and small groups of radially arranged groups of sieve tubes (1, 8). Rays dilate. Large oval, unlignified cells characterize the cortex (8). The phellem consists of 2–3 rows of irregularly formed cells (8). Intercellulars are absent. Ducts occur in the pith (1).

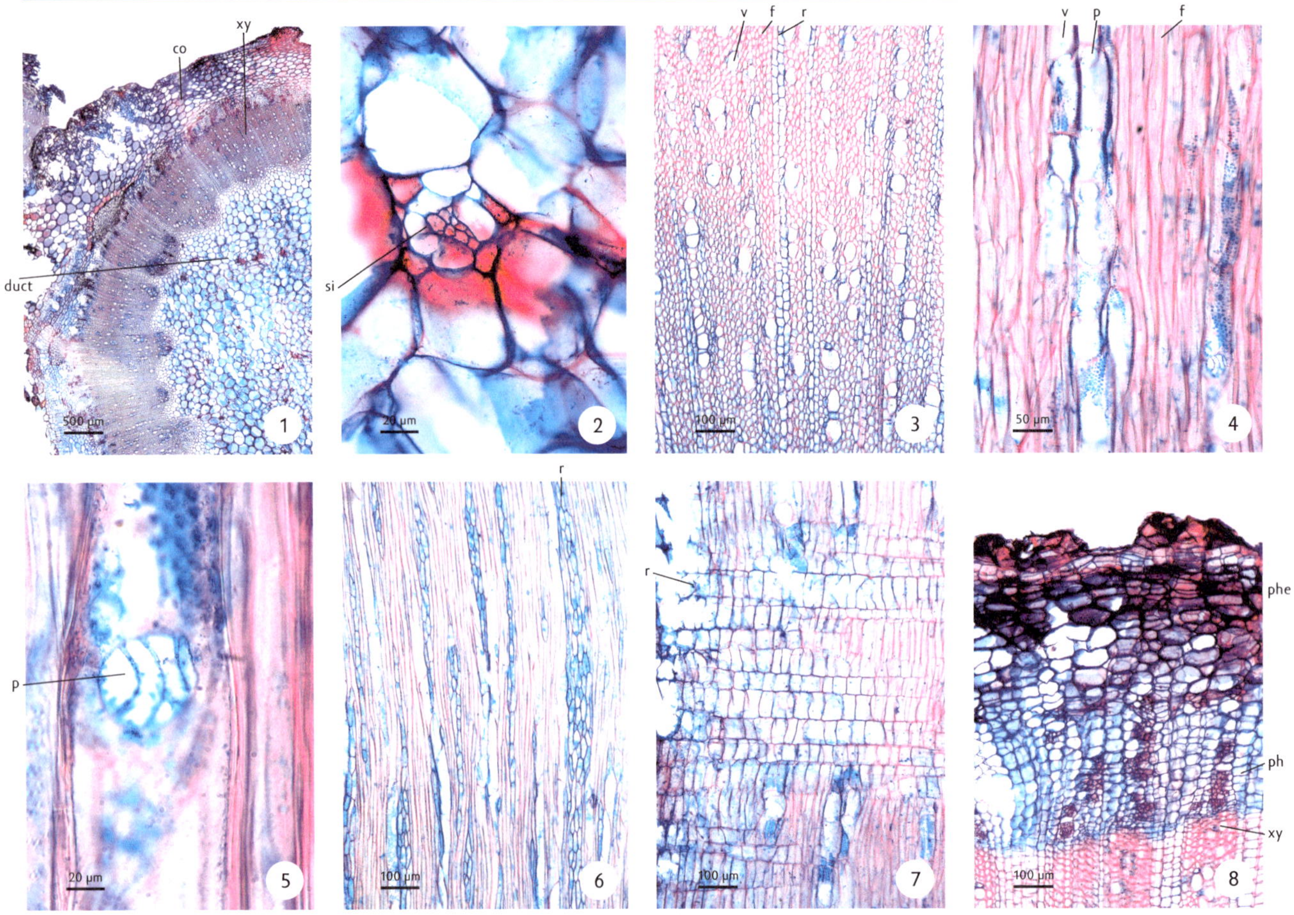

Tephroseris crispa (Jacq.) Rchb. (syn. *Senecio rivularis*)

Perennial wetland herb with short rhizomes. Stem erect, simple, grooved, hairy, 40–80 cm high. Basal leaves ovoid, serrate, with winged petiole. Yellow flower heads in an umbel-like inflorescence. Fruit is a brown achene with pappus.

Distribution and ecology

Endemic in C. Europe. Native in all CE countries except CH, mainly in foothills and mountains. Occurs in moist and wet peaty meadows, in peatbogs and springs, along stream banks, in tall-herb floodplains and alder-carrs. Prefers acid soils.

Anatomy of the rhizome

Round rhizome with a diameter of 4–5 mm. Maximum age 2 years.

The large pith of the rhizome is surrounded by solitary dilating vascular bundles, a cortex and a small phellem (1). The vascular bundles consist of vessels and paratracheal parenchyma (2). Ring boundaries are indicated by a slight semi-ring porosity (2). Vessels have a diameter of 20–30 µm (2). Vessel length is 100–120 µm (3). Vessels are solitary or arranged in small groups (2). Intervessel pits are round to oval and perforation plates are simple (3). Rays between vascular bundles consist of square and upright unlignified cells and are 6 to >10 cells (1). Rays are absent within the bundles (2). Vascular bundles form remains over two years (1).

The phloem is uniform; parenchyma cells and sieve tubes cannot be distinguished. Fibers are absent (2). Ducts occur just outside of the phloem (1, 2). Intercellulars in the cortex are very small (2). The phellem, if present, consists of a few irregularly formed cells (1).

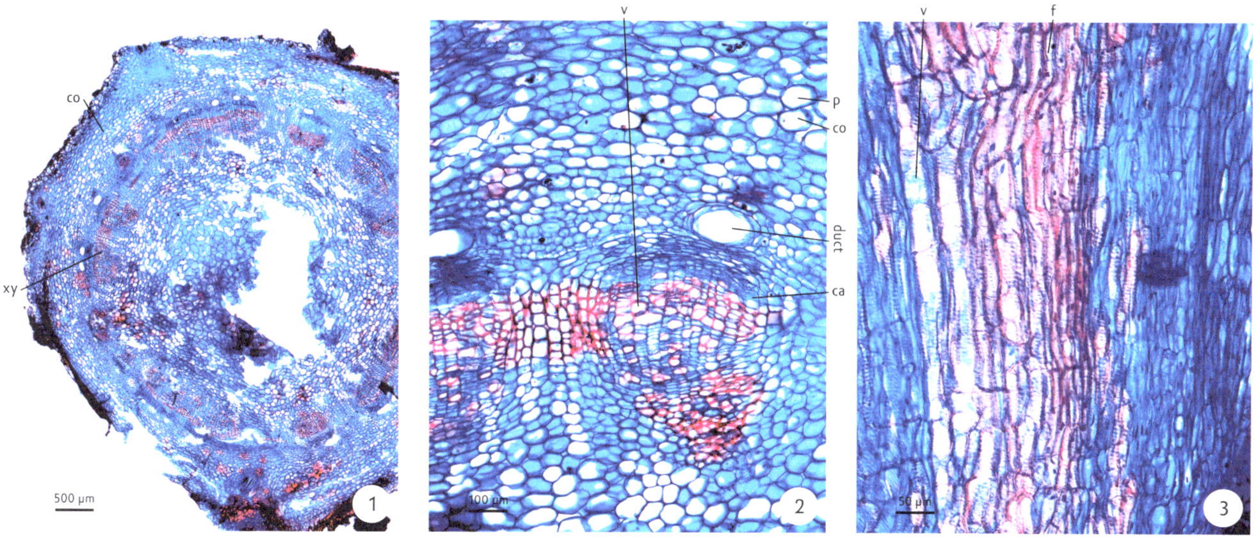

Impatiens glandulifera Royle

Indian Balsam

Balsaminaceae

Annual wetland herb. Stem erect, hollow, reddish, up to 200 cm high. Leaves lanceolate, serrate on the margin, opposite or in whorls of 3. Large flowers with short curved spurs, deep to pale pinkish-purple or white, in a raceme. Mature capsules able to shoot seeds in large distances.

Distribution and ecology

Neophyte, native of the Himalayas; widely naturalised in temperate Europe. Invasive in all CE countries. Most frequent along water banks, where it often forms continuous stands, but also established in damp woodland. Able to grow very rapidly. Attractive traits with respect to insect pollination. Self-dispersed via ballochory.

Anatomy of the basal part of the annual shoot

Round stem with a diameter of 20–60 mm. Maximum age 1 year.

A very large pith is surrounded by a belt of xylem, a small phloem and a large cortex (1, 2). The xylem is laterally continuous (1) in the upper parts of the shoot and divided in vascular bundles at the stem base (2). The cortex and the phloem are frequently separated by a thin-walled endodermis-like layer of cells (1, 2). The xylem contains solitary and radially arranged multiple vessels, thin-walled fibers and paratracheal parenchyma (1–3). Vessels have a diameter of 60–100 µm and a length of 150–200 µm. Vessel pits have simple perforation plates and oval or scalariform intervessel pits (4). The very large unlignified rays (>10 cells) are composed of very thin-walled upright cells (5) and are mostly confluent (6).

Groups of sieve tubes and parenchyma cells form the phloem (2, 3, 7). Between the radial groups are parenchyma cells, which are occasionally filled with bundles of raphids (7, 8).

The cortex and the pith consist of thin-walled parenchyma cells of very different size. The largest have a tangential diameter of 700 µm (1). The epidermis consists of quadrangular, thin-walled, unlignified cells with a slightly thicker peripheral wall (2).

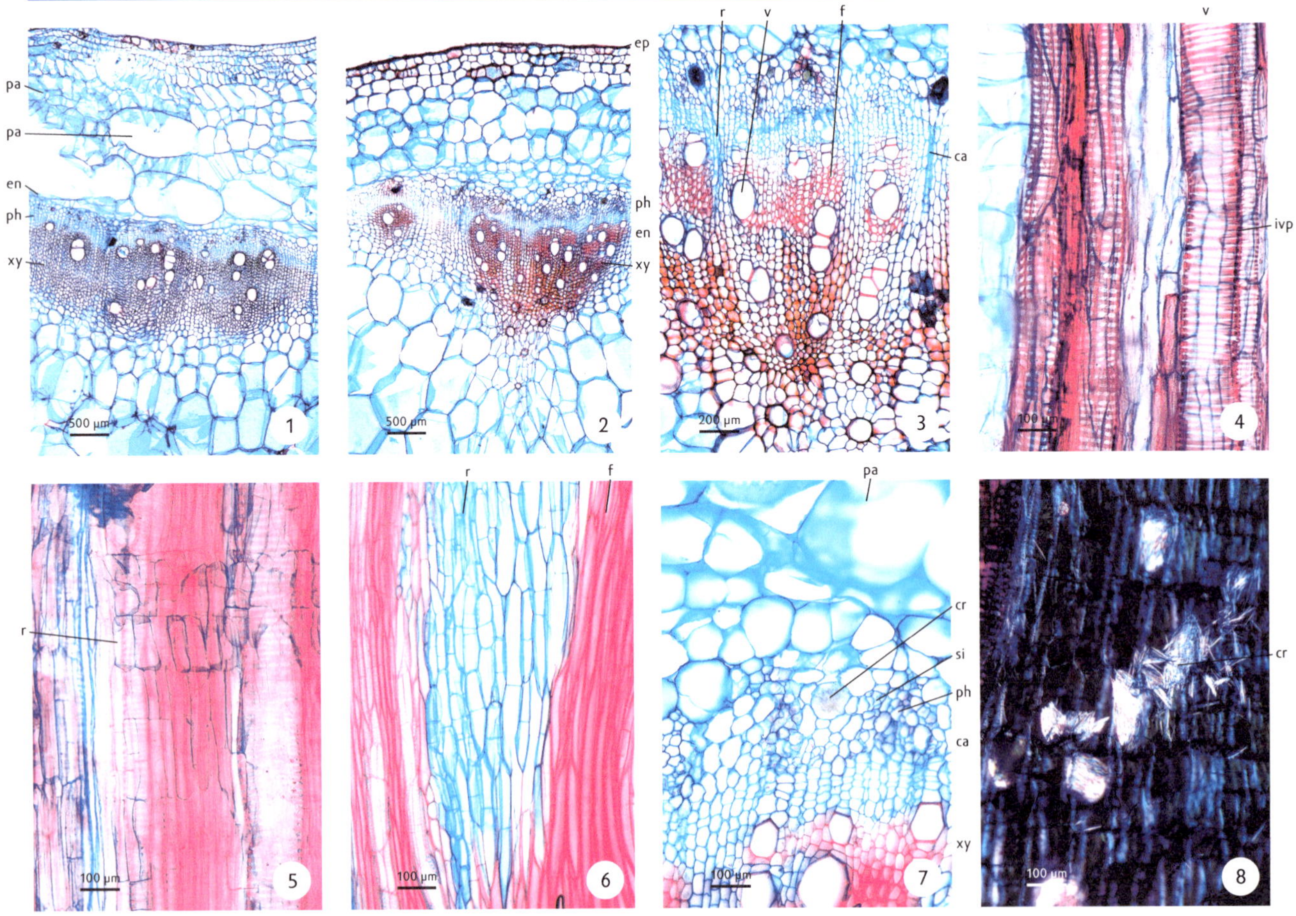

F. H. Schweingruber et al., *Anatomic Atlas of Aquatic and Wetland Plant Stems*, https://doi.org/10.1007/978-3-030-33420-8_14

Annual wetland herb. Stem erect, glabrous, translucent yellow-green, often suffused with red, becoming swollen at nodes, 20–60 cm high. Leaves opposite becoming alternate, ovate-elliptical to ovate-lanceolate or oblong, serrate. Bright yellow flowers with small reddish spots inside, with curved spurs, some cleistogamous. Fruit is an elliptic capsule.

Distribution and ecology

Eurasian temperate element. In all CE countries in wet but not waterlogged woodlands, in alluvial forests, in the surroundings of woodland springs and along shaded stream banks. Prefers wet, humic and nutrient-rich soils. Sciophyte, avoids dry and warm sites. Unable to self-pollinate, flowers poorly attractive to pollinators. Self-dispersed via ballochory.

Anatomy of the basal part of the annual shoot

Round stem with a diameter of 10–20 mm. Maximum age 1 year.

No substantial differences have been found between *Impatiens glandulifera* and *I. noli-tangere* in the anatomy of stems. That is illustrated with the distribution of pith, xylem, phloem and cortex (1) and the composition of the xylem (2), the intervessel pits (3) and the rays (4). Paratracheal parenchyma is more distinct in *Impatiens noli-tangere* than in *I. glandulifera* (2).

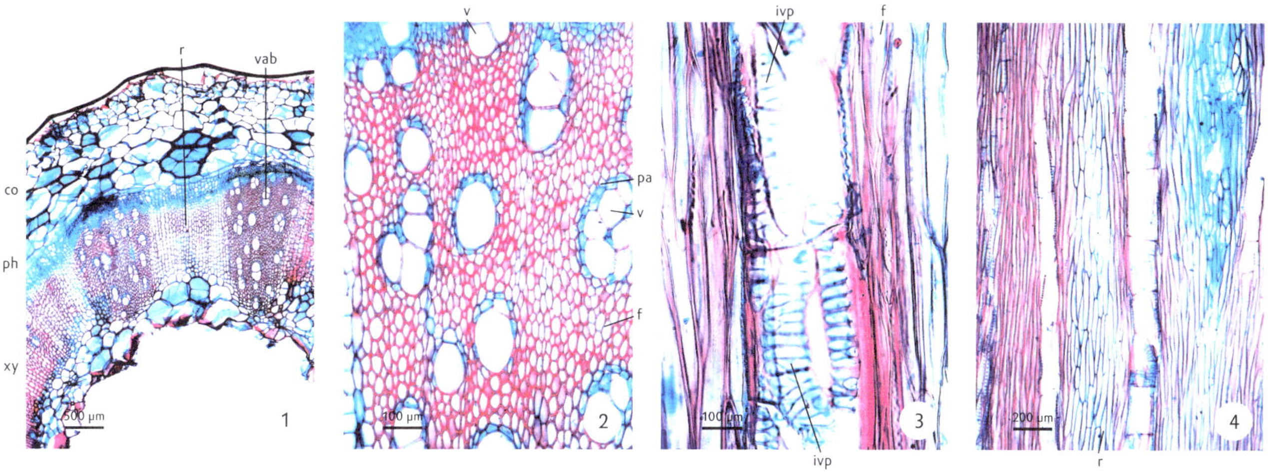

Alnus alnobetula (Ehrh.) C. Koch (syn. *Alnus viridis*)

Perennial wetland deciduous shrub. Stem up to 3 m high. Annual shoots pubescent. Leaves up to 5 cm long, irregularly biserrate, ovoid to rounded, pubescent underside. Male and female flowers in separate catkins.

Distribution and ecology

European distribution. Native in AT, CH, CZ and DE, introduced in PL and SK, absent in HU. Prefers acidic soils. Develops dense shrubby stands above tree-line, also in disused pastures, in foothills usually in wet meadows, peatbogs and in forest margins. N_2-fixing shrub currently encroaching on montane grasslands in the Alps.

Anatomy of the stem

Round stem with a diameter up to 10 cm. Xylem/bark proportion 10:1. Maximum age >40 years.

Stems consist of xylem, small phloem and phellem (1). Annual rings are marked by a row of flat latewood fibers and a discontinuous row of marginal parenchyma (2, 3) in the diffuse-porous wood. Vessels stay in short radial rows (2, 3). Vessels diameter varies from 40–50 µm (3) and vessel length from 200–300 µm. Intervessel pits are round. Perforation plates are scalariform with 10–20 bars (4). Fibers are thin- to thick-walled (3) with round pits (4). Parenchyma is apotracheal (2, 3). Rays are generally uniseriate (4) and ray cells prostrate (5). The phloem consists of tangentially layered parenchyma cells and sieve elements (6, 7). Parenchyma cells are round to oval and sieve tubes angular (7). Few rays dilated (6). Cortex consists of tangentially layered parenchyma, groups of fibers and sclereids (6, 8) and small intercellulars. The phellem consists of radially arranged thick-walled rectangular cells (6). Pith cells in the oval to triangular pith (1) appear round in cross-sections (1) and quadrangular in radial sections.

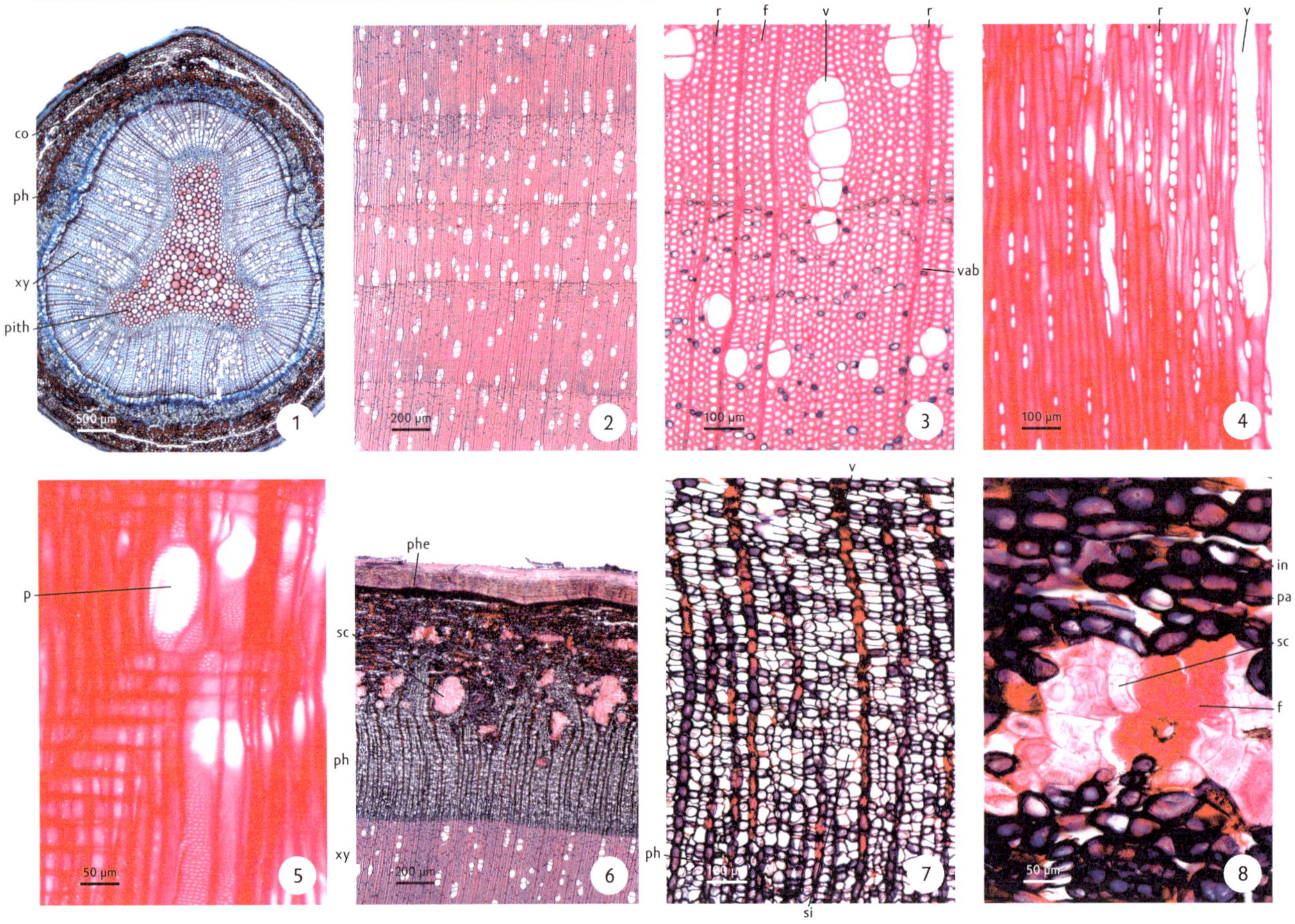

F. H. Schweingruber et al., *Anatomic Atlas of Aquatic and Wetland Plant Stems*, https://doi.org/10.1007/978-3-030-33420-8_15

Alnus glutinosa (L.) Gaertn. **Alder** Betulaceae

Perennial wetland deciduous tree. Stem up to 29 m high with dark-brown fissured bark. Habit variable from the low multiple-stemmed bush form to tall single-bole trees. Leaves obovate or elliptical, broadly obtuse to retuse at apex, irregularly shallowly biserrate, glutinous when young. Male and female flowers in separate catkins. Fruit is a brown narrowly winged achene, in compact cone-like structure.

Distribution and ecology

Euro-Siberian temperate element. Native in all CE countries, usually along rivers and streams, in floodplains, in fens, bogs and carrs. Prefers wet, basic to moderately acidic soils. Seeds are able to colonize open sites, producing even-aged stands of mature trees. Seedlings are shade-sensitive. N_2-fixing symbiotic bacteria on the roots.

Anatomy of the stem

Round stem with a diameter up to 50 cm. Maximum age >40 years.

Stems consist of a xylem, a small phloem and a rhytidome (1, 5). Annual rings are marked by flat latewood fibers and a slight semi-ring porosity (1, 2). The xylem is composed of vessels, fibers, parenchyma and rays (1, 2). The xylem is diffuse- to semi-ring porous. Vessels are solitary or in short radial rows (1, 2). Vessels diameter varies from 50–80 µm (2) and vessel length from 400–600 µm. Intervessel pits are round. Perforation plates are scalariform with 10–20 bars. Fibers are thin- to thick-walled (2) and have round pits (3). Parenchyma is apotracheal. Rays are generally uniseriate (4) and ray cells are prostrate (3). The phloem consists of round parenchyma cells, square sieve tubes and groups of sclerenchyma (5, 6). However parenchyma and sieve tubes are in detail difficult to distinguish in cross-sections (6). Intercellulars are absent in the phloem of stems (6) but present in roots (7). The phellem consists of radially arranged thin-walled rectangular cells (8).

Alnus incana

The xylem is identical to that of *A. glutinosa*. Sclereids in bark rays are radially more continuous than those in *A. glutinosa*.

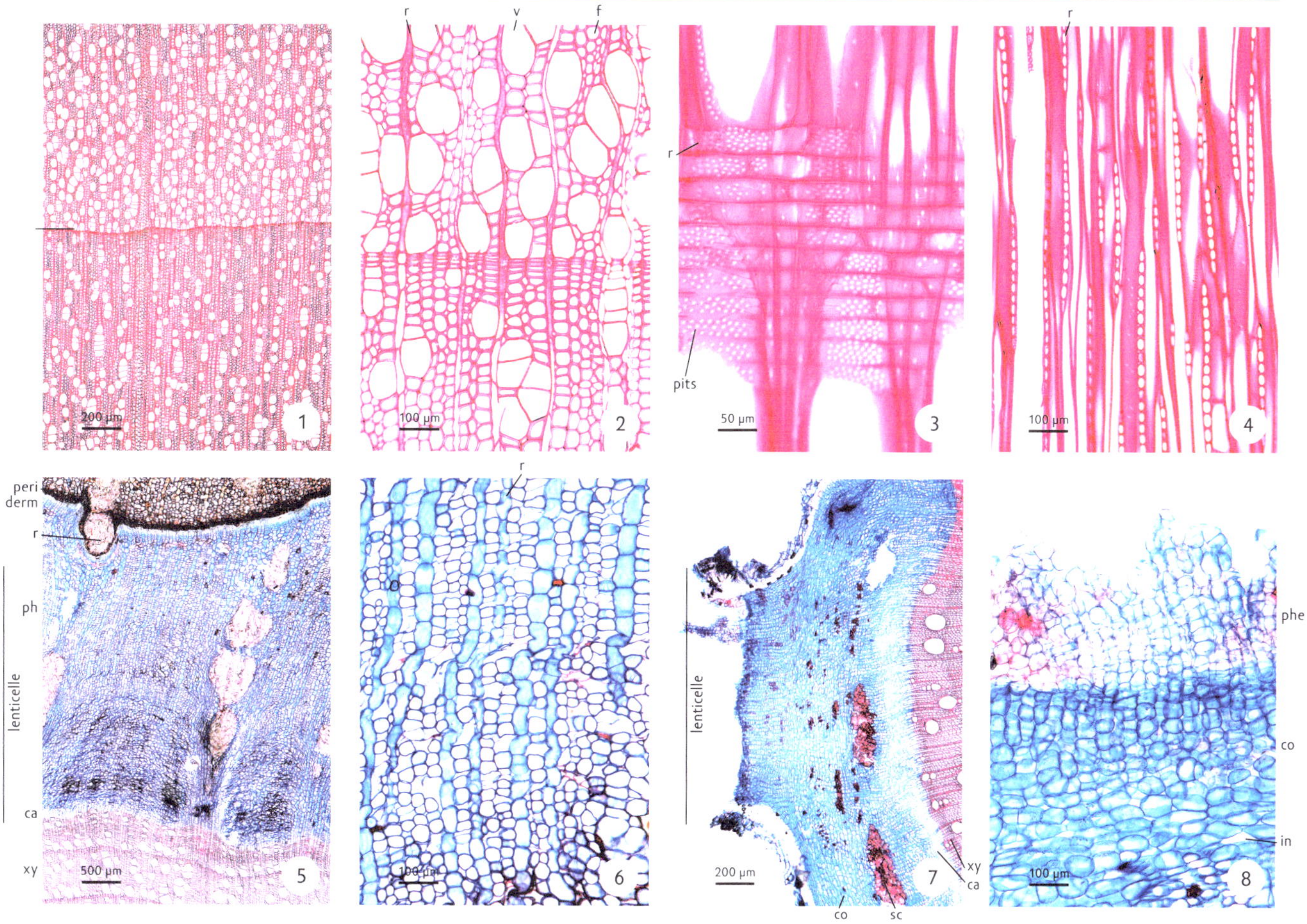

Betula nana L. **Dwarf Birch** Betulaceae

Perennial wetland deciduous shrub. Prostrate shrub up to 1 m in height. Twigs stiff, dull dark brown and pubescent. Leaves almost orbicular, rounded at both ends, crenate all around. Male and female flowers in separate catkins. Fruit is a nutlet, with very narrow wings.

Distribution and ecology

Circumpolar, boreo-arctic montane element. Native in AT, CH, CZ, DE and PL. Usually found on peatbogs and on peaty meadows from foothills to the mountains. Strictly on acidic soils, tolerant of waterlogged soils. Cold-tolerant shrub with vegetative reproduction.

Anatomy of the stem

Round stem with a diameter up to 4 cm. Xylem/bark proportion 10:1. Maximum age >40 years.

Stems consist of a xylem, a small phloem, a large cortex and a phellem (1). Annual rings are marked by a row of flat latewood fibers and a row of marginal parenchyma (1–3) in the diffuse-porous wood. Vessels are solitary or in short radial rows (2, 3). Vessel diameter varies from 30–40 µm (3) and vessel length from 150–200 µm. Intervessel pits are round. Perforation plates are scalariform with 10–20 bars (4). Fibers are thin- to thick-walled (3) with round pits (4). Parenchyma is apotracheal (1). Rays are uniseriate (5) and ray cells prostrate with very small vessel-ray pits (6).

The phloem consists of parenchyma cells, sieve elements and groups of sclereids (1, 7). Parenchyma cells are round to oval, sieve tubes are square and often compressed (7). Few rays slightly dilated (7). Cortex consists of parenchyma and small intercellulars (8). The phellem consists of radially arranged thick-walled rectangular cells (8). Pith cells in the triangular pith (1) are round in cross-section (1).

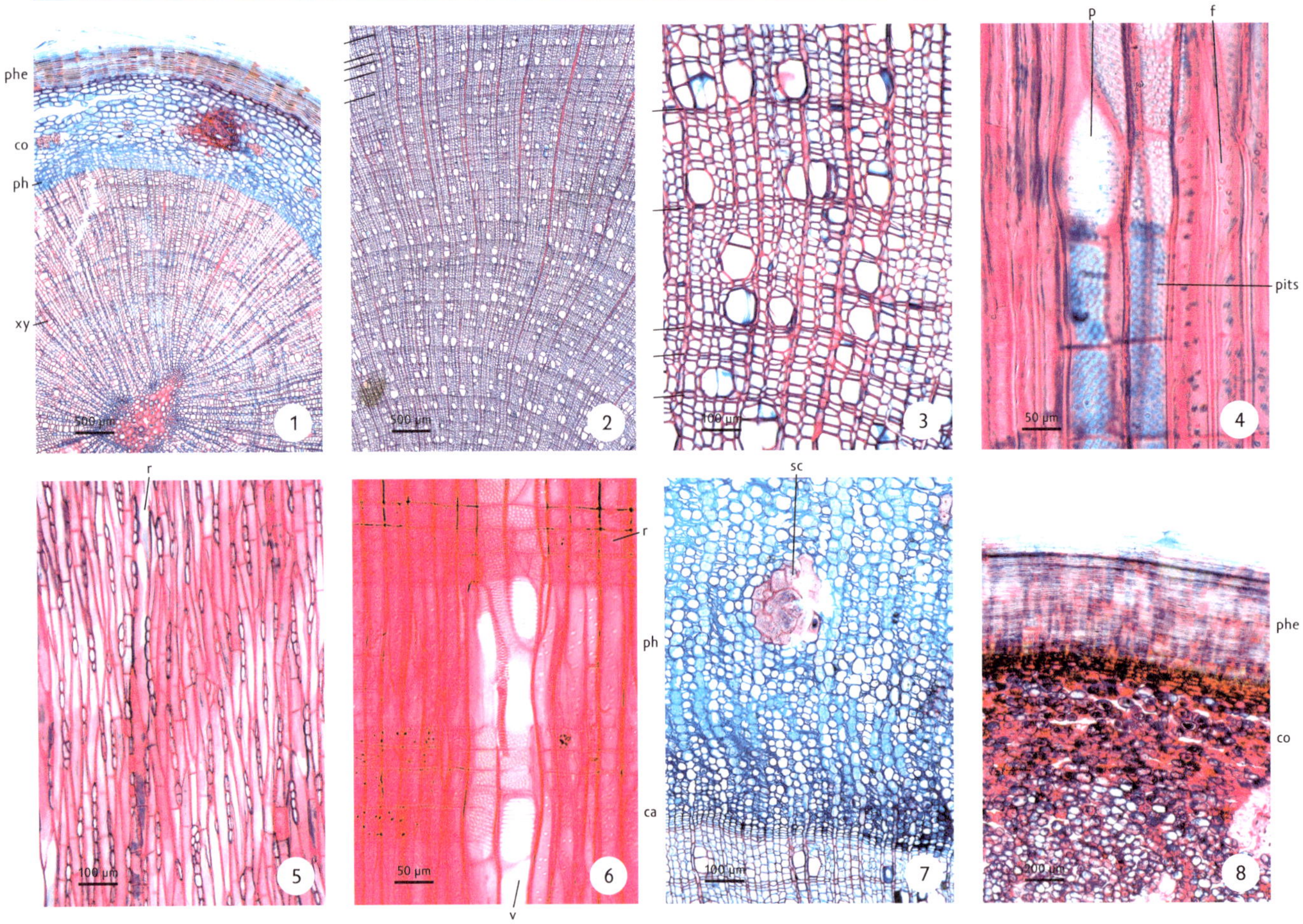

Betula pubescens Ehrh. **Downy Birch** Betulaceae

Perennial wetland deciduous tree. Stem up to 25 m high, with bark variously brown, grey or white. Twigs usually not pendent, pubescent to glabrous. Leaves ovate, singly serrate. Male and female flowers in separate catkins. Fruit is a flattened tiny nutlet with rounded wings.

Distribution and ecology

Euro-Siberian boreo-temperate element. Native in all CE countries. Prefers acidic, wet and peaty soils, especially in the uplands. It can rapidly colonise open, unshaded ground (burned areas, extracted peat bogs and fens). Heliophyte. Tolerates waterlogged soils.

Anatomy of the stem

Round stem with a diameter up to 30 cm. Xylem/bark proportion 10:1. Maximum age >40 years.

Stems consist of a xylem, a small phloem, a large cortex and a phellem (1). Annual rings are marked by a few rows of flat latewood fibers (2) and a row of marginal parenchyma (3) in the diffuse-porous wood. Vessels are solitary or in short radial rows (2, 3). Vessel diameter varies from 50–70 µm (3) and vessel length from 150–200 µm. Intervessel pits are very numerous, small and round. Perforation plates are scalariform with 10–20 bars (4) and fibers thin- to thick-walled (3). Parenchyma is apotracheal and marginal (3). Rays are 1–3-seriate (6), all ray cells are prostrate (5) with numerous very small vessel-ray pits (4).

The phloem consists of parenchyma cells, large sieve elements, small companion cells and groups of sclereids (1, 7). Sieve tubes collapse in the second year (7). The phellem consists of tangential layers of rectangular cells (1).

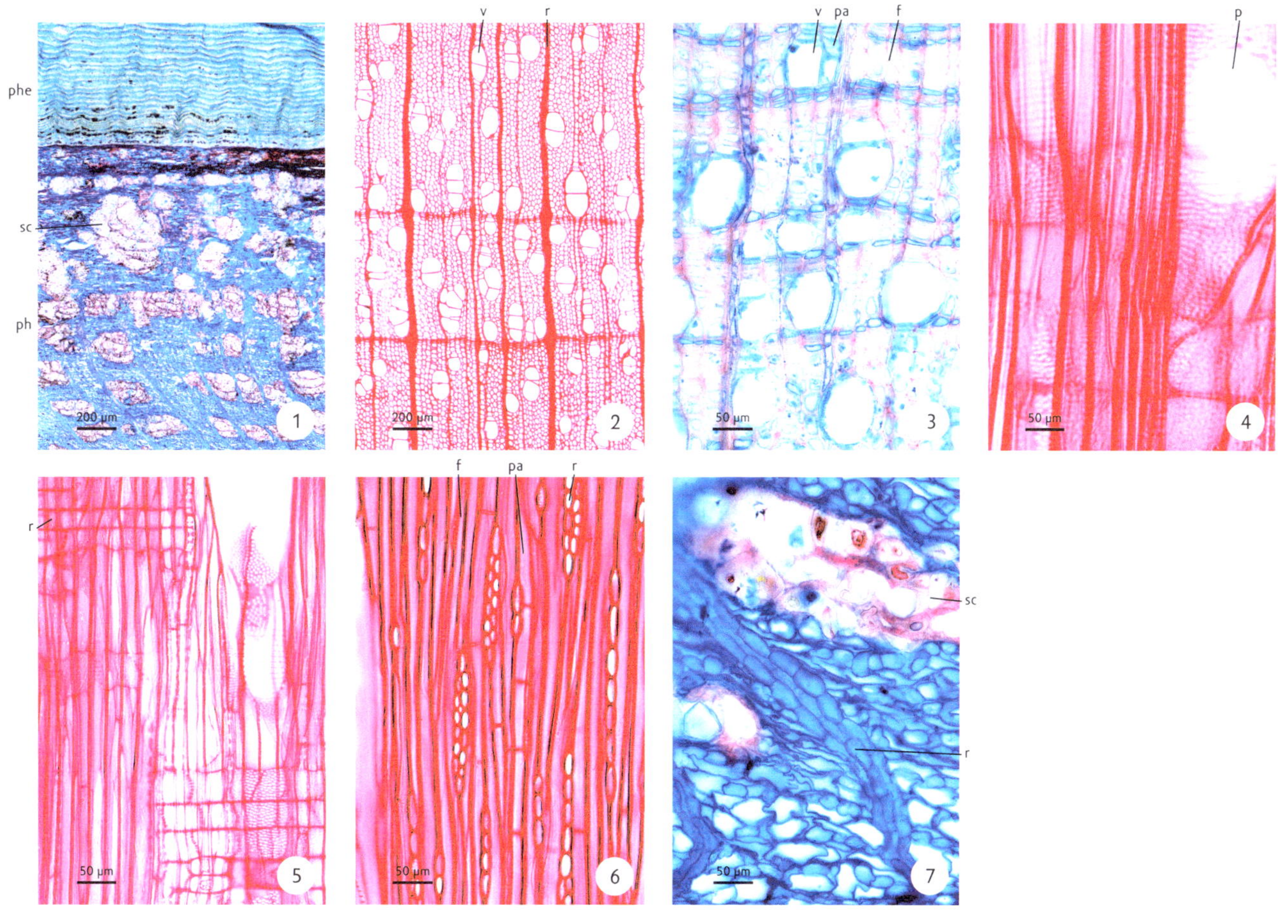

Myosotis caespitosa Schultz

Tufted Forget-me-not

Boraginaceae

(syn. *Myosotis laxa* subsp. *caespitosa*)

Annual to biennial wetland herb, often tufted. Stem erect to ascending, branched from the bottom, without runners, 20–80 cm high. Leaves elliptical with hairs orientated to leaf apex. Numerous flowers in rich terminal cymes. Wheel-shaped blue corolla, 3.5–4.5 mm across. Calyx with appressed hairs throughout, calyx teeth long, pointed. Fruit is a wide ovoid nutlet.

Distribution and ecology

Circumpolar boreo-temperate element, with a disjunct distribution. Native in all CE countries from lowlands to foothills. Occurs along banks of standing and flowing waters, in wet meadow ditches, on margins of floodplain forests and sedge stands, also in drained bottoms of fishponds. Prefers open places affected by disturbance (i.e., trampling by livestock).

Anatomy of the rhizome

Round rhizome with a diameter of 2–4 mm. Maximum age 1 year.

The pith of the rhizome is surrounded by a xylem, a phloem and a large cortex (1–3). The xylem consists of vessels and thin- to thick-walled fibers (2). Axial parenchyma is paratracheal (2). Vessels are solitary or arranged in radial rows (2), they have a diameter of 20–40 µm (2) and a length of 120–150 µm. Intervessel pits are round to oval and perforation plates are simple (4). Fiber pits are round and very small. Rays are absent (2, 5).

The phloem consists of parenchyma cells and groups of sieve tubes (6). The stele is surrounded by a poorly differentiated pericycle/endodermis zone (1–3). Intercellulars in the cortex are irregularly formed and are as large as the surrounding parenchyma cells (7). Parenchyma cells in the pith and in the cortex are filled with slime (3, 7). Prismatic crystals occur in the cortex (6). Phellem is absent (7).

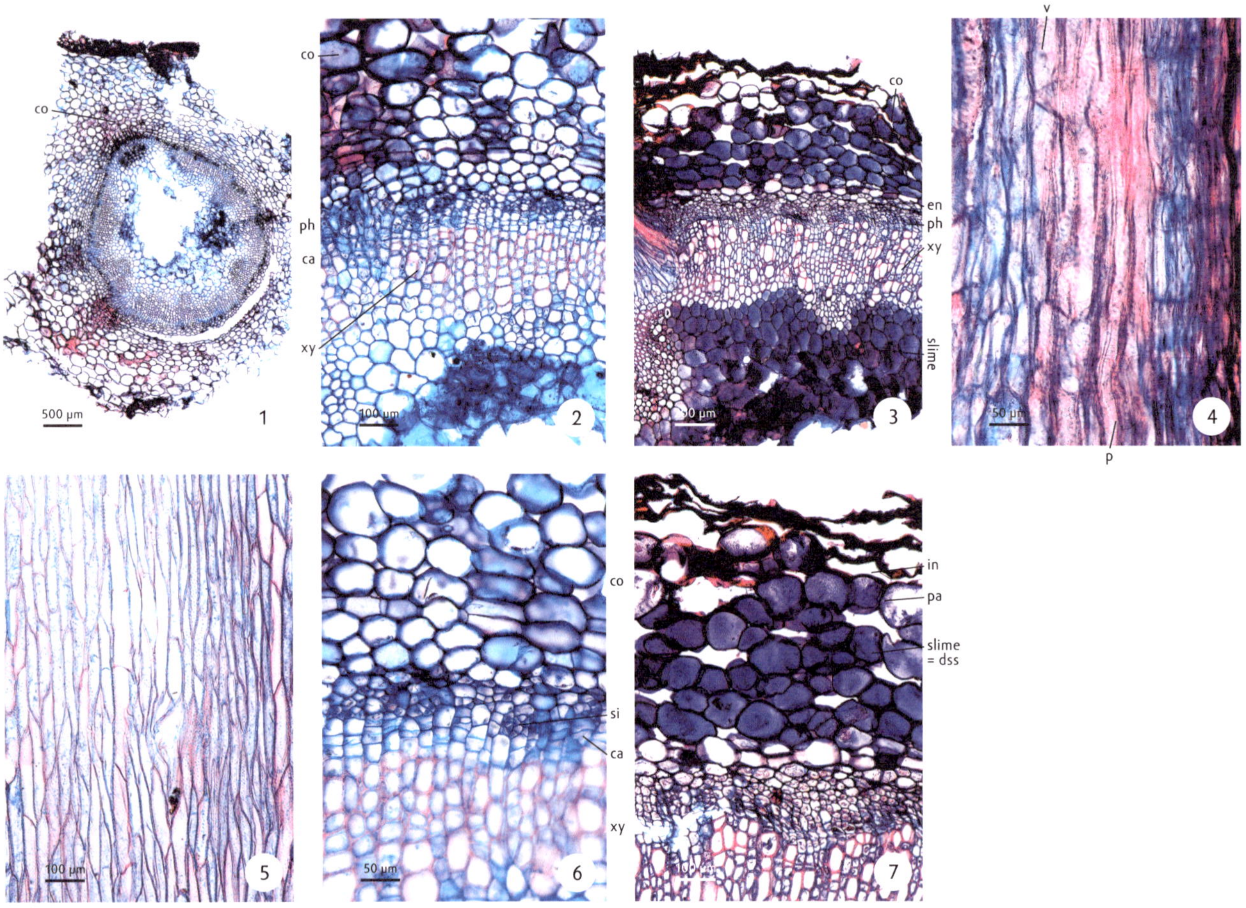

F. H. Schweingruber et al., *Anatomic Atlas of Aquatic and Wetland Plant Stems*, https://doi.org/10.1007/978-3-030-33420-8_16

Myosotis palustris (L.) L. (syn. *Myosotis scorpioides*) **Water Forget-me-not** Boraginaceae

Perennial wetland or emergent herb, usually with runners. Creeping stem with hairs spreading below, appressed above, 15–70 cm high. Oblong downy leaves, 3–5 times as long as wide. Bright blue flowers in scorpiod cymes, calyx-teeth short, flower stalks once or twice lenght of calyx. Fruit is a drupe.

Distribution and ecology

Euro-Siberian temperate element; widely naturalised outside its native range. Native in all CE countries from lowlands to subalpine zone. Occurs in muddy banks of stagnant waters, in wet meadows, floodplain forests and alder carrs.

Anatomy of a rhizome

Round rhizome with a diameter of 3–5 mm. Maximum age 2 years.

The pith of the rhizome is surrounded by a xylem, a phloem and a large cortex (1, 2). The xylem of the diffuse-porous xylem consists of vessels and thin- to thick-walled fibers (2). Axial parenchyma is marginal (3). Vessels are solitary and have a diameter of 20–40 µm (2) and a length of 120–150 µm. Intervessel pits are round to oval and perforation plates simple (3, 4). Fiber pits are round and very small. Rays are absent (2).

The phloem consists of parenchyma cells and groups of sieve tubes (6). The stele is surrounded by an endodermis (2, 5). Intercellulars in the cortex are irregularly formed (2, 6). The cortex cells contain slime. Crystals and phellem are absent.

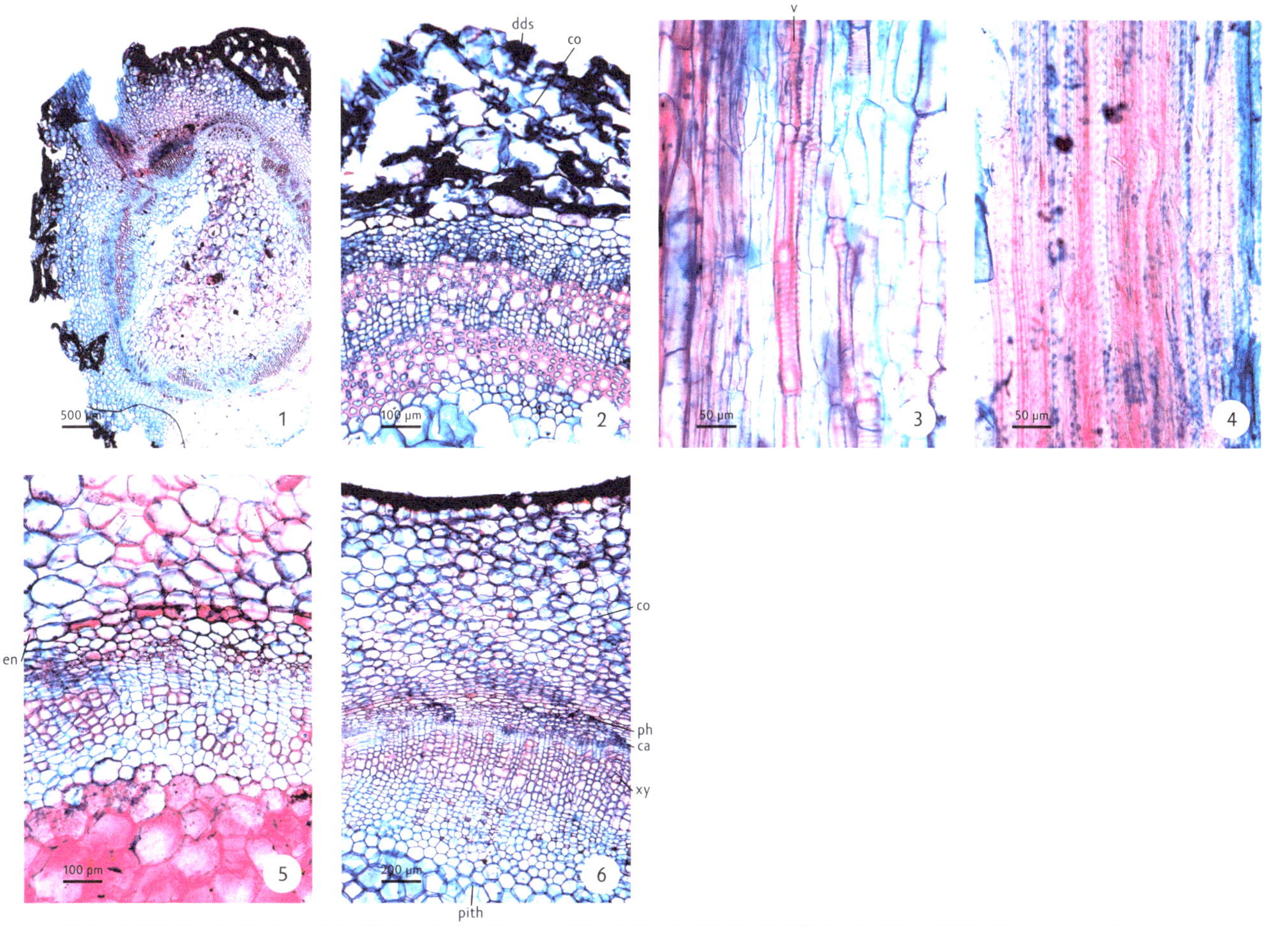

Myosotis rehsteineri Wartm. (syn. *Myosotis palustris* subsp. *caespititia*)

Perennial wetland herb with short runners, very similar to M. palustris. Stem round, with appressed hairs, only 2–10 cm high. Hairs on leaves in opposite direction to leaf apex. Wheel-shaped blue corrola 6–10 mm in diameter.

Distribution and ecology

Endemic to C. Europe. Native in AT, CH and DE, in the area of the Lake Constance. Occurs in occasionally flooded gravel banks.

Anatomy of the rhizome

Round rhizome with a diameter of 1–2 mm. Maximum age 1 year.

The pith of the rhizome is surrounded by vascular bundles with a xylem, a phloem and a large cortex (1, 2). The xylem of the diffuse-porous xylem consists of vessels and thin- to thick-walled fibers (2). Vessels are solitary and have a diameter of 15–20 µm (2) and a length of 80–100 µm. Intervessel pits are round to oval and perforation plates simple (3). Fibers are absent. Rays between the vascular bundle are 2–6-seriate. Rays are absent in vascular bundles (2).

The phloem consists of parenchyma cells and groups of sieve tubes (2). The stele is surrounded by an endodermis (2). Intercellulars in the cortex are net-like (1, 2). Crystals and phellem are absent.

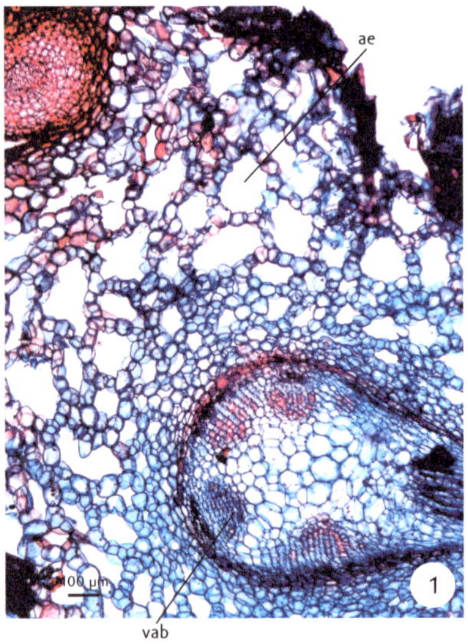

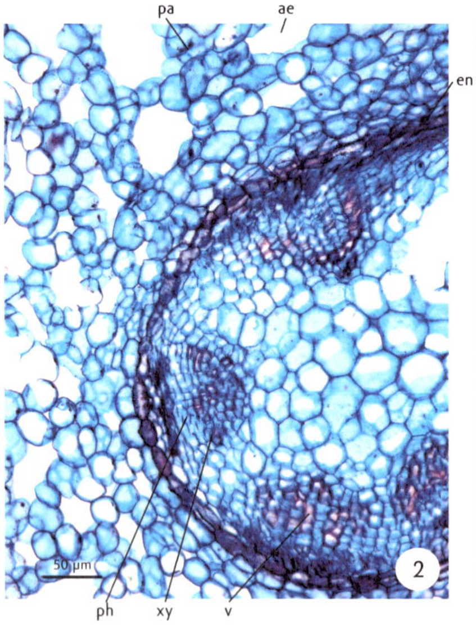

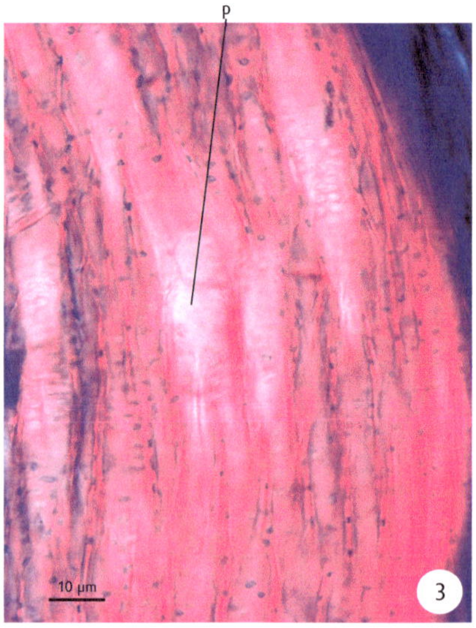

Cardamine amara L.

Large Bitter-cress Brassicaceae

Perennial wetland or emergent herb. Stem erect to ascendent, with short rhizome, little branched, 20–40 cm high. Leaves with sinuate-crenate leaflets. Flowers in several racemes, petals white or pinkish-purple. Fruit is a compressed siliqua.

Distribution and ecology

European temperate element. Common in all CE countries in forest springs, streamsides and alder carrs along small streams. Prefers moist and nutrient-rich acidic substrates, tolerant of shade. Winter-green herb.

Anatomy of the stem

Round stem with a diameter of 3 mm. Maximum age 1 year.

The pith of the stem is surrounded by a small xylem, a phloem and a large cortex (1). The xylem/phloem belt can be divided laterally into large vascular bundles or can be continuous (1, 2, 6). The xylem consists of vessels and thin- to thick-walled fibers. Axial parenchyma is pervasive near the pith (2) and paratracheal at external parts of the xylem (2, 6). Vessels have an intervessel diameter of 15–20 µm (2) and vessel length is 50–70 µm. Vessels are solitary (2, 5, 6). Pits are scalariform in the zone of pervasive parenchyma and round to oval in parts with paratracheal parenchyma (3). Perforation plates are simple (3). Fiber pits are very small. Rays are absent within the vascular bundles (2, 4, 6) and 5 to >10-seriate between vascular bundles (1, 2, 6).

The phloem consists of angular parenchyma cells, groups of sieve tubes and a few fibers at the periphery of the phloem (5, 6). The stele is surrounded by an endodermis (1, 6). Intercellulars in the cortex are irregularly formed and are as large as the surrounding parenchyma cells (6). A lignified epidermis surrounds the stem (6). The phellem is absent.

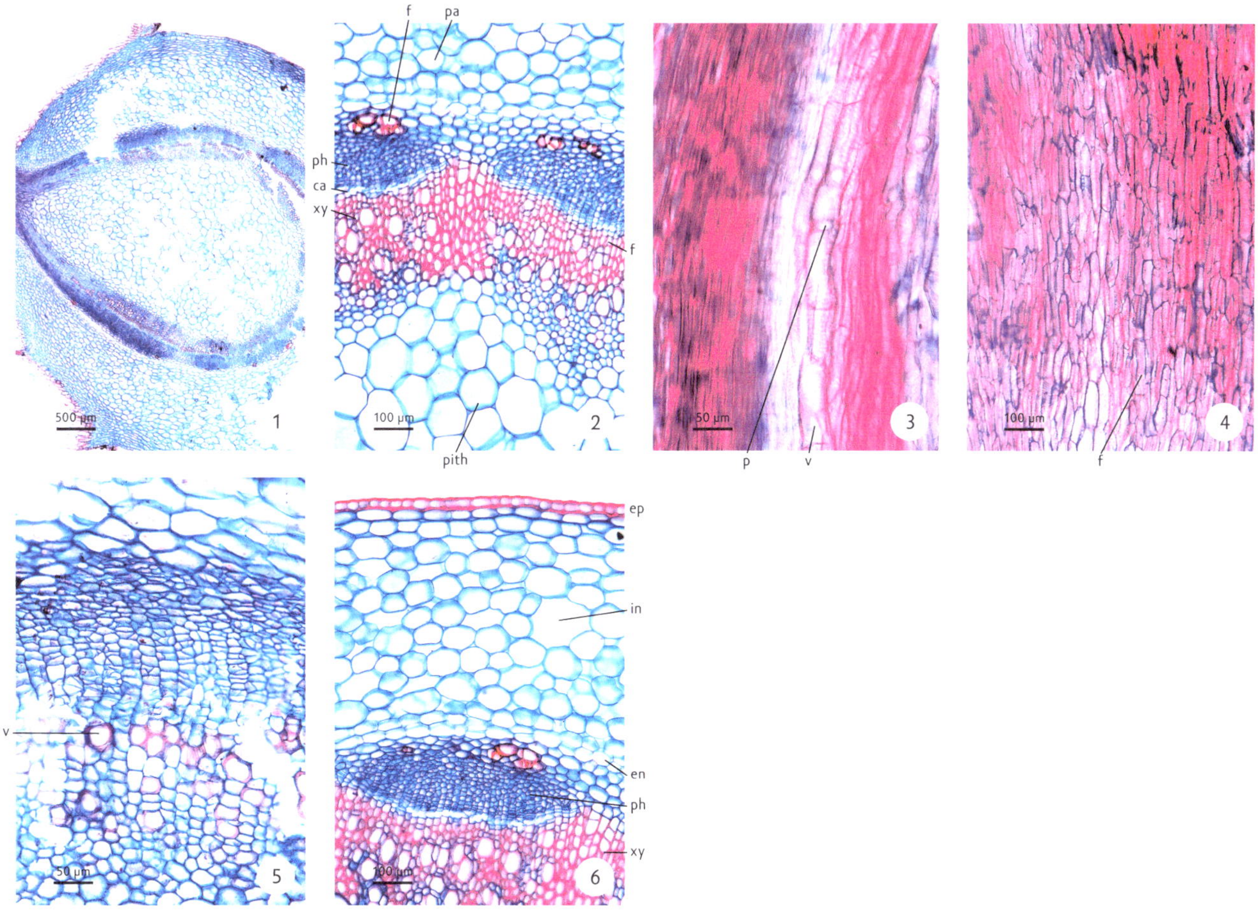

F. H. Schweingruber et al., *Anatomic Atlas of Aquatic and Wetland Plant Stems*, https://doi.org/10.1007/978-3-030-33420-8_17

Cardamine pratensis L. **Cuckooflower** Brassicaceae

Perennial wetland herb. Stem ascending to erect, up to 60 cm high. Basal leaves pinnate, with long petiole, stem leaves dissected, with short petiole. Flowers in a raceme, petals pale to deep pink, or white. Fruit is a siliqua.

Distribution

Circumpolar wide-boreal element. Common in all CE countries in moist and wet meadows, on fens, in upland areas also in pastures and bryophyte-rich springs. Prefers moderately fertile, seasonally waterlogged soils. It is morphologically and cytologically variable, many clones reproduce by rooting from the leaves. Resistant to some herbicides.

Anatomy of the stem

Base of annual shoot with a diameter of 3 mm. Maximum age 1 year.

Presented are plants from wet meadows.

A circle of collateral vascular bundles and a cortex surround a large pith. Vascular bundles in stems above the soil are laterally connected by a vessel-free belt of fibers (1, 2).

The xylem of stems above the soil consists of vessels and thin- to thick-walled fibers and paratracheal parenchyma (2). The xylem of stems in the soil is mainly composed of vessels and pervasive parenchyma (3). Vessels have a diameter of 15–20 µm (2, 3) and vessel length of 50–60 µm. Vessels are solitary (2, 3). Intervessel pits are round to oval (4, 5) and perforation plates are simple (3). Fiber pits are very small. Rays are absent within the vascular bundles (2, 3). The large confluent rays between the vascular bundles (5 to >10 cells) have upright cells.

The phloem consists of parenchyma cells and small groups of sieve tubes (2, 3). The stele is surrounded by a pericycle/endodermis zone (2, 3). The cortex contains frequently a few sclerenchyma cells (7). The frequency of intercellulars in the cortex varies, they can be almost absent (6) or fairly large and irregularly formed (2). A lignified epidermis surrounds the stem (1). The phellem is absent (1).

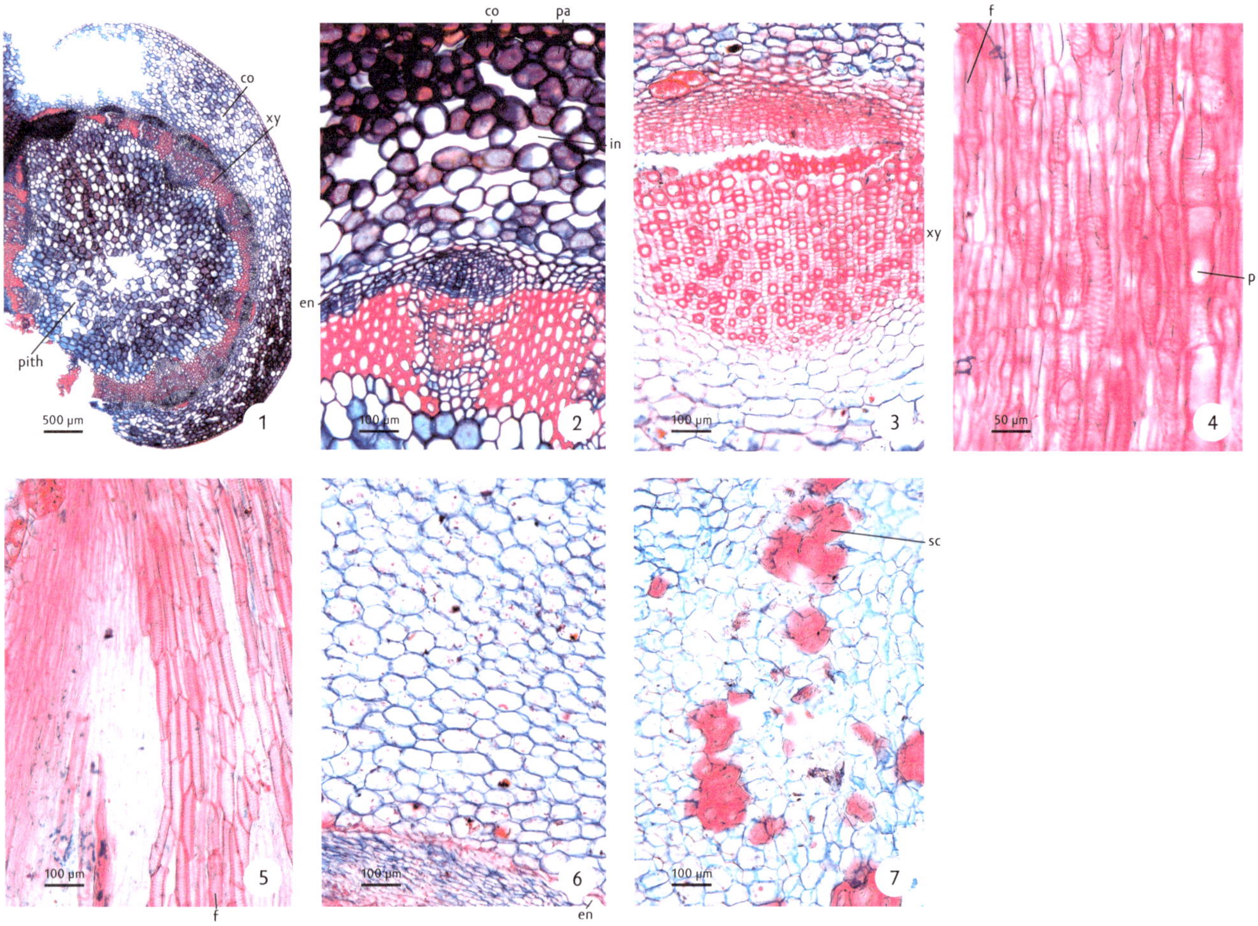

Cochlearia officinalis L.

Common Scurvygrass

Brassicaceae

Biennial or shortly perennial wetland herb. One or more ascending leafy stems, 20–40 cm high. Basal leaves in a loose rosette, heart-shaped, stem leaves ovate, sligthly dentate and shiny. Dense inflorescence of white flowers, 8–10 mm across. Fruit is a globose siliqua, narrowed into style.

Distribution and ecology

West-European distribution. Native in CH (extinct) and DE, both in lowlands and montane zone. Occurs on calcareous springs, wet basic rocks and rich fens, but also on saltmarshes and cliffs. Prefers saline or mineral-rich soils. Winter-green plant. Medicinal herb.

Anatomy of a rhizome

Rhizome with a diameter of 4 mm. Maximum age 2 years.

A continuous xylem, a large phloem and a very large cortex surround a small pith (1). The xylem is lobed and consists of thin-walled angular vessels and pervasive parenchyma (2). Vessels have a diameter of 20–30 µm (2). Vessel length is 70–90 µm (3). Intervessel pits are scalariform (3). Fibers are absent (2). Rays are absent or radially discontinuous (2).

The phloem consists of parenchyma cells and small groups of sieve tubes (4). Parenchyma zones between lobes dilate (4). Intercellulars in the cortex are small and large to net-like in the pith (5). The phellem consists of rectangular cells (1).

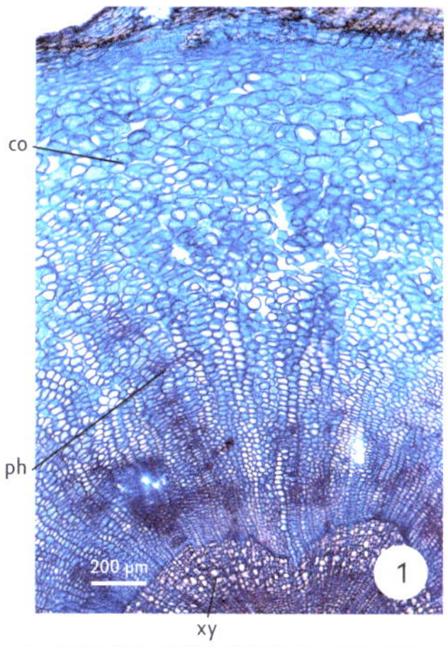

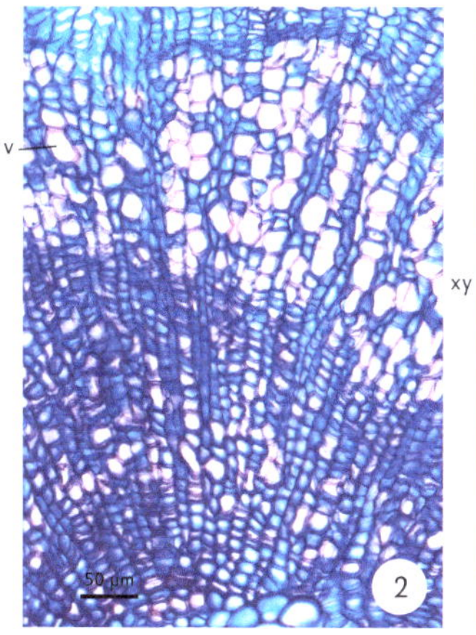

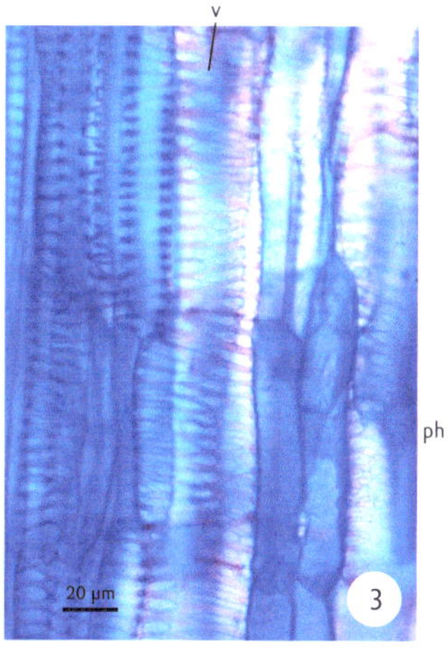

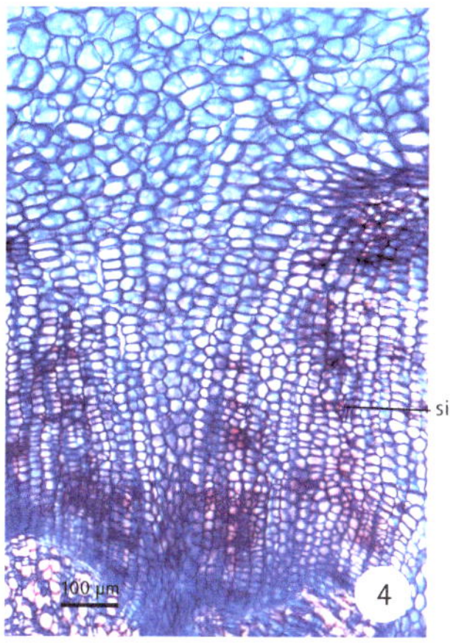

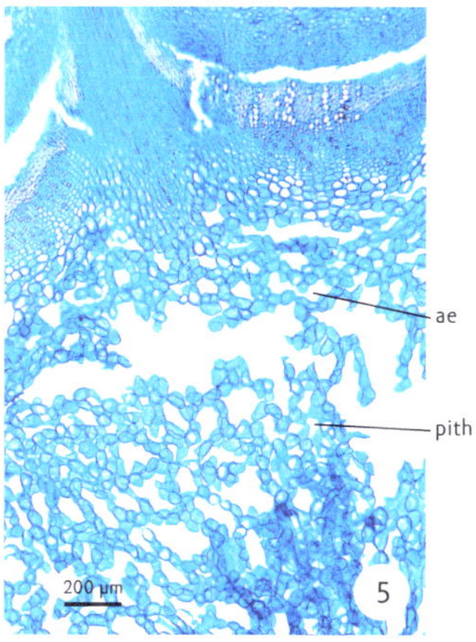

Nasturtium officinale R.Br. (syn. *Rorippa nasturtium-aquaticum*) **Water-cress** Brassicaceae

Perennial wetland or emergent herb. Procumbent to ascending stem with adventive roots, 20–80 cm high. Leaves pinnate, entire to sinuate. Flowers in a dense raceme, petals white, anthers yellow. Fruit is a cylindrical siliqua.

Distribution and ecology

Euro-Siberian southern-temperate element, but widely naturalised. Native in all CE countries, occurs in banks of clear streams, in ditches and canals, rarely beside standing waters. Prefers nutrient-rich, often organic soils, usually with higher Ca concentration.

Anatomy of the rhizome

Rhizome with a diameter of 3 mm. Maximum age 1 year. Presented is a plant from a wet meadow.

A circle of collateral vascular bundles and a cortex surrounds a large pith. Vascular bundles are laterally connected by parts with few fibers or fiber zones without vessels (1).

Most parts of the xylem consist of vessels and thin- to thick-walled fibers and paratracheal parenchyma (2). Vessels have a diameter of 15–20 µm (2) and vessel length is 70–120 µm. Vessels are primarily in small groups (2). Intervessel pits are scalariform in zones with pervasive parenchyma and round to oval in zones with paratracheal parenchyma (3). Perforation plates are simple. Fiber pits are very small. Rays are absent.

The phloem consists of parenchyma cells and small groups of sieve tubes (4). The stele is surrounded by an endodermis, which is partially sclerotized (4, 5). The frequency and size of intercellulars in the cortex varies from large to small (5, 6). The cortex contains frequently a few sclerenchyma cells (6). The phellem is absent (1, 5).

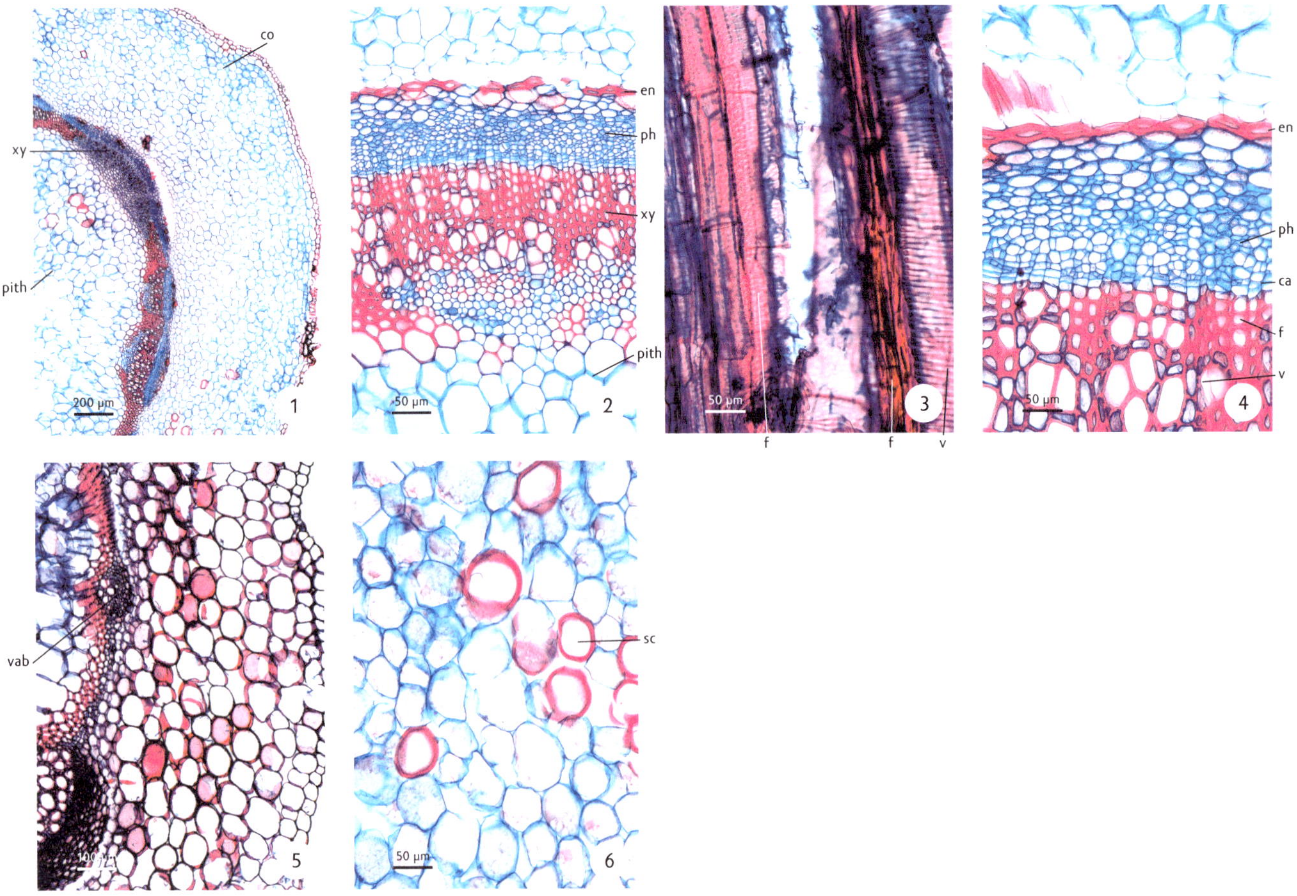

Rorippa amphibia (L.) Besser **Great Yellow-cress** Brassicaceae

Perennial wetland or emergent herb with thick rhizome. Stem ascending to erect, hollow, 0.4–1 m high. Shoots arise from axils of creeping stems. Lower stem leaves pinnately lobed, upper only toothed. Petals yellow. Fruit (siliqua) much shorter than mostly reflexed pedicels.

Distribution and ecology

Eurasian temperate element; widely naturalised outside its native range. Native in all CE countries in emergent vegetation along the edges of streams and rivers, by lakes and fishponds. Tolerates eutrophic water, fluctuating water table and periodical flooding as amphibous plant. Seed set is often poor, possibly plants are highly self-incompatible. Able to spread by fragmentation of mature plants.

Anatomy of the tuber

The intra-specific anatomy greatly varies.

Round tuber with a diameter of 12 mm and a floating stem with 3–5 mm. Secondary growth in the tuber remains for several years and only one year in the floating stem. Presented is the tuber of one plant from a backwater.

A circle of collateral vascular bundles and a cortex surround a large pith (1). The xylem consists of angular vessels, thin- to thick-walled fibers and pervasive parenchyma. Vessels have a diameter of 20–30 µm (2). Vessel length is 50–100 µm (3). Intervessel pits are scalariform in zones with pervasive parenchyma (3). Rays are absent within vascular bundles (2) and >10-seriate between the bundles (1).

The phloem consists of parenchyma cells and small groups of sieve tubes (2). The cortex in the tuber contains many intercellulars between round, thin-walled parenchyma cells (4). The phellem is absent (1).

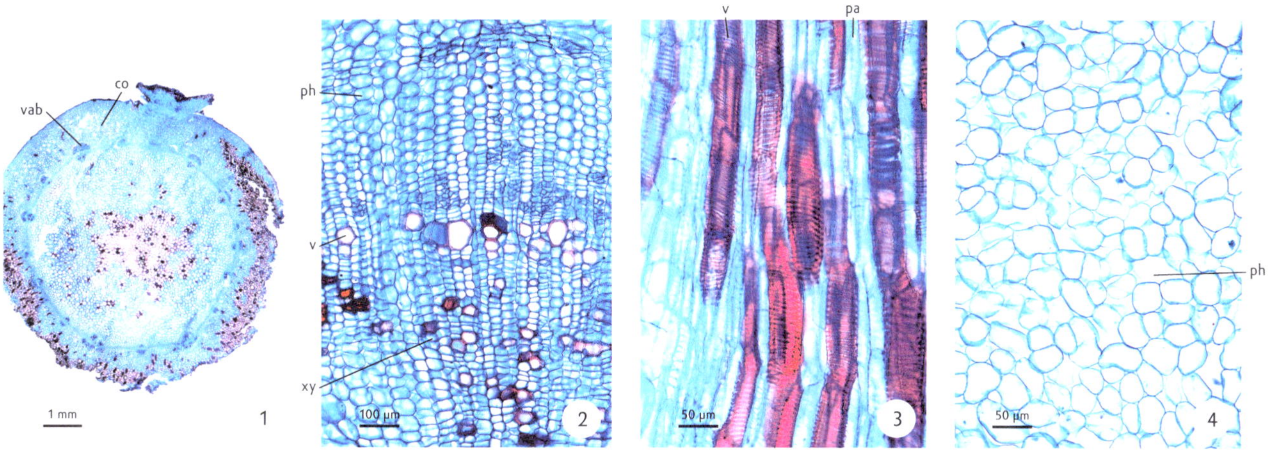

Rorippa palustris (L.) Besser

Marsh Yellow-cress Brassicaceae

Annual wetland herb, rarely short-lived perennial. Ascending to erect hollow stem, 15–60 cm high. Leaves auricled and pinnately-lobed. Petals yellow, fruit (siliqua) as long as pedicels.

Distribution and ecology

Circumpolar boreo-temperate element. Native in all CE countries. Typical of exposed drained bottoms of fishponds, exposed wet mud in lake margins or in river banks, also wet depressions in fields. Evidence for longevity of the seed bank.

Anatomy of the rhizome

Round rhizome with a diameter of 3–4 mm. Maximum age 1 year.

An almost circle of a xylem, a small phloem and a large cortex surround the large pith (1). The xylem consists of angular vessels, thin- to thick-walled fibers and paratracheal parenchyma. Vessels have a diameter of 30–40 µm (2). Vessel length is 50–100 µm (3).

Most parts of the xylem consist of vessels and thin- to thick-walled fibers and paratracheal parenchyma (2). Vessels are primarily in short radial groups (2). Intervessel pits are scalariform in zones with pervasive parenchyma and round to oval in zones with paratracheal parenchyma (3). Perforation plates are simple. Fiber pits are very small (4). Rays are absent (5).

The phloem consists of parenchyma cells and small groups of sieve tubes (6). The stele is surrounded by an endodermis, which is partially sclerotized (4, 5). Intercellulars in the cortex are small (6). A phellem is absent (1).

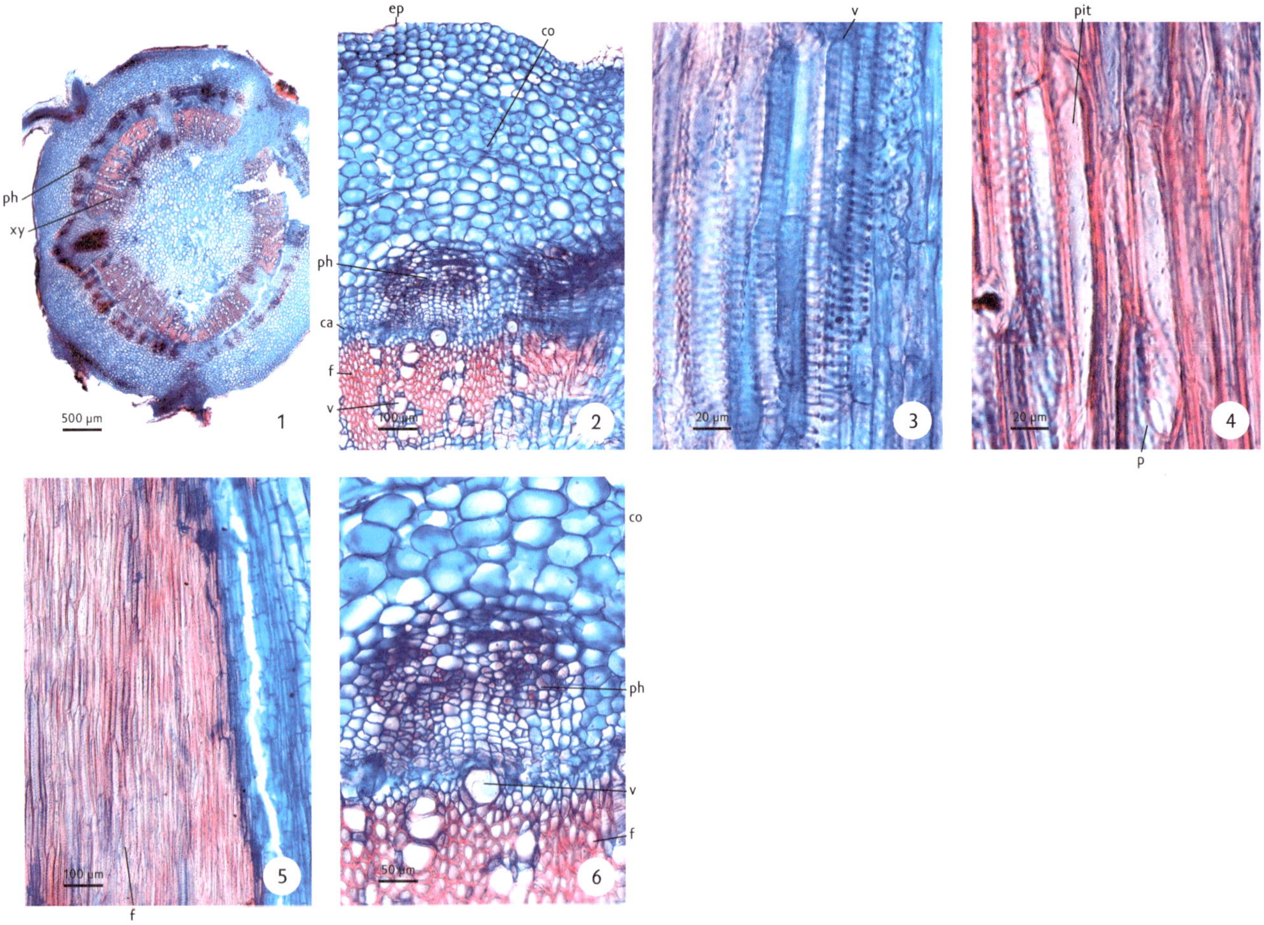

Cerastium cerastoides (L.) Britton (syn. *Cerastium trigynum*) Caryophyllaceae

Perennial wetland herb. Stem procumbent, mat-forming, 5–15 cm high. Flowering stems erect, hairy. Leaves lanceolate, ca. 1 cm long. White flowers with only 3 styles.

Distribution and ecology

Arctic-alpine element. Native in AT, CH, DE, PL and SK, in subalpine to alpine zone. Occurs often in areas of late snow-bed, on wet acidic rocks, tall-herb springs and in wet alpine pastures.

Anatomy of the root collar

Root collar with a diameter of 2–3 mm. Maximum age approximately 25 years. Rings are often indistinct. Presented is a plant from an alpine snow bed.

The small pith of the root collar is surrounded by large diffuse-porous xylem, a small phloem and cortex (1). The xylem consists of thick-walled vessels and parenchyma cells (2). Vessels are solitary with a diameter of 10–20 µm (2) and a length of 60–80 µm. Intervessel pits are scalariform and perforation plates simple (3). Parenchyma is pervasive (2). Fibers and rays are absent (2).

The phloem is uniform and consists of parenchyma (4). Intercellulars in the cortex are absent (4). The phellem consists of many layers of rectangular cells (4).

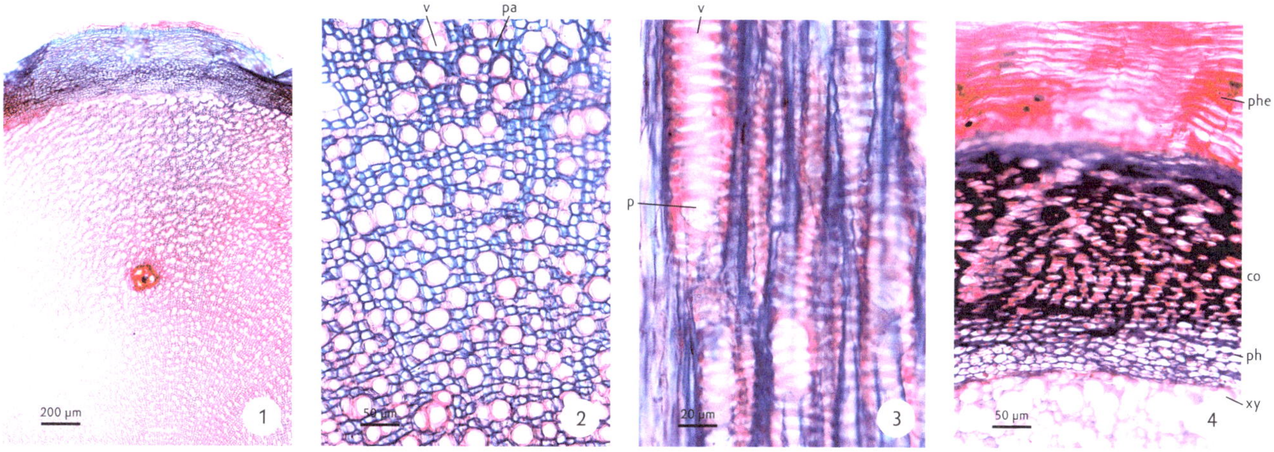

F. H. Schweingruber et al., *Anatomic Atlas of Aquatic and Wetland Plant Stems*, https://doi.org/10.1007/978-3-030-33420-8_18

Corrigiola littoralis L. **Strapwort** Caryophyllaceae

Annual wetland herb. Small prostrate herb, rarely overwintering, with shoots of 5–25 cm long. Linear leaves alternate. Small flowers of about 2 mm diameter of whitish colour, in terminal and axillary clusters. Fruit is an achene, almost not swimming.

Distribution and ecology

Subatlantic Mediterranean species, native in CH, CZ, DE and PL. It prefers damp sandy or gravely open places prone to seasonal flooding, esp. in river margins. Typical of high disturbance habitats, cattle trampling areas and winter flooding sites. Able to self pollinate via cleistogamy. Endangered by river regulation.

Anatomy of the root collar

Root collar with a diameter of 2–3 mm. Maximum age unknown. Annual rings are absent. Presented is a plant from a meadow on a sea shore.

The small pith of the root collar is surrounded by successive cambia, a few xylem and phloem belts, a cortex and a phellem (1). The xylem consists of thin-walled vessels and parenchyma cells (2). The angular vessels are solitary, they have a diameter of 20–30 µm (2) and a length of 150–200 µm (3). Intervessel pits are scalariform and perforation plates are simple (3). Parenchyma is pervasive (2, 4). Fibers and rays are absent (2).

The phloem consists of groups of sieve tubes and parenchyma cells (4). Intercellulars in the cortex are small (4). The phellem consists of many layers of irregularly formed cells (4).

ROOT COLLAR

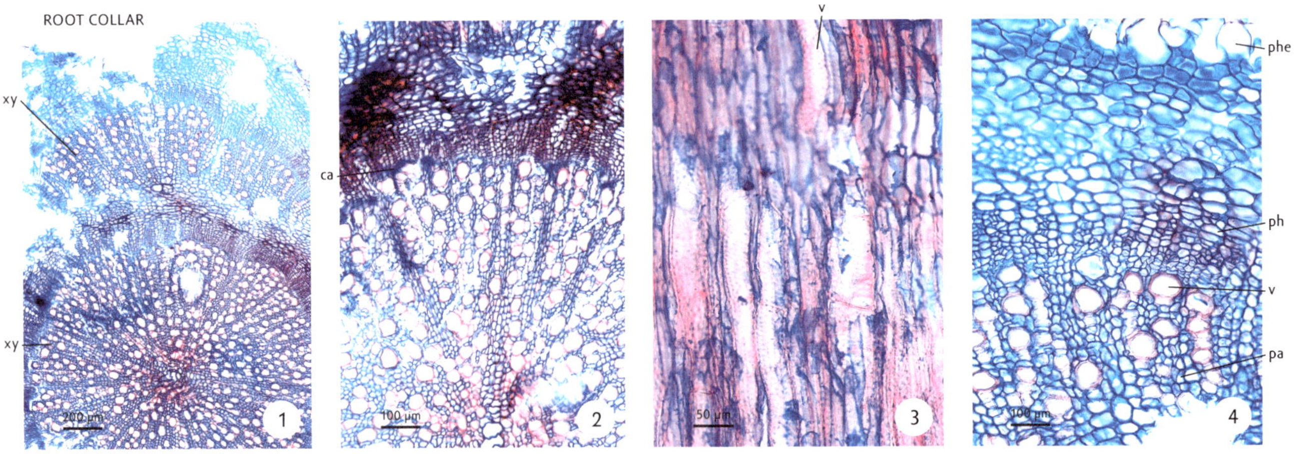

Perennial wetland herb. Much branched sprawling stem, up to 2 m long, shortly pubescent, fragile, scrambling over other plants. Leaves ovate, acuminate, with short petiole. White flowers in open dichasia, campanulate calyx. Fruit is a round black berry.

Distribution and ecology

Native of Europe, SW Asia and disjunctly in E. Asia. Native in all CE countries. Usually along large rivers in lowlands, esp. in alluvial forests, along banks of streams and rivers. Tolerates periodical flooding. Prefers wet soils.

Anatomy of the stem

Stem or a liana with a diameter of 3–4 mm. Maximum age 4 years. Presented is a plant from a riparian site.

The small pith is surrounded by a semi-ring porous xylem, a small phloem, a cortex and a phellem (1). The xylem consists of thin-walled vessels, groups of fibers and parenchyma cells (2). Vessels are solitary, have a diameter of 50–70 µm (2) and a length of 120–180 µm (3). Intervessel pits are round to slightly scalariform and perforation plates are simple (3). Parenchyma is pervasive (2, 4). Thin- to thick-walled fibers occur in groups and rays are absent (2).

The phloem consists of small groups of sieve tubes and parenchyma cells (4). Intercellulars in the cortex are absent (4). The phellem consists of some layers of rectangular cells (4).

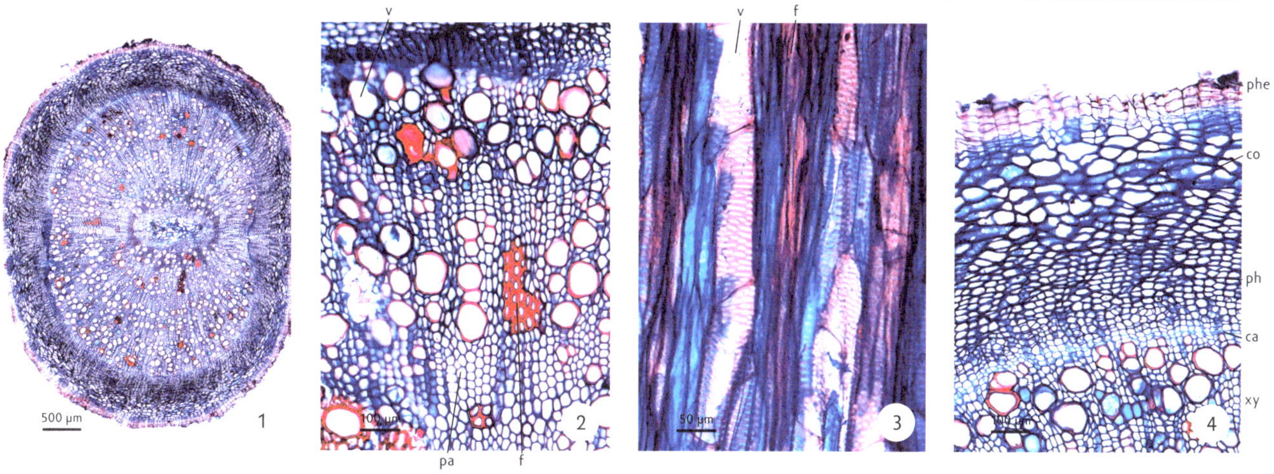

Annual wetland herb. Procumbent to decumbent stems 5–20 cm long, usually a radially spreading circular colony 5–50 cm in diameter, stems distinctly red, opposite leaves obovate, obtuse, white flowers densely clustered in leaf axils. Fruit is a capsule with one seed.

Distribution and ecology

Suboceanic southern-temperate distribution. Scattered irregularly and rarely in all CE countries except for SK and CH, in lowlands to uplands, bare acid sandy or peaty wet soils, margins of fishponds, extracted fens, fields or abandoned meadows, old sand-pits, mainly in fishpond areas. Strict heliophyte, competitively very week, pioneering species on disturbed areas.

Anatomy of the root collar

Root collar with a diameter of <1 mm. Maximum age 1 year.

The large pith is surrounded by a small xylem, a small phloem, a thick-walled endodermis and a large, thin-walled cortex (1–3). The intensively lignified xylem consists of solitary vessels and fibers which cannot be differentiated on cross-sections (3). Vessels and fibers have a diameter of 8–12 µm (2) and a length of 80–110 µm. Intervessel pits are scalariform and perforation plates are simple. Parenchyma is pervasive (3). Fibers are absent. Rays are probably 1-3-seriate but are not well differentiated from axially oriented cells (3).

The phloem is uniform, sieve tubes and parenchyma cells cannot be differentiated (3). A large thick-walled endodermis separates the stele from the cortex (2). Intercellulars in the cortex are small, but the parenchyma cells very large (40–60 µm) (1, 2). The phellem is absent (1, 2).

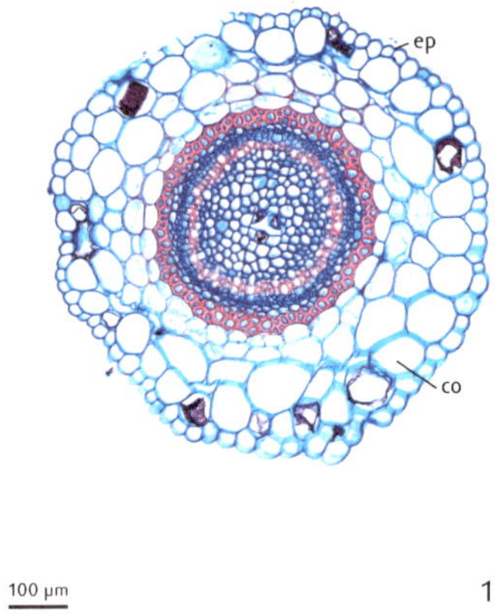

1

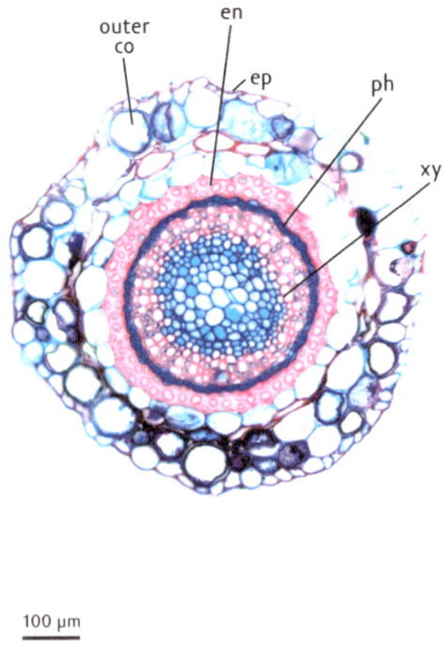

2

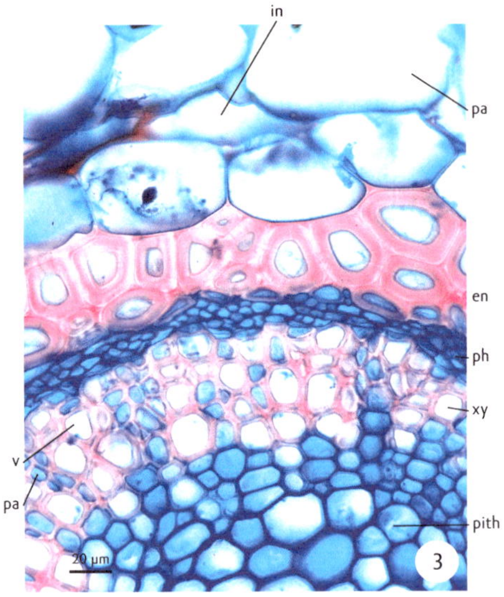

Lychnis flos-cuculi L.

Ragged Robin

Caryophyllaceae

Biennial or shortly perennial wetland herb. Stem erect, leafy, glabrous, up to 80 cm high. Leaves narrow-lanceolate, hairless. Petals rose-red, divided each into 4 narrow lobes. Fruit is a capsule.

Distribution and ecology

Euro-Siberian temperate element; widely naturalised outside its native range. Native in all CE countries from uplands to mountains. Occurs in wet pastures, wet meadows and fens, also in damp woodland margins. Prefers wet, slightly acidic soils. Persistent seed bank.

Anatomy of the root collar

Root collar with a diameter of 3–4 mm. Maximum age 3 years. Presented is a plant from a wet meadow.

The pith is surrounded by a xylem, a small phloem, an endodermis, a large, thin-walled cortex and a small phellem (1). The xylem consists of solitary vessels and parenchyma cells (3). Vessels have a diameter of 5–10 µm (2). Vessel length could not be observed. Intervessel pits are scalariform and perforation plates are simple (4). Parenchyma is pervasive (2). Fibers occur near the periphery. Rays are absent (3).

The phloem contains groups of sieve elements and parenchyma (5). Intercellulars in the cortex are absent (1, 2). The phellem, if present, consists of a few layers of irregularly formed cells (1). Very large crystal druses occur in the pith (6).

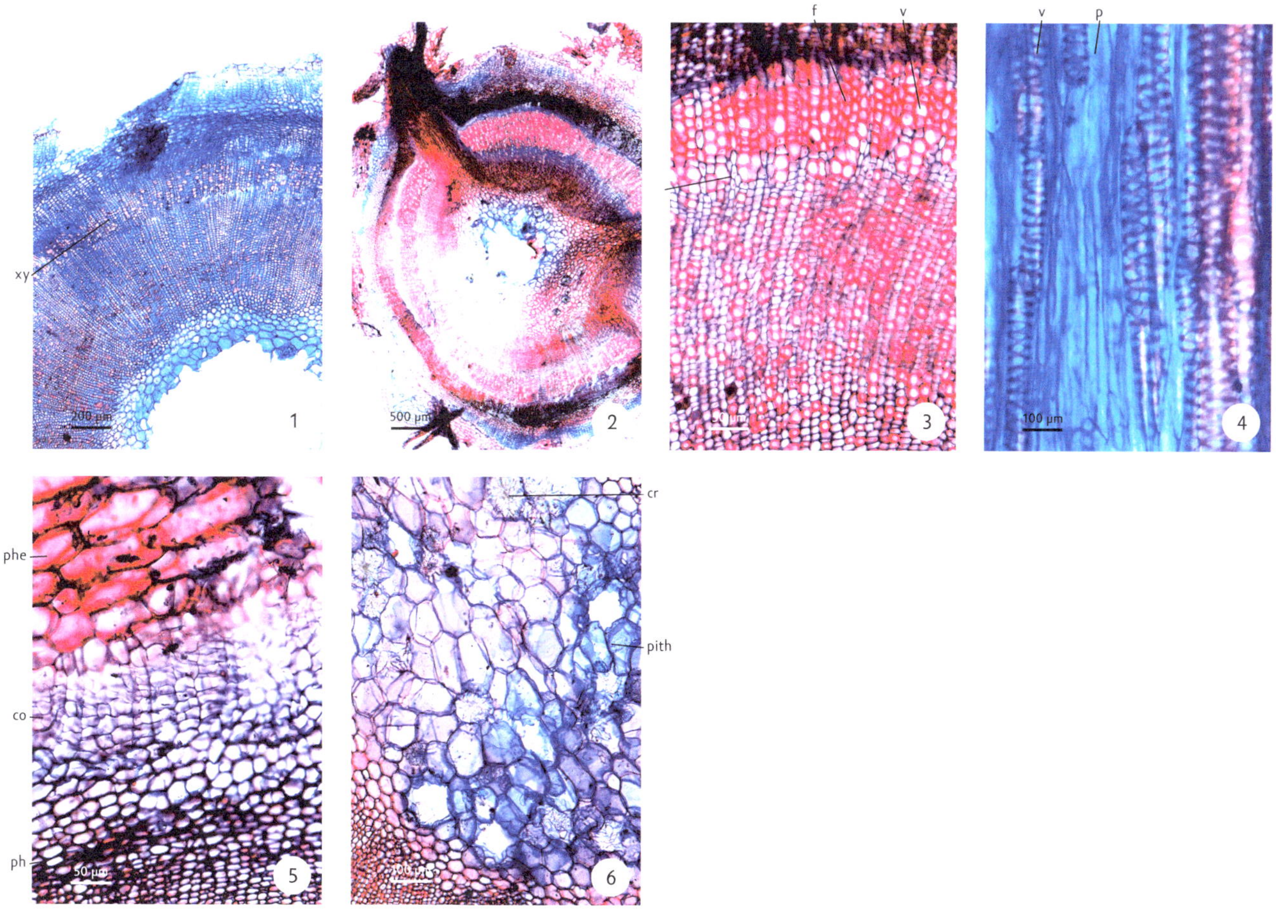

Myosoton aquaticum (L.) Moench **Water Chickweed** Caryophyllaceae

Perennial wetland herb. Stem procumbent, often branched, square-like, 20–50 cm high. Leaves oval, cordate, basal leaves with short petiole. White flowers with 10 stamens and 5 styles. Fruit is an oval capsule.

Distribution and ecology

Euro-Siberian temperate element; widely naturalised outside its native range. Native in all CE countries in damp wet habitats. Occurs on banks of rivers, streams and ditches, in alder carrs and by fishponds.

Anatomy of a basal part of the annual shoot

Shoot with a diameter of 2 mm. Maximum age 1 year. Annual rings are absent.

The central cylinder (stele) is surrounded by a large cortex and an epidermis (1). The xylem consists of solitary round vessels and parenchyma cells (2). Vessels have a diameter of 30–50 µm (2) and a length of 80–100 µm. Intervessel pits are scalariform and perforation plates are simple (3). Parenchyma is pervasive (2). Fibers and rays are absent (1, 2).

The phloem is uniform (2). The outer cortex consists of a few layers of lignified cells, a large belt of large, thin-walled, unlignified and an inner belt of small thin- to thick-walled cells around the central cylinder (2, 4). Crystals are absent. The epidermis consists of lignified, thin- to thick-walled cells (5).

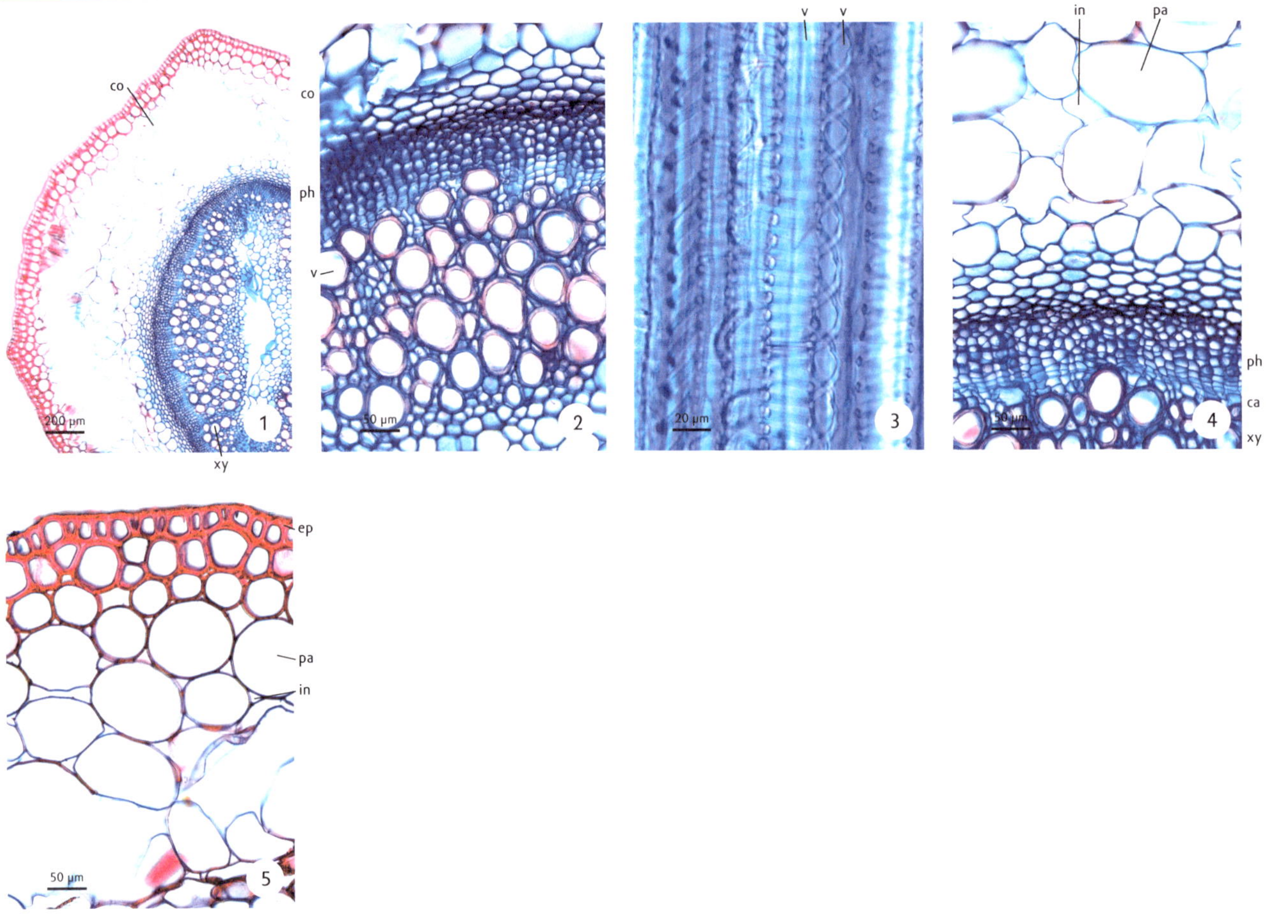

Spergularia echinosperma (Čelak.) Asch. et Graebn.

Bristleseed sandspurry

Annual wetland herb. Erect or procumbent glandulous stems. Leaves opposite, linear, stipules membraneous, triangular. Pink flowers in a sparse cymose. Fruit is a capsule with brown seeds with spiny papillas.

Distribution and ecology

Native in Central Europe, some localities in France and Morocco. Only in AT, CZ, DE, PL and SK. Typical of exposed drained bottoms of fishponds, prefers sandy, periodically flooded soils. Heliophilous, low competitive species, develops long-term seed bank.

Anatomy of the root collar

Root collar with a diameter of 2 mm. Maximum age unknown. Annual rings are absent. Presented is a plant from a wet meadow.

The root collar contains successive cambia, a few xylem and phloem belts and a cortex (1). The xylem consists of solitary vessels and parenchyma cells (2). Vessels have a diameter of 20–30 µm (2) and a length of 70–100 µm (3). Intervessel pits are large, oval to scalariform and perforation plates are simple (3). Parenchyma is pervasive (2). Fibers and rays are absent (2).

The phloem consists of groups of sieve tubes and parenchyma cells (2). Intercellulars in the cortex are very small (2). The phellem consists of several layers of rectangular cells (1).

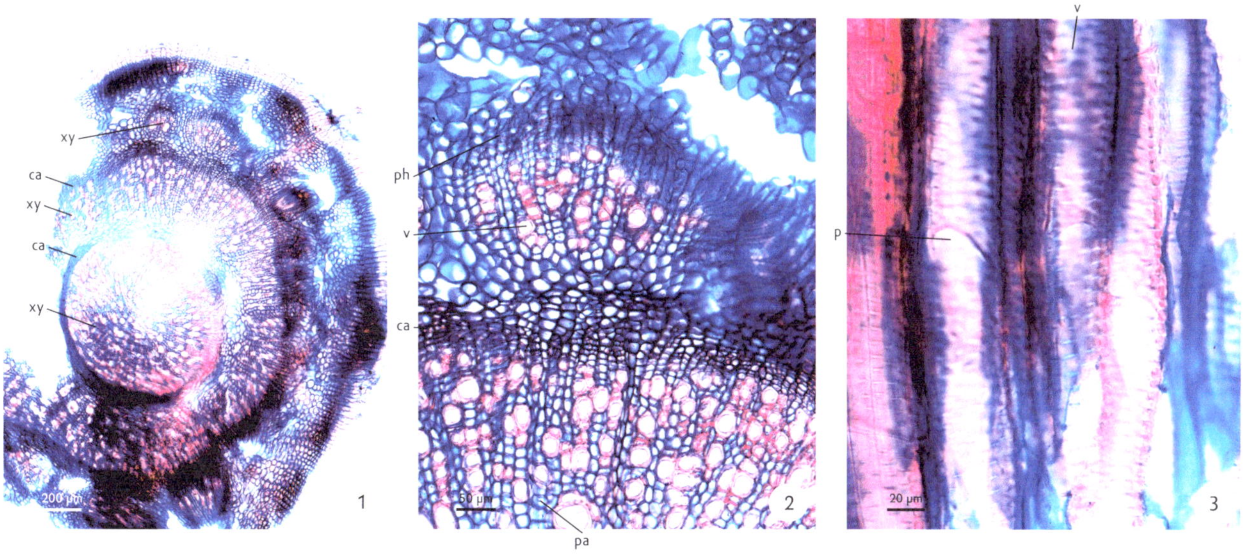

Spergularia media (L.) C. Presl (syn. *Spergularia maritima*)

Caryophyllaceae
Greater Sea-spurrey

Perennial wetland herb. Stem decumbent to ascending, in the upper part with glandulous hairs, up to 5–25 cm high. Leaves opposite, linear and fleshy. Flowers pink or white, 8–12 mm across. Fruit is a smooth capsule with a clear winged border.

Distribution and ecology

Euro-Siberian southern-temperate element; widely naturalised. Rare in AT, CZ, DE, SK, especially in lowlands. Occurs in saltmarshes and on the margins of saline ditches in coastal grazing marshes. Inland occurs rarely in salty meadows and in the surroundings of mineral springs. Obligate halophyte.

Anatomy of the root collar

Root collar with a diameter of 2 mm. Maximum age unknown. Annual rings are absent. Presented is a plant from a wet meadow.

The root collar contains successive cambia, a few xylem and phloem belts and a cortex (1). The xylem consists of solitary vessels and parenchyma cells (2). Vessels have a diameter of 20–30 µm (2) and a length of 70–100 µm (3). Intervessel pits are large, oval to scalariform and perforation plates are simple (3). Parenchyma is pervasive (2). Fibers and rays are absent (1, 2).

The parenchymatic phloem contains few small groups of sieve elements (3). Intercellulars in the cortex are very small (4). Cortex with a few unstructured crystals. The phellem consists of several layers of rectangular cells (1).

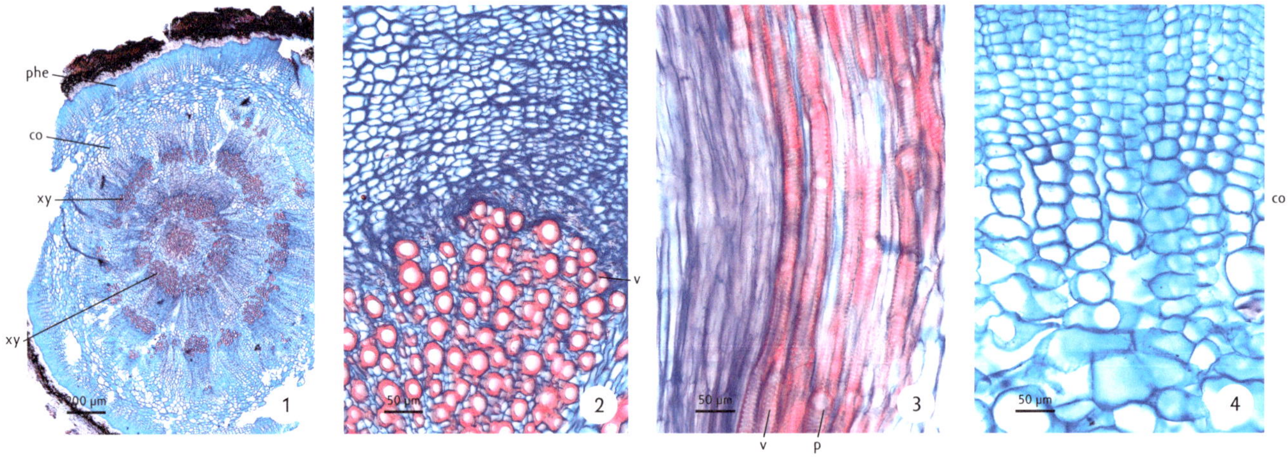

Perennial wetland herb, winter-green. Stem procumbent, square, 10–40 cm high. Leaves lanceolate to oval, smooth and grey-green. Petals white, shorter than sepals. Fruit is a capsule.

Distribution and ecology

Euro-Siberian and N American temperate distribution. Native in all CE countries from uplands to the subalpine zone. Occurs on banks of streams, in wet ditches, wet habitats in woodlands and on springs. Prefers acidic soils.

Anatomy of a basal part of the annual shoot

Shoot with a diameter of 0.7 mm. Maximum age 1 year. Annual rings are absent.

The central cylinder (stele) is surrounded by an endodermis, a large cortex and an epidermis (1, 2). The xylem consists of solitary angular vessels and parenchyma cells (2). Vessels have a diameter of 5–10 µm (2). Length of fibers is 150–180 µm. Intervessel pits are scalariform and perforation plates are simple (3, 4). Parenchyma is pervasive (2). Fibers and rays are absent (1, 2).

The phloem is uniform (2). Intercellulars in the cortex are very small (2). Crystals are absent. The epidermis consists of lignified, thin- to thick-walled cells (2).

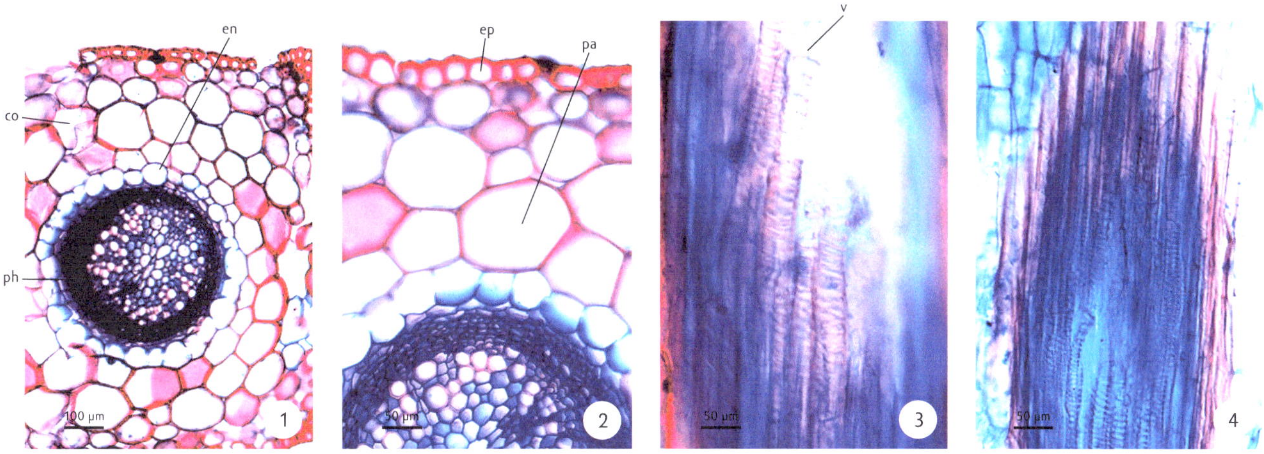

Stellaria nemorum L.

Wood Stitchwort Caryophyllaceae

Perennial wetland herb, stoloniferous. Stem decumbent to ascending, pubescent, 15–60 cm high. Lower leaves long-stalked, cordate. Petals white, mostly 2× as long as sepals. Fruit is a capsule.

Distribution and ecology

European boreo-temperate element. Native in all CE countries. Occurs mostly in wet and shaded habitats, usually by streams, on springs, also at ruderal sites in damp woods. Seeds germinate in response to addition of light and nitrates.

Anatomy of a basal part of the annual shoot

Shoot with a diameter of 3 mm. Maximum age 1 year. Annual rings are absent.

The central cylinder (stele) is surrounded by an endodermis-like zone of small cells, a large cortex and an epidermis (1). Vascular bundles are circularly arranged around the pith. The xylem consists of grouped angular vessels and parenchyma cells (2). Bundles are laterally separated by thin-walled parenchyma cells (1). Vessels have a diameter of 20–30 µm (2). Length of fibers 100–150 µm. Intervessel pits are scalariform and perforation plates are simple (3). Parenchyma is pervasive (2). Fibers and rays within vascular bundles are absent (1, 2).

The phloem is uniform (2). The outer cortex consists of large, unlignified cells with small intercellulars and the inner part consists of a belt of smaller, lignified cells (2). Crystals are absent. The epidermis consists of lignified, thin- to thick-walled cells (1).

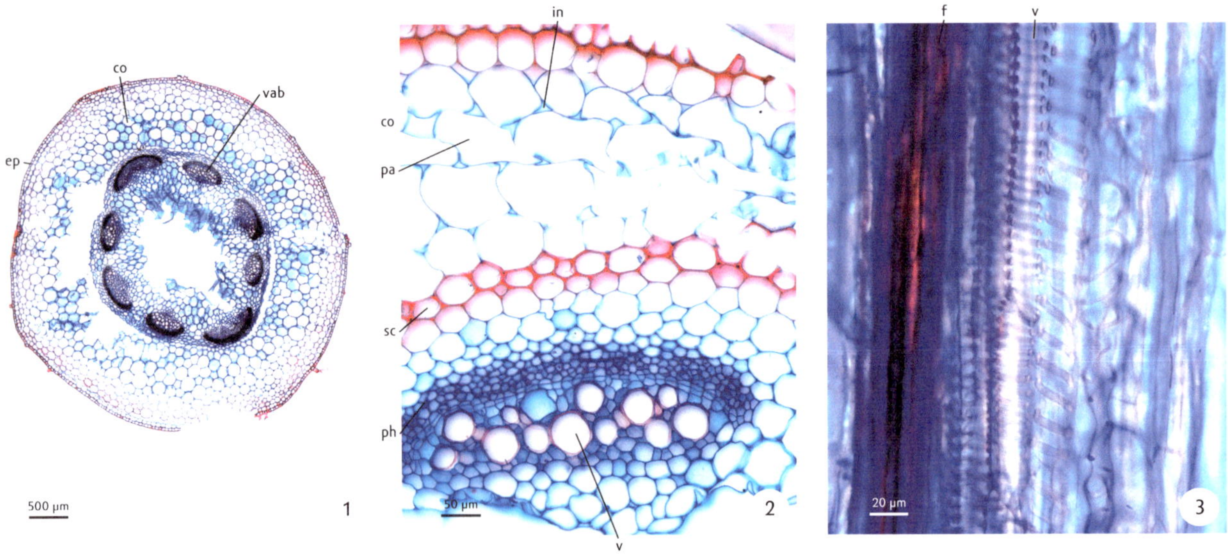

Stellaria palustris Ehrh. ex Hoffm. **Marsh Stitchwort** Caryophyllaceae

Perennial wetland herb, rhizomatous. Stem weak, ascending, glabrous, 15–45 cm high. Leaves linear-lanceolate, sessile, usually waxy grey-green. Bracts smooth and hairless. Flowers white, up to 2 cm across, in cymes, petals split to base. Fruit is a capsule.

Distribution and ecology

Eurasian boreo-temperate element. Native in all CE countries in uplands, especially in areas with standing water in winter. Occurs in fens and fen meadows, margins of eutrophic standing waters, in reeds and marshes.

Anatomy of a basal part of the annual shoot

Shoot with a diameter of 0.6 mm. Maximum age 1 year. Annual rings are absent.

The central cylinder (stele) is surrounded by an endodermis-like zone of small cells, a large cortex and an epidermis (1, 2). The xylem consists of solitary angular vessels and parenchyma cells (2). Vessels have a diameter of 7–10 µm (2). Length of fibers could not be observed. Intervessel pits are scalariform and perforation plates are simple. Parenchyma is pervasive (2). Fibers and rays are absent (1, 2).

The phloem is uniform (2). Intercellulars in the cortex are absent (2). Crystals are absent. The epidermis consists of lignified, thin- to thick-walled cells (2).

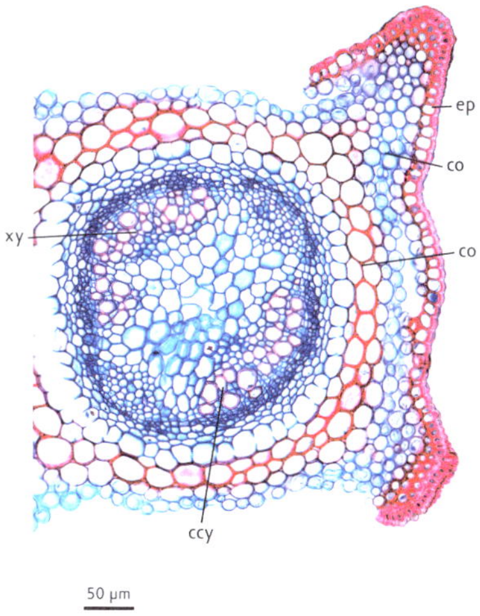

1

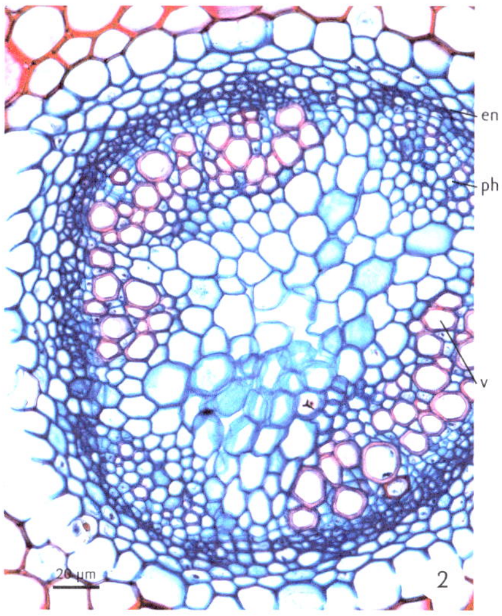

2

Parnassia palustris L. **Grass-of-Parnassus** Celastraceae

Perennial wetland plant, a few petiolated, ovate to cordate leaves in basal rosette. Flower stem with one sessile leaf 10–25 cm long, one terminal white flower. Fruit is an ovoid capsule.

Distribution and ecology

Circumpolar boreo-temperate distribution. Scattered irregularly and rarely in all CE countries, in lowlands to mountains up to 2000 m a.s.l. Wet meadows, pastures, peat bogs, fens, margins of fishponds, avoids strongly acidic soils, prefers slightly acidic to neutral, base-rich soils (pH ca. 5.5–7.5). Competitively weak species.

Anatomy of the root collar

Round transition zone between roots and the annual shoots with a diameter of 3–5 mm. Maximum age is 1 year.

Root collars consist of a small pith, a xylem, a small phloem, a large cortex and an epidermis (1, 2). The xylem is constructed of parenchyma cells, vessels and ray-like radial parenchyma zones (3). Vessels are angular and arranged solitary (2, 3). Vessel diameter varies from 10–15 µm (3) and vessel length from 80–150 µm. Intervessel pits are scalariform and reticulate with very large openings (4). Perforation plates are very small and simple. Fibers are absent (3). Parenchyma is pervasive (2). Rays are absent (2, 3).

Phloem and cortex are separated by an endodermis (1, 3). The phloem consists of parenchyma cells and small groups of sieve tubes (3). The cortex is composed of polygonal cells with irregularly formed intercellulars (5). The epidermis consists of square cells (1).

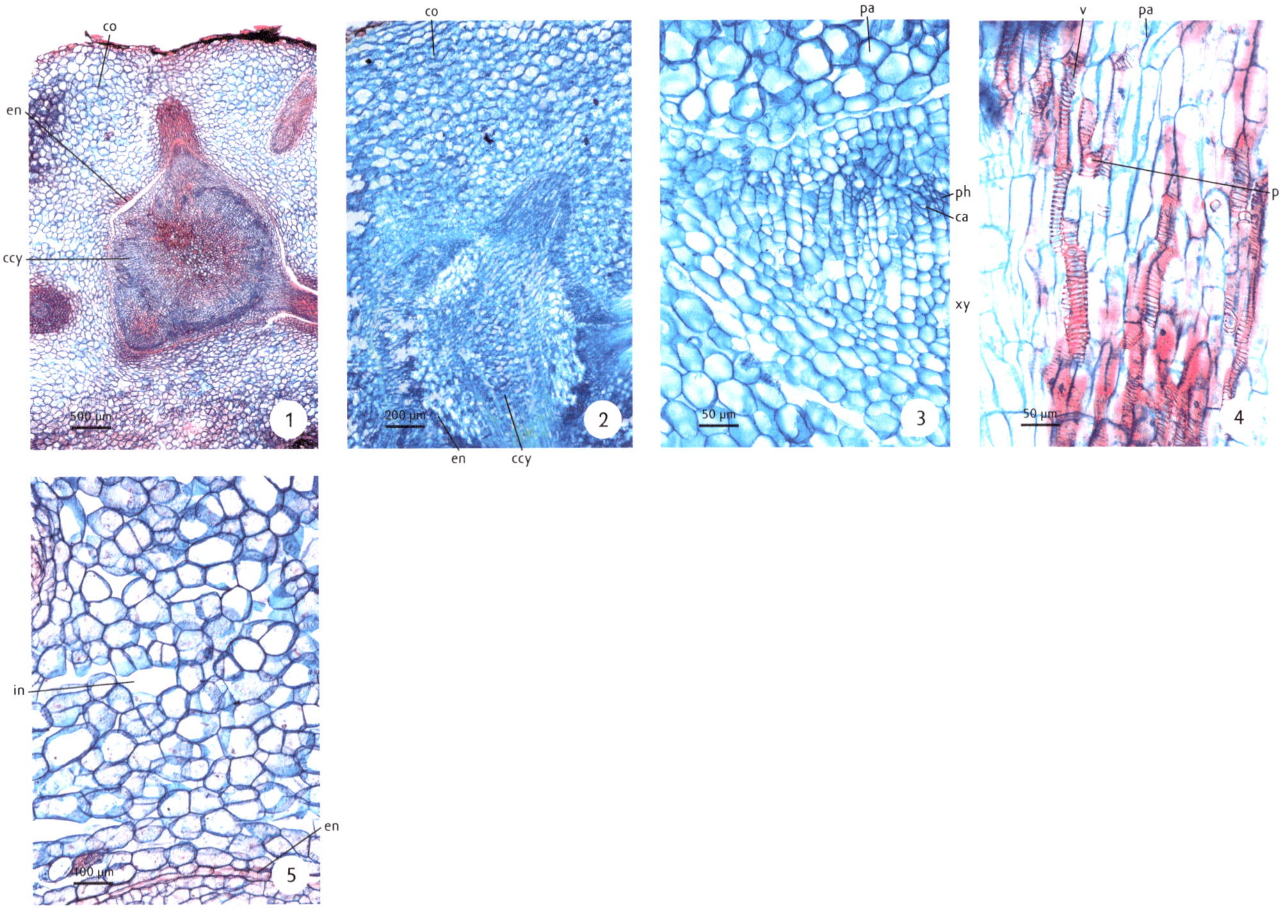

F. H. Schweingruber et al., *Anatomic Atlas of Aquatic and Wetland Plant Stems*, https://doi.org/10.1007/978-3-030-33420-8_19

Ceratophyllum demersum L.

Rigid Hornwort

Ceratophyllaceae

Submerged rootless aquatic perennial herb. Stems branched, with whorled leaves, up to 1 m long. Terrestrial form not known. Leaves dark green, rigid, 1–2 times forked and toothed. Flowers monoecious, subaquatic, sessile, solitary in leaf axils. Fruit is an achene, apiculate to long-spined at apex, often with 2 long spines at base. Overwinters in apical, loose cone-shaped turions.

Distribution and ecology

Circumpolar southern-temperate element. Native commonly in all CE countries. It grows in still or slowly flowing meso- to eutrophic waters in rivers, ditches, canals, fishponds and lakes. Prefers medium to hard (ca. 0.5–5 meq/l) and neutral to alkaline (pH ca. 7–9), nutrient-rich waters. Efficient HCO_3^- user. It can form dense stands. Vegetative reproduction is common. Tolerates strong eutrophication, competetively very strong, eurytopic species.

Anatomy of the submerged annual shoot

Round shoot with a diameter of 1 mm. Maximum age 1 year (1, 2).

The poorly differentiated central cylinder is surrounded by an endodermis, large cortex and an epidermis (1).

The pith is composed of parenchyma cells, which contain large starch grains, and a central tube (2). Cells reflect blue light in polarized light (3). The unlignified xylem/phloem complex cannot be differentiated in xylem and phloem. The belt consists of small and large cells (2). The small cells reflect green and the large cells dark blue in polarized light (3). Since longitudinal walls of both types are not structured it is impossible to classify them (4, 5).

The central cylinder is surrounded by a poorly differentiated endodermis or pericycle. The cortex consists of thin-walled, unlignified parenchyma cells and circularly arranged large intercellulars. The epidermis is composed of square, thin-walled cells and a few stomata (6).

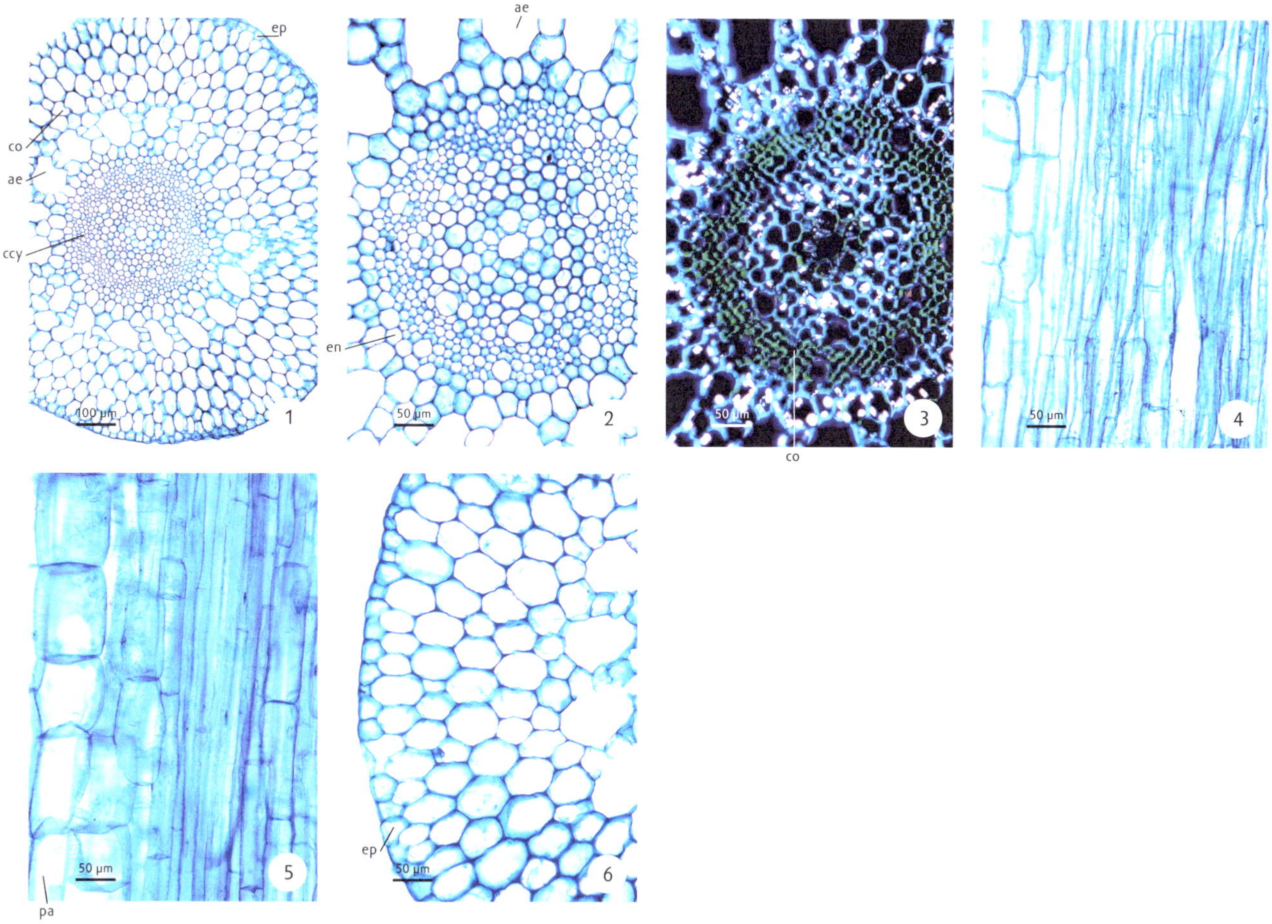

F. H. Schweingruber et al., *Anatomic Atlas of Aquatic and Wetland Plant Stems*, https://doi.org/10.1007/978-3-030-33420-8_20

Ceratophyllum submersum L.

Soft Hornwort Ceratophyllaceae

Submerged rootless aquatic perennial herb. Stems branched, with whorled leaves, up to 1 m long. Terrestrial form not known. Leaves lighter green up to reddish, less rigid, 3–4 times forked with 8 leaf filaments. Flowers monoecious, subaquatic, sessile, solitary in leaf axils. Fruit is a warty achene, apiculate to obtuse at apex, without basal spines. Overwintering in vegetative state without dormant turions.

Distribution and ecology

Euro-Siberian temperate element. Native in all CE countries. Occurs in still eutrophic or slightly brackish water in lakes, fishponds, backwater pools and ditches where it can occur in dense stands. Vegetative reproduction is common. Prefers hard (ca. 1–5 meq/l) and neutral to alkaline (pH ca. 7–9), base-rich waters, also dystrophic waters in fen lakes. Efficient HCO_3^- user. Does not tolerate strong eutrophication, competetively weaker.

Anatomy of the submerged annual shoot

Round shoot with a diameter of 1 mm. Maximum age 1 year (1, 2).

The poorly differentiated central cylinder is surrounded by an endodermis, a large cortex and an epidermis (1).

The unlignified xylem/phloem complex cannot be differentiated in xylem and phloem. The belt consists of small and large cells (2). Since longitudinal walls of both types are not structured it is impossible to classify them.

The central cylinder is surrounded by a poorly differentiated endodermis or pericycle (2). The cortex consists on thin-walled, unlignified parenchyma cells and circularly arranged large intercellulars. The epidermis is composed of square, thin-walled cells (3).

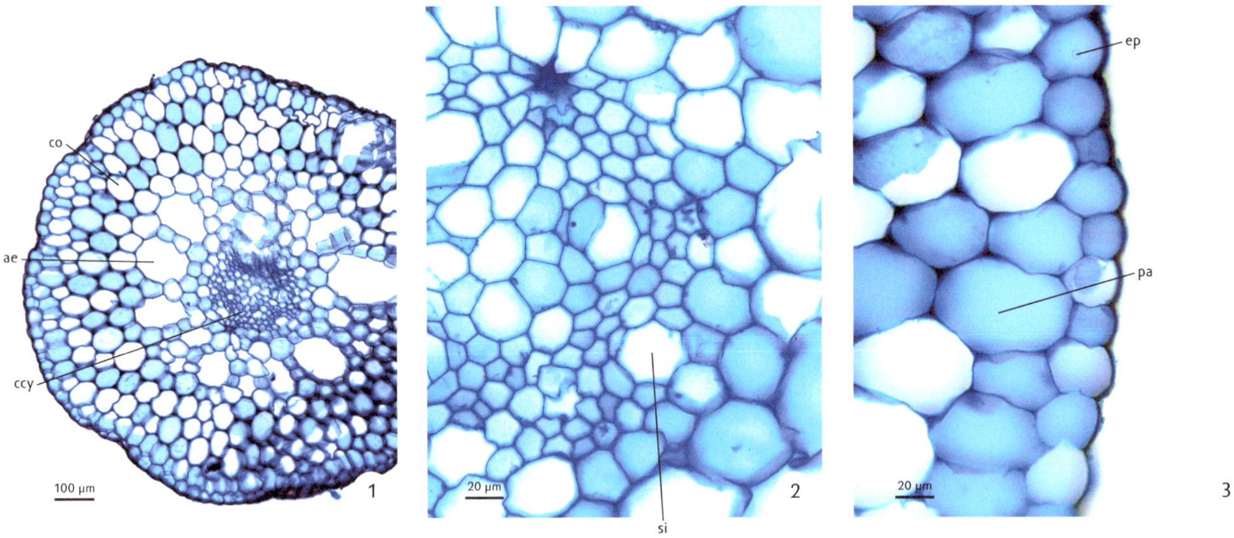

Crassula helmsii (Kirk) Cockayne

(syn. *Tillaea helmsii*)

Swamp Stonecrop Crassulaceae

Perennial amphibious or submerged aquatic herb. Stem rounded, rooting at the nodes, 10–130 cm long. Leaves narrow, opposite, sessile and succulent, 4–24 mm long. Small white flowers with four petals on long stalks arising from the upper leaf axils. The flowers are always above water. Fruits are follicles containing 2–5 elliptical and smooth seeds.

Distribution and ecology

Native of Australia and New Zealand, alien in various parts of Europe, especially NW Europe including DE. Grows in muddy margins of ponds and drainage ditches, semi-submerged in deeper water, or totally submerged up to 3 m depth forming elongated stems. Able to form dense mats. Tolerant of shade for long periods. Ability of CAM photosynthesis. Reproduces mainly through vegetative propagation.

Anatomy of the transition between the root and the annual shoot

Round stem with a diameter of 1 mm. Annual rings absent. Maximum age 1 year (1, 2).

The root collar consists of a xylem (1, 2), a small phloem and a large cortex (1, 2). The xylem contains solitary vessels and parenchyma cells. Fibers and rays are absent (2). Vessels in the center of the stem have a diameter of 10–20 μm (2) and a length of 30–50 μm. Intervessel pits are scalariform and perforation plates are simple.

The phloem is uniform. It is difficult to differentiate sieve tubes and parenchyma cells (2). The stele is surrounded by an endodermis without Casparian strips (1, 2). The cortex consists of large, round to oval, unlignified thin-walled parenchyma cells (1, 2). The phellem is absent (1).

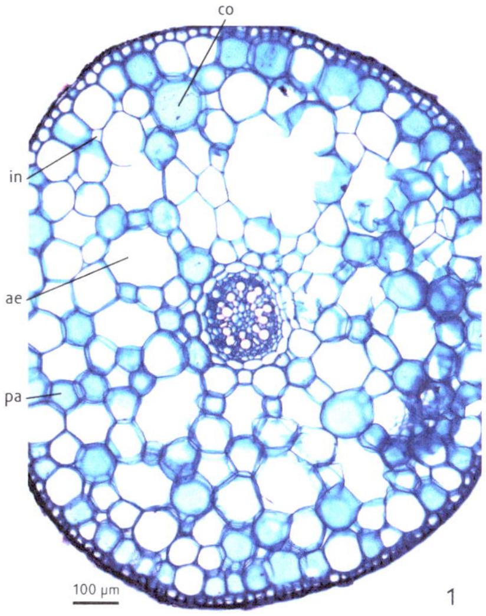

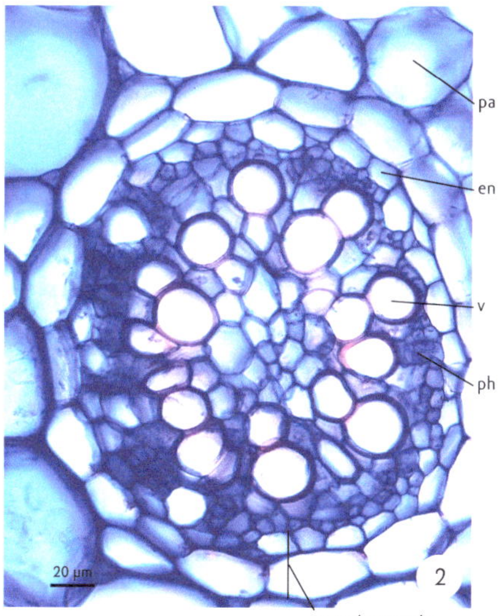

F. H. Schweingruber et al., *Anatomic Atlas of Aquatic and Wetland Plant Stems*, https://doi.org/10.1007/978-3-030-33420-8_21

Sedum villosum L.

Hairy stonecrop Crassulaceae

Biennial to perennial wetland herb. Stems erect to procumbent, rooting near base. Alternate, linear-oblong, flattened fleshy leaves often reddish and markedly pubescent, flower stem 5–15 cm long with white-pink flowers. Fruit is an utricle with numerous tiny seeds.

Distribution and ecology

European boreo-arctic montane distribution, also in N. America. Scattered irregularly and very rarely in all CE countries, in lowlands to mountains. Peaty wet meadows, fens, streamsides in hilly areas, always in base-poor, acid, barren soils. Facultative heliophyte, competitively very weak. In the last decades, it has been extinct at many CE sites, mostly in lowlands. Very sensitive to eutrophication.

Anatomy of the transition between the root and the annual shoot

Round stem with a diameter of 1 mm. Annual rings absent. Maximum age 1 year (1, 2).

The root collar consists of a pith, a dense, totally lignified xylem belt, a small phloem and a large cortex (1, 2). The xylem consists mainly of fibers. Fibers and vessels cannot be distinguished on cross-sections. The inner part of the xylem consists of unlignified vessels and the outer part primarily of thick and thin- to thick-walled living fibers (2, 3). Parenchyma and rays are absent (2, 3). Vessels in the center of the stem have a diameter of 10–20 µm and a length of 30–50 µm (4). Intervessel pits are scalariform (5) and perforation plates are simple.

The phloem is uniform. It is difficult to recognize sieve tubes (2). The central cylinder is surrounded by a poorly differentiated pericycle/endodermis zone (1, 2). The cortex consists of large, round to oval, unlignified thin-walled parenchyma cells with small intercellulars (1, 2, 6). The phellem is absent.

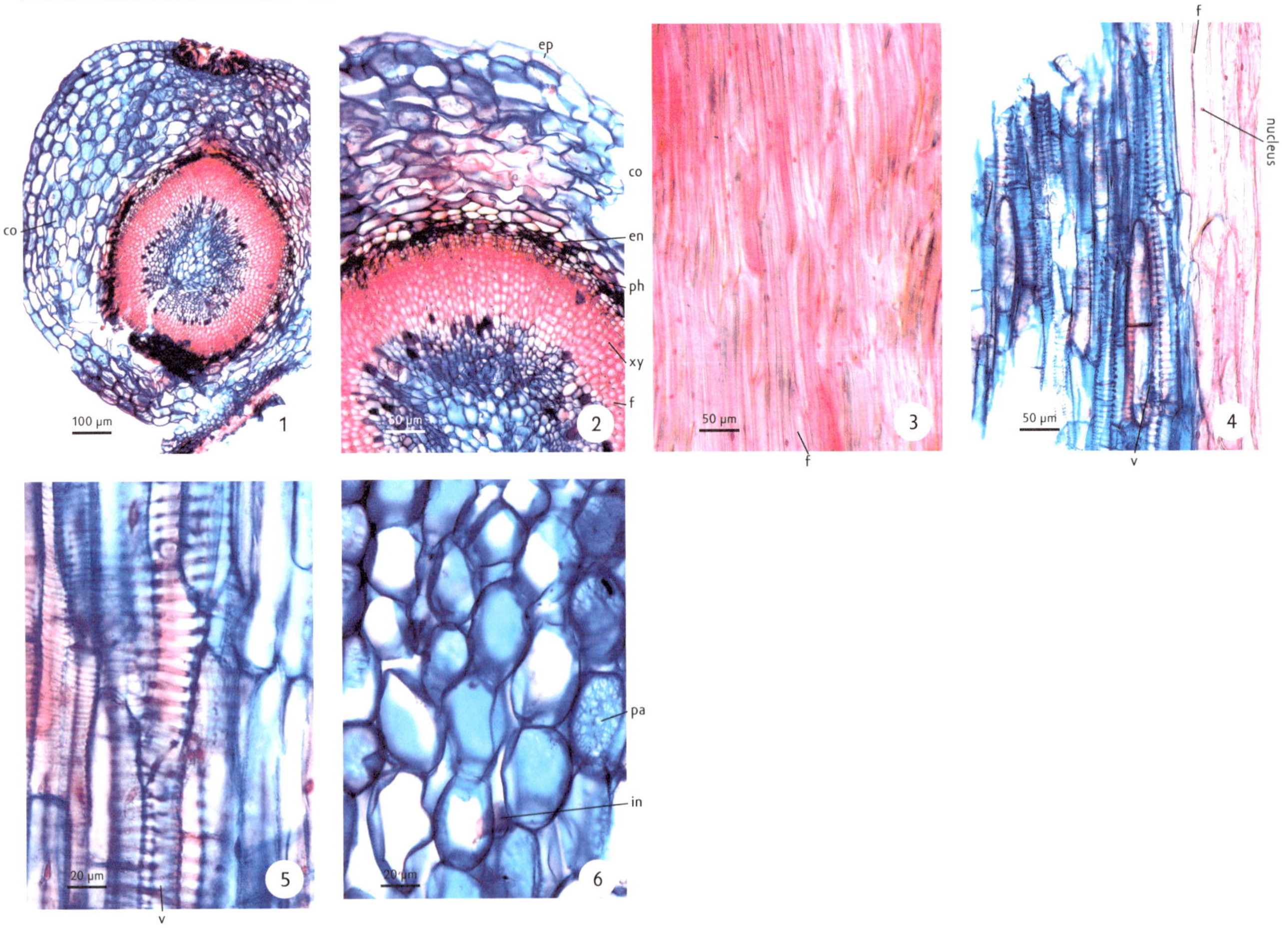

Annual wetland herb. Procumbent to decumbent stems 1–5 cm long bear succulent linear opposite leaves, shoots yellow-green, solitary flowers in leaf axils. Fruit is an utricle.

Distribution and ecology

Circumpolar temperate element, with a disjunct distribution. Scattered irregularly and rarely in all CE countries but CH, in lowlands to uplands, usually in fishpond basins. Wet denudated shores of fishponds or dam reservoirs, especially in fish stocks, muddy, silty or sandy soils. Strict heliophyte, competitively very week, pioneering species on summer-drained fishpond bottoms.

Anatomy of the transition between the root and the annual shoot

Round stem with a diameter of 1 mm. Annual rings absent. Maximum age 1 year (1, 2).

The root collar consists of a xylem (1, 2), a small phloem and a large cortex (1, 2). The xylem consists of poorly lignified, solitary vessels and parenchyma cells. Fibers and rays are absent (2). Vessels in the center of the stem have a diameter of <10 μm (2) and a length of 30–50 μm. Intervessel pits are scalariform and perforation plates are simple.

The phloem is uniform. It is difficult to differentiate sieve tubes and parenchyma cells (2). The central cylinder is surrounded by an endodermis without Casparian strips (1, 2). The cortex consists of large, round to oval, unlignified thin-walled parenchyma cells (1, 2). The phellem is absent.

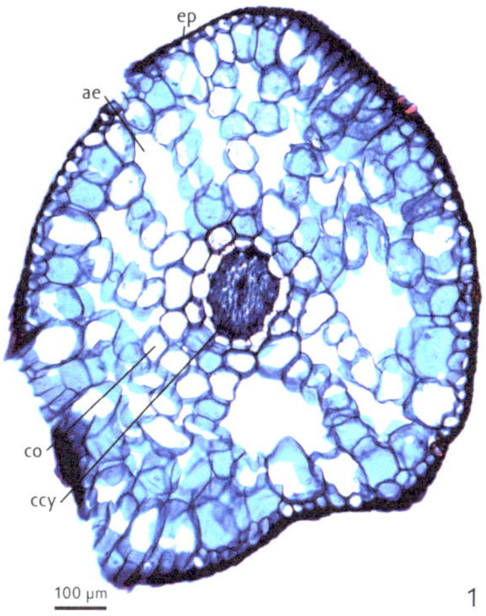

1

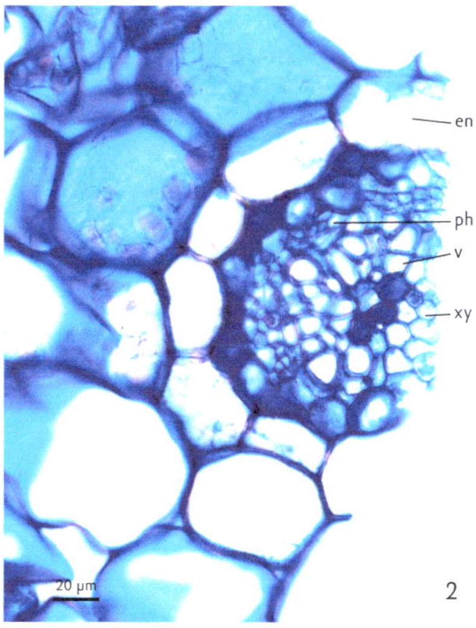

2

Succisa pratensis Moench

Devil's-bit Scabious

Perennial wetland herb with short vertical rhizome. Stem erect to ascending, bifurcately branched distally, 50 cm high. Leaves both in a basal rosette and in opposite pairs along the stem. Leaves simple, obovate-lanceolate, entire. Inflorescences are capitula growing from opposite pairs of leaf axils. Corolla violet, rarely white or pink, calyx with 5 bristle-teeth. Fruit is a 4-angled achene, densely hirsute, with persistent calyx and pappus with 5 black hairpins.

Distribution and ecology

Euro-Siberian temperate element. Native to all CE countries. Occurs in moist, wet or fen meadows, also on margins of woods. Prefers mildly acidic soils, tolerates grazing. Host plant for endangered Lepidoptera *Euphydryas aurinia.* Survival reduced by low pH in combination with high ammonium concentrations. Transient seed bank.

Anatomy of the root collar and the annual shoot

Round root collar with a diameter of 6–10 mm. Secondary growth occurs two growing seasons.

Annual shoot: The xylem consists of a continuous circular fiber/vessel zone (1, 2).

Root collar: A large pith is surrounded by a belt of large vascular bundles with a semi-ring porous xylem, a large cortex, an epidermis or a small phellem (3, 4). The xylem of the second ring contains solitary or grouped angular vessels, thin- to thick-walled fibers, paratracheal parenchyma and rays (4). Fibers are absent in the first ring (4). Vessels have a diameter of 30–40 µm and a length of 50–100 µm. Annual ring-boundaries are characterized by marginal parenchyma and semi-ring porosity (4). Vessels have oval or scalariform intervessel pits and large, simple perforation plates (5). Rays or parenchyma zones are very large (>10 cells) between vascular bundles, and 2-3-seriate within the bundles (3, 6). Rays in annual shoots are uniseriate (2).

Parenchyma cells and a few sieve tubes characterize the phloem (4). Phloem and cortex are separated by a poorly developed endodermis (7). The cortex consists of thin-walled parenchyma cells and small intercellulars (7, 8). Crystal druses occur in the cortex (7) and the pith. The epidermis is partially replaced by a small phellem (8).

ANNUAL SHOOT — ROOT COLLAR

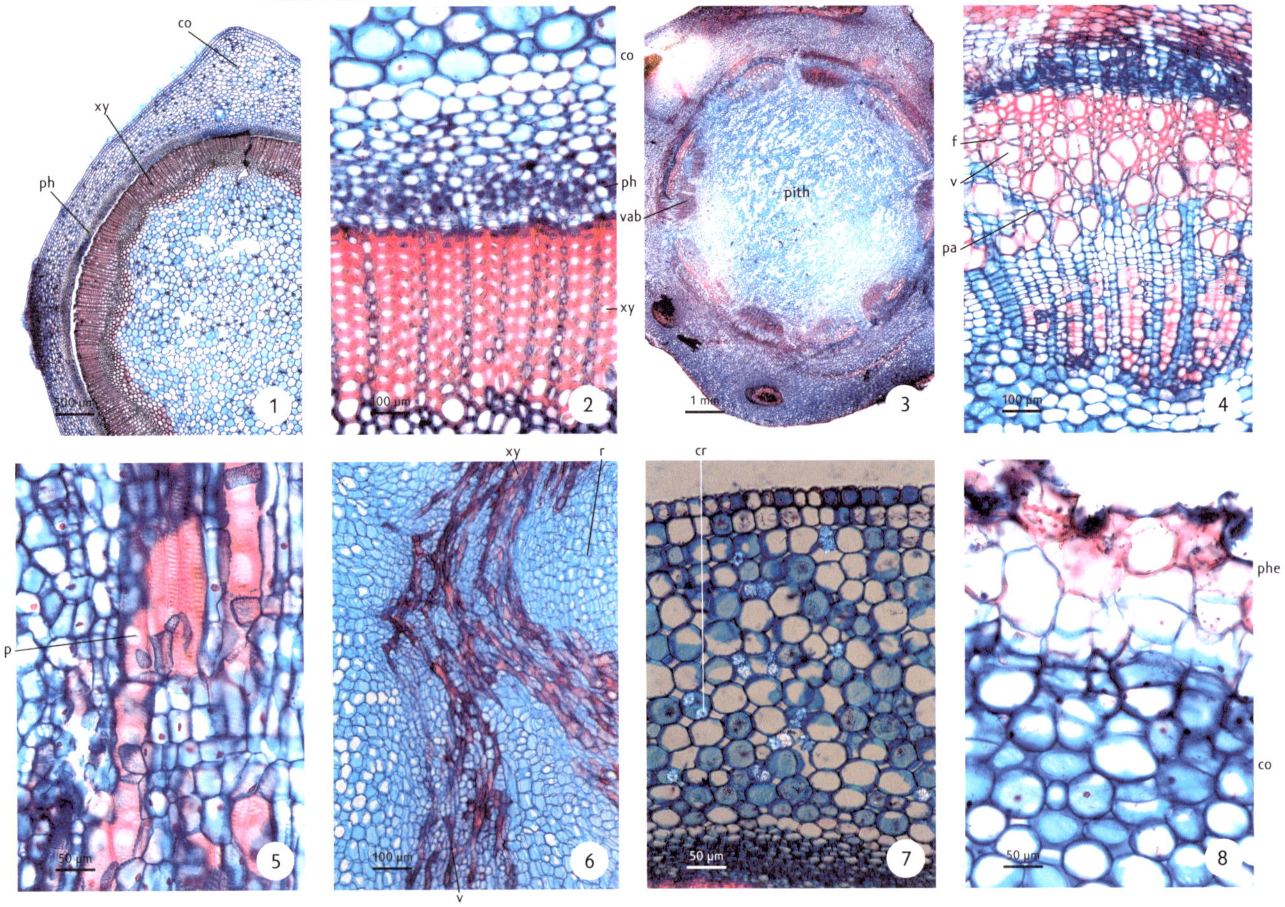

F. H. Schweingruber et al., *Anatomic Atlas of Aquatic and Wetland Plant Stems*, https://doi.org/10.1007/978-3-030-33420-8_22

Perennial, rootless submerged carnivorous plant with thin, linear, poorly branched shoot, free-floating. Shoot 8–20 cm long with regular leaf whorls, leaf with one snapping trap 3–6 mm long as two mobile lobes, fine bristles extend the trap. Solitary white flowers above water on a stalk 10–15 mm long, green fertile capsules with 1–12 seeds under water, the most capsules pale and sterile. Overwintering in 6–8 mm large, rhomboid-shaped dormant turions, actively sink down in autumn and float up in spring.

Distribution and ecology

Cosmopolitan distribution in Old World continents, introduced to the USA. Very rare natural extant distribution only in PL (ca. 8 natural sites) and HU (1–2 sites); extinct from SK, CZ, AT, DE, re-introduced to PL (ca. 10 sites), CZ (ca. 4 sites) and CH (3 sites) to potential suitable habitats. Occurs in lowland standing waters, littoral of shallow lakes, backwater pools and oxbows of larger rivers, peatbogs, fen lakes, also after peat extraction, peaty fishponds. Very stenotopic, only at limited microsites in dystrophic, oligo-mesotrophic waters, shallow loose reed or robust sedge stands. Depends strongly on prey capture and high CO_2 concentration. Under optimum conditions, the plants richly branched, flowering depends mostly on high water temperature, seed set very rare. Plants spread by water birds.

Anatomy of the submerged annual shoots

Round shoot with a diameter of <1mm. Maximum age 1 year (1, 2).

The central cylinder (stele) is surrounded by a large cortex and an epidermis (1, 2).

A round intercellular canal is located in the center of the stele. The central cylinder contains angular sieve tubes and small companion cells and parenchyma cells (3). Cell differentiation is often difficult (4).

The stele is surrounded by an indistinct endodermis-like row of cells (3). The cortex consists of thin-walled, unlignified parenchyma cells (1) or a net-like aerenchymatic tissue (2). The epidermis is composed of thin-walled cells (2). Cell walls of the parenchyma cells in the cortex contain scalariform-like thickenings (5).

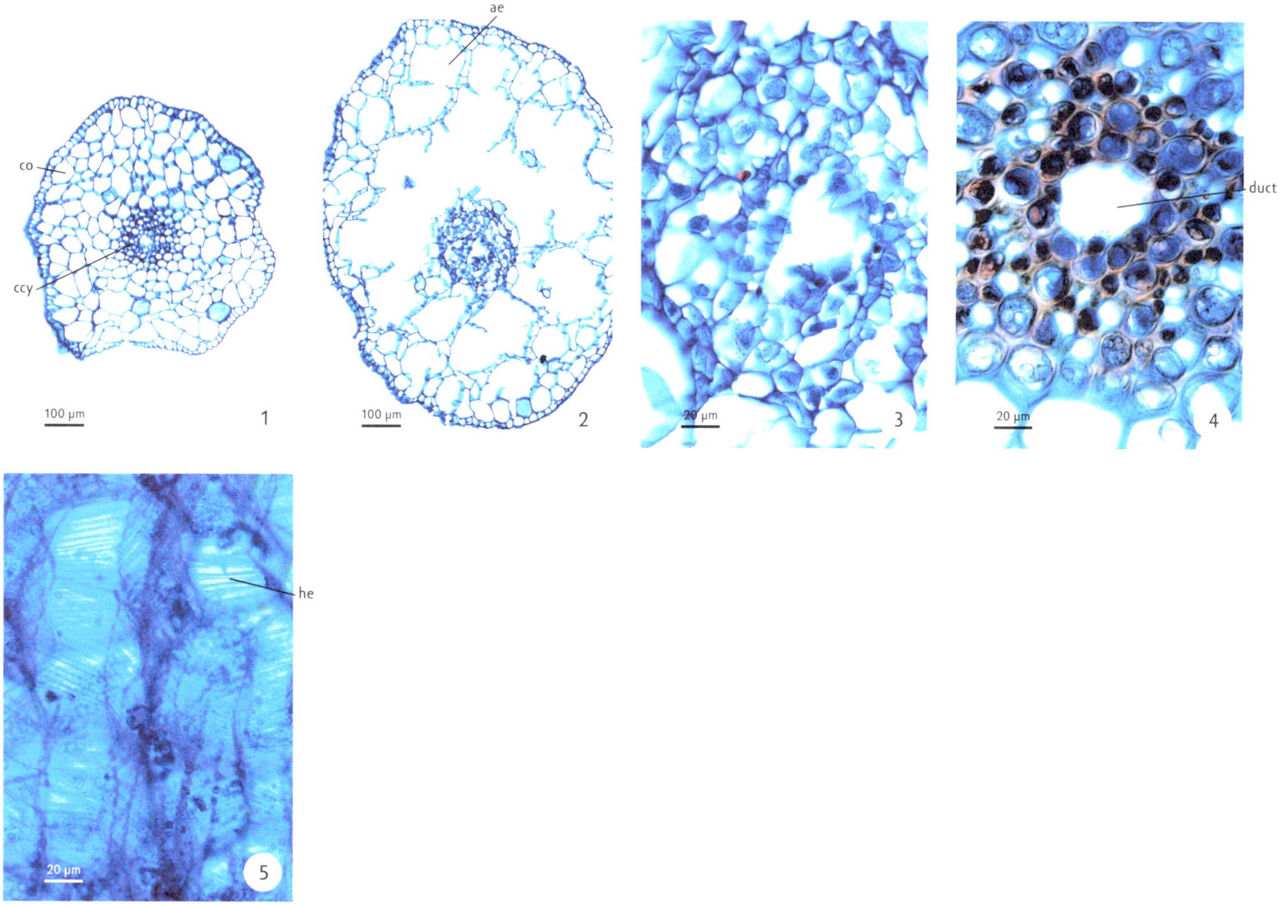

F. H. Schweingruber et al., *Anatomic Atlas of Aquatic and Wetland Plant Stems*, https://doi.org/10.1007/978-3-030-33420-8_23

Drosera anglica Huds. (syn. *D. longifolia* L.) **Great sundew** Droseraceae

Perennial wetland herb, semi-erect rosette of carnivorous leaves 4–8 cm long, rooting in the substrate. Slowly mobile leaf blade narrowly obovate to linear-oblonged, leaves bear many dozens of multicellular mobile emergences, ending with red glandular head. Non-branched thin roots 3–6 cm long with hairs. Oblonged capsules 4–5 mm long, 5–6 mm large hibernaculae.

Distribution and ecology

Circumpolar boreo-montane element. Rare in all CE countries, scattered irregularly mainly in wetland and peaty areas both in lowlands and highlands; lowland sites mostly endangered. Occurs in peatbogs, fens, lake and fishpond littorals, drainage canals in peatbogs and fens, typical microhabitats – ecotones between the depressions and *Sphagnum* hummocks. Prefers acidic to neutral substrates (pH ca. 4.5–7). Strictly heliophilous and hygrophilous, stenotopic, competitively very weak.

Anatomy of the root collar

Round root collar with a diameter of 1 mm, without secondary growth.

The transition zone between the roots and the annual stem consists of a central cylinder (stele) with vascular bundles and a large cortex (1).

The cylinder of the upper part towards the leaves contains irregularly dispersed, closed and collaterally formed vascular bundles between parenchyma cells (3, 4). Vascular bundles consist of a few polygonal cells: lignified vessels, unlignified sieve tubes and parenchyma cells (2). Fibers and rays are absent (1, 2). Vessels have a diameter of 10–20 µm and are 40–60 µm long (4). Vessels have round intervessel pits and simple perforation plates (4). Walls of parenchyma in the central cylinder and the cortex cells are net-like thickened (3).

The cortex and the stele are separated by an endodermis with Casparian strips (5). The cortex consists of thin-walled parenchyma cells and small intercellulars (1). The epidermis is thin-walled (1).

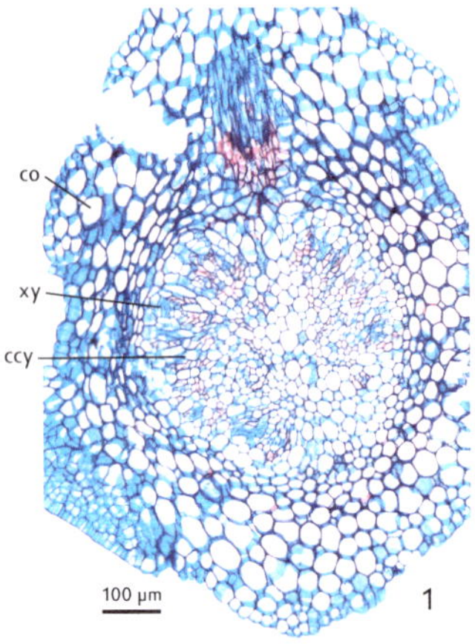

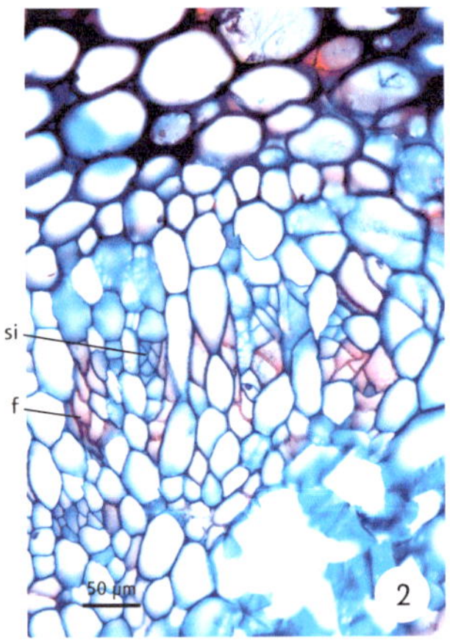

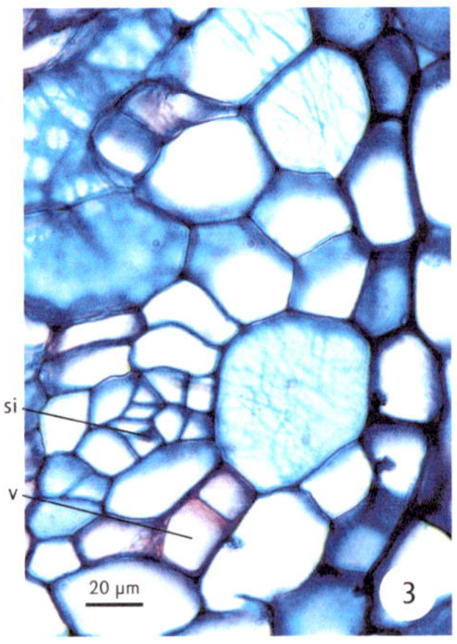

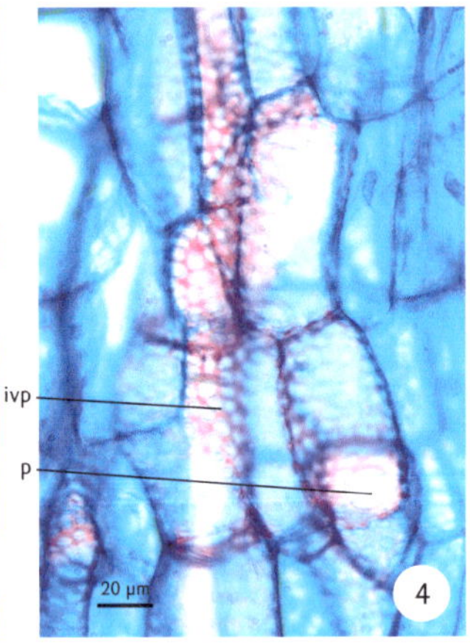

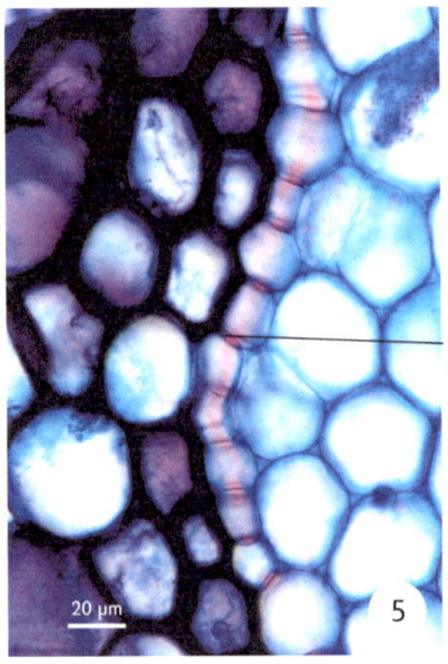

Drosera intermedia Hayne

Oblong-leaved sundew

Perennial wetland herb with semi-erect rosette of carnivorous leaves 2–5 cm long, rooting in the substrate. Slowly mobile leaf blade (narrowly) obovate, dark red-pink leaves bear many dozens of multicellular mobile emergences, ending with red glandular head, non-branched thin roots 3–5 cm long with hairs. Flower peduncle curved at base, oblonged capsules 4–5 mm long, 4–5 mm large hibernaculae (winter buds).

Distribution and ecology

Suboceanic temperate element; also in N. America and subtropical Brazil. Absent in SK and HU, rare in all other CE countries, scattered irregularly only in wetland and peaty areas. Occurs in peatbogs, fens, lake and fishpond littorals, drainage canals in peatbogs and fens, typical microhabitats – shallow depressions on the margin of *Sphagnum* hummocks. Only in acidic substrates (pH ca. 4–6), strictly heliophilous and hygrophilous, stenotopic, competitively very weak. During rainy seasons, the plants may die, survival in the form of abundant seed bank, competing wetland graminoids die.

Anatomy of the root collar and the annual shoot

Round root collar with a diameter of 1 mm without secondary growth.

The transition zone between the roots and the annual stem consists of a central cylinder (stele) with vascular bundles and a large cortex (1, 2).

Closed collateral irregularly formed vascular bundles are distributed irregularly in the basal part of the stem (1) and circular in the flower stalk (2). Vascular bundles consist of a few polygonal cells: lignified vessels, unlignified sieve tubes and parenchyma cells (2). Fibers and rays are absent (1–3). Vessels have a diameter of 10–20 μm and are 40–60 μm long (3). Vessels have round intervessel pits and simple perforation plates (3).

Cortex and stele are separated by an endodermis (see *Drosera anglica*). The cortex consists of thin-walled parenchyma cells and small intercellulars (1).

European *Drosera* species cannot be distinguished anatomically from each other. *D. rotundifolia* is identical with D. *intermedia*.

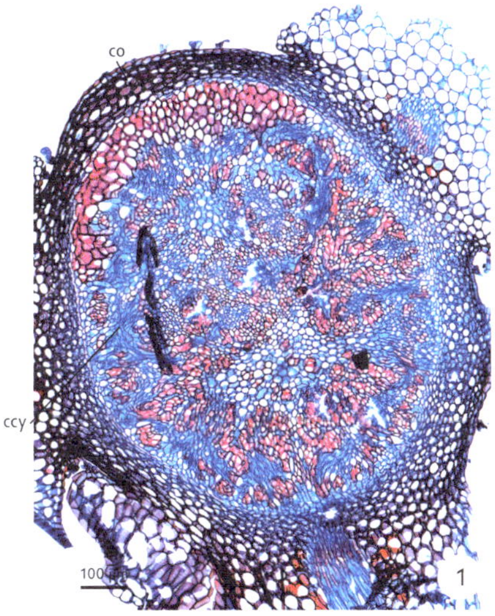

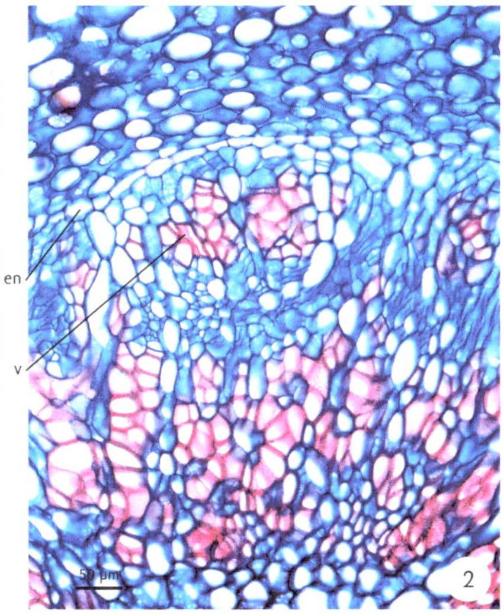

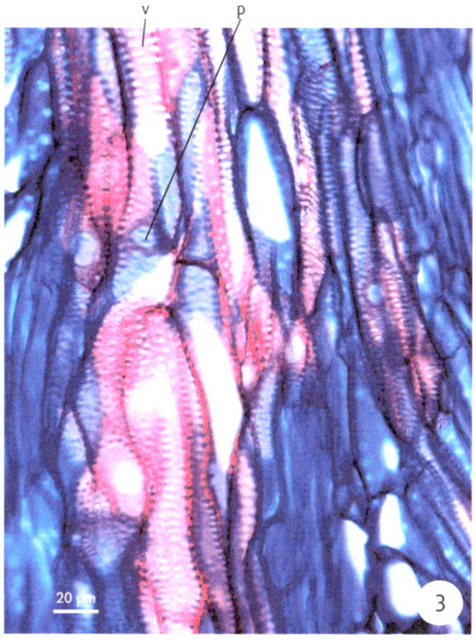

Submerged or amphibious annual herb. Stems erect or ascending, mostly simple, relatively sturdy, 2–30 cm high. Leaves in whorls of 8–17 (submerged) and of 3 (aerial). Flowers usually 3 at a node, mostly sessile or short-stalked, with four petals and sepals. Seeds cylindrical.

Distribution and ecology

Euro-Siberian sub-Mediterranean species. Native in all CE countries in periodically flooded habitats in lowlands. Terrestrial form occurs in wet depressions in cultivated fields and on wet roadsides, aquatic form in shallow stagnant waters (30–50 cm deep) in pond and lake shores, in flooded rice fields and meadows, or in drainage ditches, usually in base-rich soils. Ephemeral species with a long-term seed bank. Able of clonal propagation.

Anatomy of a basal part of the annual shoot

The round stem has a diameter of 1.3 mm (1). Maximum age 1 year.

The stem is composed of a central cylinder (stele), a large aerenchymatic cortex and an epidermis (1). The xylem of the vascular bundles consists of a circle of unlignified vessels and pervasive parenchyma (2, 3). Vessels have a diameter of 20–25 µm (2, 3) and a length of 140–170 µm. Vessel walls consist of ring-like (annular) thickenings (4, 5) and perforations are probably absent (5). Fibers and rays are absent (2, 3).

The phloem consists of large sieve tubes and groups of small companion cells (2, 6). A thin-walled endodermis-like uniseriate row of cells separates the stele from the cortex (2, 6). The outer cortex is composed of an uniseriate layer of round, thin-walled cells, the middle cortex of very large radial chambers of aerenchyma (1). The separating walls consist of a single layer of storied palisade-like parenchyma cells (7, 8) and the inner cortex is similar to that of the outer cortex. The epidermis consists of small roundish thin-walled cells. Crystals are absent.

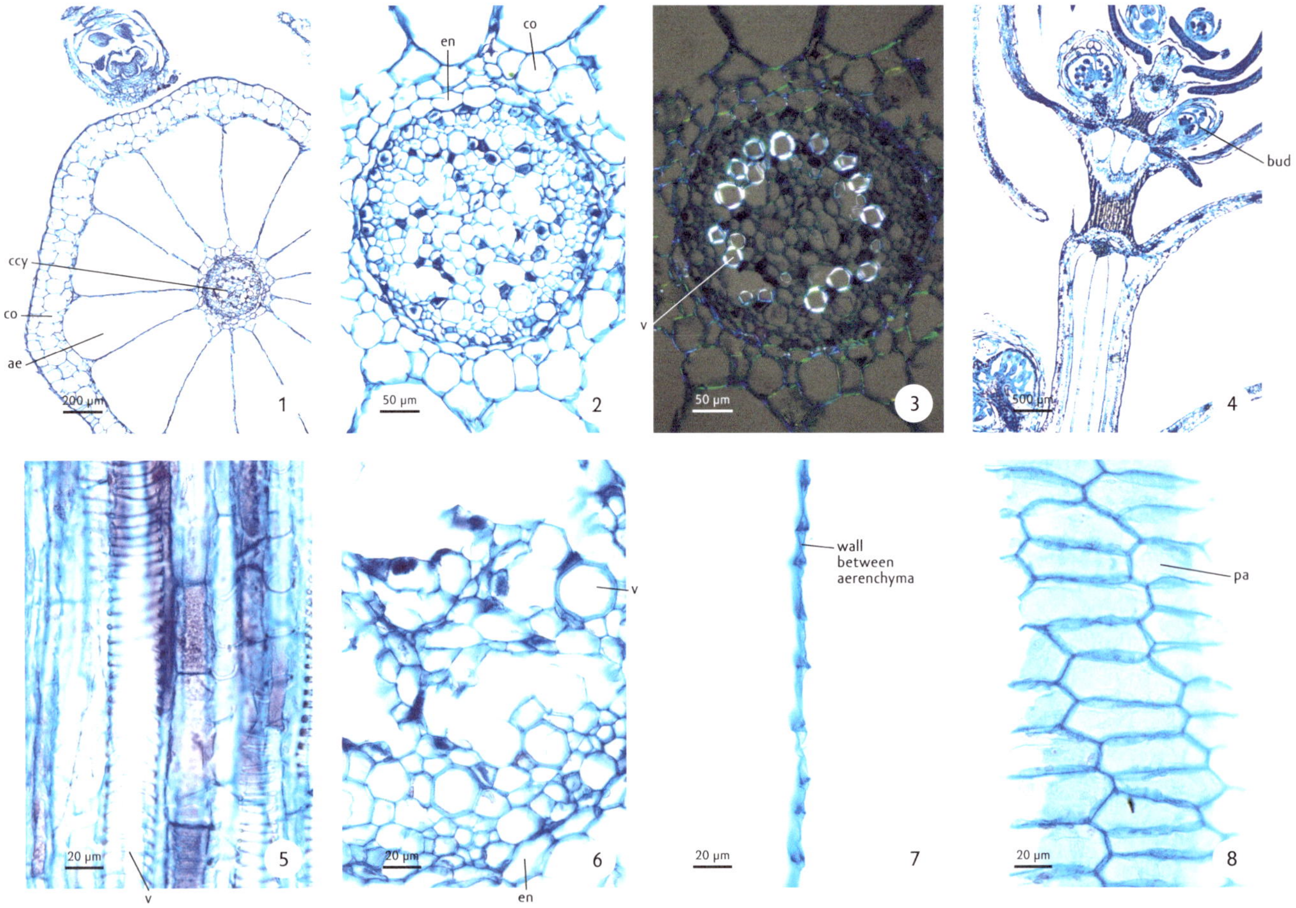

F. H. Schweingruber et al., *Anatomic Atlas of Aquatic and Wetland Plant Stems*, https://doi.org/10.1007/978-3-030-33420-8_24

Annual or short-lived perennial submerged or amphibious aquatic herb. Stems procumbent, rooting at nodes, up to 10 cm high. Leaves elliptic, in whorls or in opposite pairs. Pinkish flowers are tiny, stalked, three-petalled in leaf axils. Fruit is a stalked capsule with 3 valves.

Distribution and ecology

European temperate element. Native in all CE countries, probably absent in HU. It occurs on exposed mud of fishponds, lakes and reservoirs or in shallow sand-pits or submerged in shallow, oligotrophic to eutrophic, usually soft waters. Presence of vesicular-arbuscular mycorrhiza. Ephemeral species with a long-term seed bank.

Anatomy of a basal part of the annual shoot

The round stem has a diameter of 0.7 mm (1). Maximum age 1 year.

The stem is composed of a central cylinder (stele), an aerenchymatic large cortex and an epidermis (1). The xylem of the vascular bundles consists of a circle of unlignified vessels and pervasive parenchyma (2–4). Vessels have a diameter of 10–15 µm (2, 3). Vessel length could not be determined. Vessel walls consist of ring-like (annular) thickenings. Perforations could not be determined. Fibers and rays are absent (2–4).

The phloem consists of large sieve tubes (2–4). A thin-walled endodermis-like uniseriate row of cells separates the stele from the cortex (2, 4). The outer cortex is composed of a uniseriate layer of round, thin-walled cells, the middle cortex of large radial chambers of aerenchyma (1) and the inner cortex is similar to that of the outer cortex. The epidermis consists of small roundish thin-walled cells. Many cells contain violet-stained contents. Crystals are absent.

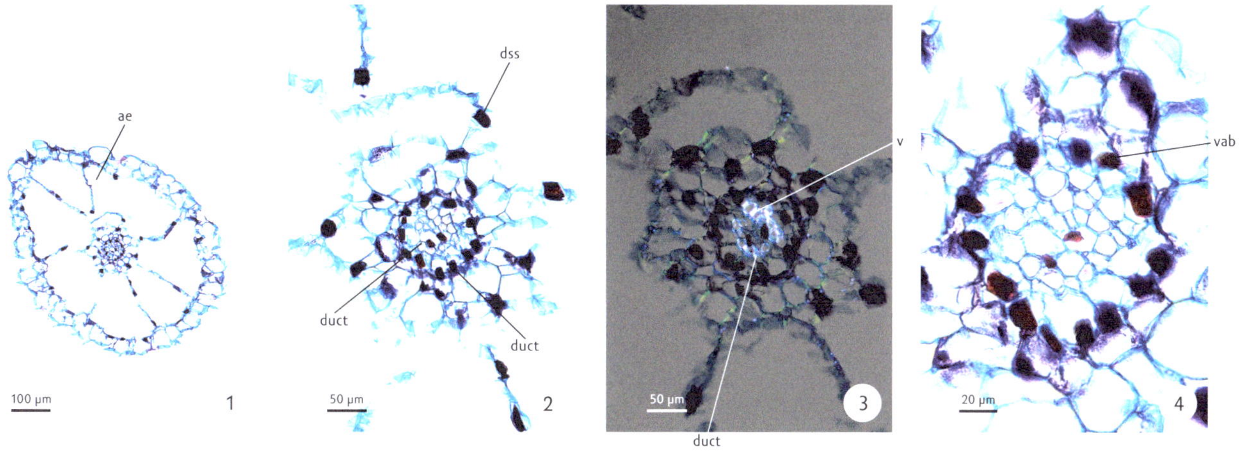

Andromeda polifolia L. **Bog-rosemary** Ericaceae

Perennial wetland evergreen shrub 15–30 cm tall. Glabrous leaves 1–4 cm long, sublinear to elliptic, revolute at margins, white-glaucous on lowerside.

Distribution and ecology

Circumpolar boreo-montane distribution. Scattered irregularly and rarely in all CE countries except HU, in lowlands to mountains, more common in peatland areas and in montane raised bogs. Mountain peat bogs, raised bogs, commonly on *Sphagnum*-dominated hummocks, strictly on acidic soils (ca. pH 3.5–5), a heliophyte.

Anatomy of the stem

Round stem with a diameter of 2–4 mm. Maximum age >30 years.

Stems consist of a xylem, a small phloem and a phellem (1). Annual rings are marked by a uniseriate layer of flat latewood fibers (2). The diffuse porous xylem is composed of solitary vessels, thin- to thick-walled fibers and rays (2). Vessel diameter varies from 15–30 µm (2) and vessel length from 150-250 µm (3). Intervessel pits are round. Perforation plates are generally scalariform with 8–15 bars, however, transitions between scalariform intervessel pits and scalariform perforation plates are frequent (3). Fibers are thin- to thick-walled (2) and have round pits. Parenchyma is absent (2). Rays are uniseriate (4) and ray cells are upright (5).

The phloem consists of thick-walled parenchyma cells and sieve tubes (6). Intercellulars are absent. The phellem consists of radially arranged thin-walled rectangular cells.

Cell dimorphism is characteristic for the pith. Very large thin-walled, pit-free, cells (80–100 µm) are between small net-like arranged, thin- to thick-walled (10–20 µm) cells (7, 8).

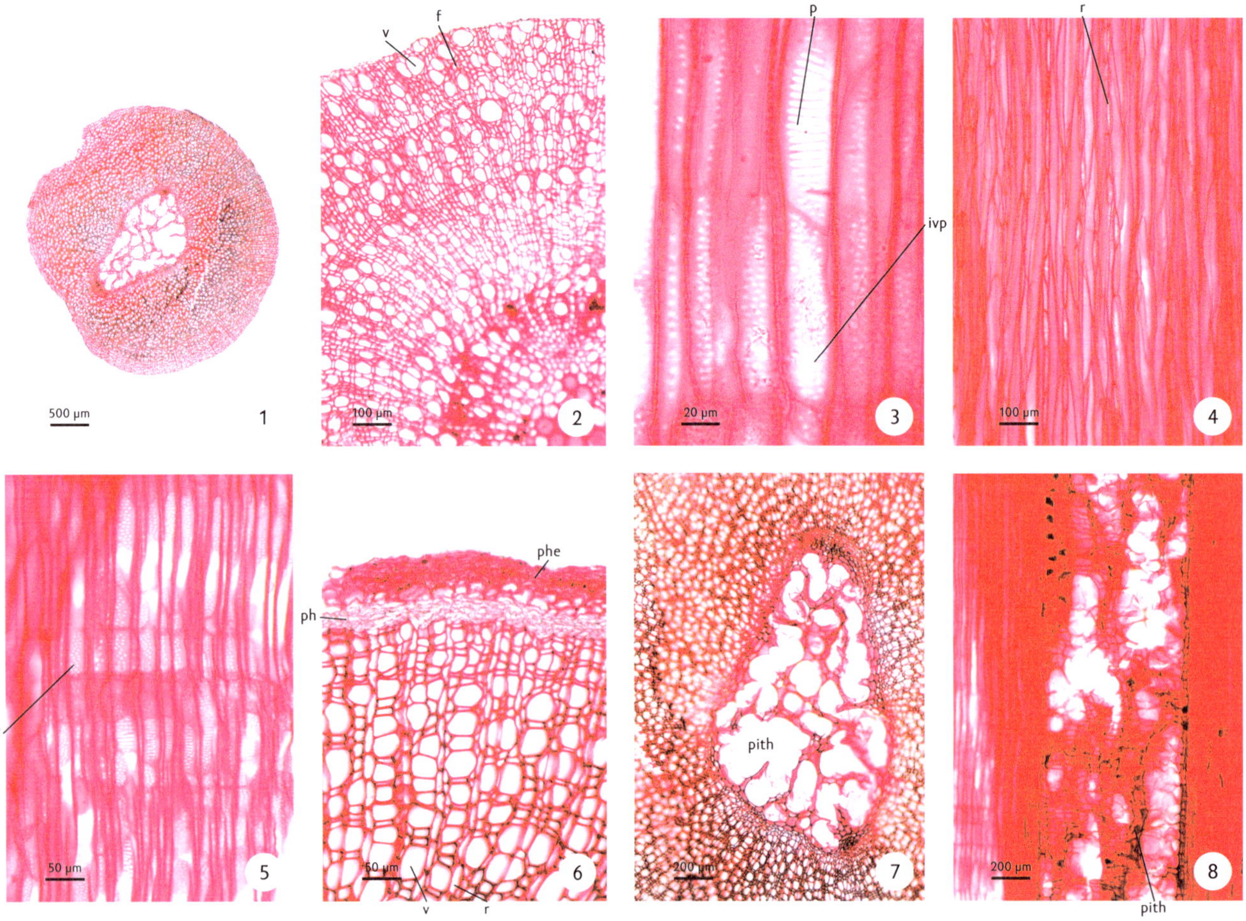

Erica tetralix L.

Cross-leaved Heath

Perennial wetland evergreening subshrub 15–60 cm tall, stem erect or procumbent, rooting along its length, bearing linear leaves in whorls of 4, grey-green leaves distinctly hairy and glandular. Flowers in umbel-like heads, oval to urn-shaped, rose-pink corolla. Fruit is a capsule.

Distribution and ecology

Suboceanic temperate distribution. Scattered irregularly and rarely in all western CE countries except for AT, HU and SK. Peat bogs, raised bogs, wet heaths, wet loose spruce forests, always acidic, base-poor soils, facultatively psychrophilous.

Anatomy of the stem

Round stem with a diameter of 4–8 mm. Maximum age >10 years.

Stems consist of a xylem, a small phloem and a phellem (1, 6). Annual rings are marked by 2-3-seriate layer of flat latewood fibers (2). The xylem is composed of vessels, thin- to thick-walled fibers and rays (2, 3). The xylem is diffuse- or semi-ring-porous with solitary vessels (2). Vessel diameter varies from 30–40 µm (3) and vessel length from 150–220 µm (4). Intervessel pits are round. Perforation plates are generally scalariform with 8–15 bars, however, transitions between scalariform intervessel pits (3) and scalariform perforation plates are frequent (4). Fibers are thin- to thick-walled (2) and have round pits. Parenchyma is absent. Rays are uniseriate (4) and ray cells are upright.

The phloem consists of thick-walled parenchyma cells and sieve tubes (5). Cell types in the phloem can be distinguished only in radial sections (5). Intercellulars are absent. The cortex is absent in older stems. The phellem consists of radially arranged thin-walled rectangular cells (5).

Pith cells appear round in cross-sections (6) and quadrangular in radial sections (7).

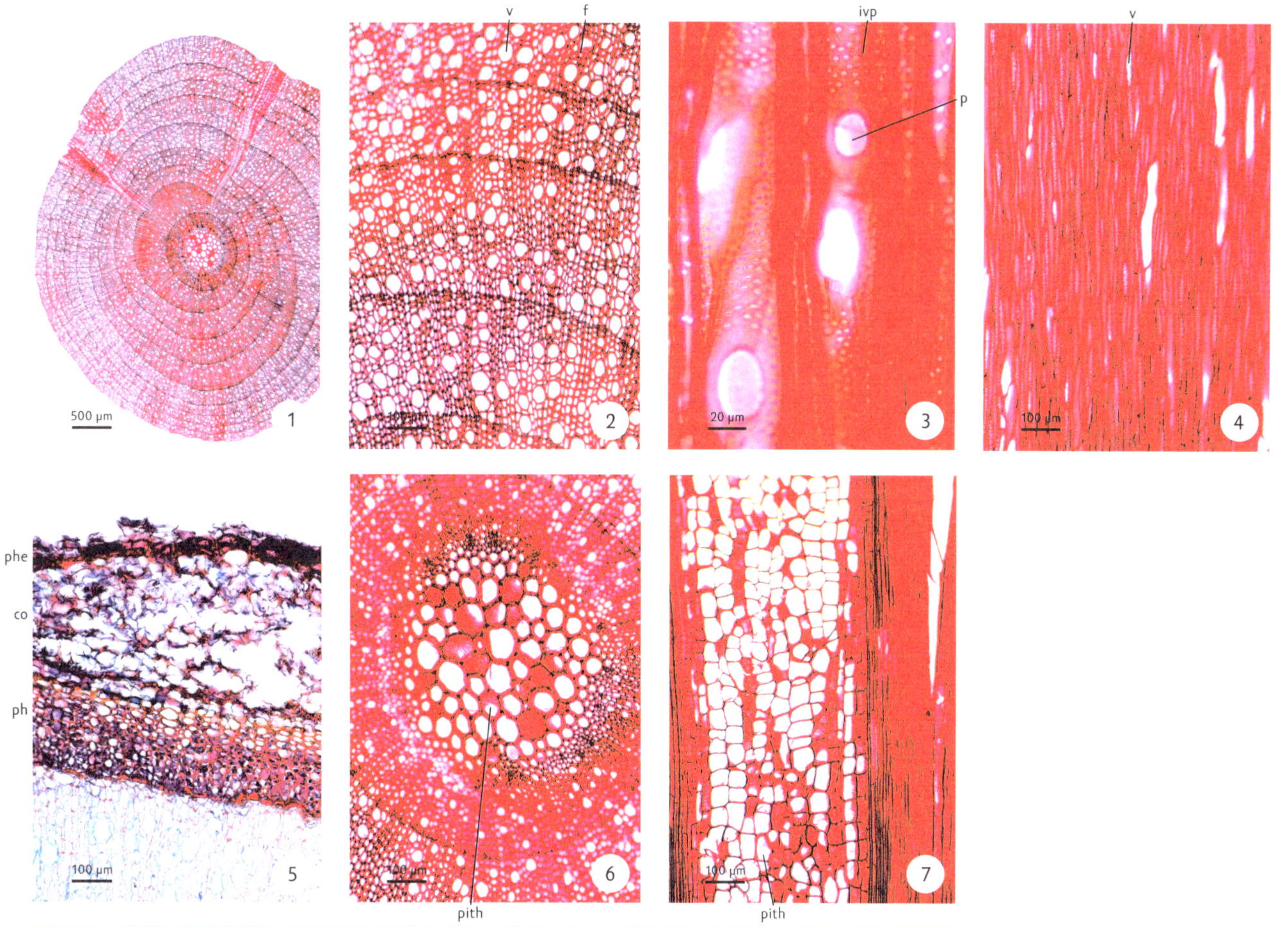

F. H. Schweingruber et al., *Anatomic Atlas of Aquatic and Wetland Plant Stems*, https://doi.org/10.1007/978-3-030-33420-8_25

Rhododendron tomentosum Harmaja **Labrador tea** Ericaceae

(syn. *Ledum palustre*)

Perennial wetland shrub 0.4–1.2 m tall. Leaves alternate, evergreen, narrowly elliptic, 2–4 cm long, with strongly revolute margins, densely tomentose on lowerside with red-brown aromatic glands. White flowers in dense terminal racemes. Fruit is a capsule with numerous tiny seeds.

Distribution and ecology

Circumpolar boreo-arctic montane distribution. Scattered irregularly in all CE countries except CH and HU, in uplands to mountains. Peaty pine-dominated wetland forests, wet sandstone rocks, always in acid, base-poor soils.

Anatomy of the stem

Round stem with a diameter of 4–10 mm. Maximum age 70 years.

Stems consist of a xylem, a small phloem and a phellem (1). Annual rings are marked by a uniseriate layer of flat latewood fibers (2). The xylem is composed of vessels, thin-walled fibers, parenchyma cells and rays (2). The xylem is diffuse- or semi-ring-porous with solitary vessels (2). Vessel diameter varies from 30–40 µm (2) and vessel length from 150–220 µm. Intervessel pits are round. Perforation plates are generally scalariform with 8–15 bars, however transitions between scalariform intervessel pits and scalariform perforation plates are frequent (3). Fibers are thin-walled (2) and have round pits. Parenchyma is paratracheal (2). Rays are uniseriate (5) and ray cells are upright (4).

The phloem consists of thick-walled parenchyma cells and sieve tubes (6). Cell types can be distinguished in radial sections with the presence of sieve plates and simple pits. Intercellulars are absent (6). The phellem consists of radially arranged thin-walled rectangular cells (6).

Cell dimorphism is characteristic for the pith. Very large thin-walled, pit-free, barrel-like cells (60–80 µm) are between small net-like arranged, thin- to thick-walled, cube-like cells (10–20 µm) (7,8).

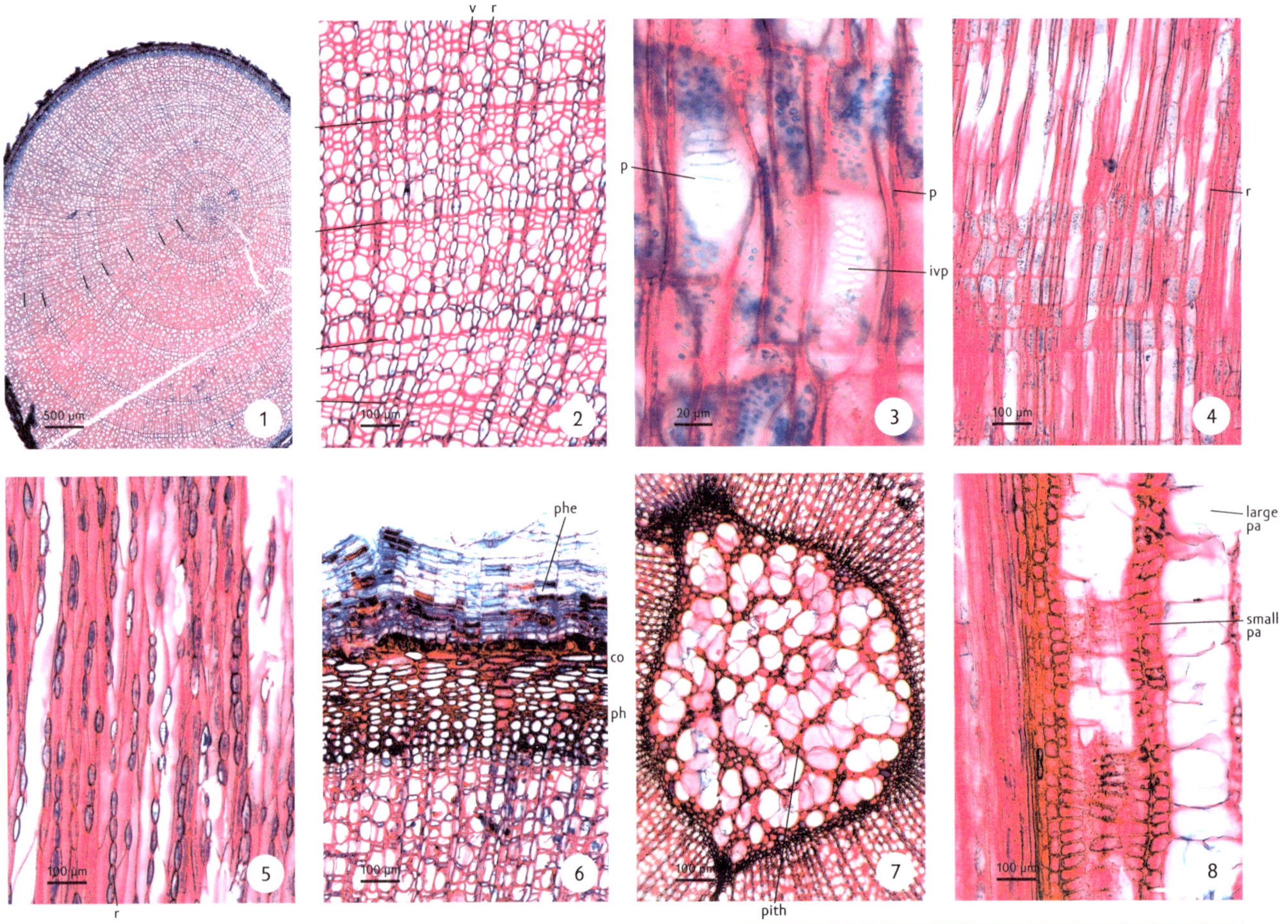

Vaccinium oxycoccos L. (syn. *Oxycoccus palustris*) **Cranberry** Ericaceae

Wetland procumbent dwarfish shrub, stems 10–80 cm long, evergreen leaves 5–12 mm long, bright pink flowers in terminal groups of 1–4, globose fruit red, 8–14 mm across. Fruit is an edible red berry.

Distribution and ecology

Circumpolar boreo-montane element. Scattered irregularly in all CE countries, in lowlands to mountains, commonly in peatland areas. Peat bogs, fens, wet spruce forests, usually on *Sphagnum*-dominated hummocks, tolerates strongly acidic to neutral soils (pH 3–7), very tolerant of irradiance.

Anatomy of the rhizome

Round stems with a diameter of 1 mm. Maximum age 4 years.

Stems consist of a xylem, a small phloem and phellem (1, 2). Annual rings are marked by a uniseriate layer of flat latewood fibers (3). The xylem is composed of vessels and thin- to thick-walled fibers and rays (2, 3). The xylem is diffuse-porous with solitary vessels (2). Vessel diameter varies from 25–30 µm (2, 3) and vessel length from 80–150 µm. Intervessel pits are round (3). Perforation plates are primarily simple and rarely scalariform with 5–12 bars (3, 4, 5). Fibers are thin- to thick-walled (2) and have round pits (6). Parenchyma is paratracheal. Rays are uniseriate (7) and ray cells are square and upright (4).

The phloem consists of parenchyma cells and sieve tubes (6). Cell types are difficult to distinguish (6). The cortex consists of fairly thick-walled oval cells. A few rays dilate. Intercellulars are absent. Few layers of rectangular cells characterize the phellem (8).

Pith cells are round in cross-sections (1).

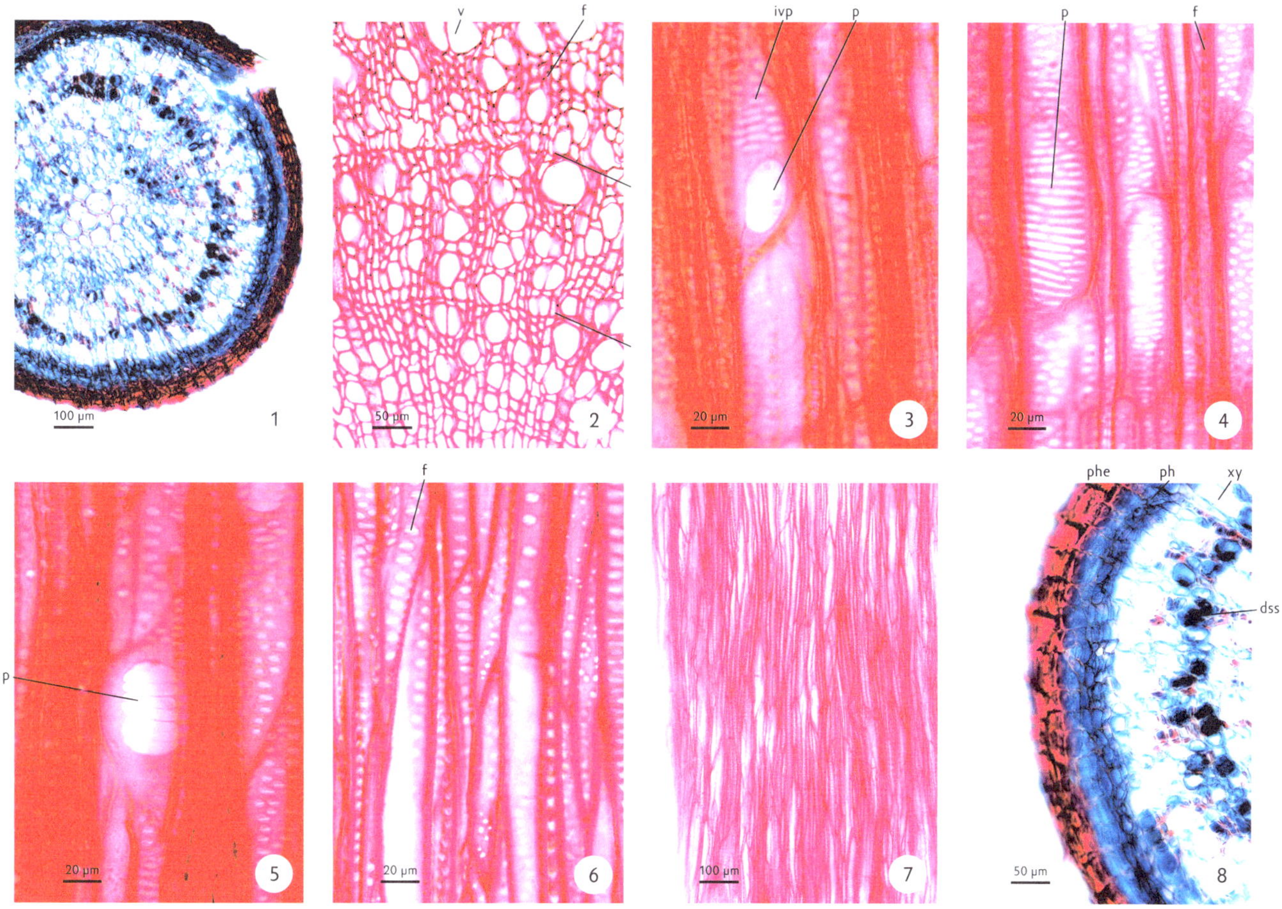

Perennial wetland deciduous subshrub, 20–50 cm high. Twigs erect to ascending, terete, glabrous, brown to red-brown. Leaves almost sessile, obovate, entire, 8–25 mm long. Flowers nodding, in axillary clusters of 1–4, petals partially fused to flask-like pale pink corolla. Fruit is a globose berry, bluish-black with whitish bloom.

Distribution

Circumpolar boreo-arctic montane element. Native to all CE countries except HU, especially from uplands to the mountains. Occurs in peatbogs, in transitional mires and peaty pine forests. Prefers moist, peaty acidic soils in mountain areas. Roots colonised by ericoid mycorrhiza. Host plant for endangered Lepidoptera *Colias palaeno* and *Vacciniina optilete*.

Anatomy of the stem

Round stem with a diameter of 4–15 mm. Maximum age 50 years.

Stems consist of a xylem, a small phloem and a phellem (1, 6). Annual rings are often marked by a uniseriate layer of flat latewood fibers (2). The xylem is composed of vessels, thin-walled fibers and parenchyma rays (2). The xylem is diffuse- or semi-ring porous with solitary angular vessels arranged in intra-annular tangential rows (2). Vessel diameter varies from 30–50 µm (2) and vessel length from 100–200 µm. Intervessel pits are round. Perforation plates are generally scalariform with 8–15 bars, however transitions between scalariform intervessel pits and scalariform perforation plates are frequent (3). Fibers are thin-walled (2) and have round pits. Parenchyma is paratracheal (2). Rays are uniseriate (5) and ray cells are upright (4).

The phloem consists of thick-walled parenchyma cells and sieve tubes (6). Cell types can be distinguished in radial sections with the presence of sieve plates and simple pits. Intercellulars are absent (6). The phellem consists of radially arranged thin-walled rectangular cells (6).

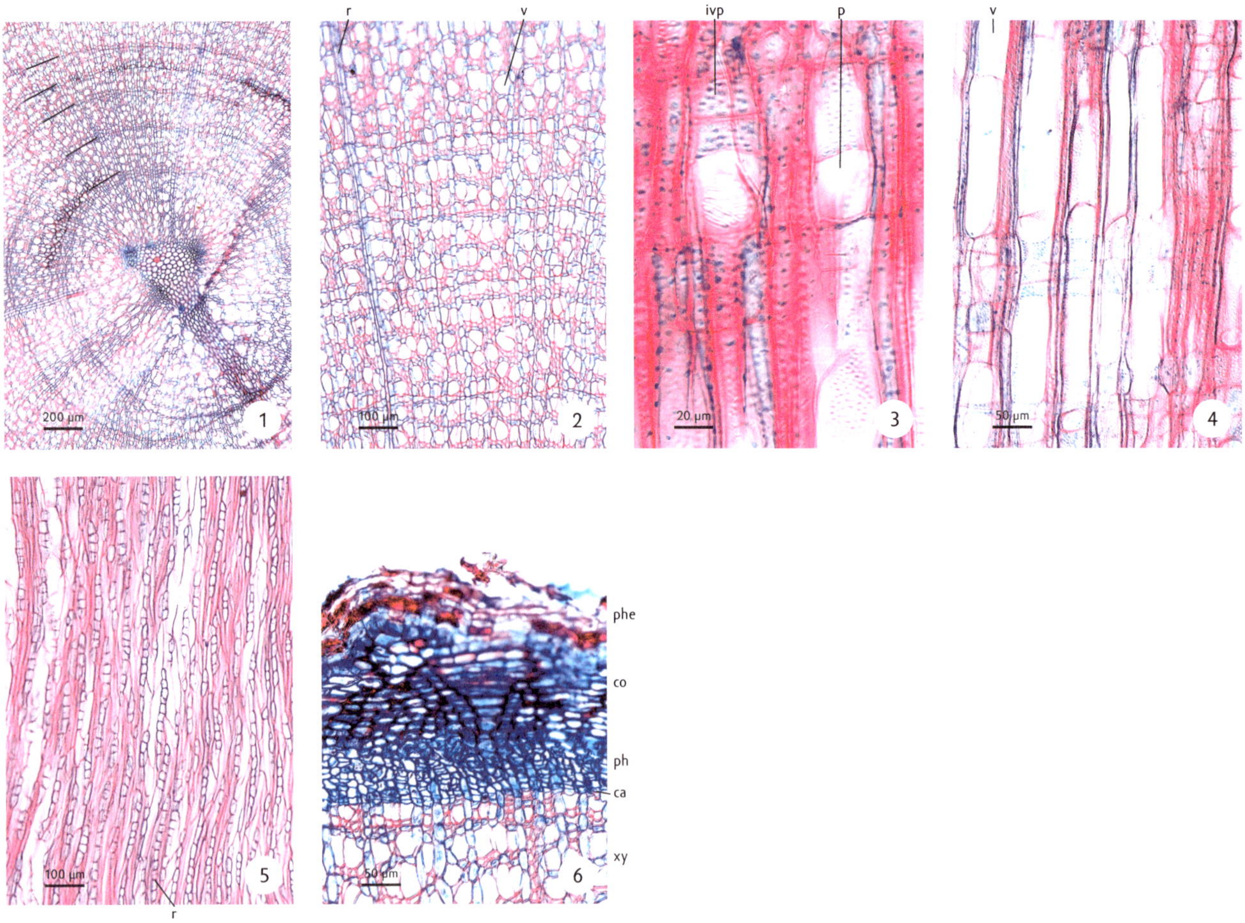

Euphorbia palustris L.

Marsh Spurge Euphorbiaceae

Perennial wetland herb, with a reddish, branched stem, resembling a small willow shrub, 80–120 cm high. Sessile leaves are bluish green, lanceolate, soft and glabrous. Highly reduced male flowers surrounding a female one in a cup-shaped cyathium, these in terminal compound umbels. Bracts elliptical, glands rounded and yellow. Fruit is a round capsule with short warts.

Distribution and ecology

Native to Europe and small parts of adjacent Western Asia, occurs in all CE countries. Typical river corridor plant confined to lowland basins of large rivers, however, occurs also under saline conditions in inland salt marshes and along the shores of the North Sea and Baltic Sea. Grows on calcareous, peaty mud and clayish soils, usually of alternating moisture. Photophilous species. Successful reproduction dependent on seeds.

Anatomy of the rhizome

Rhizome with a diameter of 6 mm. Maximum age 1 year. Presented is a plant from a wet meadow.

The pith is surrounded by a xylem, a small phloem and a large cortex (1). The xylem consists of vessels in short to long radial rows and fibers (3). Vessels have a diameter of 40–60 μm (2) and a length of 250–300 μm. Intervessel pits are round and perforation plates are simple. Parenchyma is absent (2). Fibers are square and arranged in conifer-like radial lines (2). Rays are uniseriate.

The phloem consists of groups of sieve tubes and parenchyma cells (3). Intercellulars in the cortex are very small, oval intercellulars in the cortex are tangentially enlarged (1). The phellem is absent (3). Some phellem cells (3) occur frequently below the epidermis consisting of small square cells (3). The pith contains a net-like aerenchyma (4).

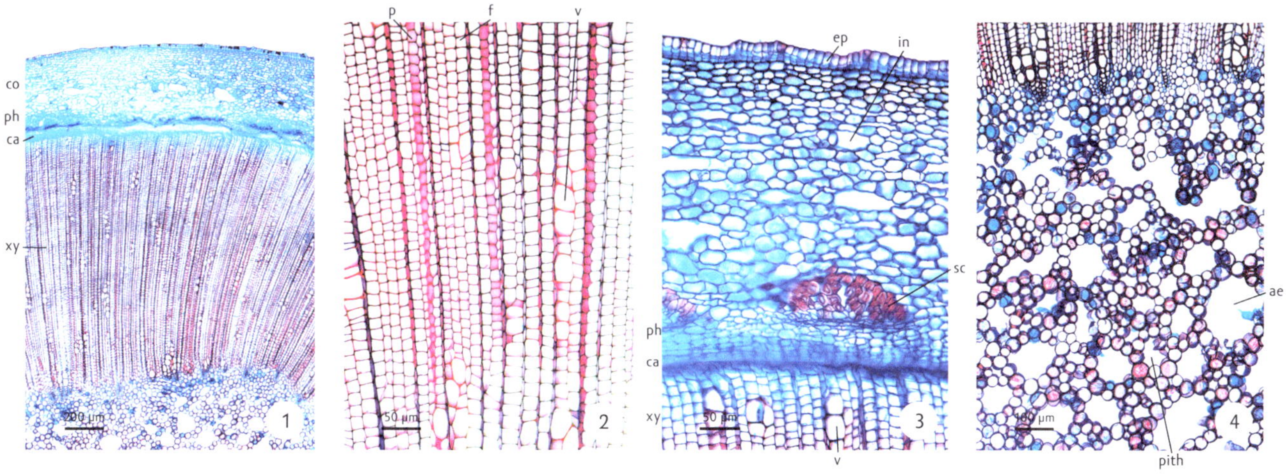

F. H. Schweingruber et al., *Anatomic Atlas of Aquatic and Wetland Plant Stems*, https://doi.org/10.1007/978-3-030-33420-8_26

Lotus maritimus L. (syn. *Tetragonolobus maritimus*) **Dragon´s-teeth** Fabaceae

Perennial wetland herb. Stem prostrate, sparsely pubescent, up to 30 cm long. Leaves with 3 leaflets, leaflets obovate to oblanceolate. Flowers solitary, on longer stalks, yellow with brown streaks. Fruit is a several-seeded legume, with four wings.

Distribution and ecology

Native of C. and S. Europe, SW. Asia and N. Africa. Native in all CE countries. Grows in both wet and dry meadows and pastures, along tracks. Facultative halophyte, photophilous species.

Anatomy of the root collar

Stem with a diameter of 4–5 mm. Maximum age 5 years.

The root collar consists of a pith, a xylem, a large phloem, a small cortex and a distinct phellem (1). The angular vessels in the semi-ring porous xylem occur in groups. They have a diameter of 40–60 µm (2) and a length of 100–150 µm. Intervessel pits are round to scalariform and perforation plates are simple (3). Fibers are thin- to thick-walled (2). Parenchyma is paratracheal and partially pervasive (2). Rays are 2- to 10-seriate (1, 2, 4) with upright cells.

Groups of sieve tubes, sclereids, parenchyma cells and dilating rays form the phloem (5, 6). The cortex consists of roundish parenchyma cells with small intercellulars (6). Many layers of rectangular cells form the phellem (5).

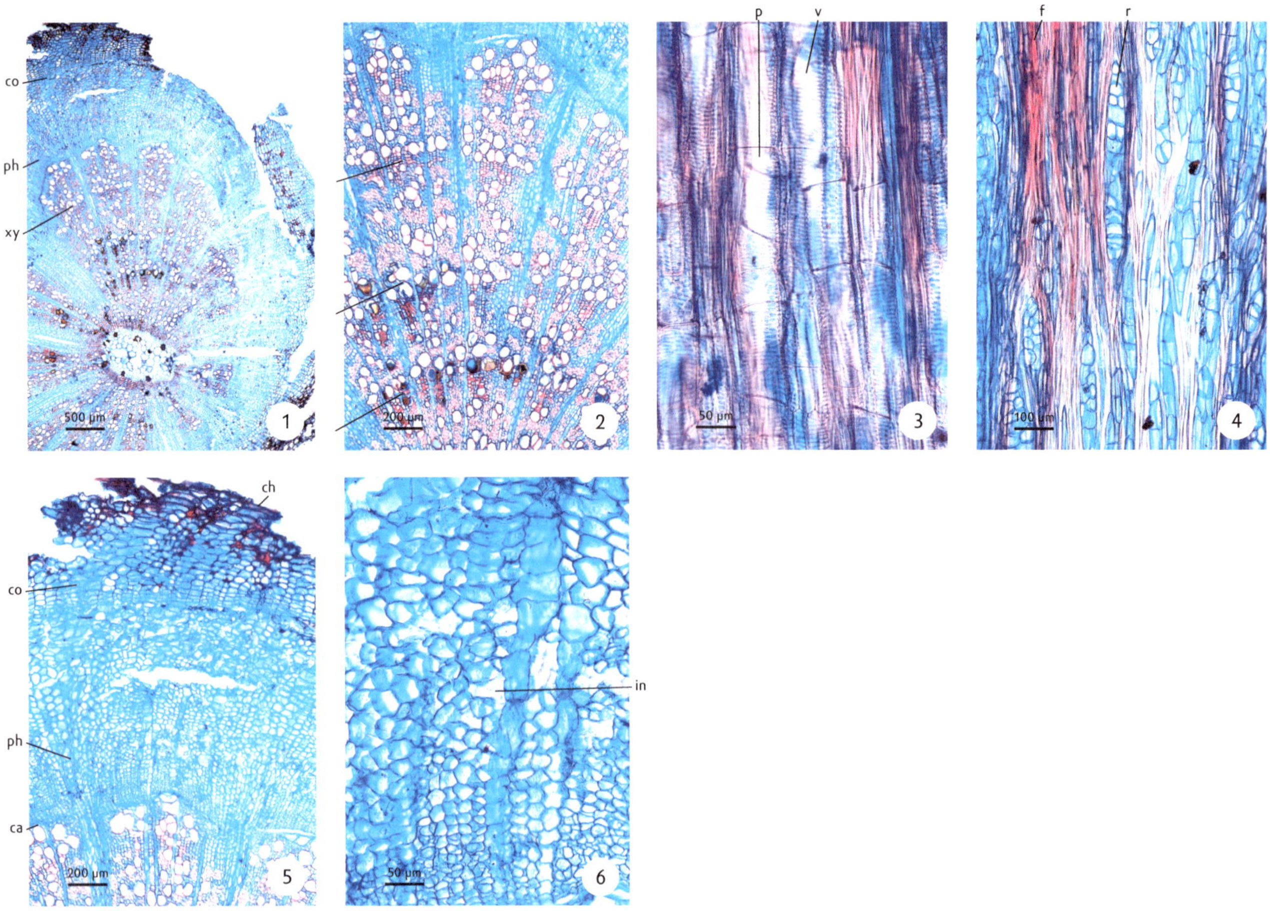

F. H. Schweingruber et al., *Anatomic Atlas of Aquatic and Wetland Plant Stems*, https://doi.org/10.1007/978-3-030-33420-8_27

Lotus pedunculatus Cav. (syn. *Lotus uliginosus*) **Greater Bird's-foot-trefoil** Fabaceae

Perennial wetland herb. Stem cylindrical and hollow, erect to ascending, up to 60 cm high. Leaves with 5 leaflets, the lowest two at the base of the petiole, leaflets ovate to obovate. Yellow flowers in one-sided heads on long stalks in leaf axils. Fruit is a several-seeded legume.

Distribution and ecology

European temperate element; widely naturalised outside its native range. Native in all CE countries except HU. Grows in wet peaty meadows, marshes and fens, also along the banks of fishponds, ditches and streams. Prefers humic, mineral-rich soils. The flooded stems develop extensive aerenchyma and adventitious roots. N_2 fixing bacteroids on root nodules.

Anatomy of the root collar

Stem with a diameter of 3–4 mm. Maximum age 6 years.

The root collar consists of a pith, a xylem, a large phloem, a cortex and a distinct phellem (1). The angular vessels in the semi-ring porous xylem are solitary or in groups. They have a diameter of 50–70 µm (2) and a length of 120–160 µm. Intervessel pits are round and perforation plates are simple (5). Fibers are thick-walled (2). Parenchyma is paratracheal (2). Vascular bundles remain (1). Rays between the bundles are 6- to 10-seriate (3, 4) with upright cells. Sheath cells are distinct (3, 4).

The phloem is homogeneous, sieve elements and parenchyma cells are not differentiated (5, 6). Rays dilate (6). The cortex consists of roundish parenchyma cells with small intercellulars and a few groups of fibers (6, 7). Some parenchyma cells are filled with blue staining substances with many styloids (7). Many layers of rectangular cells form the phellem (7).

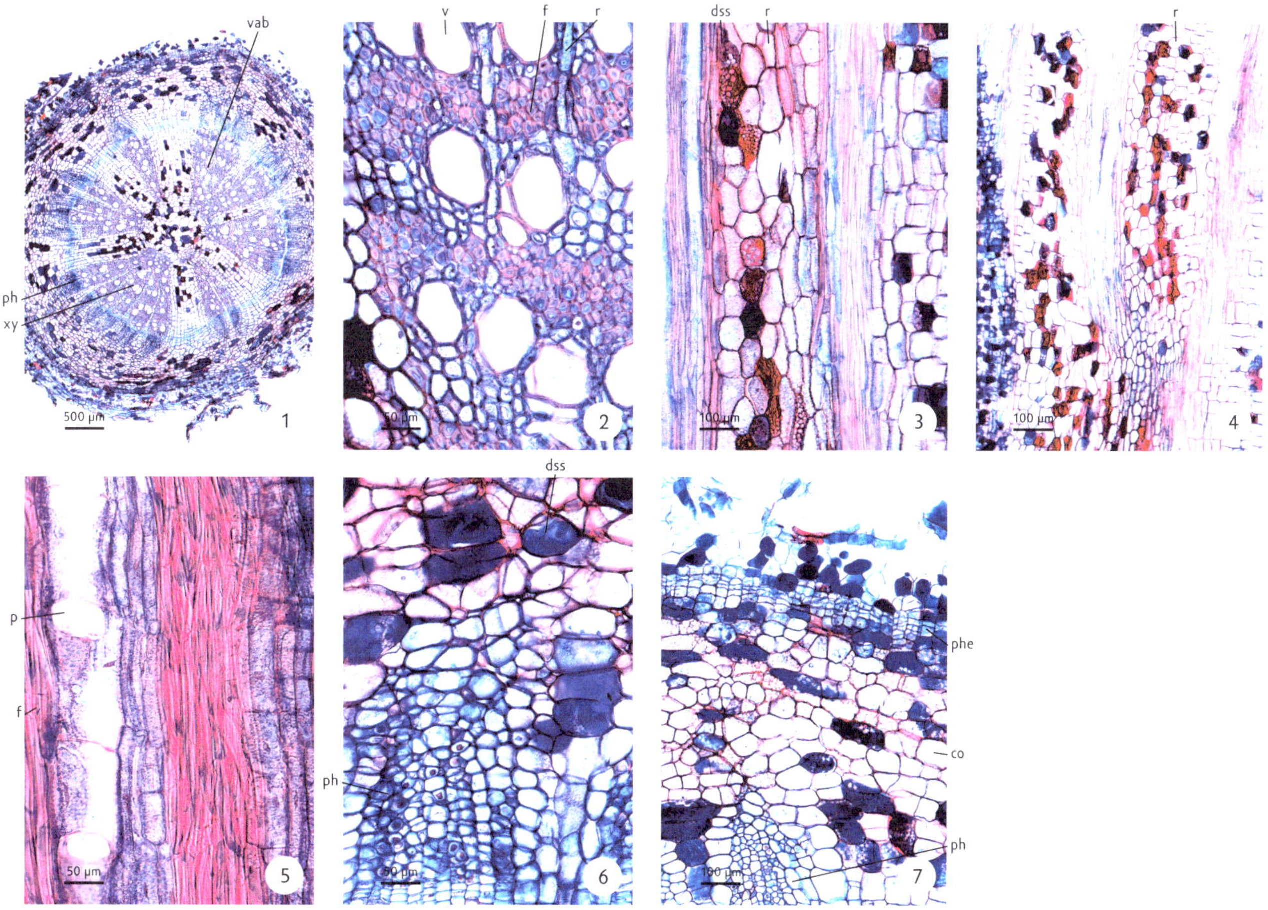

Centaurium pulchellum (Sw.) Druce **Lesser Centaury** Gentianaceae

Annual or biennial wetland herb. Stem erect, without basal leaf rosette, 3–15 cm high. Leaves opposite, glabrous, entire and all elliptical. Cymose inflorescence, corolla dark pink, 5-lobed. Fruit is a capsule.

Distribution and ecology

Euro-Siberian southern-temperate element. Native in all CE countries. Inland it grows in fen meadows and along wet tracks. On the coast it grows on open sandy and muddy habitats, often in sand dunes and in upper saltmarshes. Prefers disturbed ground and wet, mildly acidic to calcareous soils. Able of autonomous selfing.

Anatomy of the annual shoot

Round annual shoot and root collar with a diameter of 1 mm. Maximum age 1 year.

The pith is surrounded by a circle of vascular bundles, a small cortex and an epidermis (1, 2). The xylem of the vascular bundles contains a few solitary vessels. Rays appear as blue-stained cells in cross-sections (2). Axial parenchyma is absent (2). Vessels have a diameter of 10–15 µm. Vessel length cannot be evaluated. Vessels have round intervessel pits and simple perforation plates (3). Fibers are thin- to thick-walled (4).

The very small phloem contains a few sieve elements within a parenchymatic tissue. Groups of sieve tubes within a parenchymatic tissue characterize the cortex (4, 5), which consists of thin-walled parenchyma (4, 5). The phellem is absent (1, 4). Medullary phloem groups characterize the periphery of the pith (1).

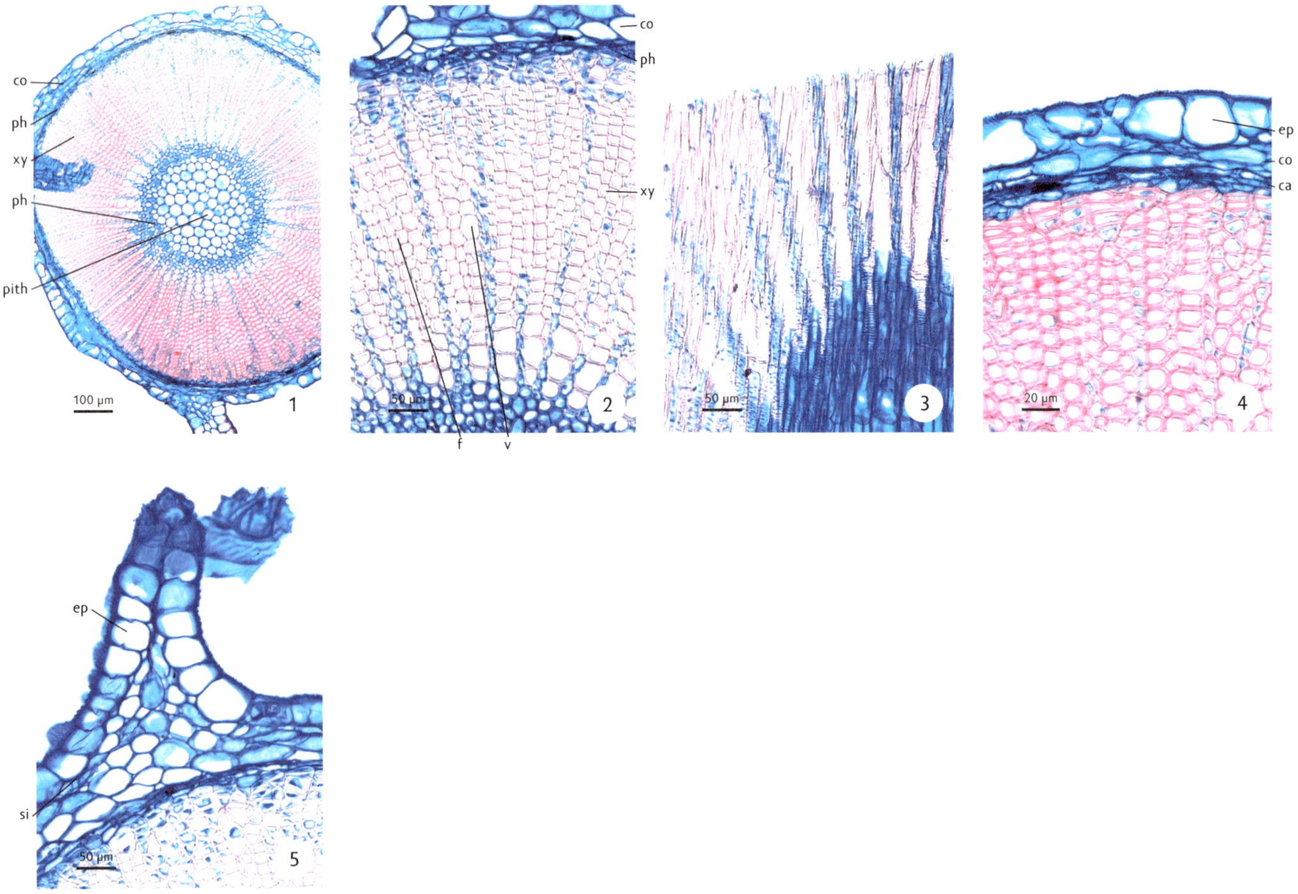

F. H. Schweingruber et al., *Anatomic Atlas of Aquatic and Wetland Plant Stems*, https://doi.org/10.1007/978-3-030-33420-8_28

Swertia perennis L.

Perennial wetland herb with a whitish rhizome. Flowering stem erect, 15–60 cm high. Basal rosette with petiolate leaves, stem leaves sessile, glabrous, opposite and ovate. Inflorescence from 10–20 blue-violet pentamerous flowers. Two fringed nectary glands occur at the base of each petal. Fruit is an ovate capsule with numerous seeds.

Distribution and ecology

Discontinuous distribution from C. Europe through Asia to western N. America, with a centre in the European Alps. Native in all CE countries except HU. Typical plant of calcareous pre-alpine fens. Grows also in mountain springs, peaty meadows and wet rocks.

Anatomy of the root collar and the annual shoot

Round annual shoot and root collar with a diameter of 3 mm. Maximum age 1 year.

Annual shoot: The large pith is surrounded by a vessel-free belt of fibers, a small phloem, a large cortex and an epidermis (1). The xylem belt is vessel- and ray-free and the phloem consists of sieve tubes and parenchyma cells (2). Special is the presence of medullary phloem (1, 2).

The large pith is surrounded by a circle of large concentric vascular bundles, a large cortex and an epidermis (3). The xylem of the vascular bundles contains unlignified, solitary vessels and pervasive parenchyma (4). Vessels have a diameter of 10–15 µm and a lenght of 80–100 µm. Vessels have scalariform intervessel pits with large openings and simple perforation plates (5). Fibers are absent (3). Rays are represented by radial groups of parenchyma cells between vascular bundles (1, 2).

Groups of sieve tubes within a parenchymatic tissue characterize the cortex (3) which consists of thin-walled parenchyma cells and small intercellulars (6). The epidermis is partially replaced by a small phellem (7). Medullary groups of phloem characterize the periphery of the pith (2, 8).

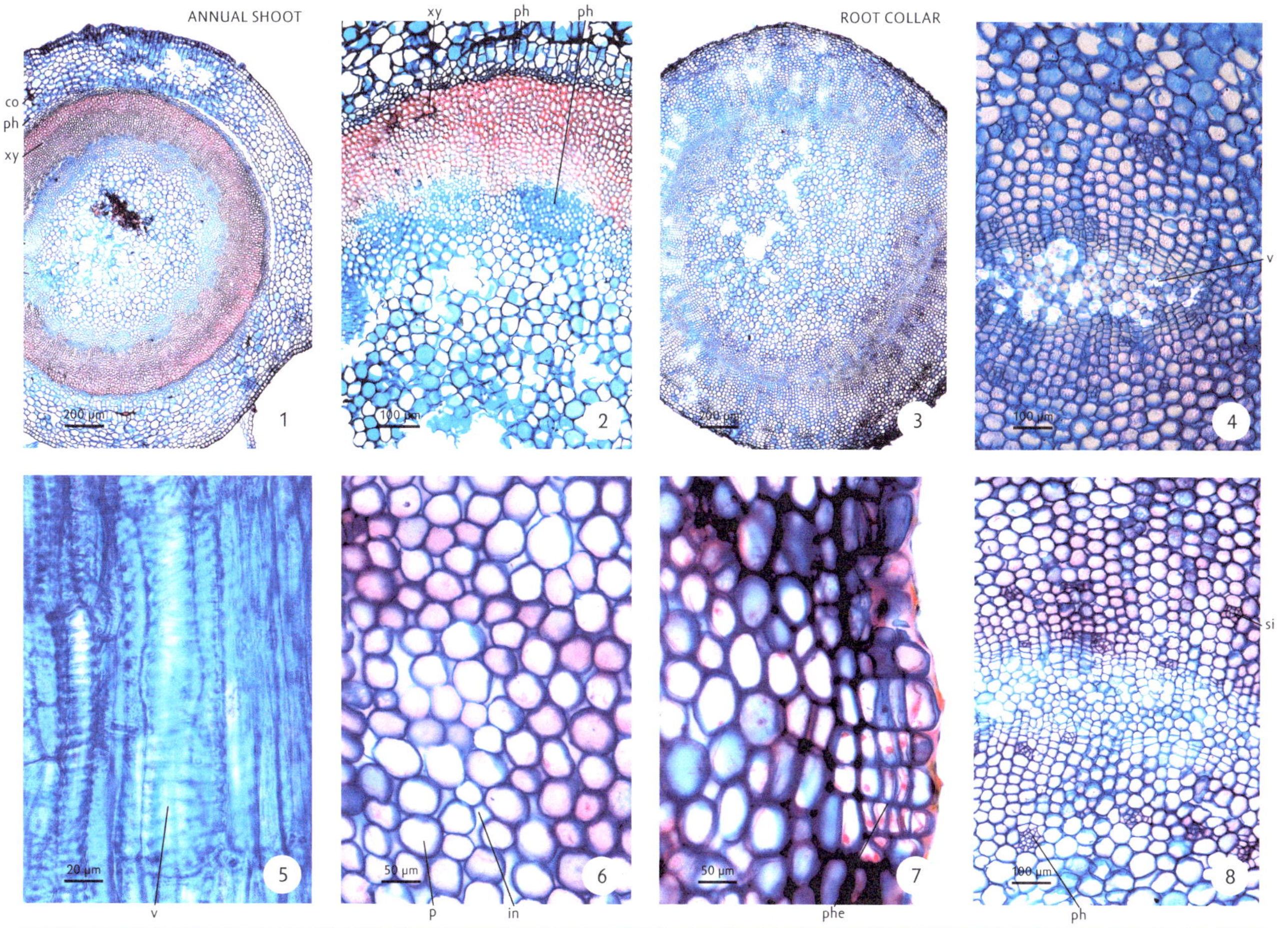

Geranium palustre L.

Perennial wetland herb with a thick rhizome. Stem erect to ascending, bristly, with thick nodes, 30–70 cm high. Leaves opposite, lower leaves 5–7 times palmately-lobed with long petiole, upper leaves 3–5 times palmately lobed with short petiole, lobes obtuse. Pentamerous flowers in pairs, petals purple-red. Fruit is a dry fruit with beak, splitting into watch-sping-coiled sections (mericarps).

Distribution and ecology

European and West Asian distribution. Native in all CE countries in waterlogged mineral-rich soils. Occurs on wet meadows, on banks of small streams and ponds, also in floodplain forests and carrs. Prefers mineral-rich, humic soils. Tolerates partial shade. The ability of ballistic expulsion of the seeds from the mericarp.

Anatomy of the rhizome

Round rhizome with a diameter of 7–12 mm. Rings indistinct.

The large pith is surrounded by a circle of large vascular bundles and a large cortex (1–3). The xylem of the vascular bundles contains angular solitary vessels and pervasive parenchyma (4). Vessels have a diameter of 40–50 μm and are 50–70 μm long. Vessels have round to scalariform intervessel pits and simple perforation plates (5). Fibers are absent (3). Rays are represented by radial groups of parenchyma cells between vascular bundles (1, 2). Xylem and bark contain much slime (2).

The phloem consists of groups of sieve elements within a parenchymatic tissue (6). Intercellulars in the cortex are small. The epidermis is partially replaced by a small phellem, which consists of a few layers of rectangular cells (7).

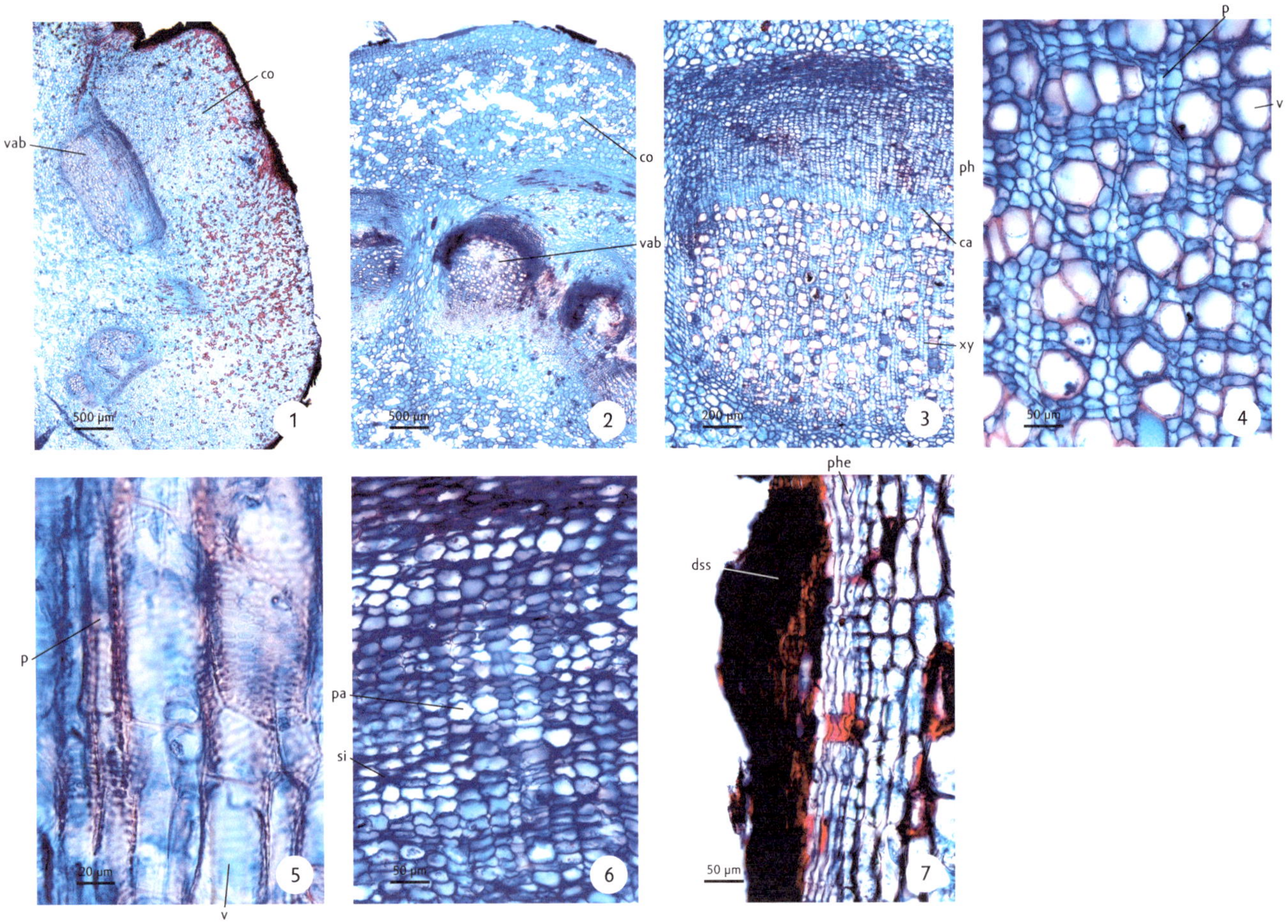

F. H. Schweingruber et al., *Anatomic Atlas of Aquatic and Wetland Plant Stems*, https://doi.org/10.1007/978-3-030-33420-8_29

Myriophyllum alterniflorum DC.

Alternate water-milfoil Haloragaceae

Perennial submerged herb with thin, cylindric, fragile, poorly branched porous stem, usually 4 leaves in regular whorls, rooted, terrestrial form unknown. Shoot 20–120 cm long, unbranched fibrous roots 10–20 cm long, adventitious roots formed at shoot bases, finely pinnate leaves with leaf filaments in one plane, leaf filaments circular in cross section. Pendent or erect flower spikes float above the surface, no overwintering organs.

Distribution and ecology

Suboceanic boreo-temperate element; also in N. America. Lacking in SK and HU, rarely scattered in other CE countries, mainly in base-poor streaming waters mostly in uplands. Occurs in slowly to fast streaming rivers, rarely standing waters, shallow lakes (usually northern CE). Prefers softwaters (total alkalinity 0.3–1 meq/l), clear to slightly dystrophic waters. HCO_3^- user.

Anatomy of a basal part of the stem

The round basal part of the plant with a diameter of 1–2 mm (1). Absent annual rings, age unknown.

The stem is composed of a circle of xylem and phloem, a large cortex and an epidermis (1, 2). The xylem consists of angular, grouped, unlignified vessels and pervasive parenchyma (2). Vessels have a diameter of 20–30 µm (2) and a length of 80–100 µm. Intervessel pits are scalariform (3) and perforations are absent or could not be observed. Some vessel walls have annular thickenings (3). Fibers and rays are absent (1, 2).

The phloem consists of sieve tubes and groups of small companion cells (2). The stele is surrounded by an endodermis (2). The large cortex is characterized by an aerenchyma in which chambers are separated by radial cell strips (1). Inside of the epidermis are a few layers of parenchyma cells without intercellulars. Crystal druses occur in the cortex (4).

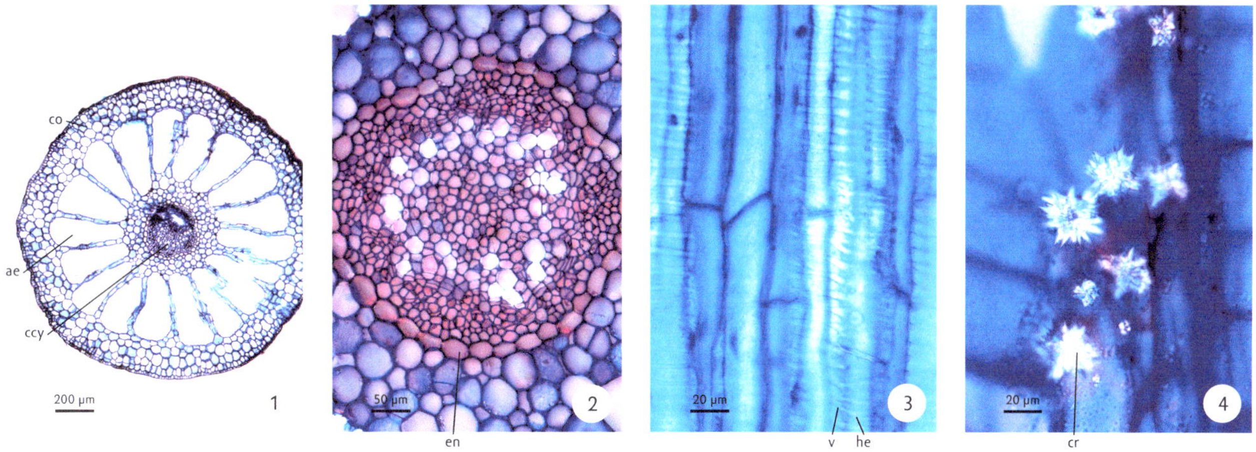

F. H. Schweingruber et al., *Anatomic Atlas of Aquatic and Wetland Plant Stems*, https://doi.org/10.1007/978-3-030-33420-8_30

Myriophyllum heterophyllum Michx. **Variable-leaf water-milfoil** Haloragaceae

Perennial evergreen submerged herb with heterophylly, rhizomatous. Stem stout, up to 3 mm in diameter and up to 100 cm long. Leaves in whorls of 4–6, submerged leaves feather-like, pinnate, 2–5 cm long and 2–4 cm wide. Emergent leaves stiff, entire or serrated to lobed along the margins, occur only during the summer months. A short, emergent floral spike above the water surface with conspicuous floral bracts. Reddish, tiny flowers in whorls of 4–6 in leaf axils, they have four petals and four stamens. Fruit is a group of nutlets and splits open when ripe.

Distribution and ecology

Native in eastern N. America. As neophyte in AT, CH and DE. Occurs in a variety of different aquatic habitats, such as lakes, ponds, canals, ditches, rivers and mudflats, in water up to 1.8 m deep. It can form dense monospecific beds with large biomass (>3 kg of dry mass per m^2). Strict CO_2 user, aerial shoots provide access to atmospheric CO_2. Spread rapidly by fragmentation and by seeds.

Anatomy of the shoot

The specimen from a sea shore. The round basal part of the plant with a diameter of 1–2 mm (1). Absent annual rings, age unknown.

The stem is composed of a circle of xylem and phloem, a large cortex and an epidermis (1, 2). The xylem consists of a few solitary vessels and pervasive parenchyma (2). Vessels have a diameter of 15–20 µm (2) and a length of <100 µm. Intervessel pits are scalariform (3). Vessel walls have annular thickenings (3). Perforations are absent or could not be observed.

The phloem consists of sieve tubes and parenchyma cells (2). The stele is surrounded by an endodermis (2). The large cortex is characterized by round intercellular chambers (1). Inside the epidermis there are a few layers of parenchyma cells without intercellulars. Crystal druses are very frequent in the cortex (4).

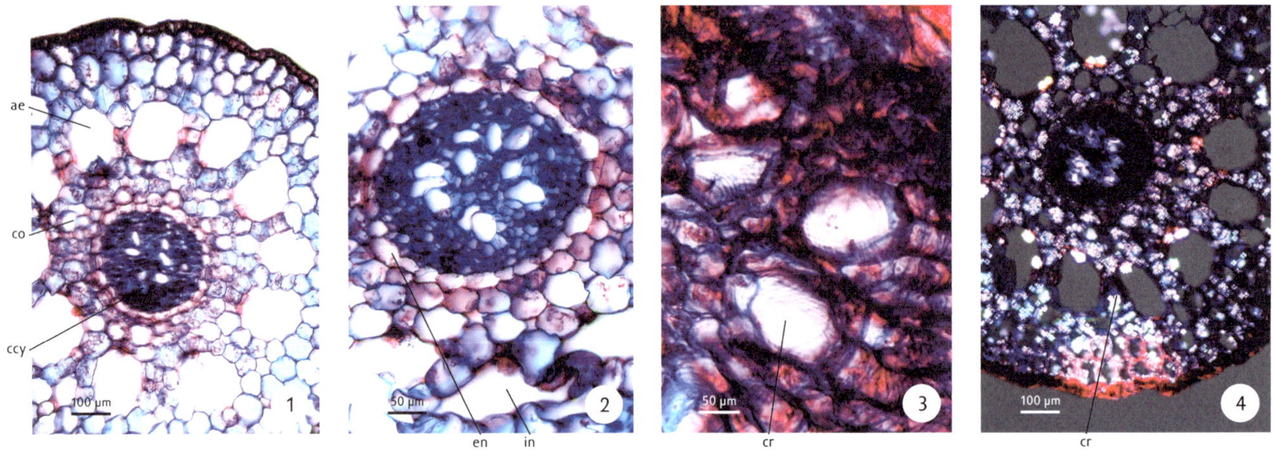

Myriophyllum spicatum L.   **Spiked (Eurasian) water-milfoil**   Haloragaceae

Perennial summer-green submerged herb with thin, cylindric, fragile, poorly branched porous stem, with 4 leaves in regular whorls, rooted. Shoot 30–250 cm long, unbranched fibrous roots 10–40 cm long, adventitious roots formed at shoot bases or in fragments of shoots as plant propagation. Finely pinnate leaves with leaf filaments in one plane, leaf filaments circular in cross section. Dwarfish terrestrial form known. Red erect flower spikes 4–16 cm long above the surface, overwintering in buds in rooted basal stem segments.

Distribution and ecology

Eurasian temperate element, naturalised in N. America. Common in all 7 CE countries, mainly in lowlands and hilly areas, retreat from strongly eutrophicated habitats (*e.g.*, CZ, SK). Occurs in shallow standing waters in water depth 0.3–3 m, in fishponds, lakes, dam reservoirs, backwater pools and oxbows, canals, sand-pits, also in rivers. Prefers meso- to mildly eutrophic waters with high water transparency (>0.5 m, clear-water phase, facultative heliophyte) and organic, nutrient-rich sediments. Tolerates medium hard to hard waters (total alkalinity 1–5 meq/l). Very efficient HCO_3^- use. Seed set for seed bank, mainly vegetative propagation, very dense stands to 1–2 kg dry weight/m^2, highly invasive species, very sensitive to winter water drawdown and frosts.

Anatomy of the shoot

The round basal part of the plant with a diameter of 1–2 mm (1). Absent annual rings, age unknown. The structure is very similar to that of *Myriophyllum alterniflorum* and *M. verticillatum*.

The stem is composed of a circle of xylem and phloem, a large cortex and an epidermis (1, 2). The xylem consists of a few angular, solitary vessels and pervasive parenchyma (2). Vessels have a diameter of 20–30 µm (2) and a length of 80–100 µm. Intervessel pits are scalariform (3) and perforations are simple (3). Vessel walls have annular thickenings (3, 4). Fibers and rays are absent (1, 2).

The phloem consists of sieve tubes and parenchymatic cells (2). The stele is surrounded by a starch-free endodermis (2). The large cortex is characterized by an aerenchyma in which chambers are separated by radial cell strips (1). Inside of the epidermis, there are a few layers of parenchymatic cells without intercellulars. Crystal druses occur in the cortex (5).

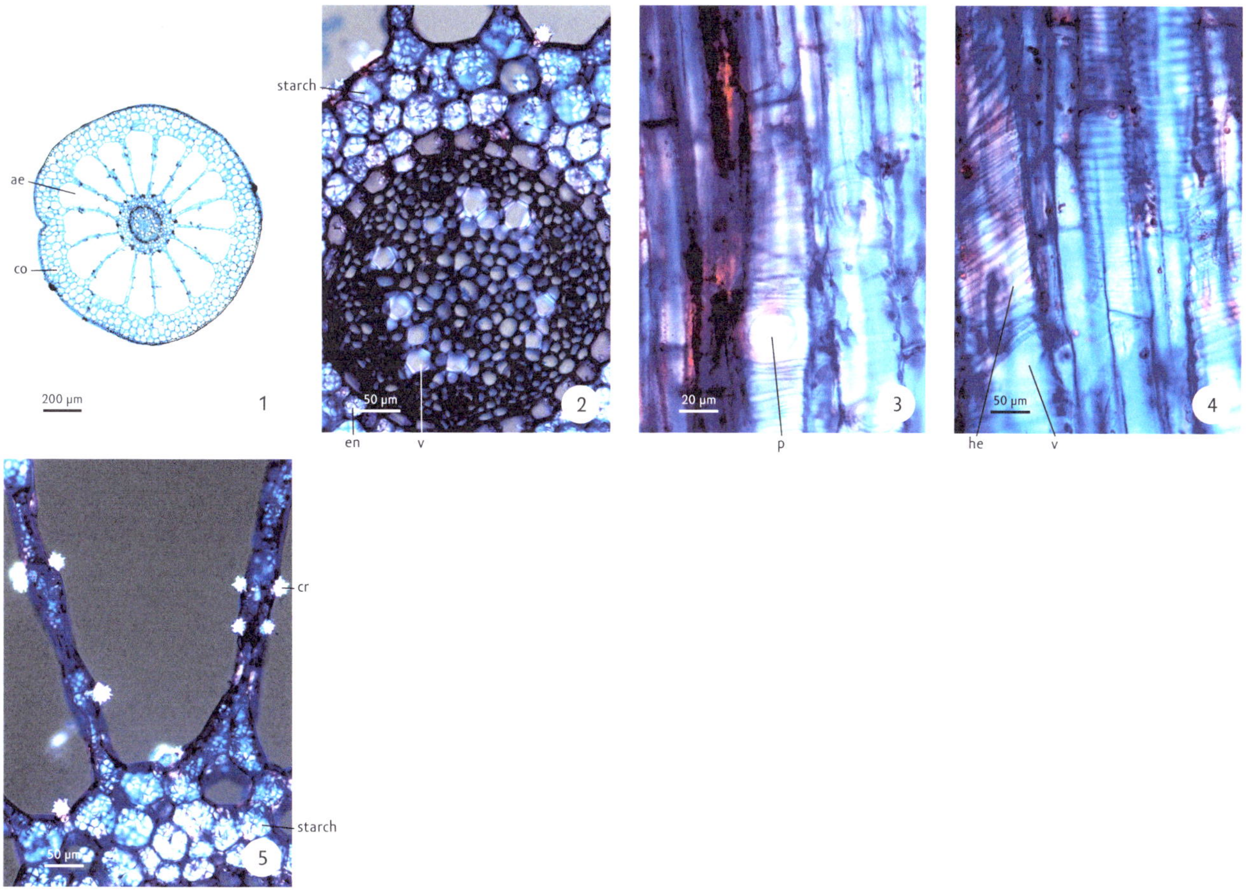

Myriophyllum verticillatum L. **Whorled water-milfoil** Haloragaceae

Perennial submerged plant with thin, cylindric, fragile, poorly branched porous stem, usually 5 leaves in regular whorls, rooted. Low terrestrial form (10–15 cm) with thicker dissected leaves known. Shoot 20–250 cm long, unbranched fibrous roots 10–30 cm long, adventitious roots formed at shoot bases, finely pinnate leaves with leaf filaments in one plane. Erect flower spikes 4–16 cm long above the surface. Dormant, compact elongated turions (2–5 cm), overwinter and sprout at the bottom.

Distribution and ecology

Circumpolar temperate element. Scattered distribution, less common to rare in all 7 CE countries, mainly in lowland and hardwaters, retreat from eutrophicated waters. Occurs in shallow standing waters in water depth 0.2–3 m, in fishponds, lakes, backwater pools and oxbows, canals, slow rivers and fen lakes. Prefers clear to mildly dystrophic waters, meso- to mildly eutrophic waters with high water transparency >0.5 m and organic, nutrient-rich sediments. Strictly requires organic bottom. Tolerates partly shaded habitats (facultative sciophyte). Tolerates medium hard to hard waters (total alkalinity 1-5 meq/l). Strict CO_2 user, prefers CO_2-rich waters. Mainly vegetative propagation by fragments of shoots.

Anatomy of the shoot

The round basal part of the plant with a diameter of 1–2 mm (1). Absent annual rings, age unknown. The structure is very similar to that of *Myriophyllum alterniflorum* and *M. spicatum*.

The stem is composed of a circle of xylem and phloem, a large cortex and an epidermis (1, 2). The xylem consists of angular, solitary vessels and pervasive parenchyma (2). Vessels have a diameter of 20–30 µm (2) and a length of 80–100 µm. Intervessel pits are scalariform with wide openings (3) and perforations are simple (3). Vessel walls have annular thickenings (3, 4). Fibers and rays are absent (1, 2).

The phloem consists of sieve tubes and parenchyma cells (2). The stele is surrounded by a starch free endodermis (2). The large cortex is characterized by an aerenchyma in which chambers are separated by radial cell strips (1). Inside of the epidermis there are a few layers of parenchyma cells without intercellulars. Crystal druses occur in the cortex.

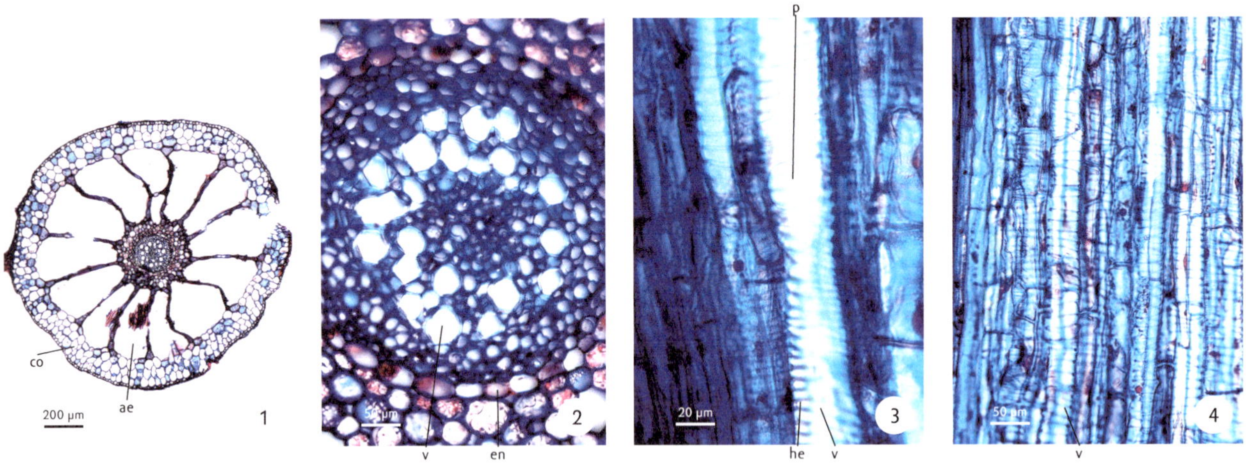

Hypericum tetrapterum Fr.

Square-stalked St John's-wort Hypericaceae

Perennial wetland herb, rhizomatous. Stem erect, glabrous, square in section with distinct wings, 20–80 cm high. Leaves opposite, sessile, elliptic, with many translucent glands. Pentamerous flowers in branched cymes, 10–15 mm across, petals pale yellow. Fruit is a capsule containing narrow cylindric seeds, with fine reticulate surface.

Distribution and ecology

European temperate element. Native to all CE countries from lowlands to uplands. Occurs in wet meadows, along streams and ditches, also in springs. Prefers nutrient-rich soils. Medicinal plant.

Anatomy of the stem

Round stem with a diameter of 3–4 mm. Maximum age 1 year.

Stems and rhizomes consist of a xylem, a small phloem, a cortex and a large phellem (1). The xylem is composed of vessels, thin- to thick-walled fibers and rays (2). Axial parenchyma is absent (2). Vessels are arranged solitary and in short radial rows. Vessel diameter varies from 30 to 40 µm (2) and vessel length from 100 to 200 µm. Intervessel pits are round and vessel-ray pits are round to horizontal-oval. Perforation plates are generally simple. Plates with 1–5 bars are rare (3, 4). Fibers are thin- to thick-walled (2). Tension wood occurs frequently (2). Rays are normally uniseriate, rarely biseriate (5) and ray cells are square and upright (3).

The phloem consists of polygonal parenchyma cells and sieve elements (6), it contains a few oil cells (7). Rays dilate (6). Intercellulars are absent in the phloem and the cortex (6, 7). The phellem consists of radially arranged rectangular cells (7).

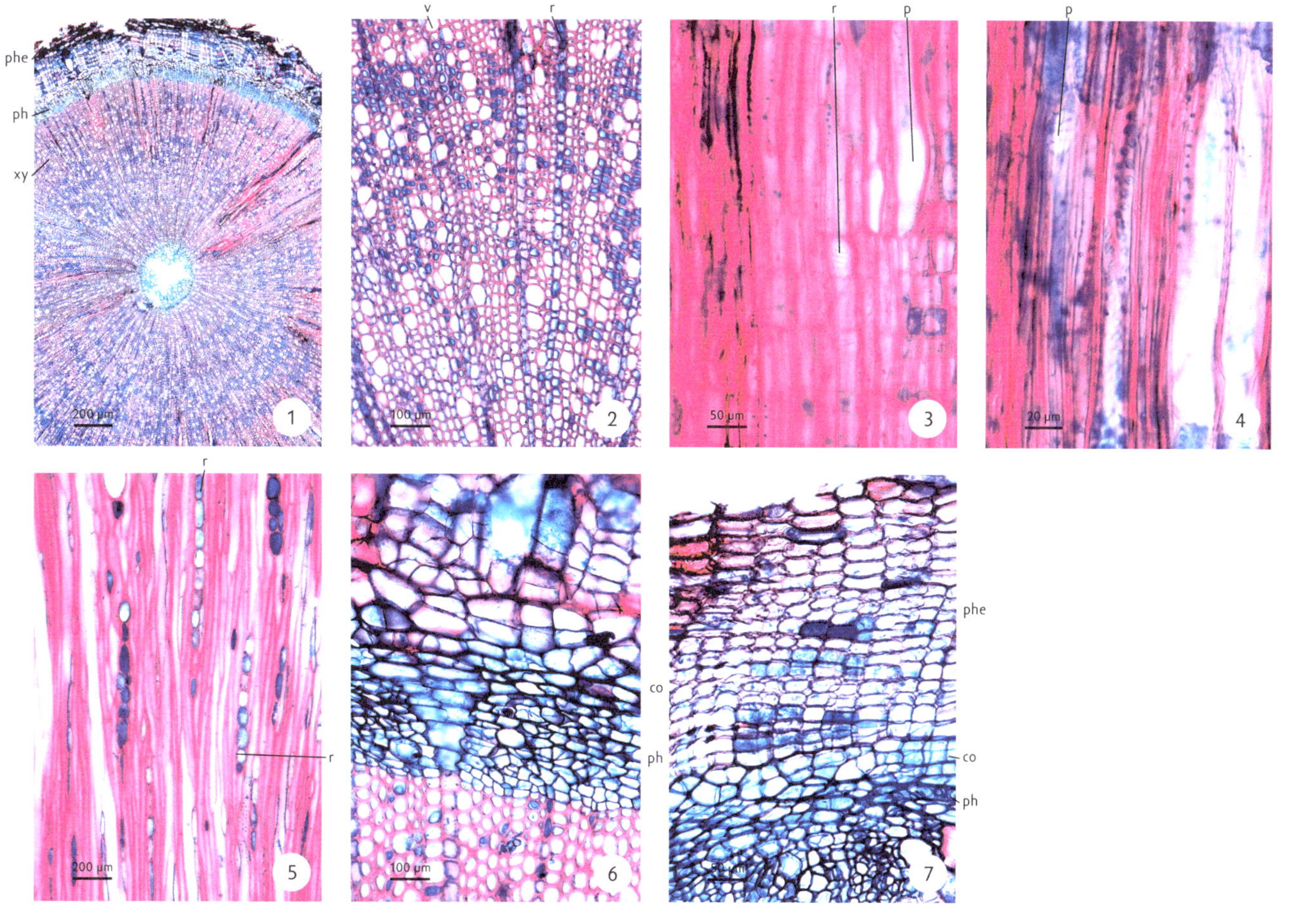

F. H. Schweingruber et al., *Anatomic Atlas of Aquatic and Wetland Plant Stems*, https://doi.org/10.1007/978-3-030-33420-8_31

Lycopus europaeus L. **Gipsywort** Lamiaceae

Perennial wetland or emergent herb, rhizomatous. Stem erect, with ascending branches, 30–100 cm high. Leaves elliptical-lanceolate, with toothed margin. Flowers in dense whorls in leaf axils. Calyx bell-shaped, corolla white, lower lip purple-dotted. Fruit consists of four nutlets.

Distribution and ecology

Euro-Siberian temperate element; widely naturalised outside its native range. Native in all CE countries. Grows in banks of lakes, fishponds, rivers, streams and ditches, also in fens. Able to colonise exposed mud. Tolerant of temporary flooding. Germination depends on both diurnal temperature fluctuation and light. Pharmacologically active substances.

Anatomy of the root collar

Square stem with a diameter of 3–4 mm. Maximum age 2 years.

A central part consists of a large pith and a continuous xylem-phloem-belt. A large cortex surrounds this part (1, 6). Vascular bundles remain in some individuals. The diffuse-porous xylem contains vessels with a diameter of 40–60 μm (2) and a length of 200–250 μm. Vessels are arranged solitary or in short radial rows (2). Intervessel pits are small and round perforation plates are simple (3). Some vessels contain tyloses (2). Fibers are thin- to thick-walled (2). Septate fibers are present (4). Parenchyma is scanty paratracheal or absent (2). Rays are 1–4 seriate and often confluent with upright cells (2, 5).

The phloem consists of parenchyma cells and groups of sieve tubes and dilating rays (6). Phloem and cortex are separated by an endodermis (6). The cortex contains isolated fibers and small intercellulars (6). Crystals are absent.

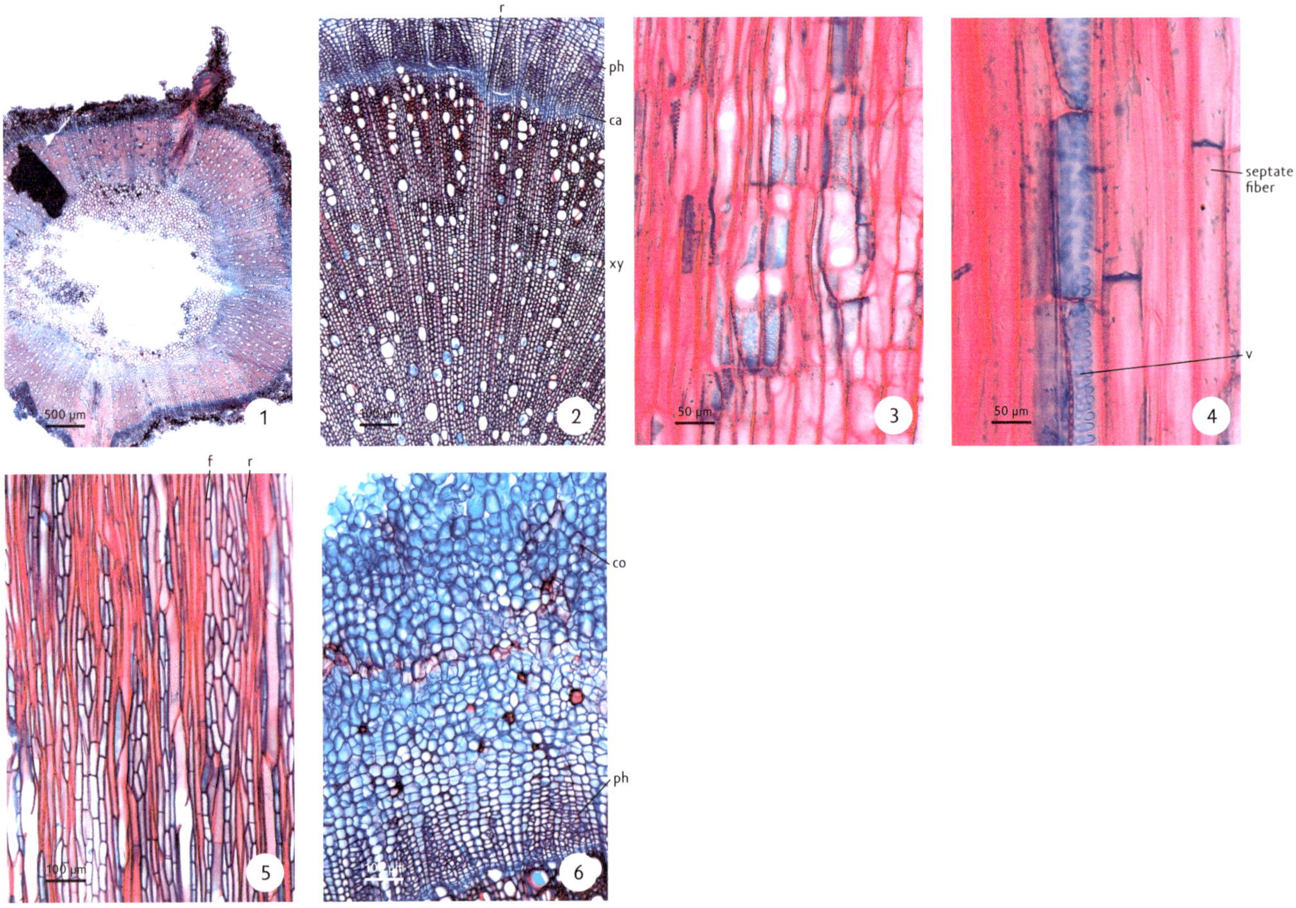

F. H. Schweingruber et al., *Anatomic Atlas of Aquatic and Wetland Plant Stems*, https://doi.org/10.1007/978-3-030-33420-8_32

Lycopus exaltatus L. f. **Cut Leaved Gypsywort** Lamiaceae

Perennial wetland herb. Stem erect, lignifying, usually unbranched, 70–150 cm high. Leaves deeply pinnately-lobed, lobes shortly pointed. White flowers in well-spaced whorls in leaf axils. Fruit consists of four nutlets rounded on the top.

Distribution and ecology

Eurasian subcontinental distribution. Native in CZ, HU and SK, probably introduced to PL, extinct in DE. Occurs in floodplain forests, alluvial meadows and ditches. Prefers wet, nutrient- and mineral-rich soils. Tolerates both flooding and drying of soils during summer.

Anatomy of the root collar

Round stem with a diameter of 4–5 mm. Maximum age 1 year.

A central part consists of a large pith and a continuous xylem-phloem-belt. A large cortex surrounds this part (1). Vascular bundles remain. The xylem contains vessels with a diameter of 30–50 µm (2) and a length of 100–120 µm. Vessels are arranged solitary or in short radial rows (2). Intervessel pits are small and round perforation plates are simple (3). Fibers are thin- to thick-walled (2). Septate fibers are absent (3). Parenchyma is scanty paratracheal or absent (2). Rays are 1–6-seriate and often confluent with upright cells (2, 5).

The phloem consists of parenchyma cells and groups of sieve tubes. Rays do not dilate. Phloem and cortex are separated by an endodermis (4). The cortex contains isolated fibers and small intercellulars. Crystals are absent.

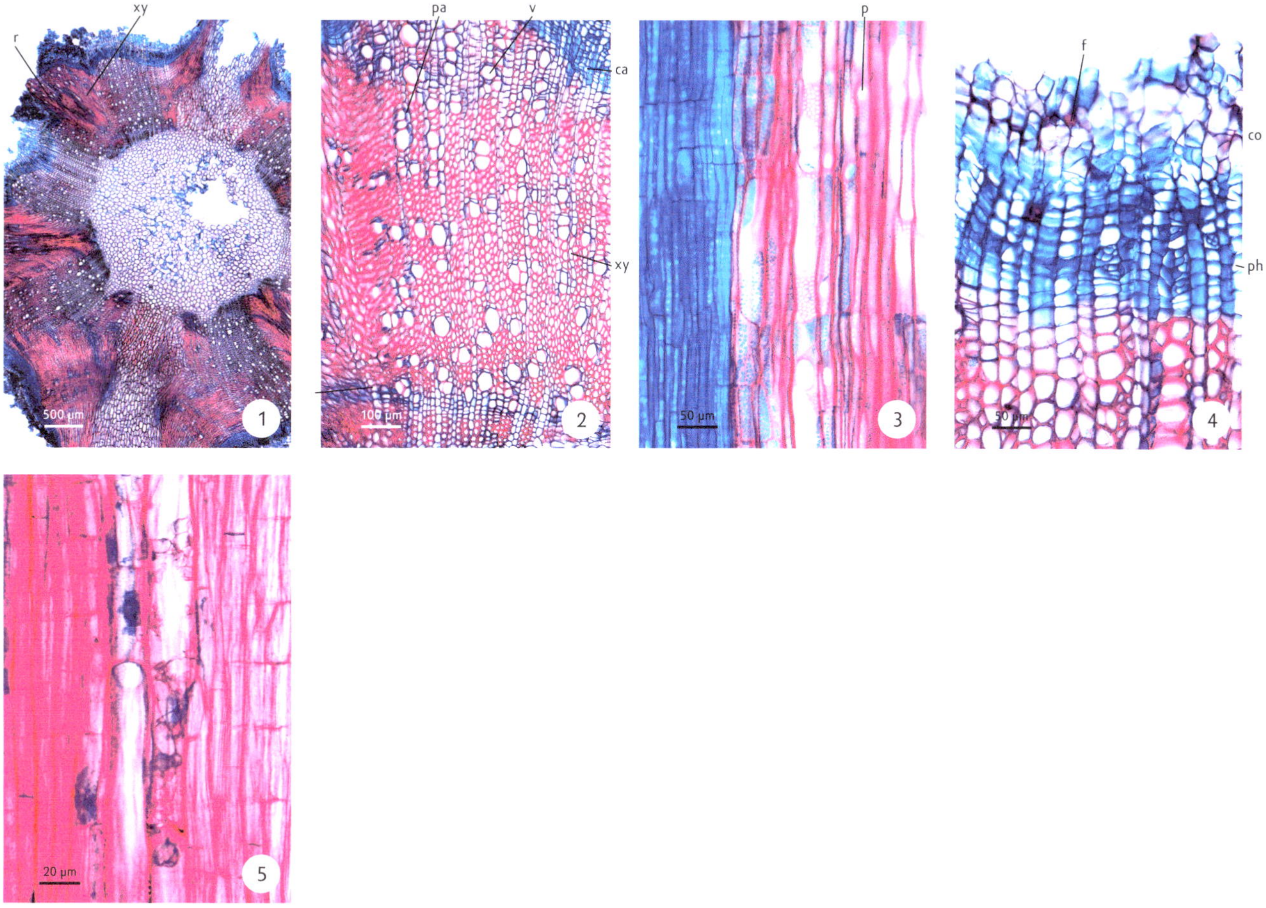

Mentha aquatica L. — **Water Mint** — Lamiaceae

Perennial wetland herb, rhizomatous. Stem erect, 15–60 cm high. Leaves opposite, oval, with blunt tips and teeth. Inflorescence in a terminal rounded head, corolla lilac, stamens projecting from flowers. Fruit consists of four nutlets.

Distribution and ecology

European temperate element; widely naturalised outside its native range. Native in all CE countries on banks of ditches, ponds and rivers, in wet pastures and fens. Prefers mildly acidic to calcareous, mineral-rich to peaty soils. A high degree of phenotypic plasticity in response to flooding. Self-incompatible, frequently spreads by rhizomes and rhizome fragments. Aromatic and medicinal plant. Hybrids with other *Mentha* species very common.

Anatomy of a basal part of the annual shoot

Round stem with a diameter of 2-3 mm. Maximum age 1 year.

The central part consists of a large pith, a continuous xylem- and a phloem-belt and a large cortex (1). The xylem consists of solitary round vessels with a diameter of 30–50 µm (2) and a length of 150–250 µm. Intervessel pits are small and round perforation plates are simple (3). Fibers are thin- to thick-walled (2). Septate fibers occur. Pits in fibers are rare or absent. Parenchyma is scanty paratracheal (2). Rays are uniseriate (4) with upright cells. Primary vascular bundles surround the arcade-like pith (1). A few of them remain in the adult phase.

The fairly homogeneous phloem consists mainly of round parenchyma cells and few smaller sieve tubes (2). The phloem and the cortex are separated by a thin-walled endodermis (2). The pericycle of some individuals contains fibers (5). The cortex contains thin-walled parenchyma with net-like intercellulars between parenchyma cells (1). The epidermis consists of small, unlignified thin-walled square cells (1).

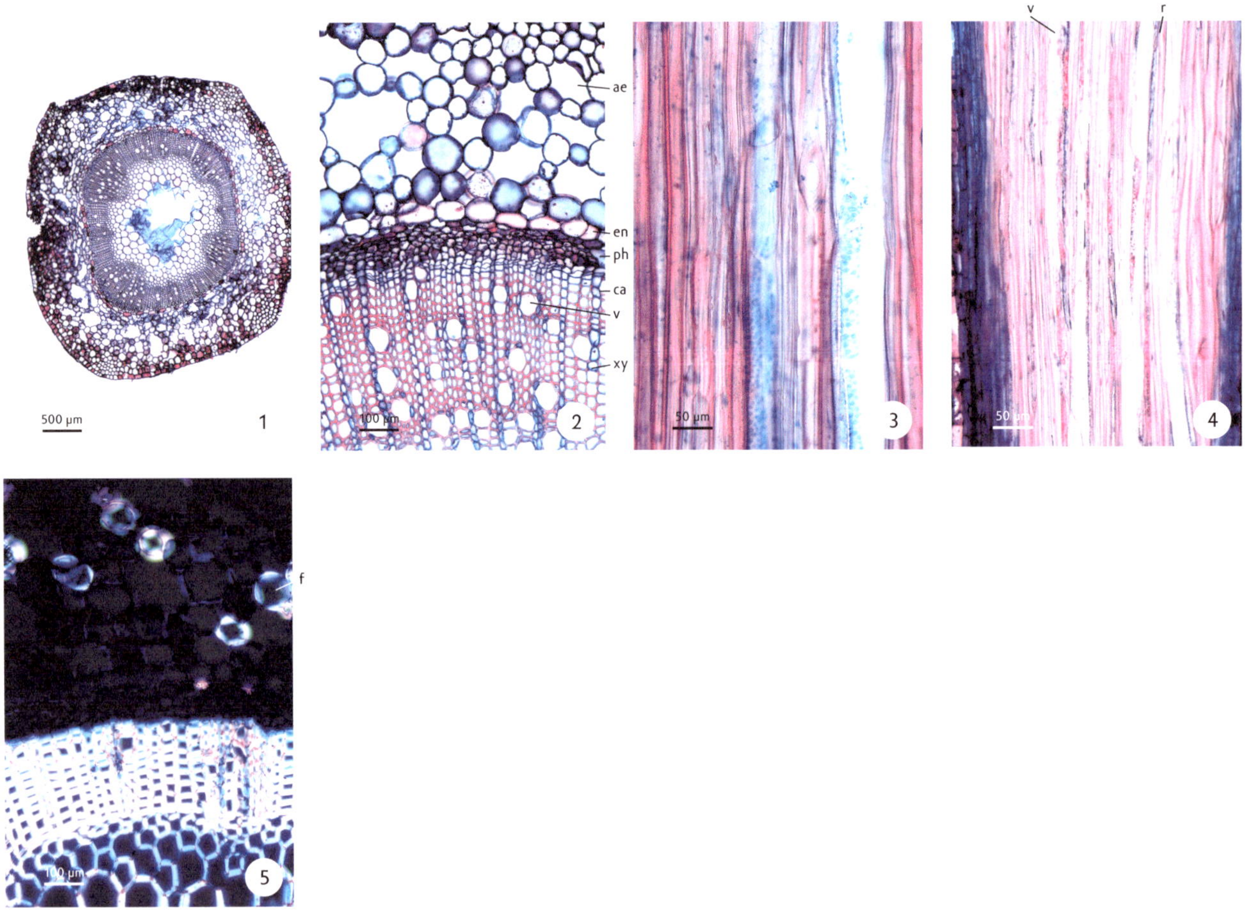

Mentha longifolia (L.) Huds. **Horse Mint** Lamiaceae

Perennial wetland herb, with creeping rhizome. Stem erect, 60–90 cm high. Leaves lanceolate, stalkless, toothed, hairy above, grey-woolly below. Inflorescence is a terminal spike, corolla lilac and hairy. Fruit consists of four brown nutlets.

Distribution and ecology

Native to Europe, western and central Asia and extratropical Africa. In all CE countries close to rivers and streams, in ditches and wet meadows. Prefers mineral-rich substrates. Usually forms clonal colonies by rhizomes. Medicinal and aromatic plant with unpleasant scent.

Anatomy of a basal part of the annual shoot

Round stem with a diameter of 2–3 mm. Maximum age 3 years.

A central part consists of a large pith and a continuous xylem-phloem-belt. A large cortex surrounds this part (1). The xylem consists mainly of solitary round vessels with a diameter of 30–50 µm (2) and a length of 150–250 µm. Intervessel pits are small and round perforation plates are simple. Fibers are thin- to thick-walled (2). Septate fibers are frequent (3). Pits in fibers are rare. Parenchyma is scanty paratracheal, however often indistinct (2, 5). Rays are uni- and biseriate with upright cells (4).

The phloem is fairly homogeneous, consisting mainly of round parenchyma cells and few smaller sieve tubes (5). The phloem is externally delimited by a layer of cells containing dark staining substances (5). The cortex contains net-like intercellulars between parenchyma cells and single fibers (5). The epidermis consists of small, unlignified thin-walled square cells. Spots of mucilage, which appear in polarized light, occur in the phloem and cortex (6).

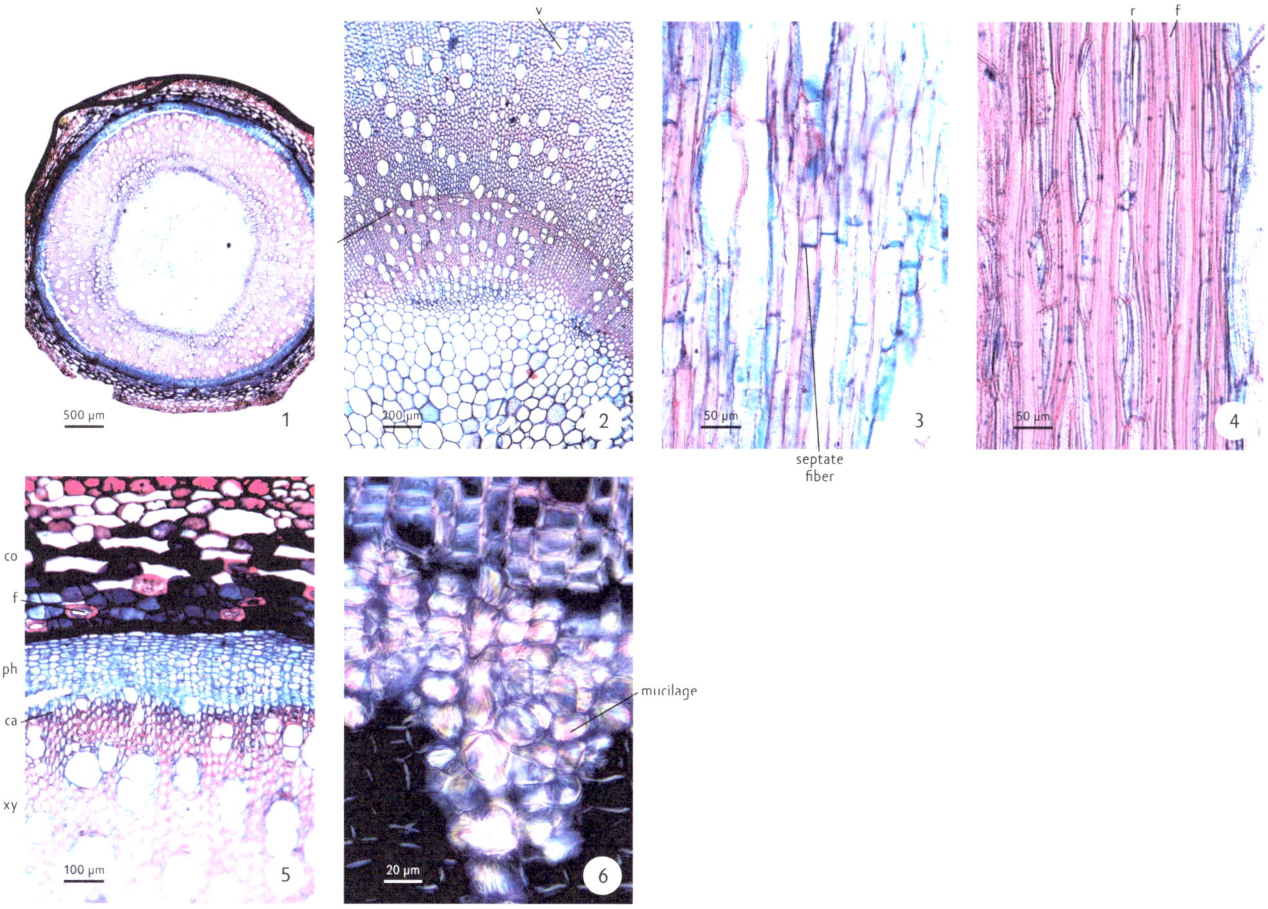

Perennial wetland herb, short-lived. Stem prostrate, creeping, glabrous, to 30 cm long. Leaves oval, short-stalked and toothed. Flowers in distant whorls in leaf axils, calyx downy, corolla lilac. Fruit consists of four dry nutlets.

Distribution and ecology

European southern-temperate element. Native in all CE countries in lowlands. Grows in wet meadows and pastures, water banks, also on heavily grazed village greens. Prefers seasonally inundated grasslands with nutrient-rich soil. Facultative halophyte. Able to growth under water. Aromatic and medicinal plant.

Anatomy of a basal part of the annual shoot

Round stem with a diameter of 2–3 mm. Maximum age 1 year.

A central part consists of a large pith and a continuous xylem-phloem-belt. A large cortex surrounds this part (1). The xylem consists mainly of solitary round vessels with a diameter of 40–70 µm (2) and a length of 100–150 µm. Intervessel pits are small and round perforation plates are simple (3). Fibers are thin- to thick-walled (2). Septate fibers are frequent (3). Pits in fibers are rare. Parenchyma is scanty paratracheal, however, often indistinct (2). The majority of rays are uni- and biseriate with upright cells (2, 4).

The phloem is fairly homogeneous, consisting mainly of round parenchyma cells and few smaller sieve tubes (2). Phloem and cortex are separated by an endodermis (2). The cortex contains large irregularly formed intercellulars between parenchyma cells. Fibers are absent (2). The epidermis consists of small, unlignified thin-walled square cells.

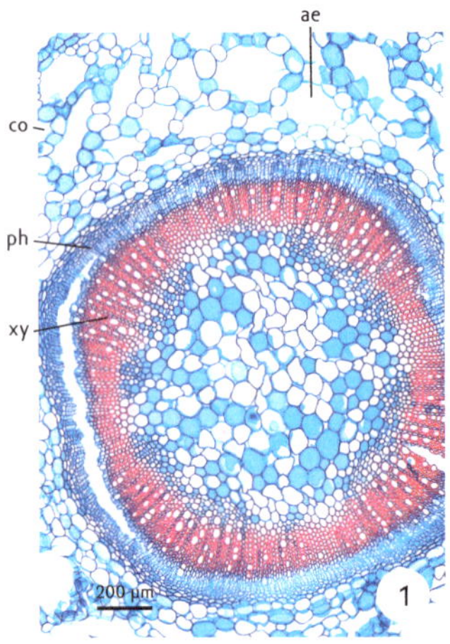

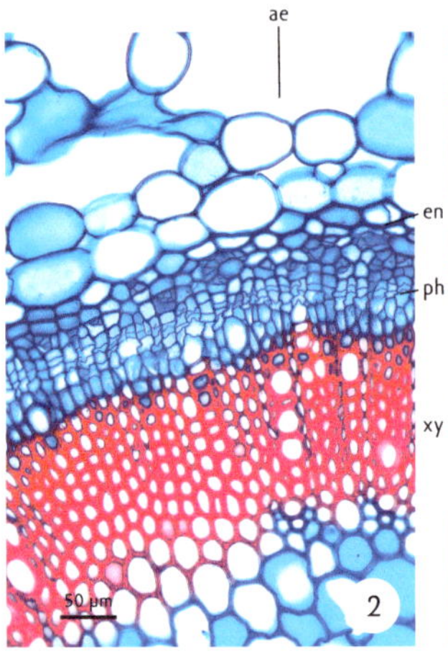

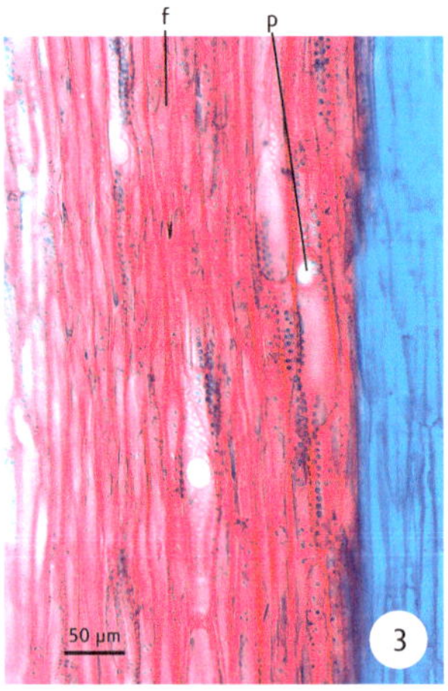

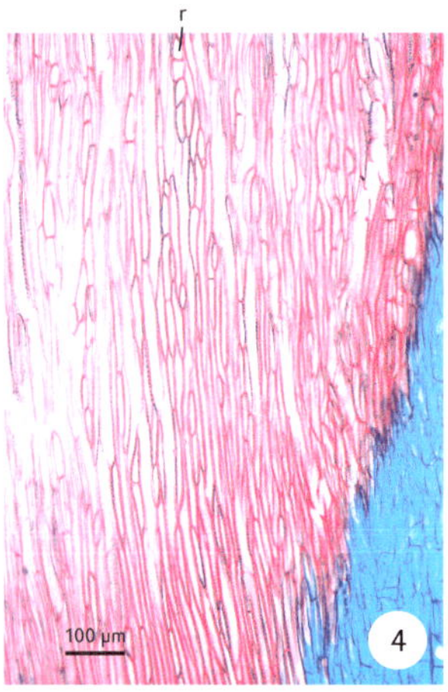

Mentha spicata L.

Spear Mint

Lamiaceae

Perennial wetland herb with characteristic spearmint scent, rhizomatous. Stem erect, glabrous or tomentose, 45–110 cm high. Leaves sessile, lanceolate to ovate, with sharp, forward-directed teeth. Flowers in a spike-like head. Calyx bell-shaped, corolla white to pink or lilac. Fruit is an ovate to elliptic nutlet with large hilum at base. Very variable taxon, susceptible to hybridization with other Mentha *spp.*

Distribution and ecology

Archaeophyte, origin unknown; derived from *M. longifolia* and *M. suaveolens* by hybridisation and chromosome doubling. Naturalised in all CE countries. Occurs in different wet habitats, usually close to human settlements, on waste ground.

Anatomy of a basal part of the annual shoot

Round stem with a diameter of 2–3 mm. Maximum age 2 years.

A central part consists of a large pith and a continuous xylem-phloem belt. A large cortex surrounds this part (1). The xylem contains solitary round vessels with a diameter of 20–30 µm (2) and a length of 120–170 µm. Intervessel pits are small and round perforation plates are simple (3). Fibers are thin- to thick-walled (2). Septate fibers are absent (3). Pits in fibers are rare. Parenchyma is scanty paratracheal, however often indistinct (2). The majority of rays is uni- and biseriate with upright cells (4).

The phloem is fairly homogeneous, consisting mainly of round parenchyma cells and few smaller sieve tubes (5). Phloem and cortex are separated by an endodermis (5). The cortex contains large irregularly formed intercellulars.

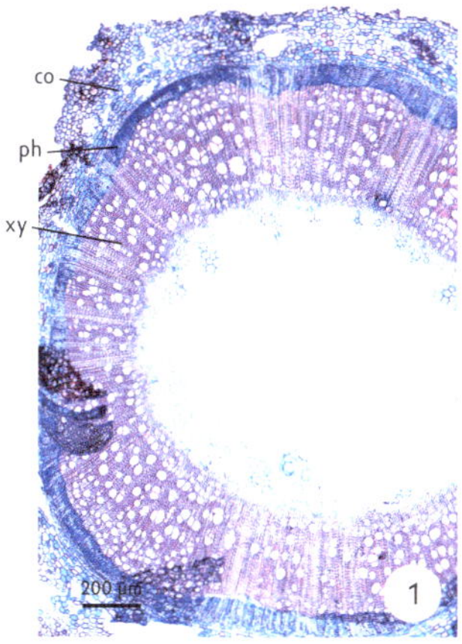

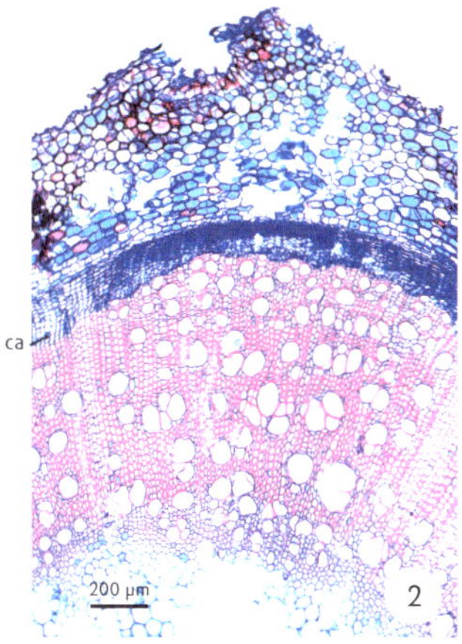

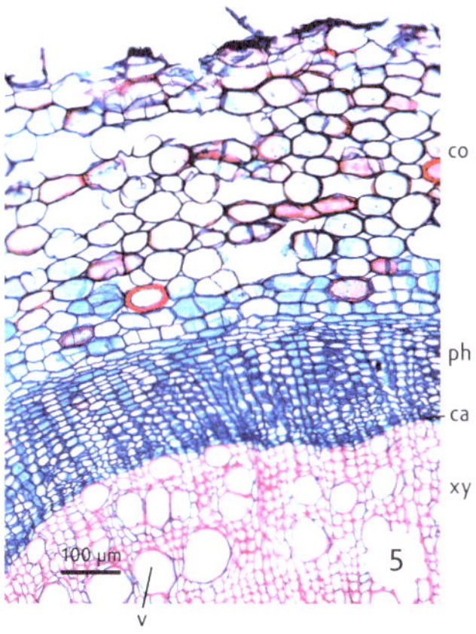

Mentha × verticillata L. = *Mentha aquatica × arvensis*

Lamiaceae
Whorled Mint

Perennial wetland herb, rhizomatous. Stem erect, 20–90 cm high. Plant usually sterile, pubescent, similar to M. aquatica in habit and scent but more robust. Sharp leaf teeth, calyx mostly <3.5 mm, stamens usually included.

Distribution and ecology

Frequent in mainland Europe, although rare in the Mediterranean region. Native in all CE countries on wet meadows and pastures, banks of rivers, ponds and ditches. Prefers disturbed ground. Aromatic and medicinal plant.

Anatomy of a basal part of the annual shoot

Round stem with a diameter of 2–3 mm. Maximum age 1 year.

A central part consists of a large pith and a continuous xylem-phloem-belt. A large cortex surrounds this part (1, 2). The xylem contains solitary round vessels with a diameter of 25–40 µm (2) and a length of 150–180 µm. Intervessel pits are small and round perforation plates are simple (3). Fibers are thin- to thick-walled (2). Septate fibers are absent (3). Pits in fibers are rare. Parenchyma is scanty paratracheal (2). Rays are uni- and biseriate (2) with upright cells.

The phloem is fairly homogeneous, consisting mainly of round parenchyma cells and few smaller sieve tubes (4). Phloem and cortex are separated by an endodermis (4). The cortex contains small groups of fibers and large irregularly formed intercellulars (4). The endodermis contains single prismatic crystals.

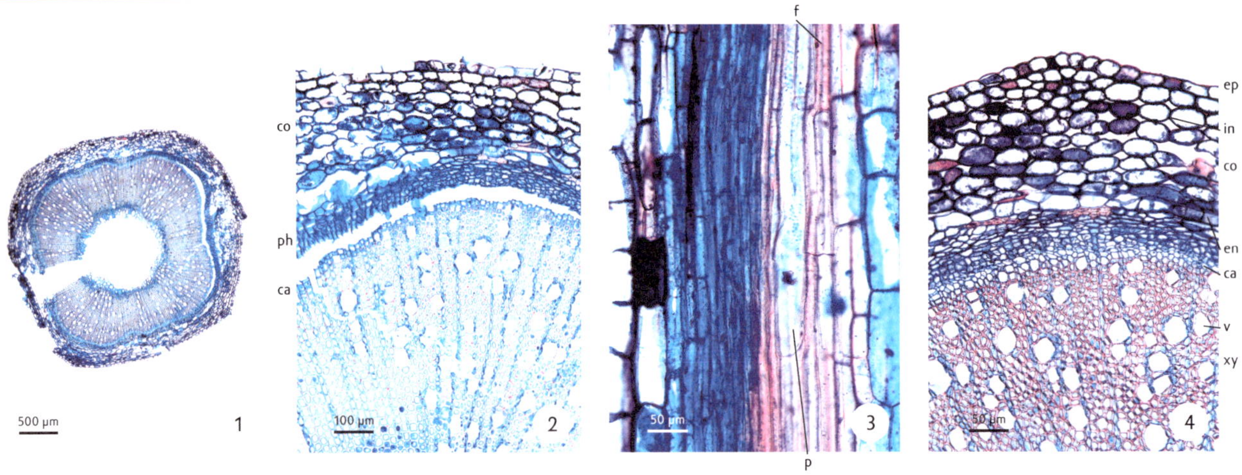

Perennial wetland herb. Stem ascending, feeble, usually branched, 20–70 cm high. Leaves oval-lanceolate, cordate-based, scalloped on the margin. Flowers in pairs in leaf axils. Calyx two-lipped, lips untoothed. Corolla two-lipped and blue-violet, corolla-tube long and slightly curved. Fruit consists of four dry nutlets.

Distribution and ecology

Euro-Siberian boreo-temperate element; also in N. America. Native in all CE countries. Typical of margins of standing waters, reeds, alder carrs, fen meadows and woodlands. Tolerates short-term flooding.

Anatomy of the annual shoot and the root collar

The annual shoot is quadrangular in the cross-section (1). The basal part of the stem has a diameter of 2–3 mm. Maximum age is up to 5 years (2).

The basal part of the stem consists of a large hollow pith, a small xylem and phloem and a large cortex (3). Ring boundaries are characterized by semi-porosity (2). The xylem contains vessels, thin- to thick-walled fibers, parenchyma cells and rays. Vessels are angular, arranged solitary and have a diameter of 40–60 µm and a length of 80–150 µm (5). Vessel-walls have round intervessel pits (5) and simple perforation plates (4). Fibers are thin- to thick-walled (4). Axial parenchyma is paratracheal (4). Rays are indistinct, 1–4-seriate (2, 3, 5) and consist of upright cells (5, 6).

Cortex and phloem of the annual shoot are separated by two layers of a poorly differentiated endodermis (2). Parenchyma cells and small groups of sieve tubes between dilating rays characterize the phloem (4, 7, 8). The inner, multilayered zone contains occasionally small thick-walled, lignified fibers (7). The cortex is composed of round and oval parenchyma cells and intercellulars (3, 7). The periphery of the annual shoot consists of an epidermis (8).

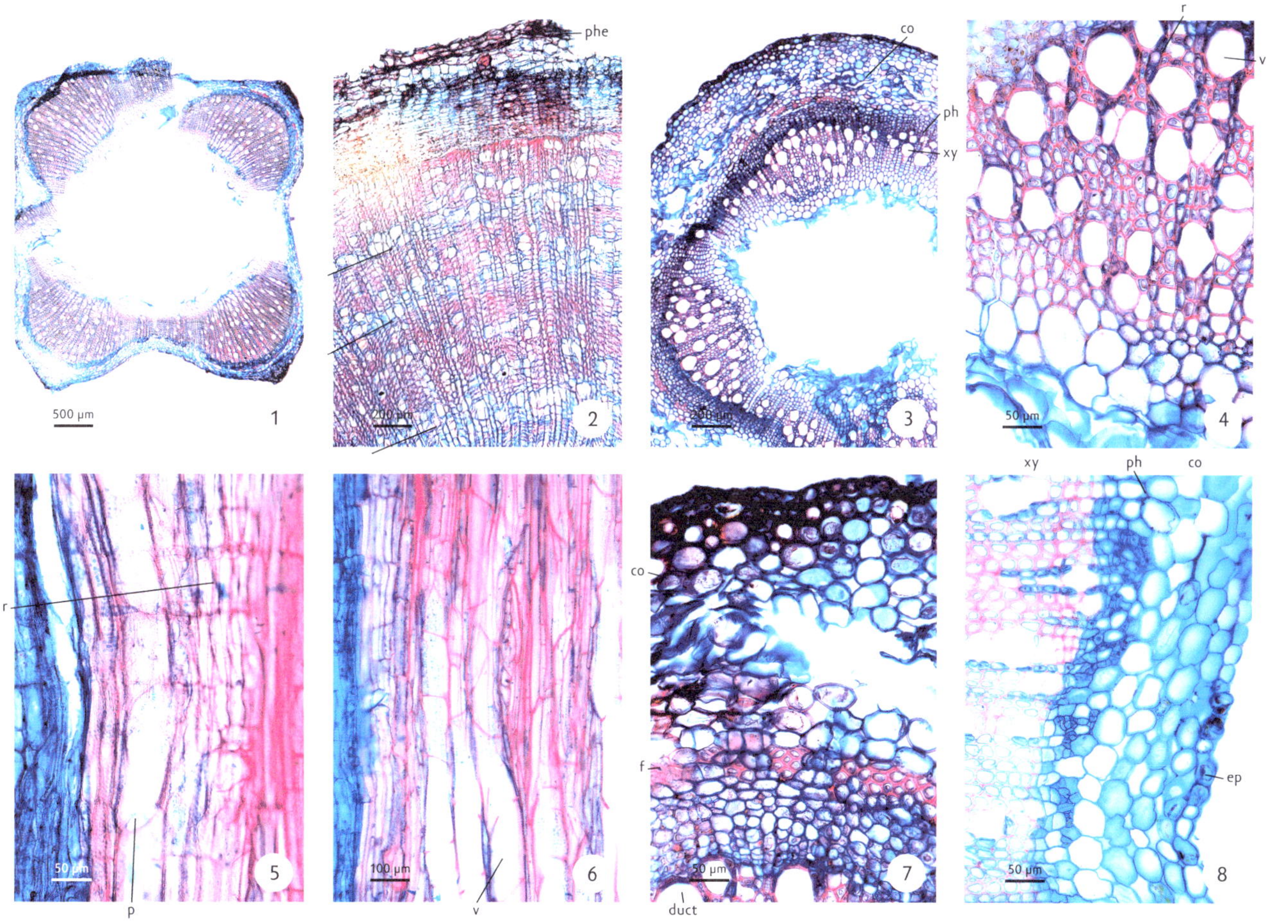

Stachys palustris L.

Marsh Woundwort

Perennial wetland herb with long, creeping subsurface rhizomes producing small tubers at the apex in autumn. Stems erect, sparsely to densely hairy, 30–120 cm high. Leaves opposite; oblong-ovate, acute at apex, cordate at base, appressed-pubescent on both surfaces. Inflorescence is a terminal spike, flowers in whorls. Corolla purplish red or purple, mottled with light and dark tones, bilabiate, lower lip 3-lobed. Fruit consists of four ellipsoid dry nutlets, rounded at apex.

Distribution and ecology

Circumpolar boreo-temperate element. Native in all CE countries. Typical of disturbed wetland habitats like banks of streams and rivers, by ponds and ditches, in fens and marshes, occasionally in cultivated fields. Tolerates short-term flooding. Releases an unpleasant odour when bruised.

Anatomy of the root collar

Square stem with a diameter 5–6 mm. Maximum age 1 year.

A large cortex surrounds this part (1). Vascular bundles remain. The xylem contains vessels with a diameter of 35–50 µm (2) and a length of 150–170 µm. Vessels are arranged solitary (2). Intervessel pits are small and round perforation plates are simple (3). Fibers are thin- to thick-walled (2). Septate fibers are present (3). Parenchyma is scanty paratracheal (2). Rays are uniseriate with upright cells within vascular bundles (4) and >10-seriate between them. Large rays are often confluent (1).

The phloem consists of parenchyma cells and groups of sieve tubes. Rays do not dilate (1). Phloem and cortex are separated by an indistinct endodermis (5). Fibers are absent (5, 6). Irregularly formed intercellulars characterise the cortex (5, 6). Short needle-shaped crystals are present on the cortex.

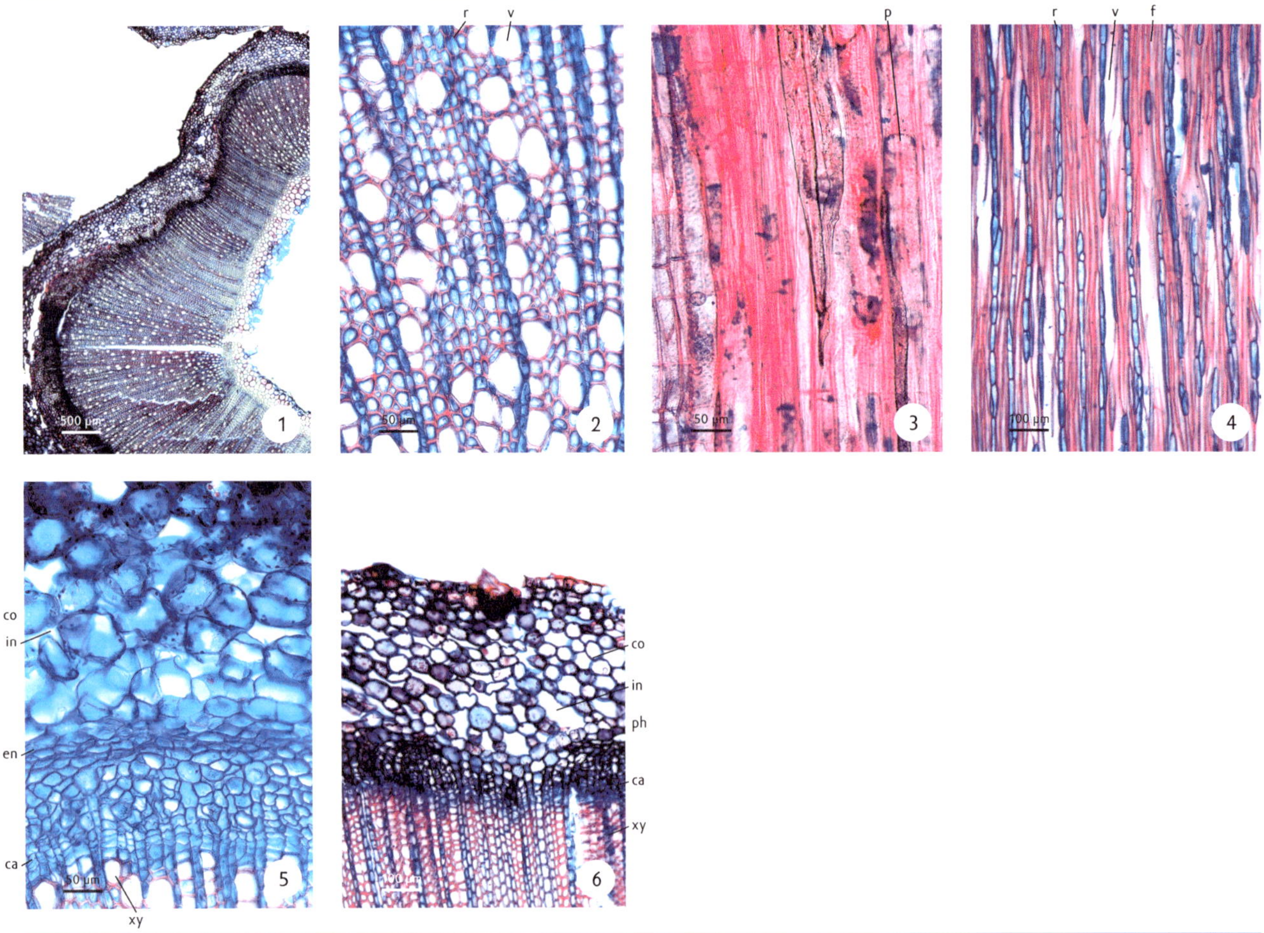

Pinguicula alpina L.

Alpine butterwort

Lentibulariaceae

Perennial carnivorous wetland herb, slimy leaves in basal rosette, 2–5 cm long, elliptic-oblong, with two types of glands: sessile and stalked, leaves usually yellow-green but some Alpine populations with red coloration of anthocyanins. Non-branched fibrous roots 4–10 cm long survive winter period. Solitary white zygomorphic flower with yellow spots on erect pedicel 5–10 cm long, ovate capsule 3 mm large. Fleshy and starchy hibernaculae (4–10 mm large) partly sunken to the ground by perennial roots.

Distribution and ecology

Euro-Siberian arctic-montane element. All CE countries except for CZ, extinct in HU. It grows very locally only in calcareous mountains over ca. 1000–1200 to 2000 m a.s.l. in the Carpathians and the Alps; here very common, abundant populations. Wet mountainous meadows, dripping rocks, always neutral calcareous soils or limestone rocks, facultative heliophyte and hygrophyte, tolerates drier substrates, very weak competitor.

Anatomy of the root collar

Irregularly formed 3 mm wide root collar. Maximum age 1 year.

The central part consists of a large pith and a continuous xylem-phloem-belt and large cortex (1). The structure of xylem is often difficult to observe because the longitudinal axis is twisted.

The xylem is constructed of parenchyma cells and lignified vessels (2, 3). Vessels are angular and arranged in small groups (2–4). Vessel diameter varies from 15–25 µm (2) and vessels are not longer than 50 µm (4). Intervessel pits are scalariform and reticulate with very large openings (4). Perforation plates are very small and simple (4). Fibers and rays are absent (3). Parenchyma is pervasive (3).

The phloem consists of small groups sieve tubes (2). Phloem and cortex are separated by an endodermis (2). The cortex is composed of large, thin-walled, roundish parenchyma. Intercellulars are absent (1–3). Walls of parenchyma cells in the cortex are net-like thickened (3). Medullary phloem is absent (2, 3).

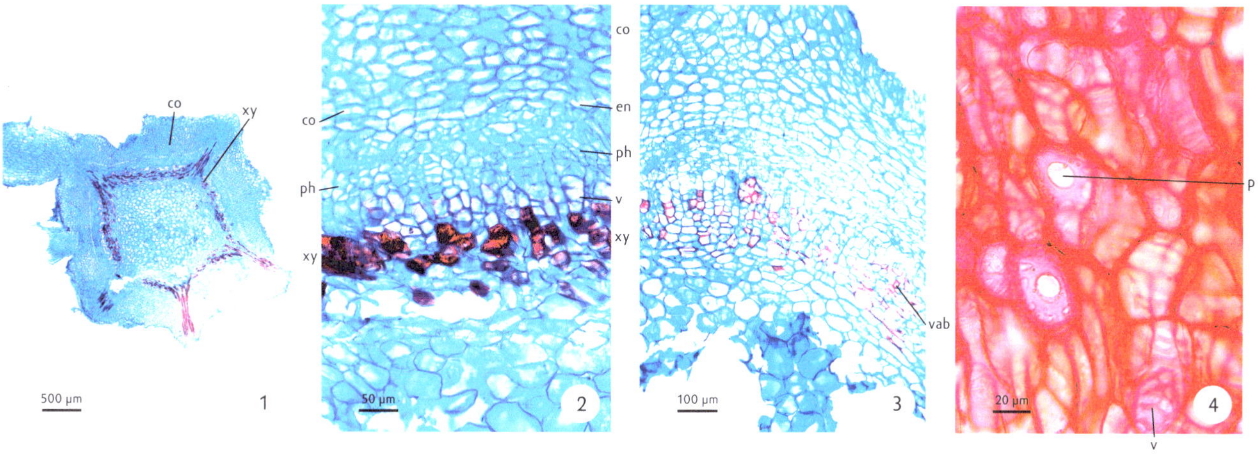

F. H. Schweingruber et al., *Anatomic Atlas of Aquatic and Wetland Plant Stems*, https://doi.org/10.1007/978-3-030-33420-8_33

Pinguicula vulgaris L.   **Common butterwort**   Lentibulariaceae

Perennial carnivorous wetland herb, leaves in basal rosette, 2–8 cm long slimy leaves ovate-oblong with two types of glands: sessile and stalked. Non-branched fibrous roots 3–5 cm long, solitary violet zygomorphic flower with whitish throat on erect pedicel 5–18 cm long, ovate capsule 3–4 mm large, 4–10 mm large, fleshy and starchy hibernaculae partly sunken to the ground by roots.

Distribution and ecology

Circumpolar boreo-montane element, with a disjunct distribution. Rare in all 7 countries, scattered irregularly only in wetland and peaty areas, more common in highlands. Peatbogs, fens, mossy springs, dripping mossy rocks, wet peaty meadows in lowlands and highlands, lowland sites usually strongly endangered by eutrophication and overgrowing by tall vegetation. Relatively eurytopic, organic substrates and wet sand, tolerates strongly acidic and neutral substrates (pH 3.5–7) and different level of shading. Vegetative propagation by small daughter hibernaculae prevails over the reproduction by seed.

Anatomy of the root collar

Irregularly formed 2 mm wide root collar. Maximum age 1 year.

The central part consists of a large pith and a continuous xylem-phloem-belt and large cortex (1). The structure of xylem is often difficult to observe because the longitudinal axis is twisted.

The xylem is constructed of parenchyma cells and poorly lignified vessels (2). Vessels are angular and arranged solitary and in small groups (2, 3). Vessel diameter varies from 15–25 µm (2) and vessels are not longer than 50 µm (4). Intervessel pits are scalariform and reticulate with very large openings (3). Perforation plates are very small and simple (4). Fibers are absent. Parenchyma is pervasive (3). Rays are absent (2).

The phloem consists of small groups of sieve tubes (3). Equal groups occur just inside the xylem (medullary phloem) (5). Phloem and cortex are separated by an endodermis (3). The cortex is composed of large, thin-walled unlignified, roundish parenchyma. Intercellulars are absent (6). Walls of parenchyma cells in the cortex are net-like thickened (6).

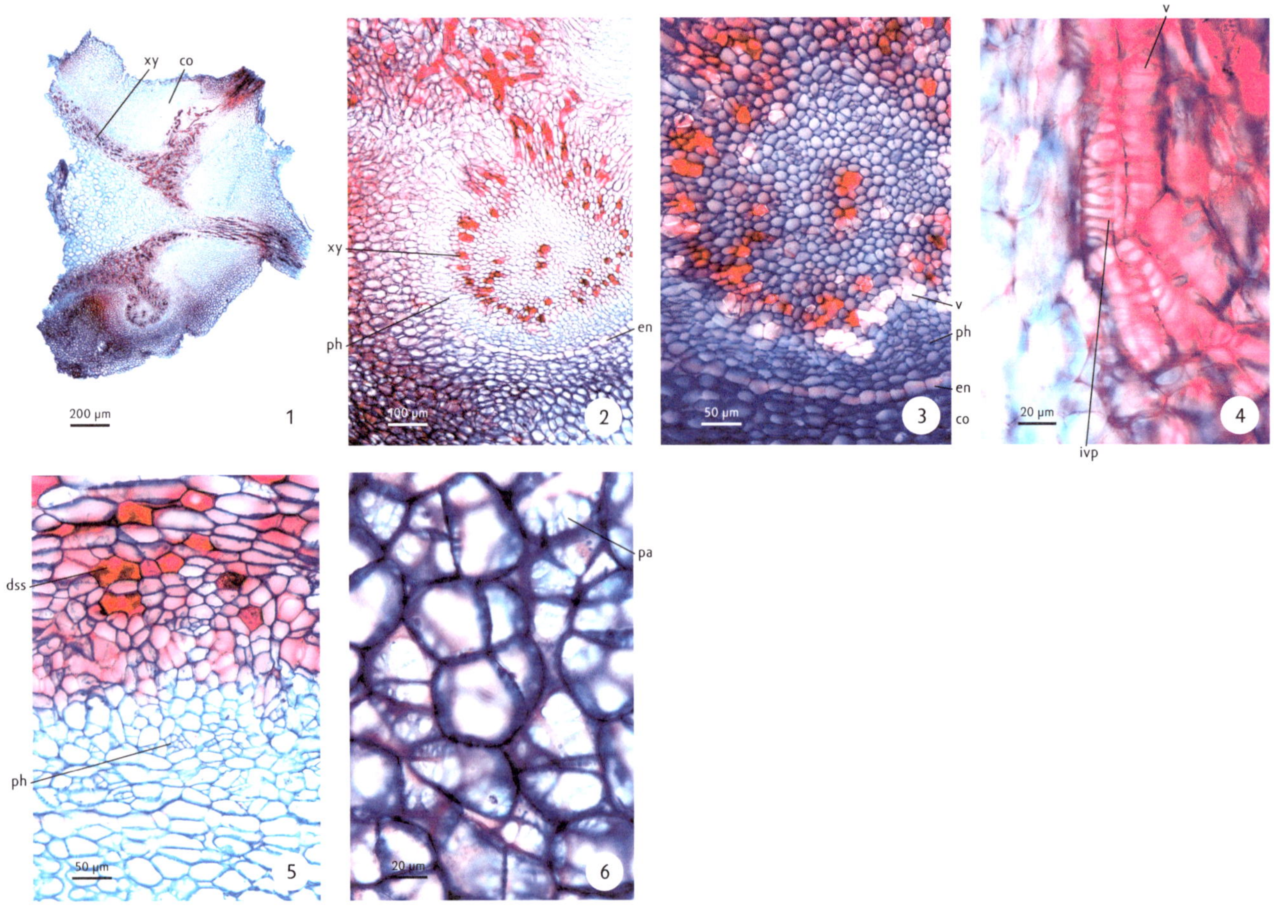

Utricularia australis R.Br. (syn. *U. neglecta*) **Southern bladderwort** Lentibulariaceae

Perennial, rootless submerged carnivorous plant with thin, linear, poorly branched shoot, free-floating. Shoots 8–150 cm long, regular leaf nodes, thin, branched filamentous leaves 1.5–4 cm long. Dozens of suction traps 1–4 mm large of foliar origin, sometimes very thin 'air shoots' with stomata. Inflorescence with yellow flowers 15–20 mm large above water on flower peduncle, all capsules sterile. Overwintering in 2–12 mm large, circular-shaped dormant turions.

Distribution and ecology

Eurasian boreo-temperate element. Cosmopolitan spread, Old World continents, relatively very common in all 7 countries except for highlands above ca. 700 m a.s.l. or warm lowlands with hard water. Occurs in lakes, fishponds, dam reservoirs, backwater pools and oxbows, peatbogs, fen lakes, sand-pits, usually standing shallow, slightly or strongly dystrophic waters, littorals in loose reed or sedge stands, dwarfish form survives terrestrial conditions. Prefers soft to medium hard waters (pH 5.5–9, total alkalinity 0.2–3 meq/l), from oligo- to strong eutrophy. Ecologically extremely plastic, very tolerant of irradiation, water chemistry, transparency and water depth. Strict CO_2 user. Spread commonly by water birds to new sites.

Flower stalk and submerged shoot

Round flower stalk (1, 2) and submerged shoot with diameters of 2 mm (3, 6). Maximum age 1 year.

Flower stalk: The large pith contains small vascular bundles. It is surrounded by a small continuous lignified fiber belt, a large aerenchymatic cortex and an epidermis (1). Single irregularly formed vascular bundles are in the pith (2). Xylem and phloem are difficult to distinguish. Xylem with few 15–20 µm wide, unlignified angular vessels. Homogeneous phloem (2). Cortex with large radial, oval aerenchymatic intercellulars.

Shoot: Xylem and phloem within the central cylinder (stele) are difficult to distinguish (3–5). Xylem with few, isolated unlignified vessels (4) with annular thickenings or scalariform pits with very wide openings (6). Vessel diameter varies from 15–20 µm (4). Perforations absent (6). Parenchyma pervasive (4). Rays absent (4). Phloem and cortex are separated by an endodermis without Casparian strips (4). Cortex with large radial, oval aerenchymatic intercellulars (3).

FLOWER STALK

SUBMERGED SHOOT

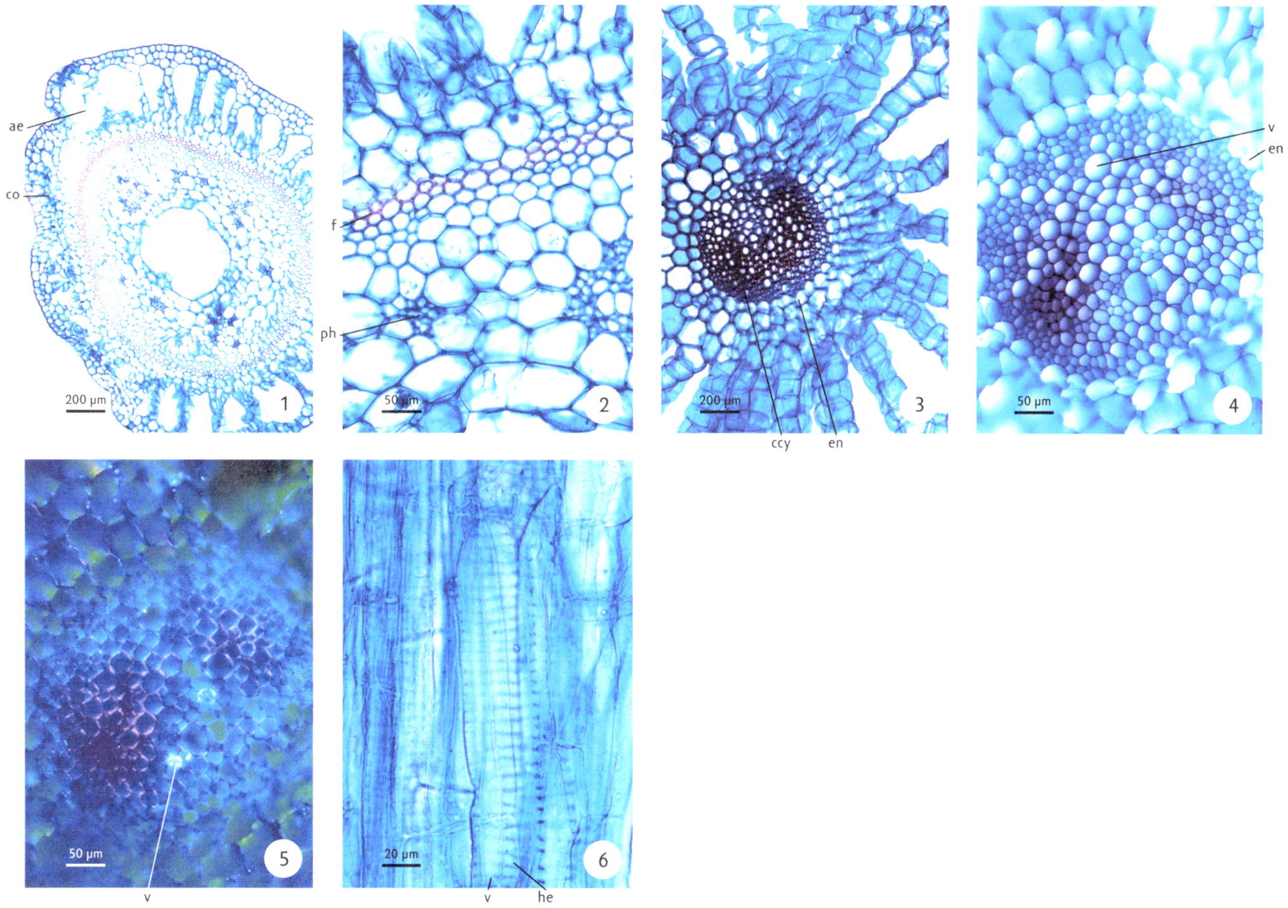

Utricularia bremii Heer

Perennial, rootless amphibious aquatic carnivorous plant with thin, linear, poorly branched shoot, free-floating or weakly attached to bottom, shoots also terrestrial on wet substrates. Shoot 10–50 cm long, regular leaf nodes, thin, branched filamentous leaves 4–14 mm long. Several suction traps 1–3 mm large. Light-yellow flowers 8–10 mm large above water on flower peduncle. Capsules sterile (at least in CE). Overwintering in 1–4 mm large turions.

Distribution and ecology

European endemite. Very rare, irregular spread in CE countries, a few sites in CZ, SK, PL, DE and CH; probably absent in AT and HU. In CZ, successfully introduced to 3 suitable sites in sand-pits. Occurs in lakes, peaty fishponds, peatbogs, fen lakes, forest pools, shallow sand-pits, standing very shallow, slightly or strongly dystrophic waters, littorals in loose reed or sedge stands or sandy/clayish bottom, also terrestrial form on very wet substrate. Prefers very soft to medium soft, acidic, oligo-mesotrophic waters (pH 5-7, total alkalinity 0.0–0.6 meq/l). Ecology mostly unknown, very tolerant of irradiance, strict CO_2 user, flowering rather rare.

Flower stalk and submerged shoot

Round flower stalk (1, 2) and submerged shoot with diameters of <1 mm (3–6). Maximum age 1 year.

Flower stalk: The large pith contains small vascular bundles. It is surrounded by a biseriate continuous lignified fiber belt, an aerenchymatic cortex and an epidermis (1, 2). Single irregularly formed vascular bundles are in the pith. Xylem and phloem are difficult to distinguish. Vessels probably absent (2). Homogeneous phloem (2). Cortex with tangential, oval aerenchymatic intercellulars.

Shoot: Xylem and phloem within the central cylinder (stele) are difficult to distinguish (3–5). Xylem with few, isolated unlignified vessels (4, 5) with annular thickenings or scalariform pits with very wide openings (6). Perforations absent (6). Vessel diameter varies from 10–14 µm (4). Parenchyma pervasive (4). Rays absent (4). Phloem and cortex are separated by an endodermis without Casparian strips (3, 4). Cortex with large aerenchymatic intercellulars (3).

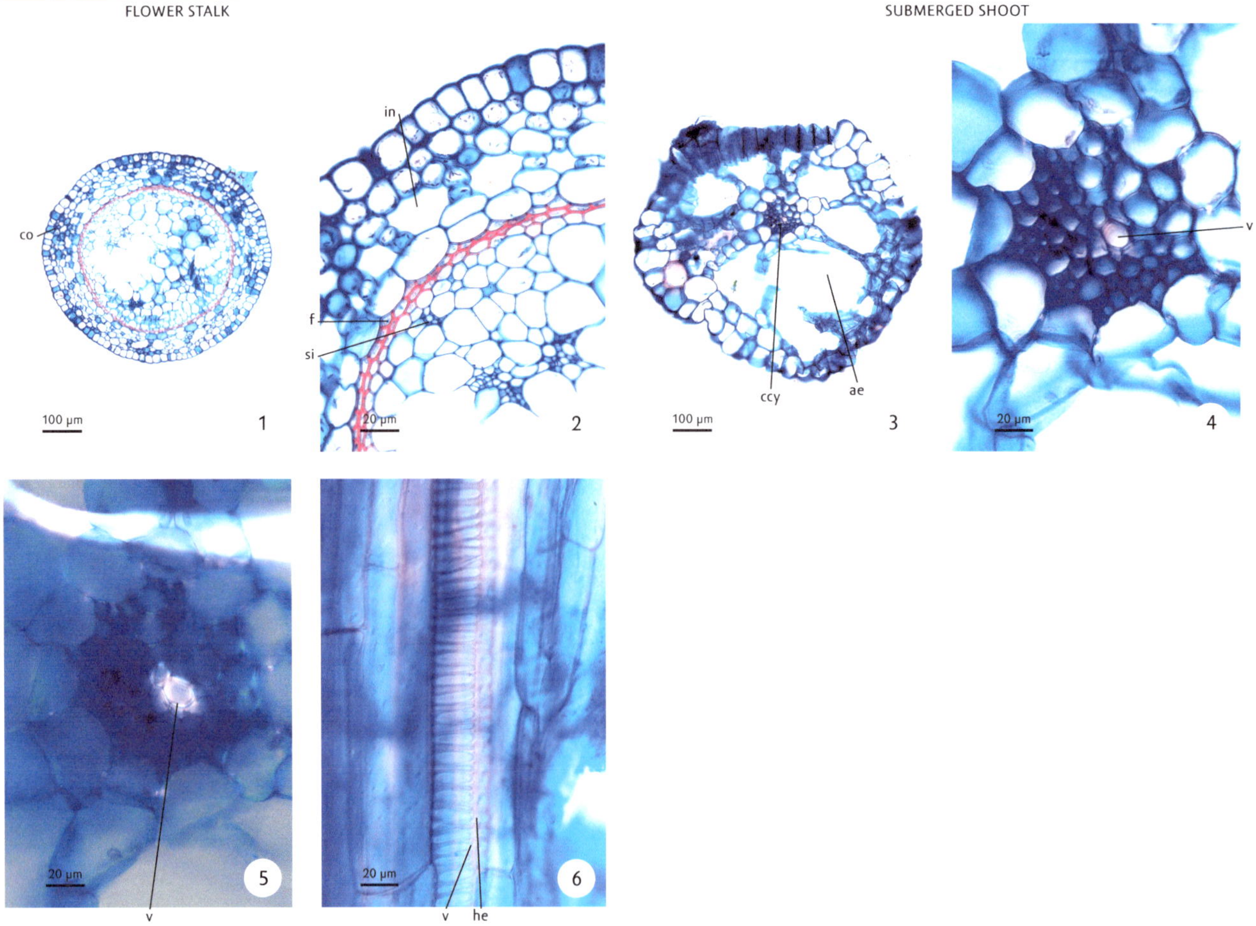

Utricularia intermedia Hayne

Intermediate bladderwort Lentibulariaceae

Perennial, rootless amphibious aquatic carnivorous plant with thin, linear, poorly branched shoot, free-floating or weakly attached to bottom or terrestrial on wet substrates. Shoot 20–40 cm long, regular leaf nodes, shoots distinctly differentiated in green photosynthetic ones without traps and pale carnivorous shoots with numerous traps 2–5 mm long. Inflorescence with a few yellow flowers 12–15 mm large above water on flower peduncle, capsule ca. 3 mm large. Overwintering in 3–7 mm large, oval or oblonged turions with flat leaflets.

Distribution and ecology

Circumpolar boreo-montane element. Very irregular and rare spread in all CE countries, only in areas with peatbogs or acidic fens, usually only a few sites in each country; lacking in SK, HU, more common in PL and CH. Occurs in lakes, peaty fishponds, peatbogs, fen lakes, standing very shallow, dystrophic waters, littorals in loose reed or sedge stands or *Sphagnum* hummocks. Prefers shaded microhabitats, very soft to slightly hard, always acidic (pH 4–6), oligo- to slightly eutrophic waters with high CO_2 concentration. Stenotopic, strict CO_2 user. Flowering and seed set common.

Flower stalk and submerged shoot

Round flowering stalk (1, 2) and submerged shoot with diameters of <1 mm (3–6). Maximum age 1 year.

Flower stalk: The large pith contains small vascular bundles. It is surrounded by a 2–3-seriate continuous lignified fiber belt, an aerenchymatic cortex and an epidermis (1, 2). Single irregularly formed vascular bundles are in the pith. Xylem and phloem are difficult to distinguish. Vessels unlignified, thin-walled (2). Homogeneous phloem (2). Cortex with large aerenchymatic intercellulars (1, 2).

Shoot: Xylem and phloem within the central cylinder (stele) are difficult to distinguish (3–5). Xylem with a single, isolated unlignified vessel (4, 5) with annular thickenings or scalariform pits with very wide openings (6). Perforations absent (6). Vessel diameter varies from 15–20 µm (4). Parenchyma pervasive (4). Rays absent (4). Phloem and cortex are separated by an endodermis without Casparian strips (3, 4). Cortex with large radial, oval aerenchymatic intercellulars and many prismatic crystals (3).

FLOWER STALK SUBMERGED SHOOT

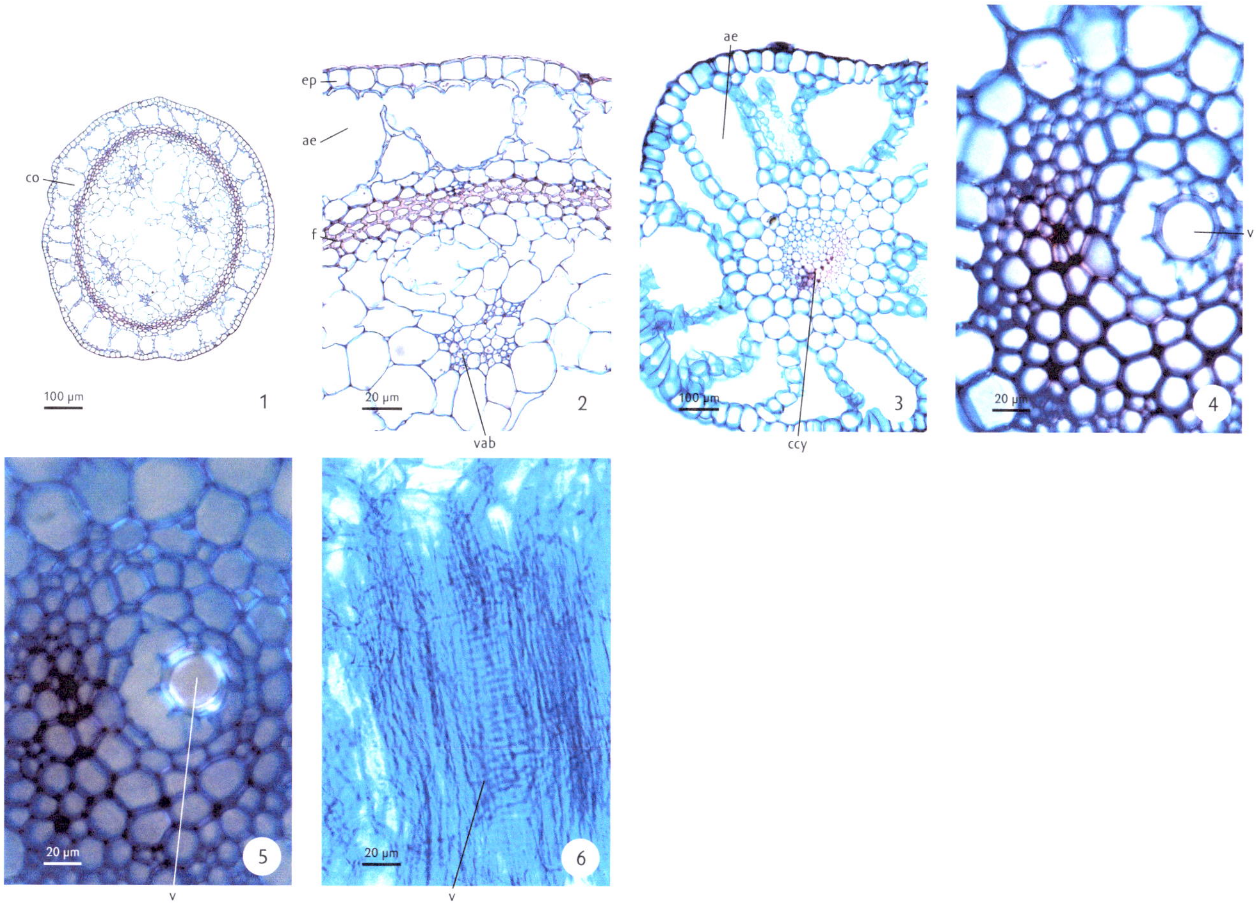

Utricularia minor L.

Lesser bladderwort

Perennial, rootless amphibious aquatic carnivorous plant with thin, linear, poorly branched shoot, free-floating or weakly attached to the bottom. Shoot also terrestrial on wet substrates, 8–40 cm long, regular leaf nodes, thin, branched filamentous leaves 3–12 mm long with several suction traps 1–2.5 mm large. Inflorescence with a few light-yellow flowers 8 mm large above water on flower peduncle. Overwintering in turions 1–4 mm large.

Distribution and ecology

Circumpolar boreo-temperate element. Scattered, irregular spread in all CE countries, usually dozens of sites in each country, rarer in HU, AT and DE. Occurs in lakes, peaty fishponds, peatbogs, fen lakes, forest pools, shallow sand-pits, standing very shallow, slightly or strongly dystrophic waters in littorals, loose reed or sedge stands or sandy/clayish bottom. Prefers very soft to medium hard, acidic to alkaline (pH 3.5–9.2, total alkalinity 0.0–2 meq/l), oligo- to slightly eutrophic waters. Slightly eurytopic, tolerates very wide amplitudes of habitat factors, strict CO_2 user.

Flower stalk and submerged shoot

Round flower stalk (1, 2) and submerged shoot with diameters of <1 mm (3). Maximum age 1 year.

Flower stalk: The large pith contains small vascular bundles. It is surrounded by a 2-3-seriate continuous lignified fiber belt, an aerenchymatic cortex and an epidermis (1). Single irregularly formed vascular bundles are in the pith (1). Xylem and phloem are difficult to distinguish. Vessels unlignified, thin-walled (2). Homogeneous phloem (1). Cortex with aerenchymatic intercellulars (1).

Shoot: Xylem and phloem within the central cylinder (stele) are difficult to distinguish (3). Xylem with unlignified vessels with annular thickenings or scalariform pits with very wide openings. Perforations absent. Vessel diameter varies from 10–15 µm. Parenchyma pervasive. Rays absent. Phloem and cortex are separated by an endodermis without Casparian strips (3). Cortex with large radial, oval aerenchymatic intercellulars (3).

FLOWER STALK

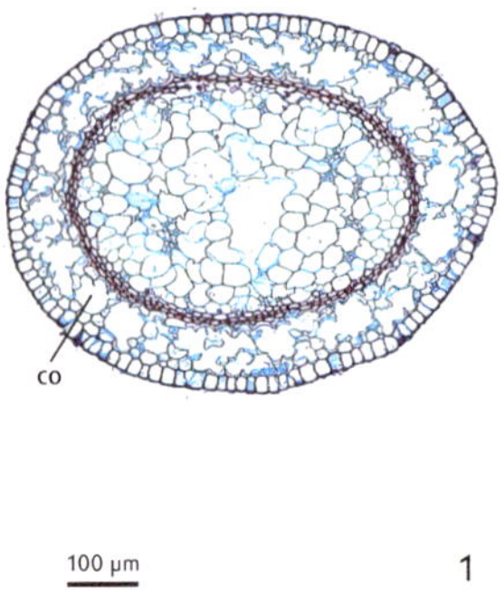

1

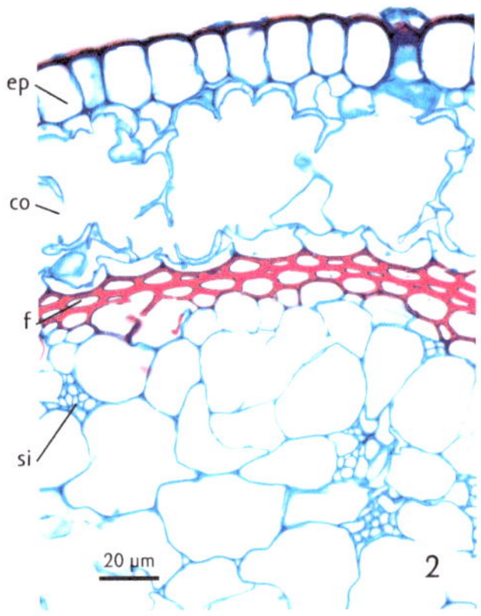

2

SUBMERGED SHOOT

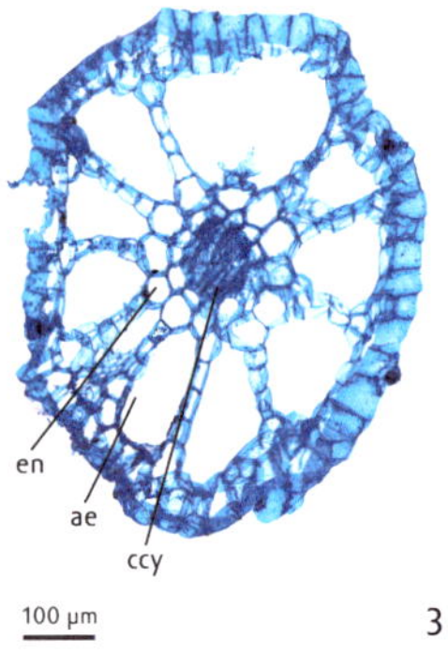

3

Utricularia ochroleuca Hartm. **Pale bladderwort** Lentibulariaceae

Perennial, rootless amphibious aquatic carnivorous plant with thin, linear, poorly branched shoot, free-floating, weakly attached to bottom or commonly terrestrial on wet substrates. Shoots 20–50 cm long, regular leaf nodes, distinctly differentiated in green photosynthetic with only a few traps, and pale carnivorous shoots with numerous traps 2–5 mm long. Inflorescence with a few yellow flowers 10–12 mm large above water on flower peduncle, plants sterile. Overwintering in oval or eliptic turions 2–8 mm large with flat leaflets.

Distribution and ecology

Species differentiated from very similar *U. stygia* since the end of the 1990's, still the exact distribution partly unknown, much rarer than *U. stygia*. Circumpolar boreo-montane element. Irregular and extremely rare spread in all CE countries, only areas with peatbogs or acidic fens, only a very few sites in each country; lacking in SK and HU, 4 small sites in CZ. Occurs in lakes, peaty fishponds, peatbogs, fen lakes, pools after peat extraction, standing very shallow (0–20 cm) dystrophic waters in littorals, loose reed or sedge stands or on *Sphagnum*. Prefers very soft to slightly hard, acidic to neutral (pH 4–7), oligo- to slightly eutrophic waters with high CO_2 concentration. Stenotopic, strict CO_2 user. Flowering common, plants sterile due to pollen malformation.

Flower stalk and submerged shoot

Round flower stalk and submerged shoot with diameters of <1 mm. Maximum age 1 year.

Flower stalk: The large pith contains small vascular bundles. It is surrounded by a 2–3-seriate continuous lignified fiber belt, an aerenchymatic cortex and an epidermis (1). Single irregularly formed vascular bundles are in the pith (1). Xylem and phloem are difficult to distinguish. Vessels unlignified, thin-walled. Homogeneous phloem. Cortex with large aerenchymatic intercellulars (1).

Shoot: Xylem and phloem within the central cylinder (stele) are difficult to distinguish (2, 3). Xylem with a few isolated poorly lignified angular vessels (3) with annular thickenings or scalariform pits with very wide openings (4). Perforations absent (4). Vessel diameter varies from 15–20 µm (3). Parenchyma pervasive (4). Rays absent (4). Phloem and cortex are separated by an endodermis without Casparian strips (2). Cortex with large, oval aerenchymatic intercellulars (3).

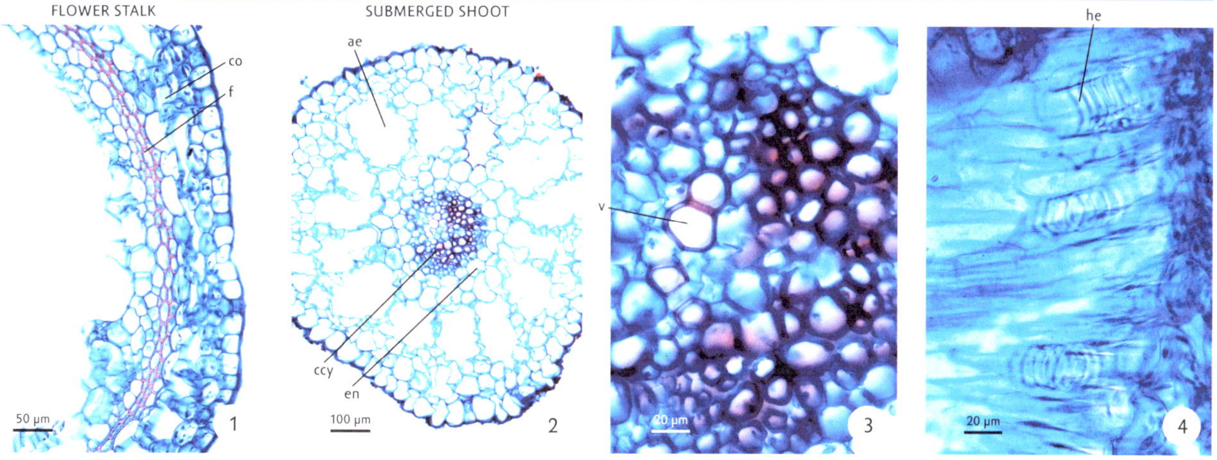

Utricularia stygia Thor (syn. *U. ochroleuca* s. lato) **Nordic bladderwort** Lentibulariaceae

Perennial, rootless amphibious aquatic carnivorous plant with thin, linear, poorly branched shoot, free-floating, weakly attached to bottom or commonly terrestrial on wet substrates. Shoots 20–50 cm long, regular leaf nodes, distinctly differentiated in green photosynthetic with only a few traps, and pale carnivorous shoots with numerous traps 2–5 mm long. Inflorescence with a few yellow flowers 12–14 mm large above water on flower peduncle, plants sterile. Overwintering in oval or eliptic turions 2–8 mm large with flat leaflets.

Distribution and ecology

Species differentiated from very similar *U. ochroleuca* s. str. since the end of the 1990's, still the exact distribution partly unknown, more common than *U. ochroleuca*. Circumpolar boreo-montane element. Irregular and very rare spread in all CE countries, only areas with peatbogs or acidic fens, only a very few sites in each country; lacking in SK and HU, 8 sites in CZ. Occurs in lakes, peaty fishponds, peatbogs, fen lakes, pools after peat extraction, standing very shallow (0–20 cm) dystrophic waters in littorals, loose reed or sedge stands or on *Sphagnum*. Prefers very soft to slightly hard, acidic to neutral (pH 4–7), oligo- to slightly eutrophic waters with high CO_2 concentration. Stenotopic, strict CO_2 user. Flowering common, plants sterile due to pollen malformation.

Flower stalk and submerged shoot

Round flower stalk and submerged shoot with diameters of <1 mm. Maximum age 1 year.

Flower stalk: The large pith contains small vascular bundles. It is surrounded by a 2–3-seriate continuous lignified fiber belt, an aerenchymatic cortex and an epidermis (1, 2). Single irregularly formed vascular bundles are in the pith (1, 2). Xylem and phloem are difficult to distinguish. Vessels are probably absent (2). Homogeneous phloem (2). Cortex with tangential oval aerenchymatic intercellulars (2).

Shoot: Xylem and phloem within the central cylinder (stele) are difficult to distinguish (3, 4). Xylem with a few, isolated unlignified vessel (3, 4) with annular thickenings or scalariform pits with very wide openings. Perforations absent. Vessel diameter varies from 15–20 µm (4), with thin- to thick-walled fibers. Parenchyma pervasive (4). Rays absent (4). Phloem and cortex are separated by an endodermis without Casparian strips (3, 4). Cortex with large, oval aerenchymatic intercellulars (3).

FLOWER STALK — SUBMERGED SHOOT

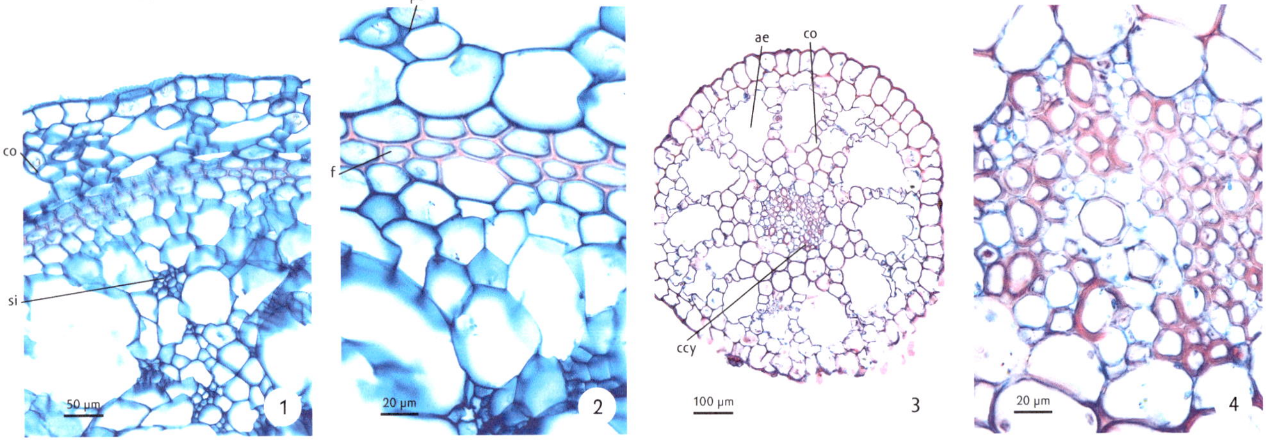

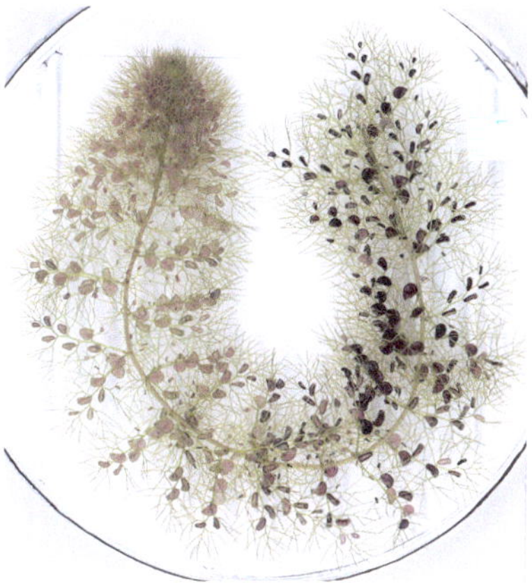

Perennial, rootless submerged carnivorous plant with thin, linear, poorly branched shoot, free-floating. Shoot 50–250 cm long, regular leaf nodes, thin, branched filamentous leaves 2–5 cm long. Dozens of suction traps 1–5 mm large, very thin 'air shoots' with stomata on scales formed at O_2 shortage. Inflorescence with yellow flowers 15–20 mm large above water on flower peduncle, oval capsules 4 mm large, circular-shaped. Overwintering in slimy dormant turions 2–25 mm large.

Distribution and ecology

Euro-Siberian temperate element. Common to rare in all 7 countries, mainly warm lowlands with hard water, number of sites decreasing in agricultural areas. Occurs in lakes, fishponds, dam reservoirs, backwater pools and oxbows of big rivers, peatbogs, fen lakes, usually in standing shallow or deeper (or slowly streaming), slightly or strongly dystrophic waters in littorals. Rather stenotopic, thermophilous, requires strictly medium to strongly hard, neutral or slightly alkaline, base-rich waters (pH 7–8, total alkalinity 1.5–6 meq/l), from meso- to slight eutrophy. Strict CO_2 user. Poorly tolerates intensive fishpond management and strong eutrophication. Flowering and seed set very common.

Flower stalk and submerged shoot

Round flower stalk and submerged shoot with diameters of 1–2 mm. Maximum age 1 year.

Flower stalk: The large pith contains small vascular bundles. It is surrounded by a 2–3-seriate continuous lignified fiber belt, an aerenchymatic cortex and an epidermis (1, 2). Single irregularly formed vascular bundles in the pith (1, 2). Xylem and phloem are difficult to distinguish. Vessels are probably absent (2). Homogeneous phloem (2). Cortex with large aerenchymatic intercellulars (1).

Shoot: Xylem and phloem within the central cylinder (stele) are difficult to distinguish (4, 5). Xylem with a circularly arranged, solitary unlignified vessel (4, 5) with annular thickenings or scalariform to reticulate pits with very wide openings (6, 7). Perforations absent (6). Vessel diameter varies from 10–15 µm (4). Fibers absent (5). Parenchyma pervasive (4). Rays absent (4). Phloem and cortex are separated by an endodermis without Casparian strips (3, 4). Cortex with irregularly formed aerenchymatic intercellulars and few prismatic crystals (3).

FLOWER STALK SUBMERGED SHOOT

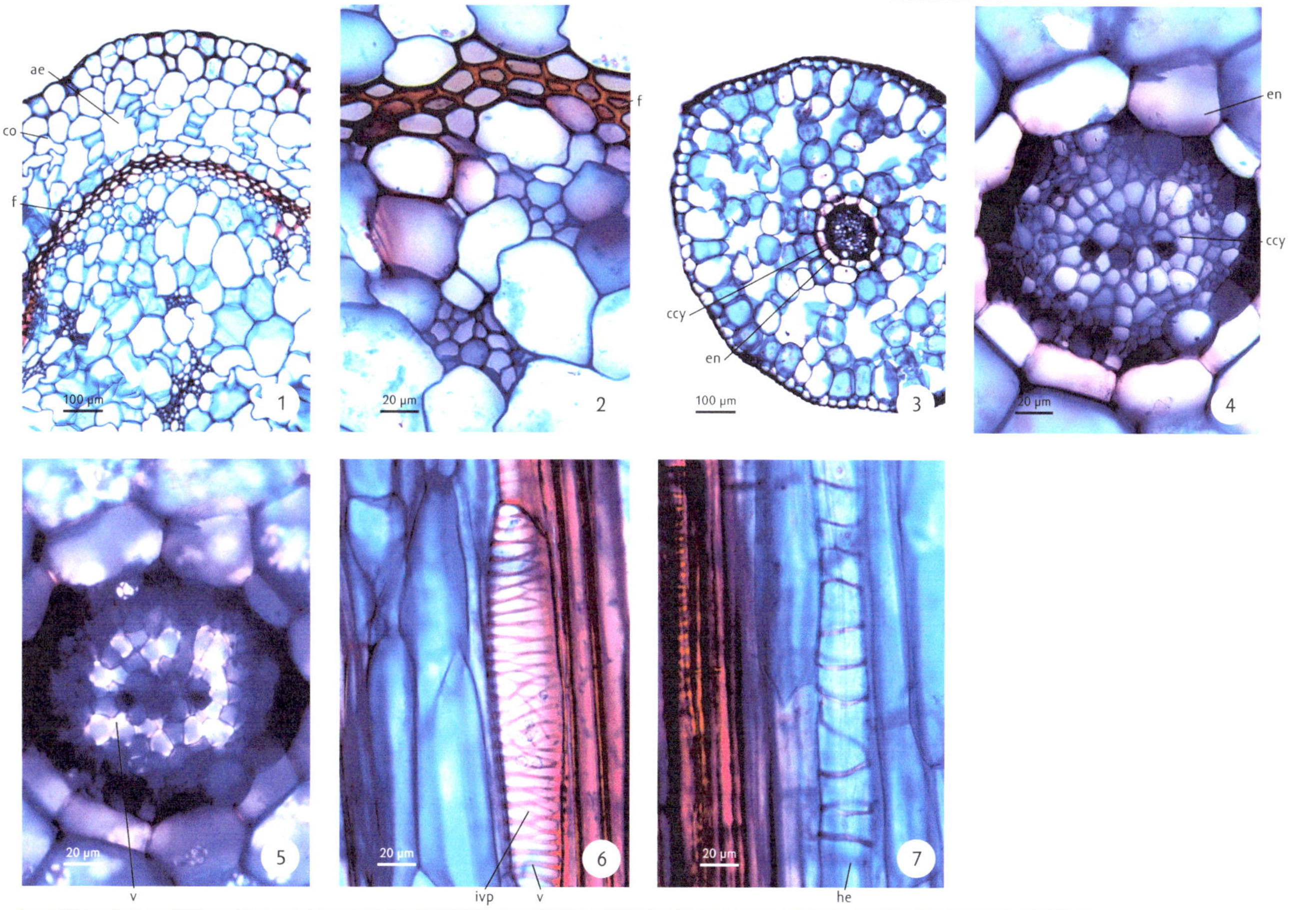

Annual wetland miniature herb. Extremely slender stems (2–6 cm long) much branched, opposite ovate to elliptic leaves. Numerous tiny flowers with white corolla. Fruit is a rounded capsule.

Distribution and ecology

European temperate distribution. Scattered irregularly to rarely in all CE countries except CH, in lowlands to uplands. Bare acid sandy or peaty wet soils, margins of fishponds, fields or abandoned meadows, old sand-pits, woodland rides, mainly in fishpond areas. Strict heliophyte, competitively very week, pioneering species on disturbed areas.

Anatomy of the root collar

Round stem with a diameter of <1mm. Maximum age 1 year.

The root collar consists of poorly lignified pith, a dense totally lignified xylem belt, a small phloem and a large cortex (1, 2, 5). Vessels are arranged solitary and in short radial rows. They have a diameter of 10–15 μm and a length of approximately 100 μm (3). Intervessel pits are small, round (4) and perforation plates are simple (4). Fibers are living (5) and thick-walled (2) with septate fibers. Parenchyma is paratracheal and apotracheal (5). Rays are absent (2, 5).

Phloem and cortex are separated by a small endodermis (5). The phloem consists of groups of sieve tubes, parenchyma cells and a few solitary fibers (5). The large cortex consists of large round to oval, unlignified thin-walled parenchyma cells (5). The phellem is absent (1).

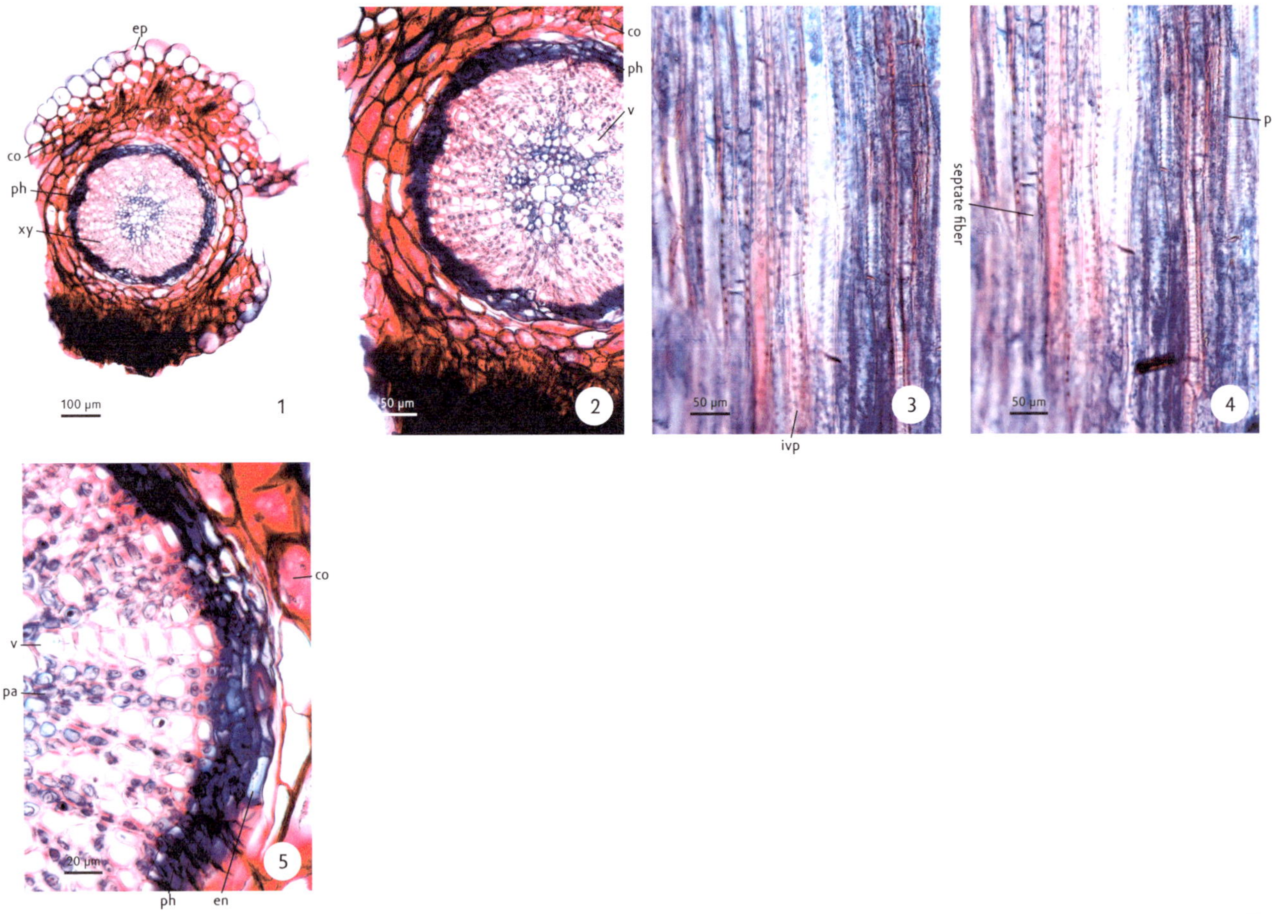

F. H. Schweingruber et al., *Anatomic Atlas of Aquatic and Wetland Plant Stems*, https://doi.org/10.1007/978-3-030-33420-8_34

Lindernia procumbens (Krock.) Borbás

Annual wetland herb. Erect to procumbent stems 5–20 cm long, ovoid opposite leaves 10–20 mm long bear single flower stalks in leaf axils. Corolla is white or pinkish. Fruit is a capsule.

Distribution and ecology

Eurasian distribution, secondarily in N. America and Australia. Scattered irregularly and rarely in all CE countries, in lowlands to uplands, wet denudated shores of standing and streaming waters, usually in fishpond basins or floodplains of big rivers, muddy or sandy soils. Strict heliophyte, competitively very week, pioneering species on summer-drained fishpond bottoms.

Anatomy of the flower stalk and the root collar

Round flower stalk with a diameter of 1 mm. Maximum age 1 year (1).

The basal part of the stem consists of a pith, a xylem, a small phloem, a large cortex and an epidermis (2).

The xylem contains vessels, thin- to thick-walled fibers, parenchyma cells and rays (2). Vessels are angular and are arranged in radial rows (2). Vessels have a diameter of 15–20 µm and length of 60–100 µm. Intervessel pits are round (3) and a perforation simple (4). Axial parenchyma is paratracheal (2). Rays are absent (2, 5).

Cortex and phloem are separated by two layers of a poorly differentiated endodermis. Parenchyma cells and small groups of sieve tubes characterize the phloem (2, 6). Few fibers occur at the peripheral side of the phloem (6). The cortex and the pith represent a net-like aerenchyma with large intercellulars (1). The periphery of the stem consists of an epidermis (1).

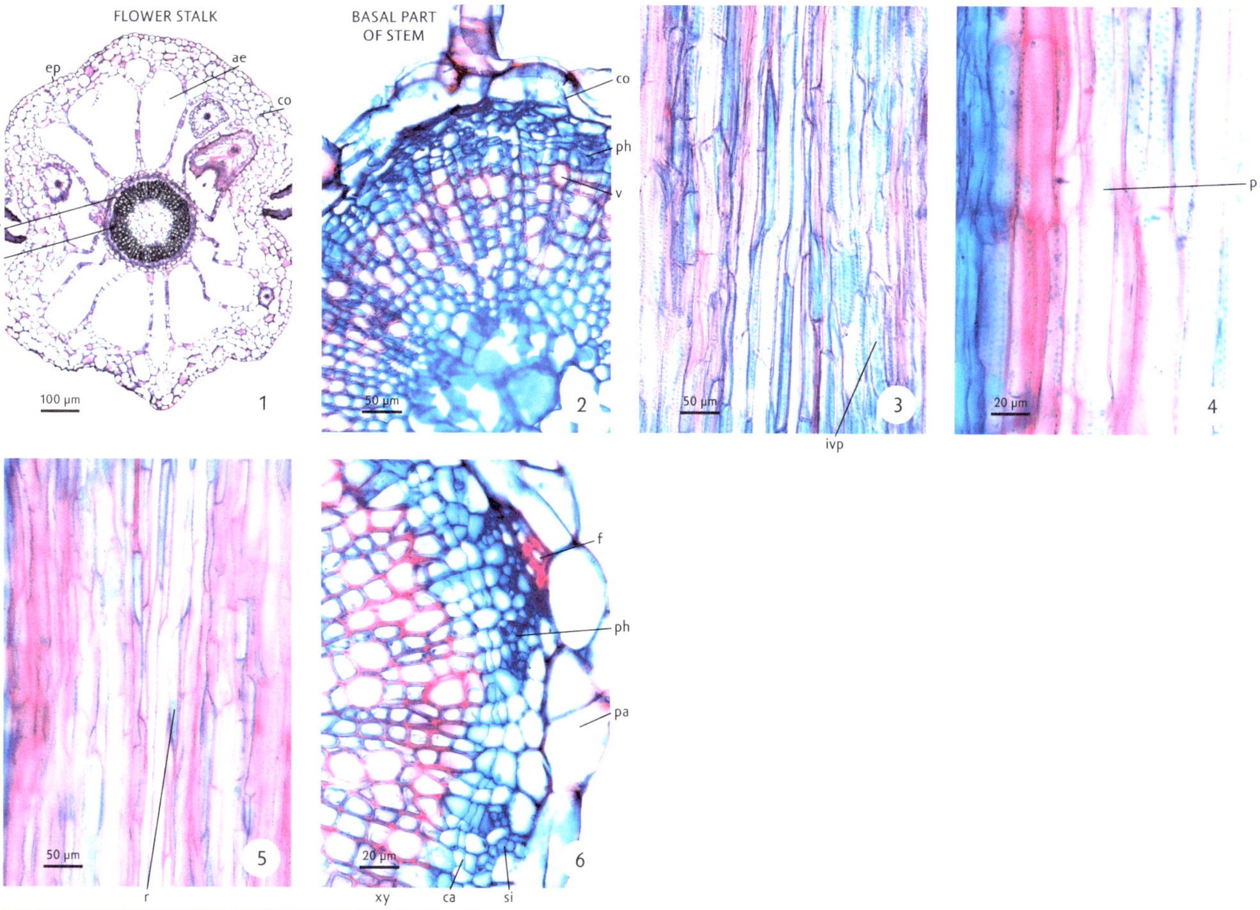

F. H. Schweingruber et al., *Anatomic Atlas of Aquatic and Wetland Plant Stems*, https://doi.org/10.1007/978-3-030-33420-8_35

Lythrum hyssopifolia L. (syn. *Lythrum hyssopifolium*) **Grass-poly** Lythraceae

Annual wetland herb. Stem erect to ascendent, branched, 5–50 cm high. Leaves sessile, opposite and narrow linear. Flowers solitary or in pairs in bract axils in the upper part of stem, corolla pink or pale pink. Fruit is a cylindrical capsule.

Distribution and ecology

Euro-Siberian southern-temperate distribution. Native in all CE countries in salty or fen meadows, damp pastures, drained bottoms of fishponds and ditches, also in depressions in arable fields. Prefers disturbed mineral-rich soils which are flooded in winter. Facultative halophyte. The ability to develop extensive aerenchymatic phellem in flooded plants. Contains chemicals toxic for cattle and sheeps.

Anatomy of the root collar

Round root collar with a diameter up to 4 mm. Maximum age 1 year.

Stem consists of a xylem, a small phloem and large phellem (1). Annual rings are marked by a slight semi-ring porosity (1, 2). The xylem is composed of vessels, thin- to thick-walled fibers and rays (1). Vessels are arranged solitary and in radial rows (2). Earlywood vessels diameter varies from 20–30 µm (2) and vessel length from 150–200 µm. Intervessel pits are round (2) and perforation plates are simple (3). Fibers are thin- to thick-walled (1). Some of them contain septate fibers and gelatinous fibers (tension wood) (2). Parenchyma is absent. Rays are uniseriate (4) and ray cells are upright (3). Crystals are absent in the xylem.

The phloem consists of groups of sieve tubes and parenchyma cells (5). The cortex is composed of unlignified thin-walled parenchyma (6). The phellem consists of several layers of rectangular cells (5). Pith with medullary phloem (6).

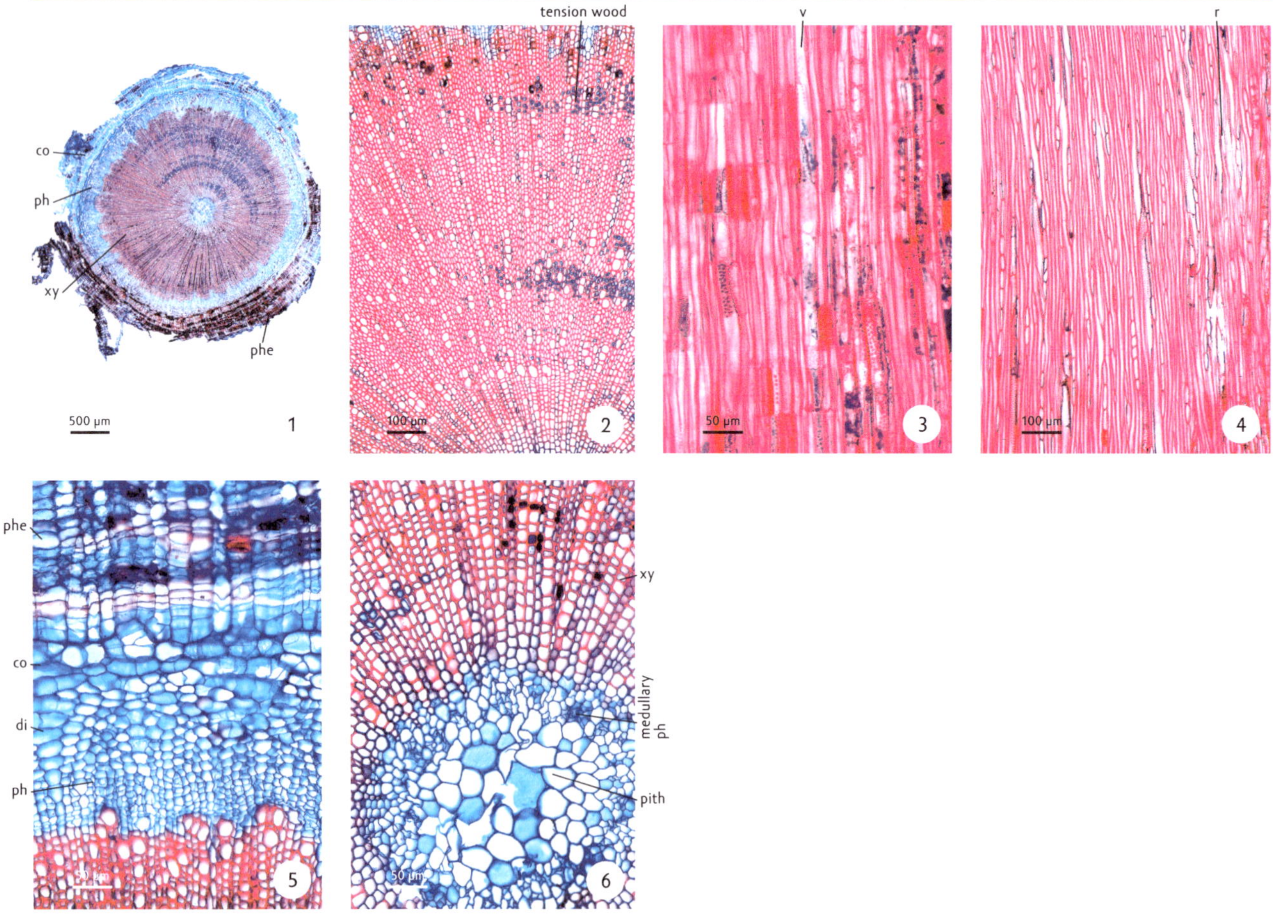

F. H. Schweingruber et al., *Anatomic Atlas of Aquatic and Wetland Plant Stems*, https://doi.org/10.1007/978-3-030-33420-8_36

Perennial wetland herb. Stem erect, downy, 30–120 cm high. Leaves sessile, oval-lanceolate, untoothed, in opposite pairs or in whorls of 3. Inflorescence dense, spike-like, flowers in whorls in bract axils with 6 red-purple petals. Fruit is a small capsule containing numerous minute seeds.

Distribution and ecology

Eurasian temperate element, invasive in N. America. Native in all CE countries on banks of lakes, fishponds, slow-flowing rivers, also in tall-herb fens and willow carrs. Prefers wet or periodically inundated fertile soils. Wide ranges in diurnal temperature fluctuations conducive to maximum seed germination.

Anatomy of a basal part of the annual shoot

Round stem with a diameter up to 10 mm. Maximum age 14 years.

Stem consists of a large xylem, a small phloem and phellem (1). Annual rings are marked by a few rows of flat latewood fibers and a few vessels (2, 3). The xylem is composed of vessels, thin- to thick-walled fibers, parenchyma and rays (2, 3). The xylem is diffuse-porous to semi-ring porous (2, 3). Vessels are angular and solitary and occur frequently primarily in the earlywood (2, 3). Vessel diameter varies from 30–50 µm (3) and vessel length from 150–250 µm. Intervessel pits are round and perforation plates are simple (4). Fibers are thin- to thick-walled (2, 3) and contain locally gelatinous fibers (tension wood) (3). Parenchyma, if present, is paratracheal (3, 5). Rays are prostrate (6) and generally 1–4-seriate with sheath cells (5). Large prismatic crystals and druses occur in axial parenchyma cells along rays (7).

The phloem consists of parenchyma cells and few sieve tubes (8) with dilating rays. Parenchyma cells in the cortex are oval and contain crystals. Intercellulars are absent. The phellem consists of uniseriate, tangentially layered rectangular cells (8).

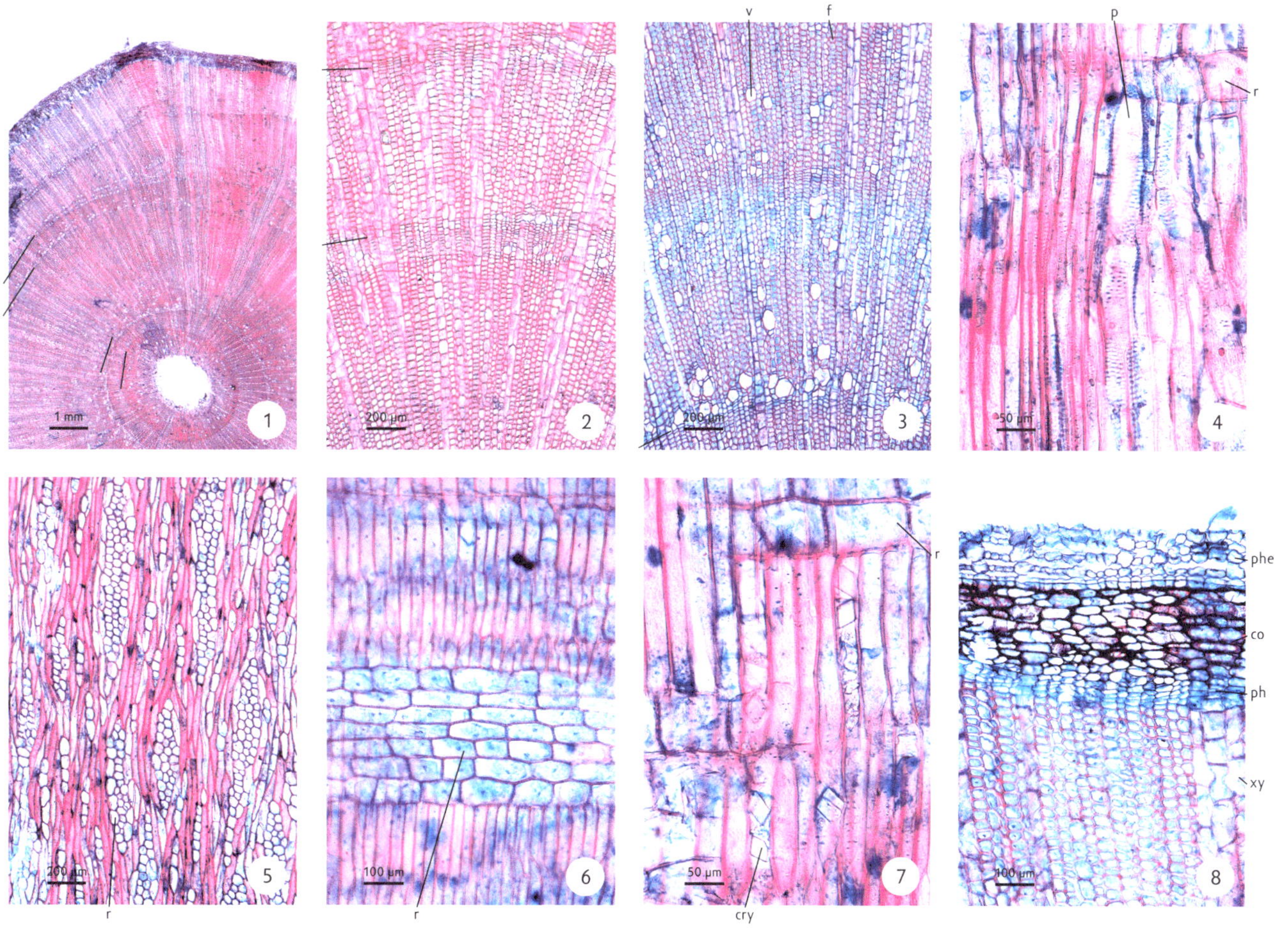

Perennial wetland herb with woody rhizome. Stem erect, glabrous, richly branched in the upper part, 50–120 cm high. Leaves alternate, sessile, lanceolate to narrow-lanceolate. Purpur-red flowers in whorls in leaf or bract axils. Fruit is an ovoid capsule.

Distribution and ecology

Eurasian continental distribution, secondary in N. America. Native in AT, CZ, SK and HU. Occurs on wet and periodically flooded meadows in floodplains of large rivers, margins of floodplain woodlands, in ditches and wet arable fields.

Anatomy of the root collar

Round root collar with a diameter up to 10 mm. Maximum age 1 year.

Stem consists of a xylem, a small phloem and phellem (1). The xylem is composed of vessels, thin- to thick-walled fibers, parenchyma and rays (1). Vessels are arranged in short radial rows (1). Vessel diameter varies from 30–50 µm (3) and vessel length from 150–200 µm. Intervessel pits are round (2) and perforation plates are simple (3). Tyloses occur locally (4). Fibers are thin- to thick-walled (1). Some of them contain septate fibers (5) and gelatinous fibers (tension wood) (2). Parenchyma has not been observed. Rays are uniseriate (6) and ray cells are upright (4). Crystals are absent in the xylem.

The phloem consists of uniform cells (7). The cortex is composed of unlignified thin-walled parenchyma and lignified, thick-walled fibers and intercellulars (7). Crystal druses occur in the thin-walled cells. The phellem consists of several layers of rectangular cells. The pith contains medullary phloem (1).

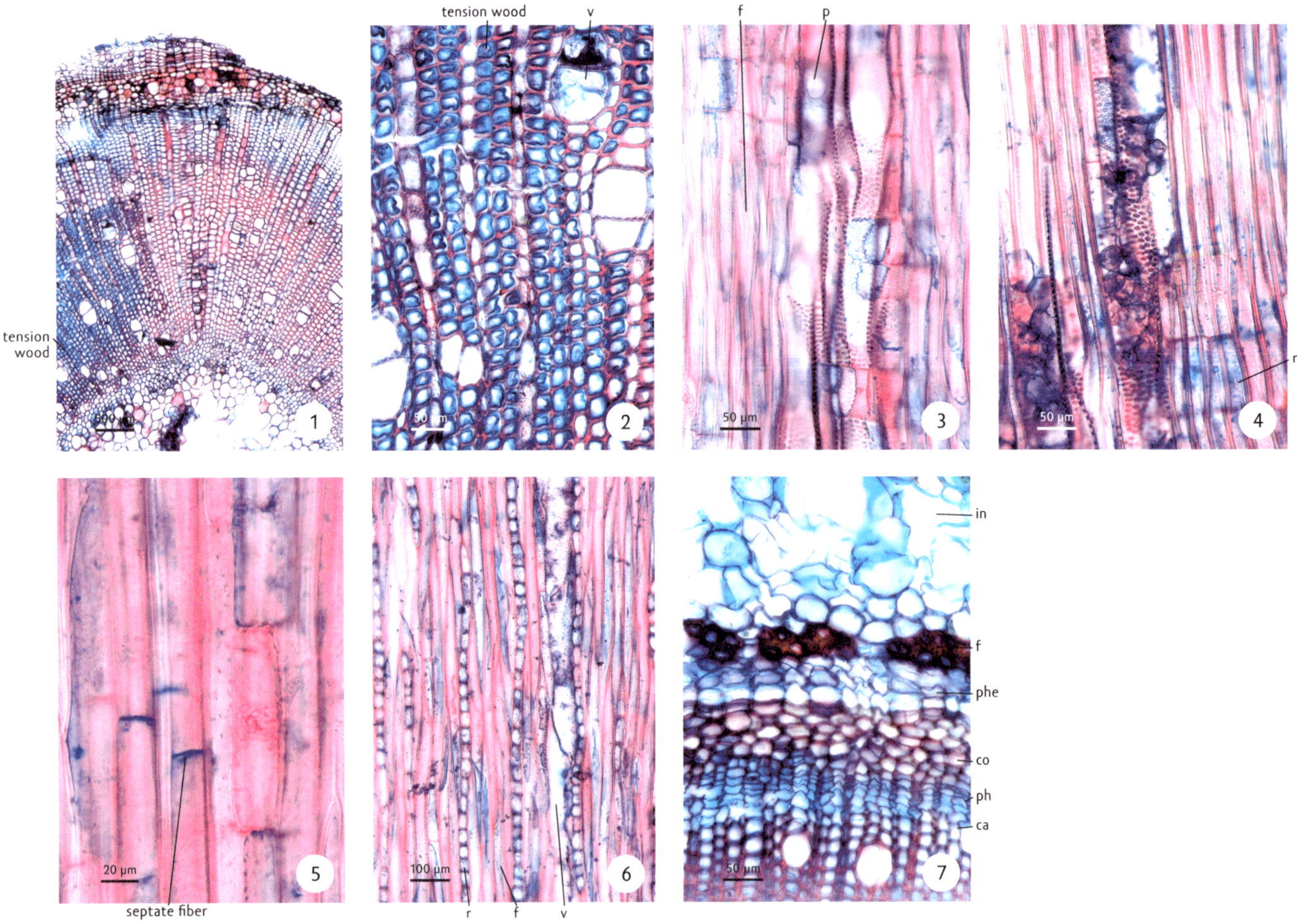

Usually annual wetland or amphibious herb. Procumbent stems 4–25 cm long, rooting at nodes. Leaves petiolate, obovate, opposite. Solitery flowers in leaf axils, submerged stems much elongated. Fruit is a reddish rounded capsule.

Distribution and ecology

European temperate distribution. Scattered irregularly and locally frequently in all CE countries, from lowlands to mountains. Sandy or clayish shorelines of fishponds and lakes, denudated fishpond bottoms, margins of wet meadows and fields, damp trackways, sand-pits, usually wet barren, sandy or clayish soils or base-poor waters, facultatively photophilous.

Anatomy of the basal part of the shoot

The round basal part of the plant has a diameter of 1.2–1.5 mm (1). Maximum age 1 year.

The stem is composed of a circle of xylem and phloem, a large aerenchymatic cortex and an epidermis (1). The xylem of the stem consists of solitary vessels and pervasive parenchyma (2). Vessels outline is angular (3). They have a diameter of 20–30 µm (2, 3) and a length of <100 µm. Intervessel pits are round and scalariform (4). Vessel walls have annular thickenings (2). Perforations are simple (4).

The phloem consists of groups of sieve tubes and parenchyma cells (3). The large cortex is characterized by large, round to radial intercellulars (1). There are a few layers of parenchyma cells without intercellulars inside of the epidermis. The pith contains medullary phloem (3) and crystal druses.

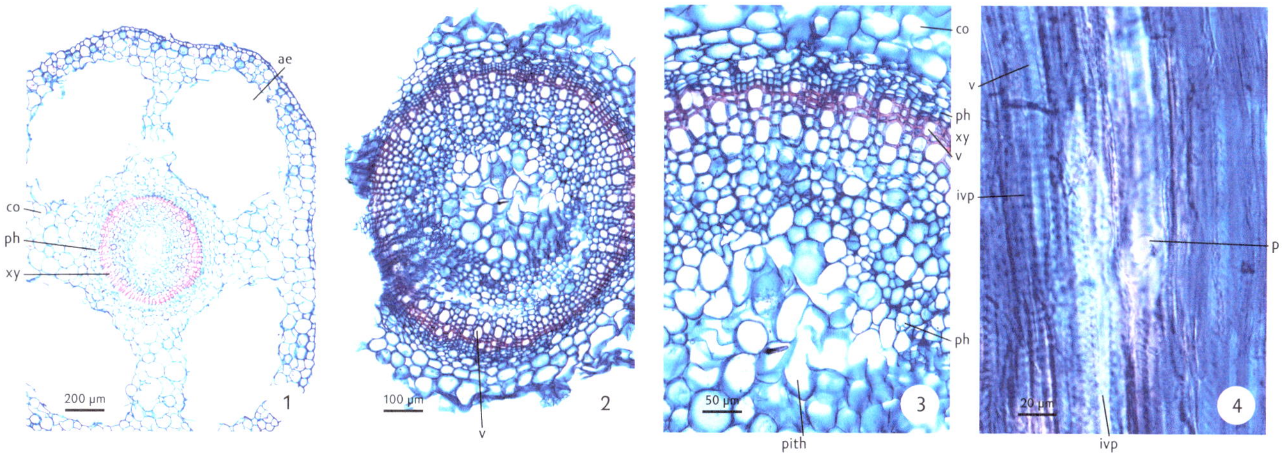

Trapa natans L.

Water Chestnut

Lythraceae

Annual floating-leaved herb with heterophylly. Stem thin, branched, 50–200 cm long, anchored into the mud by very fine roots. Floating rhomboid leaves with saw-tooth edges on inflated petioles form a leaf rosette on the water surface, submerged leaves finely divided, feather-like along the submerged part of the stem. Small four-petalled white flowers. Fruit is a black nut with 2–4 long triangular sharp spines.

Distribution and ecology

Temperate Eurasia and N. Africa, naturalized in N. America and Australia. Native in all CE countries. Prefers standing or slowly flowing waters with high nutrient content, prefers high water temperatures, tolerates water table fluctuation. Heliophilous species. Seeds can germinate in strict anoxia, seedlings are also highly tolerant of anoxia.

Anatomy of a basal part of the whole rosette shoot and the rhizome

Basal part: Round stem with a diameter of 7 mm. Rhizome with a diameter of 3 mm, tuber with a diameter of 10 mm. Maximum age 1 year (1, 2).

The large pith is surrounded by a xylem, a small phloem and a large cortex (1, 2). The xylem consists of large vessels and pervasive parenchyma (3). Vessels are unlignified (4) and are surrounded by one layer of small square parenchyma cells (3). Vessels have annular (4) and helical wall-thickenings (5). Vessels have a diameter of 50–90 µm (2) and a length of 200–250 µm. Fibers and rays are absent (3). Large ducts occur within the xylem (3).

The phloem of the principal stem and the rhizome consists of parenchyma cells and sieve tubes (6). Groups of sieve tubes (6) occur inside (medullary phloem) and outside of the xylem (1, 2, 3). Phloem and xylem are separated by an endodermis (3). The cortex contains net-like intercellulars (1, 2). Crystal druses occur in the inner (7) and small druses in the outer cortex (8).

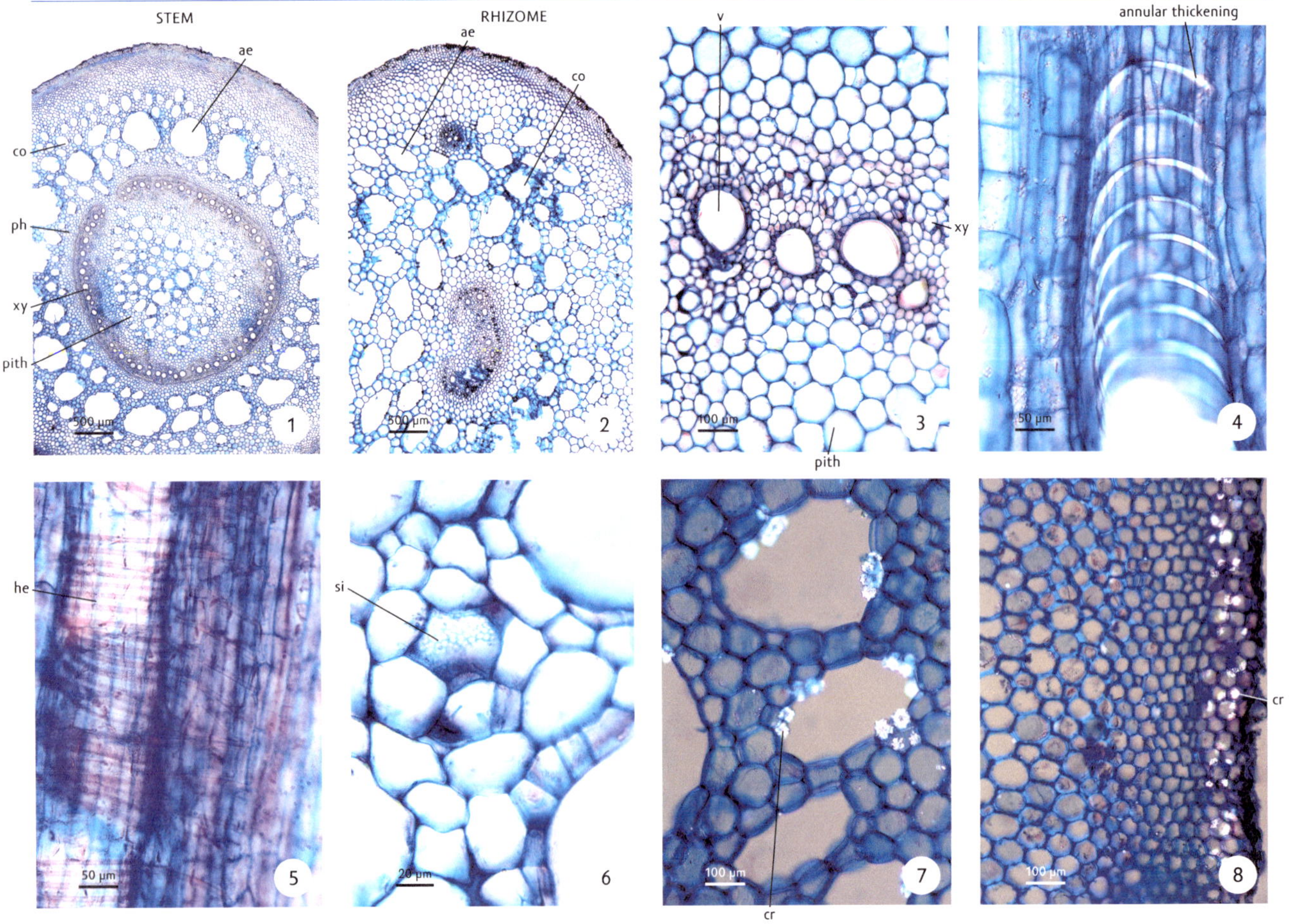

Perennial wetland or emergent herb with creeping rhizome bearing the tuft of trifoliate leaves at apex. The glabrous leaflets are obovate or elliptic, 3–7 cm long. The flowering stem is 12–35 cm high. The racemose inflorescence arises above water surface. Flowers are heterostylous dimorphic. Corolla five lobed, broadly campanulate, pink outside, white inside, 15 mm across. Fruit is a capsule, up to 10 mm long.

Distribution and ecology

Circumpolar boreo-temperate element. Native in all CE countries. Occurs in fens, shallow shores of lakes, pools and ponds, also on fen meadows. Prefers peaty soils and bottoms. Heliophilous species. Self-incompatible species with extensive clonal propagation. Medicinal herb.

Anatomy of the rhizome

The round basal part of the plant has a diameter of 5–7 mm (3). Maximum age 1 year.

The stele contains several collateral, open vascular bundles and is surrounded by a large cortex and an epidermis (1, 2). The xylem of the vascular bundles consists of radially grouped, lignified vessels and pervasive parenchyma (2, 3). Vessels have a diameter of 50–70 µm (2) and a length of 180–250 µm. Intervessel pits are scalariform (4). Special are the scalariform and perforation plates with 20 to >30 bars (4). Few fibers occur in the endodermis but are absent within vascular bundles (2, 3). Large, confluent rays exist between vascular bundles (5).

The phloem consists of parenchyma cells and groups of sieve tubes (6). An endodermis layer of thin-walled cells surrounds the stele. Their cells contain Casparian strips (7). The large cortex in the annual shoot and the rhizome are characterized by a net-like aerenchyma with well-delimited air cavities (1, 2). The epidermis consists of small roundish, thin-walled unlignified cells. Inside of the epidermis, there are 3–4 layers of small round parenchyma cells (8). Spiculate hairs do not exist.

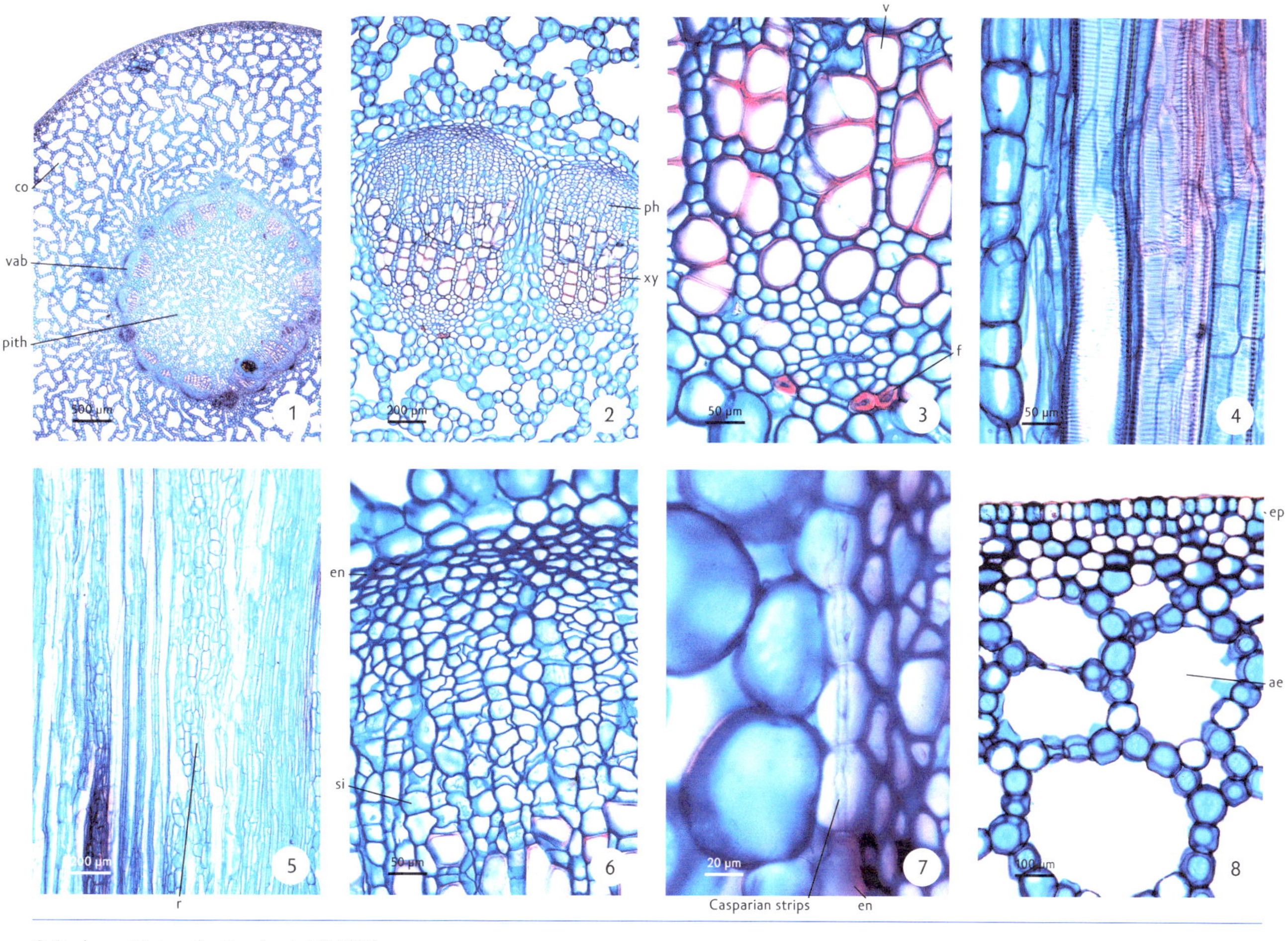

F. H. Schweingruber et al., *Anatomic Atlas of Aquatic and Wetland Plant Stems*, https://doi.org/10.1007/978-3-030-33420-8_37

Nymphoides peltata (S. G. Gmelin) O. Kuntze

Menyanthaceae
Fringed Water-lily

Perennial floating-leaved herb with creeping rhizoms. Submerged stems 80–150 cm long. Leaves round to kidney-shape, purple below, with long petioles. Flowers heterostylous dimorphic, in leaf axils, flowers above water. Corolla 3 cm across, golden-yellow, with five fringed lobes. Fruit is a capsule with several black disc-like seeds with many cilia on the border.

Distribution and ecology

Eurasian temperate element; widely naturalised outside its native range. Naturalized in CH, native in other CE countries. Occurs in water 0.5–2 m deep in lakes, fishponds and oxbows or slowly flowing rivers and canals. Prefers eutrophic, well-buffered alkaline waters. No use of HCO_3^-. Thermo-osmotic gas transport. Persistent seed bank, seeds do not germinate under hypoxic conditions.

Anatomy of the annual shoot and the rhizome

Round annual shoot with a diameter of 2 mm (1) and a rhizome with a diameter of 5–7 mm (5). Maximum age 1 year.

Annual shoot: The stele with 4 collateral, open vascular bundles is surrounded by a large cortex and an endodermis (2). The xylem consists of thin-walled solitary vessels and pervasive parenchyma (2). The unlignified vessels have a diameter of 50–70 µm (2). Cortex with radially oval aerenchymatic intercellulars. Special for the cortex of the annual shoot is the presence of lignified, thick-walled spiculate hairs (3), which occur also in *Nymphoides aquatica* (4).

Rhizome: The stele is surrounded by a large xylem and a small phloem, a very large cortex with a small additional stele and an epidermis (5). The xylem of the vascular bundles consists of a few solitary, angular lignified vessels and pervasive parenchyma (6). Vessels have a diameter of 30–40 µm (7) and a length of 100–150 µm (7). Intervessel pits are scalariform and perforation plates are simple (7). Fibers and rays are absent within vascular bundles (7).

The phloem consists of parenchyma cells and groups of sieve tubes (6). Cortex and phloem are separated by an endodermis without Casparian stripes (2, 4). The large cortex in the annual shoot and the rhizome are characterized by a net-like aerenchyma with well-delimited air cavities (8). The epidermis consists of small tetrangular thin-walled unlignified cells (8). Spiculate hairs are absent.

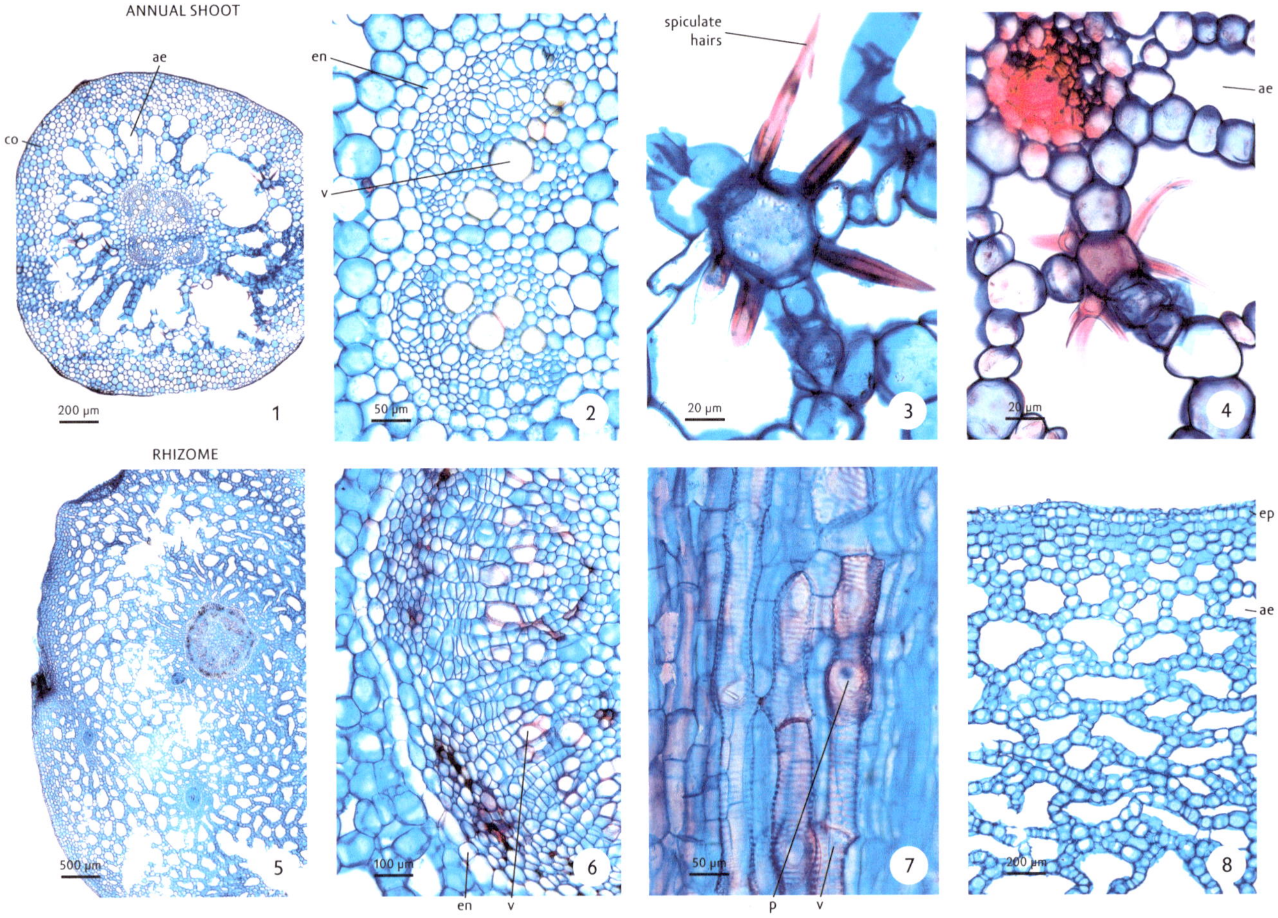

Montia fontana L. subsp. *fontana*

Blinks Montiaceae

Annual to perennial wetland plant. Stems 3–20 cm long, branched, erect to procumbent or floating. Oblonged leaves opposite, adventive roots formed in leaf axils, white single flowers. Fruit is a 3-seeded capsule.

Distribution and ecology

European boreo-temperate distribution; also in C. and E. Asia and N. America. Scattered rarely in all CE countries. Shallow wetlands in springs, bogs, fens, ditches, drainage canals, always in calcium-poor, acidic waters, barren organic soils. Psychrophyte, prefers wet and cold habitats mainly in uplands and highlands.

Anatomy of the root collar

Root collar with a diameter of <1 mm. Maximum age 1 year.

Presented is a plant from an alpine lake shore.

The large pith is surrounded by a small xylem in vascular bundles, a small phloem and a triple-layered cortex (1, 2). The hardly lignified xylem consists of hardly lignified groups of angular vessels with a diameter of 5–8 µm (2) and a length of vessel 60–80 µm. Intervessel pits are scalariform and plates are simple. Parenchyma is pervasive (3). Rays are absent (1, 2). Vascular bundles are laterally separated by unlignified parenchyma.

The cortex consists of an innermost layer of unlignified, a middle layer of sclerenchyma, and an outermost layer of unlignified large thin-walled parenchyma cells (1, 2). Intercellulars in the cortex are small and occur in the outermost layer (1). The phellem is absent (1).

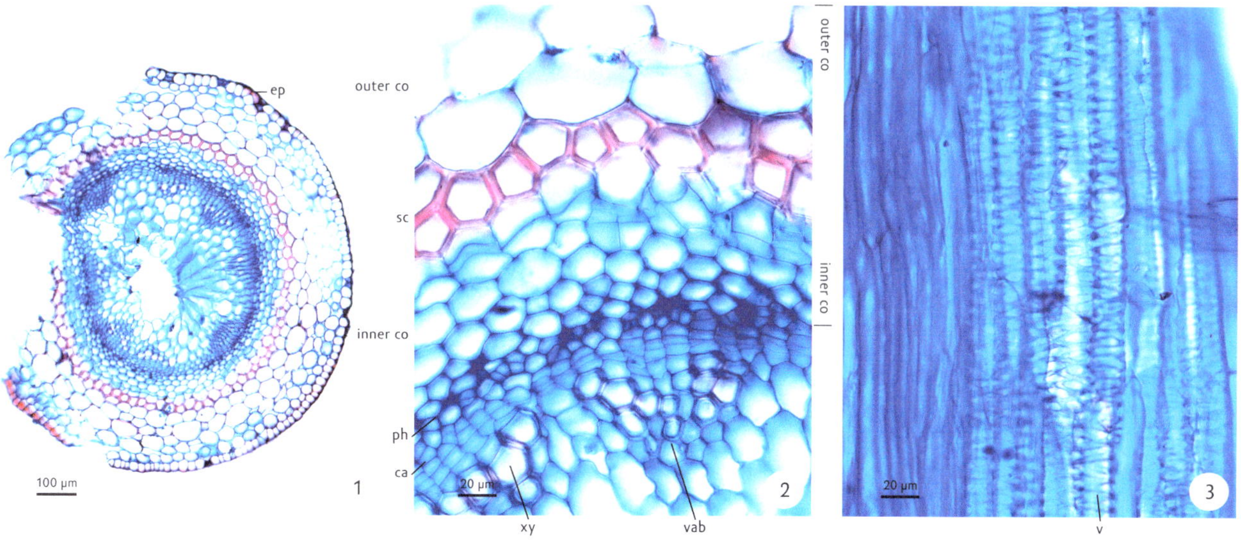

F. H. Schweingruber et al., *Anatomic Atlas of Aquatic and Wetland Plant Stems*, https://doi.org/10.1007/978-3-030-33420-8_38

Nuphar lutea (L.) Sm.

Yellow Water-lily

Perennial floating-leaved herb with thick rhizome rooting in sediment. Floating leaves cordate, leathery, on 50–250 cm long elastic petioles, stomata only on the upper leaf side. Submerged leaves thinner with undulate margins and on shorter petiole. Flowers solitary above water, 4–6 cm in diameter, on long pedicels, both sepals and petals yellow. Fruit is a flask-shaped capsule ripening above water. Terrestrial form known.

Distribution and ecology

Euro-Siberian boreo-temperate element. Native in all CE countries. Occurs in mesotrophic or eutrophic water of lakes, ponds and oxbows or in slowly flowing rivers, canals and ditches. Inability to utilize HCO_3^-. Flowers self-compatible and protogynous, preventing automatic selfing. Transient seed banks, seeds germinate under hypoxic conditions.

Anatomy of the flower stalk and petiole

Round stalk with a diameter of 5–10 mm. Maximum age 1 year (1).

Closed collateral vascular bundles are irregularly distributed within a net-like parenchymatic tissue in stalks (1) and petioles (2). The round vascular bundles are partially surrounded by an endodermis (3, 4). The vascular bundles of the flower stalk contain only a phloem consisting of sieve tubes and accompanying cells (3, 4). The xylem is absent in the flower stalk (3). A xylem with annular thickenings or scalariform intervessel pits with large openings occur only in the stalk (5). The radial ends of the vascular bundles are separated by a round air duct (3).

Between the thin-walled parenchyma cells, many aerenchymatic, net-like intercellulars exist. Many of the surrounding cells of the very large air cavities contain thick-walled, spiculate hairs with small prismatic crystals (calcium oxalate) (6). The collenchymatic peripheral zone does not contain thick-walled hairs (7).

Hybrid *Nuphar × spenneriana* is anatomically identical with *Nuphar lutea*.

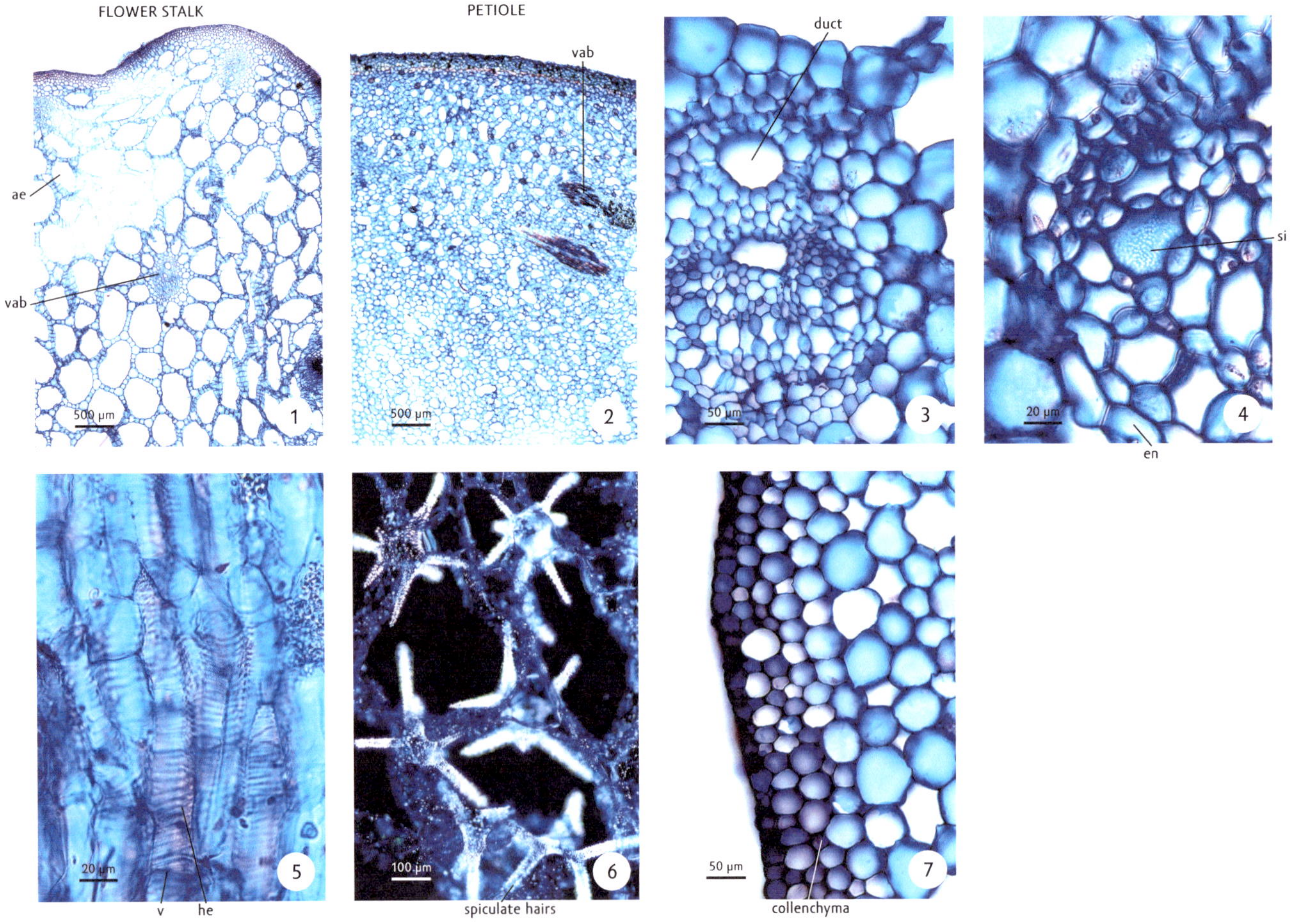

F. H. Schweingruber et al., *Anatomic Atlas of Aquatic and Wetland Plant Stems*, https://doi.org/10.1007/978-3-030-33420-8_39

Nuphar pumila (Timm) DC.

Least Water-lily

Nymphaeaceae

Perennial floating-leaved herb with thick rhizome rooting in sediment. Floating leaves wide elliptic to circular, leathery, on long elastic petioles, stomata only on the upper leaf side. Submerged leaves thinner and on shorter petiole. Flowers solitary above water, 2–3 cm in diameter, on long pedicels, petals yellow. Fruit is an ovoid capsule ripening above water.

Distribution and ecology

Circumpolar boreal element. Native in all CE countries except HU and SK. Climate relicts in Europe with few isolated populations in the Alpine arc with ongoing hybridization between *N. pumila* and *N. lutea*. Occurs in bog pools, oxbows and fishponds with muddy or peaty sediments in water 0.4–1 m deep. Prefers soft oligo- to mesotrophic water poor in Ca^{2+} and Mg^{2+}, but rich in total Fe, P and N. Overwintering in rhizomes.

Anatomy of the flower stalk

Round flower stalk with a diameter of 4–6 mm. Maximum age 1 year (1).

Closed collateral vascular bundles are irregularly distributed within a net-like parenchymatic tissue in culms (1). The vascular bundles of the flower stalk contain only a phloem consisting of sieve tubes and accompanying cells and unlignified, thin-walled vessels (2). The central radial ends of the vascular bundles contain round air ducts (2).

Between the thin-walled parenchyma cells, many net-like intercellulars exist. Many of the surrounding cells of the very large air cavities contain thick-walled, spiculate hairs with small prismatic crystals (calcium oxalate) (3).

Longitudinal section missing.

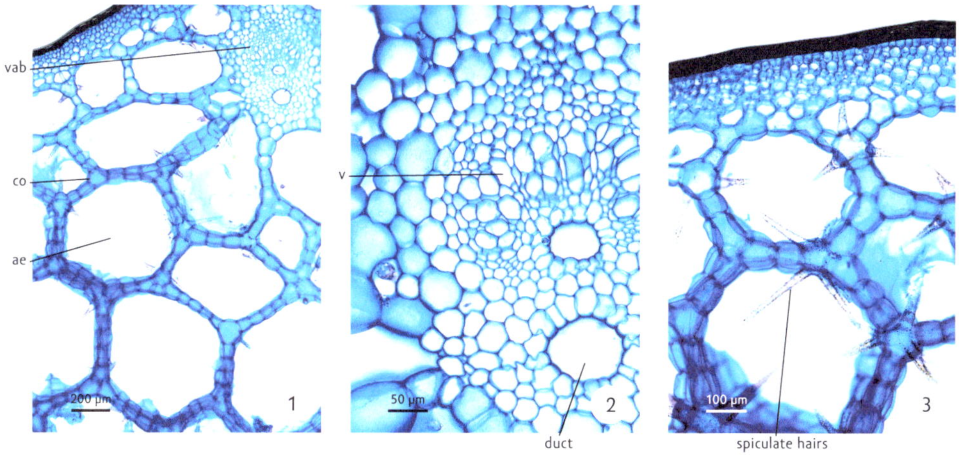

Nymphaea alba (L.) **White Water-lily** Nymphaeaceae

Perennial floating-leaved herb with arm-thick rhizome rooting in sediment. Floating oval leaves with long flexible petiole, petioles 30–250 cm long, leaf veins with reticulum within the leaf lamina, stomata only on the upper leaf side. White flowers solitary above water, on long pedicels, 9–12 cm in diameter. Filaments of inner stamens linear, central stigma projection short and hemispherical. Fruit is a globular capsule ripening under water.

Distribution and ecology

European temperate element. Scattered commonly to rarely in all CE countries, mostly in lowlands and fishpond basins in base-rich soils. Standing waters, fishponds, lakes, dam reservoirs, backwater pools and oxbows, slow rivers, fen lakes, water depth 0.4–3 m. Tolerates neutral to strongly alkaline, medium hard to hard water (pH 6.5–9, total alkalinity 1.5–5 meq/l), organic, sandy or clayish bottom, prefers mesotrophic to eutrophic waters, neutral to alkaline soils. Overwintering in rhizomes. Terrestrial form known.

Anatomy of the flower stalk

Round stalk with a diameter of 5–10 mm (1). Secondary growth is absent.

Closed collateral vascular bundles (primarily arranged in pairs) and round air cavities are irregularly distributed within a parenchymatic tissue (1, 2). The oval vascular bundles are surrounded by an endodermis-like layer of cells (3). At one small side occur one or two unlignified vessels with scalariform pits (4). Transverse walls and perforation plates seem to be absent. At the other end of the vascular bundles are small groups of sieve tubes and accompanying cells (3).

Small parenchyma cells and probably larger air cavities of different size occupy the large central part (3). The smaller ends of the vascular bundles are separated by a round air duct, which is surrounded by one row of parenchyma cells (3). Between the thin-walled parenchyma cells, many intercellulars exist. Some of the surrounding cells of the very large air cavities contain thick-walled, furcate hairs with small prismatic crystals (calcium oxalate) (2) between the unlignified thin-walled primary and the lignified, thick-walled secondary wall (5). The hairs inside in the collenchymatic peripheral zone are spiculate, not branched (6).

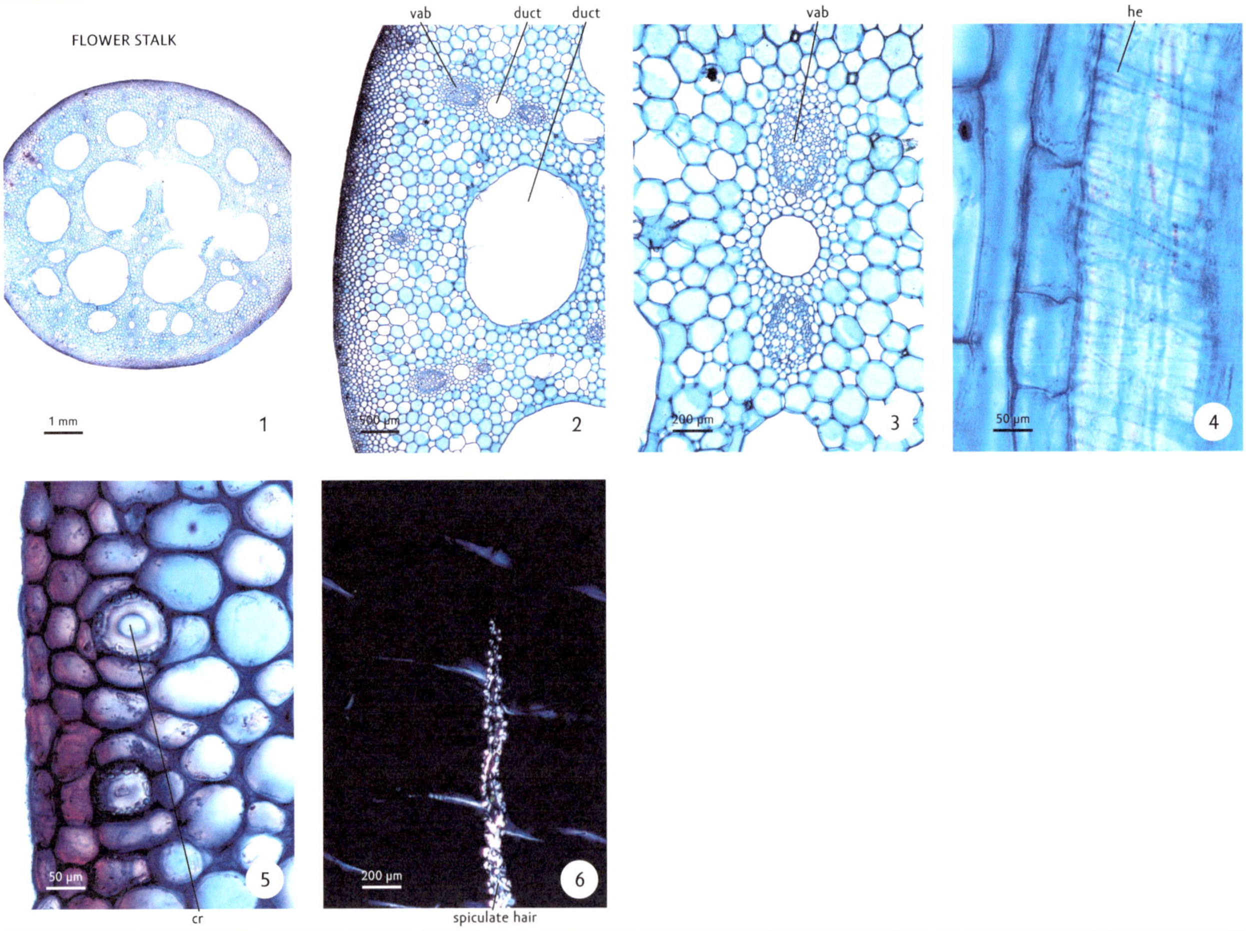

Nymphaea candida J. Presl

Perennial floating-leaved herb with arm-thick rhizome rooting in sediment. Floating oval leaves have long flexible petioles, leaf blade 20–30 cm long, 17–27 cm wide, leaf veins with reticulum within the leaf lamina. White flowers solitary above water, on long pedicels. Filaments of inner stamens lanceolate or round, central stigma projection long and conical. Fruit is a globular or ovoid capsule ripening under water. An allopolyploid originated by hybridization between the mainly European N. alba lineage and the mainly Asian N. tetragona lineage.

Distribution and ecology

Euro-Siberian distribution. Native in all CE countries except HU and SK. Occurs in standing waters as fishponds, fen lakes, backwater pools and oxbows, also in slow rivers, in water depth of 0.7–1.7 m. Prefers meso- and oligotrophic waters, with a pH ranging from weakly acidic to neutral (pH 5.5–7.1), usually on peaty or muddy sediments. Overwintering in rhizomes.

Anatomy of the flower stalk

Round stalk with a diameter of 4–6 mm. Maximum age 1 year (1).

Closed collateral vascular bundles (primarily arranged in pairs) and round air cavities are irregularly distributed within a parenchymatic tissue (1, 2). The vascular bundles contain only a phloem consisting of sieve tubes and accompanying cells. The xylem is absent (2). In the smaller ends of the vascular bundles, round air ducts occur, which are surrounded by one row of parenchyma cells (2). Between the thin-walled parenchyma cells, many intercellulars exist.

The cortex contains few spiculate hairs (3). Some cells of the epidermis have a thick-walled, externally red stained cell walls (4).

Longitudinal section is missing.

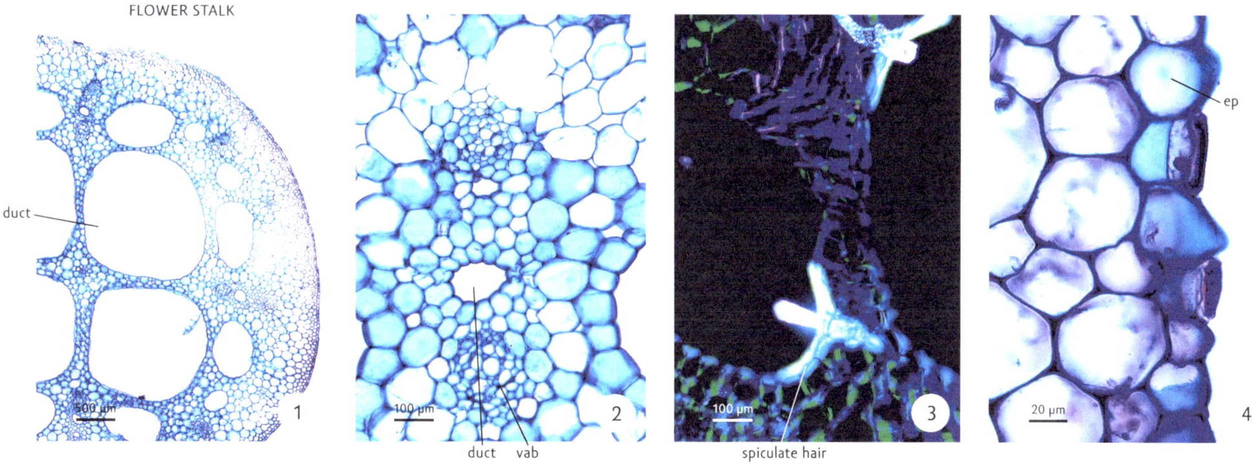

Epilobium adenocaulon Hausskn. (syn. *E. ciliatum*) **American Willowherb** Onagraceae

Perennial wetland herb overwintering as sessile green leaf rosette. Stem erect to ascendent, with patent glandular and appressed non-glandular hairs, 20–100 cm high. Leaves opposite, oblong-lanceolate, toothed. Flowers in a loose terminal raceme, petals pinkish-purple, stigma clavate. Fruit is a linear capsule containing tiny seeds with long apical hairs.

Distribution and ecology

Native of N. America. Widely naturalised in C. and N. Europe, common in all CE countries. Pioneer species of open habitats (waste ground, woodland clearings), occurs also in disturbed ground in semi-natural habitats like wetland margins, shorelines and stream banks. Able to colonize newly disturbed sites by light wind-dispersed seeds.

Anatomy of the stem

Round stem with a diameter of 4–5 mm. Maximum age 2 years (1).

The stem consists of a xylem, a small phloem and cortex and a large phellem (1). The diffuse-porous xylem is composed of angular vessels and thin- to thick-walled fibers (2). Vessels are arranged solitary (1, 2). Vessel diameter varies from 30–50 µm and vessel length from 150–200 µm. Intervessel pits are round (3) and perforation plates are simple (4). Fibers are thin- to thick-walled (2). Locally they contain gelatinous secondary walls (tension wood) (2). Parenchyma cannot be recognized on cross-sections. Rays are 2–3-seriate (5). Crystals are absent in the xylem.

The phloem contains few small groups of sieve tubes and raphids in idioblasts (5, 6). The cortex is composed of unlignified thin-walled round parenchyma and intercellulars. Groups of sieve tubes occur at the periphery of the pith (medullary phloem) (7).

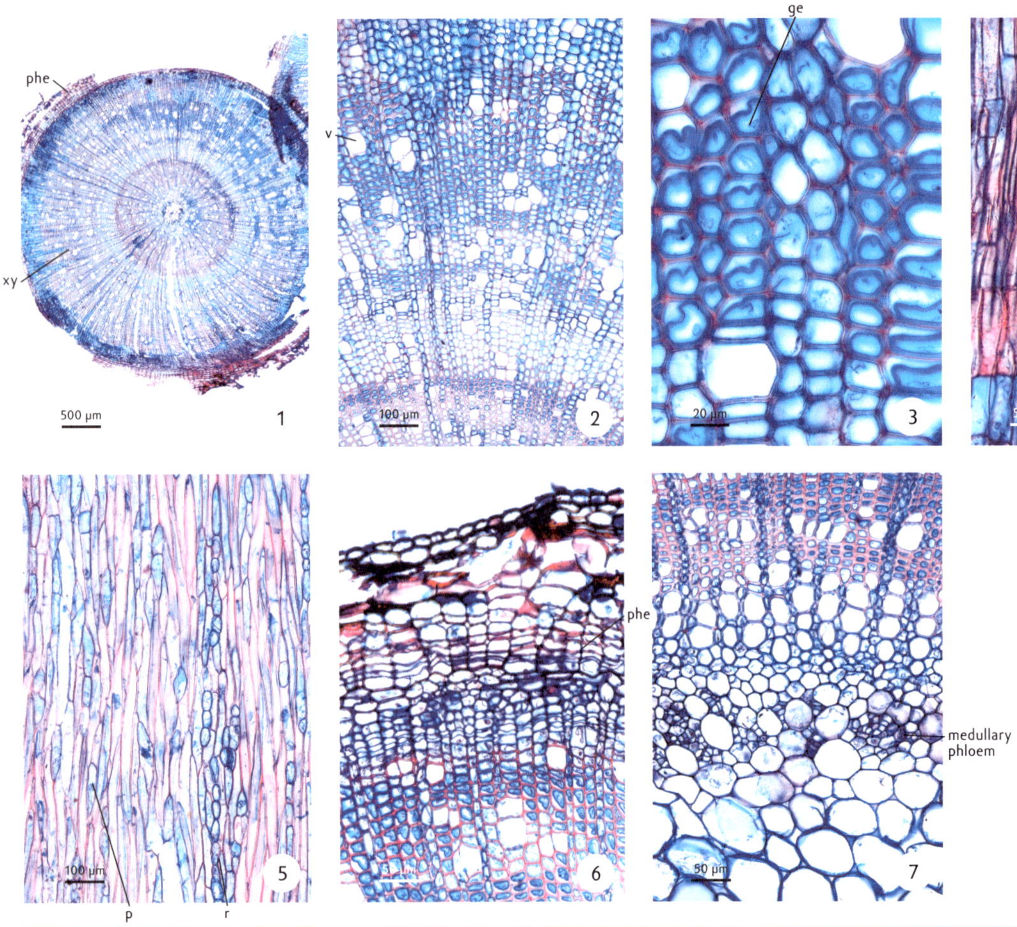

F. H. Schweingruber et al., *Anatomic Atlas of Aquatic and Wetland Plant Stems*, https://doi.org/10.1007/978-3-030-33420-8_40

Epilobium alsinifolium Vill. **Chickweed Willowherb** Onagraceae

Perennial wetland herb with subterranean stolons. Stem ascending to decumbent, up to 30 cm high. Leaves ovate to narrowly ovate, distinctly dentate, pointed. Petals 7–12 mm long, pale-purplish, stigma clavate, hypanthium without appressed non-glandular hairs, often reddish. Fruit is a linear capsule.

Distribution and ecology

European arctic-montane element; also in Greenland. Native in all CE countries except HU. Occurs by mountain springs or streams, rarely on wet screes or springs in open places in mountain forests. Tolerates eutrophic conditions.

Anatomy of the stem

Round stem with a diameter up to 0.5–1 mm. Maximum age 1 year (1).

Stem consists of a xylem, a small phloem, a large cortex and an epidermis (1). The xylem is composed of vessels and thin- to thick-walled fibers (2). Vessels are arranged solitary or in short radial rows (1, 2). Vessel diameter varies from 15–25 µm and vessel length from 50–80 µm. Intervessel pits are round and scalariform (3, 4) and perforation plates simple (3). Fibers are thin- to thick-walled (2). Parenchyma, rays and crystals are absent (2, 4).

The phloem consists of parenchyma cells and groups of sieve tubes (5). Both inner and outer cortex are separated by a layer with dark staining substances (endodermis) (5). The cortex is composed of unlignified thin-walled parenchyma. Intercellulars are rare and small (5, 6). The cortex contains raphids in idioblasts. Groups of sieve tubes occur at the periphery of the pith (medullary phloem) (7).

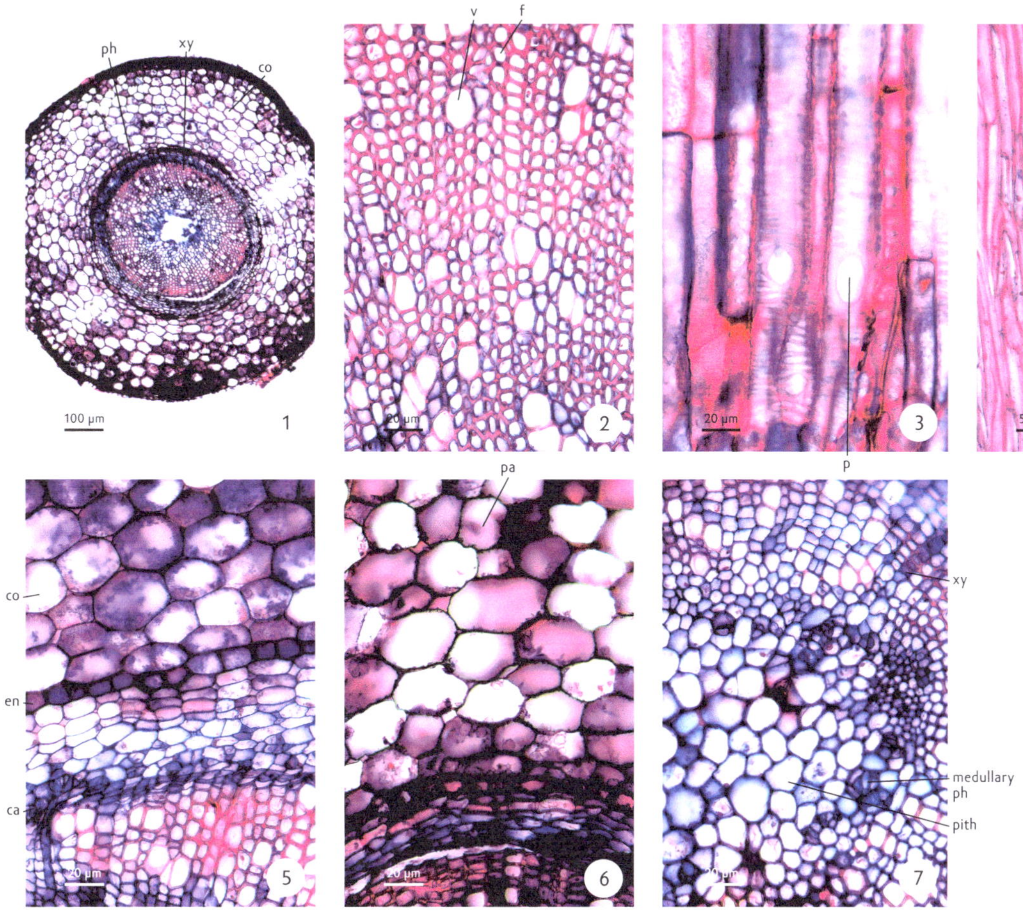

Epilobium hirsutum L.

Great Willowherb Onagraceae

Perennial wetland herb with subterranean rhizomes. Stem erect, rounded, densely pubescent with non-glandular and glandular hairs, 80–150 cm high. Leaves densely pubescent, opposite, lanceolate and sessile. Corolla 10–20 mm across, petals usually bright pinkish-purple, stigma with 4 stout lobes. Fruit is a long capsule.

Distribution and ecology

Eurasian southern-temperate element; widely naturalised outside its native range. Native in all CE countries. Occurs in open wet habitats, along banks of river and streams, in edges of fishponds, ditches and damp ground. Prefers fertile, neutral to basic soils with low iron content. It is able to form adventitious roots and stem hypertrophy under waterlogging conditions. Germinates at low temperatures and short days.

Anatomy of the stem

Round stem with a diameter up to 5–15 mm. Maximum age 1 year (1).

Stem consists of a xylem, a small phloem and cortex phellem (1, 6). The xylem is composed of vessels and thin- to thick-walled fibers (2). Vessels are arranged solitary or in short radial rows (1, 2). Vessel diameter varies from 40–50 µm and vessel length from 150–200 µm. Intervessel pits are round (3) and perforation plates are simple (3). Fibers with small pits are thin- to thick-walled (2–4). Locally they contain gelatinous secondary walls (tension wood) (2, 6). Parenchyma cannot be recognized on cross-sections. Rays are uni- to biseriate (4). Ray-vessel-pits are horizontally enlarged (5) with intra-xylary phloem (3). Crystals are absent in the xylem.

The phloem contains small groups of sieve tubes. Raphids in large idioblasts occur in the phloem and the cortex. Intercellulars are absent. Groups of sieve tubes occur at the periphery of the pith (medullary phloem) (7).

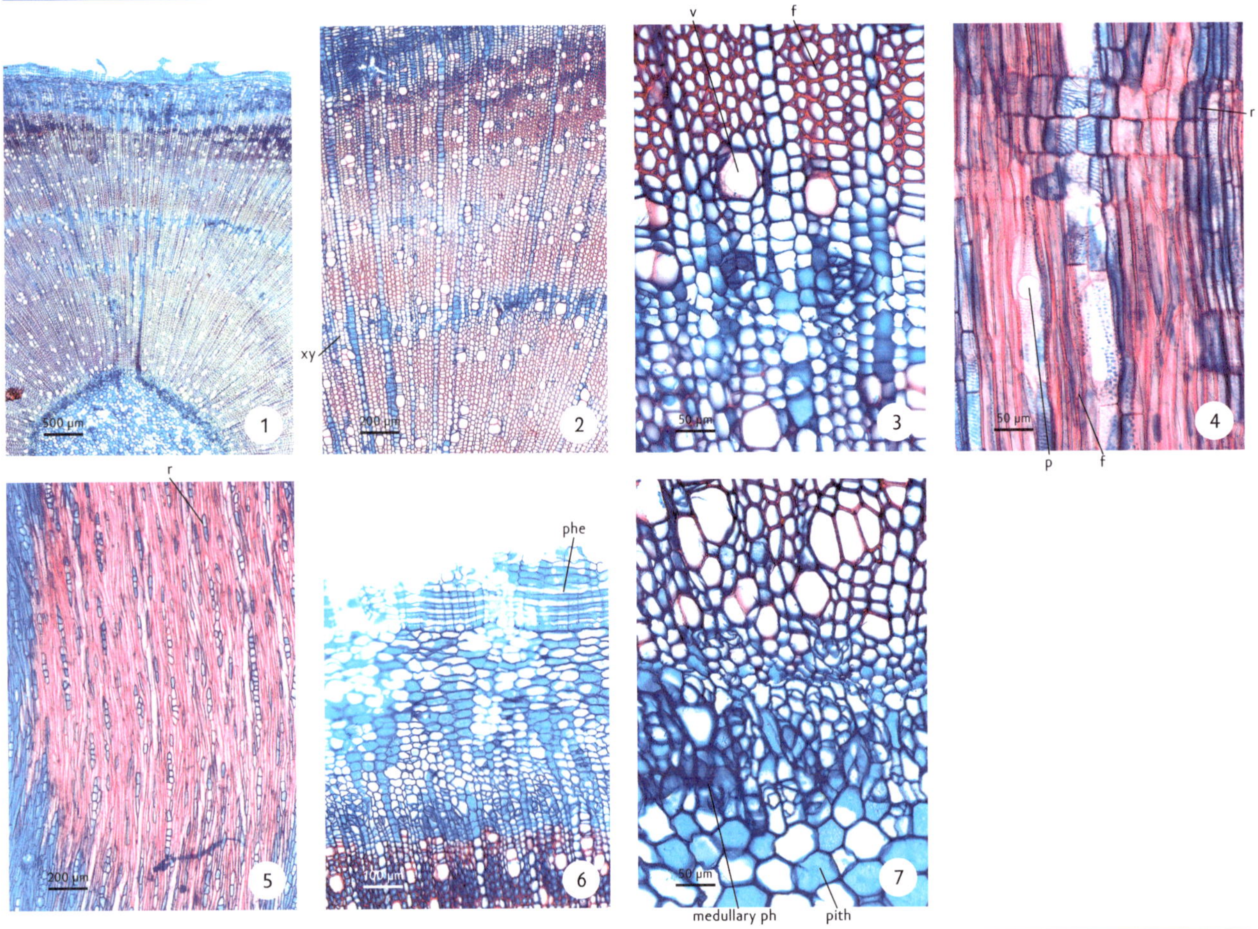

Epilobium nutans F. W. Schmidt

Perennial wetland herb with very thin creeping stolons. Stem erect, unbranched, 8–25 cm high, upper part pubescent. Leaves glabrous, usually sessile, ovate to lanceolate, usually with fine teeth. Inflorescence from 2–4 pendant flowers with pale violet petals, stigma clavate. Fruit is a 3.5–4 cm long capsule.

Distribution and ecology

C. and S. Europe. Native in all CE countries except HU, very rare in SK. Grows in partially shaded springs, only in the mountains.

Anatomy of a basal part of the annual shoot

Round stem with a diameter up to 0.5–1 mm. Maximum age 1 year (1).

Stem consists of a xylem, a small phloem, a large cortex and an epidermis (1). The xylem is composed of vessels and thin- to thick-walled fibers (2). Vessels are arranged solitary (1, 2). Vessel diameter varies from 15–20 µm and vessel length from 60–80 µm. Intervessel pits are round and scalariform (3, 4). Perforation plates are generally simple, however, a few are scalariform with a very few bars (3). Fibers are thin- to thick-walled (2). Parenchyma and rays cannot be distinguished in radial sections (3). Uniseriate rays appear in cross-sections (2) in radial rows but cannot be distinguished from axial parenchyma cells in radial sections. Crystals are absent in the xylem.

The phloem consists of uniform cells (4). Cortex and phloem are separated by few layers of a poorly differentiated endodermis/pericycle zone (1, 2, 5), which is characterized by lignified strips in the edges (4). The cortex is composed of unlignified thin-walled, round parenchyma cells and small intercellulars (4, 5). Groups of sieve tubes occur at the periphery of the pith (medullary phloem) (2). The piths contain raphids in idioblasts (6).

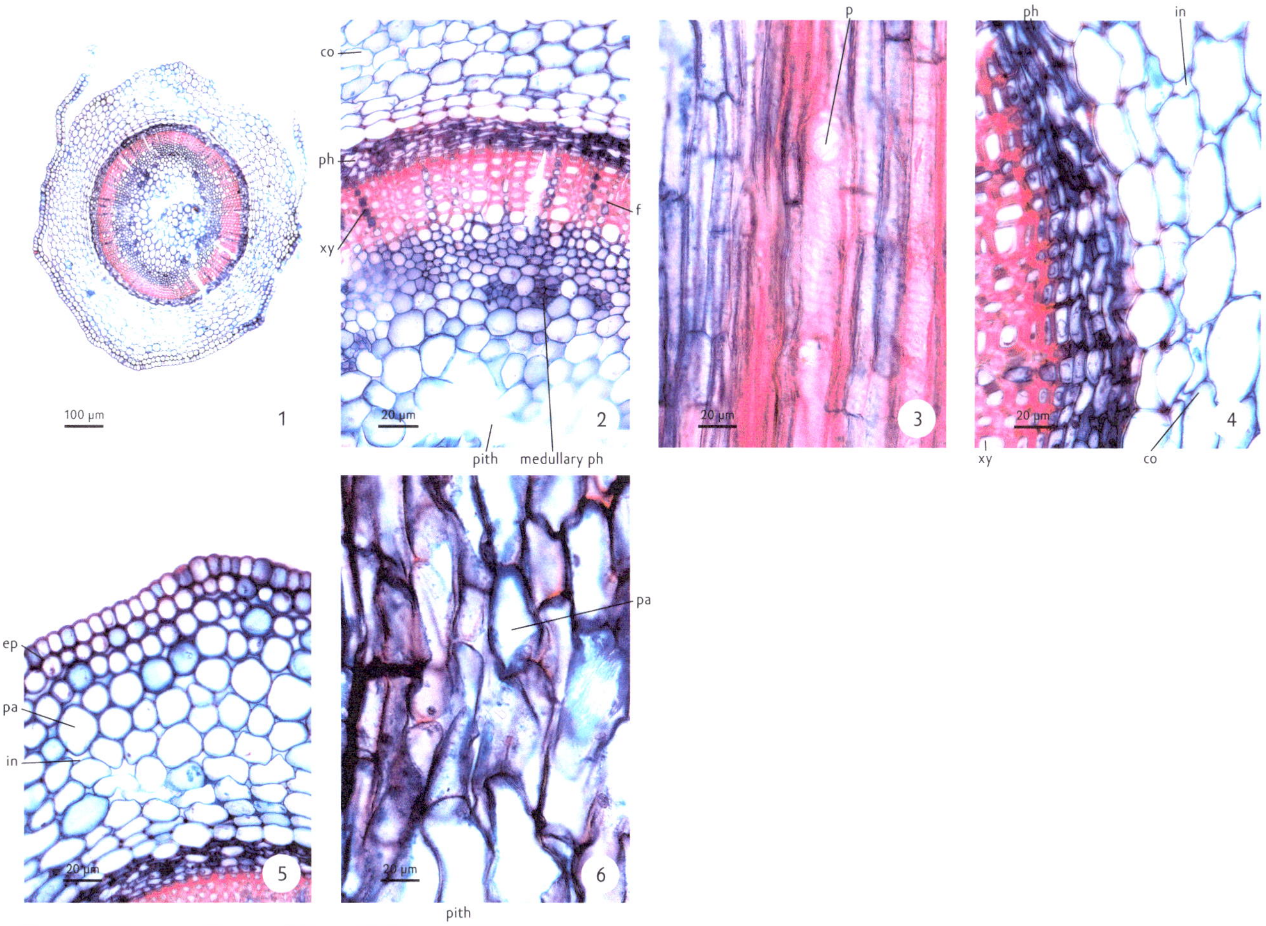

Epilobium palustre L.

Marsh Willowherb

Onagraceae

Perennial wetland herb, with long filiform rhizomes. Stem erect, 10–70 cm high, unbranched, rounded, with sparse appressed non-glandular hairs and fewer patent glandular hairs near top. Leaves sessile or with short petiole (up to 4 mm), lanceolate to linear-lanceolate. Flowers in a loose terminal raceme, petals pale pink, stigma clavate. Fruit is a linear capsule containing tiny seeds with long apical hairs.

Distribution and ecology

Circumpolar boreo-temperate element. Native in all CE countries. Occurs on fens, peaty meadows, peaty margins of fishponds and meadow springs. Prefers wet acidic soils. Overwinters by tubers which form at the ends of the stolons. It spreads by wind-dispersed seeds.

Anatomy of the stem

Round stem with a diameter up to 1–1.5 mm. Maximum age 1 year (1).

The stem consists of a xylem, a small phloem and a large cortex and an epidermis (1). The xylem is composed of vessels and thin- to thick-walled fibers (2). Vessels are arranged solitary or in short radial rows (1, 2). Vessel diameter varies from 15–25 μm and vessel length from 50–80 μm. Intervessel pits are generally round, rarely scalariform (3) and perforation plates are simple (3). Fibers are thin- to thick-walled (2). Locally they contain gelatinous secondary walls (tension wood). Parenchyma and rays cannot be distinguished (2, 4). Crystals are absent in the xylem.

The phloem consists of parenchyma cells and groups of sieve tubes (5). The inner and outer cortex is separated by a layer of rectangular cells (endodermis?) (1). The cortex is composed of unlignified thin-walled parenchyma and intercellulars (1, 6). The cortex contains raphids in idioblasts (7). Groups of sieve tubes occur at the periphery of the pith (medullary phloem) (2).

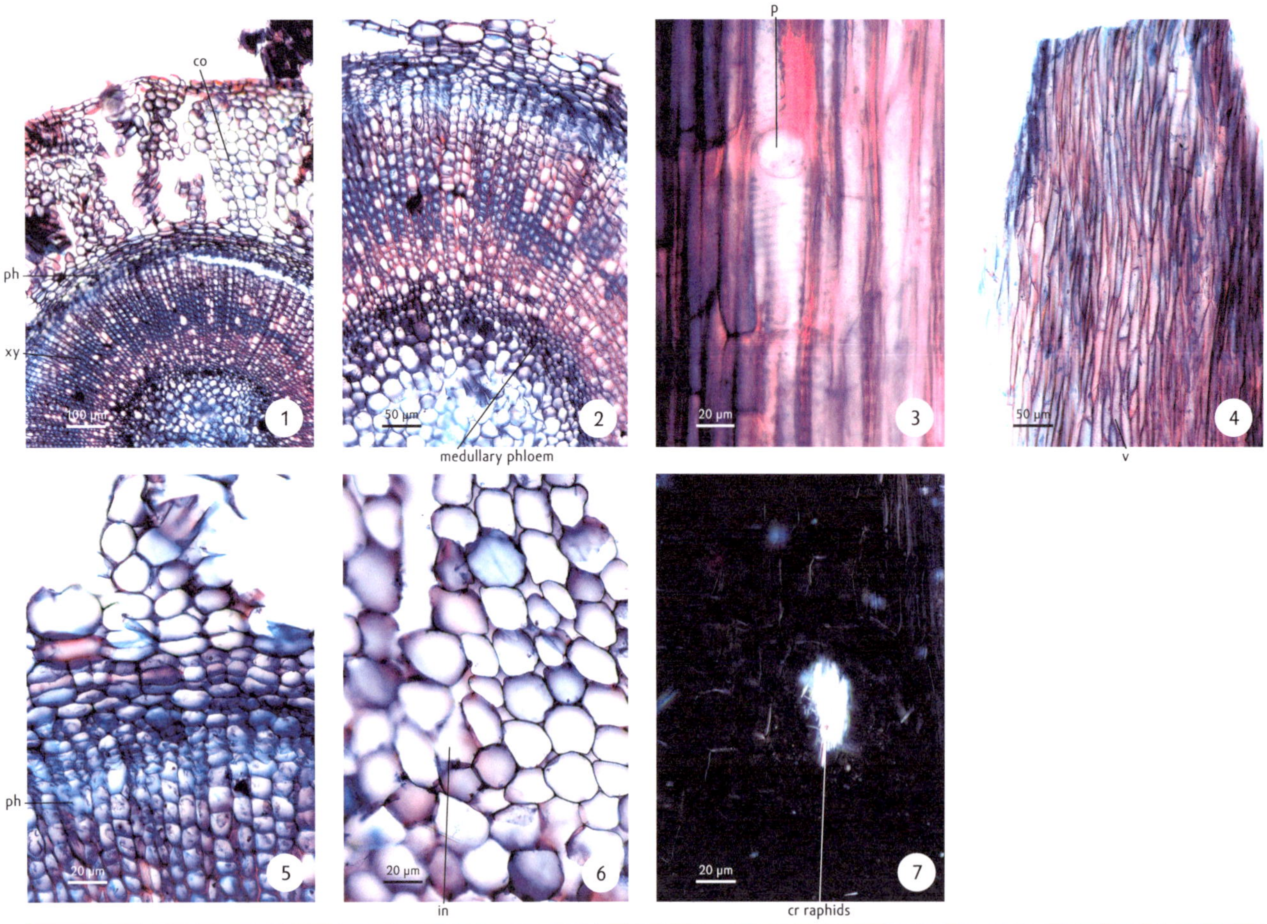

Epilobium parviflorum Schreb. **Hoary Willowherb** Onagraceae

Perennial wetland herb with short leafy stolons or perennating leaf rosette. Stem erect, rounded, densely pubescent, 30–60 cm high. Leaves oblong-lanceolate to narrowly lanceolate, pubescent, opposite, rounded at base and not clasping the stem. Petals <1 cm, pale-pink, four-lobed stigma. Fruit is a linear capsule.

Distribution and ecology

Eurasian temperate element. Native in all CE countries. Occurs along banks of streams and ditches, in marshes, in wet pastures or in other disturbed wet sites. Prefers periodically flooded or waterlogged clayish or loamy soils. Medicinal herb.

Anatomy of the root collar

Round stem with a diameter of 5–8 mm. Maximum age 2 years (1).

Stem consists of a xylem, a small phloem, a cortex and a large phellem (1). The diffuse-porous xylem is composed of vessels and thin- to thick-walled fibers (2). Vessels are arranged solitary or in short radial rows (1, 2). Vessel diameter varies from 40–50 µm and vessel length from 150–200 µm. Intervessel pits are round (3) and perforation plates are simple (3). Fibers with small pits are thin- to thick-walled (2). Locally they contain gelatinous secondary walls (tension wood) (2). Parenchyma cannot be recognized on cross-sections. Rays are uni- to biseriate (4). Crystals are absent in the xylem.

The phloem contains small groups of sieve tubes. Raphids in large idioblasts occur in the phloem and the cortex (5, 6). Intercellulars are absent. Groups of sieve tubes occur at the periphery of the pith (medullary phloem) (7).

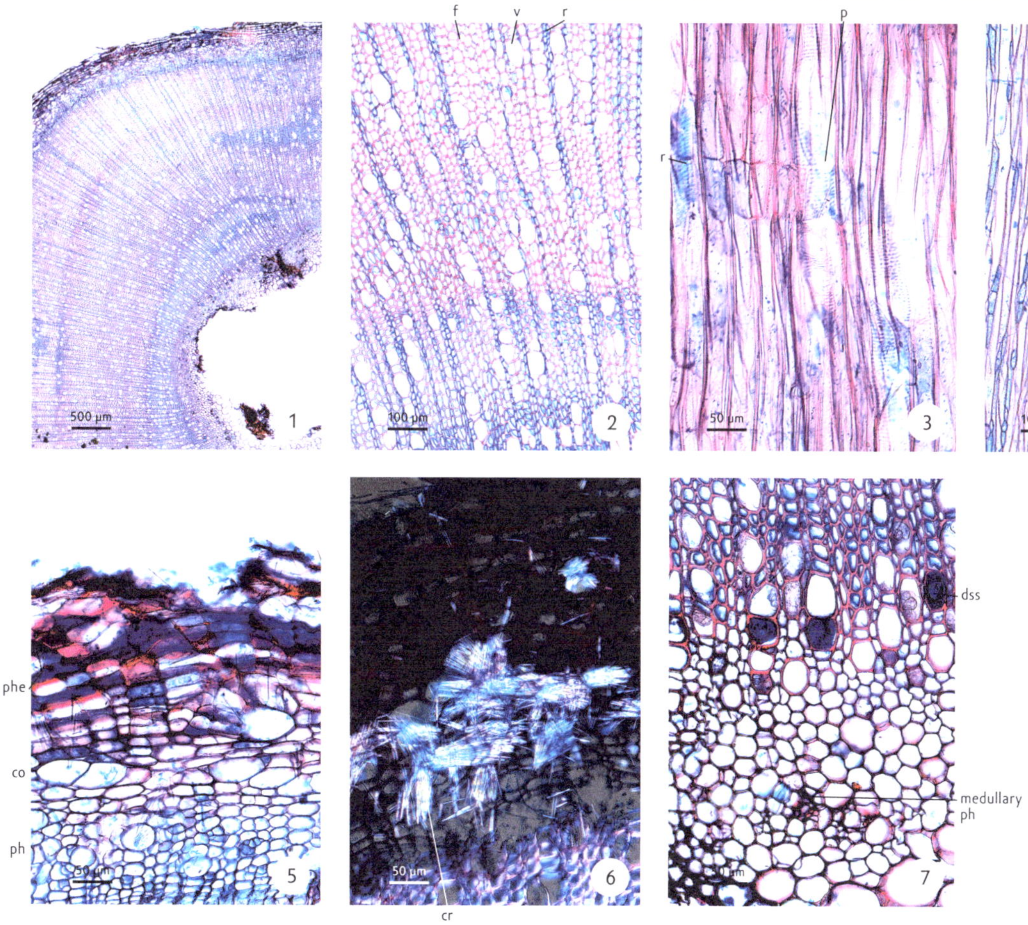

Ludwigia grandiflora (Michx.) Greuter & Burdet

Onagraceae
Water primrose

Perennial emergent or floating-leaved herb with white spongy roots. Two growth stages, in the first, the stem can grow horizontally over the soil or water, rooting at nodes; in the second growth stage, shoots can grow vertically up to 1 m and flower. Leaves glabrous, alternate and have petioles. Flowers on solitary stalks (2.5 cm long) with 5 obovate, bright golden-yellow petals with a darker spot at the base. Fruit is a pubescent light-brown capsule, with seeds embedded in a woody endocarp.

Distribution and ecology

Native to South and Central America and parts of the USA, introduced to France. As neophyte in CH. Grows in a variety of habitats, including wet meadows, river banks and still and slow-flowing waters. Prefers high nutrient concentrations, high temperatures and high irradiance. Propagation by small plant fragments. Allelopathic effects on native plants.

Anatomy of the stem

Round stem with a diameter up to 2–4 mm. Maximum age 1 year (1).

Stem consists of a large pith, a small xylem and phloem, a large cortex and an epidermis (1). The peripheral xylem is composed of vessels and thin- to thick-walled fibers (2) and the internal xylem of vessels and unlignified pervasive parenchyma (2). Vessels are arranged solitary (1, 2). Vessel diameter varies from 50–80 µm and vessel length from 300–500 µm. Intervessel pits are round (3) and perforation plates are simple (3) and scalariform (4, 5). Fibers are thin- to thick-walled in the outer part of the xylem and absent in the central part (2). Parenchyma and rays cannot be distinguished (2, 6). Rays are absent (4). Crystals are absent in the xylem.

The phloem consists of parenchyma cells, single sieve tubes and a few isolated fibers (7). The phloem and cortex are separated by an endodermis (1). The cortex represents a net-like aerenchyma with well-delimited air cavities (1). The cortex and the pith contain raphids in idioblasts (8). Groups of sieve tubes occur at the periphery and in the center of the pith (medullary phloem) (2).

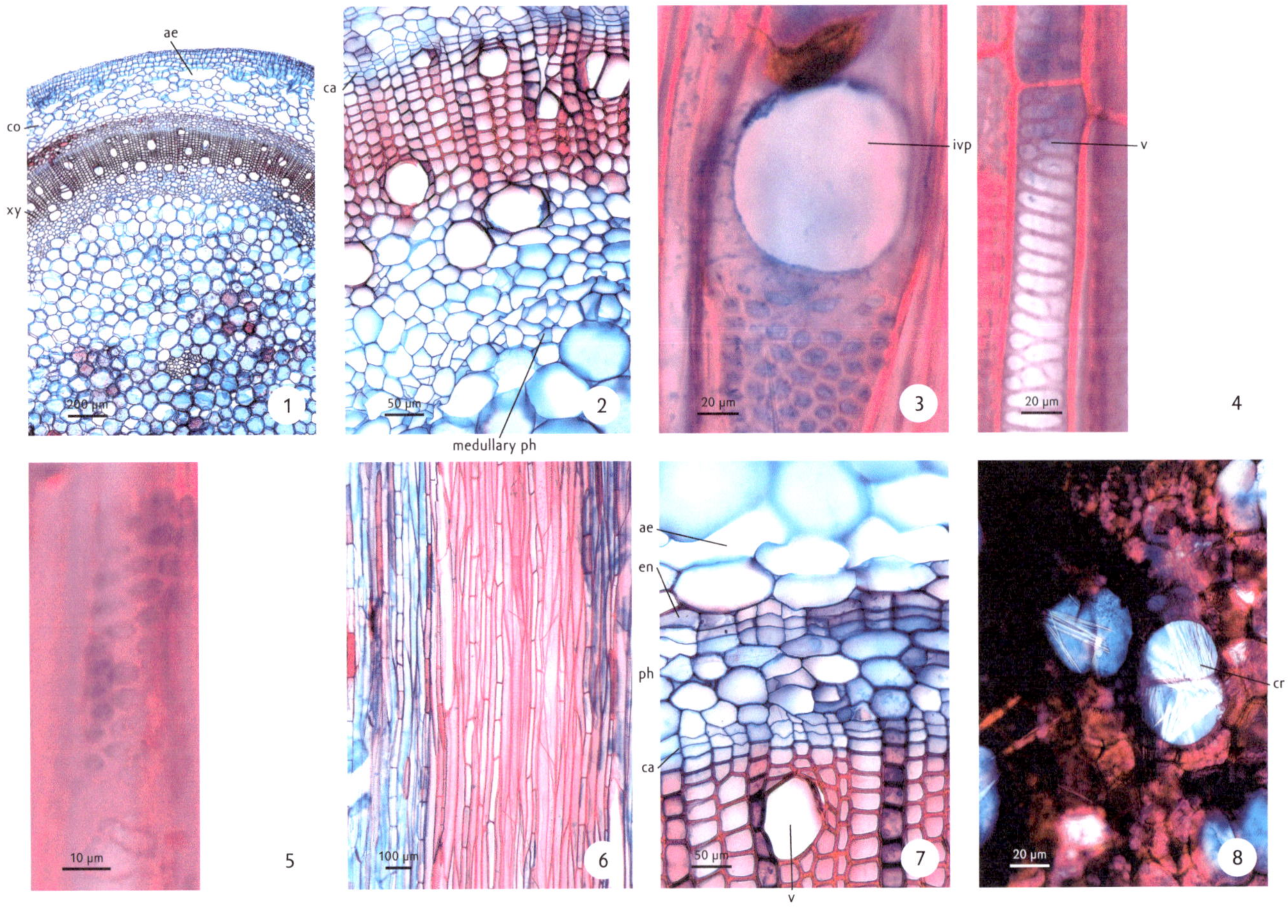

Ludwigia palustris (L.) Elliott

Hampshire-purslane

Onagraceae

Annual or perennial rooted floating-leaved herb. Stem glabrous, creeping, rooted in nodes, usually reddish-suffused, 10–50 cm long. Leaves in opposite pairs, glabrous, oval to elliptical, short-stalked. Small square flowers solitary in leaf axils, 3 mm across, with a four-lobed stigma. Fruit is a 4–5 mm long capsule.

Distribution and ecology

Subatlantic-submediterrean-circumpolar element; secondarily in New Zealand and Australia. Native in CH, DE, HU, very rare in AT and SK, probably extinct in PL. Grows in shallow water bodies, in seasonally-inundated pools and ditches. Prefers muddy and nutrient-rich sediments.

Anatomy of the stem

Round stem with a diameter up to 1–2 mm. Maximum age 1 year (1).

Stem consists of a large pith, a small xylem and phloem, a large cortex and an epidermis (1). The peripheral xylem is composed of vessels and thin- to thick-walled fibers (2) and the internal xylem of vessels and unlignified pervasive parenchyma (2). Vessels are angular and arranged solitary (1, 2). Vessel diameter varies from 30–40 µm and vessel length from 150–200 µm. Intervessel pits are round (3) and perforation plates are simple (3). Fibers are thin- to thick-walled in the outer part of the xylem and absent in the central part (2). Parenchyma and rays cannot be distinguished in the outer part of the xylem (2). Rays and crystals are absent in the xylem (4).

The phloem consists of parenchyma cells and small groups of sieve tubes (2). The phloem and cortex are separated by an endodermis (1). The cortex represents a net-like aerenchyma with well-delimited air cavities (1, 5). The cortex and the pith contain raphids in idioblasts. Groups of sieve tubes occur at the periphery and in the center of the pith (medullary phloem) (2).

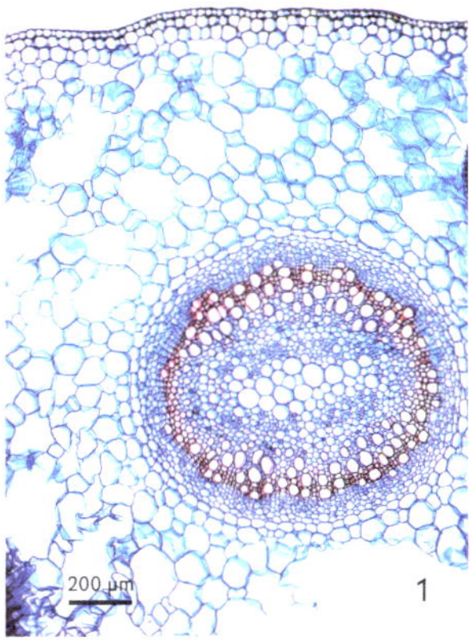

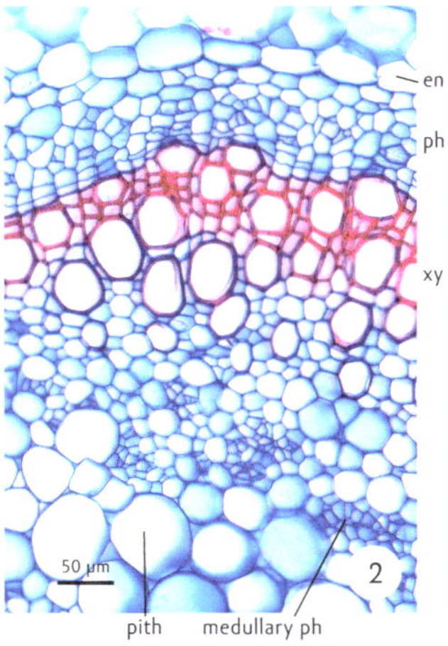

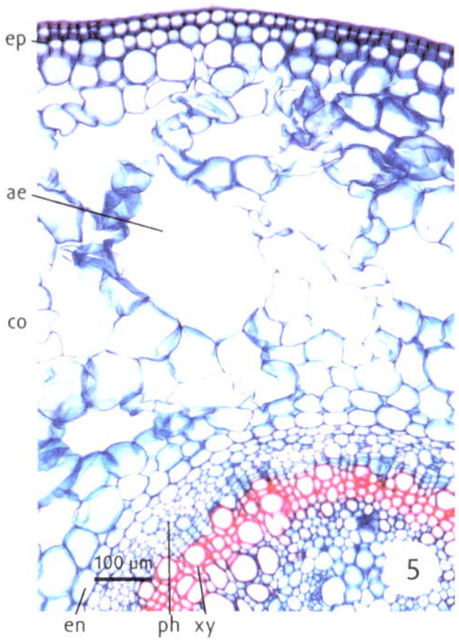

Pedicularis palustris L.

Marsh Lousewort

Orobanchaceae

Biennial to short-lived wetland herb. Stem erect, single, branched, up to 60 cm high. Leaves 2–6 cm long, deeple pinnatified, lobes toothed. Flowers in loose leafy raceme, calyx tubular, inflated below, corolla pink-purple, two-lobed. Fruit is a capsule longer than calyx.

Distribution and ecology

European boreo-temperate element. Native in all CE countries. Prefers nutrient-rich organic soils with high groundwater table. Occurs in a wide range of base-rich to acidic soils on peaty and fen meadows, in the margins of peatbogs and oligotrophic fishponds. Root-hemiparasite. Pollen limitation in small populations.

Anatomy of the annual shoot and the root collar

The basal part of the stem (root collar) has a diameter of 3–4 mm (1, 2). Maximum age 2 years (2).

The annual shoot has a large hollow pith, a large xylem, a small phloem and a large cortex (1). The root collar consists of a large xylem, a large phloem and a large cortex. (2) Annual rings are indicated by semi-ring-porosity (2). The xylem contains vessels, thin- to thick-walled fibers and axial parenchyma cells. Vessels are angular and are arranged solitary and in short radial rows (2, 3). Earlywood vessels have a diameter of 50–70 µm and a length of 80–150 µm. Vessel-walls have round intervessel pits (4) and simple perforation plates (4). Rays are absent (5). Fibers are thin- to thick-walled (3). Most of them contain nuclei. Parenchyma is paratracheal (3). Rays are absent (2, 3, 5).

Parenchyma cells and small groups of sieve tubes characterize the phloem (6). The cortex consists of oval parenchyma cells and irregularly formed, large intercellulars (1, 2, 7). The periphery of the annual shoot consists of an epidermis (1, 7).

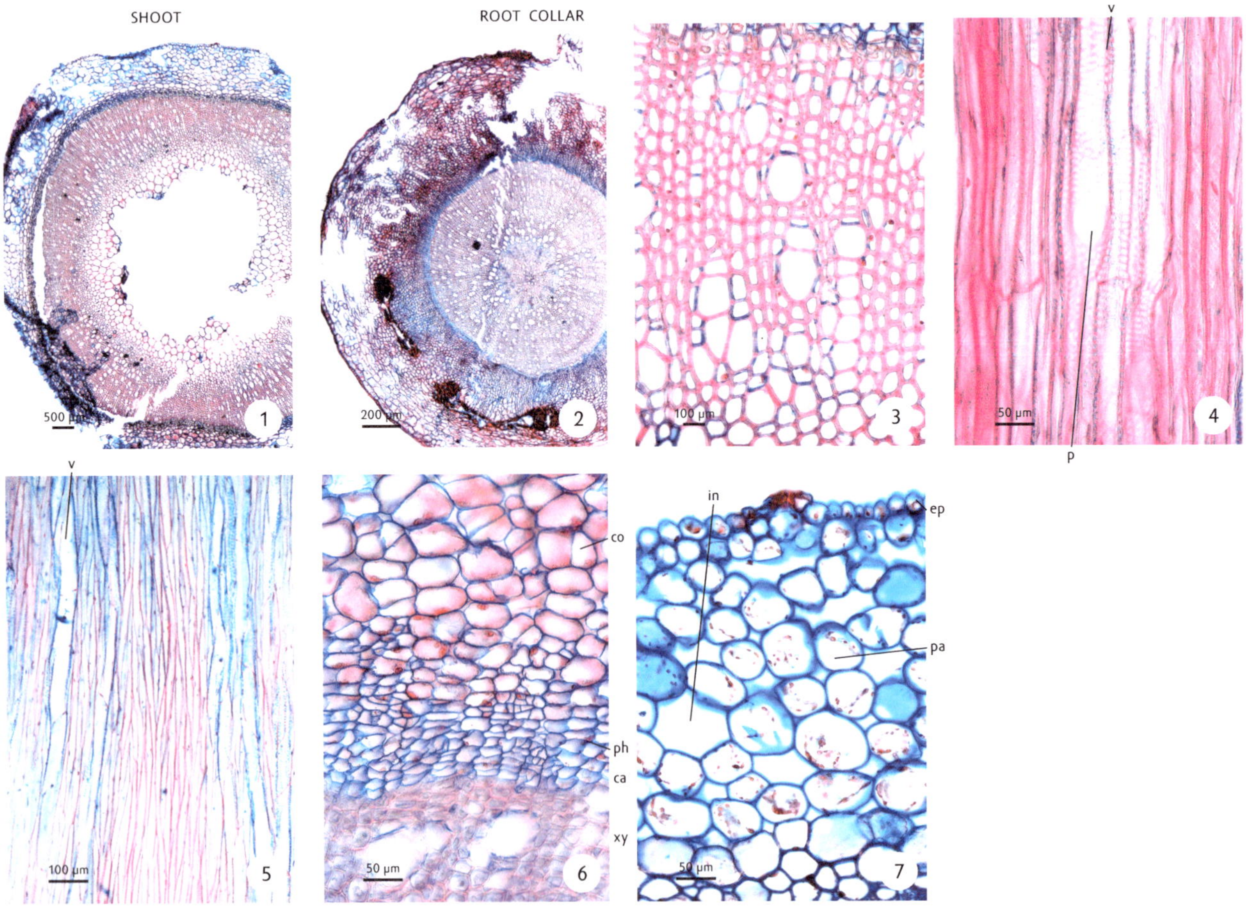

F. H. Schweingruber et al., *Anatomic Atlas of Aquatic and Wetland Plant Stems*, https://doi.org/10.1007/978-3-030-33420-8_41

Pedicularis sylvatica L.

Lousewort

Orobanchaceae

Biennial to short-lived wetland herb. Stems several, spreading from base, ascending to procumbent, only 10–20 cm high. Leaves 2–3 cm long, deeply pinnately lobed, lobes toothed. Flowers in loose leafy raceme, calyx inflated, five-angled, corolla pink, upper lips two-toothed. Fruit is an ellipsoid capsule shorter than calyx.

Distribution and ecology

European temperate element. Native in all CE countries except HU. Prefers acidic, nutrient-poor soils, occurs on peaty meadows, wet pastures and heathlands with high water table. Heliophilous, competitively weak species. Root-hemiparasite.

Anatomy of the annual shoot

The basal part of the stem (root collar) has a diameter of 3–4 mm (1, 2). Maximum age 2 years (2).

The root collar consists of a large xylem, a large phloem, a large cortex and a small phellem (2). Annual rings are indicated by semi-ring-porosity (1, 2). The xylem contains vessels and thin- to thick-walled fibers and axial parenchyma cells. Vessels are angular and are arranged solitary (2). Earlywood vessels have a diameter of 40–60 μm and a length of 80–100 μm. Vessel walls have round intervessel pits (3) and simple perforation plates (3). Vascular bundles remain and are separated by large rays (1, 2). Rays within vascular bundles are absent (2, 4). Fibers are thin- to thick-walled (3). Parenchyma is pervasive (3).

Parenchyma cells and groups of sieve tubes characterize the phloem (5). The cortex consists of oval parenchyma cells and irregularly formed, large intercellulars (1). The periphery of the annual shoot consists of a phellem containing dark stained substances (1). Central parts are bent due to contractile vessels (6).

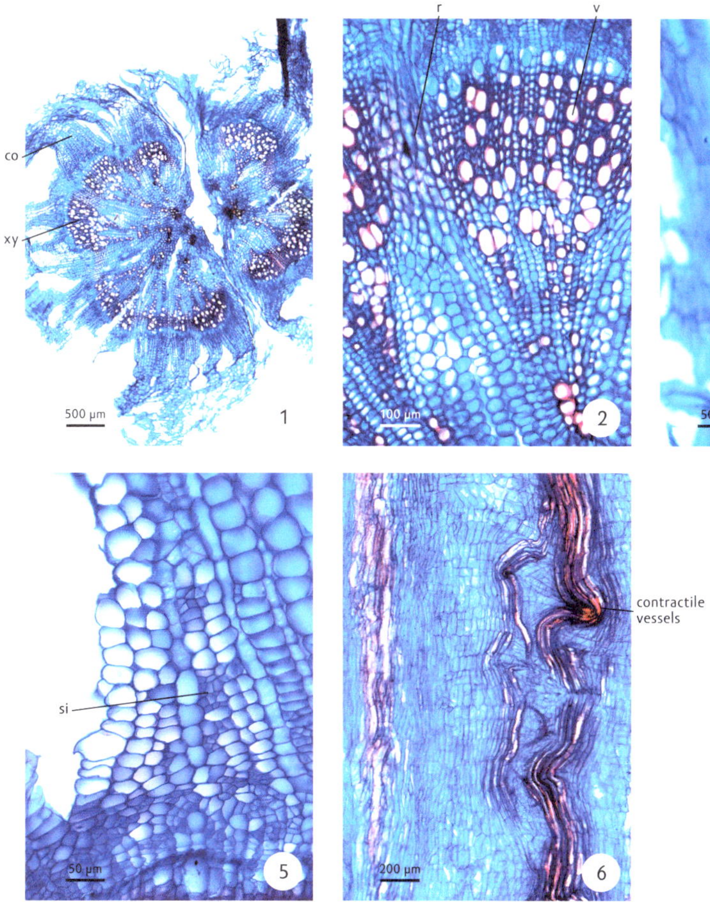

Mimulus guttatus DC. (syn. *Erythranthe guttata*) **Monkeyflower** Phrymaceae

Perennial wetland herb. Stem hollow, ascending to erect, branched, glabrous below, glandular-pubescent above, 25–50 cm high. Leaves opposite, glabrous, broadly ovate, margin dentate. Inflorescence a raceme, flowers with upper and lower lips, corolla deep yellow or with red spots on lower lip, up to 4 cm long. Seeds ellipsoid, base narrowed with hilum.

Distribution and ecology

Native to western N. America. Neophyte in all CE countries, rare in HU. Occurs along streams, rivers and ditches, also in damp meadows. Prefers partly shaded and wet, oligotrophic sandy or stony soils. Able to spread vegetatively by rooting from the nodes.

Anatomy of the root collar

Stem base with a diameter of 2–6 mm. Maximum age 1 year.

The basal part of the stem consists of a large pith, a small xylem and a phloem, and a large cortex (1).

The xylem contains solitary vessels, thin- to thick-walled fibers, parenchyma cells and rays (2). Vessels are arranged solitary or in short radial rows and have a diameter of 50–80 µm (2) and a length of 100–200 µm (3). Vessel-walls have round intervessel pits (4) and simple perforation plates (3). Axial parenchyma is paratracheal (2). Rays are absent (2, 4).

Cortex and phloem are separated by two layers of a poorly differentiated endodermis (5). Parenchyma cells and small groups of sieve tubes characterize the phloem (6). The inner one contains small thick-walled, lignified cells (6). The cortex and the pith are composed of angular parenchyma cells almost without intercellulars (1, 7). The phellem is absent.

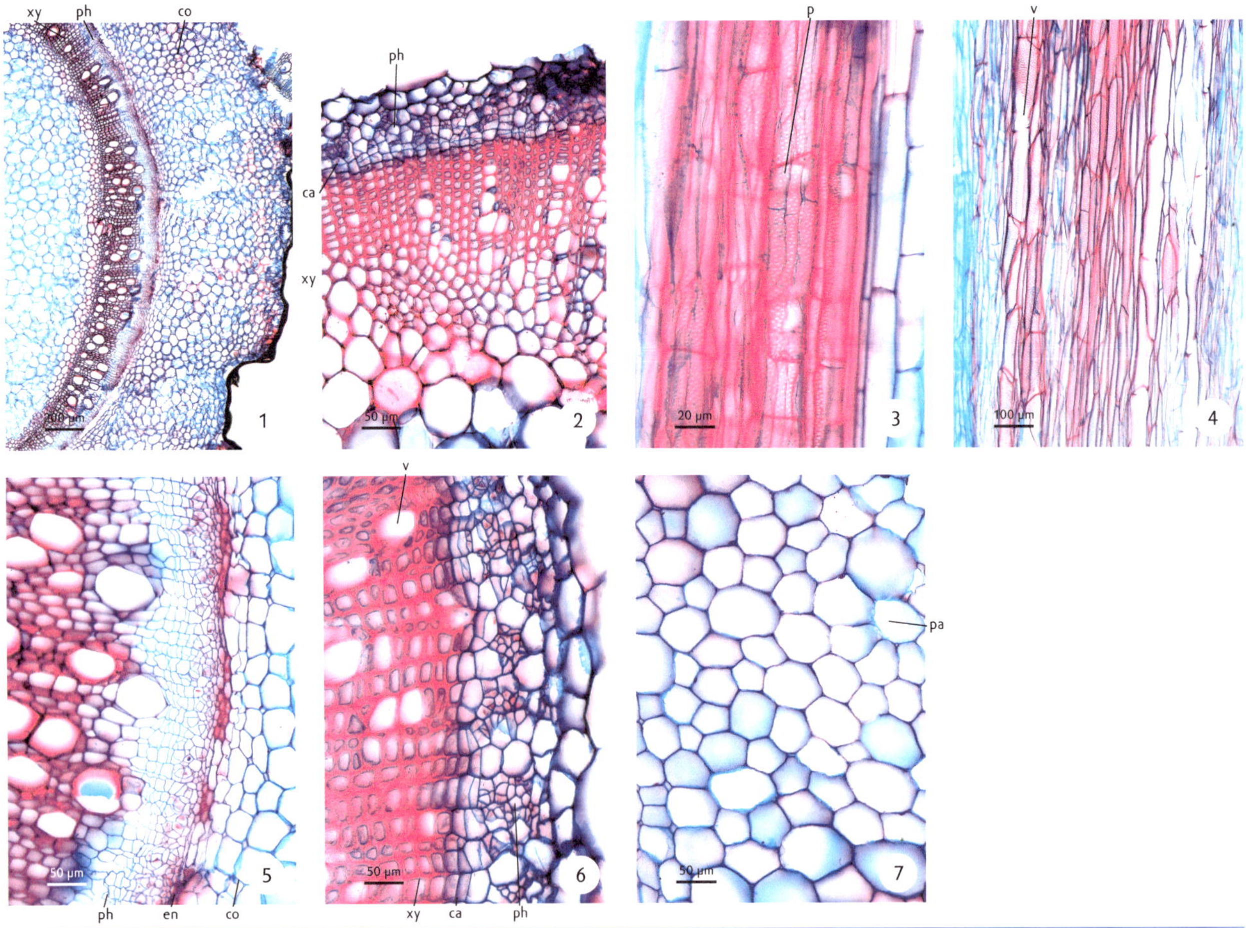

F. H. Schweingruber et al., *Anatomic Atlas of Aquatic and Wetland Plant Stems*, https://doi.org/10.1007/978-3-030-33420-8_42

Callitriche hamulata Kütz ex W. D. J. Koch

Intermediate Water-starwort

Submerged, floating or terrestrial annual or perennial herb with heterophylly. Branched shoots up to 100 cm long with leaf rosettes floating on the water surface. Submerged leaves thin and long, with bicycle-spanner-like leaf apex, floating leaves elliptic. Fruits always sessile, dark brown, orbicular, with narrow wings.

Distribution and ecology

European temperate distribution. Native in all CE countries except HU. Prefers rapidly flowing, well oxygenated, meso-eutrophic streams and rivers, usually poor in calcium. Occurs also in softwater lakes and ponds with sufficient nutrient and C availability. Strict CO_2 user.

Anatomy of the submerged shoot

Round shoot with a diameter <1mm. Maximum age approximately 1 year.

The small stele is surrounded by a very large cortex and an epidermis (1). The stele consists of a few unlignified vessels and parenchyma cells. Xylem and phloem cannot be distinguished. Vessels have a diameter of 5–10 µm. They have scalariform to reticulate intervessel pits. Perforation plates are probably absent. Fibers and rays are absent. Parenchyma is pervasive (2, 3).

The stele and the cortex are surrounded by an endodermis. The cortex consists of large radial aerenchymatic intercellulars (1, 2).

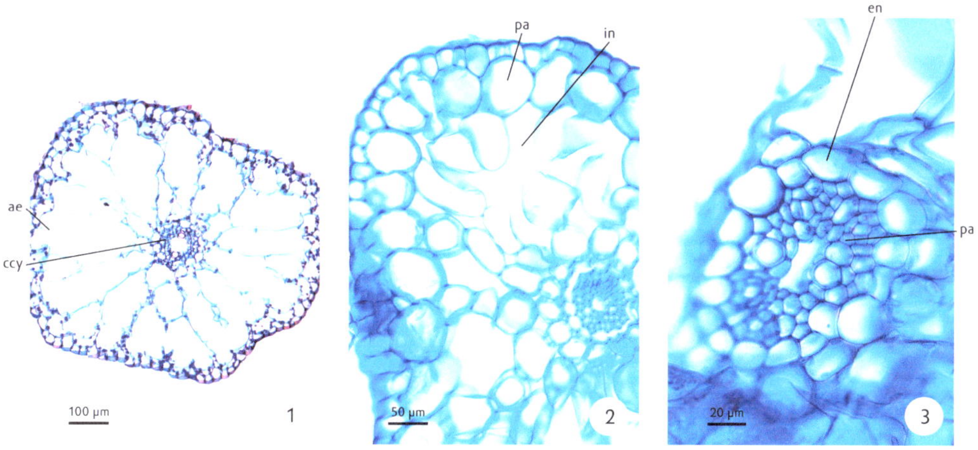

F. H. Schweingruber et al., *Anatomic Atlas of Aquatic and Wetland Plant Stems*, https://doi.org/10.1007/978-3-030-33420-8_43

Callitriche palustris L.

Vernal Water-starwort

Perennial submerged to biennial terrestrial wetland herb. Floating submerged stems up to 40 cm long, fragile. Rosette-leaves rounded, submerged leaves narrower. Fruits are blackish schizocarps, each of 4 nutlets obovate to elliptic, mericarp winged only at apex.

Distribution and ecology

Circumpolar boreal-temperate element. Native in all CE countries, especially in ephemeral water bodies, in lakes with marked water level fluctuations, temporary pools and rutted tracks. Prefers meso-eutrophic freshwaters. Frequent also as a terrestrial form on wet mud, may dominate on the drained fishpond bottoms.

Anatomy of the submerged shoot

Round shoot with a diameter < 1mm. Maximum age approximately 1 year.

The small stele is surrounded by a very large cortex and an epidermis (1, 2). The stele consists of a few solitary vessels and parenchyma cells. Vessels have a diameter of 5–10 µm. They have annular thickenings with scalariform to reticulate intervessel pits. Perforation plates are probably absent. Fibers and rays are absent. Parenchyma is pervasive (2, 3).

The stele and the cortex are surrounded by two rows of endodermis-like cells (4). The phloem consists of groups with a few sieve tubes. The cortex consists of large radial aerenchymatic intercellulars (1, 2, 4).

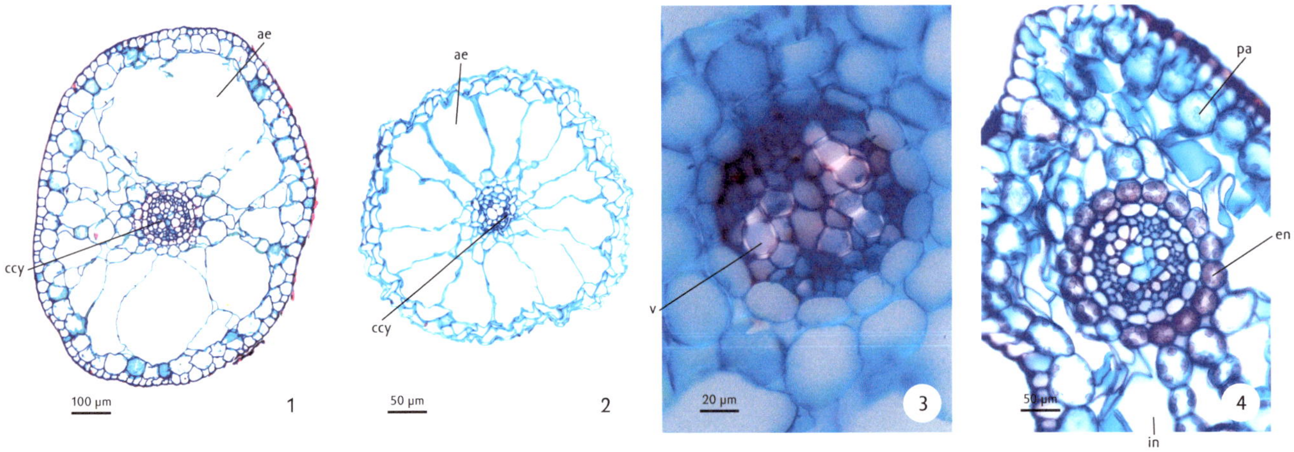

Gratiola officinalis L.

Hedge Hyssop Plantaginaceae

Perennial wetland herb with underground creeping rhizome. Stem erect to ascendent, hollow, 4-angular, 25–40 cm high. Leaves sessile, opposite, lanceolate and serrate on the margin. Flowers in leaf axils, corolla two-lipped, white. Fruit is an orbicular capsule.

Distribution and ecology

Eurasian distribution. Native in all CE countries. Endangered river corridor plant of Central European lowlands, grows on subcontinental flooded meadows, on muddy banks of fishpond, oxbows and ditches. Typical pioneer species with absolute light requirement for germination. Medicinal plant.

Anatomy of the root collar

Round stem with a diameter of 3–5 mm. Maximum age 2 years.

The basal part of the stem consists of a large pith, a small xylem and phloem, a large cortex and an epidermis (1). Annual ring-boundaries are characterized by semi-ring porosity (2). The xylem contains solitary vessels, thin- to thick-walled fibers, parenchyma cells and rays (2). Vessels have a diameter of 30–50 µm and a length of 80–120 µm. Vessel walls have round intervessel pits and simple perforation plates (3). Axial parenchyma is paratracheal (2). Rays appear uniseriate in cross-sections (2) but are not recognized in tangential sections (4).

Cortex and phloem are separated by an endodermis (2, 5). Parenchyma cells and small groups of sieve tubes characterize the phloem (5). Few isolated fibers occur at the outer border of the phloem (5). The cortex and the pith are represented by a net-like aerenchyma with large intercellulars (6, 8). Special is the presence of ring-like lignified cell-wall thickenings (7). The periphery consists of an epidermis and a hypodermis (1).

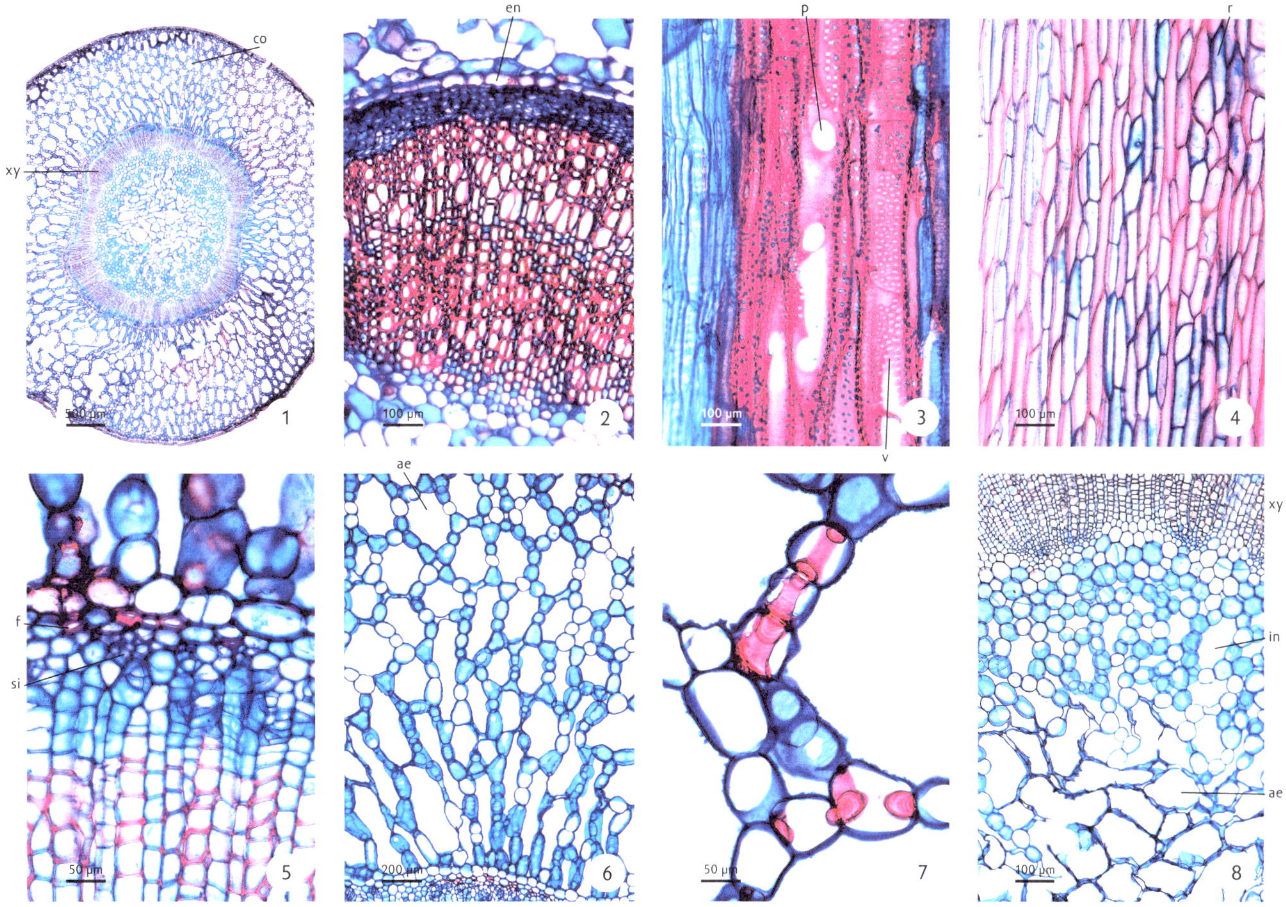

Hippuris vulgaris L.

Mare's tail Plantaginaceae

Perennial amphibious plant with heterophylly. Submerged shoots 30–100 cm long bear whorls of thin, up to 8 cm long linear membraneous leaves, emergent shoots 20–40 cm long with whorls of thick and up to 22 mm long leathery leaves. Greatly reduced flowers are in axils of leaves. Fruit is an achene.

Distribution and ecology

Circumpolar boreo-temperate element. Scattered rarely and irregularly in all CE countries, from lowlands to uplands. Shores of fishponds and lakes, backwater pools, ditches, canals, slow-flowing rivers. Prefers hard waters and base-rich soils. Facultative halophyte, tolerates even brackish waters.

Anatomy of a basal part of the stem

The round basal part of the stem has a diameter of 5–6 mm (1). Annual rings absent.

The stem is composed of a circle of xylem and phloem, a large cortex and an epidermis (1, 2). The xylem consists of angular, grouped, unlignified vessels and pervasive parenchyma (3). Vessels have a diameter of 20–30 µm (3) and a length of 200–250 µm. Intervessel pits are scalariform (4) and perforations are simple. Fibers and rays are absent (3).

The phloem consists of large angular sieve tubes and groups of small companion cells (3). An endodermis surrounds the stele (3). The large cortex is characterized by a net-like aerenchyma with well-delimited net-like cavities (1, 2, 5). The epidermis consists of small roundish thin-walled cells. Crystals are almost absent.

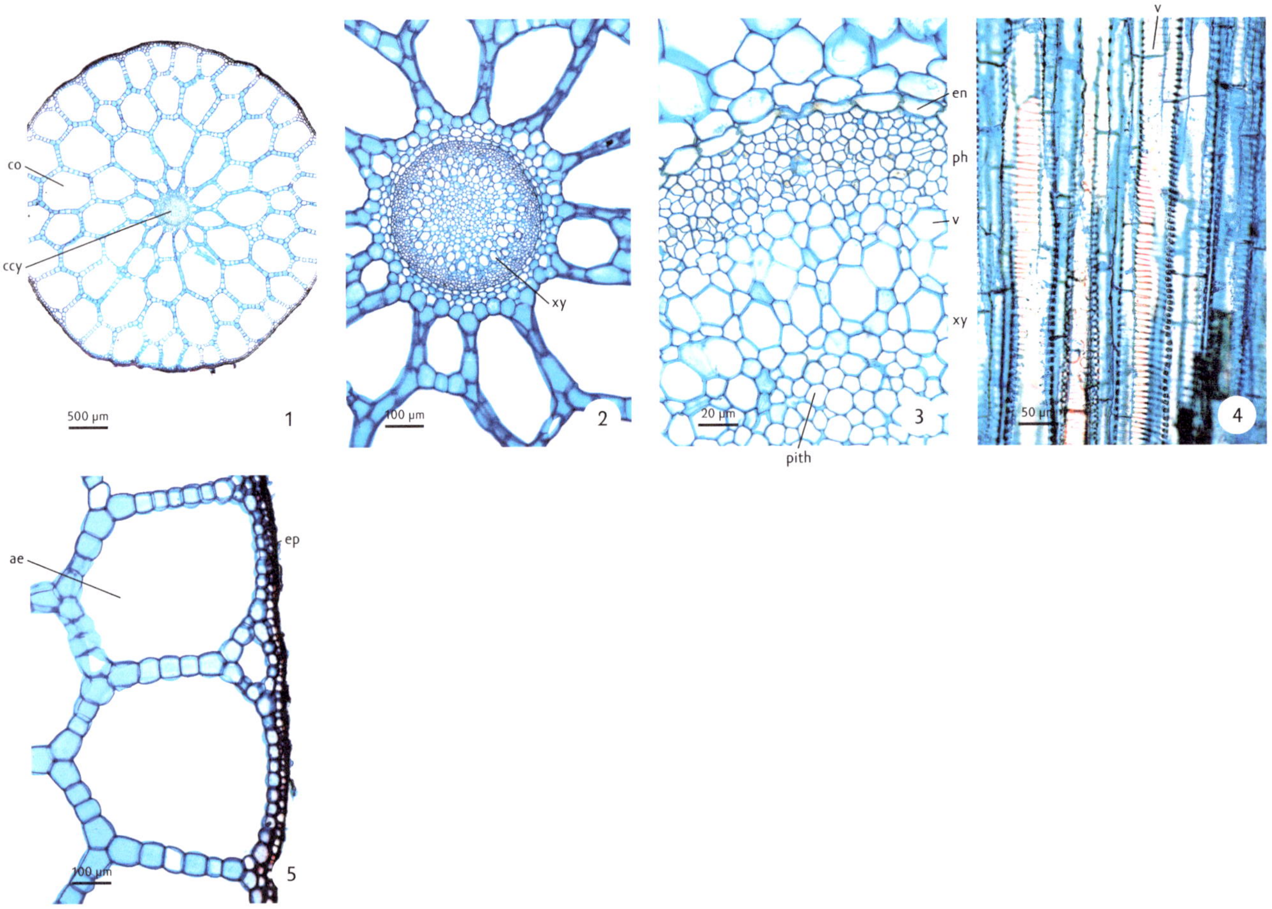

Littorella uniflora (L.) Asch. (syn. *Plantago uniflora*) **Shoreweed** Plantaginaceae

Perennial, evergreen aquatic amphibious plant. Basal rosette, stiff, subulate, septate leaves 2–15 cm long, numerous fasciated hollow roots, thin rhizome with numerous adventitious roots, rise to daughter plants. Submerged and aerial leaves with different adaptive morphology and anatomy – submerged longer, more fragile, 2–3 times thicker than the aerial ones, lack stomata, with larger lacunal space, covered with thick cuticle. Submerged plants sterile, reproduce by runners. Flowers unisexual, fruit is a hard achene.

Distribution and ecology

Suboceanic temperate element. Scattered rarely in all CE countries except for SK and HU. Standing shallow waters, lakes, fishponds, old shallow sand-pits, water depth 0–100 cm. Oligotrophic, very soft waters (pH 6.0–7.5, total alkalinity 0.0–1 meq/l), sandy bottom, clear water, sandy littorals, permanent submerged or terrestrial growth. Submerged plants exhibit typical ecophysiological traits of isoetids: root CO_2 uptake, CAM photosynthesis. Strict heliophyte, strict CO_2 user, low competitive ability, sensitive to eutrophication.

Anatomy of the root collar

Round stem with a diameter of 7 mm. Secondary growth occurs for several growing seasons.

A central part consists of a xylem and a phloem (1), which is surrounded by a cortex. The xylem consists of radially arranged rows of lignified, thick-walled secondary vessels with a diameter of 10–15 µm (2) and a length of 40–60 µm. They have scalariform to reticulate intervessel pits and simple perforation plates (3, 4). Vessels and fibers are absent. Parenchyma cells are thin-walled and unlignified. Distinct rays are absent (2).

Irregularly arranged groups of sieve tubes and parenchyma cells form the phloem (2).

The cortex consists of a net-like aerenchyma with small intercellulars (5). The stele is surrounded by a poorly differentiated pericycle/endodermis zone. The parenchyma cells of the aerenchyma have very large pits with net-like primary walls (5). The phellem consists of very thin-walled pentangular, unlignified cells (5). Air cavities in the pith are absent (1).

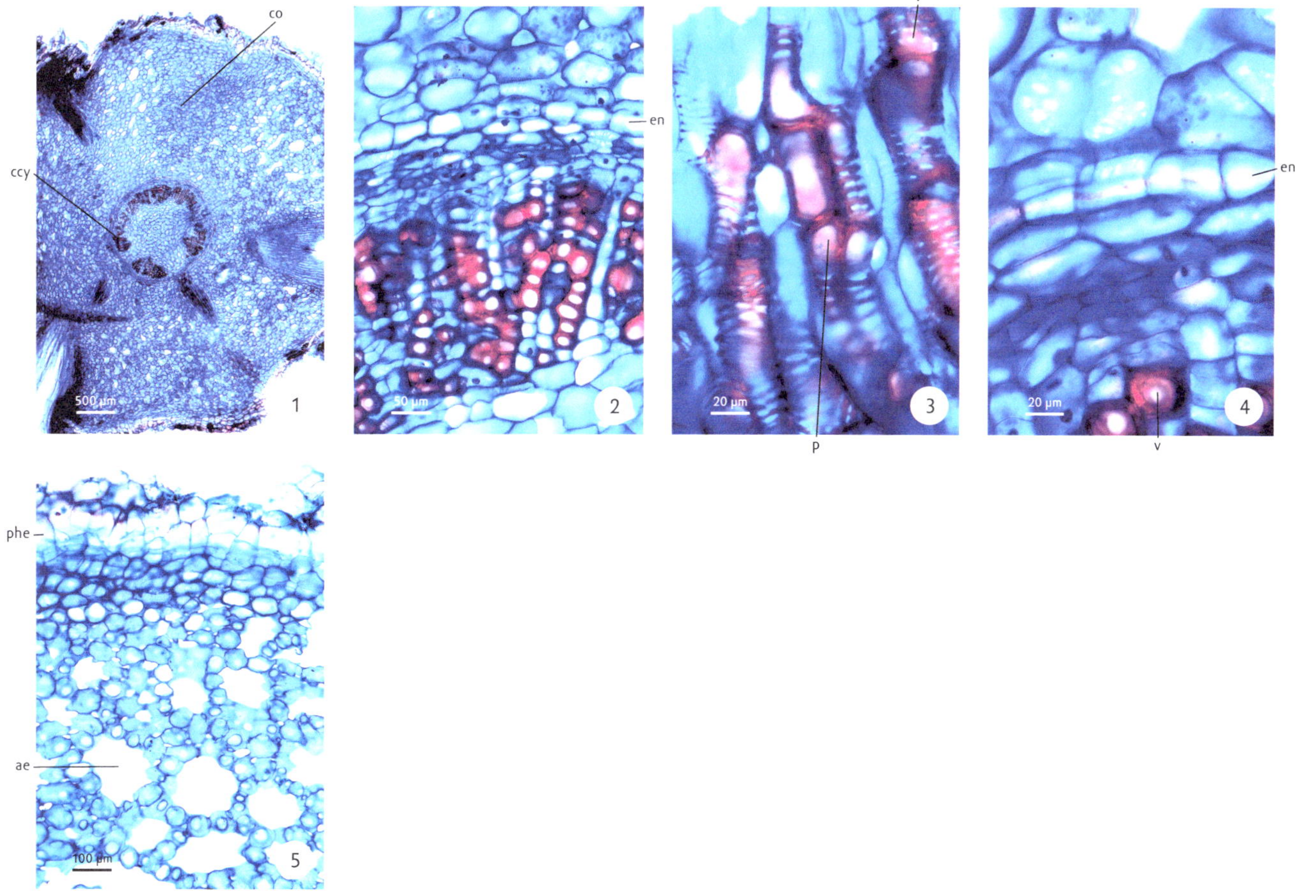

Plantago maritima L. **Sea Plantain** Plantaginaceae

Perennial wetland herb with woody base and ramified rhizome. Basal leaf rosette of narrow-linear, glabrous and fleshy leaves, 5–20 cm long. Inflorescence spike 2–6 cm long, on erect flower stalk 15–45 cm high. Flowers 3 mm long with brown corolla. Fruit is an ovoid capsule usually with two seeds.

Distribution and ecology

Euro-Siberian wide-boreal element; also in N. America. Native in all CE countries except CH. Typical of the middle and upper zones of saltmarshes, on coastal heaths and occasionally on inland saltmarshes. Facultative halophyte, accumulates high levels of sorbitol, glycine, betaine and proline in roots in response to increasing salt concentration. Very low floating capacity of seeds. Gynodioecious species.

Anatomy of the root collar

The anatomical variability is very large. Here plants from a site with a short vegetation period in northern Iceland are described. Maximum age approximately 12 years.

The xylem and the phloem are surrounded by a large cortex and a phellem (1). The semi-ring porous xylem consists of solitary lignified angular vessels with a diameter of 40–50 µm (2) and a length of 120–200 µm (3). They have scalariform to reticulate intervessel pits and simple perforation plates (3). Fibers and rays are absent (2). Parenchyma is pervasive (2). Small groups of sieve tubes (inter-xylary phloem) are characteristic (2).

Small bands of phellem separate the cortex from the phloem. Radial lines of parenchyma cells with enclosures of sieve-tube groups form the phloem (4). The cortex contains a net-like aerenchymatic tissue (4). The phellem consists of very thin-walled rectangular, unlignified cells (1).

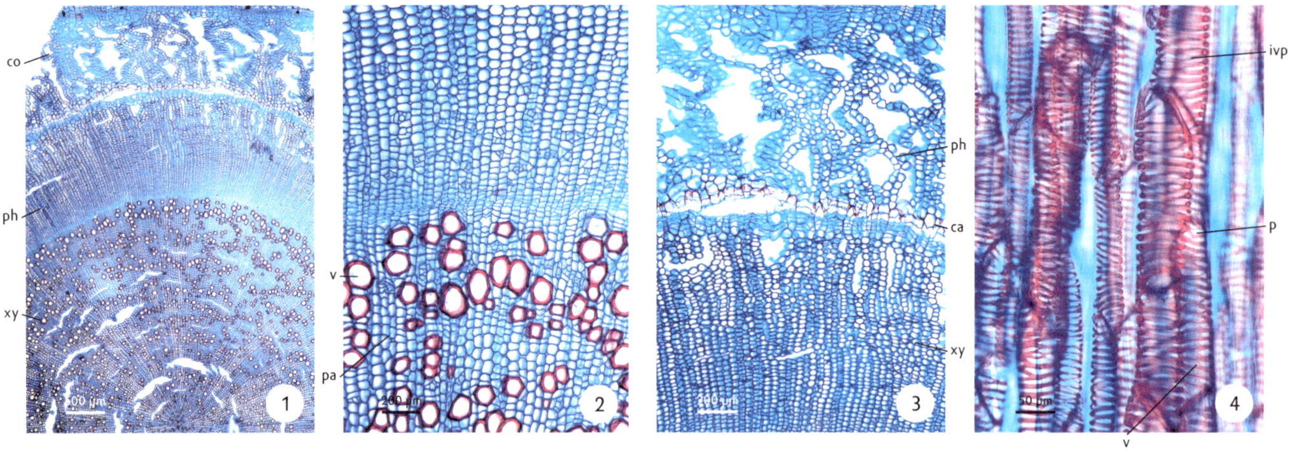

Veronica anagallis-aquatica L. **Blue Water-Speedwell** Plantaginaceae

Annual or perennial (only in the southern part of its distribution) amphibious herb with short rhizome. Stem erect to ascending, four-sided, hollow, 20–70 cm high. Leaves opposite, oval-lanceolate, pointed, scarcely-toothed, sessile above. Flowers in axillary racemes. Wheel-shaped corolla white or pale blue with purple veins, 4–8 mm across. Fruit is a glabrous capsule with calyx longer than the capsule at fruiting. V. anagallis-aquatica, V. anagalloides, and V. catenata have more or less the same seed features and are also taxonomically very relative.

Distribution and ecology

Eurasian southern-temperate element; widely naturalised outside its native range. Native in all CE countries. Prefers nutrient-rich soils in margins of rivers, streams, ditches and ponds. Ability to grow submerged in shallow water. High morphological plasticity according to the current velocity. Strict CO_2 user in submerged conditions. Vegetative reproduction by rooted stem fragments.

Anatomy of the root collar

The basal part of the stem has a diameter of 2–3 mm. Maximum age 1 year (1, 2).

The basal part of the stem consists of a large pith, a small xylem and phloem, and a large cortex (1, 2). The xylem contains vessels and thin- to thick-walled fibers (5). Axial parenchyma cells and rays are absent (1, 2, 5). Vessels are angular, arranged in short radial rows and have a diameter of 15–20 µm and a length of 120–180 µm. Vessel walls have round intervessel pits and simple perforation plates (3). Fibers are partially septate. Rays are absent (4).

Parenchyma cells and small groups of sieve tubes characterize the phloem (5). Cortex and phloem are separated by an endodermis (5, 6). The inner, multilayered zone contains occasionally small thick-walled, lignified cells. The cortex represents partially a net-like aerenchyma with large intercellulars (1, 2, 6). Intercellulars are absent in the pith (1). The periphery of the annual shoot consists of an epidermis (1, 2, 6).

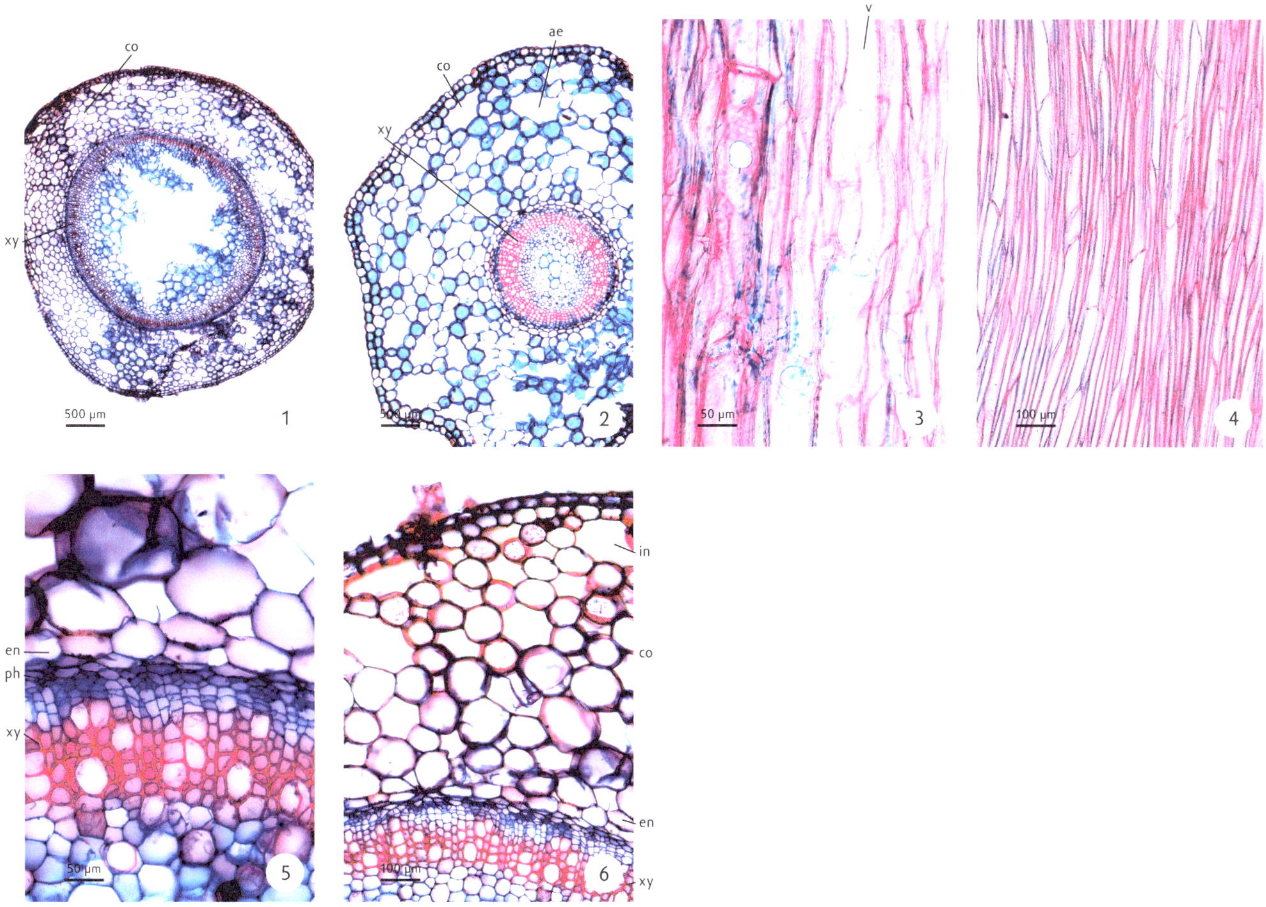

Veronica anagalloides Guss.

Annual wetland herb with short creeping rhizome. Stem full, glabrous, usually reddish, 10–30 cm high. Leaves opposite, sessile and lanceolate. Flowers in axillary racemes. Wheel-shaped corolla whitish with dark purple veins, only 2–4 mm across. Fruit is a glandulous capsule with calyx shorter than the capsule at fruiting.

Distribution and ecology

Eurasian distribution, also N. Africa. Native in all CE countries except PL. Grows in muddy banks of standing or slowly flowing water, on drained bottoms, on wet depressions in fields or wet meadows. Facultative halophyte.

Anatomy of the root collar

The basal part of the stem has a diameter of 2–3 mm. Maximum age 1 year (1).

The basal part of the stem consists of a large pith, a small xylem and phloem, and a large cortex (1). The xylem contains vessels and thin- to thick-walled fibers (2). Axial parenchyma cells and rays are absent (2). Vessels are angular and arranged solitary and have a diameter of 15–20 µm (2) and a length of 120–180 µm. Vessel walls have round intervessel pits (3) and simple perforation plates (3). Fibers are partially septate (3).

Cortex and phloem are separated by an endodermis (4). Parenchyma cells and small groups of sieve tubes characterize the phloem (2). The cortex represents partially a net-like aerenchyma with large intercellulars (1, 2). Intercellulars are absent in the pith (1). The periphery of the annual shoot consists of an epidermis (1).

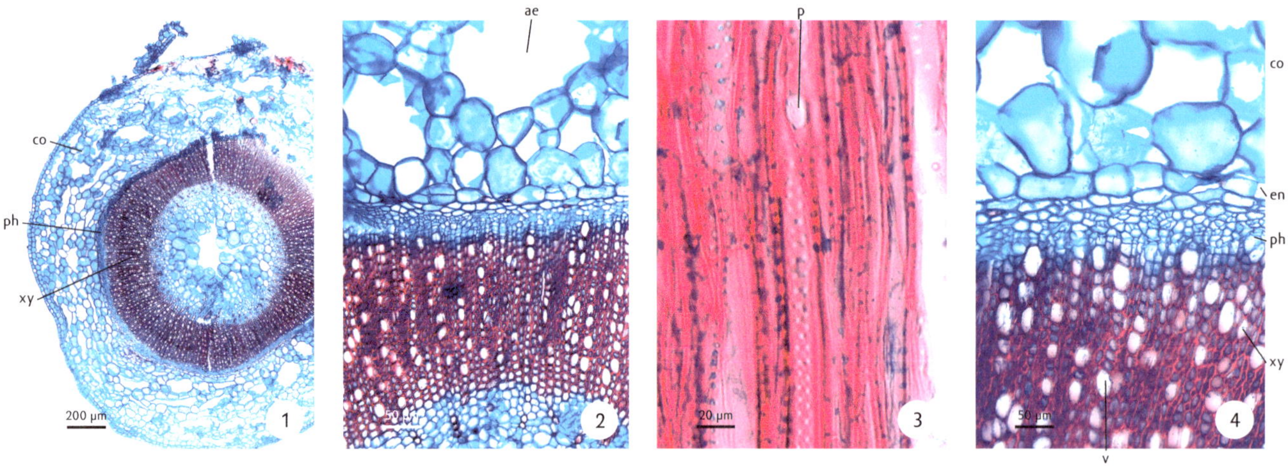

Veronica beccabunga L.

Brooklime

Plantaginaceae

Perennial amphibious herb with creeping, partially aboveground rhizome. Stem full, fleshy, creeping, rooting and ascending, up to 60 cm high. Leaves thick, ovate-oblong to elliptic, shallow-toothed and short petiolate. Inflorescence is a loose many-flowered raceme in leaf axils. Bright blue corollas are 7–8 mm wide. Fruit is a rounded capsule.

Distribution and ecology

Euro-Siberian temperate element; widely naturalised outside its native range. Native in all CE countries. Prefers wet, slightly acidic soils and open habitats. Occurs on banks of rivers and streams, on springs, in ditches and ponds. Strict CO_2 user. Able to propagate vegetatively from rooted stems.

Anatomy of the root collar

The basal part of the stem has a diameter of 3-5 mm. Maximum age 1 year (1).

The basal part of the stem consists of a large hollow pith, a xylem, a small phloem and a large cortex (1). The xylem contains vessels and thin- to thick-walled fibers: Axial parenchyma cells and rays are absent or difficult to recognize (2). Vessels are polygonal, are arranged solitary and have a diameter of 30–50 µm and a length of 80–120 µm. Vessel-walls have round intervessel pits (3) and simple perforation plates (3). Rays are absent (4).

Parenchyma cells and small groups of sieve tubes characterize the phloem (5). Cortex and phloem are separated by two layers of a poorly differentiated endodermis/pericycle zone (2, 5). The cortex and the pith are composed of round and oval parenchyma cells and small intercellulars (5, 6). A very small phellem is located between the inner and outer cortex (5).

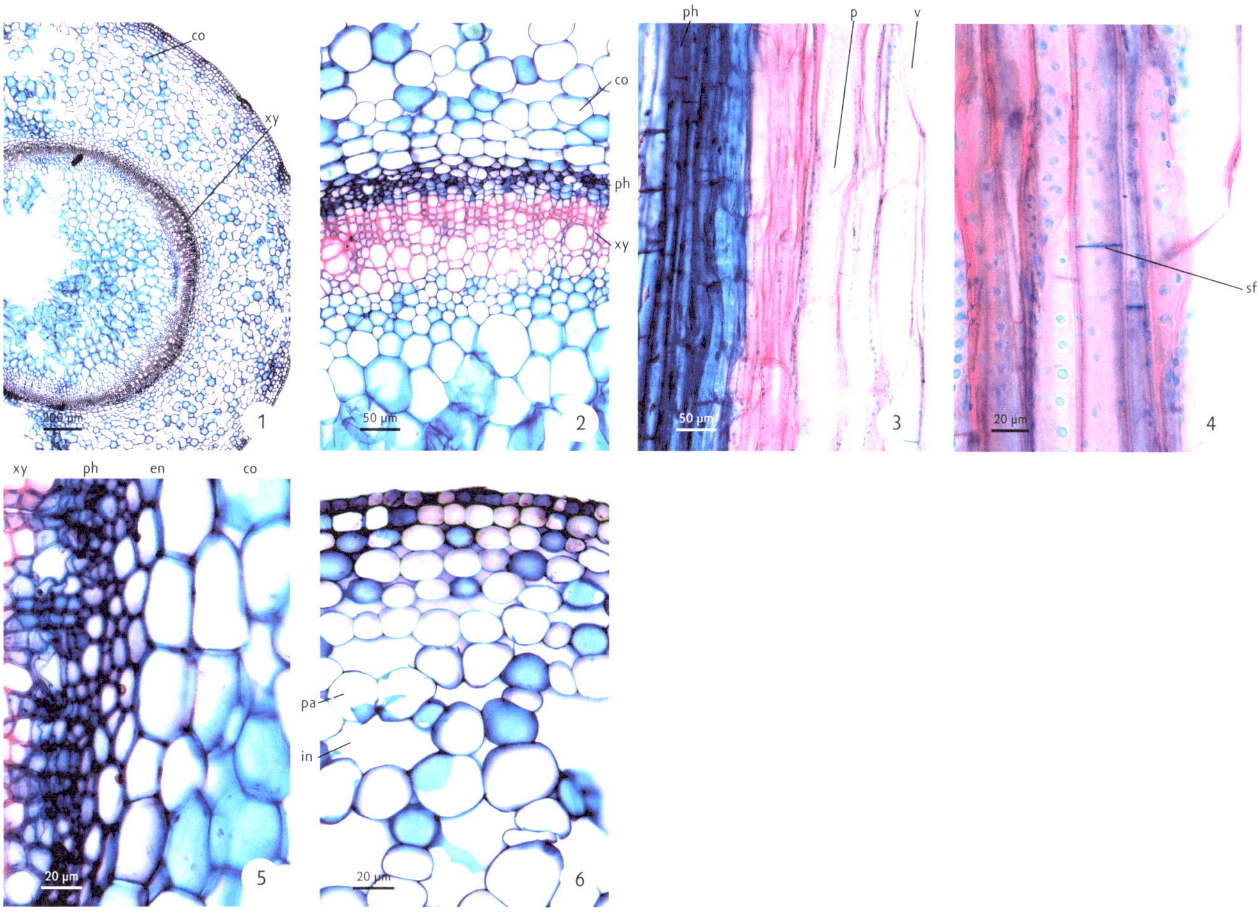

Annual, monocarpic or perennial (only in the southern part of its distribution) wetland herb with creeping rhizome. Stem erect to ascendent, usually hollow, often reddish, 15–50 cm high. Leaves opposite, sessile, narrow lanceolate to wide-linear, glabrous. Flowers in axillary racemes. Wheel-shaped corolla whitish or pinkish, only 4–5.5 mm across. Fruit is a rounded capsule.

Distribution and ecology

Circumpolar temperate element, with a disjunct distribution. Native in all CE countries, especially in lowlands. Typical of banks of shallow waters and on the muddy edges of rivers or streams, also in dune-slacks or wet depressions in fields. Prefers calcareous clayish soils.

Anatomy of the root collar

The basal part of the stem has a diameter of 2–3 mm. Maximum age 1 year (1).

The basal part of the stem consists of a large pith, a small xylem and phloem, and a large cortex (1). The xylem contains vessels and thin- to thick-walled fibers (2). Vessels are angular and arranged solitary and have a diameter of 15–20 µm (2) and a length of 100–120 µm. Vessel walls have round intervessel pits (3) and simple perforation plates (3). Axial parenchyma cells and rays are absent (2).

Cortex and phloem are separated by an endodermis (2, 4). Parenchyma cells and small groups of sieve tubes characterize the phloem (2). The cortex represents partially a net-like aerenchyma with large intercellulars (1, 2). The periphery of the annual shoot consists of an epidermis (2).

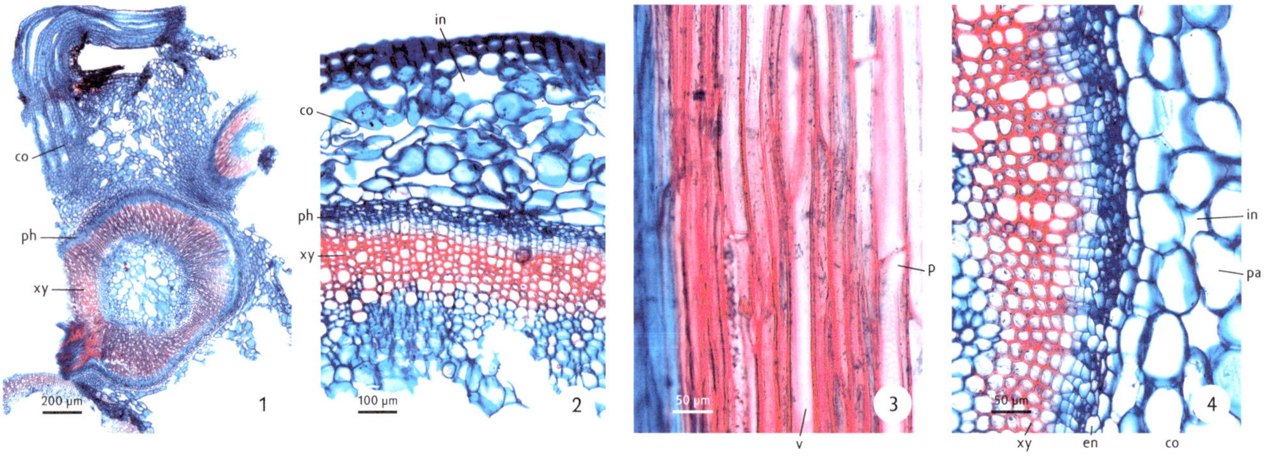

Veronica maritima L. (syn. *Pseudolysimachion maritimum*)

Perennial wetland herb with short underground rhizome. Stem branched in the upper part, 60–150 cm high. Leaves in whorls, lanceolate, toothed or serrate on the margin. Flowers in long dense terminal racemes, corolla purplish blue. Fruit is a glabrous ovoid capsule.

Distribution and ecology

European distribution, also in W. Asia. Adventive in CH, native in other CE countries especially in lowlands. Occurs in regularly flooded alluvial meadows, on banks of rivers and oxbows. Prefers nutrient-rich, neutral to slightly alkaline wet soils. Heliophilous, river corridor species.

Anatomy of the root collar

The basal part of the stem has a diameter of 3–4 mm. Maximum age 1 year (1, 2).

The basal part of the stem consists of a large pith, a small xylem and phloem, a large cortex and a small phellem (1, 5). Vessels are arranged solitary. They have a diameter of 15–20 µm (2) and a length of 200–300 µm (4). Vessel-walls have round intervessel pits (3) and simple perforation plates (3). Fibers are thin- to thick-walled (2). Axial parenchyma cells and rays are absent (2, 4).

Parenchyma cells and small groups of sieve elements characterize the phloem (5). Cortex and phloem are separated by an endodermis-like row of round cells (5). The phloem contains small thick-walled fibers. The cortex and the pith consists of round cells and small, irregularly formed intercellulars and the phellem of a few layers of square cells and an epidermis (5).

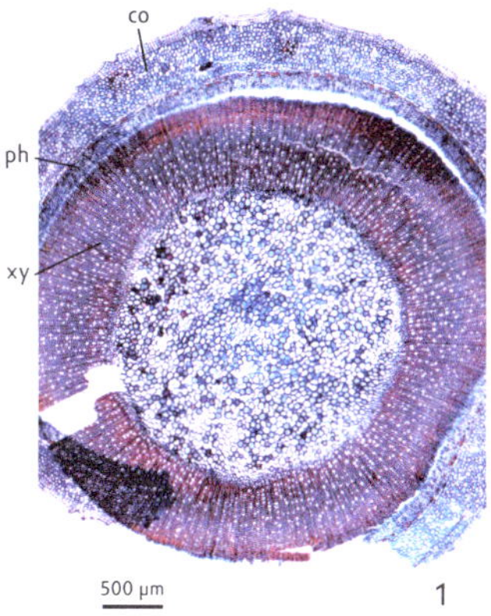

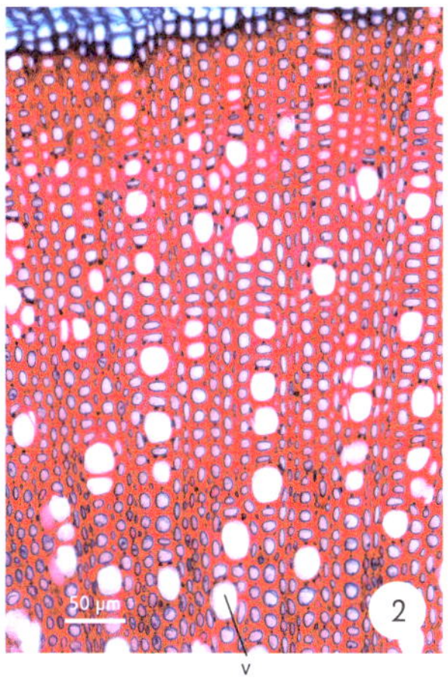

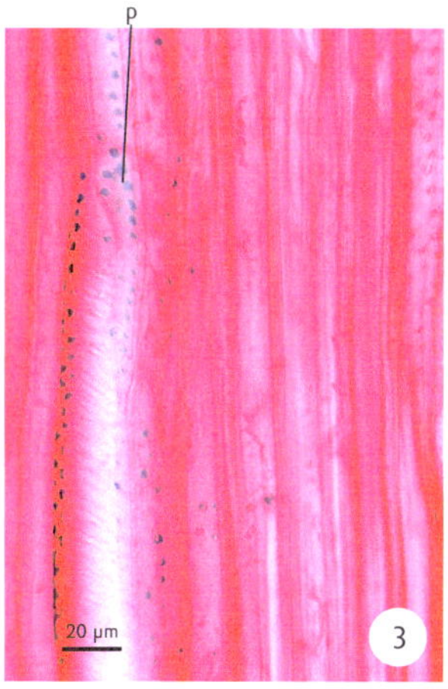

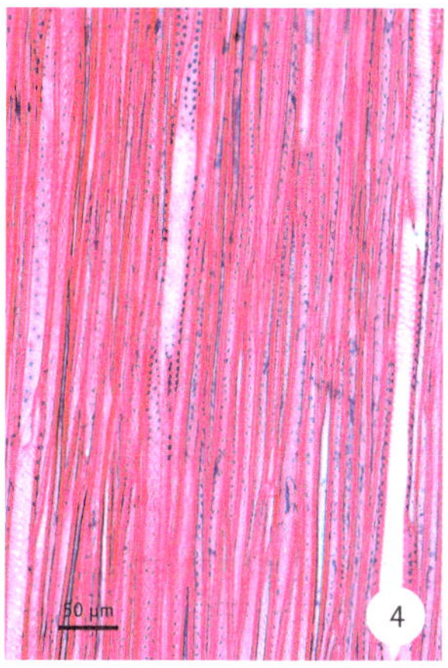

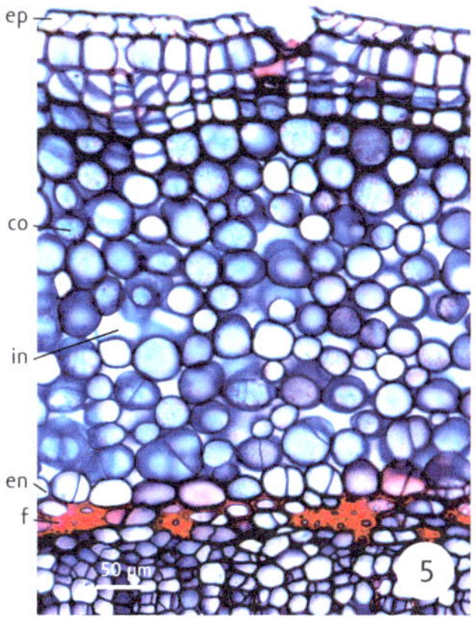

Veronica scutellata L.

Marsh Speedwell Plantaginaceae

Perennial wetland herb with horizontal rhizome. Stem slender, creeping, 15–50 cm high. Leaves opposite, linear-oblong to lanceolate, sessile. Inflorescense is a loose few-flowered raceme in leaf axil, flower stalk long. White or pale blue corolla, 4–6 mm across. Fruit is a reniform capsule.

Distribution and ecology

Euro-Siberian boreo-temperate element; also in N. America. Native in all CE countries. Occurs in margins of lakes, fishponds and ditches, also in fen-meadows and amongst tall reeds and sedges. Prefers slightly acidic, mesotrophic soils with high water table. Heliophilous species.

Anatomy of the root collar

The basal part of the stem (root collar) has a diameter of 1–2 mm. Maximum age 1 year (1).

The basal part of the stem consists of a large pith, a small xylem and phloem and a very large cortex (1). The xylem contains vessels and parenchyma cells. The vessels are polygonal, are arranged solitary or in groups and have a diameter of 15–20 μm and a length of 120–180 μm. Intervessel pits are scalariform (3), perforations are simple (3). Fibers and rays are absent (2). Axial parenchyma is paratracheal and pervasive (2).

Cortex and phloem are separated by a poorly differentiated pericycle/endodermis zone (1, 2). Parenchyma cells and small groups of sieve tubes characterize the phloem (2). The inner, multilayered zone contains occasionally small, thick-walled, lignified cells (2, 4). The cortex represents a net-like aerenchyma with large intercellulars (1, 2). Intercellulars are absent in the pith (2). The periphery of the annual shoot consists of an epidermis (1).

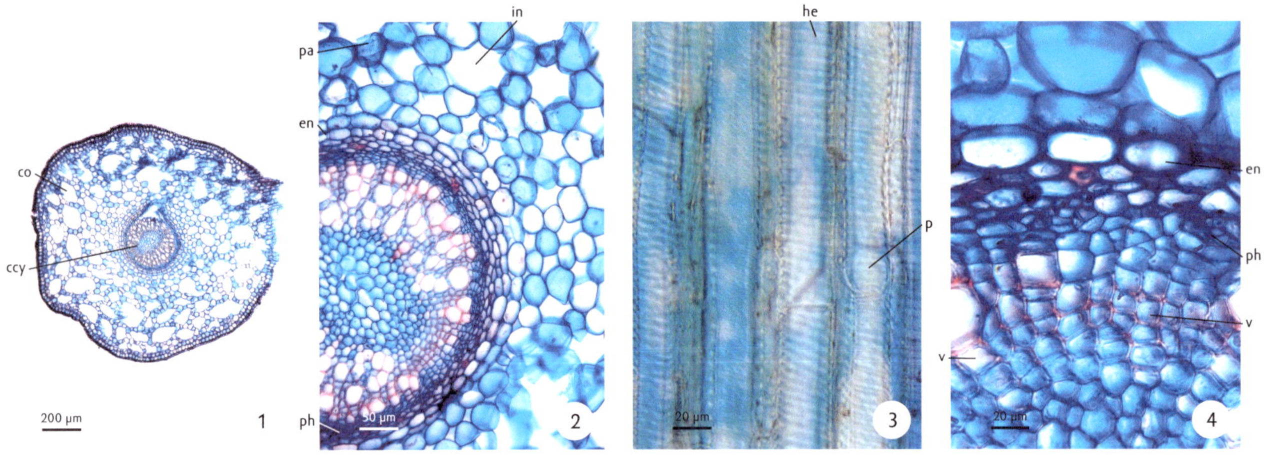

Bistorta officinalis Delarbre

(syn. *Bistorta major, Polygonum bistorta, Persicaria bistorta*)

Perennial wetland herb with fleshy rhizome. Stem erect to ascendent, glabrous, 35–100 cm high. Basal leaves with cordate bases, stem leaves petiolate, triangular and pointed. Inflorescence is a terminal dense spike of pink flowers, 4–5 mm across. Fruit is a shiny brown achene.

Distribution and ecology

Eurasian boreo-temperate element. Native in all CE countries, rare in HU. Occurs on wet mountain meadows, in damp pastures, also in tall-herb vegetation along mountain streams and rivers. Prefers acidic soils with high water table. High nitrogen content in its rhizomes. Medicinal plant.

Anatomy of the rhizome

Round rhizome with a diameter of 10–14 mm. Maximum age 4 years.

The large pith of the rhizome is surrounded by single vascular bundles, a cortex and a small phellem (1, 2). The xylem of the vascular bundles contains angular vessels, thin- to thick-walled fibers and paratracheal parenchyma (3). Ring boundaries are mainly indicated by a slight semi-ring porosity (3, 4). Earlywood-vessels have a diameter of 40–60 µm (3) and vessel length is 100–150 µm. Vessels are solitary or in short radial groups (3, 4). Intervessel pits are large (1–1.5 µm) and round and perforation plates are simple (5). Fibers occur laterally of the vascular bundles (3). Large rays separate vascular bundles laterally (1–3), are >10 cells wide (6) and consist of square and upright cells (5).

The phloem is composed of parenchyma cells and groups of sieve tubes (3). The cortex consists of round and oval parenchyma cells with large intercellulars (7). Crystal druses occur in the pith, the inter-vascular tissue and the cortex (8). The phellem is very small and indicated by a few rectangular cells containing dark staining substances (7).

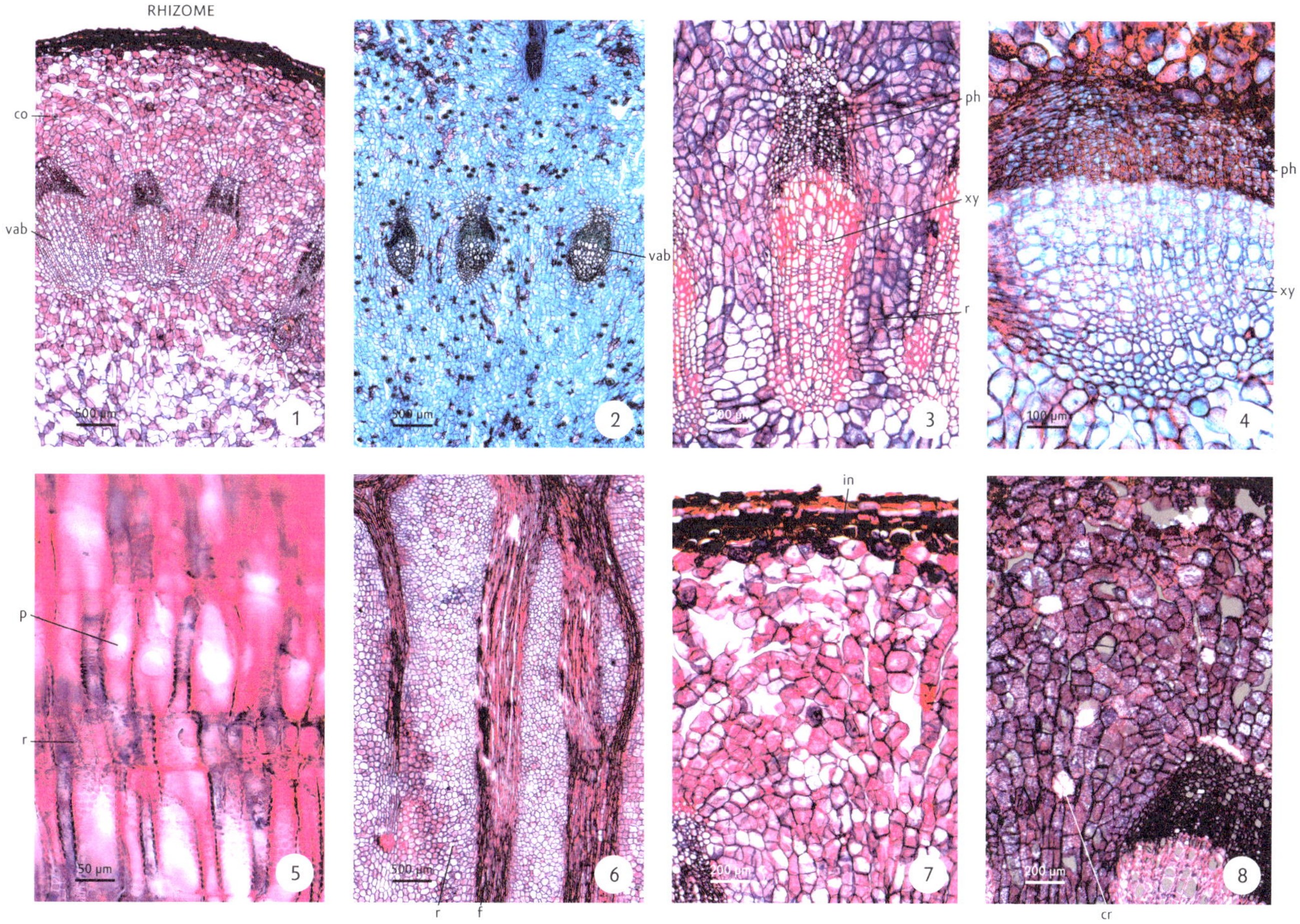

F. H. Schweingruber et al., *Anatomic Atlas of Aquatic and Wetland Plant Stems*, https://doi.org/10.1007/978-3-030-33420-8_44

Persicaria amphibia (L.) Delarbre

(syn. *Polygonum amphibium*)

Polygonaceae
Amphibious Bistort

Amphibious perennial herb with very dissimilar aquatic and terrestrial forms, rhizomatous. Aquatic form has sparsely branching underwater stem with glabrous alternating floating leaves, lanceolate or narrow-ovate, 4–10 cm long, 2–4 cm wide. Terrestrial form has procumbent to erect stem, with pubescent lanceolate leaves, 0.4–1 m high. Infloresence is a dense spike of pink flowers, rarely produced by the terrestrial form. Fruit is a dark dry achene enclosed in persistent perianth.

Distribution and ecology

Circumpolar boreo-temperate element; widely naturalised outside its native range. Native in all CE countries. Aquatic form occurs in lakes, fishponds, also in slowly flowing rivers and ditches. Terrestrial form is typical of wet disturbed habitats in marshes or wet meadows. Well adapted to changing waterlevels. Rapid vegetative spread by rhizome fragments. Male-fertile or male-sterile clones.

Anatomy of the annual shoot

Round shoot with a diameter of 3–4 mm. Maximum age 1 year (1–4).

The large hollow pith of the basal part is surrounded by a zone of fibers and single vascular bundles, a phloem, a cortex and a small phellem (1, 2). Vascular bundles are laterally connected by a vessel-free fiber zone (2).

The major part of the xylem of the vascular bundles consists of vessels, thin- to thick-walled fibers and parenchyma (2). Vessels have a diameter of 50–60 µm and a length of 150–200 µm. Vessels are angular and solitary (2). Intervessel pits are large (1–1.5 µm) and round and perforation plates are simple (3). Parenchyma is paratracheal with septate fibers (4). Large rays (inter-vascular tissue) separate vascular bundles laterally (1, 2) but rays within vascular bundles are absent.

The phloem consists of parenchyma cells and small groups of sieve tubes (5). Phloem and cortex are separated by a small discontinuous fiber zone (5).

The cortex consists of round and oval parenchyma cells and large intercellulars (5). Crystal druses occur in the pith, in the inter-vascular tissue and in the cortex (6). A uniseriate layer forms the epidermis and a biseriate layer the hypodermis (7). Intraxylary cork has been observed at the periphery of the pith (8).

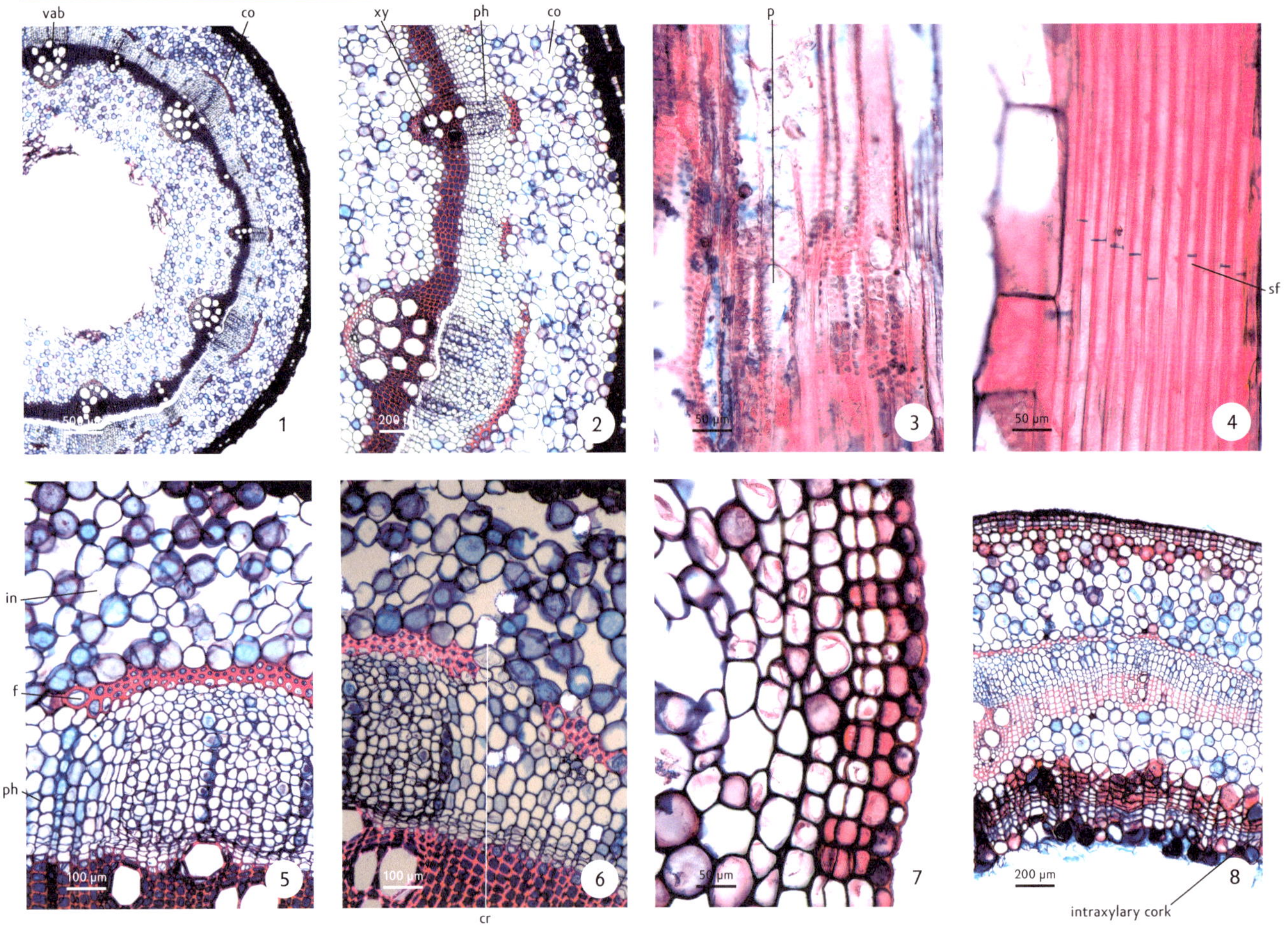

Persicaria hydropiper Delarbre (syn. *Polygonum hydropiper*) **Water-pepper** Polygonaceae

Annual wetland herb. Stem erect, branched, glabrous, often reddish, 25–60 cm high. Leaves lanceolate with sharp peppery taste. Inflorescence lax, the flowers mostly separated, perianth pinkish or greenish-white. Fruit is an oval dry achene enclosed in persistent perianth.

Distribution and ecology

Circumpolar temperate element; widely naturalised outside its native range. Native in all CE countries. Occurs on margins of lakes, ponds, rivers and streams, also on drained fishpond bottoms, wet depressions in wet meadows or fields. Prefers base-poor muddy soils, tolerates partial shade. Develops persistent seed-bank, requires moist chilling to break seed dormancy.

Anatomy of the annual root collar

Round shoot with a diameter of 4–6 mm. Maximum age 1 year (1).

The central part is surrounded by a xylem, a phloem and a cortex (1). Vessels are solitary and in short radial multiples (1, 2). They have a diameter of 30–50 µm and a length of 100–150 µm. Intervessel pits are round and stay in alternating position (4). Perforation plates are simple (3). Parenchyma is paratracheal (2). Rays are uniseriate, rare or absent (5).

The phloem contains groups of sieve elements (6). The cortex consists of large cells of different forms and a few isolated fibers. Crystal druses occur in the cortex. Intercellulars are absent (6). The phellem is small and consists of a few layers of rectangular cells (6).

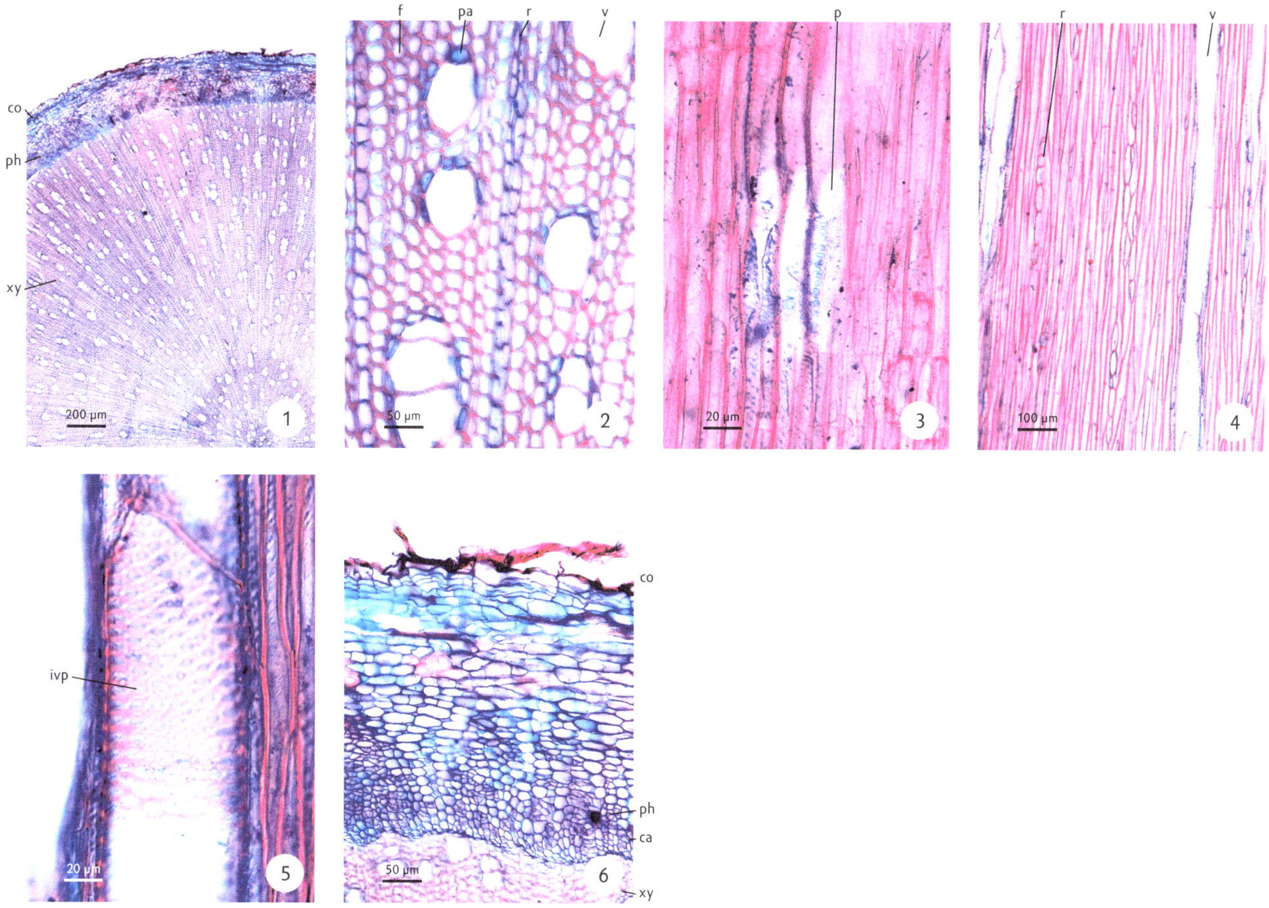

Persicaria maculosa Gray

Redshank

Polygonaceae

(syn. *Persicaria maculata, Polygonum maculatum, P. persicaria*)

Annual wetland herb. Stem branched, erect to ascendent, glabrous, reddish, often swollen above the leaf-bases, 20–60 cm high. Leaves elliptic to lanceolate, often black-blotched, 5–10 cm long with short petiole. Inflorescence is a dense spike, 1–3 cm long, perianth whitish or pink. Fruit is a glabrous dry achene enclosed in persistent perianth.

Distribution and ecology

European distribution, but widely naturalised outside its native range. Native in all CE countries. Occurs in a wide range of disturbed habitats, on drained fishpond bottoms, along rivers, streams and ditches, also in wet depressions in fields or ruderal habitats. Broad ecological amplitude. Seasonal changes in dormancy.

Anatomy of the annual root collar

Round shoot with a diameter of 3–4 mm. Maximum age 1 year (1).

The central part is surrounded by a xylem, a phloem and a cortex (1). Vessels are solitary in short radial multiples (1, 2). They have a diameter of 40–50 µm and a length of 100–150 µm. Intervessel pits are round and stay in alternating position (3). Perforation plates are simple (3). Parenchyma is paratracheal (2). Rays are uniseriate, rare or absent (4).

The phloem contains groups of sieve elements and companion cells (5). The cortex consists of large cells of different forms and a few isolated fibers (5). Crystal druses occur in the cortex (6). Intercellulars are absent (5). The phellem consists of a few layers of rectangular cells (5).

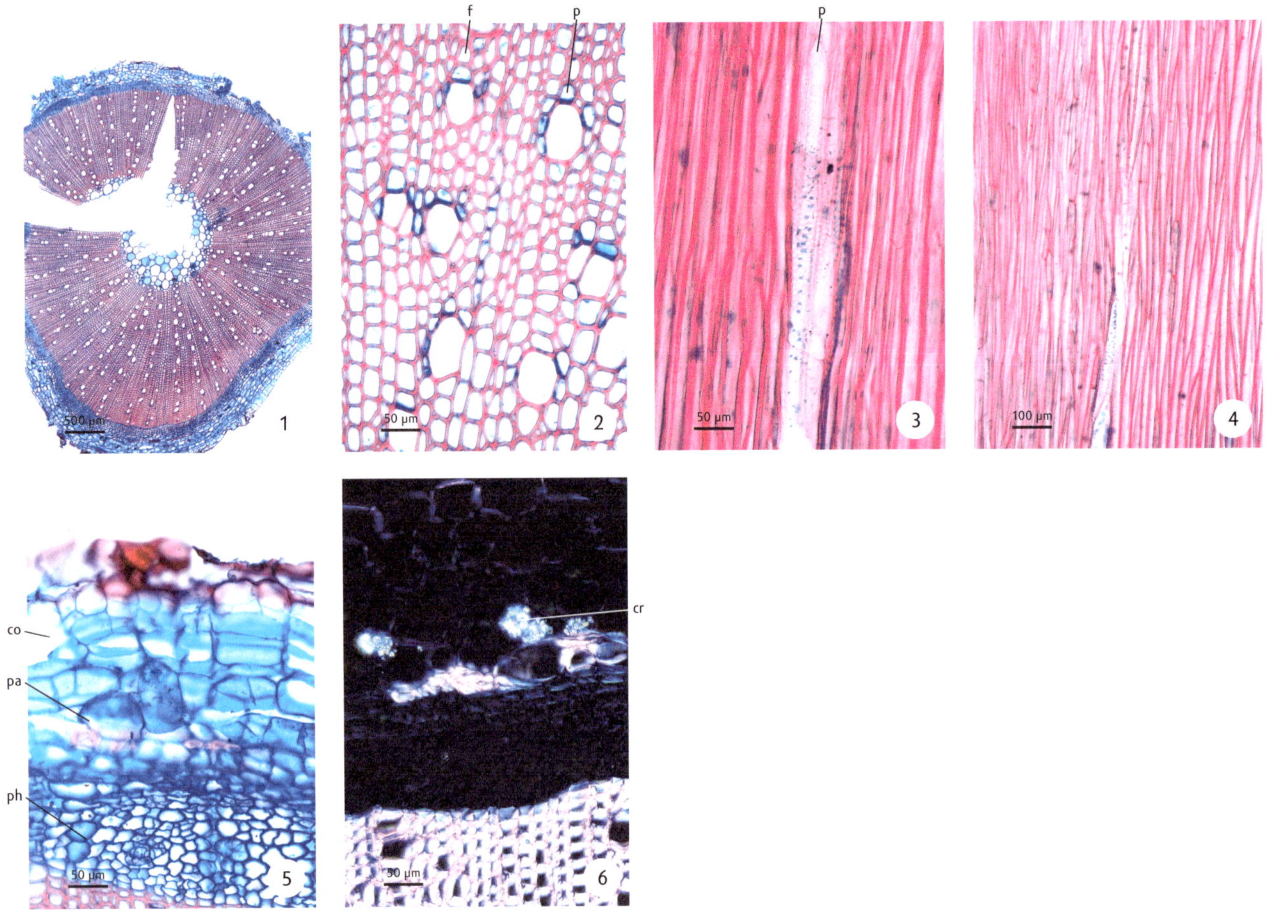

Persicaria minor (Huds.) Opiz (syn. *Polygonum minus*) **Small Water-pepper** Polygonaceae

Annual wetland herb. Stem decumbent to ascending, with thin lateral branches, only slightly swollen in nodes, 10–20 cm high. Leaves linear-lanceolate to linear, only 4–8 mm wide. Inflorescense lax, the flowers mostly separated, perianth greenish. Fruit is a lenticular dry achene enclosed in persistent perianth.

Polygonaceae

Distribution and ecology

Eurasian temperate element. Native in all CE countries. Occurs on a wide range of wet disturbed soils close to lakes, fishponds and ditches, also on wet depressions in pastures.

Anatomy of annual shoots

Round shoot with a diameter of 3–4 mm. Maximum age 1 year (1–4).

The large hollow pith of the basal part is surrounded by a xylem, a phloem and a cortex (1, 5). Vessels are solitary and have a diameter of 50–60 µm and a length of 100–150 µm. Intervessel pits are large and round. Perforation plates are simple (4). Parenchyma is paratracheal (2, 5). Rays are uniseriate, very rare or absent (3).

The phloem is uniform. Sieve elements and parenchyma cannot be differentiated (5). The cortex consists of large cells of different forms and a few isolated fibers (5) and contains a few crystal druses. Intercellulars are absent (5). The phellem is only partially present (1).

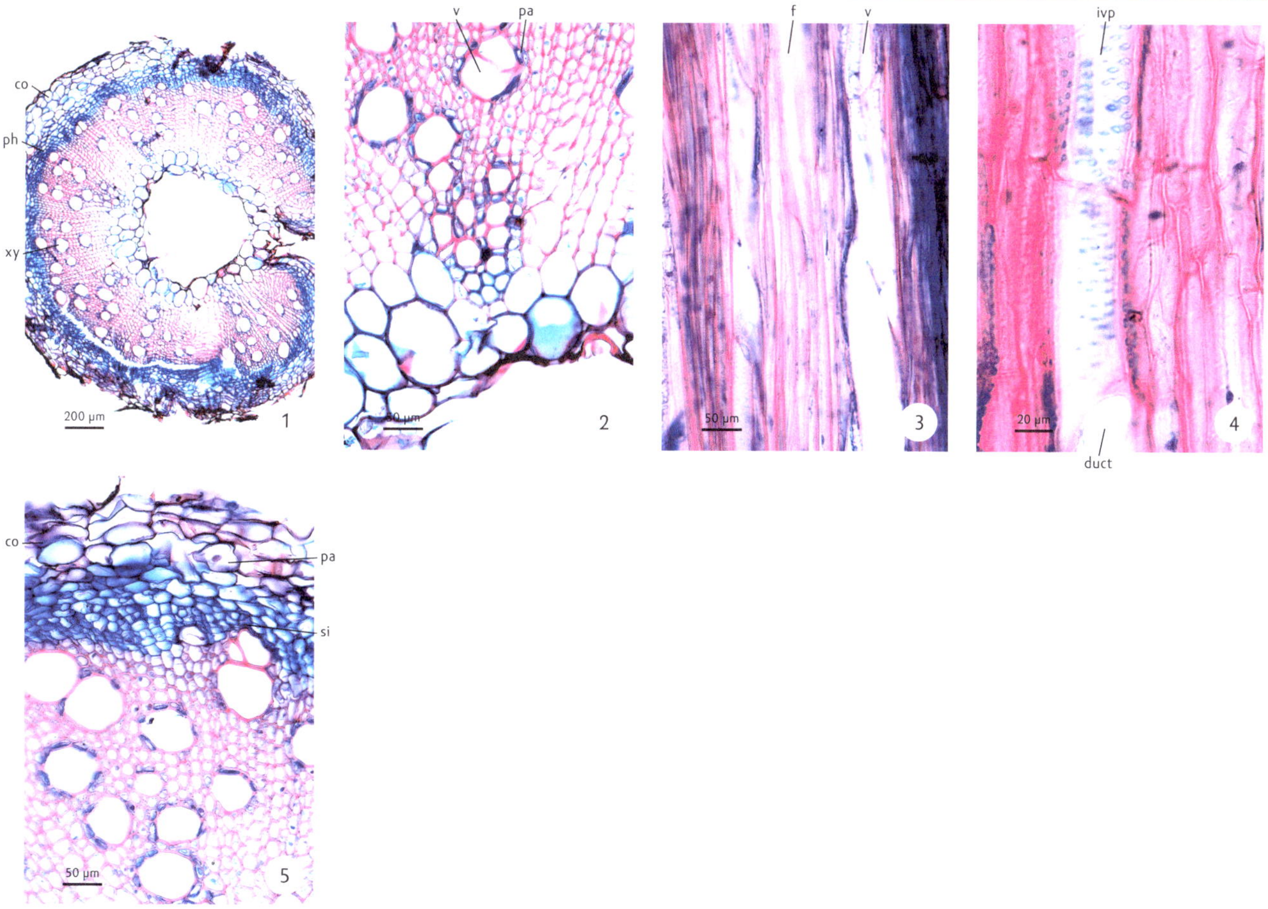

Persicaria mitis (Schrank) Assenov (syn. *Polygonum mite*)

Annual wetland herb. Stem erect to ascendent, glabrous, usually branched, only slightly swollen in nodes, 15–60 cm high. Leaves linear-lanceolate, 10–15 mm wide. Inflorescense sparse, lax, the flowers mostly separated, perianth greenish. Fruit is a three-sided black dry achene enclosed in persistent perianth.

Distribution and ecology

European temperate distribution, also in W. Asia. Native in all CE countries except HU. Occurs on wet disturbed sites close to streams, lakes, fishponds and ditches, also on wet depressions in fields or at ruderal sites. Prefers heavy soils, avoids soils with high nitrogen content.

Anatomy of the annual root collar

Round shoot with a diameter of 3–5 mm. Maximum age 1 year (1).

The central part is surrounded by a xylem, a phloem and a cortex (1). Vessels are primarily solitary in short radial multiples (1, 2). They have a diameter of 30–50 µm and a length of 120–150 µm. Intervessel pits are round and stay in alternating position (3). Perforation plates are simple (3). Parenchyma is paratracheal (2). Rays are uni- to biseriate, rare or absent (4).

The phloem contains groups of sieve elements and companion cells (5, 6). The cortex consists of large cells of different forms and a few isolated fibers (5). Crystals are absent. Intercellulars are absent (6). The phellem consists of a few layers of rectangular cells (6).

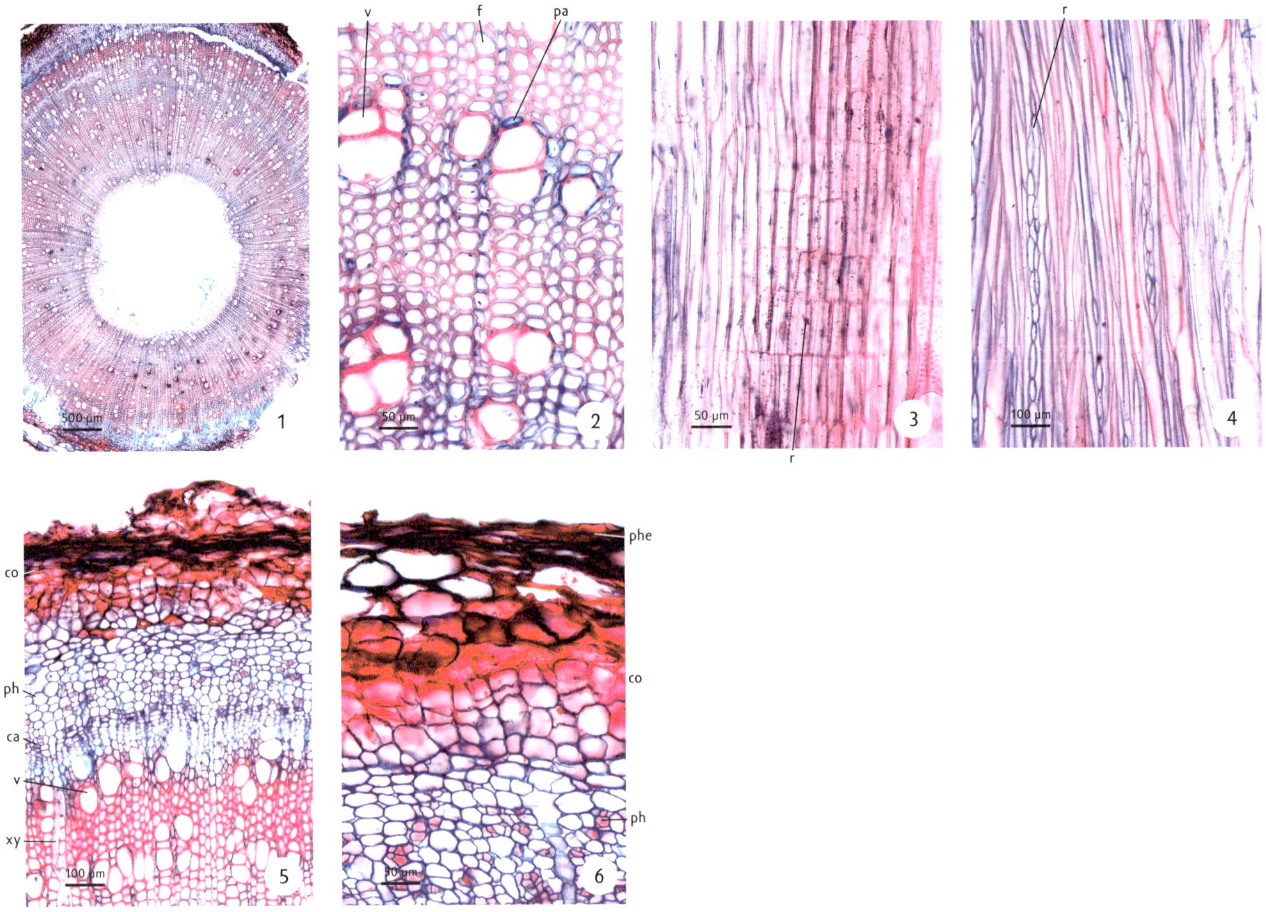

Perennial wetland herb with creeping horizontal rhizomes. Fertile stalks erect, 50–120 cm high, bearing only small leaves. Above-ground vegetative shoots with 3–5 leaves, broadly ovate, cordate, 20–40 cm long and wide. Inflorescence dense, with erect branches, tepals narrowly ovate-triangular, entire, without tubercles. Fruit is a trigonous achene enclosed by the persistent perianth.

Distribution and ecology

Eurasian distribution. Native in all CE countries except HU, most abundant in montane, subalpine and alpine zones. Occurs on stream banks, spring areas, nutrient-rich and mesic to wet abandoned pastures and meadows, also at anthropogenic sites, such as surroundings of sheep shelters, in the vicinity of mountain chalets, resting places of domestic animals. Strictly nitrophilous and heliophilous.

Anatomy of the rhizome

Round rhizome with a diameter of 15–30 mm. Maximum age 6 years.

The large pith of the rhizome is surrounded by single vascular bundles (1–3), a cortex and a phellem. The xylem contains vessels, parenchyma and fibers (3). Ring boundaries are indicated by a slight semi-ring porosity and tangentially arranged groups of fibers (3). Earlywood vessels are angular and have a diameter of 50–70 μm (3) and a length of 50–180 μm. Vessels are arranged generally in irregular groups (3). Intervessel pits are large (1–1.5 μm), round and horizontally enlarged. Perforation plates are simple (4). Fibers occur in groups. Parenchyma is paratracheal and pervasive. Rays separate vascular bundles laterally (2, 3) and are >10 cells wide (5). The center of most of the large rays contains groups of very small cells (4). All ray cells consist of square and upright cells (5).

The phloem is composed of parenchyma cells and groups of sieve tubes (6). Each vascular bundle is surrounded by a group of fibers (7). The cortex consists of round and oval parenchyma cells with large, often net-like intercellulars (1, 7). Crystal druses occur in the pith, the inter-vascular tissue and the cortex (7). The phellem is small and indicated by small irregularly formed rectangular cells containing dark staining substances (1).

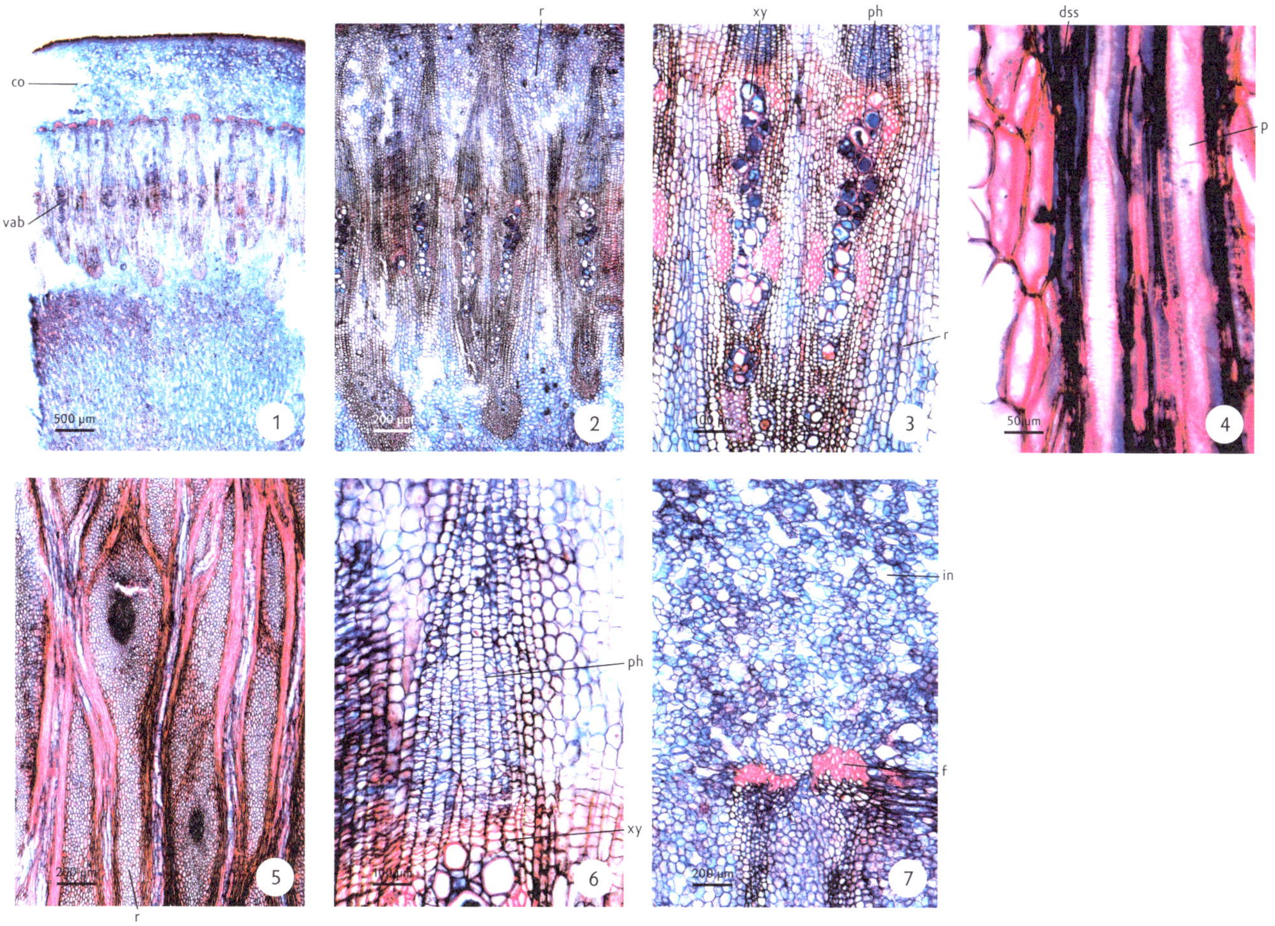

Rumex hydrolapathum Huds. **Water Dock** Polygonaceae

Emergent or perennial wetland robust herb. Stem erect, angular, up to 200 cm high. Leaves lanceolate to narrowly ovate, up to 100 cm long and 20 cm wide. Inflorescence large, with many branches, tepals ovate-triangular, entire, all with elongated smooth tubercle. Fruit is a trigonous achene enclosed by the persistent perianth.

Distribution and ecology

European temperate element. Native in all CE countries. Occurs on banks of slow-flowing rivers and streams, by ponds, canals and in ditches and oxbows, also in fens. Prefers calcium-rich, permanently waterlogged soils. Seeds are rather short-lived, germinate only in waterlogged soils.

Anatomy of the root collar

Round root colar with a diameter of 15–30 mm. Maximum age 4 years.

The large pith of the root collar is surrounded by single vascular bundles, a cortex and a small phellem (1). The xylem consists of vessels, pervasive parenchyma and groups of fibers (1, 2). Ring boundaries are indicated by semi-ring porosity and tangentially arranged groups of fibers (1, 3). Earlywood vessels have a diameter of 50–70 µm (3) and vessel length of 90–110 µm. Vessels are generally arranged solitary or in irregular groups (2, 3). Intervessel pits are scalariform with large openings (4). Perforation plates are simple (4). Vessel-free parenchyma zones separate vascular bundles laterally (1, 3) and are >10 cells wide (5). The center of most of the large rays contains groups of very small cells (5). All ray cells consist of upright cells.

The phloem is composed of parenchyma cells and tangential layers of groups of sieve tubes (6). The cortex consists of round and oval parenchyma cells with large, often net-like intercellulars (7, 8). The presence of large, very thick-walled fibers is special (1). They are concentrated just below the phellem (7). Crystal druses occur in the pith, in the inter-vascular tissue and in the cortex. The phellem is small and represents a few layers of small rectangular cells (1). Large thin-walled pith cells (100–150 µm) are arranged around large irregularly formed intercellulars (1).

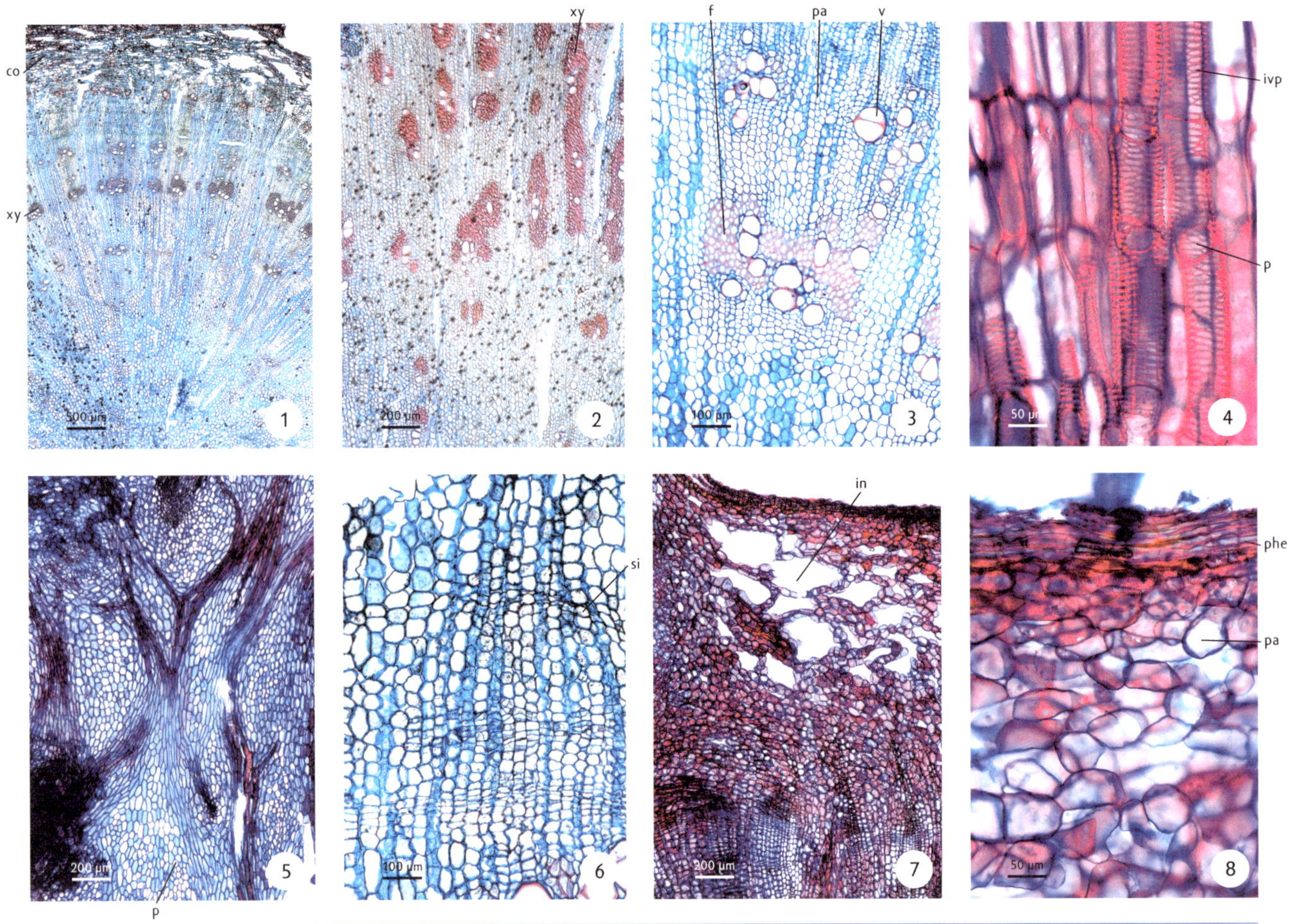

Rumex maritimus L. **Golden Dock** Polygonaceae

Annual to biennial wetland herb. Stem 10–70 cm high, golden-brown in fruiting. Basal leaves petiolate, narrow-lanceolate, stem leaves smaller. Inflorescence with many spreading leafy branches. Tepals triangular with very fine teeth and narrow tubercle. Fruit is a brown-green achene enclosed in persistent perianth.

Distribution and ecology

Circumpolar temperate element. Native in all CE countries. Grows on wet muddy soils in margins of rivers, pools and ditches, also in fishpond drained bottoms. Prefers nutrient-rich soils, tolerates sligthly saline soils. Long-lived seeds undergo a seasonal dormancy cycle. Flooding-tolerant plant.

Anatomy of the root collar

Round root collar with a diameter of 5–10 mm. Maximum age 1 year.

The small pith of the root collar is surrounded by a continuous belt of xylem and phloem, a cortex and a small phellem (1). The xylem contains vessels, a parenchyma and fibers (1, 2). Earlywood vessels have a diameter of 40–60 µm (2) and vessel length of 100–140 µm. Vessels are generally arranged solitary in short radial rows or in irregular groups (1, 2). Intervessel pits are round (3). Perforation plates are simple. Fibers are thin- to thick-walled (2) with septate fibers (4). Parenchyma is paratracheal (2). Ray width varies between 3–6 cells and contains sheath cells (5). All ray cells consist of upright cells.

The phloem is composed of parenchyma groups of sieve tubes and dilating rays (6). The cortex consists of round and oval parenchyma cells and irregular groups of fibers (6). Intercellulars are very small and contain crystal druses. The phellem is small and indicates a few layers of small rectangular cells (6).

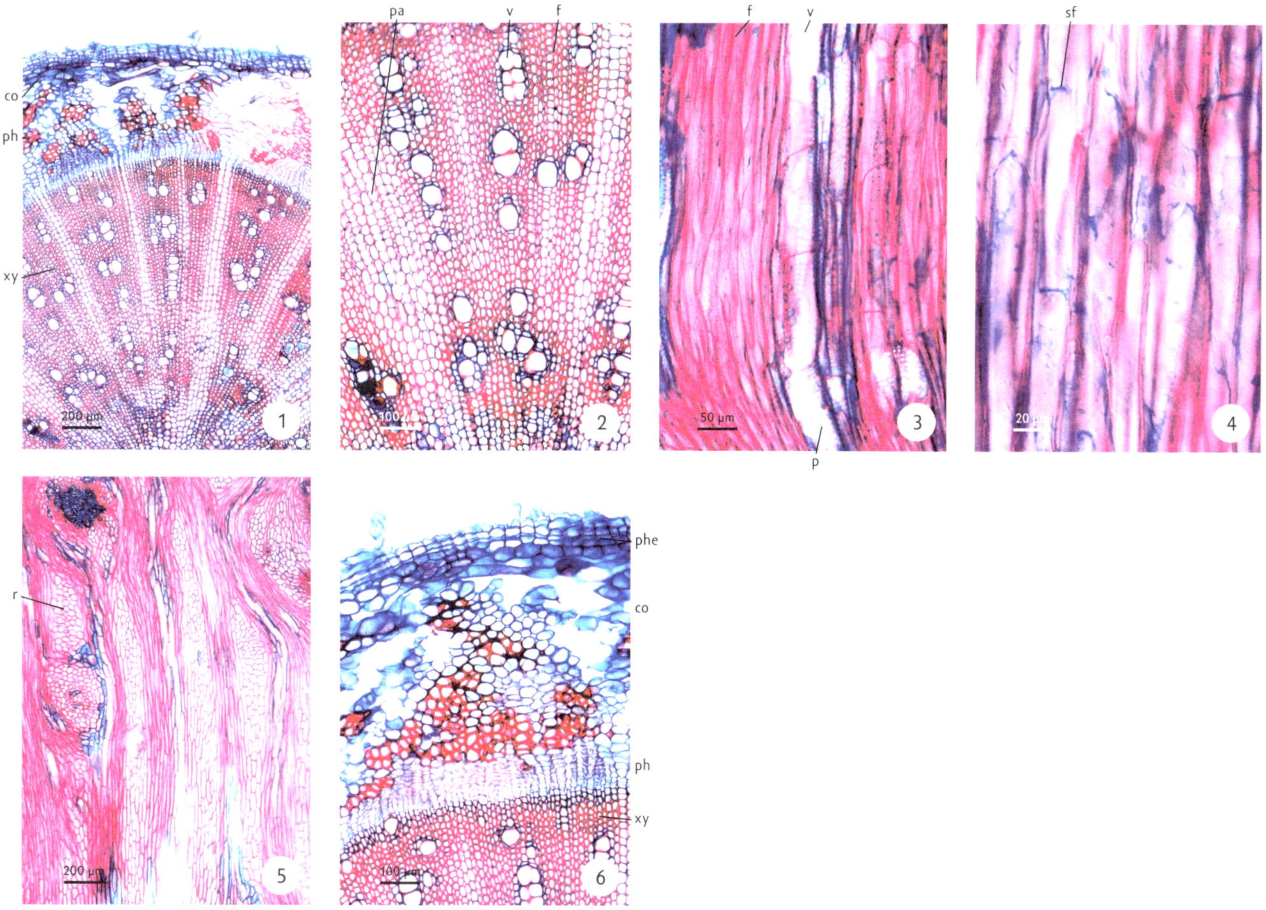

Centunculus minimus L.

Chaffweed Primulaceae

Annual wetland plant. Erect to decumbent stems only 2–6 cm long, alternating cordate leaves 3–5 mm long. Flowers solitary in leaf axils with white to pink corolla. Fruit is a rounded capsule.

Distribution and ecology

Extensive distribution from Europe to N. Africa, both N. and S. America. Scattered irregularly to rarely in all CE countries, in lowlands to uplands. Bare sandy or clayish wet soils, margins of fishponds or fields, abandoned meadows, old sand-pits, mainly in fishpond areas. Strict heliophyte, competitively very week, pioneering species on disturbed areas.

Anatomy of the shoot

Round flower stalk with a diameter of approximately <1 mm. Maximum age 1 year.

The central cylinder is surrounded by a large parenchymatic cortex and an epidermis (1). The pith is surrounded by a 2–3 cell wide fiber-free belt of xylem (2). It consists of grouped, angular and thin-walled angular vessels with a diameter of 8–12 µm (2) and a length 50–80 µm. Vessels have scalariform intervessel pits and simple perforation plates (4). Fibers and rays are absent. Parenchyma is pervasive (3).

The central cylinder is surrounded by an endodermis (1, 3). The 3–5 cell wide phloem consists of round parenchyma cells and small groups of sieve tubes (3). The cortex consists of large cells and small intercellulars. Dark staining substances fill endodermis and cortex cells (1).

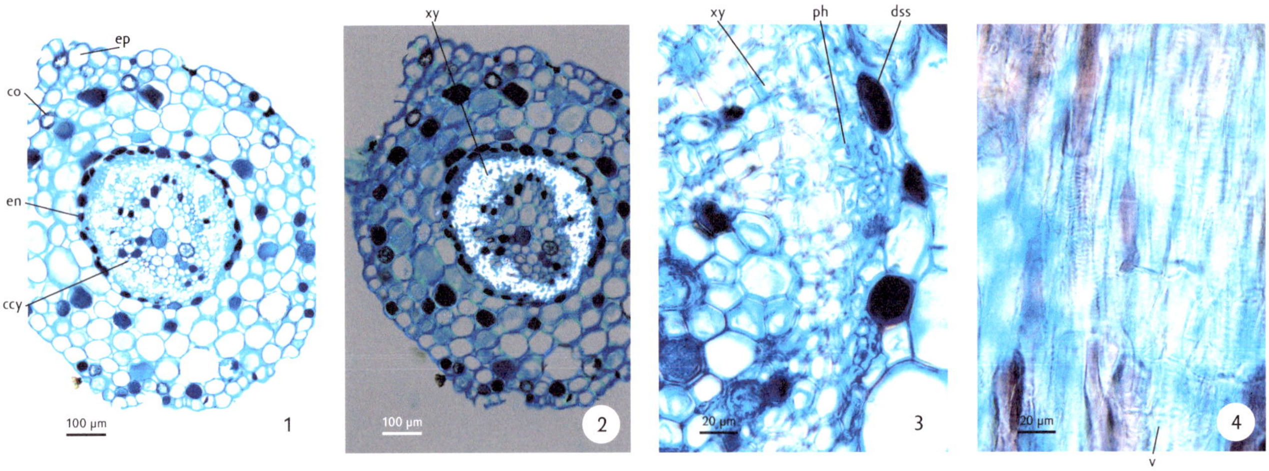

F. H. Schweingruber et al., *Anatomic Atlas of Aquatic and Wetland Plant Stems*, https://doi.org/10.1007/978-3-030-33420-8_45

Glaux maritima L.

Sea-milkwort

Perennial wetland herb with runners. Stem ascending, branched at base, glabrous, 3–25 cm high. Leaves mostly opposite, stalkless, lowest leaves scaly and brown, upper leaves ovately lanceolate to elliptic, fleshy, glabrous and bluish green. Flowers solitary in leaf axils, corolla lacking, corolla-like calyx campanulate, light red to pink and dark-spotted. Fruit is a spherical capsule.

Distribution and ecology

Circumpolar boreo-temperate element. Native in all CE countries except CH, rare in AT, CZ, SK and HU. Occurs in saltmarshes, brackish dune-slacks, strandlines, very rarely also inland in saline wetlands. Prefers open communities, forms dense stands. Heliophyte, halophyte, poor competitor. Specialized hibernaculae for overwintering.

Anatomy of the basal part of the shoot

Round stalk with a diameter of approximately 1–3 mm. Maximum age 1 year.

A small stele (central cylinder) is surrounded by a large aerenchymatic cortex and an epidermis (1). The pith is surrounded by a fiber-free belt of xylem (2). Vessels are arranged solitary or in short radial rows. Vessels are angular and have a diameter of 10–20 µm (2) and a length of 100–130 µm. They have round to scalariform intervessel pits (3) and simple perforation plates (4). Parenchyma is paratracheal and pervasive (2). Fibers and rays are absent (2).

The stele is surrounded by a poorly differentiated endodermis (2), which is lignified in some specimens (5). The phloem is special. It consists of groups of sieve tubes, parenchyma cells and large (20 µm) laticifer-like tubes.

The cortex represents a net-like aerenchyma with well-delimited air cavities (1). The epidermis consists of small square thin-walled unlignified cells (1).

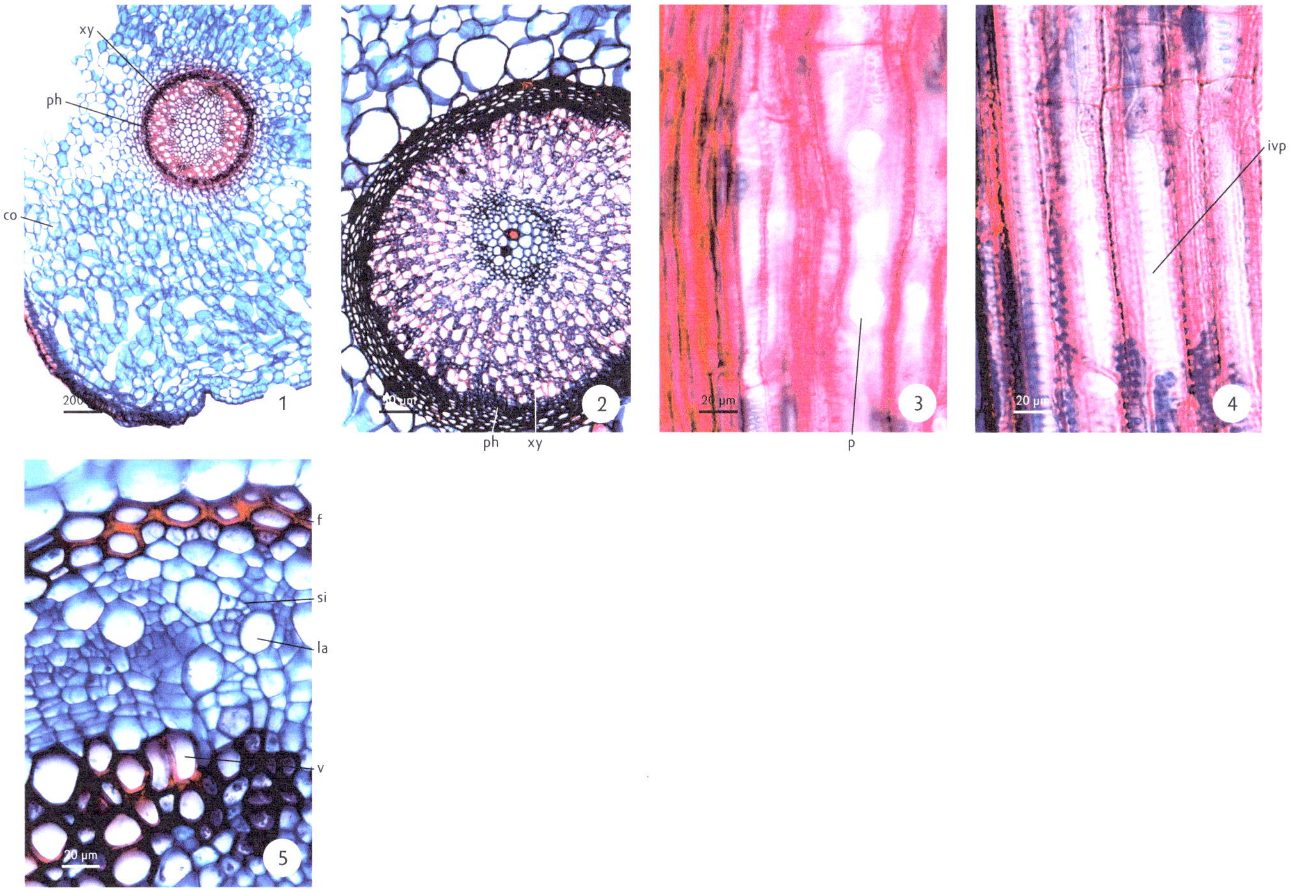

Perennial herb rooting densely in sediment, shoots 10-60 cm long, amphibious habit: in the spring usually submerged, only apical rosette of leaves floating, later emergent or terrestrial. Leaves whorled, pinnate, adventitious roots at nodes, flowers on tiered whorls, single flowers 15–25 mm across, inflorescence emergent, fruit is dry capsule.

Distribution and ecology

European temperate element. Scattered commonly to rarely in all CE countries, from lowlands to highlands. Shallow fishponds, lakes, ditches, backwater pools and oxbows, peat bogs and fen pools after peat extraction, slow rivers and canals, sandy, clayish or organic (peaty) sediments, avoids strongly acidic sediments, water depth 10–50 cm. Prefers mesotrophic, soft to hard waters (pH 6.0–8.5, total alkalinity 0.5–5 meq/l), nutrient-poor to rich soils, clear to slightly dystrophic waters, tolerant of periodical drying, intolerant of eutrophication, competitively weak, facultative sciophyte, probably strict CO_2 user.

Anatomy of the flower stalk

Round stalk with a diameter of approximately 2 mm. Secondary growth is absent or occurs during a very short time.

A small central cylinder is surrounded by a large aerenchymatic cortex (1, 3). The pith is surrounded by a 2–3 cell-wide fiber-free xylem (2). Solitary or irregularly grouped vessels with a diameter of 20–30 µm are unlignified and thin-walled (2). Vessels with a length of 200–300 µm have scalariform intervessel pits and simple perforation plates (4).

The 3–5 cell-wide phloem consists of round parenchyma cells and small angular sieve tubes with a diameter of approximately 4 and 2 µm, respectively (2). The stele is surrounded by a poorly differentiated pericycle or endodermis (2). The cortex represents a net-like aerenchyma with well-delimited air cavities (4). The epidermis consists of small tetrangular thin-walled unlignified cells.

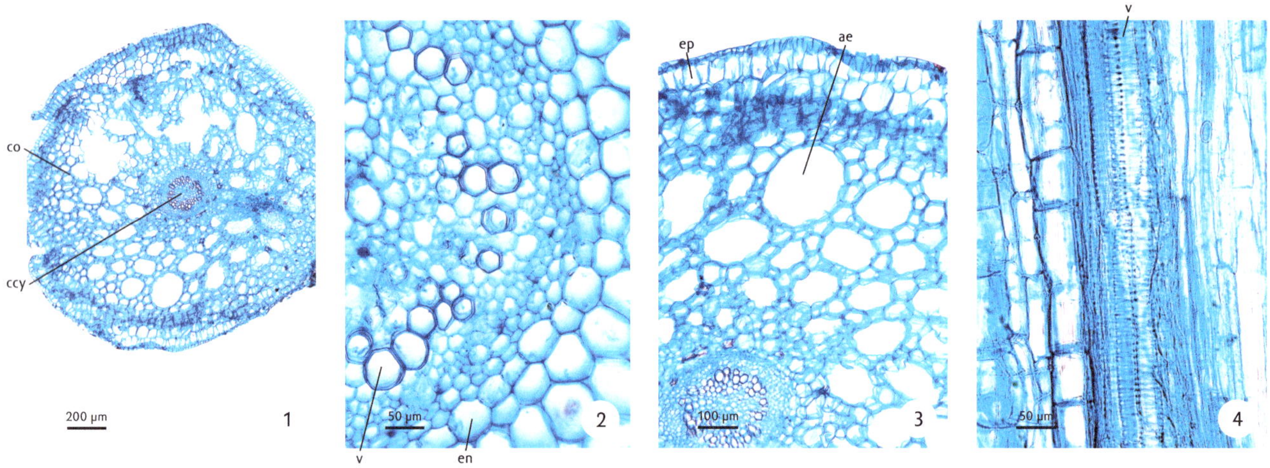

Lysimachia thyrsiflora L. (syn. *Naumburgia thyrsiflora*) **Tufted Loosestrife** Primulaceae

Perennial wetland or emergent herb. Stem erect, 25–60 cm high. Leaves opposite, linear-lanceolate to narrowly elliptic, glabrous. Flowers in rounded stalked racemes, 5–10 cm long, in axils of mid-stem leaves. Corrola yellow with black glandular dots. Fruit is a capsule with rose-brown seeds.

Distribution and ecology

Circumpolar boreo-montane element. Native but endangered in all CE countries, extinct in HU. Occurs in fens, fen meadows, margins of pools and ditches, also in shallow water. Prefers permanently wet and slightly acidic to neutral soils.

Anatomy of a basal part of the annual shoot

Round stem with a diameter of 5 mm. Maximum age 1 year.

The basal part consists of a large pith, a xylem, a phloem and an epidermis (1). Vascular bundles are located within a continuous lignified fiber belt (1). Vessel-free fiber zones delimit the bundles laterally (2).

The xylem consists mainly of fibers and solitary, angular vessels with a diameter of 20–30 µm (3) and a length of 100–200 µm. They have small, round intervessel pits and simple perforation plates (4). Fibers are thin- to thick-walled. Parenchyma is paratracheal and pervasive and rays within vascular bundles are absent (3).

Groups of sieve tubes and parenchyma cells form the phloem (3). The stele is surrounded by an endodermis (2, 3). The cortex and the pith consist of net-like intercellulars (aerenchyma) (2, 3). The epidermis consists of small, square, thin-walled, unlignified cells with slightly thicker peripheral walls.

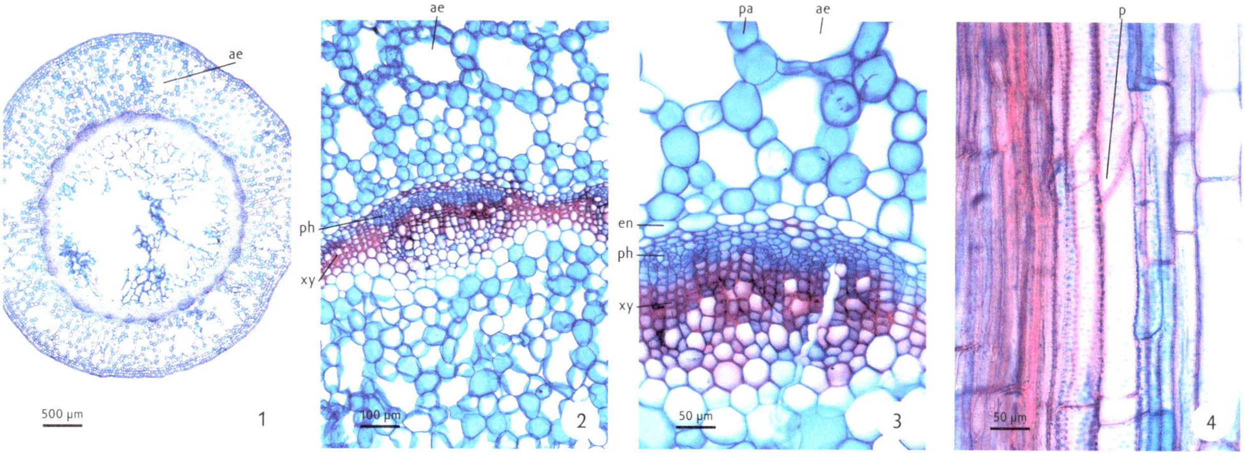

Samolus valerandi L.

Brookweed

Short-lived perennial wetland herb. Stem erect, glabrous, 5–50 cm high. Obovate, untoothed leaves in basal rosette with short petiole and alternating up the stem. Inflorescence is a raceme, white bell-shaped corolla with 5 lobes, longer than calyx-tube. Usually monocarpic, fruit is a capsule opening by 5 teeth.

Distribution and ecology

Subcosmopolitan temperate element. Native but endangered in all CE countries, extinct in SK. Occurs in dune slacks, lagoons, seasonally flooded springs, inland in saline meadows, ditches or drained fishpond bottoms. Prefers permanently wet calcareous or saline soils. Heliophyte and facultative halophyte.

Anatomy of a basal part of the annual shoot

Round stem with a diameter of 5–8 mm. Maximum age 1 year.

A central part consists of a large pith, a continuous xylem- and a phloem-belt. A large cortex surrounds this part (1). The xylem consists of angular vessels in short and long radial rows with a diameter of 30–40 µm (2) and a length of 150–200 µm. Intervessel pits are small and round. Perforation plates are simple (3). Fibers and rays are absent (4). Parenchyma is pervasive (2).

The stele is surrounded by an endodermis (2). The phloem consists of isolated groups of sieve tubes and parenchyma cells (2, 5). The cortex is composed of parenchyma cells with large intercellulars (1).

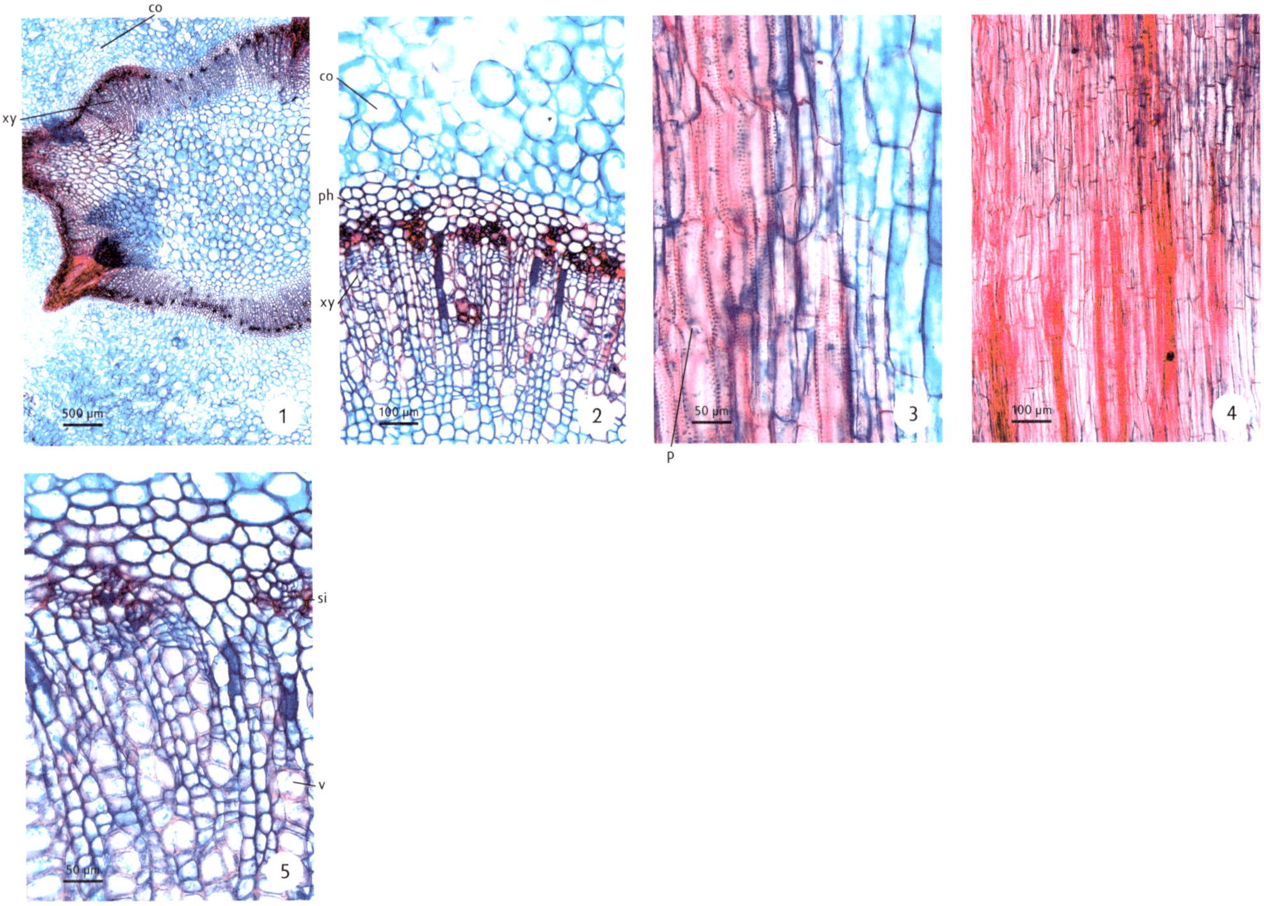

Aconitum napellus agg.

Monk's-hood Ranunculaceae

Perennial wetland herb with tuberous tap roots. Stem erect, up to 1.5 m high. Leaves deeply palmately lobed with narrow, deeply divided segments. Flowers in terminal racemes, petaloid sepals dark blue to purple, the upper forming an elongated hood, petals in the form of nectaries. Fruit is an aggregate of follicles. Very variable species with several subspecies.

Distribution and ecology

European temperate element, probably native and endemic to western and central Europe. Native to AT, CH and DE. Occurs on wet banks along streams, alpine-montane tall herb vegetation, in mountain and upland meadows and pastures and in open woodlands. Prefers moist, calcareous to slightly acidic soils. Reproduces both sexually and asexually through the formation of rhizomes.

Anatomy of the root collar

Root collars have a diameter of 10–15 mm. Secondary growth remains one growing season.

The parenchymatic center is surrounded by circularly arranged, open collateral vascular bundles, which are embedded in a thin-walled parenchymatic tissue (1, 2). The xylem is composed of radial groups of angular vessels and paratracheal and pervasive parenchyma (2). Vessel diameter is 15–25 µm and vessel length 80–100 µm. Intervessel pits are round to scalariform. Intervessel pits and perforation plates are simple (3, 4). Fibers are absent (2).

The phloem is composed of parenchyma cells and small isolated groups of sieve tubes (5). The cortex consists of unlignified thin-walled parenchyma cells without intercellulars (1).

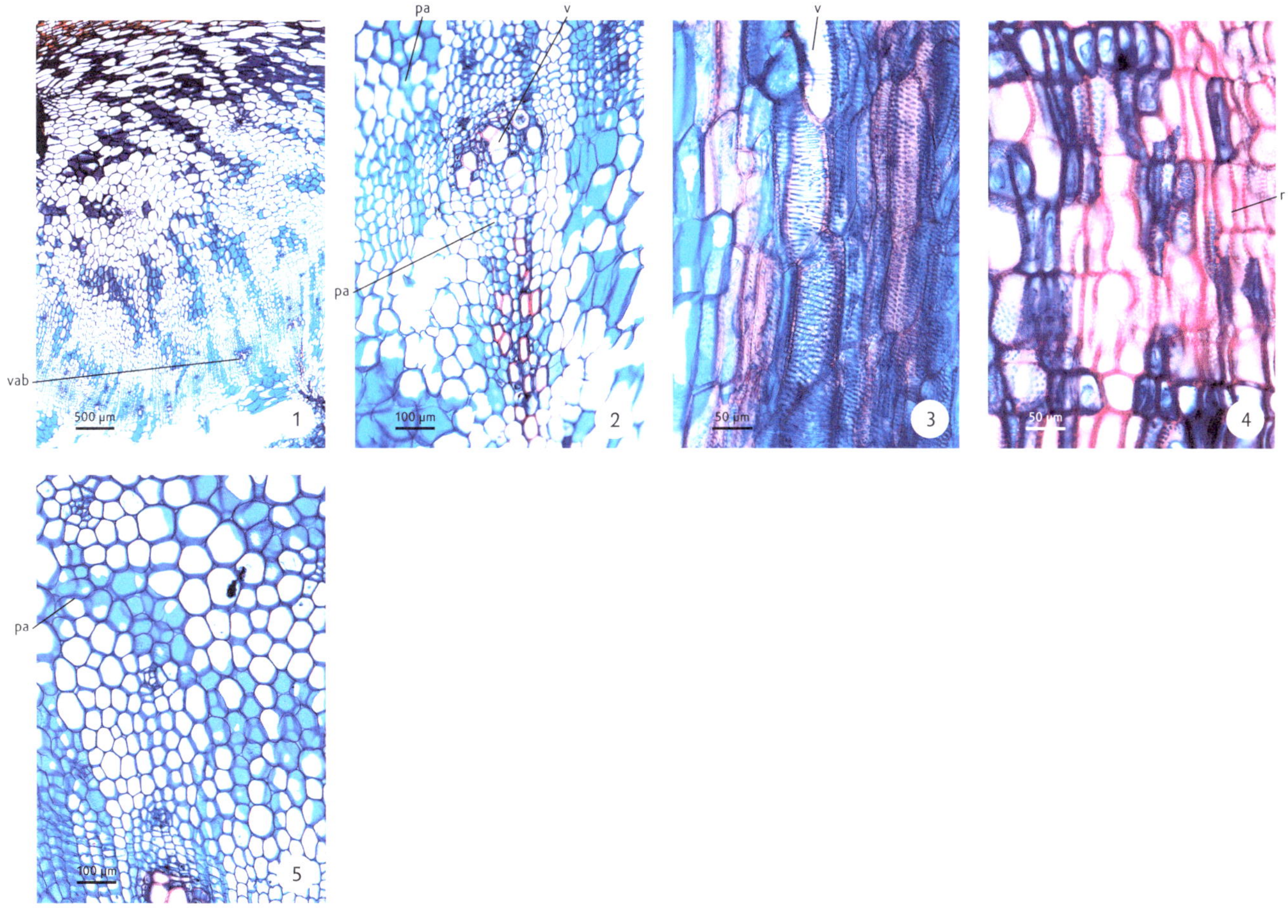

F. H. Schweingruber et al., *Anatomic Atlas of Aquatic and Wetland Plant Stems*, https://doi.org/10.1007/978-3-030-33420-8_46

Batrachium circinatum (Sibth.) Spach

(syn. *Ranunculus circinatus*)

Ranunculaceae
Fan-leaved Water-crowfoot

Perennial, sometimes annual submerged herb, homophyllous. Stem usually unbranched, glabrous, up to 250 cm long. All leaves very finely divided, circular, with short rigid segments lying on one plane. Single flowers in leaf axils on flower stalks longer than leaves, petals white, with lunate nectar-pit. Fruit is an achene, reniform-elliptic, with small beak. Terrestrial form known.

Distribution and ecology

Eurasian temperate element. Native to all CE countries. Grows in still or very slowly flowing freshwaters, both shallow and deeper than 1.5 m, mainly in pools, ponds, oxbows, ditches and channels, also in lakes and flooded gravel-pits. Prefers base-rich meso-eutrophic to eutrophic water, avoids shallow water which can dry up for a longer period.

Anatomy of the submerged annual shoot

Round shoot with a diameter of 1–2 mm. Maximum age 1 year.

A hollow shoot center is surrounded by a few open collateral vascular bundles, which are embedded in a thin-walled parenchymatic tissue (1). Some bundles are surrounded by an endodermis (2) with Casparian strips (3). The endodermis is sometimes absent (4). The xylem of the vascular bundles is composed of few primary unlignified vessels without secondary walls and a group of unlignified secondary angular vessels with secondary walls with a diameter of 10–15 μm. Parenchyma is pervasive (2). Xylem and phloem are anatomically separated by a few cambium cells (2). Few bundles miss the cambium (4). Fibers are absent.

The phloem consists of large sieve tubes and smaller accompanying cells (2, 4). The cortex consists of thin-walled, unlignified parenchyma cells and large, partially net-like intercellulars. The epidermis is composed of square, thin-walled cells (5).

Longitudinal section is missing.

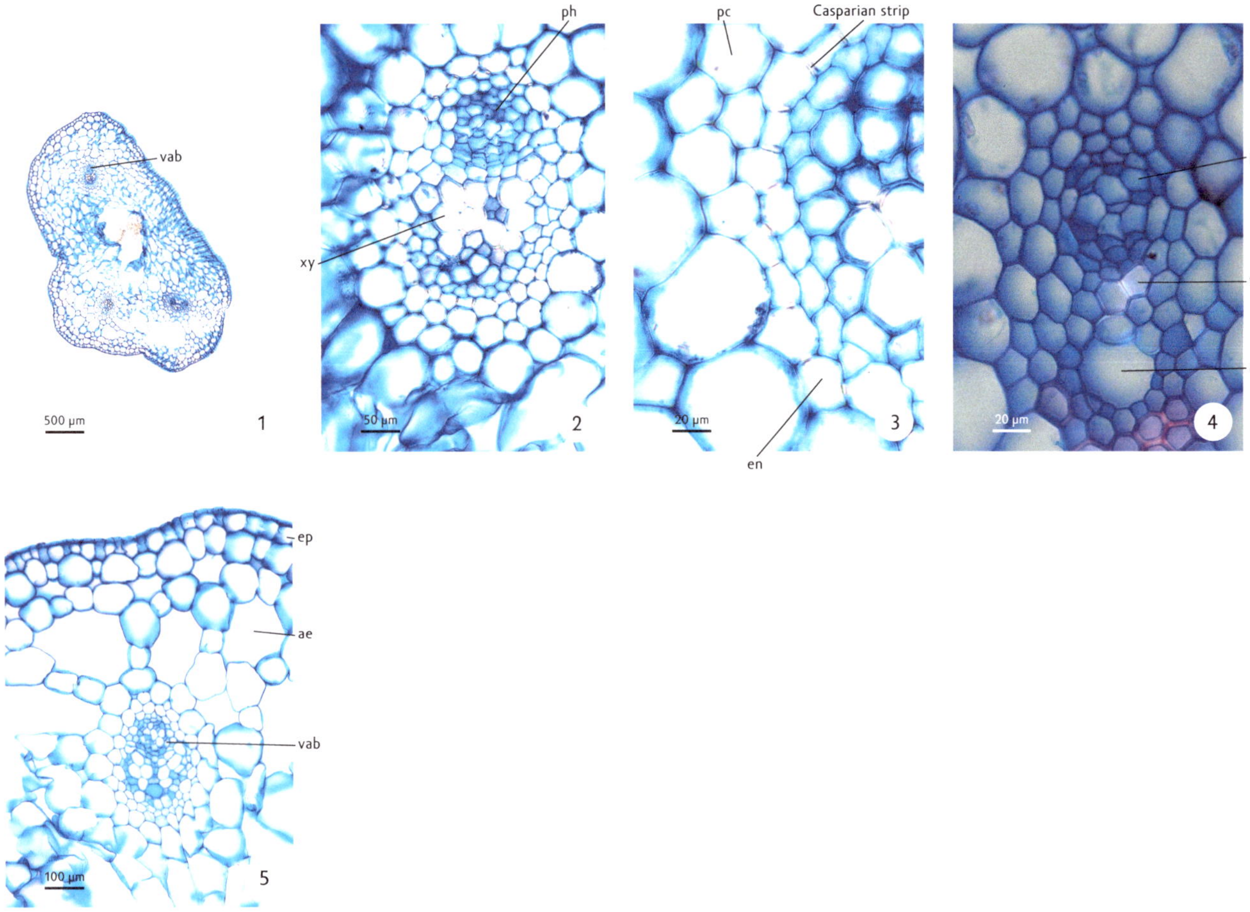

Batrachium fluitans (Lam.) Wimm.

(syn. *Ranunculus fluitans*)

Ranunculaceae
River Water-crowfoot

Perennial submerged herb with submerged leaves, terrestrial form known. Stem usually unbranched, glabrous, up to 6 m long. All leaves very finely divided, usually longer than adjacent stem internode. Single flowers in leaf axils on flower stalks usually shorter than leaves, petals white with pear-shaped nectar-pit. Fruit is an achene, ovate-elliptic, apex with short and blunt beak, fruits mature very rarely.

Distribution and ecology

European temperate element. Native to all CE countries. Occurs in rapidly flowing rivers, usually with gravel or stony stable sediments, from very shallow to 3 m deep water. Prefers base-rich meso-eutrophic water, tolerant of mild eutrophication. Stem has a low rigidity which helps to withstand drag from fast-flowing water. The ability to use HCO_3- for photosynthesis. Mainly vegetative propagation by stem fragments.

Anatomy of the submerged annual shoot

Round shoot with a diameter of 1.5–2.5 mm. Maximum age 1 year.

Few open circularly arranged collateral vascular bundles are embedded in a thin-walled parenchymatic tissue (1). The bundles are centripetally surrounded by a lignified sclerenchymatic sheath (1–3). The xylem of the vascular bundles consists of a few angular, unlignified vessels with a diameter of 15–30 µm. Parenchyma is pervasive (2). Fibers are absent. Perforation plates are net-like (5) and intervessel pits are scalariform. The xylem and phloem are anatomically separated by a few cambium cells (2).

The phloem consists of large sieve tubes and smaller accompanying cells (2, 4). The cortex consists of thin-walled, unlignified parenchyma cells and large irregularly formed intercellulars. The epidermis is composed of square, thin-walled cells and a few stomata (1).

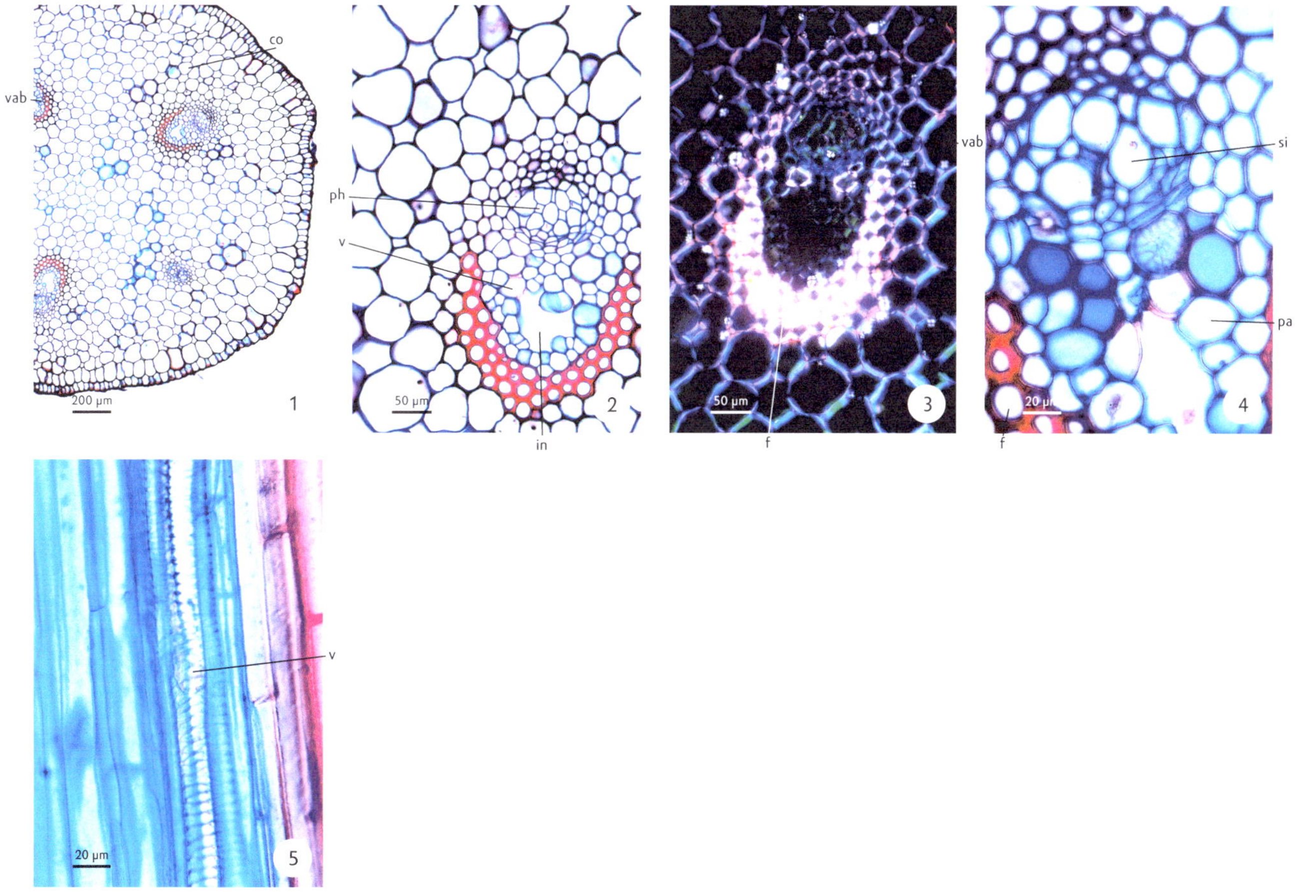

Batrachium peltatum (Schrank) J. Presl

(syn. *Ranunculus peltatus*)

Ranunculaceae

Pond Water-crowfoot

Annual to perennial submerged herb, heterophyllous or with finely divided leaves only. Stem branched, glabrous, 30–150 cm long. Floating leaves palmately-lobed, kidney-shaped, up to 1/3 divided. Flowers single in leaf axils, flower stalks longer than 5 cm and longer than leaf petiole, petals white, 9–17 mm long, with pear-shaped nectar-pit. Fruit is an achene, obovate-reniform, short bristly. Terrestrial form known.

Distribution and ecology

European wide-temperate element. Native to all CE countries except HU. Occurs in ponds, pools, ditches and slowly flowing streams from lowlands to submontane areas. Tolerates both calcareous and base-poor water, prefers meso-eutrophic water. Submerged leaves able to use HCO_{3^-}. Production of flowers is enhanced by low P concentrations in water.

Anatomy of the submerged annual shoot

Round stem with a diameter up to 1–1.5 mm. Maximum age 1 year (1).

A hollow shoot center is surrounded by a few open collateral vascular bundles, which are embedded in a thin-walled parenchymatic tissue (1). The bundles are surrounded by a partially or completely lignified sclerenchymatic sheath (2, 3). The xylem of the vascular bundles consists of few primary vessels and a group of secondary lignified vessels with a diameter of 10–15 µm. Fibers are absent. Parenchyma is pervasive (2). The xylem and phloem are anatomically separated by a few cambium cells (2).

The phloem consists of large sieve tubes and smaller accompanying cells (3, 4). The cortex consists of thin-walled, unlignified parenchyma cells and large intercellulars. The epidermis is composed of square, thin-walled cells and a few stomata (4).

Longitudinal section is missing.

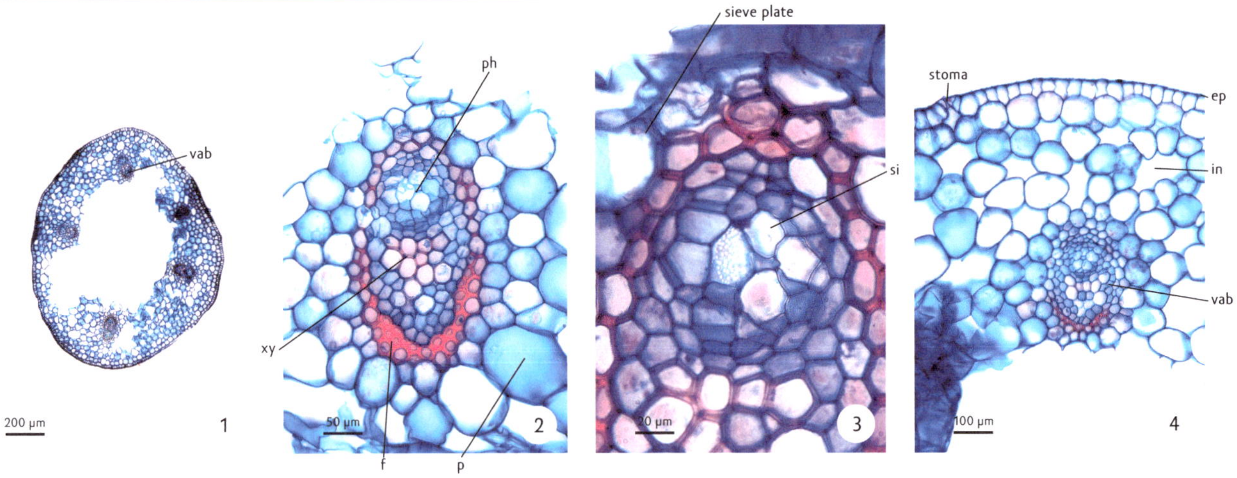

Batrachium trichophyllum (Chaix) F. W. Schultz

(syn. *Ranunculus trichophyllus*)

Ranunculaceae

Thread-leaved Water-crowfoot

Annual or perennial submerged herb with submerged leaves only. Stem branched, usually glabrous, up to 100 cm long. Leaves all finely divided with short flaccid segments. Flowers single in leaf axils, petal white, 3.5–6 mm across, with lacunate nectar-pit, sepals reflexed. Fruit is an achene, obovate-elliptic, dorsal part bristly. Terrestrial form known.

Distribution and ecology

Circumpolar wide-boreal element. Native to all CE countries. Occurs in ponds, pools, oxbows and drainage ditches, also in slowly flowing rivers. Terrestrial form occurs on drained bottoms. Prefers mesotrophic to eutrophic water with sandy bottom. Photosynthesis mainly in the epidermis cells, HCO_3^- user. Cleistogamous flowers, flowers also under water.

Anatomy of the submerged annual shoot

Round shoot with a diameter of 1.5–2.5 mm. Maximum age 1 year.

A hollow shoot center is surrounded by a few open collateral vascular bundles, which are embedded in a thin-walled parenchymatic tissue (1). The bundles are surrounded by a partially or completely lignified sclerenchymatic sheath (1, 2). The xylem of the vascular bundles consists of round, unlignified vessels with a diameter of 10–15 µm with spiral thickenings (5). Parenchyma is pervasive (2). The xylem and phloem are anatomically separated by a few cambium cells (2).

The phloem consists of large sieve tubes and smaller accompanying cells (3, 4). The cortex consists of thin-walled, unlignified parenchyma cells and large intercellulars. The epidermis is composed of square, thin-walled cells and a few stomata (1).

Longitudinal section is missing.

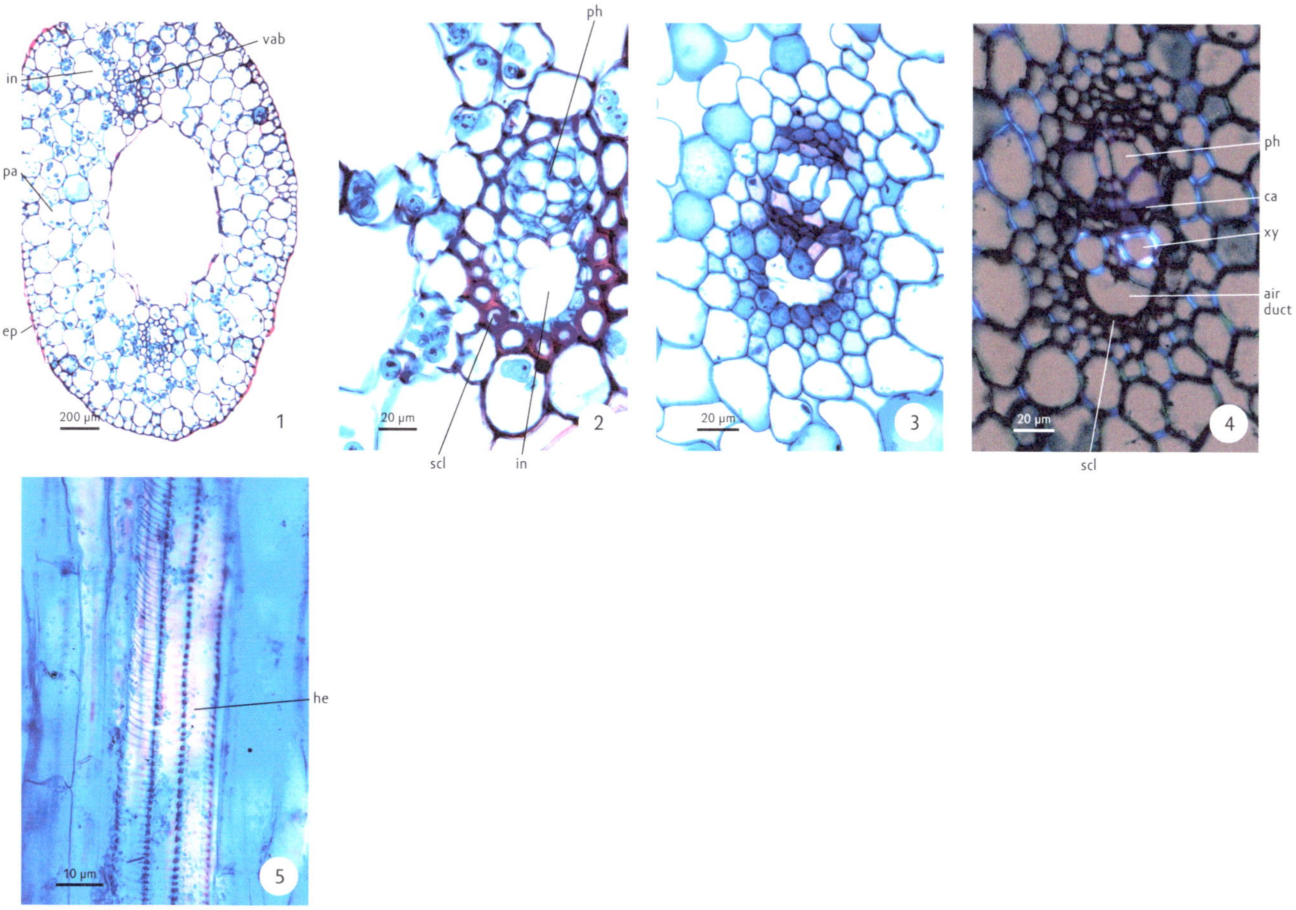

Caltha palustris L.

Marsh-marigold Ranunculaceae

Perennial wetland herb, rhizomatous. Flower stalks erect to ascending, usually unbranched, glabrous, 30–50 cm high. Basal leaves long-petiolate, cordate, dark green. Flowers usually solitary, 2.5–3.5 cm across, petaloid sepals golden-yellow. Fruit is an aggregate of follicles.

Distribution and ecology

Circumpolar wide-boreal element. Native to all CE countries. Occurs along wet banks of streams, ponds and lakes, in reeds, in alder carrs, also in wet meadows and pastures from lowlands up to the mountains. Prefers wet, neutral to base-rich soils.

Anatomy of the root collar

Root collar have a diameter of 5–10 mm. Maximum age 1 year.

The center is surrounded by a few circularly arranged circles of open collateral vascular bundles, which are embedded in a thin-walled parenchymatic tissue (1). Some of them are laterally linked with an fascicular cambium (2). The xylem is composed of grouped lignified vessels and paratracheal and pervasive parenchyma (3). Vessel diameter is 30–40 µm and a vessel length 80–100 µm. Intervessel pits are scalariform and perforation plates are simple (4). Fibers are absent (4).

The phloem is primarily composed of parenchyma cells (3). The cortex consists of unlignified thin-walled parenchyma cells and small intercellulars (5).

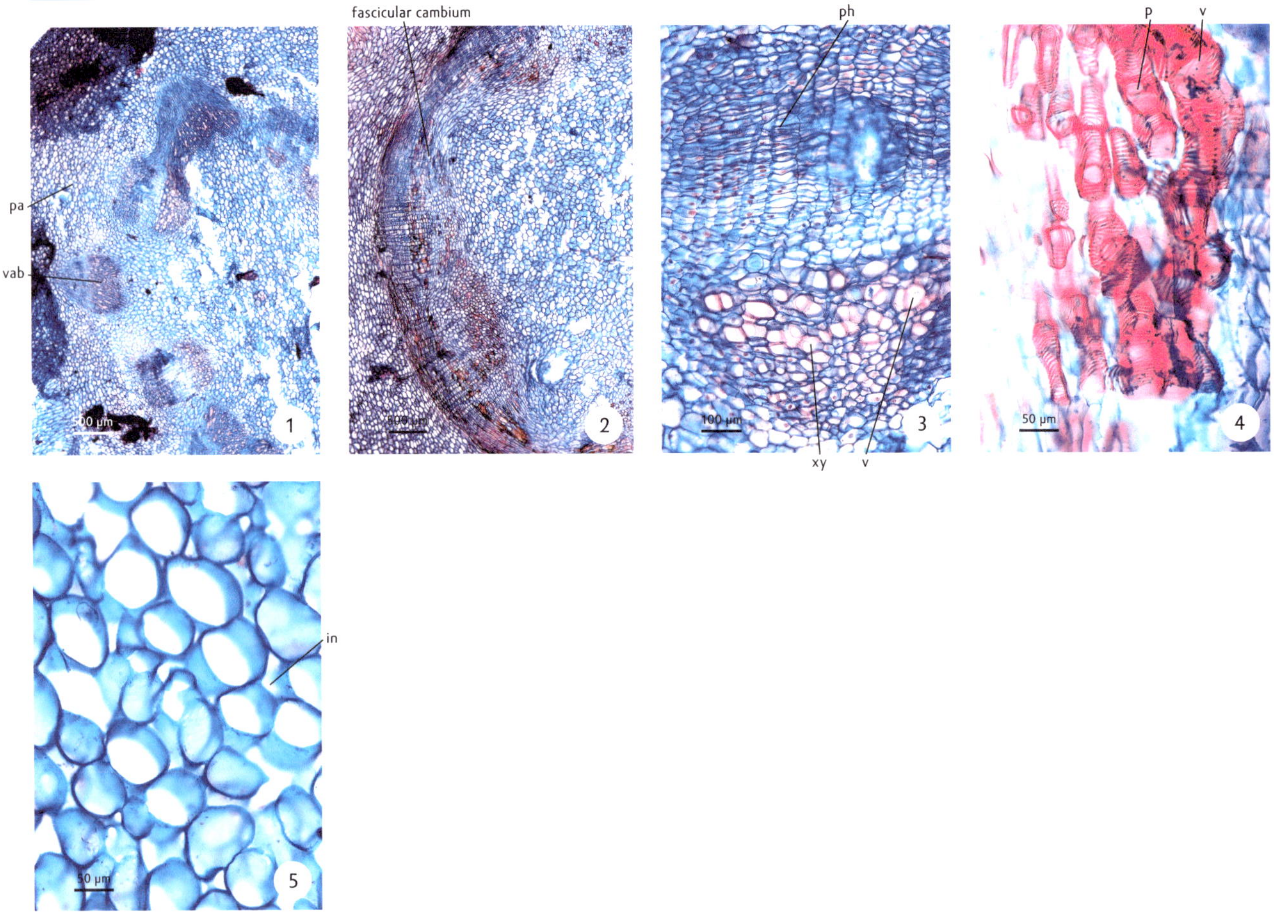

Clematis integrifolia L. **Solitary Clematis** Ranunculaceae

Perennial wetland herb, rhizomatous. Stem erect, tough, often purplish brown, 30–60 cm high. Leaves opposite, sessile, ovate to wide lanceolate, entire. Flowers usually solitary, terminal or in axillary cymes, 4 petaloid sepals ovate, 3.5–4.5 cm long, dark violet or blue. Fruit is an achene, hairy, the style becoming elongated and feathery.

Distribution and ecology

Eastern Europe to western Asia. Native to AT, HU and SK, almost extinct in CZ. Occurs in wet meadows in river floodplains. Endosperm dormancy, chilling requirements for seed germination.

Anatomy of a basal part of the stem

Polygonal stem with a diameter of 5–8 mm. Maximum age 2 years.

Annual rings are marked by a slight semi-ring porosity (3). The lobed pith is surrounded by a xylem, a phloem, a cortex and a phellem (1, 2). The xylem consists of radially arranged groups of solitary vessels with a diameter of 40–60 µm (2, 3) and a length of 200–300 µm. Vessels have round intervessel pits and simple perforation plates (4). Fibers are thin- to thick-walled (2, 3). Parenchyma is paratracheal (3). Vascular bundles remain (2), large vessel-free zones (>10-seriate rays) (1, 2) occur between zones with vessels.

The phloem is uniform, parenchyma cells and sieve elements cannot be distinguished (5, 6). The cortex consists of large parenchyma cells and small, irregularly formed intercellulars and a thin-walled phellem (2).

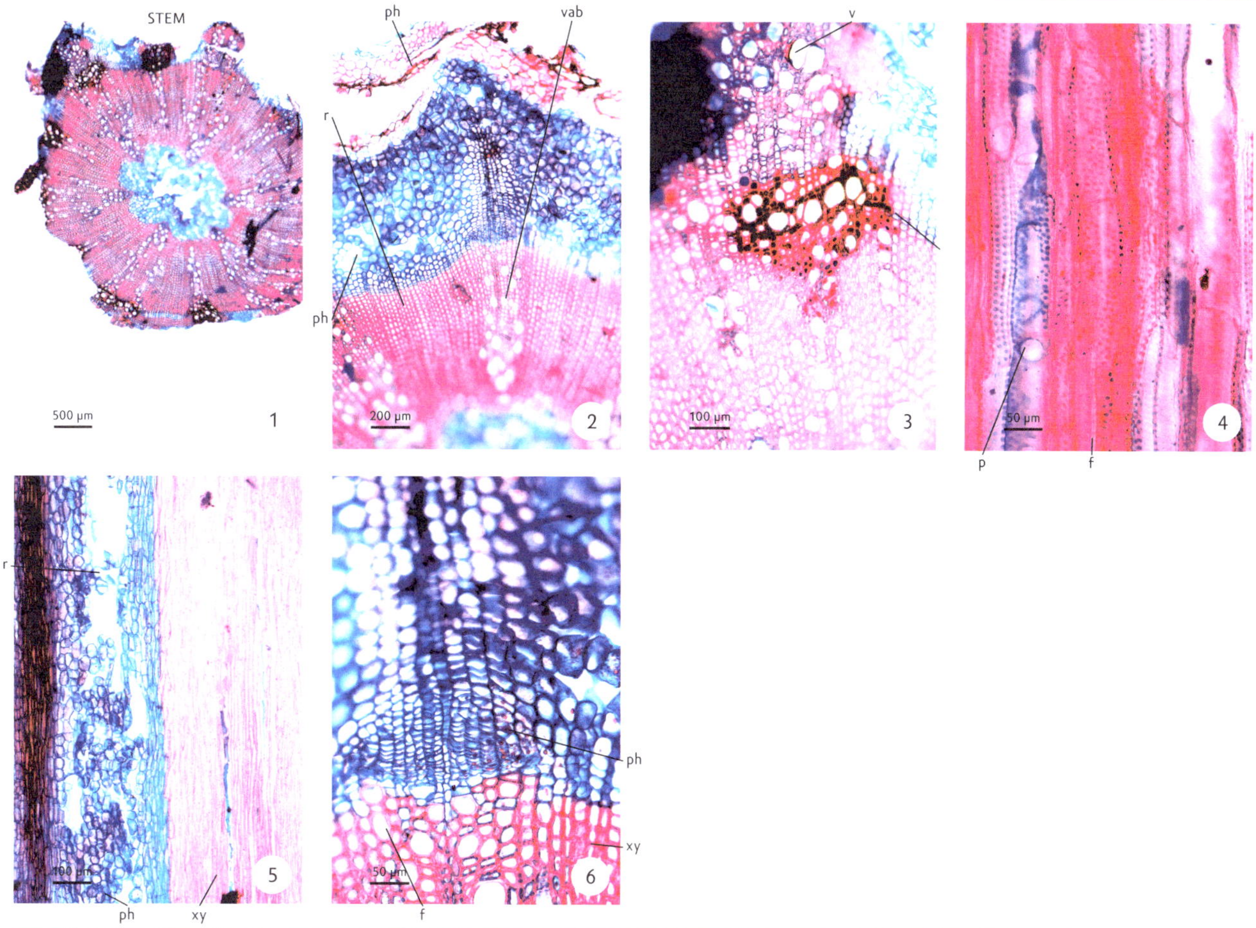

Myosurus minimus L.

Mousetail

Annual wetland herb. Stem erect to ascendent, glabrous, 10–16 cm high. Leaves in a basal leaf rosette, simple, linear, 2–6 cm long, 0.5–1.0 cm wide. Flowers terminal, 5 petals 3–4 mm long, greenish, inconspicuous, the flower bed is conspicuously elongated when fruiting. Fruit is an elliptic achene, with lustrous surface.

Distribution and ecology

European temperate element; also in N. America and Australasia and widely naturalised elsewhere. Native to all CE countries. Occurs in temporarily flooded arable fields or in drained fishpond bottoms. Prefers regularly disturbed wet sandy or clayish soils, both acid and slightly alkaline, tolerates also slightly salty soils. Long-lived seeds.

Anatomy of the root collar

Round stem with a diameter of 2 mm. Maximum age 1 year (1, 2).

The root collar consists of a lignified xylem belt, a small phloem and a large cortex (1). The xylem consists of vessels and of fibers (2). The inner part of the xylem consists of unlignified vessels and the outer part primarily of thick and thin- to thick-walled living fibers (2). Parenchyma is paratracheal and pervasive, rays are absent (2). Vessels in the center of the stem have a diameter of 10–15 µm and a length of <50 µm (4). Intervessel pits are round (3) and perforation plates are simple (3).

The stele is surrounded by an endodermis (1, 2). The phloem consists of small groups of sieve tubes and parenchyma cells (5). It is difficult to recognize sieve tubes (2). The cortex consists of large, round to oval, unlignified thin-walled parenchyma cells and intercellulars (1, 5). The phellem is absent.

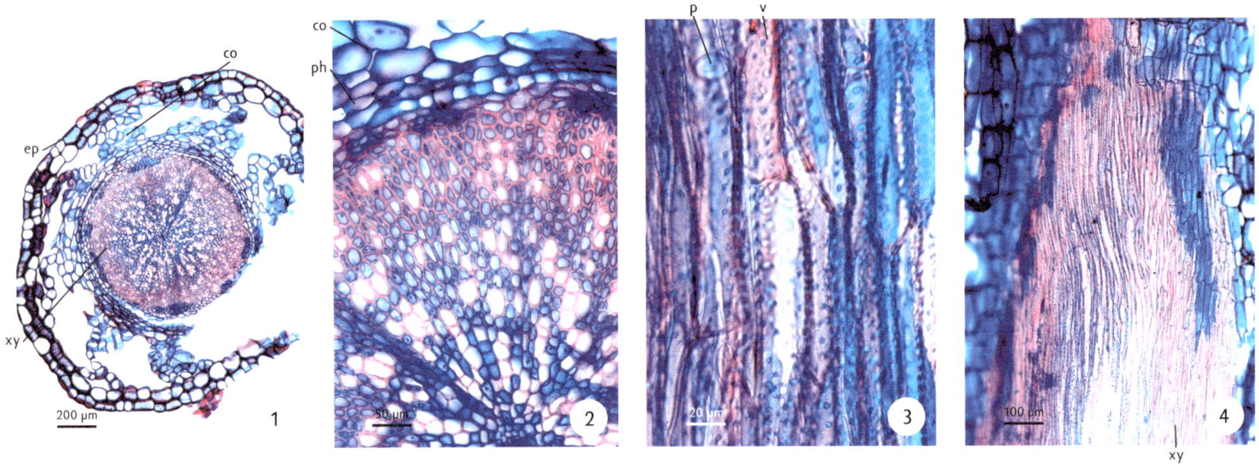

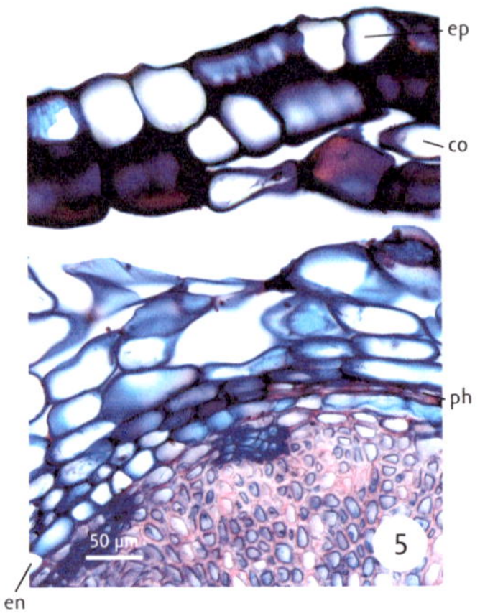

Ranunculus aconitifolius L.

Aconite-leaved Buttercup Ranunculaceae

Perennial wetland herb with short rhizome. Stem erect, richly branched, glabrous, basal part sulcate, 40–70 cm high. Lowest leaves long-petiolate, deeply palmately 3–6-lobed. Flower stalks sulcate and pubescent, flowers 10–25 mm across, petals ovate, white. Fruit is a glabrous achene, nearly roundish, with short curved beak.

Distribution and ecology

Endemic to C. Europe. Native to AT, CH, CZ and DE. Occurs in wet banks along mountain streams, in alpine-montane tall herb vegetation, in mountain pastures and in springs. Prefers silicate bedrocks.

Anatomy of the annual shoot and the root collar

Annual shoot: Round shoot with a diameter of 2–4 mm (1). Maximum age 1 year.

The center is surrounded by a few open collateral vascular bundles, which are embedded in a thin-walled parenchymatic tissue with small intercellulars (1, 3). The bundles are surrounded by a dense, lignified thick-walled sclerenchymatic sheath cells (2, 3). The xylem of the vascular bundles is composed of a group of lignified vessels, which are surrounded by unlignified parenchyma cells. Vessels have a diameter of 15–35 µm. Fibers are absent. Parenchyma is pervasive (3).

The phloem consists of sieve tubes and smaller accompanying cells (2). The cortex consists of thin-walled, unlignified parenchyma cells and small intercellulars. The epidermis is composed of square, thin-walled cells (1).

Root collars have a diameter of 8–12 mm (4, 5). Secondary growth remains one growing season.

The center is surrounded by a few, circularly arranged, solitary, open collateral vascular bundles, which are embedded in a thin-walled parenchymatic tissue (4). Vascular bundles are surrounded by a thin-walled endodermis (5, 6). The xylem of the vascular bundles is composed of lignified vessels with a diameter of 15–25 µm. They are surrounded by unlignified parenchyma cells (5, 6). Vessel length varies from 50 to 70 µm. Perforation plates are simple (7). Fibers are absent.

The phloem is composed of parenchyma cells and small groups of sieve tubes (5). The cortex consists of unlignified parenchyma cells and small intercellulars (8).

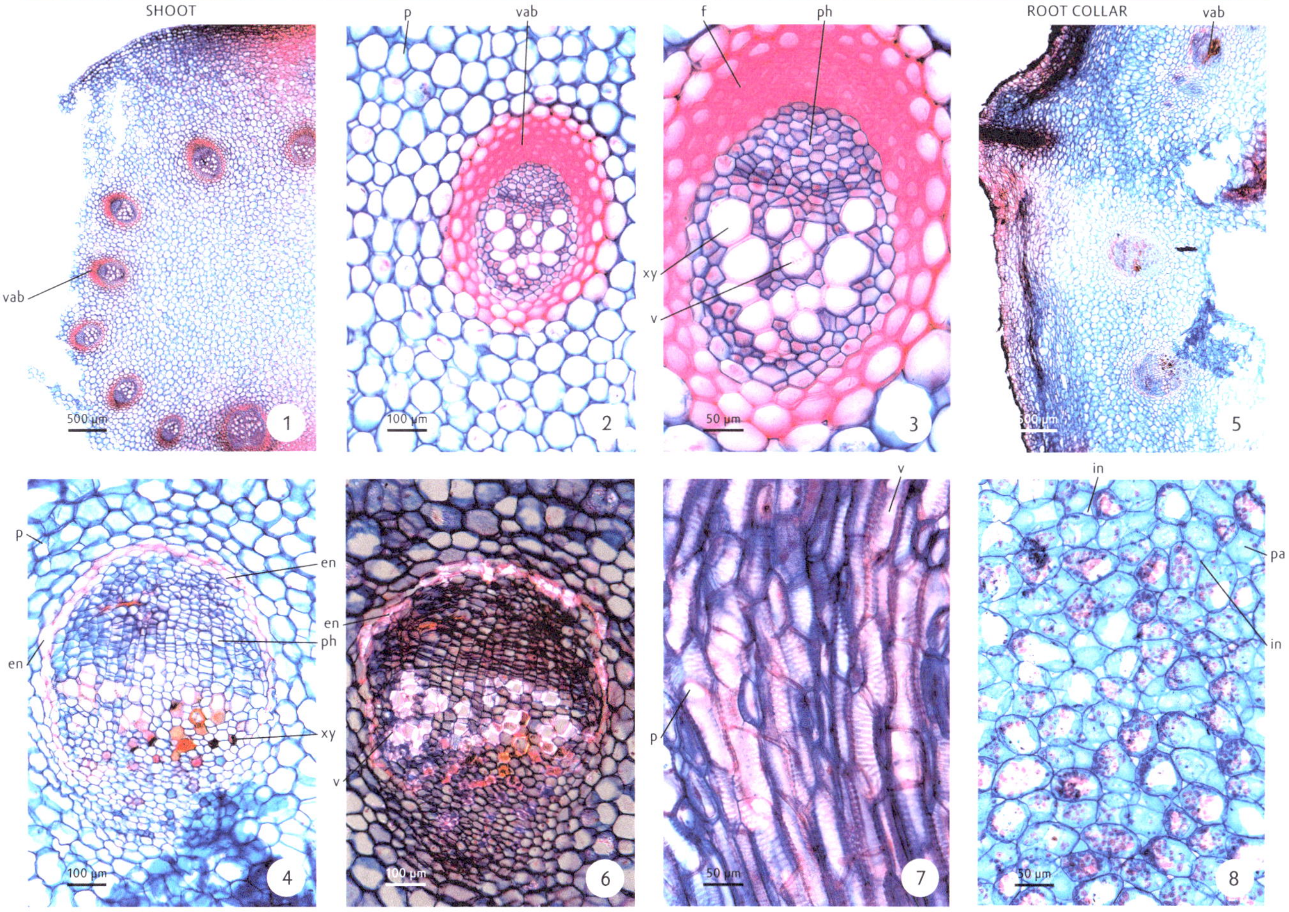

Ranunculus flammula L.

Lesser Spearwort Ranunculaceae

Perennial amphibious herb with short rhizome and heterophylly. Stem erect to procumbent, branched, 20–50 cm high. Basal leaves glabrous, long-petiolate, lanceolate to oblanceolate, stem leaves sessile, linear. Flowers 7–15 mm across, petals yellow and shiny. Fruit is a glabrous achene, with short curved beak.

Distribution and ecology

European temperate element. Native to all CE countries. Occurs on wet soils or in shallow still water with seasonal water level fluctuations, along banks of ponds and lakes, in ditches, in wet meadows and pastures. Prefers oligo- to mesotrophic water and acid to neutral substrates. Roots are colonised by vesicular-arbuscular mycorrhiza.

Anatomy of the annual shoot and the root collar

Annual shoot: Round shoot with a diameter of 2–3 mm. Maximum age 1 year.

The center is surrounded by a few open collateral vascular bundles, which are embedded in a thin-walled parenchymatic tissue with small intercellulars (1). The bundles are surrounded by a partially or completely lignified thick-walled sclerenchymatic sheath (2). The xylem of the vascular bundles is composed of a few primary unlignified vessels without secondary walls and a group of lignified secondary vessels with secondary walls, which are surrounded by unlignified parenchyma cells (2).

The phloem consists of large sieve tubes and smaller accompanying cells (2). The cortex consists of thin-walled, unlignified parenchyma cells and large intercellulars (3).

Root collars have a diameter of 3–4 mm. Maximum age 1 year. The center is surrounded by a few, circularly arranged open collateral vascular bundles, which are embedded in a thin-walled parenchymatic tissue (4). Vascular bundles are surrounded by a thin-walled endodermis (5). The xylem of the vascular bundles is composed of lignified angular vessels with a diameter of 15–20 µm. Vessel length varies from 50 to 60 µm. Perforation plates are simple (6). Fibers are absent.

The phloem is composed of parenchyma cells and small groups of sieve tubes (5). The cortex consists of unlignified parenchyma cells and small intercellulars (7).

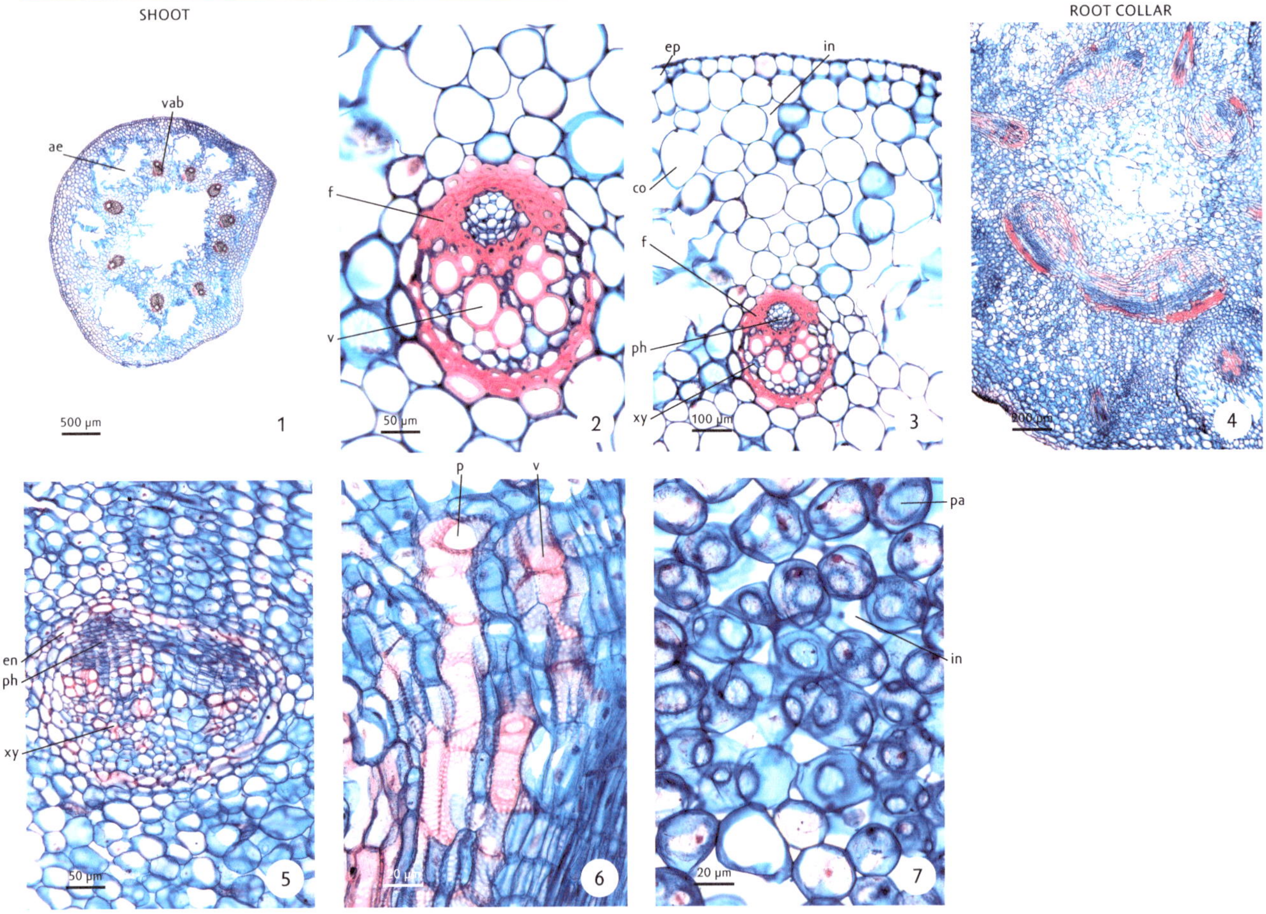

Ranunculus lanuginosus L.

Perennial wetland herb with short rhizome. Stem erect, branched, lanate, 30–50 cm high. Basal leaves long-petiolate, 3–5-lobed, leaf lobes broad ovoid, stem leaves rather sessile with lanceolate leaf lobes. Flowers on 5–10 cm long hairy stalks, 5 petals glabrous and ochre-yellow. Fruit is an obovate achene with a spiral-curved beak.

Distribution and ecology

European temperate element. Native to all CE countries from lowlands to uplands. Occurs on humid soils in shaded environments such as floodplain forests or gorges.

Anatomy of the basal root collar

Root collars have a diameter of 5–7 mm. Secondary growth remains one growing season.

The center is surrounded by a few, circularly arranged open collateral vascular bundles, which are embedded in a thin-walled parenchymatic tissue (1). An inter-fascicular cambium is absent (1). Vascular bundles are surrounded by an endodermis-like layer of lignified fibers (2). The xylem of the vascular bundles is composed of a group of lignified vessels with a diameter of 25–35 µm and a length of 120–150 µm. They are accompanied by unlignified parenchyma cells (2). Intervessel pits are round to scalariform and perforation plates are simple (3). Fibers are absent.

The phloem is fairly homogeneous and composed primarily of parenchyma cells (2). The cortex consists of unlignified thin-walled parenchyma cells and small intercellulars (4). They appear roundish in cross-sections (4).

Sections of annual shoots are missing.

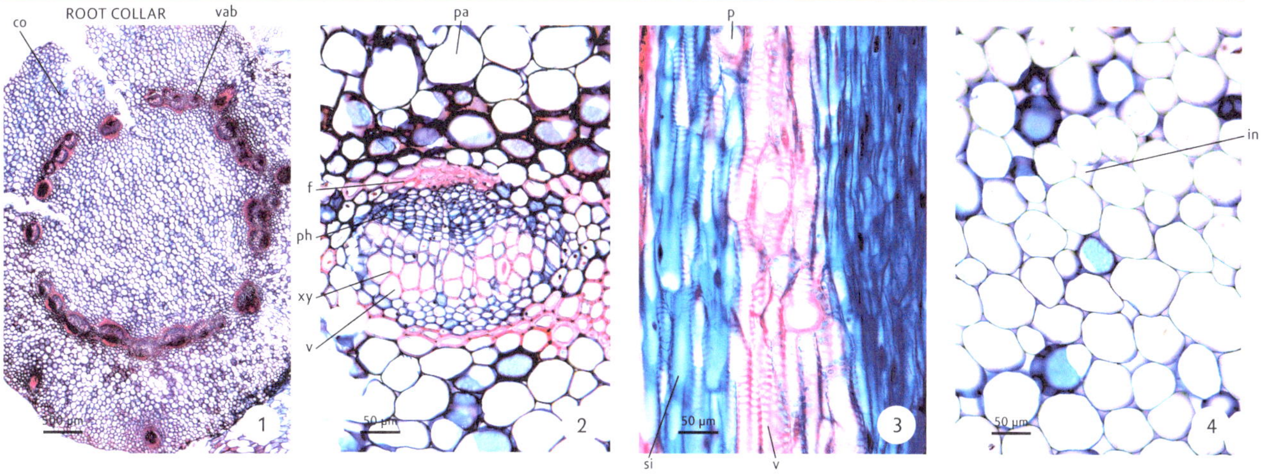

Ranunculus lingua L. **Greater Spearwort** Ranunculaceae

Perennial amphibious aquatic herb, stoloniferous. Stem erect, poorly branched, 60–120 cm high. Lowermost leaves entire, shortly petiolate, broadly lanceolate, stem-leaves sessile, lanceolate to linear lanceolate, toothed and glabrous. Flowers 20–40 mm across, petals glabrous and golden-yellow. Fruit is a glabrous ovoid achene with short curved beak.

Distribution and ecology

Euro-Siberian temperate element. Native to all CE countries. Occurs in swamps, fens, also in shallow water at the edges of lakes, ponds, rivers or ditches. Prefers base-rich, still or slowly flowing water and substrate rich in humus and nutrients. Dispersal mainly by floating (buoyant) rhizome fragments. Roots are colonised by arbuscular mycorrhiza.

Anatomy of the annual shoot

Round shoot with a diameter of 2.5–3.5 mm. Maximum age 1 year.

A hollow shoot center is surrounded by a few open collateral vascular bundles, which are embedded in a thin-walled parenchymatic tissue (1, 3). The bundles are surrounded by a partially or completely lignified sclerenchyma-sheath (2, 5). The xylem of the vascular bundles consists of a group of lignified vessels with a diameter of 50–60 µm. Vessels are surrounded by paratracheal parenchyma (2, 5). Fibers are absent (2, 5). Intervessel pits are round to scalariform and perforation plates are simple (5). The xylem and phloem are anatomically separated by a few cambium cells (2, 5).

The phloem consists of large sieve tubes and smaller accompanying cells (2, 5) and the cortex of thin-walled, unlignified parenchyma cells and large intercellulars (4). The epidermis is composed of square, thin-walled cells (4).

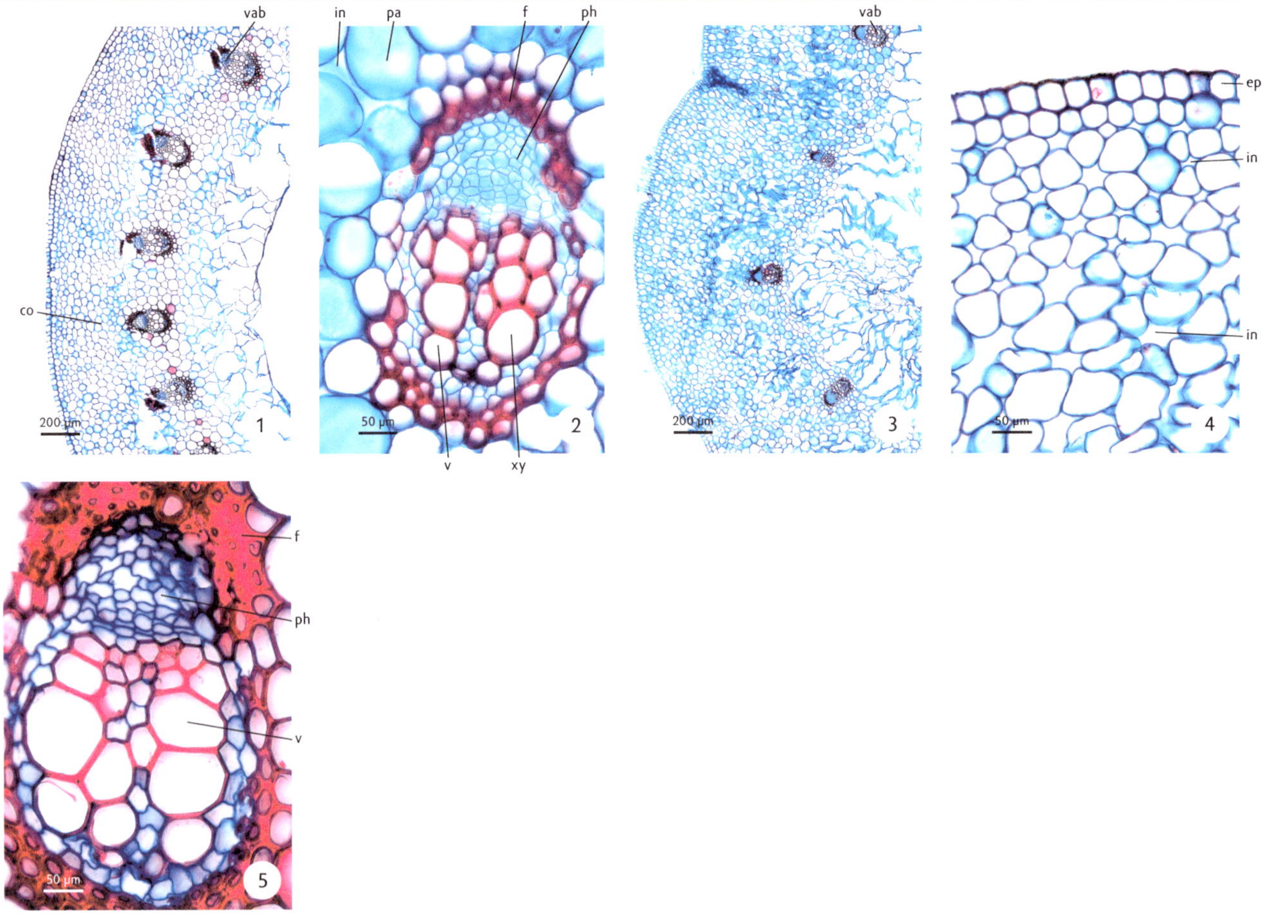

Ranunculus pygmaeus Wahl.

Dwarf Buttercup Ranunculaceae

Perennial wetland herb. Stem erect to ascendent, pubescent, 1–5 cm high. Basal leaves 3–5-lobed, glabrous, stem-leaves deeply 3–5-lobed, glabrous. Single flower 0.5–1 cm across, petals light yellow. Fruit is a cylindrical to ovoid achene with curved beak.

Distribution and ecology

Circumarctic-alpine species with disjunct distribution: very rare in the Alps, more widespread in the Arctic. Native to AT and CH. Occurs in snow fields or at glacier edges in small isolated relic populations in the central Alps, especially on silicate bedrocks. Submergence promotes the elongation of petioles. Propagates only by seeds.

Anatomy of the basal rootstock

Basal rootstocks have a diameter of 4–5 mm. Secondary growth remains one growing season.

The pith is surrounded by a few, circularly arranged circles of open collateral vascular bundles, which are embedded in a thin-walled parenchymatic tissue (1–3). In contrast to all other analyzed *Ranunculus* species, the bundles are laterally linked with an inter-fascicular cambium (3). The xylem is composed of grouped lignified vessels and pervasive parenchyma (2). A vessel diameter is 15–20 µm and a vessel length 40–50 µm. Intervessel pits are round to scalariform and perforation plates are simple (4). Fibers are absent.

The phloem is primarily composed of parenchyma cells (3, 4). The cortex consists of unlignified thin-walled parenchyma cells and small intercellulars (5).

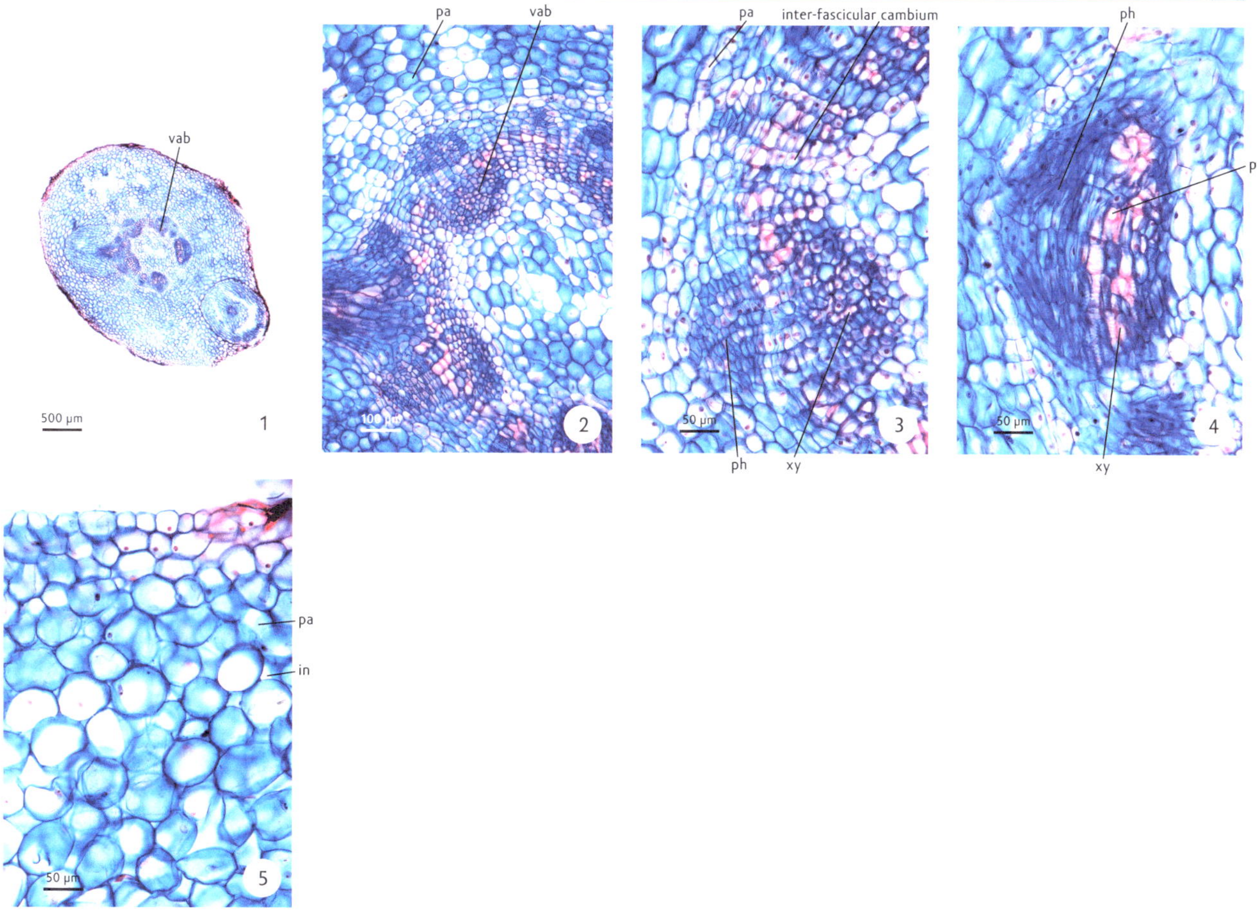

Ranunculus sceleratus L.

Celery-leaved Buttercup

Annual to biennial (overwintering) wetland to amphibious herb. Stem erect, sulcate, hollow, richly branched, 20–50 cm high. Basal leaves long-petiolate, deeply 3-lobed, glabrous and shiny. Corolla 5–10 mm across, petals yellow. Fruit is a glabrous ovoid achene with a very short beak.

Distribution and ecology

Circumpolar boreo-temperate element; widely naturalised outside its native range. Native to all CE countries. Prefers wet, disturbed soils rich in nutrients. Occurs on mud at the edges of streams, rivers, ponds and ditches. Salt-tolerant, very resistant to waterlogging and shallow floods due to its ability to greatly increase the elongation rate of petioles under water and a high root porosity. Seeds are long-lived.

Anatomy of the annual shoot and the root collar

Annual shoot: Round shoot with a diameter of 2–3 mm. Maximum age 1 year.

The center is surrounded by a few open collateral vascular bundles, which are embedded in a thin-walled parenchymatic tissue with small intercellulars (1). The bundles are surrounded by a partially lignified thin-walled sclerenchymatic sheath (2). The xylem of the vascular bundles is composed of lignified vessels (2). The phloem consists of large sieve tubes and smaller companion cells (2). The cortex consists of thin-walled, unlignified parenchyma cells and large, partially net-like intercellulars. The epidermis is composed of square, thin-walled cells (1).

Root collars have a diameter of 3–4 mm. Annual rings are absent.

The center is surrounded by a few, circularly arranged open collateral vascular bundles, which are embedded in a thin-walled parenchymatic tissue (3, 5). The xylem of the vascular bundles is composed of lignified angular vessels with a diameter of 15–20 µm. They are surrounded by unlignified parenchyma cells (4, 6). Vessel length varies from 50 to 60 µm. Intervessel pits are round and perforation plates are simple (7). Fibers are absent (4, 6).

The phloem is composed of parenchyma cells and small groups of sieve tubes (4). The cortex consists of unlignified parenchyma cells and small intercellulars (3).

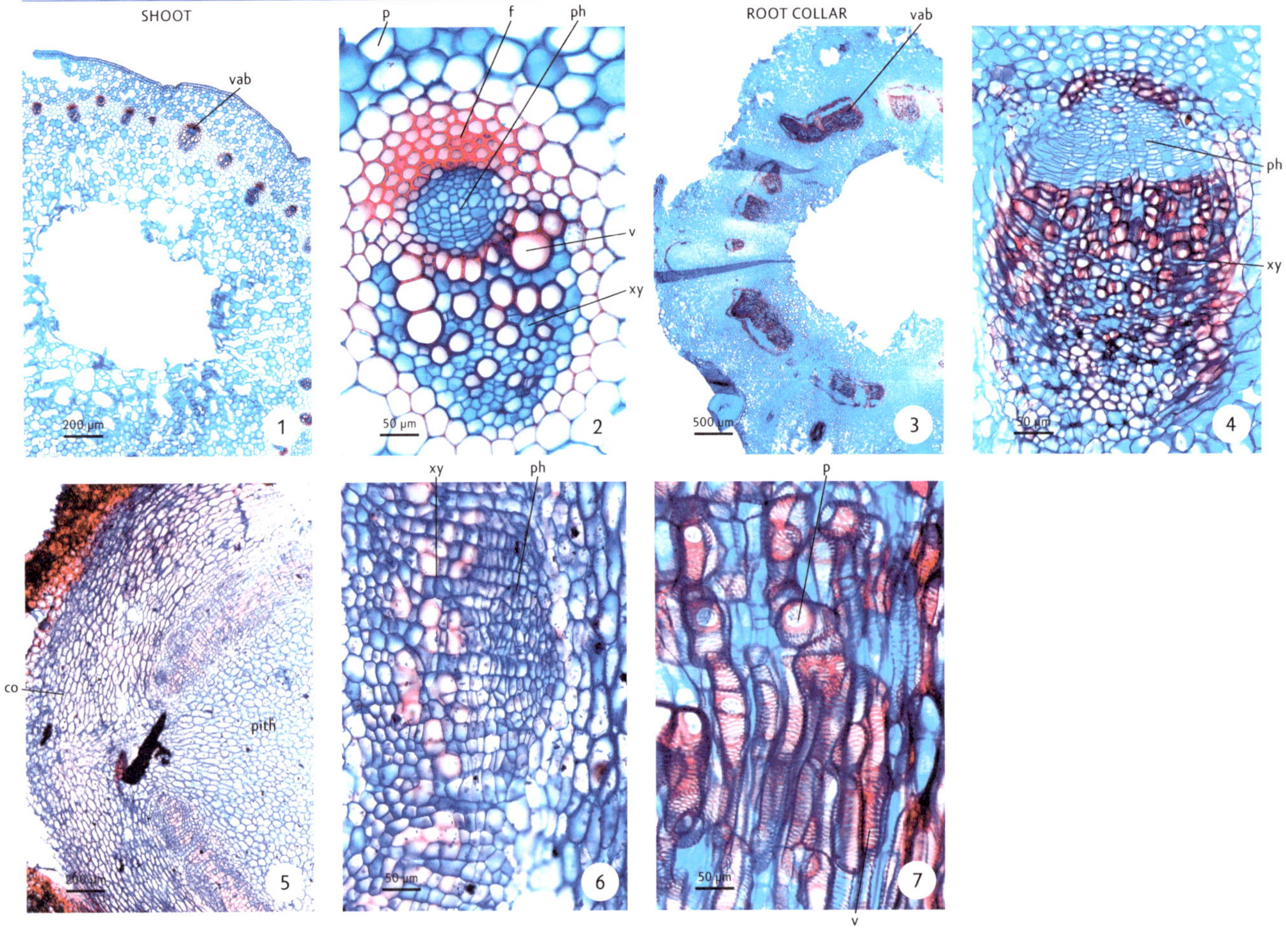

Perennial wetland herb, shortly rhizomatous. Stem erect, usually simple, glabrous, 40–100 cm high. Leaves alternate, lowermost petiolate, upper sessile, blade triangular, 2-3-ternate. Flowers in dense clusters, 10–15 mm across, petaloid sepals whitish to lilac or pink. Fruit is a long-stalked achene with 4 ridges.

Distribution and ecology

Native to Eurasia, with a disjunct distribution (C. and E. Europe, E. Asia). Native to all CE countries. Occurs in alpine tall herb vegetation, in floodplain forests and riparian shrubs, also in floodplain meadows. Prefers moist or wet soils rich in nutrients. Ornamental plant.

Anatomy of the rhizome

Rhizomes have a diameter of 10–15 mm and are occupied by many roots. Maximum age 2 years (1, 2).

The very large pith is surrounded by large and laterally extended vascular bundles, which consist of a xylem, a phloem and a cortex (1, 2). The semi-ring porous xylem is composed of radial groups of angular vessels, paratracheal and pervasive parenchyma and irregular groups of sclereids (2). Vessel diameter is 30–40 µm and a vessel length 80–100 µm. Intervessel pits are round and perforation plates are simple (3, 4). Fibers are thin- to thick-walled (2) and parenchyma is pervasive (2). Rays between vascular bundles are >10-seriate and often unlignified (5).

The phloem is composed of parenchyma cells and small groups of sieve tubes (6). The cortex consists of unlignified thin-walled parenchyma cells without intercellulars (1). Irregular groups of sclereids are frequent in the pith.

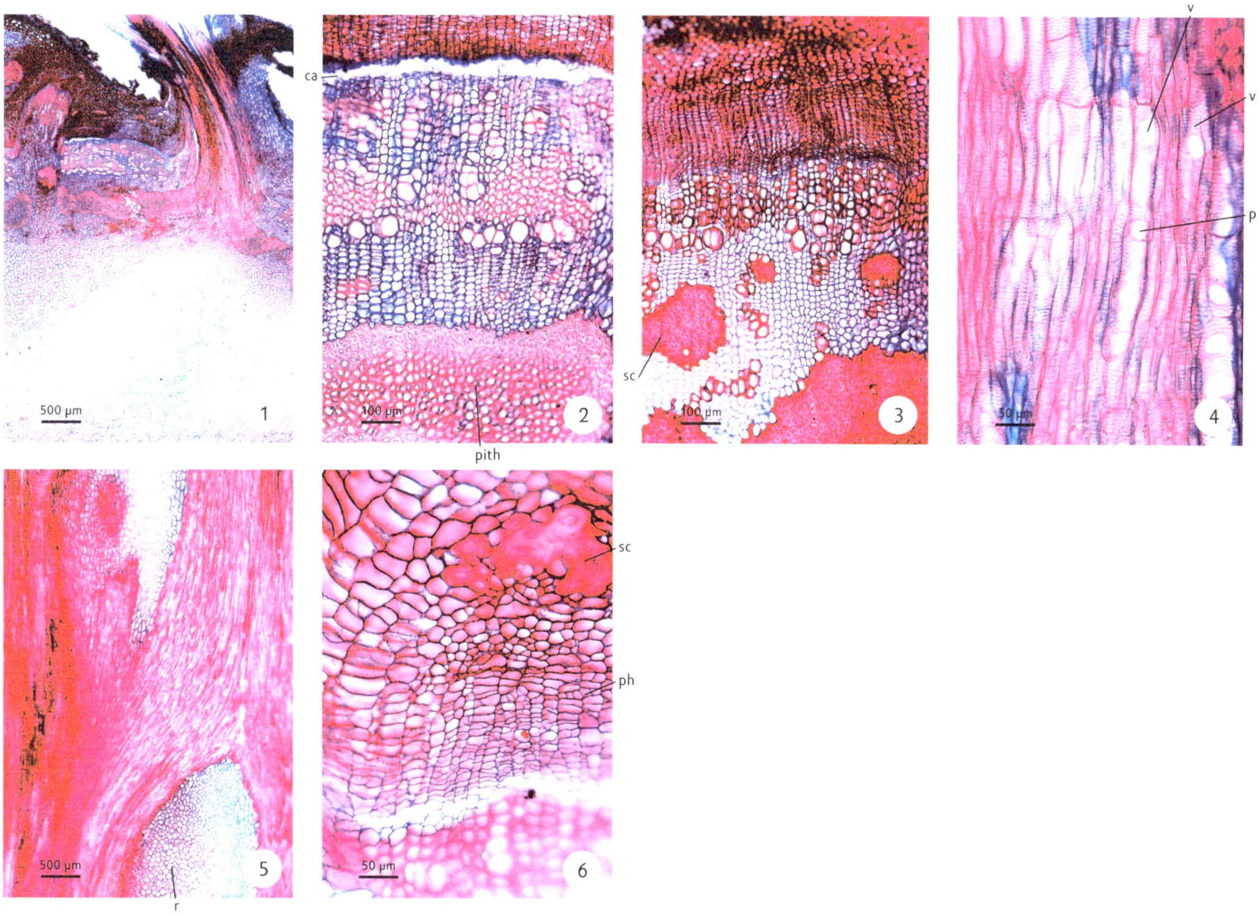

Thalictrum flavum L. **Common Meadow Rue** Ranunculaceae

Perennial wetland herb, with creeping rhizomes. Stem erect, simple or little-branched, glabrous, 50–150 cm high. Leaf blade triangular, 2-3-pinnate, dark green and shiny. Aromatic flowers in dense panicle, petaloid sepals whitish. Fruit is a sessile achene with 6–8 longitudinal ribs.

Distribution and ecology

Euro-Siberian boreo-temperate element. Native to all CE countries. Occurs in wet fen meadows, in tall vegetation along floodplain forests, at the edges of streams, ditches or in open fen carrs. Prefers permanently wet, nutrient-rich, neutral to base-rich soils.

Anatomy of the rhizome

Rhizomes have a diameter of 10–15 mm and are occupied by many roots. Maximum age 2 years (1).

The very large pith is surrounded by large and laterally extended vascular bundles, which consist of a xylem and a phloem, and a cortex (1). The xylem is composed of radial groups of vessels and paratracheal parenchyma (2) and irregular groups of sclereids (1). A vessel diameter is 30–40 μm and a vessel length 80–100 μm. Intervessel pits are round and perforation plates are simple (3). Fibers are thick-walled (3, 4). Rays are uniseriate within vascular bundles (5) and >10-seriate between the bundles (2).

The phloem is composed of parenchyma cells and groups of sieve tubes (5). The cortex consists of thin-walled parenchyma cells without intercellulars and groups of sclereids (6). Irregular groups of sclereids are frequent around the pith (1).

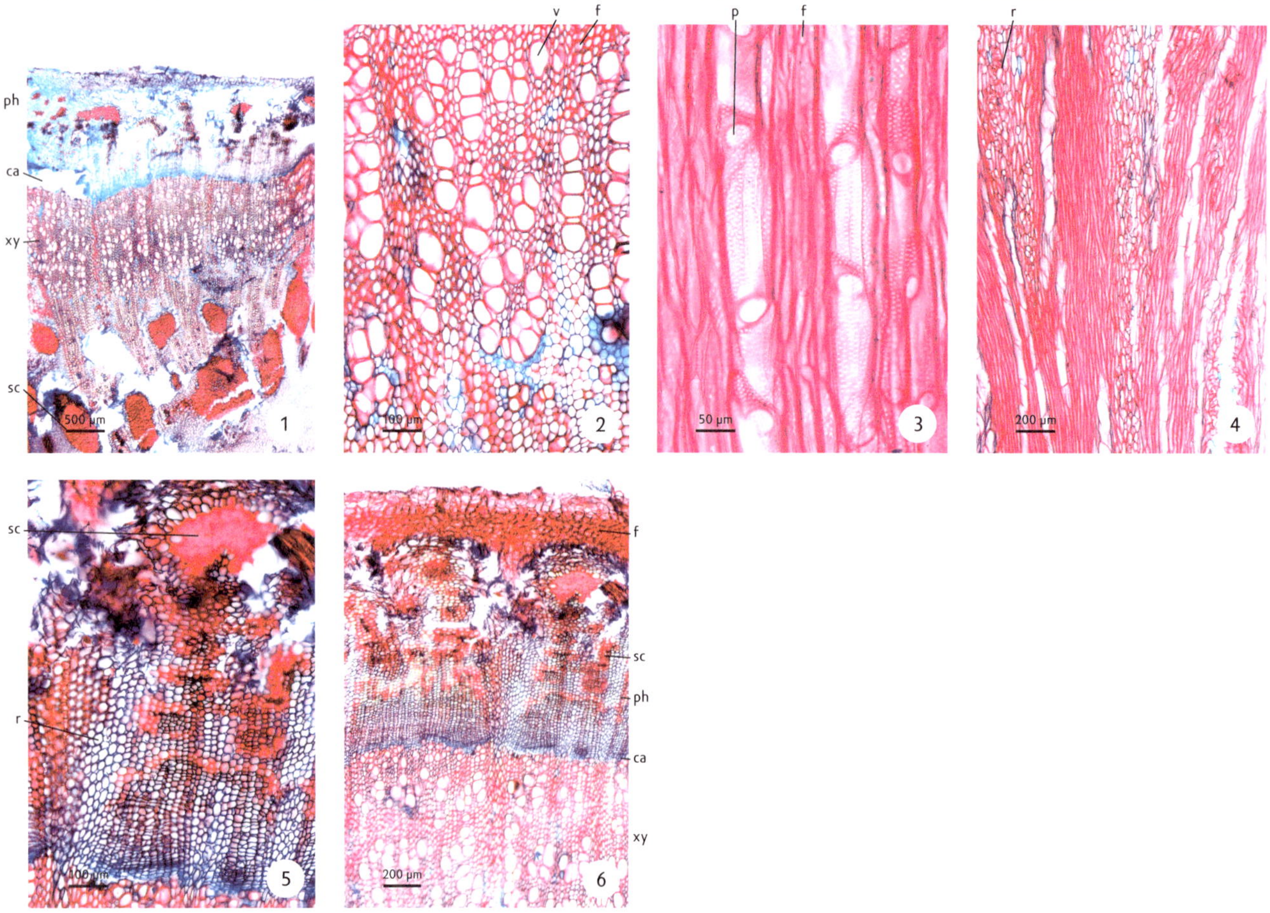

Perennial wetland herb with short rhizomes. Stem erect, hollow, glabrous, almost unbranched, 15–120 cm high. Old leaf sheaths common in basal part of the stem. Lower leaves petiolate, upper leaves sessile, the blade triangular, 2–3-pinnate. Flowers in dense panicle, petaloid sepals yellowish green. Fruit is an ovoid achene with 10 longitudinal ribs.

Distribution and ecology

European temperate element. Native to all CE countries. Occurs in wet alluvial meadows and floodplain forests, also in peaty meadows or in riparian shrubs. Prefers permanently wet, nutrient-rich soils.

Anatomy of the rhizome

Rhizomes have a diameter of 10–15 mm and are occupied by many roots. Maximum 2 years (1, 2).

The very large pith is surrounded by large and laterally extended vascular bundles, which consist of a xylem and a phloem, and a cortex (1, 2). The xylem is composed of radial groups of vessels, paratracheal parenchyma (3) and irregular groups of sclereids (1, 2). A vessel diameter is 30–40 µm and a vessel length 80–100 µm. Intervessel pits are round and perforation plates are simple (4). Fibers are thin- to thick-walled (2) and parenchyma is paratracheal (2). Rays between vascular bundles are >10-seriate and often unlignified (5).

The phloem is composed of parenchyma cells and groups of sieve tubes (5). The cortex consists of unlignified thin-walled parenchyma cells without intercellulars and sclereids (1). Irregular groups of sclereids are frequent around the pith (1).

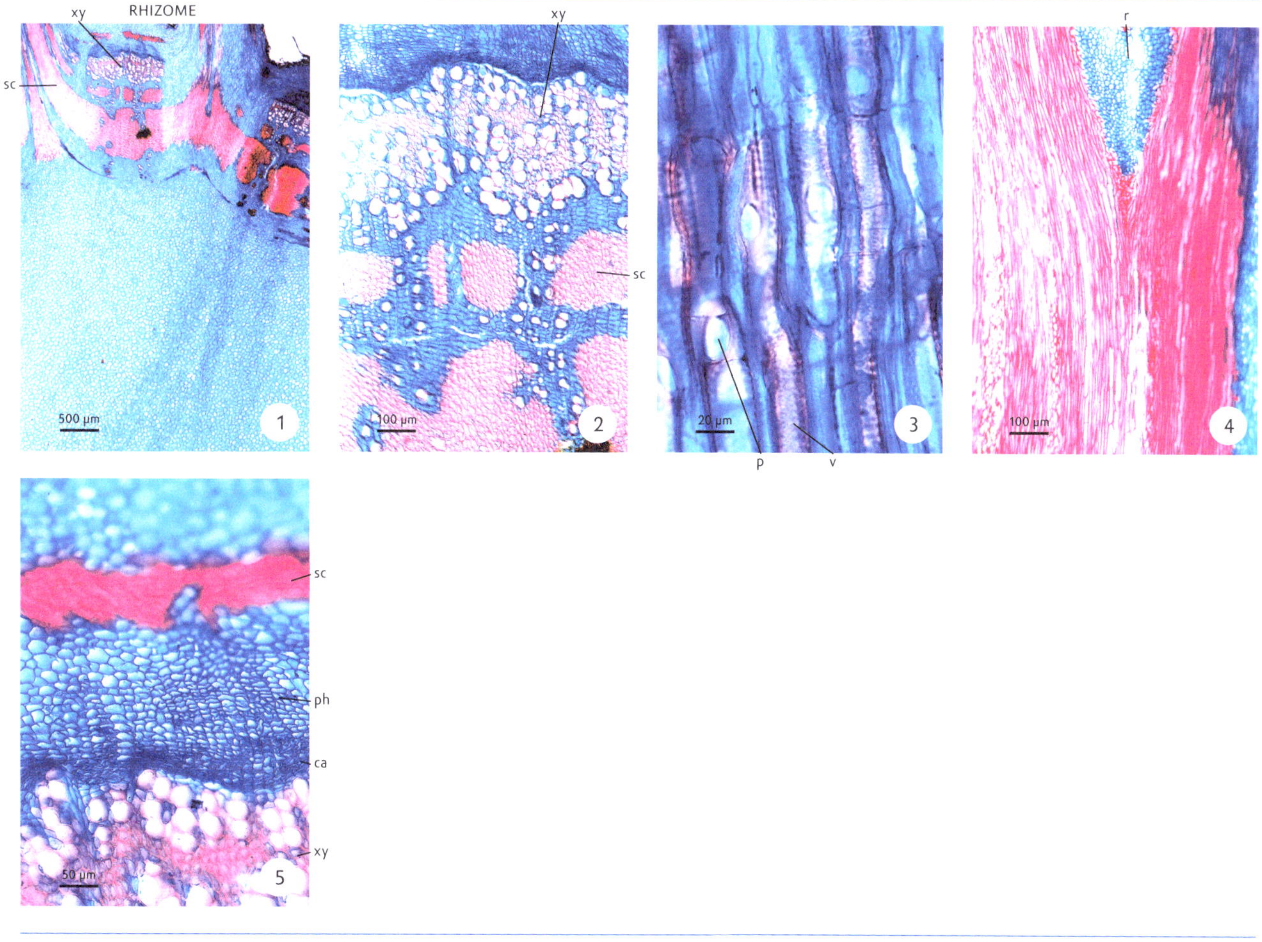

Trollius europaeus L. incl. *T. altissimus* Crantz

Ranunculaceae
Globeflower

Perennial wetland herb. Stem erect, glabrous, usually unbranched, arising from stout stock, 30–60 cm high. Basal leaves long-petioled, 5-palmately lobed, lobes obovate, toothed. Stem leaves 3-lobed, glabrous, dark green and shiny. Terminal flowers solitary, 1.8–3 cm across, globose, petaloid sepals pale yellow. Fruit is a glabrous follicle with a short beak.

Distribution and ecology

European boreo-montane element. Native to all CE countries. Occurs in alpine tall herb vegetation, in mountain spring areas, along mountain streams and in wet meadows. Prefers basic soils and is often associated with limestone. Sensitive to grazing.

Anatomy of the root collar

Root collars have a diameter of 2–3 mm. Secondary growth remains one growing season.

The center is surrounded by a few, circularly arranged, open collateral vascular bundles, which are embedded in a thin-walled parenchymatic tissue (1, 2). A few bundles are laterally linked with an inter-fascicular cambium (2, 4). The xylem is composed of radial (4) and irregular groups (3) of vessels and paratracheal and pervasive parenchyma (3, 4). Vessel diameter is 15–25 µm and a vessel length 80–100 µm. Intervessel pits are round to scalariform and perforation plates are simple (5). Fibers are absent (4).

The phloem is primarily composed of parenchyma cells (3). Special are the groups of very thick-walled fibers at the peripheral tips of vascular bundles (3). The cortex consists of unlignified thin-walled parenchyma cells and small intercellulars (6).

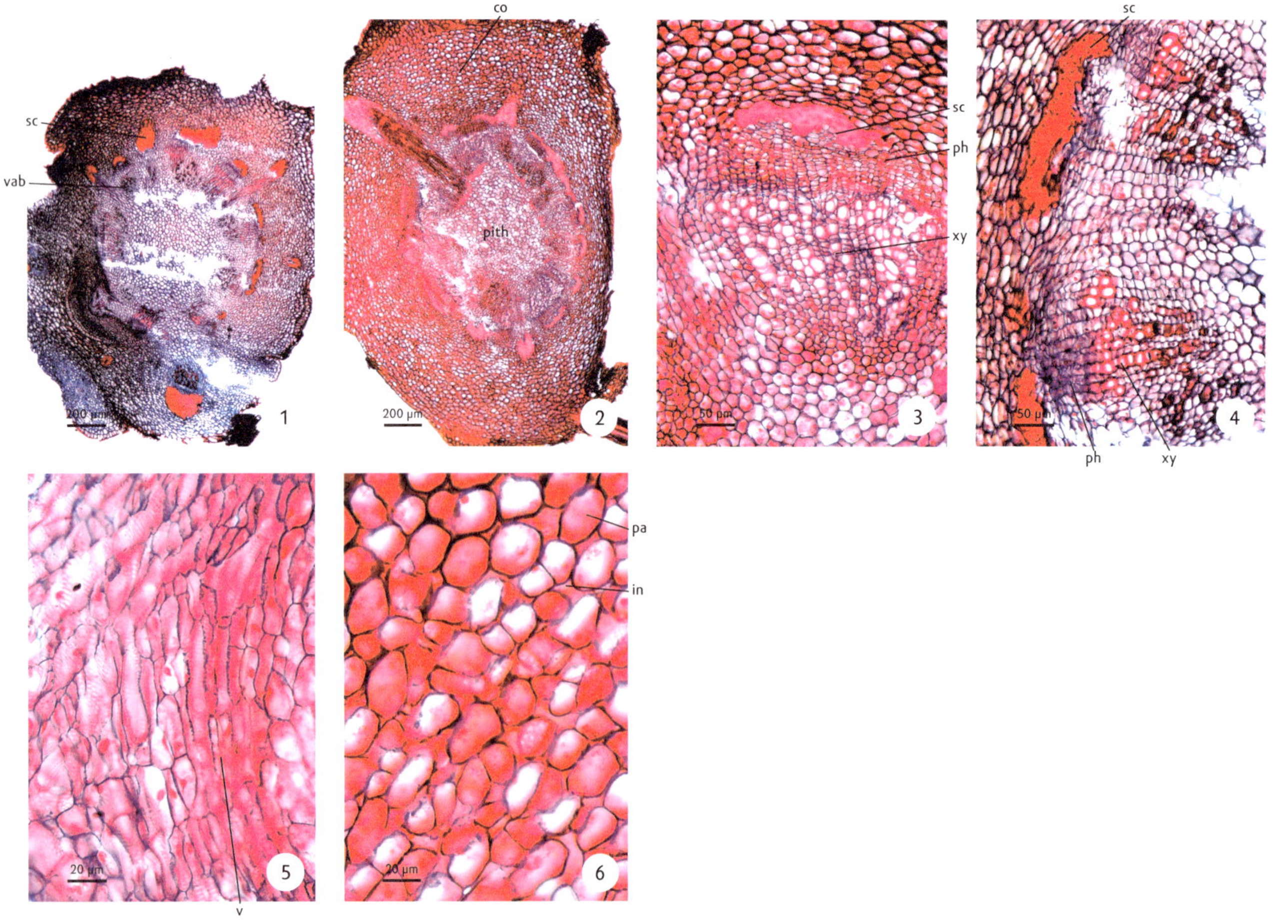

Frangula alnus Mill.

Deciduous wetland shrub. Stem 4–5 m high, sparsely branched, bark of twigs smooth. Leaves alternate, entire, petiolate, 2–7 cm long, with conspicuous veins. Flowers are pentamerous, small, white-greenish, entomophilous. Fruit is an obovoid violet-black drupe with 2–3 seeds.

Distribution and ecology

Euro-Siberian temperate element; naturalised outside its native range. Native to all CE countries. Occurs in fens, on the edges of raised bogs, in wet open woodlands. Prefers wet soils but avoids permanently waterlogged sites. Ability to regenerate after cutting and burning. Medicinal plant.

Anatomy of the stem

Round stem with a diameter of 10 cm. Maximum age >20 years.

Stems consist of a pith, a large xylem, a small phloem, a cortex and a phellem.

Annual ring-boundaries are characterized primarily by semi-ring porosity (1). The xylem contains angular, solitary vessels (1), thin- to thick-walled fibers (1), paratracheal parenchyma and rays (2). Vessels have a diameter of 50–100 μm and a length of 300–400 μm. Vessel-walls have round intervessel pits (3), simple perforation plates (3) and helical thickenings (3, 4). Fibers are thin- to thick-walled (2). Fibers with gelatinous fibers (tension wood) are frequent (2). Parenchyma is paratracheal (2). Rays are biseriate with prostrate cells (3).

The phloem is characterized by tangentially arranged groups of sieve tubes and groups of fibers (5, 6). Prismatic crystals are arranged around fiber groups and crystal druses occur in all other parenchyma cells (7). The cortex consists of oval parenchyma cells. Intercellulars are almost absent (7, 8). The phellem is characterized by radially orientated rectangular cells (8).

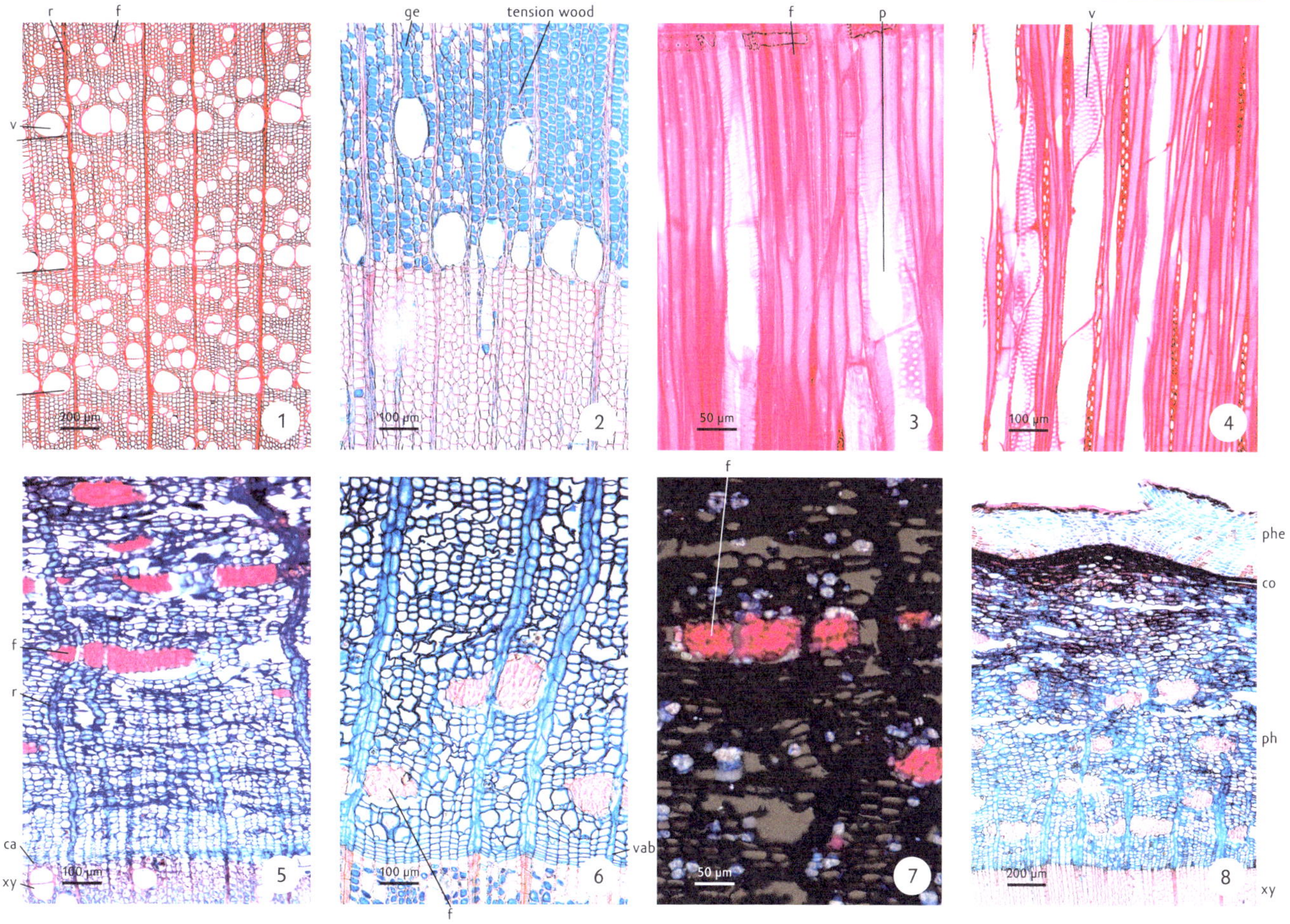

F. H. Schweingruber et al., *Anatomic Atlas of Aquatic and Wetland Plant Stems*, https://doi.org/10.1007/978-3-030-33420-8_47

Alchemilla fissa Günther et Schummel

Perennial wetland herb, tufted. Stem glabrous, bluish green, thin, feeble, ascending, 5–35 cm long. Basal leaves rounded to reniform, usually with 7 lobes, lobes dentate on the margin. Inflorescence is a rather lax compound cyme, tetramerous flowers relatively large (5–6 mm in diameter), no petals, hypanthium shortly campanulate. Fruit is an ovoid achene, smooth and brown.

Distribution and ecology

Central and Southern Europe. Native to the Alps (AT, CH and DE), very rarely in CZ and PL (The Giant Mts.). Occurs in alpine meadows on wet screes, in alpine springs and along alpine streams, usually in late-melting snow-beds. Prefers calcareous soils, but tolerates acidic substrates.

Anatomy of the root collar

Round root collar with a diameter of 2–3 mm. Perennial, maximum age unknown.

A large pith is surrounded by a belt of large vascular bundles, a large cortex and a small phellem (1). The xylem of the vascular bundles contains solitary or grouped angular vessels, thin- to thick-walled fibers, paratracheal parenchyma and rays (2). Vessels have a diameter of 15–25 µm and are 50–80 µm long. Annual ring-boundaries are absent. Vessels have oval intervessel pits (3) and simple perforation plates (3). Rays within the bundles are 2-3-seriate, those between them >10-seriate.

Parenchyma cells and a few sieve tubes characterize the phloem (4). The cortex consists of thin-walled parenchyma cells. Intercellulars are absent (4). The phellem is initiated immediately outside of the phloem and is characterized by a large zone of thin-walled, rectangular cells (4).

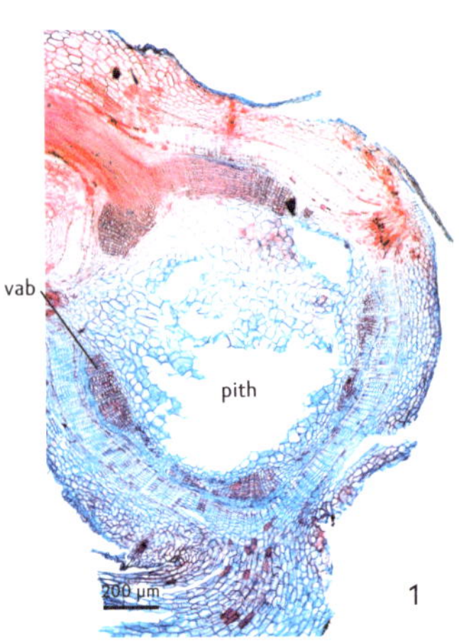

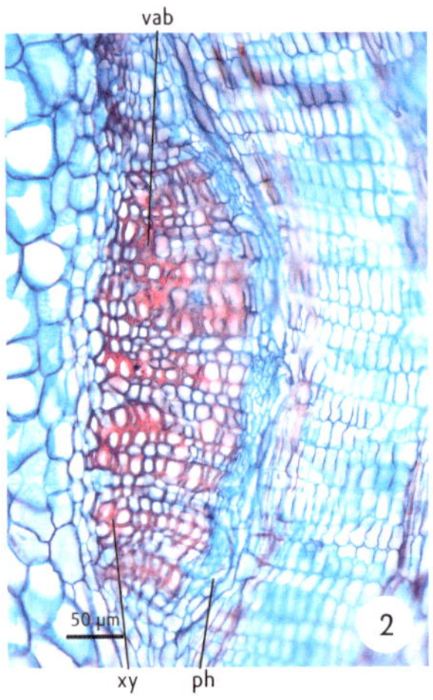

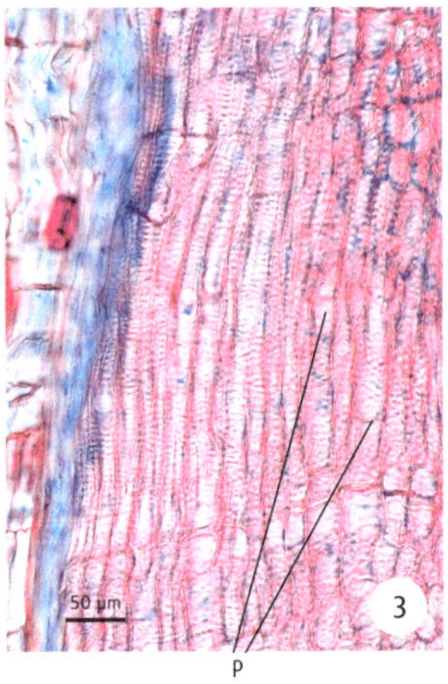

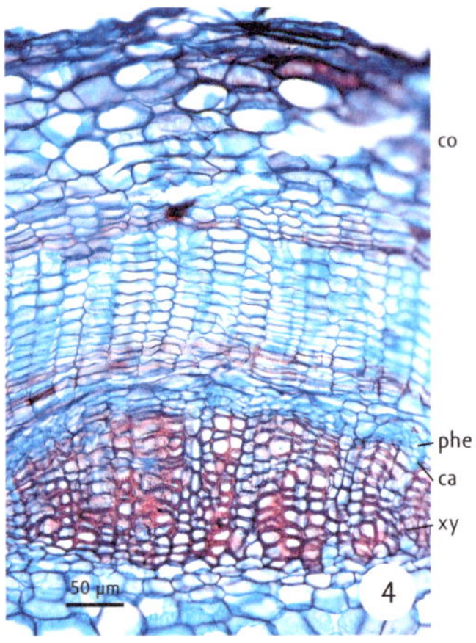

F. H. Schweingruber et al., *Anatomic Atlas of Aquatic and Wetland Plant Stems*, https://doi.org/10.1007/978-3-030-33420-8_48

Alchemilla pentaphyllea L.

Rosaceae

Perennial wetland herb. Stem thin, usually red, creeping, rooting at nodes, 5–7 cm high. Leaves rounded to reniform in the outline, palmatisect, lobes 5, ovate. Flower tetramerous, hypanthium campanulate, sepals yellowish green. Fruit is an achene.

Distribution and ecology

Mountain ranges of W. and C. Europe. Native to AT and CH. Occurs in wet alpine meadows on wet screes and moraines and at sites with long snow bed between 2300–3200 m. a.s.l. Prefers non-calcareous substrate.

Anatomy of the root collar (stained only by safranine)

Round root collar with a diameter of 3–5 mm. Maximum age 3 years.

A large pith is surrounded by a belt of large vascular bundles, a small cortex and a phellem (1). Annual ring-boundaries are characterized by a slight semi-ring porosity (2). The xylem contains angular solitary vessels, thin- to thick-walled fibers and rays (2). Vessels have a diameter of 20–30 μm and are 50–70 μm long. Vessel walls have oval intervessel pits (3) and simple perforation plates (3). Rays within the bundles are 2–3-seriate. Parenchyma is probably absent. Rays within vascular bundles are uniseriate with upright cells (4), those between them 10-seriate.

Parenchyma cells and a few sieve tubes characterize the phloem. The cortex consists of thin-walled parenchyma cells. Intercellulars are almost absent (5). The phellem is located inside the cortex (6) and is characterized by a large zone of thin-walled, rectangular cells.

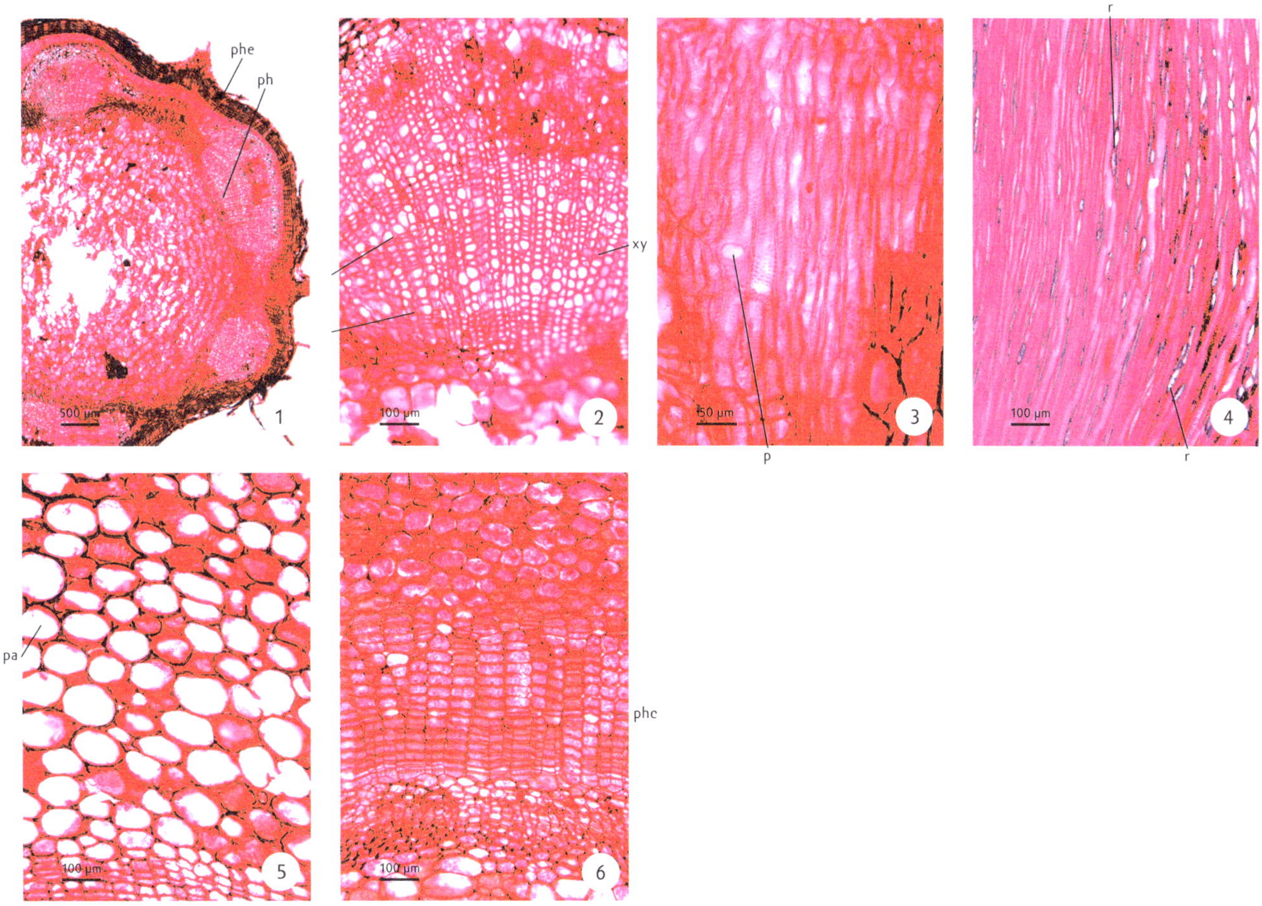

Comarum palustre L. (syn. *Potentilla palustris*) **Marsh Cinquefoil** Rosaceae

Perennial amphibious herb, with creeping woody rhizomes. Leafy rosette and ascending flowering stem up to 50 cm long. Leaves pinnate, with 5–7 leaflets, guttating leaf teeth reddish-brown. Few nectariferous flowers in a terminal cymes, petals purple, much shorter than sepals. Fruit is a rounded ovoid glabrous achene.

Distribution and ecology

Circumpolar boreo-temperate element. Native to all CE countries. Occurs in acidic transitional mires, in fens and in fen meadows, also at the edges of lakes, ponds, bog pools and on floating rafts of vegetation. Prefers nutrient-poor, slightly to moderately base-rich water. High phenotypic plasticity, self-compatible.

Anatomy of a basal part of the prostrate stem

Round stem with a diameter of 4–8 mm. Maximum age 8 years.

Stems consist of a large hollow pith, a large xylem and a small phloem, a cortex and a phellem (1). Annual ring-boundaries are characterized primarily by a row of flat latewood-fibers (2) or/and a slight semi-ring porosity (3). The xylem contains angular solitary vessels, thin- to thick-walled fibers, parenchyma cells and rays (2, 3). Vessels have a diameter of 30–50 µm and are 150–200 µm long. Vessel-walls have round intervessel pits (4) and generally simple perforation plates (4), however, a few perforation plates contain a few bars (5). Axial parenchyma is paratracheal (2, 3). Rays are uniseriate (6) with upright cells.

Parenchyma cells and small groups of sieve tubes characterize the phloem (7). The cortex consists of oval parenchyma cells. Intercellulars are absent (7, 8). The phellem is characterized by tangential layers of relatively thick-walled, rectangular cells (8).

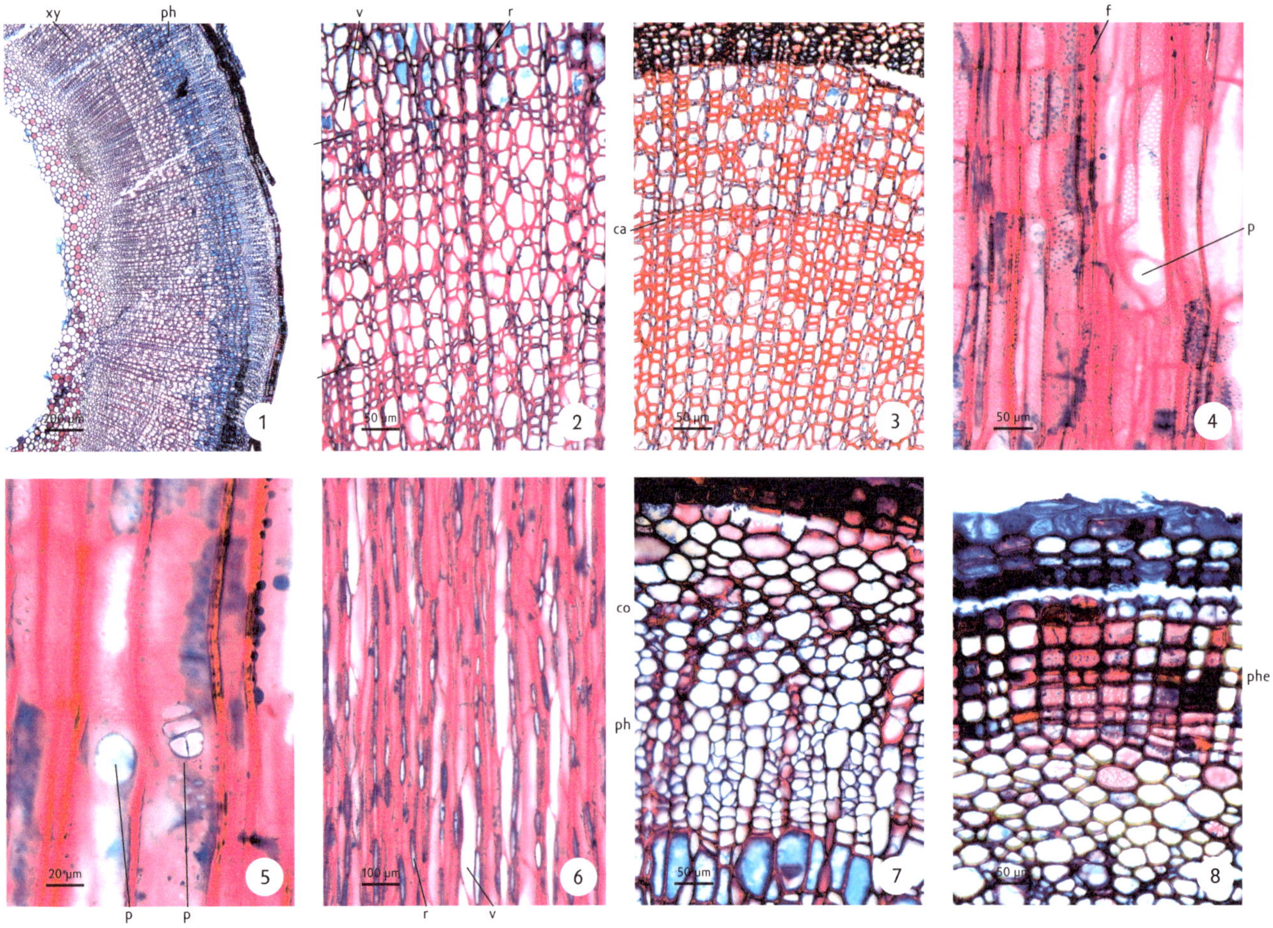

Filipendula ulmaria (L.) Maxim. **Meadowsweet** Rosaceae

Perennial wetland herb, rhizomatous. Stem erect, angular, simple or branched distally, 60–200 cm high. Leaves pinnate with 2–5 pairs of main leaflets and smaller ones between. Flowers in terminal panicle. Flowers 5–6-merous, white. Fruit is a head of reniform glabrous achenes.

Distribution and ecology

Eurasian boreo-temperate element. Native to all CE countries. Occurs in tall-herb fens, wet woodlands, damp meadows, in banks of pools and ponds, also in montane tall-herb stands. Prefers sites where water level fluctuates, and moderately fertile soils. It is intolerant of grazing and rarely flowers in shade. Foliar 'rust' fungus *Triphragmium ulmariae* with bright orange spores have severe effect on the survival of seedlings.

Anatomy of the rhizome

Round rhizome with a diameter of 5–10 mm. Parts with secondary growth remain 3 growing seasons.

The large pith of the rhizome is surrounded by single vascular bundles, a cortex and a small phellem (1). The major part of the xylem of the vascular bundles consists of vessels, thick-walled fibers and paratracheal parenchyma (2). A large zone of fibers in the latewood (1, 2) mainly indicates ring boundaries of the semi-ring porous xylem. Vessels have a diameter of 30–50 µm (3) and vessel length is 200–400 µm. Vessels are angular and arranged mainly in irregular groups (2). Intervessel pits are round and perforation plates are simple (3). Rays between vascular bundles are >10 cells wide with laterally separated sheet cells (4). Rays consist of square and upright unlignified cells (5). Rays within vascular bundles are uni- to biseriate (2) and those between vascular bundles >1-seriate (1, 2).

The phloem is located far inside the cortex and is composed of parenchyma cells, groups of sieve tubes (5) and dilating rays (1). The cortex consists of oval parenchyma cells and small intercellulars (6) and the phellem of few rows of rectangular cells (5). Crystal druses occur in the rays between vascular bundles and the pith.

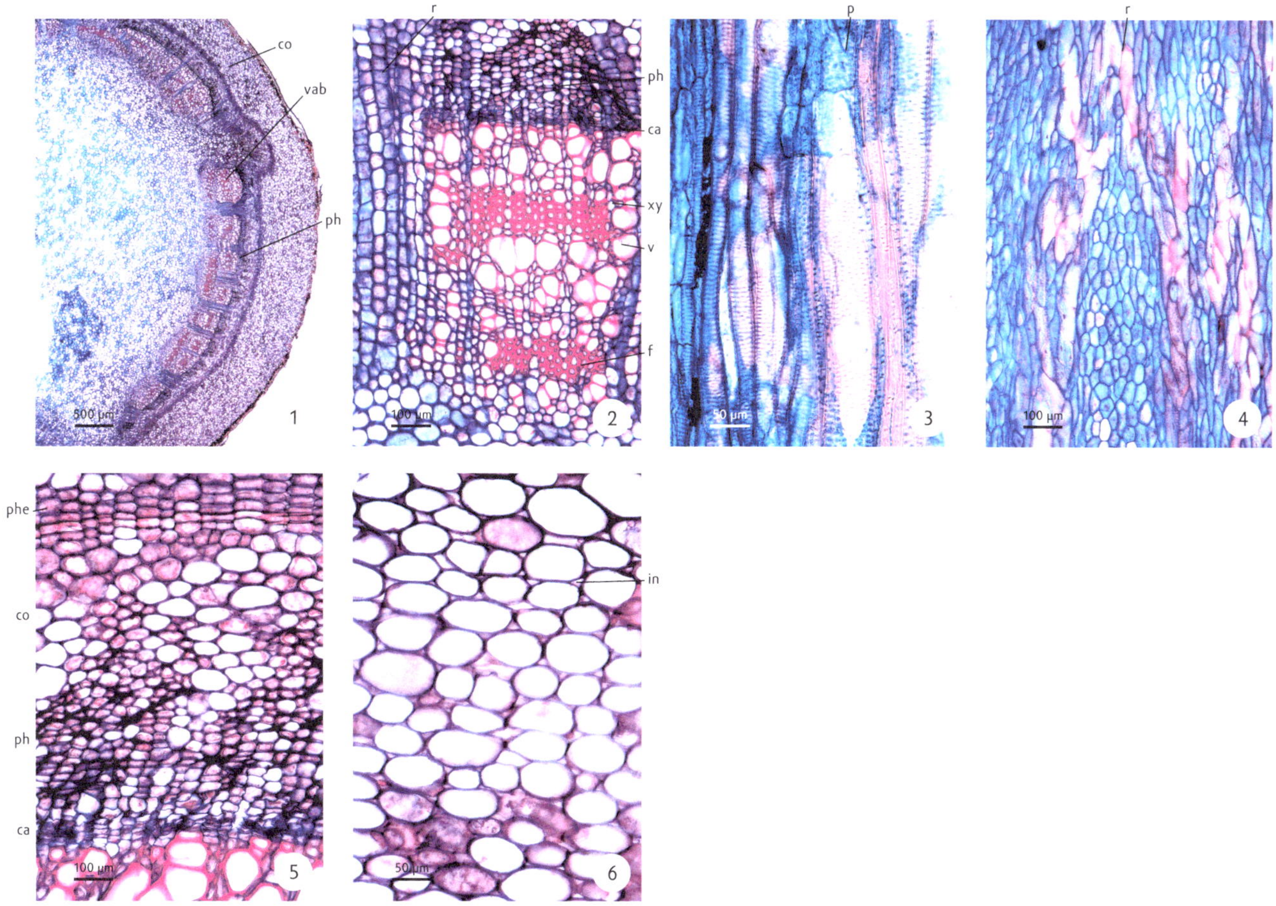

Perennial wetland herb with thick rhizome. Stem erect, pubescent, red-coloured, arising from basal leaf rosette, 15–60 cm high. Basal leaves pinnate, with 3–6 pairs of unequal lateral leaflets and large suborbicular, lobed terminal leaflet. Stem leaves small, with a few very small lateral leaflets. Inflorescence is a narrow cyme, 2–5-flowered; flowers nodding, campanulate. Calyx dark-brownish purple; petals cream-pink to pinkish purple, as long as sepals. Fruit is a hairy achene with a strong hooked awn.

Distribution and ecology

Euro-Siberian boreo-temperate element; also in N. America. Native to all CE countries, very rare in HU. Occurs on stream banks, in herb-rich meadows, also in flushes in deciduous woodland and in montane willow scrubs. Prefers mildly acidic to calcareous, nutrient-rich, wet soils. Dense populations can develop as a result of rhizomatous spread. Vesicular-arbuscular mycorrhizal colonization.

Anatomy of a basal part of the stem

Round stem with a diameter of 4–6 mm. Maximum age 5 years.

Stems consist of a large pith, a small xylem, phloem, cortex and phellem (1).

Annual ring-boundaries are characterized primarily by semi-ring porosity (2). The xylem contains solitary or grouped vessels, thin- to thick-walled fibers, parenchyma cells and rays (2). Vessels have a diameter of 20–30 µm and are 100–120 µm long. Vessel-walls have round intervessel pits and generally simple perforation plates. Axial parenchyma is paratracheal at external rings and paratracheal in the center (2). Rays are confluent and 5-to >10-seriate (3) with upright cells.

Parenchyma cells and small groups of sieve tubes characterize the phloem (3). The cortex consists of oval parenchyma cells. Prismatic crystals occur in the cortex. Intercellulars are absent. The phellem is composed of rectangular cells and occurs far inside of the epidermis (4).

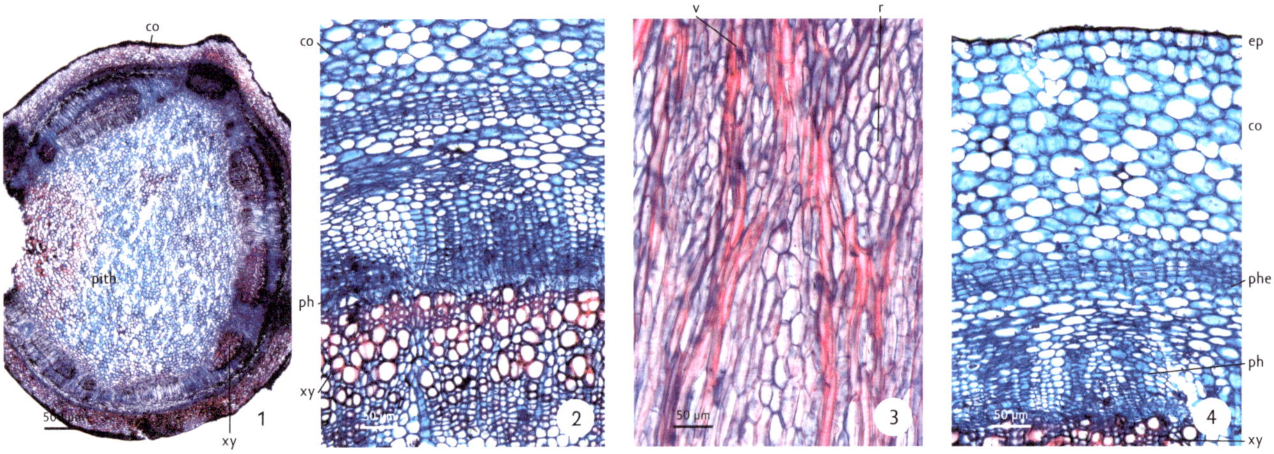

Rubus chamaemorus L.

Cloudberry Rosaceae

Perennial wetland dioecious herb with extensive underground rhizome system. Stem erect, unbranched, glandular, without prickles, 5–20 cm high. Leaves simple, orbicular, palmately 5–7-lobed. Flowers unisexual, terminal, white, 2–3 cm across. Fruit is a orange drupe with ovoid seeds.

Rosaceae

Distribution and ecology

Circumpolar boreal-montane element. Native to DE and PL, very rare in CZ as a glacial relict. Occurs in peatbogs, in peaty stands of dwarf pine. Prefers acidic peaty soils. Vegetative propagation prevails.

Anatomy of a basal part of the prostrate stem

Round stem with a diameter of 3–5 mm. Maximum age >30 years.

Stems consist of a small pith, a large xylem, a small phloem, a cortex and a phellem (1).

Annual ring-boundaries are characterized primarily by semi-ring porosity (2). The xylem contains angular solitary vessels, thin-walled, often unlignified fibers, paratracheal parenchyma and rays (3). Vessels have a diameter of 40–50 µm and are 120–150 µm long. Vessel walls have round intervessel pits (4) and simple and scalariform perforation plates (3, 4). Rays are uniseriate (5) with upright cells (6).

Parenchyma cells and a few sieve tubes characterize the phloem (7). The cortex consists of oval parenchyma cells (7) and contains crystal druses. Intercellulars are absent (7).

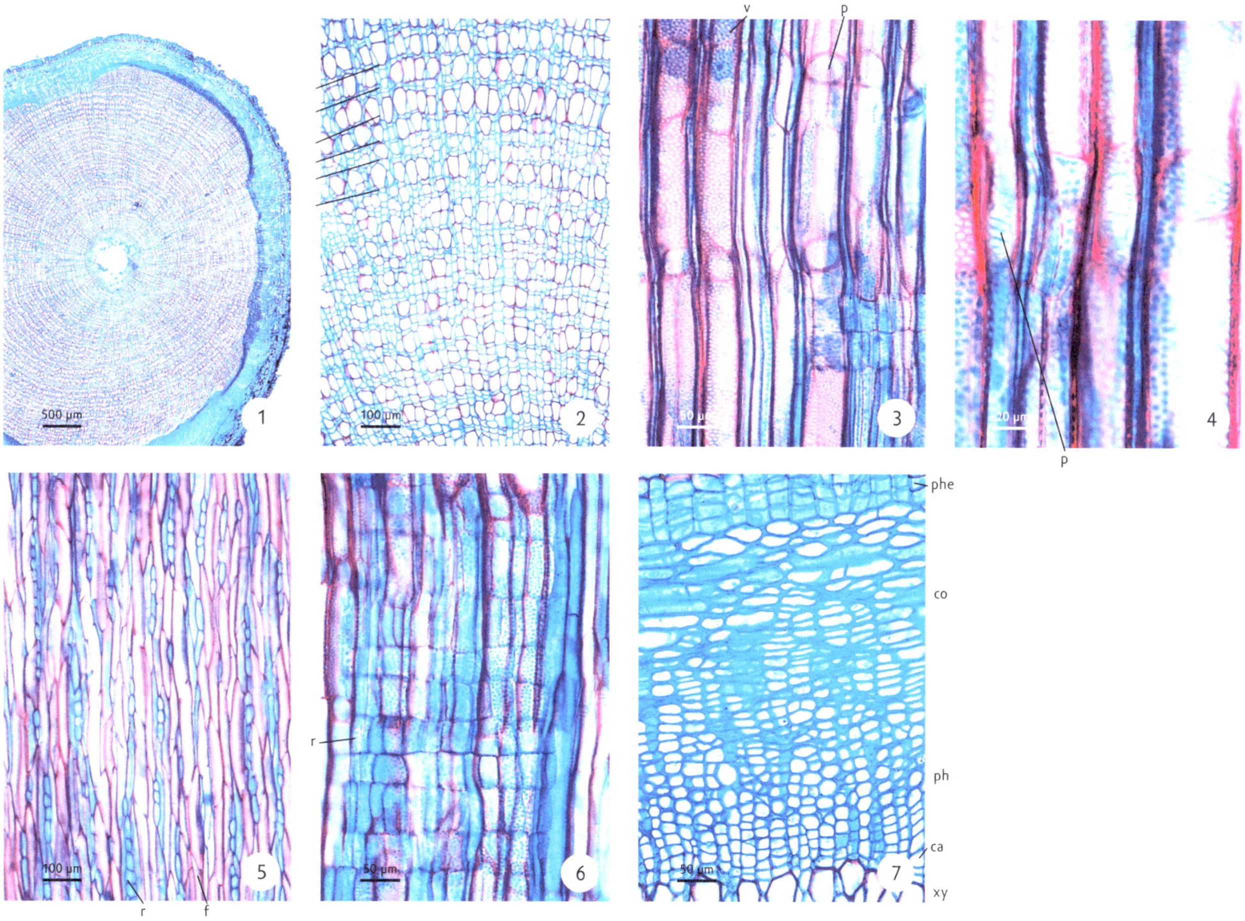

Sanguisorba officinalis L. **Great Burnet** Rosaceae

Perennial wetland herb with a long horizontal rhizome. Stem erect, hollow, branched, glabrous, 30–120 cm high. Leaves pinnate, basal leaves with 3–7 pairs of leaflets. Flowers in terminal subglobose spikes, hypanthium deeply concave, petals absent, sepals and stamens dull purplish. Fruit is an achene, sclerous, dry, with winged angles.

Distribution and ecology

Circumpolar boreo-temperate element, widely naturalised outside its native range. Native to all CE countries. Occurs on river banks, in wet meadows and pastures from lowlands to the mountains. Prefers deep, wet, clayish or peaty soils. Heliophyte. River corridor plant with transient seed bank. Medicinal herb.

Anatomy of the rhizome

Round rhizome with a diameter of 10–12 mm. Maximum age 10 years.

The pith of older rhizomes is surrounded by radial vessel/parenchyma strips, and a small phellem. The vessel/parenchyma strips consist primarily of lignified oval vessels and unlignified axial parenchyma cells (1).

Ring boundaries are characterized by semi-ring porosity (2). Earlywood-vessels have a diameter of 50–70 µm (3) and a length of 180–300 µm (4). Vessels are solitary or in irregular groups (2, 3). Intervessel pits are round and scalariform and perforation plates are simple (4). Thick-walled fiber-groups occur rarely in the latewood (3). Vessels are surrounded by pervasive parenchyma (2). Rays between vascular bundles are very large (5) and consist of procumbent, square and upright unlignified cells, which are mixed throughout the ray (8). Rays are laterally indistinctly limited by sheath cells towards the axial parenchyma (5).

The phloem consists of parenchyma cells and groups of sieve tubes and dilating rays (6). The phellem occurs at the periphery and within the xylem (interxylary cork) (7). Crystal druses are very frequent and located in all types of ray cells in axial parenchyma (8) and pith cells. The pith consists of thin-walled, unlignified oval parenchyma cells and intercellulars.

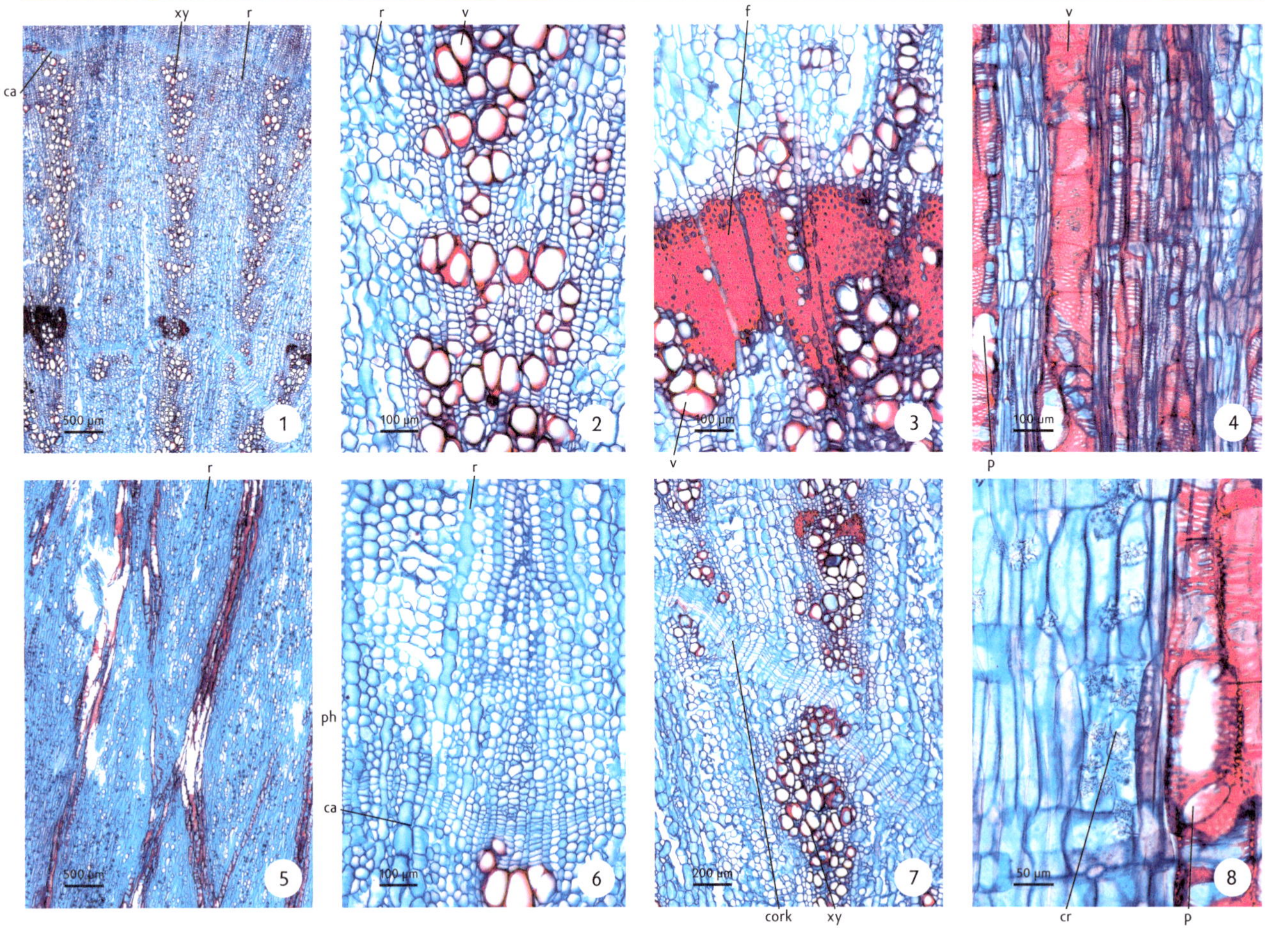

Sibbaldia procumbens L. (syn. *Potentilla sibbaldia*)

Perennial wetland herb, tufted. Flowering stems axillary, leafless, 1–3 cm high, basal rosettes of leaves grow from woody rootstock, which is dark brown and much branched. Leaves ternate, petioles often longer than obovate-cuneiform leaflets with tridentate apex. Leaves usually bluish- or glaucous-green. Flowers few in a rather dense cyme, 7 mm in diameter, with prominent epicalyx. Petals five, yellowish-green or yellow. Fruit is an ovoid achene, indistinctly rugose-ribbed.

Distribution and ecology

Circumpolar arctic-montane element, with a disjunct distribution. Native to AT, CH and DE, very rare in PL and SK. Occurs on sites of late snow-melt in the mountains. Prefers mesotrophic moist soils, avoids permanently waterlogged sites. Chionophilous heliophyte. Reproduction by seeds and by fragmentation of branched rootstock.

Anatomy of the basal part of the prostrate stem

Round stem with a diameter of 4–6 mm. The pith is eccentric (1, 2). Maximum age >20 years.

Stems consist of a pith, a large xylem, a small phloem, a cortex and a large phellem (1, 2).

Annual ring-boundaries are characterized primarily by a slight semi-ring porosity (3). The xylem contains solitary vessels, thick-walled fibers, paratracheal parenchyma and rays (3). Vessels have a diameter of 30–50 µm and are 80–100 µm long. Vessel-walls have round intervessel pits (4) and simple perforation plates (4). Rays within vascular bundles are uniseriate with upright cells (5, 6) and those between them are 10-seriate and mostly unlignified (2).

Parenchyma cells and a few sieve tubes characterize the phloem (7). The cortex consists of oval parenchyma cells. Intercellulars are almost absent (7, 8). The phellem is characterized by tangential packages of thin-walled, rectangular cells, typical for Rosaceae (7).

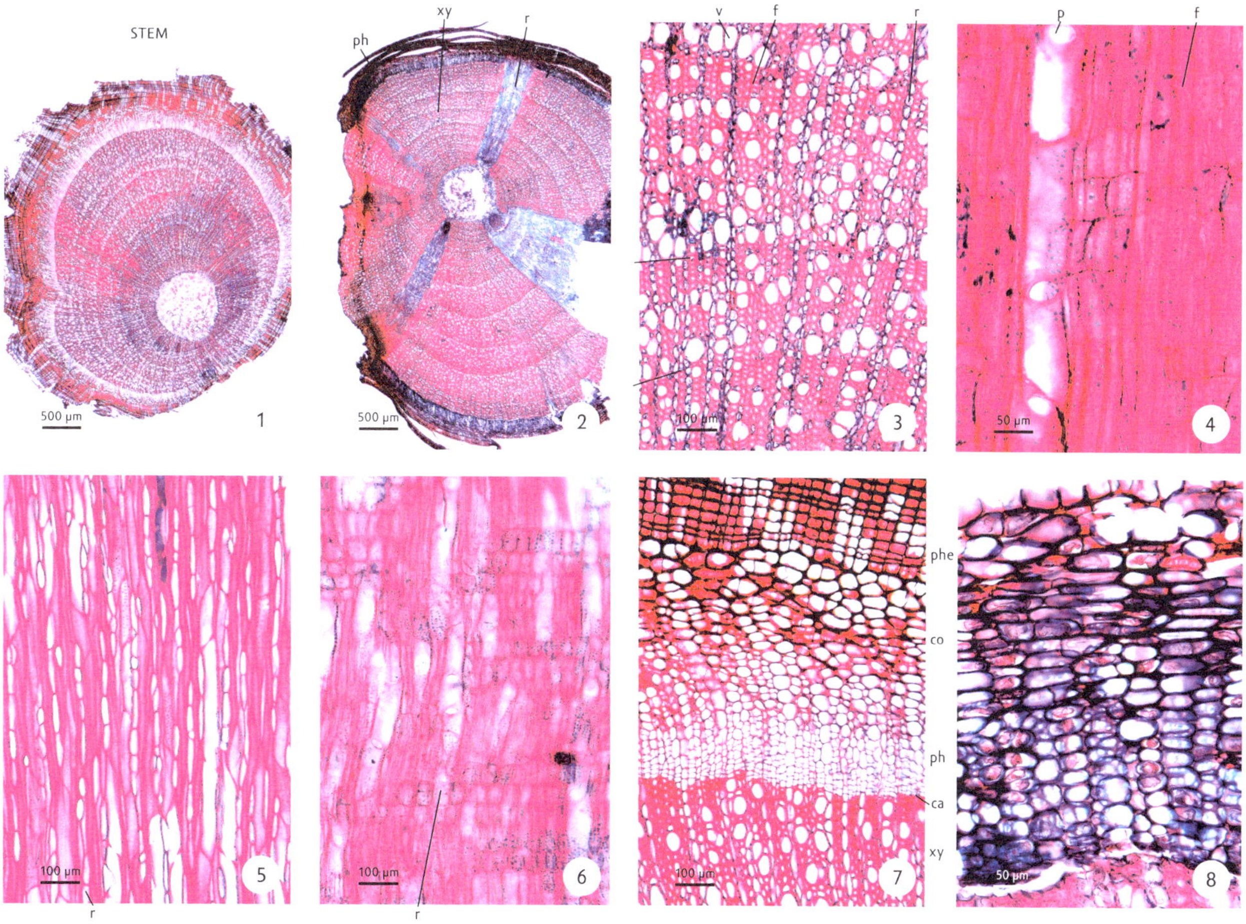

Spiraea salicifolia L. **Bridewort** Rosaceae

Perennial wetland deciduous shrub. Stem erect, 1–2 m high, twigs erect, yellowish-brown, downy when young, glabrous later on. Leaves elliptic-oblong to lanceolate, sharply serrate, glabrous. Inflorescence is a cylindrical panicle, petals bright pink. Fruit is a follicle with numerous oblong oval, narrow winged seeds.

Distribution and ecology

Native to Eurasia, from C. and E. Europe eastwards to Japan. Native to CZ, PL and SK, rare in HU, introduced to AT, CH and DE. Occurs along river banks, in alder and willow carrs, in peatbogs, also at margins of fishponds and oxbows. Prefers acidic peaty soils. Tolerates partial shadow.

Anatomy of the stem

Round stem with a diameter of 5–10 mm. Maximum age >5 years.

Stems consist of a small pith, a large xylem, a small phloem, a cortex and a phellem (1).

Annual ring-boundaries are characterized primarily by semi-ring porosity and enlarged by large rays (2). The xylem contains solitary vessels, thin-walled fibers, paratracheal parenchyma and rays (2, 3). Vessels have a diameter of 25–35 µm and are 100–160 µm long. Vessel walls have round intervessel pits (4) and simple and rarely net-like perforation plates (6, 7). Rays are uniseriate (6) and multiseriate with prostrate cells (5). A few crystal druses occur in vessels.

Parenchyma cells and a few sieve tubes and many prismatic crystals characterize the phloem (8). The cortex consists of a dense belt of sclerenchyma (8). Intercellulars are absent (8).

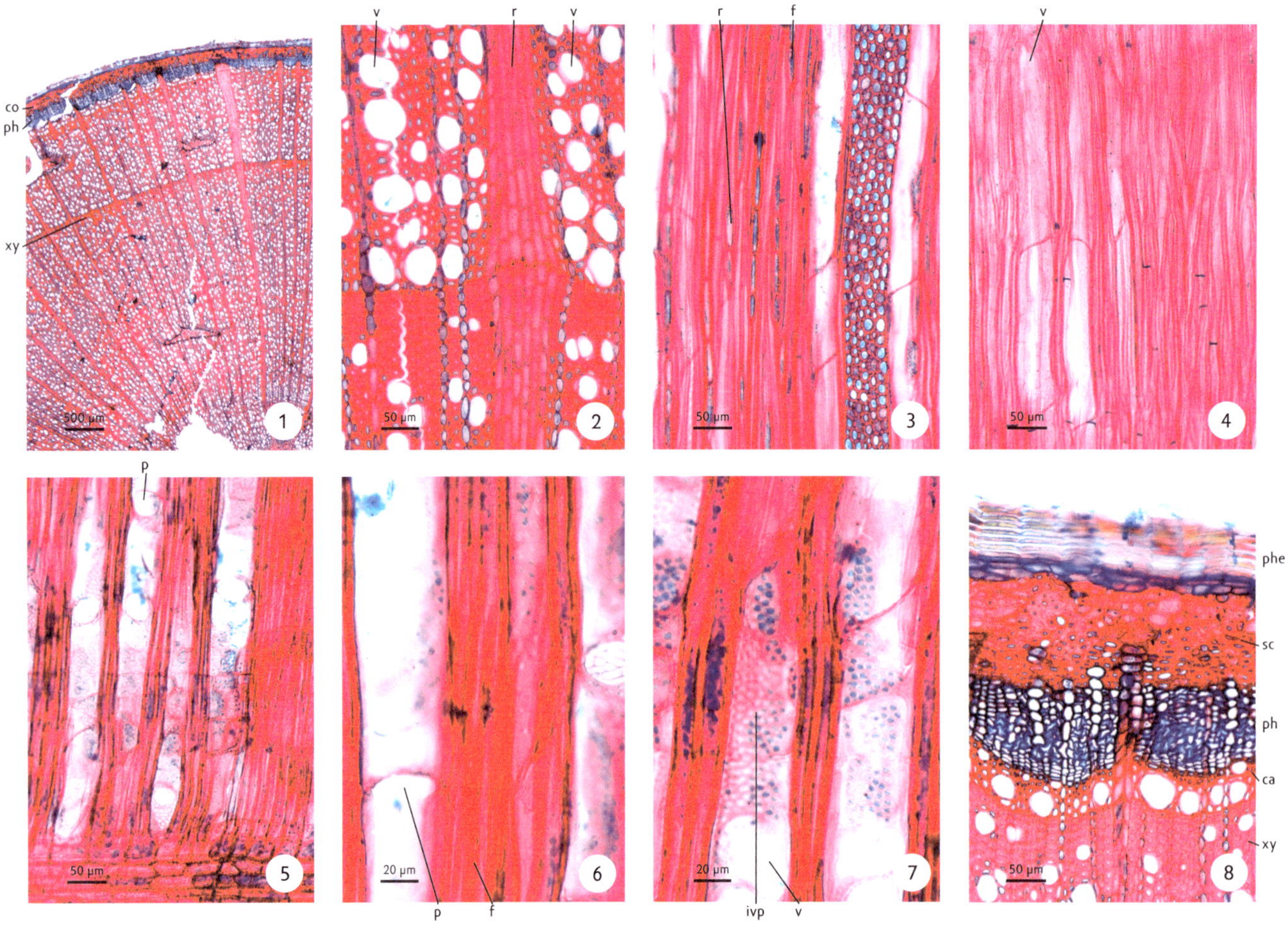

Galium palustre L.

Common Marsh-bedstraw Rubiaceae

Perennial wetland herb with thin, richly branched rhizomes. Stem slender, decumbent to ascending, smooth, 20–60 cm high. Leaves 4–6 per whorl, linear-oblong to narrowly elliptic, rounded at apex. Flowers in terminal panicles, corolla white, 4-lobed. Fruit is a pair of black nutlets, with wrinkly surface.

Distribution and ecology

Euro-Siberian boreo-temperate element; also in N. America. Native to all CE countries. Occurs in fens, in peaty and waterlogged meadows, in muddy banks of streams, ditches, oxbows and ponds. Prefers non-calcareous substrates, seasonally or permanently wet soils rich in nutrients.

Anatomy of the annual shoot and the rhizome

Square annual shoot (1) and round rhizome (3) with a diameter of 1–1.5 mm. Annual rings are absent. Maximum age unknown.

Annual shoot: The large pith is surrounded by a small xylem and a phloem, a large cortex and an epidermis (1). The xylem consists of angular, thin-walled vessels and fibers and paratracheal parenchyma. (2). Vessels have a diameter of 30–50 μm (2). Vessel length is 200–300 μm. Vessels are arranged in groups in a small belt of xylem (2).

Rhizome: The small pith is surrounded by a large xylem and a small phloem, a cortex and a large phellem (3). The xylem consists of vessels, thin- to thick-walled fibers and paratracheal and pervasive parenchyma (4). Vessels have a diameter of 30–40 μm (2) and a length of 100– 200 μm (5). Vessels are arranged solitary (4), intervessel pits are round with slit-like apertures (5) and perforation plates are simple (5). Fibers are thin- to thick-walled (4). Rays are absent or have a form like axial parenchyma cells (6).

The phloem consists of parenchyma cells and a few sieve tubes (2, 7). Cortex and phloem are separated by an endodermis (1). The cortex in the annual shoot is characterized by large unlignified parenchyma cells and at the periphery by an epidermis, and a zone of fairly thick-walled cells (collenchyma) (1) at the edges of the shoot. The cortex in the rhizome is small but the phellem is well developed (7). The cortex contains long idioblasts with raphids (8).

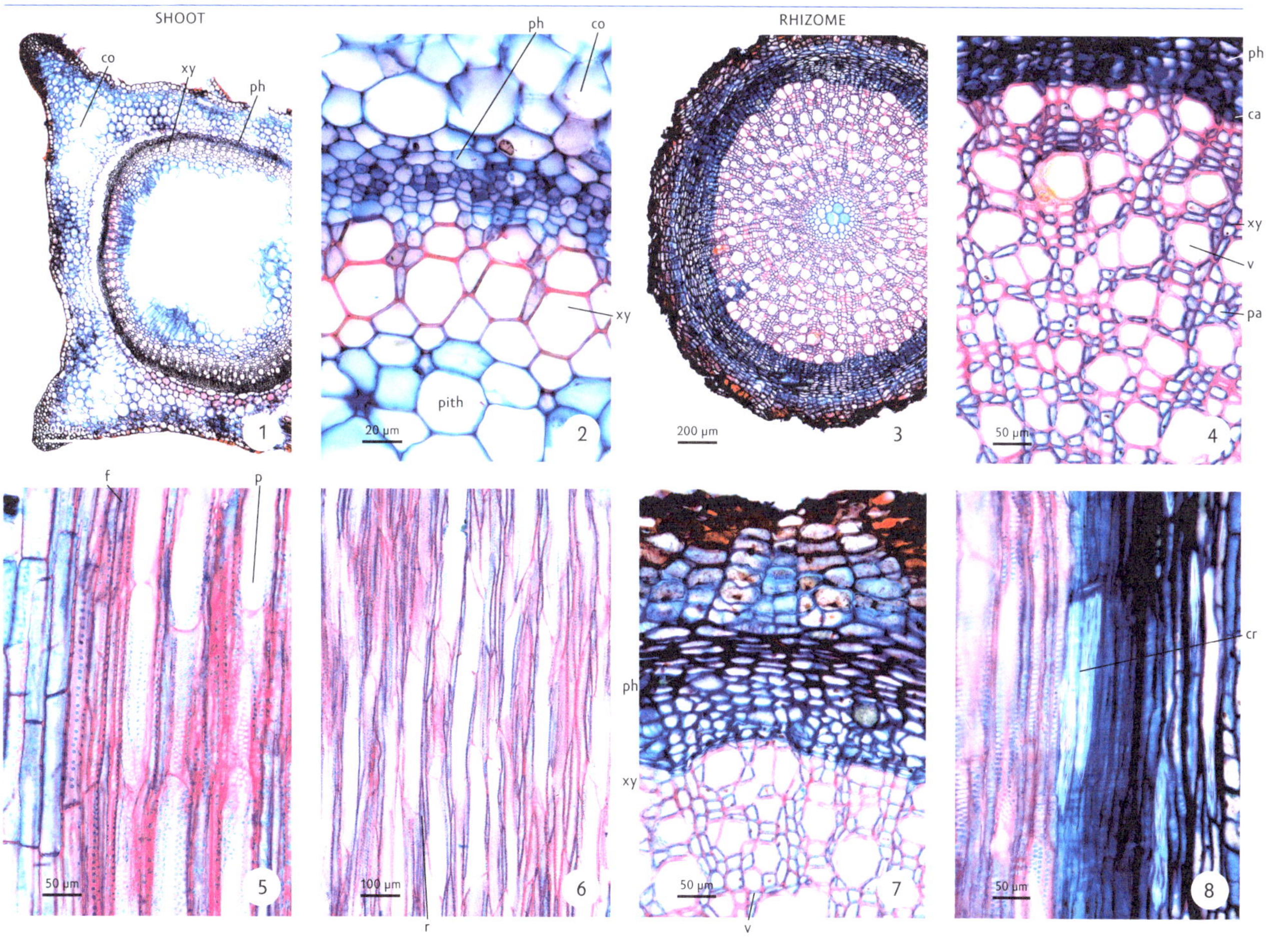

F. H. Schweingruber et al., *Anatomic Atlas of Aquatic and Wetland Plant Stems*, https://doi.org/10.1007/978-3-030-33420-8_49

Perennial wetland dwarf shrub, dioecious and deciduous, with branched, underground rhizomes. Prostrate stem partially underground, forming loose, flattened mats. Aerial twigs few, short (up to 5 cm above the soil surface), shining dark brown or reddish. Leaves 6–20 mm long, orbicular or broadly ovate, venation prominently reticulate on both sides. Catkins appearing with the leaves, terminal on short branches. Fruit is a reddish purple capsule, with cylindrical seeds embedded in white down.

Distribution and ecology

European arctic-montane element; also in N. America. Native to all CE countries except HU. Occurs in the mountains in late-melting snow-beds, also in conditions of extreme exposure on hill slopes and stony-gravel plains. Prefers open communities on acidic soils with regular disturbance (e.g., solifluction). Heliophyte with ectomycorrhiza.

Anatomy of the rhizome

Round rhizome with a diameter up to 7 mm. Maximum age >40 years.

Stems and rhizomes consist of a xylem, a small phloem, a large cortex and a small phellem (1). Annual rings are marked by one to two rows of flat latewood fibers (2, 3). The diffuse-porous xylem is composed of vessels, thin- to thick-walled fibers, parenchyma and rays (2, 3). Vessels are solitary or arranged in short radial groups (2, 3). Vessel diameter varies from 60–100 µm and vessel length from 150–250 µm. Intervessel pits are large, round or angular (5) and perforation plates are simple (5). Vessel ray-pits are extremely large (characteristic for the genus *Salix*) (4). Fibers are thin- to thick-walled (2, 3). Parenchyma is difficult to recognize (2, 3). Rays are uniseriate (4). Central ray cells are prostrate and marginal cells upright (4).

The phloem is composed of rectangular sieve tubes, companion cells and round parenchyma cells (6). The cortex consists of round to oval parenchyma cells and intercellulars (7). A few cortex cells contain crystal druses (7). The phellem is very dense and consists of tangentially layered rectangular cells, which are filled with dark substances (1).

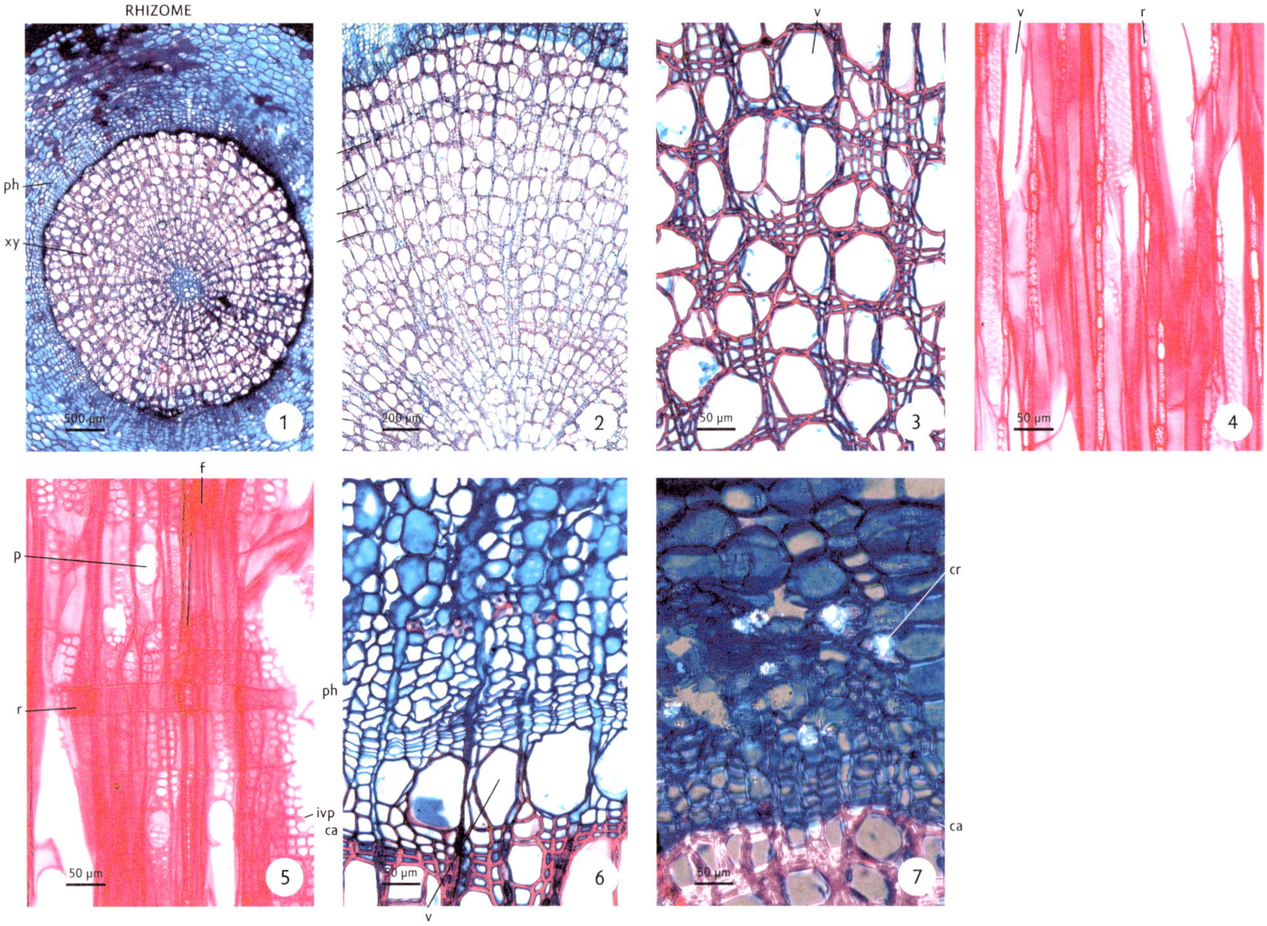

F. H. Schweingruber et al., *Anatomic Atlas of Aquatic and Wetland Plant Stems*, https://doi.org/10.1007/978-3-030-33420-8_50

Salix myrsinifolia Salisb. (syn. *Salix nigricans*) **Dark-leaved Willow** Salicaceae

Perennial wetland dioecious and deciduous shrub, 2–4 m high, richly branched. Twigs densely pubescent at first, reddish brown to greenish. Leaves alternate, usually elliptic or obovoid, 3.5–5.5 cm long and 1.5–3 cm wide, dark green and shiny upside. Catkins sessile, 1.5–3 cm long. Fruit is a capsule.

Distribution and ecology

Euro-Siberian boreo-montane element. Native to all CE countries, very rare in HU. Occurs at the edges of peatbogs, in marshes, at margins of fishponds. Prefers wet and waterlogged sites on mineral-rich soils.

Anatomy of the stem

Round stem with a diameter up to 8 cm. Maximum age >40 years.

Stems and rhizomes consist of a xylem, a small phloem, a large cortex and a small phellem. Annual rings are marked by one to two rows of flat latewood fibers and a row of marginal parenchyma (1, 2). The xylem is composed of vessels, thin- to thick-walled fibers, parenchyma and rays (1, 2). The xylem is diffuse-porous to slightly semi-ring porous (1, 2). Vessels are usually solitary (1). Vessel diameter varies from 60–90 µm and vessel length from 200–300 µm. Intervessel pits are large, round or angular (3) and perforation plates are simple (3). Vessel ray-pits are extremely large. Fibers are thin- to thick-walled (2, 3) and contain often gelatinous fibers (tension wood) (2). Parenchyma is difficult to recognize (2, 3). In the present material, one marginal row of parenchyma cells exists. Rays are uniseriate (5). Central ray cells are prostrate and marginal cells upright (4, 5).

The phloem is composed of rectangular sieve tubes, companion cells and round parenchyma cells (6). Tangentially arranged, thin-walled sieve-tube/parenchyma and thick-walled fiber bands (5, 6) are characteristic. No ray-initiated dilatations occur in the outer part of the phloem (5). The cortex consists of tangentially layered, round to oval parenchyma cells and intercellulars (7, 8). A few cortex cells contain crystal druses and prismatic crystals (8). The phellem is very dense and consists of tangentially layered rectangular cells, which are filled with dark substances (7).

The structure of this species is identical to that of many other *Salix* species, e.g. *S. alba*, *S. aurita*, *S. cinerea*, *S. fragilis*, *S. pentandra*, *S. repens*, *S. triandra*, *S. viminalis*.

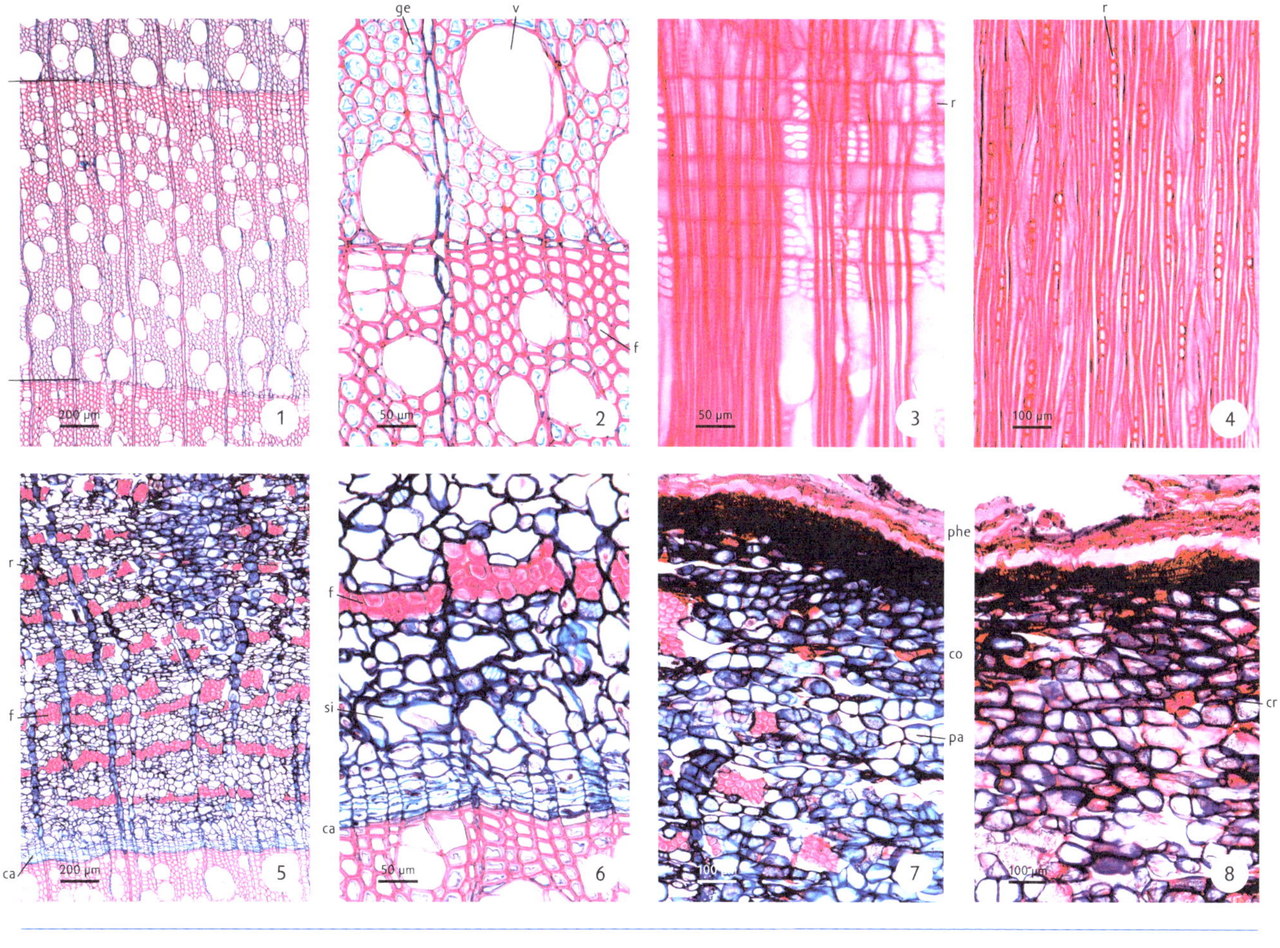

Salix purpurea L.

Purple Willow

Salicaceae

Perennial wetland dioecious and deciduous shrub, 1.5–5 m high. Twigs flexible when young, glabrous, yellowish to purplish-brown, phloem conspicuously yellow. Leaves alternate, but opposite on the basal part of the annual shoots, oblanceolate, glabrous, 3–12 cm long, 0.8–1.2 cm wide, greyish below. Catkins cylindric, 2–3 cm long, appearing before the leaves. Fruit is an ovoid capsule with cylindric seeds embedded in white down.

Distribution and ecology

Euro-Siberian temperate element. Native to all CE countries. Occurs by rivers and streams in wood margins, on river gravel deposits and along ditches. Prefers mineral-rich, stony soils. Tolerant of flooding. Medicinal plant.

Anatomy of the stem

Round stem with a diameter up to 20 cm. Maximum age >40 years.

Stems and rhizomes consist of a xylem, a small phloem, a large cortex and a small phellem. Annual rings are marked by one to two rows of flat latewood fibers and a row of marginal parenchyma (1, 2). The xylem is composed of vessels, thin- to thick-walled fibers, parenchyma and rays (1, 2). The xylem is diffuse-porous to slightly semi-ring porous (1, 2). Vessels are usually solitary (1). Vessel diameter varies from 60–90 µm and vessel length from 200–300 µm. Intervessel pits are large, round or angular and perforation plates are simple (3). Vessel ray-pits are extremely large (3). Fibers are thin- to thick-walled (2) and contain often gelatinous fibers (tension wood) (2). Parenchyma is paratracheal (2, 3) and marginal (2). Rays are uniseriate (5). Central ray cells are prostrate and marginal cells upright (4, 5).

The phloem is composed of rectangular sieve tubes, companion cells and round parenchyma cells (5). Tangentially arranged thin-walled sieve-tube/parenchyma and thick-walled fiber bands (5, 6) are characteristic. No ray-initiated dilatations occur in the outer part of the phloem (6). The cortex consists of tangentially layered, round to oval parenchyma cells (6). A few cortex cells contain crystal druses and prismatic crystals. The phellem is very dense and consists of tangentially layered rectangular cells, which are filled with dark substances (6).

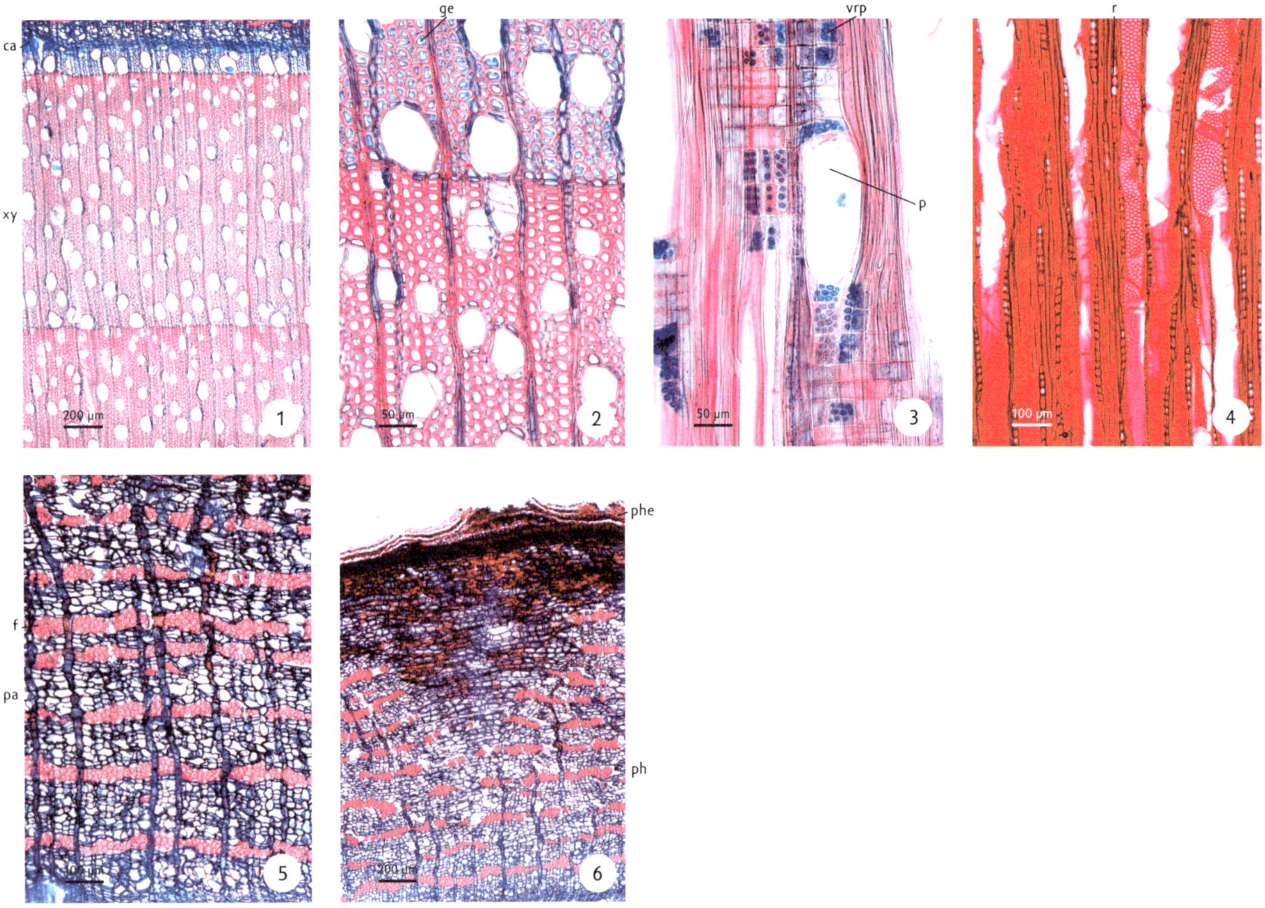

Chrysosplenium alternifolium L. **Alternate-leaved Golden-saxifrage** Saxifragaceae

Perennial wetland herb, stoloniferous. Sterile stem procumbent with only scale-leaves, flowering stem erect, fragile, 5–20 cm high, 3-angled. Leaves alternate, orbicular, cordate at base, the lowest with petiole much longer than blade. Inflorescence of dichotomous cymes, tetramerous flowers without petals golden-yellow. Seeds ellipsoid, dark purple-brown.

Distribution and ecology

Circumpolar wide-boreal element. Native to all CE countries. Occurs by streams, in wet woods and springs from lowlands to the mountains. It can form extensive stands. Sciophyte.

Anatomy of a basal part of the shoot

Triangular shoot with a diameter of 2 mm. Maximum age 1 year.

The small center is surrounded by a few, circularly arranged, open collateral vascular bundles, which are embedded in a thin-walled parenchymatic tissue (1, 2). The xylem is composed of groups of poorly lignified, angular vessels and paratracheal parenchyma (3). Vessel diameter is 15–20 µm and vessel length 80–120 µm. Intervessel pits are scalariform and perforation plates are simple (4). Fibers are absent.

The phloem is composed of sieve tubes and smaller accompanying cells (3). Cortex and phloem are separated by an endodermis (2, 3). The cortex consists of unlignified thin-walled, large parenchyma cells (150–200 µm) and small intercellulars (5). The epidermis consists of thin-walled square cells (5).

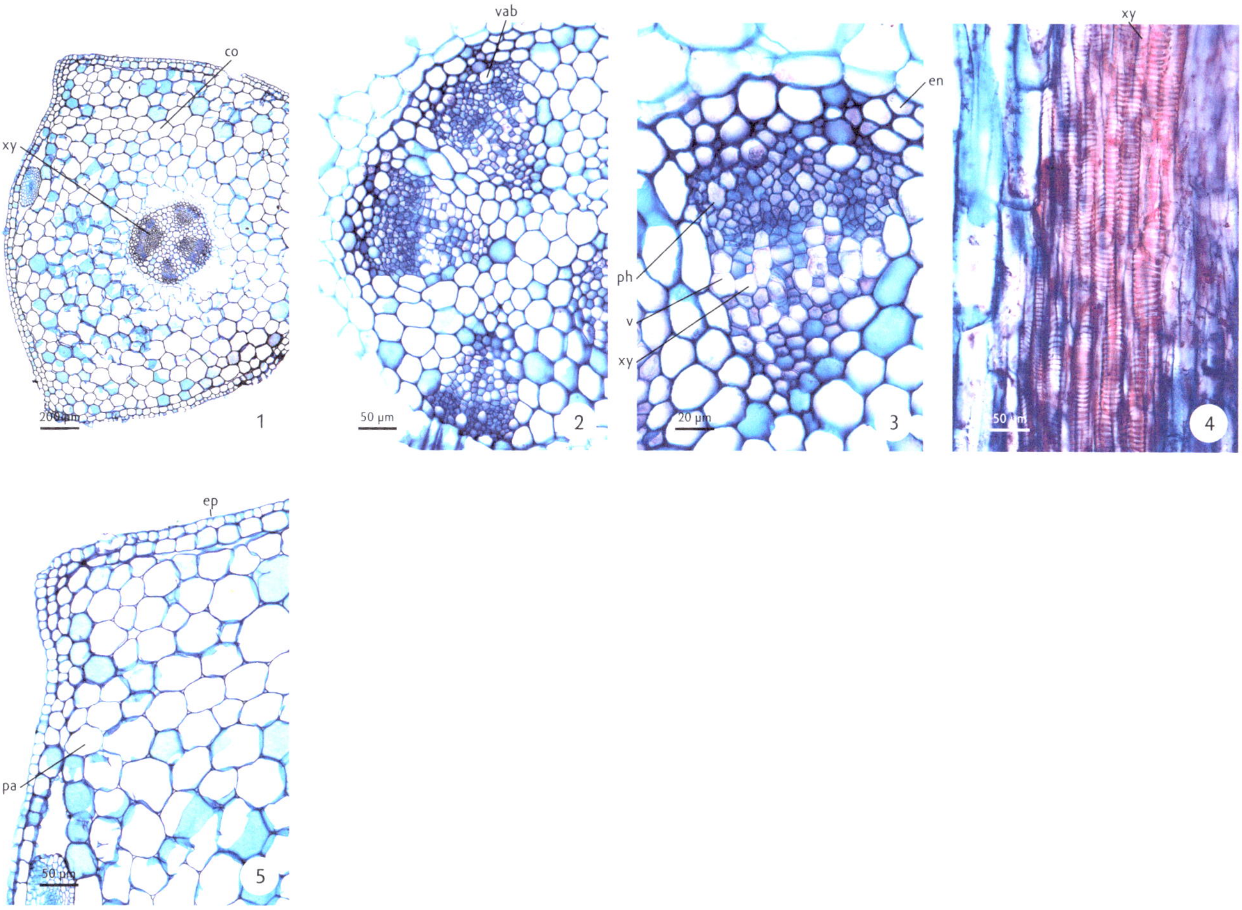

F. H. Schweingruber et al., *Anatomic Atlas of Aquatic and Wetland Plant Stems*, https://doi.org/10.1007/978-3-030-33420-8_51

Chrysosplenium oppositifolium L. **Opposite-leaved Golden-saxifrage** Saxifragaceae

Perennial wetland herb, stoloniferous. Sterile shoots leafy, flowering stem decumbent, fragile, 5–20 cm high, sharply 4-angular. Leaves opposite, dark-green, blade reniform, with petiole as long as blade. Inflorescence of dichotomous cymes, tetramerous flowers yellow-green, without petals. Fruit is a capsule with ellipsoid seeds, purple-brown to dark-brown.

Distribution and ecology

Suboceanic temperate element. Native to CH, CZ, DE and PL. Occurs in very wet sites on forest springs, in alder carrs or by streams. Tolerates more acidic soils than *C. alternifolium*. Usually forms extensive stands. Sciophyte.

Anatomy of a basal part of the shoot

Triangular shoot with a diameter of 1 mm. Maximum age 1 year.

The small center is surrounded by a belt of a xylem, a phloem, a large cortex and an epidermis (1). The xylem consists of unlignified, angular vessels and paratracheal parenchyma (2, 3). Vessel diameter is 15–20 µm and vessel length 80–100 µm. Intervessel pits are scalariform and perforation plates are simple (4). Fibers are absent (2, 3).

Cortex and phloem are separated by an endodermis (2). The phloem is composed of sieve tubes and smaller accompanying cells (3). The cortex consists of unlignified thin-walled, large parenchyma cells (150–200 µm) and small intercellulars (6). Some cells are filled with dark staining substances (5, 6). The epidermis consists of thin-walled square cells (6).

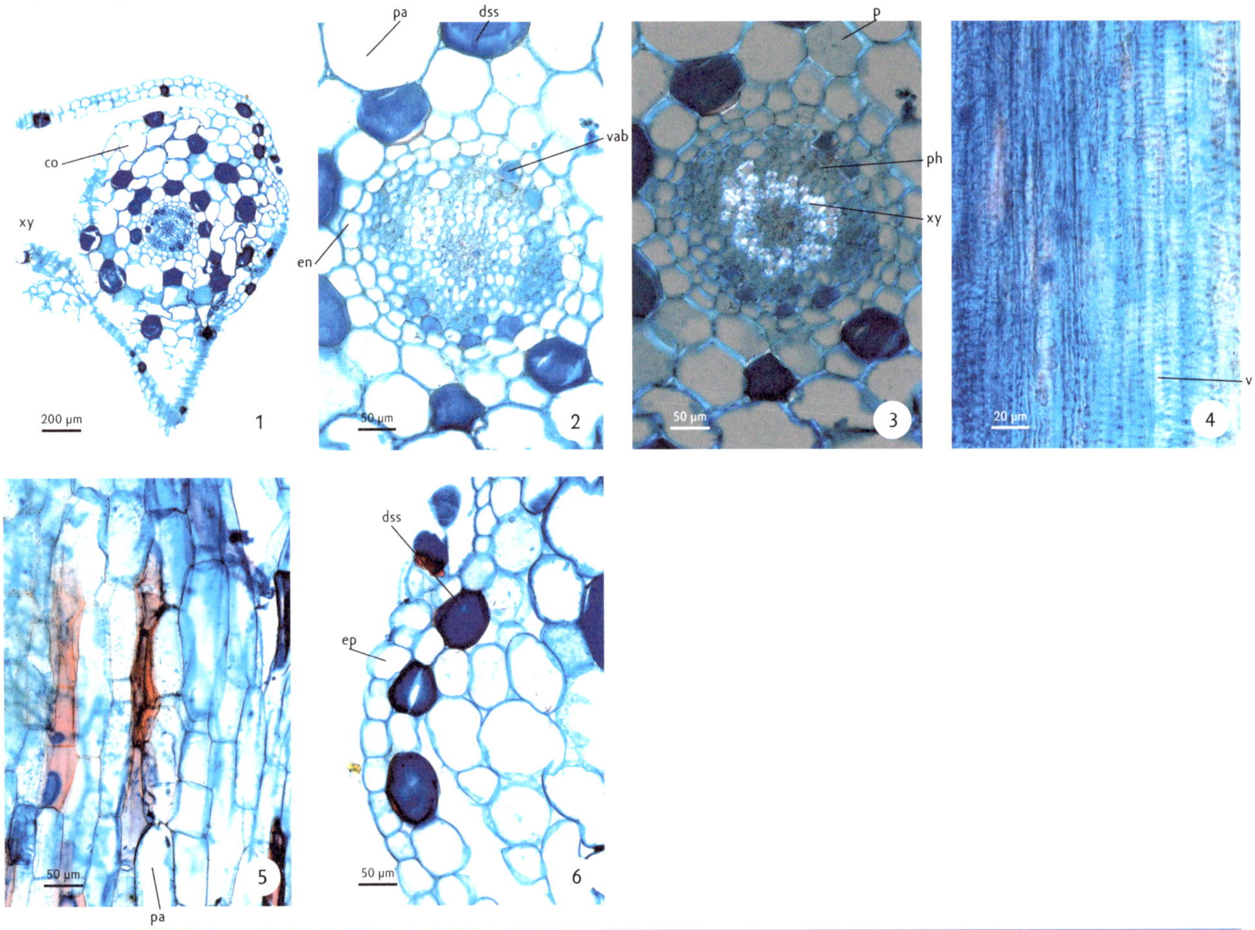

Annual wetland amphibious herb, creeping by runners. Oblanceolate or narrowly elliptic leaves with long petioles in basal rosette, leaves totally 3–15 cm long. Single flowers born on long pedicels. Corolla star-like, white or liliac. Submerged form with very enlarged petioles known, roots with large air spaces.

Distribution and ecology

Circumpolar boreo-temperate distribution. Scattered irregularly and rarely in all CE countries, in lowlands to uplands, usually in fishpond basins or floodplains of big rivers. Wet denudated shores of standing and streaming waters, muddy or sandy soils, in fishponds, wet fields, ditches. Strict heliophyte, competitively very week, pioneering species on summer-drained fishpond bottoms.

Anatomy of the root collar

The basal part of the stem has a diameter of 2–3 mm. Maximum age 1 year (1).

The basal part of the stem consists of a large pith, a small xylem, a small phloem and a large cortex (1). The xylem contains vessels and parenchyma cells (2, 4). Rays are absent (2, 4). Vessels are angular and arranged in radial rows. They have a diameter of 15–20 µm (2) and a length of 80–100 µm. Vessel walls have scalariform intervessel pits (3) and simple perforation plates.

Cortex and phloem are separated by an endodermis (2, 4). Parenchyma cells and small groups of sieve tubes characterize the phloem (4). The cortex represents a net-like aerenchyma with large intercellulars (1, 2, 4). Intercellulars are absent in the pith (1).

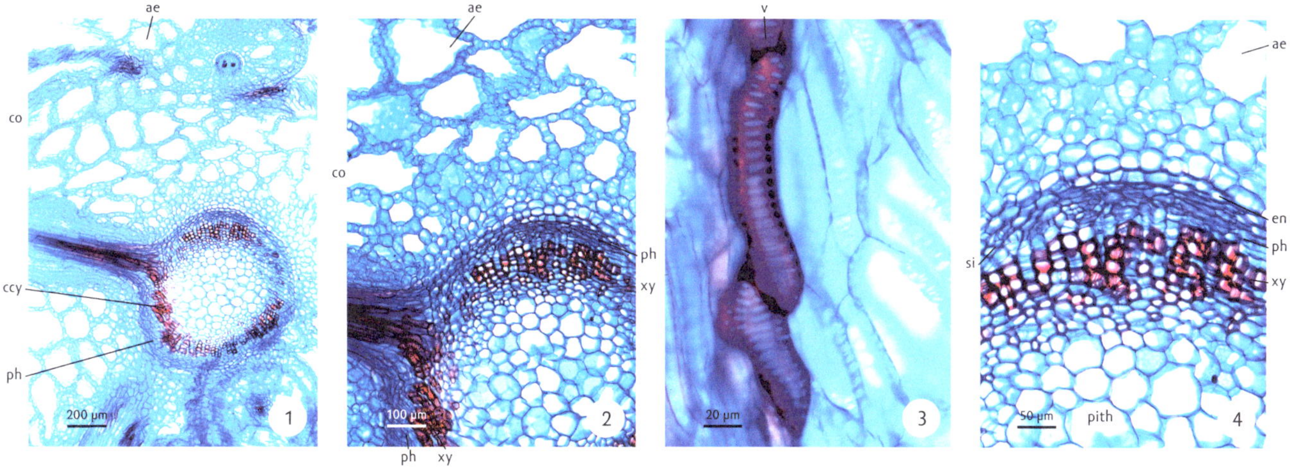

F. H. Schweingruber et al., *Anatomic Atlas of Aquatic and Wetland Plant Stems*, https://doi.org/10.1007/978-3-030-33420-8_52

Scrophularia umbrosa Dumort. **Green Figwort** Scrophulariaceae

Perennial wetland herb, rhizomatous. Stem erect to ascending, branched above, glabrous, 4-angular, broadly winged, 60–100 cm high. Leaves opposite, petiolate, ovate to elliptic, serrate. Corolla greenish to brown-reddish, with 2-lobed upper and 3-lobed lower lips and a short tube. Fruit is a glabrous globular capsule.

Distribution and ecology

Euro-Siberian temperate element. Native to all CE countries from lowlands to uplands. Occurs along streams, rivers and ditches, in wet meadows, also on wet abandoned sites. Prefers fertile soils, tolerates partial shade. Wasp-pollinated.

Anatomy of the annual shoot and the root collar

The annual shoot has a winged, square stem (1).

The basal part of the stem has a diameter of 10–15 mm. Maximum age 1 to 2 years (2).

The basal part of the stem consists of a large hollow pith, a large xylem, a small phloem, and a large cortex (1, 2). Ring boundaries are indistinct or characterized by a zone of flat cells (2).

The xylem contains vessels, thin-walled fibers, parenchyma cells and rays (3). Vessels are angular, arranged solitary or in short radial rows (3) and have a diameter of 40–60 µm and a length of 80–120 µm. Vessel-walls have round intervessel pits (4) and simple perforation plates (4). Fibers are thin-walled (3). Axial parenchyma is rare and difficult to recognize. If present, it is paratracheal (3). Rays are 1–4-seriate (5). Larger rays consist of small procumbent and large upright cells.

The cortex and the phloem of the annual shoot are separated by two layers of a poorly differentiated endodermis/pericycle zone (2, 6, 7). The inner one contains occasionally small, thick-walled, lignified cells (6). Parenchyma cells and small groups of sieve tubes between dilating rays characterize the phloem (6). The cortex and the pith are composed of round and oval parenchyma cells and intercellulars (2, 8). The periphery of the annual shoot consists of an epidermis and a hypodermis (8).

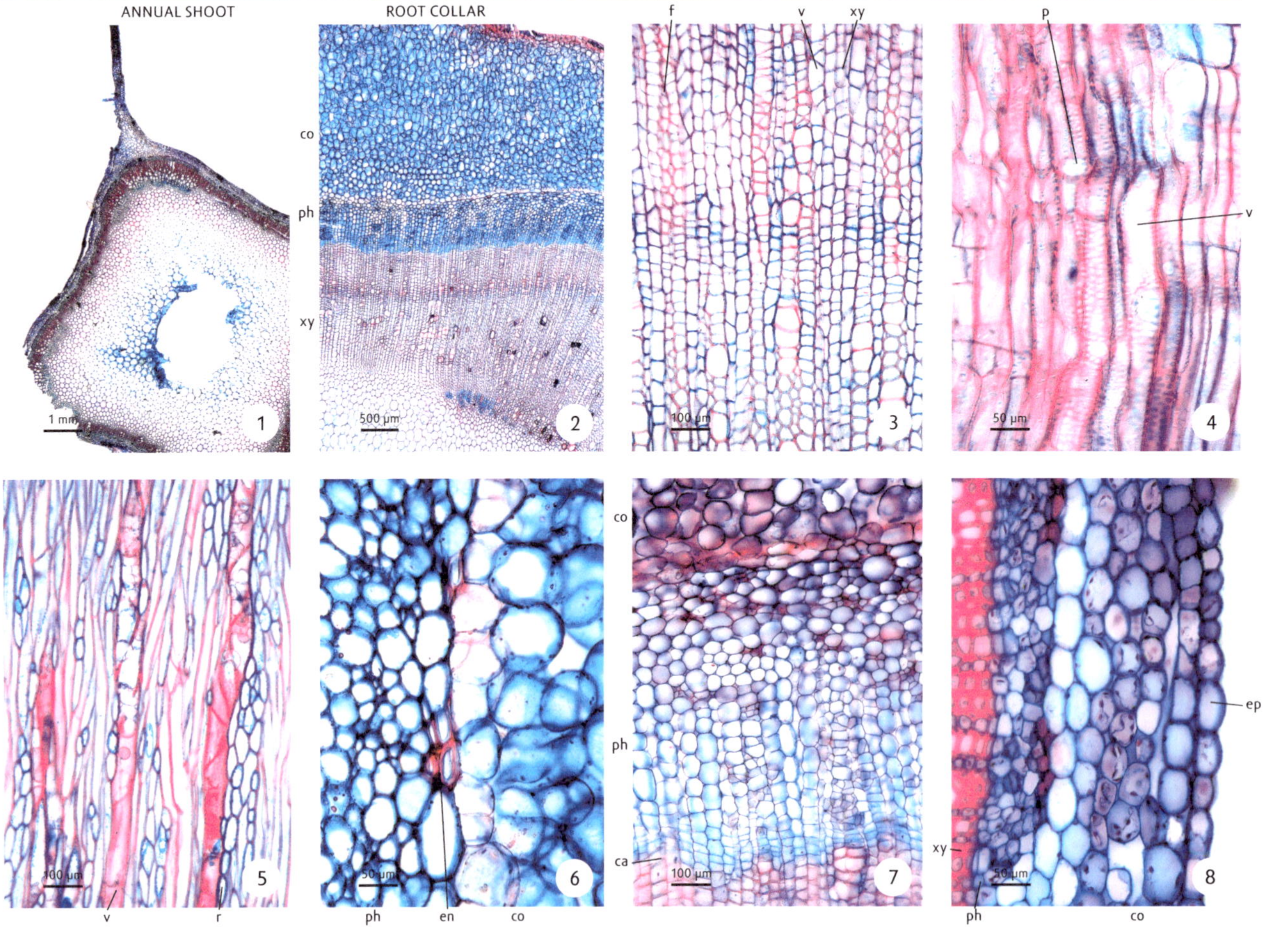

Myricaria germanica (L.) Desv. **Tamariske** Tamaricaceae

Perennial wetland deciduous shrub. Twigs erect, 60–250 cm high, annual shoots yellowish. Leaves linear, boat-like, 1–5 mm long, glabrous, grey-green. Pentamerous tiny flowers in dense terminal racemes, corolla whitish or pinkish. Fruit is a capsule with irregular-cylindric seeds with feathery appendage.

Distribution and ecology

Eurasian distribution. Native to all CE countries, very rare in CZ and HU. Occurs on mesic, montane to subalpine-flooded riverine areas on non-saline limestone and dolomite soils. Secretes large amounts of Ca and Mg salts. Tolerates flooding. Heliophyte. Anemochorous and hydrochorous. Medicinal and toxic plant.

Anatomy of the stem

Round stem with a diameter up to 4 cm. Xylem/bark proportion 10:1. Maximum age >40 years.

Stems consist of a xylem, a large phloem, a small cortex and a phellem (1). Annual rings are marked by ring porosity (1–4). Earlywood vessels are arranged solitary or in small groups (2–4). Earlywood vessel diameter varies from 100–120 µm and latewood from 20–30 µm (4). Vessel length varies from 80–100 µm. Intervessel pits round (5). Perforation plates are simple (5). Heartwood vessels contain dark staining substances (3). Fibers are thin- to thick-walled (4) with round pits (5). Parenchyma is paratracheal and storied (6). Rays 4–6-seriate (6) with upright sheath cells and prostrate central cells (5, 6).

The phloem consists of groups of sieve tubes, parenchyma cells and tangential arranged groups of fibers (7, 8). Sieve tubes are compressed in older parts of the phloem (7). Rays dilate. Crystals are absent. The phellem consists of radially arranged thick-walled rectangular cells.

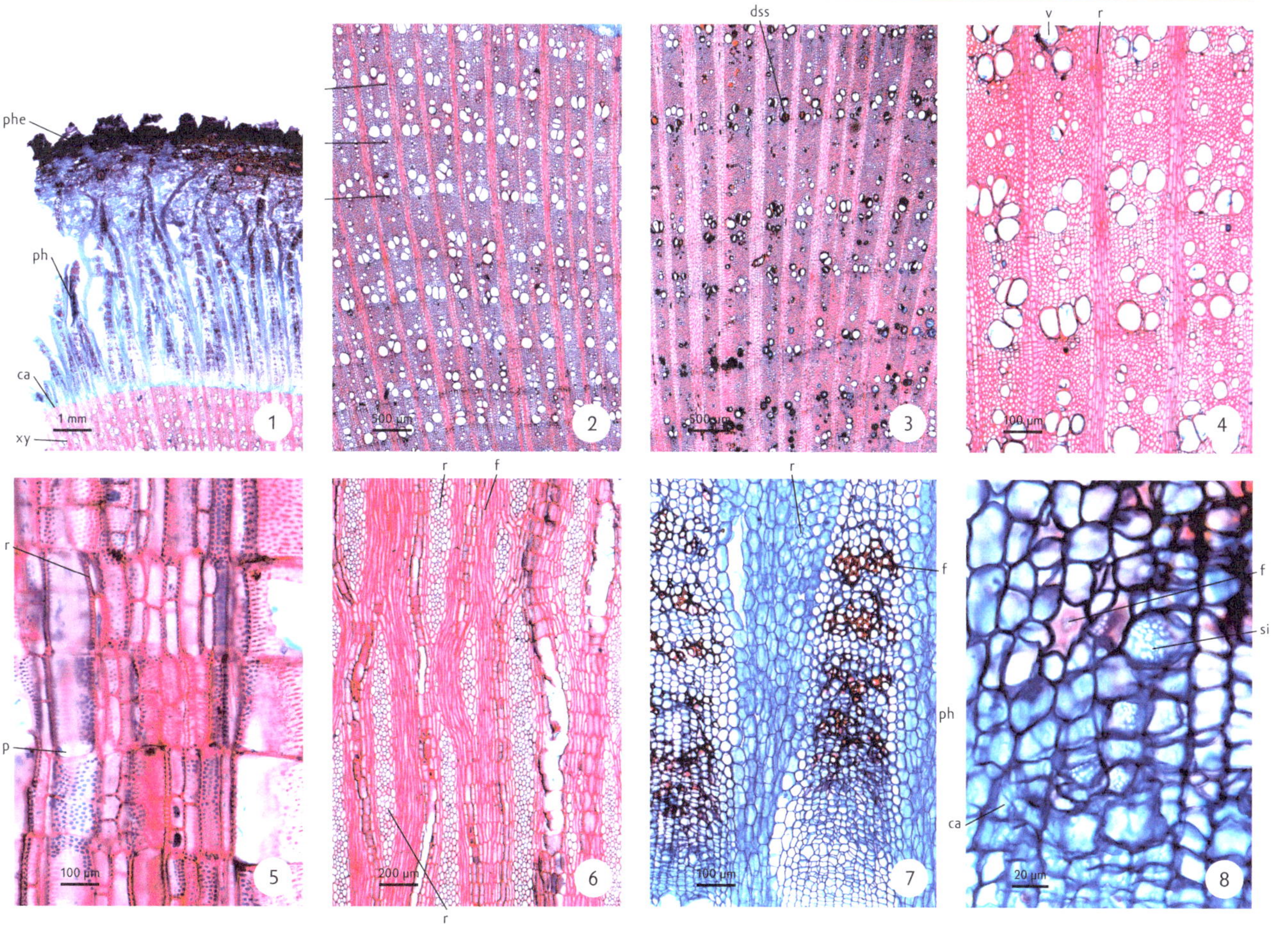

F. H. Schweingruber et al., *Anatomic Atlas of Aquatic and Wetland Plant Stems*, https://doi.org/10.1007/978-3-030-33420-8_53

Valeriana dioica L.

Marsh Valerian

Valerianaceae

Perennial wetland dioecious herb, stoloniferous. Stem erect, simple, 15–30 cm high. Basal leaves simple, ovate, entire. Stem-leaves pinnate, with entire to slightly dentate leaflets. Unisexual flowers in dense compound cymes, corolla white to pink. Fruit is an oblong-ovoid achene, with hairy bristles at the apex forming the crest.

Distribution and ecology

European temperate element; also in C. Asia and N. America. Native to all CE countries. Occurs on calcareous fens, peaty meadows, in *Alnus* or *Salix* carrs. Tolerates partial shade.

Anatomy of the root collar

Round root collar with a diameter up to 2–3 mm. Maximum age 1 year.

Stems consist of a xylem, a small phloem and a large cortex (1). The xylem is composed of angular vessels and thin- to thick-walled fibers (1). Vessels are arranged in radial rows (1, 4). Vessel diameter varies from 30–50 µm (2) and vessel length from 150–250 µm. Intervessel pits are round (3) and perforation plates are simple (3). Fibers are thin- to thick-walled and contain nuclei (3). Parenchyma is pervasive (1, 2). Rays are absent. Crystals are absent.

The phloem consists of groups of sieve tubes and parenchyma cells (2). The cortex is composed of unlignified thin-walled parenchyma containing small intercellulars (4). The phellem is absent (1).

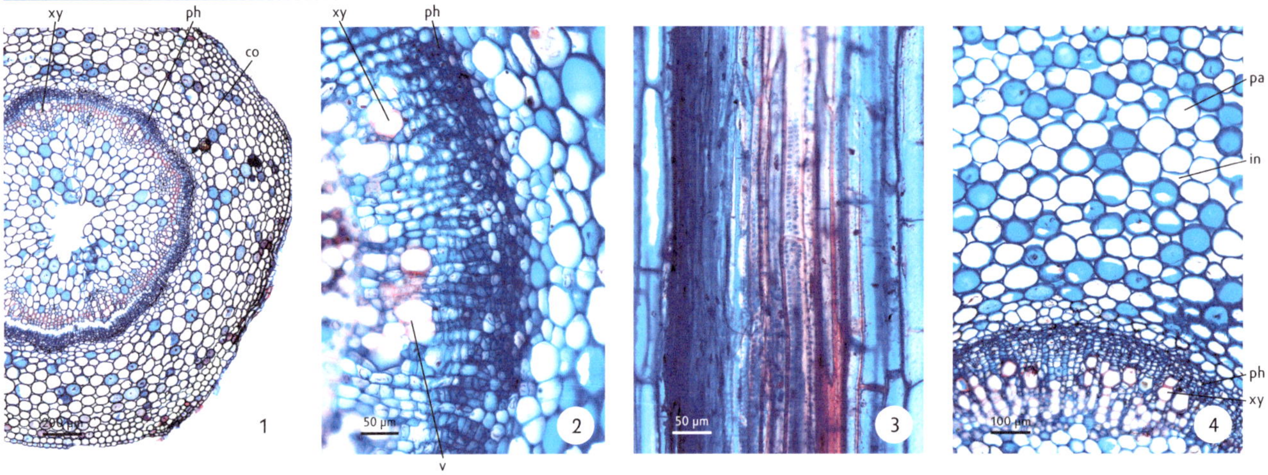

F. H. Schweingruber et al., *Anatomic Atlas of Aquatic and Wetland Plant Stems*, https://doi.org/10.1007/978-3-030-33420-8_54

Perennial wetland herb, stoloniferous. Stem erect, unbranched, glabrous, 60–200 cm high. All leaves pinnate, leaflets linear-lanceolate with entire to serrate margins. Bisexual flowers in dense compound cymes, funnel-shaped corolla white or pinkish. Fruit is a prolonged ovoid achene with hairy bristles at the apex forming the crest. Very variable polyploid complex/taxon.

Distribution and ecology

Eurasian boreo-temperate element; widely naturalised outside its native range. Native to all CE countries. Occurs along streams, rivers and ditches, in tall-herb vegetation along water margins, in wet meadows, also in damp synanthropic habitats. Medicinal herb with a typical odour.

Anatomy of the root collar

Round root collar with a diameter up to 10–15 mm. Maximum age 1 year.

Stems consist of a xylem, a small phloem and a large cortex (1). The xylem is composed of vessels, thin- to thick-walled fibers and a few rays (2). Vessels are arranged solitary and in short radial rows (2). Vessel diameter varies from 50–70 µm (2) and vessel length from 150–250 µm. Intervessel pits are round (3) and perforation plates are simple (3). Fibers are thin- to thick-walled (2). Parenchyma is absent. Rays are rare if present, uni-biseriate (4) and ray cells are upright. Crystals are absent in the xylem.

The phloem consists of groups of sieve tubes and parenchyma cells (5). The cortex is composed of unlignified thin-walled parenchyma (6) with solitary, round crystals. The phellem is very small (1).

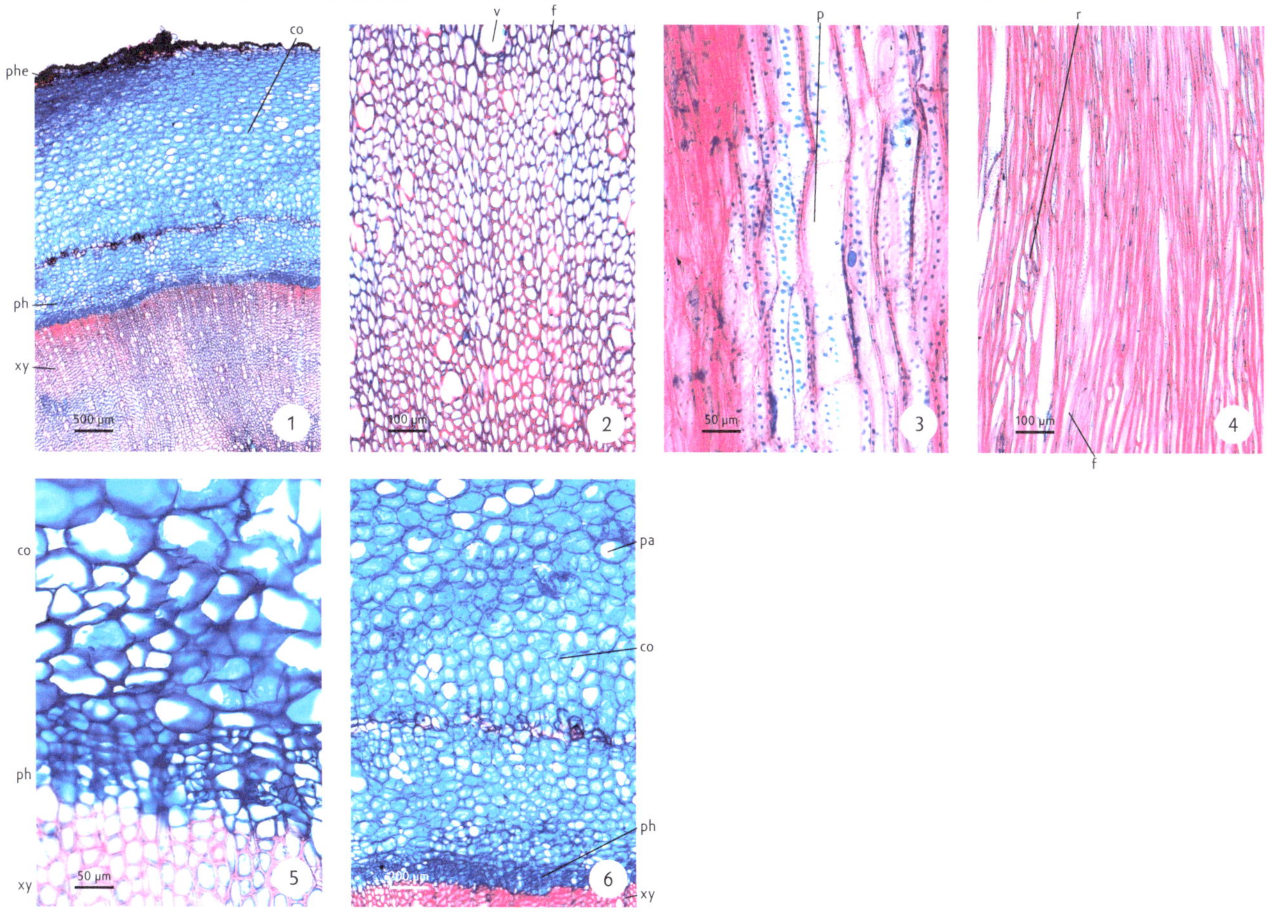

Viola biflora L.

Alpine yellow-violet

Perennial wetland herb, rhizomatous. Stem erect or ascending from basal leaf rosette, rounded, glabrous, 10–15 cm high. Leaves reniform, petiolate, serrate. Flowers solitary in axils, nodding, fragrant. Corolla yellow, the lower petal with three violet lines, the lateral petals directed askew upwards. Fruit is a broadly ovoid capsule containing obovate seeds with elaiosomes.

Distribution and ecology

Eurasia and western N. America. Native to all CE countries in the mountains, very rare in HU. Occurs in mountain springs, along mountain streams and at snow-bed sites. Tolerates partial shade in the northern part of its distribution.

Anatomy of the rhizome

Irregularly formed rhizome with a diameter of 2–3 mm. Maximum age 2 years (1).

The large pith is surrounded by a xylem, a small phloem and a large cortex (1). The xylem consists of angular vessels and paratracheal parenchyma (2). Vessels have a diameter of 15–25 µm (2) and a length of 50–100 µm. Vessels are arranged in solitary or in short radial rows (2). Intervessel pits are round and perforation plates are simple (3). Rays are uni- to biseriate (4). Fibers are absent (2).

The phloem consists of parenchyma cells and groups of sieve tubes (5). The cortex consists of large parenchyma cells (5). Intercellulars are absent (5). Crystals and phellem are absent.

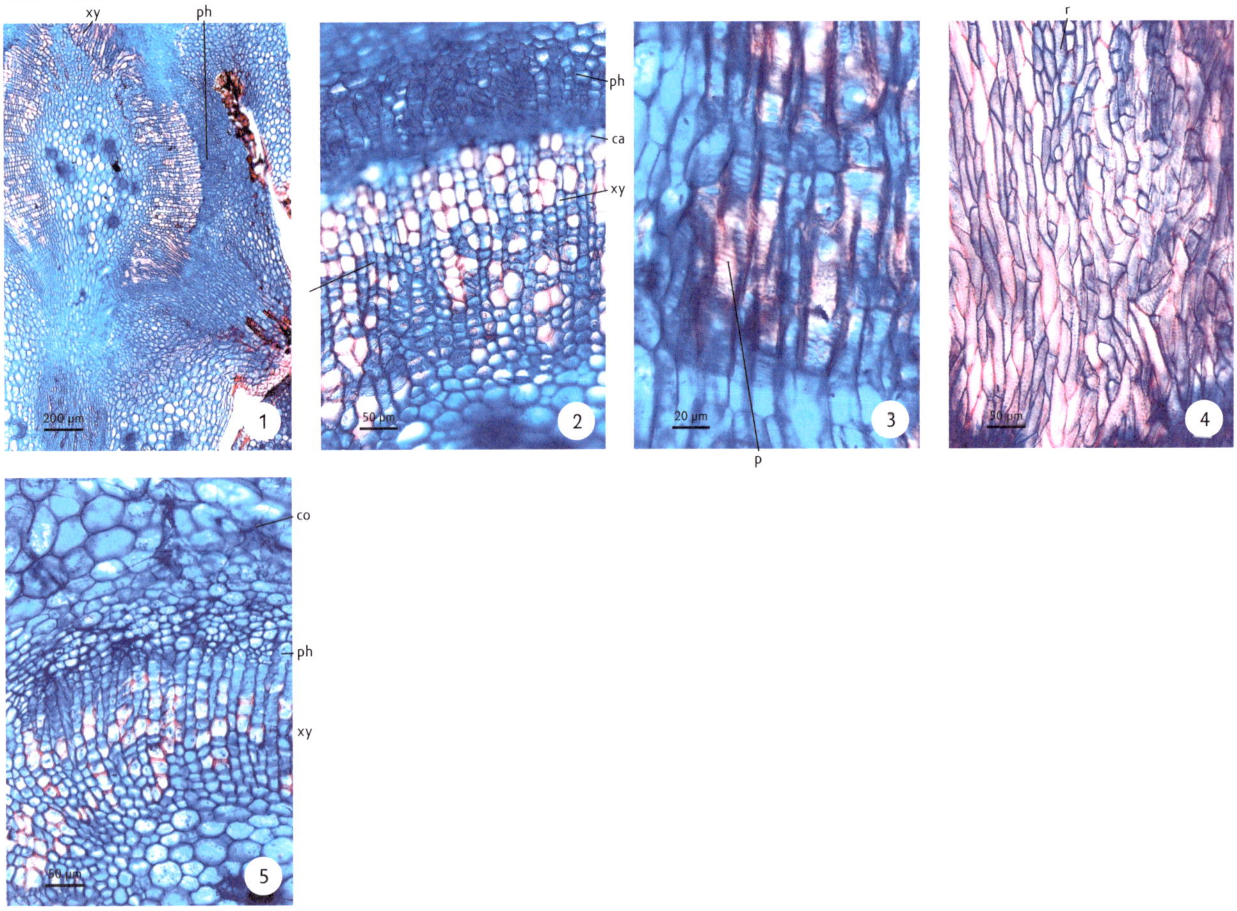

F. H. Schweingruber et al., *Anatomic Atlas of Aquatic and Wetland Plant Stems*, https://doi.org/10.1007/978-3-030-33420-8_55

Viola elatior Fries

Marsh Willowherb Violaceae

Perennial wetland herb, rhizomatous. Stem erect, rounded, shortly hirsute, 20–35 cm high. Basal leaf rosette absent, stem leaves lanceolate, up to 9 cm long, 2.5–4 × longer than wide, serrated. Stipules narrowly eliptic to lanceolate, markedly large, up to 6 cm long. Flowers solitary in leaf axils, nodding. Corolla iris-blue to blue-violet, with blue-violet venation. Fruit is an elongated ovoid capsule containing obovate seeds with elaiosomes.

Distribution and ecology

Eurasian distribution. Native to all CE countries, rarely in lowlands. Occurs on the margins of floodplain forests and shrubs, also in fens and floodplain meadows. Prefers base-rich sites, wet or flooded during spring. Tolerates long periods of summer drought and disturbance by management or flooding. Endangered river corridor plant.

Anatomy of a basal part of the annual shoot

Round stem with a diameter of 1–3 mm. Maximum age 2 years (2).

The large pith is surrounded by a xylem, a small phloem, a large cortex and an epidermis (1). The xylem consists of vessels, thin- to thick-walled fibers and parenchyma (2). Vessels are angular and have a diameter of 15–25 µm (2). Vessel length in both parts is 150–200 µm. Vessels are arranged solitary or in short radial rows (2). Intervessel pits are round and perforation plates are simple (3). Parenchyma is paratracheal and pervasive (1, 2). Rays are uniseriate (2, 4), they are difficult to recognize in tangential sections (4). Fibers are thin- to thick-walled (2) with round pits (3).

The phloem consists of parenchyma cells and groups of sieve tubes (5). The cortex consists of large parenchyma cells and small intercellulars (5). The phellem is absent. Crystal druses occur in the pith and the cortex (5). The epidermis consists of round cells (5).

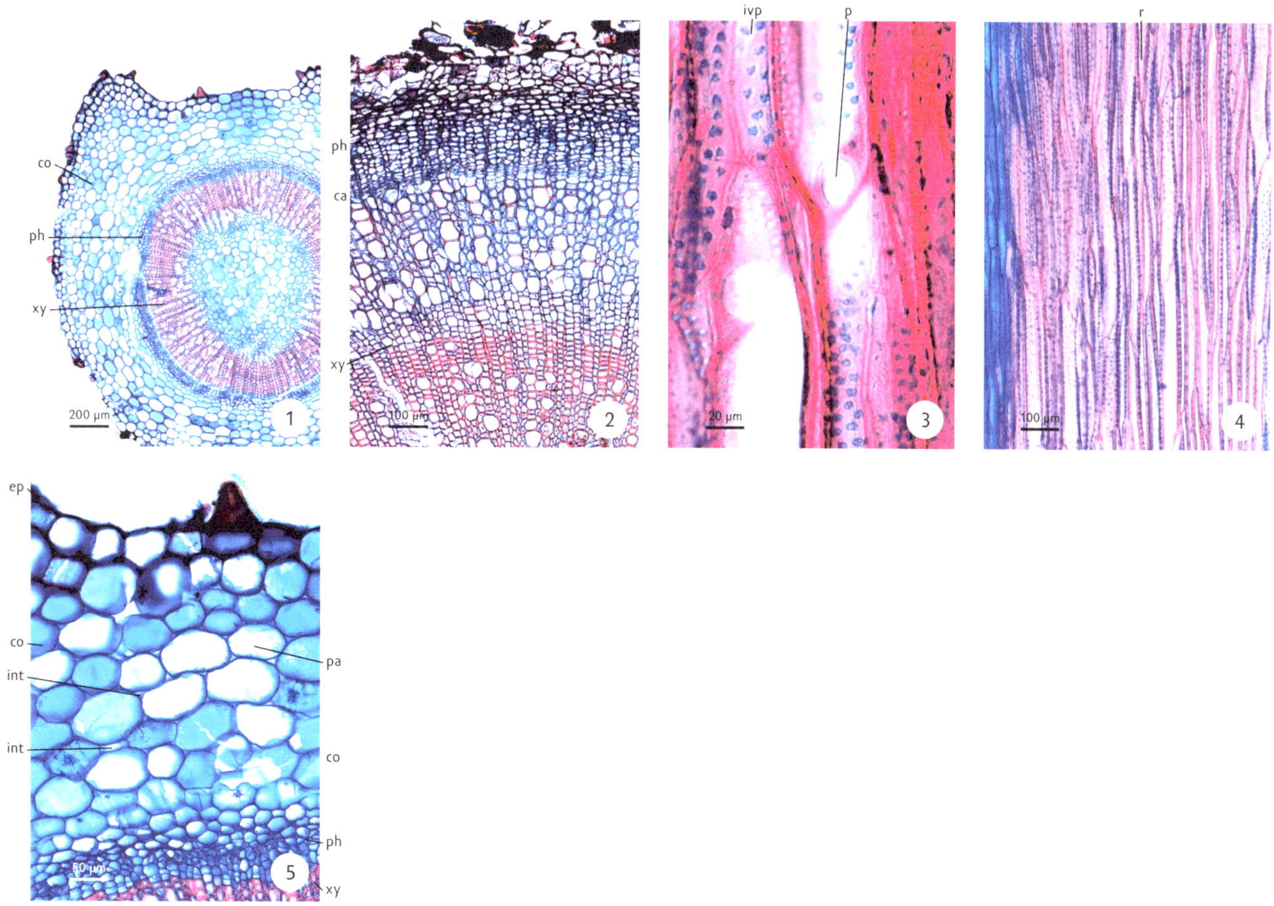

Perennial wetland herb with thin creeping rhizomes. Stem 5–10 cm high, leaves in basal rosette. Leaves rounded to reniform, 2–5 cm wide, always glabrous above. Flowers solitary, nodding, scentless. Corolla and spur pale violet. Fruit is an ovoid capsule containing obovoid seeds with very small elaiosome.

Distribution and ecology

European boreo-temperate element; also in N. America. Native to all CE countries except HU. Occurs on peatbogs and acidic fens, in peaty meadows, also in *Alnus* and *Salix* carrs. Prefers wet acidic soils, often associated with *Sphagnum*. Chasmogamous flowering in early spring, cleistogamous flowers throughout the summer.

Anatomy of a basal part of the annual shoot

Round stem with a diameter of 1–2.5 mm. Maximum age 1 year (1, 2).

The large pith is surrounded by a xylem, a small phloem, a large cortex and an epidermis (1). The xylem consists of vessels, thin- to thick-walled fibers and parenchyma (2). Vessels are angular and have a diameter of 15–25 µm (2). Parenchyma is paratracheal and pervasive (1, 2). Rays are uniseriate (2). Fibers are living and thin- to thick-walled (2).

Cortex and phloem are separated by an endodermis (3, 4). The phloem consists of parenchyma cells and groups of sieve tubes (2). The cortex consists of large parenchyma cells and small intercellulars (3). The phellem is absent. Crystal druses occur in the pith and the cortex (4).

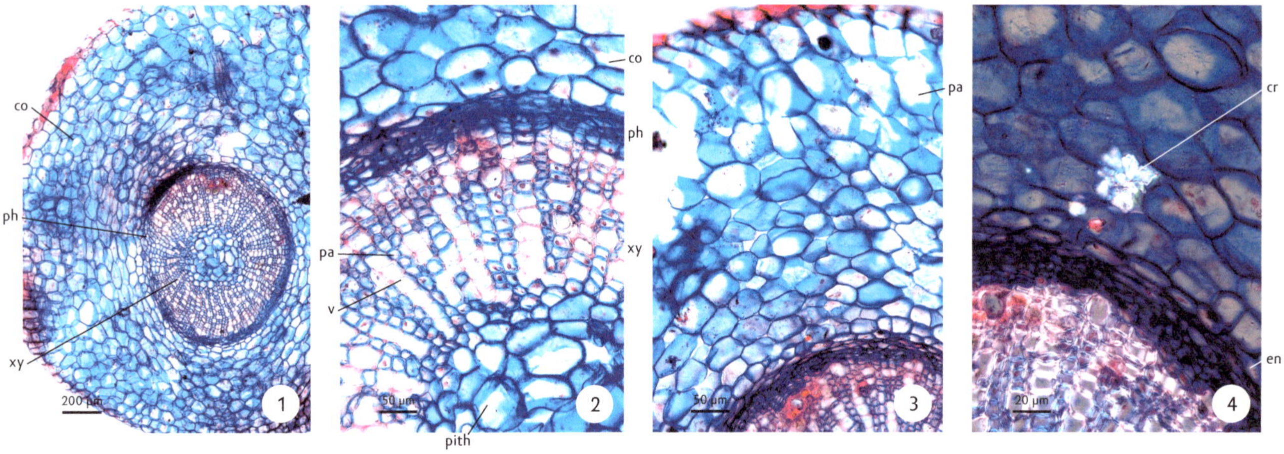

Perennial wetland herb with a many-headed rootstock. Stem erect or shortly ascending, glabrous or papillose, branched, 12–20 cm high, without basal leaf rosette. Stem leaves lanceolate to narrowly triangulate, 2.3–3.5 × longer than wide, serrate. Flowers scentless, petals milk-white or slightly lilac, with dark veins. Fruit is a narrowly ovoid capsule, containing obovate seeds with very small elaiosome.

Distribution and ecology

Eurasian element. Native to all CE countries. Occurs on marshes and meadows in floodplains, also in river banks but also in weakly to strongly acidic fens and at gley sites with a more constant ground-water level outside floodplains. Develops both chasmogamous and cleistogamous flowers, and a persistent seed bank. Endangered river corridor plant.

Anatomy of the basal part of the annual shoot

Round stem with a diameter of 2 mm. Maximum age 1 year (2).

The large pith is surrounded by a xylem, a small phloem, a large cortex and an epidermis (1). The xylem consists of vessels, thin- to thick-walled fibers and parenchyma (2). Vessels are angular and have a diameter of 15–25 µm (2). Vessel length in both parts is 150–200 µm. Vessels are arranged solitary or in short radial rows (1). Intervessel pits are round and perforation plates are simple (3). Parenchyma is paratracheal (1, 2). Rays are uniseriate (2, 3), they are difficult to recognize in tangential sections (3). Fibers are living and thin- to thick-walled (2) with round pits (fiber tracheids) (2).

The phloem consists of parenchyma cells and groups of sieve tubes (2). The cortex consists of large parenchyma cells and small intercellulars (5). The phellem is absent. Crystal druses occur in the pith and the cortex (4). The epidermis consists of round cells (5).

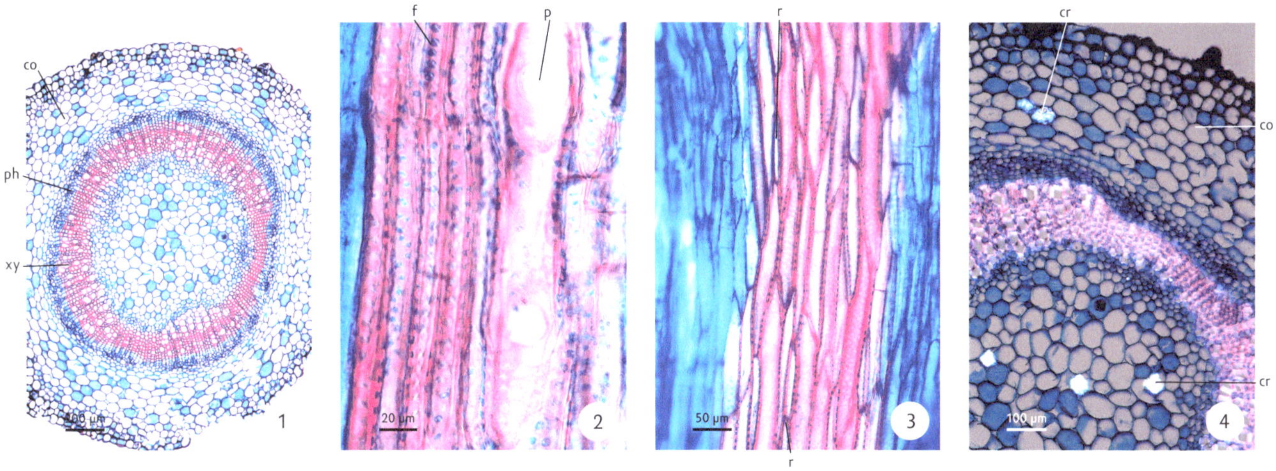

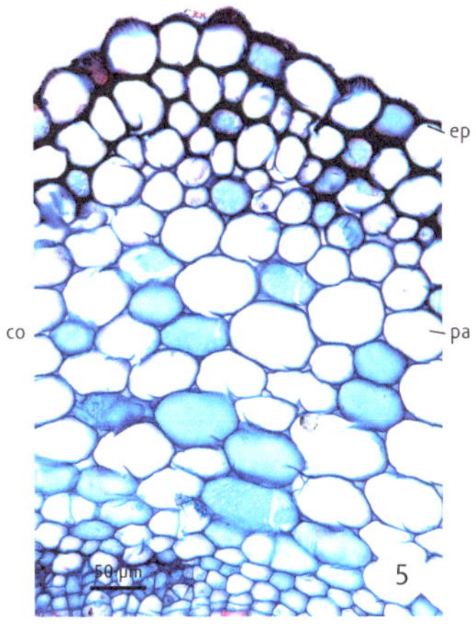

Anatomical descriptions of Monocotyledons

Acorus calamus L.

Sweet-flag

Perennial emergent herb, stem erect, triangular, 50–150 cm high, without stalks, leaves linear, 20–90 cm long, sword-shaped. Dioecious, inflorescence yellowish-green spadix, often not flowering. Aromatic medicinal plant.

Distribution and ecology

Native to subtropical Asia, in Europe cultivated as a medicinal plant, naturalized though fruits sterile, in all CE countries. Native plants are diploid (Siberia, N America) or tetraploid (S and E Asia); the origin of the sterile triploid is unknown but is naturalized in Europe and W Asia, the Himalayas and eastern N America. On the margins of streams, canals, fishponds and lakes in shallow, nutrient-rich water.

Anatomy

Triangular flower stalk with a diameter of 6–8 mm (1) and a round-shaped rhizome with 10–15 mm in diameter (5).

The flower stalk consists of an aerenchymatic tissue with regularly distributed vascular bundles. A cortex is absent (1). Vascular bundles consist of a centripetal air duct. External of an arc of poorly lignified vessels is a large group of sieve tubes with companion cells (2). Small rudimentary vascular bundles occur just inside the epidermis (3). The epidermis is characterized by external tooth-like wall-thickenings (4).

The rhizome consists primarily of an aerenchymatic reticulate tissue (5). An endodermis separates the central part from the cortex (6). A circle of concentric vascular bundles surrounds an almost vascular bundle-free central part (6). Some few fibers occur at the periphery of some vascular bundles. The vessel walls have scalariform intervessel pits or ring-like structures. Crystals are absent.

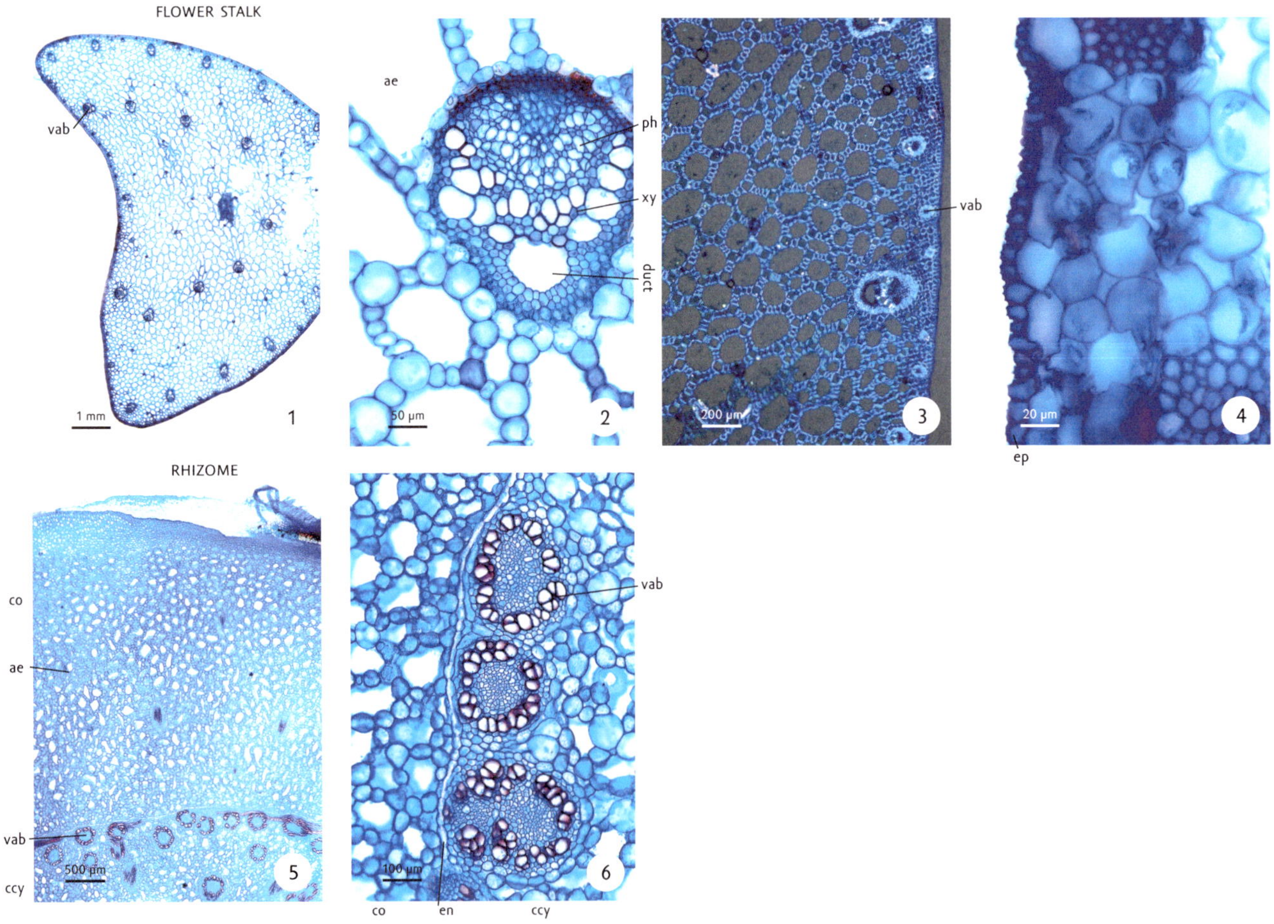

F. H. Schweingruber et al., *Anatomic Atlas of Aquatic and Wetland Plant Stems*, https://doi.org/10.1007/978-3-030-33420-8_56

Alisma gramineum Lej.

Ribbon-leaved water-plantain

Perennial emergent, fully amphibious plant, annual shoots, leaves from a shortened stem base. Ribbon-like or very narrowly elliptic leaves (10–50 cm long) with wide, non-differentiated petioles. Unbranched, fibrous porous roots 10–20 cm long, erect flower stems 0.1–0.5 m high, whorled panicles, flowers 5–8 mm across, fruits small achenes, overwintering only in underground organs.

Distribution and ecology

Circumpolar temperate element. Scattered rarely in all 7 CE countries, more common in lowlands and fishpond basins. Shallow standing waters, banks of fishponds, lakes, dam reservoirs, banks of slow-streaming rivers, denudated fishpond bottoms. Small plants or seedlings submerged or natant, adult flowering plants emergent or terrestrial, great leaf shape plasticity. Organic, sandy or clayish bottoms, mostly in slightly calcareous, neutral, mineral-rich soils, in soft to hard waters (total alkalinity 0.5–4 meq/l), prefers meso- to eutrophy, facultative heliophyte, probably strict CO_2 user.

Anatomy

Circular flower stalk with a diameter of 6–8 mm (1).

The flower stalk consists of an aerenchymatic cortex containing vascular bundles (1).

Vascular bundles contain a large air duct. An arc of round vessels and polygonal sieve tube exist externally (2).

Rhizome with a diameter of 10–20 mm (3): The central part and the cortex are separated by an endodermis (3). Concentric vascular bundles surround the central part of the rhizome (3, 4). An arc of lignified vessels surrounds a group of sieve tubes (4). Vessels walls have scalariform intervessel pits or ring-like structures. Crystals are absent.

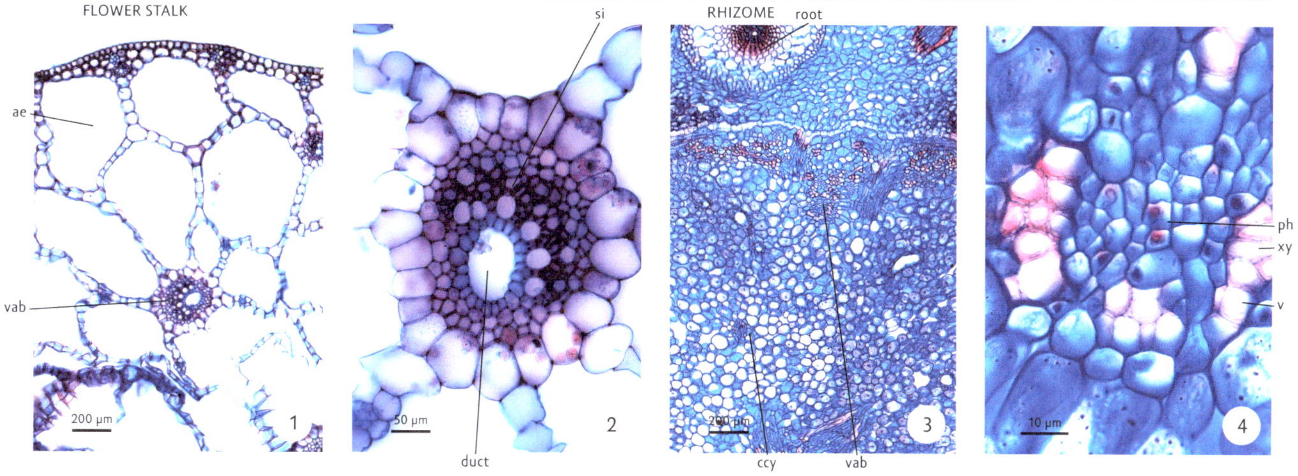

F. H. Schweingruber et al., *Anatomic Atlas of Aquatic and Wetland Plant Stems*, https://doi.org/10.1007/978-3-030-33420-8_57

Alisma plantago-aquatica L.

Water-plantain Alismataceae

Perennial emergent, amphibious plant, annual shoots, leaves from shortened stem base. Leaves (10–80 cm long) with long petioles elliptic to ovate elliptic, leaf shape very plastic: seedlings and submerged leaves fragile, narrow and ribbon-like to oblong-ovate, emergent or natant leathery leaves tougher and subcordate or elliptic. Unbranched, fibrous porous roots 10–30 cm long. Erect flower stems 0.2–1 m high, inflorescence in whorled panicles, fruits small achenes. Overwintering only in underground organs (rhizomes, roots).

Distribution and ecology

Circumpolar wide-temperate element; widely naturalized outside its native range. In all 7 CE countries, very common from lowlands to highlands. Shallow standing waters. Smaller plants or seedlings usually submerged or natant, adult plants emergent or terrestrial. Organic, sandy or clayish bottoms, mostly infertile, acidic or neutral, non-calcareous soils or softwaters, soft- to medium hard waters (total alkalinity 0.1–2 meq/l), absent from hardwaters, very eurytopic species, tolerates oligo- to hypertrophy, strict CO_2 user.

Anatomy

Circular flower stalk with a diameter of 6–8 mm (1).

The center of the shoot is characterized by large net-like intercellulars (1). A compact large belt of thin-walled, angular parenchyma cells forms the middle part.

A sheath of thick-walled lignified fibers surrounds the inner part of the shoot and all vascular bundles (2, 3). Large centripetal air ducts are surrounded by unlignified parenchyma cells in each vascular bundle. The xylem/phloem complex in vascular bundles is difficult to interpret. It consists of large round, large angular, small polygonal cells and small thick-walled fibers (2, 3).

Rhizome: Central part with a diameter of 10–20 mm (4).

Many often non-axially directed concentric vascular bundles form the inner part of the stem (4, 5). Groups of poorly lignified vessels without air ducts and unlignified parenchyma cells surround groups of sieve tubes (5). Vessel walls have scalariform intervessel pits or ring-like structures (6). An endodermis/pericycle zone separates the cortex from the central part (7). Special is the presence of secretory canals (8) within the net-like central aerenchyma. Crystals are absent.

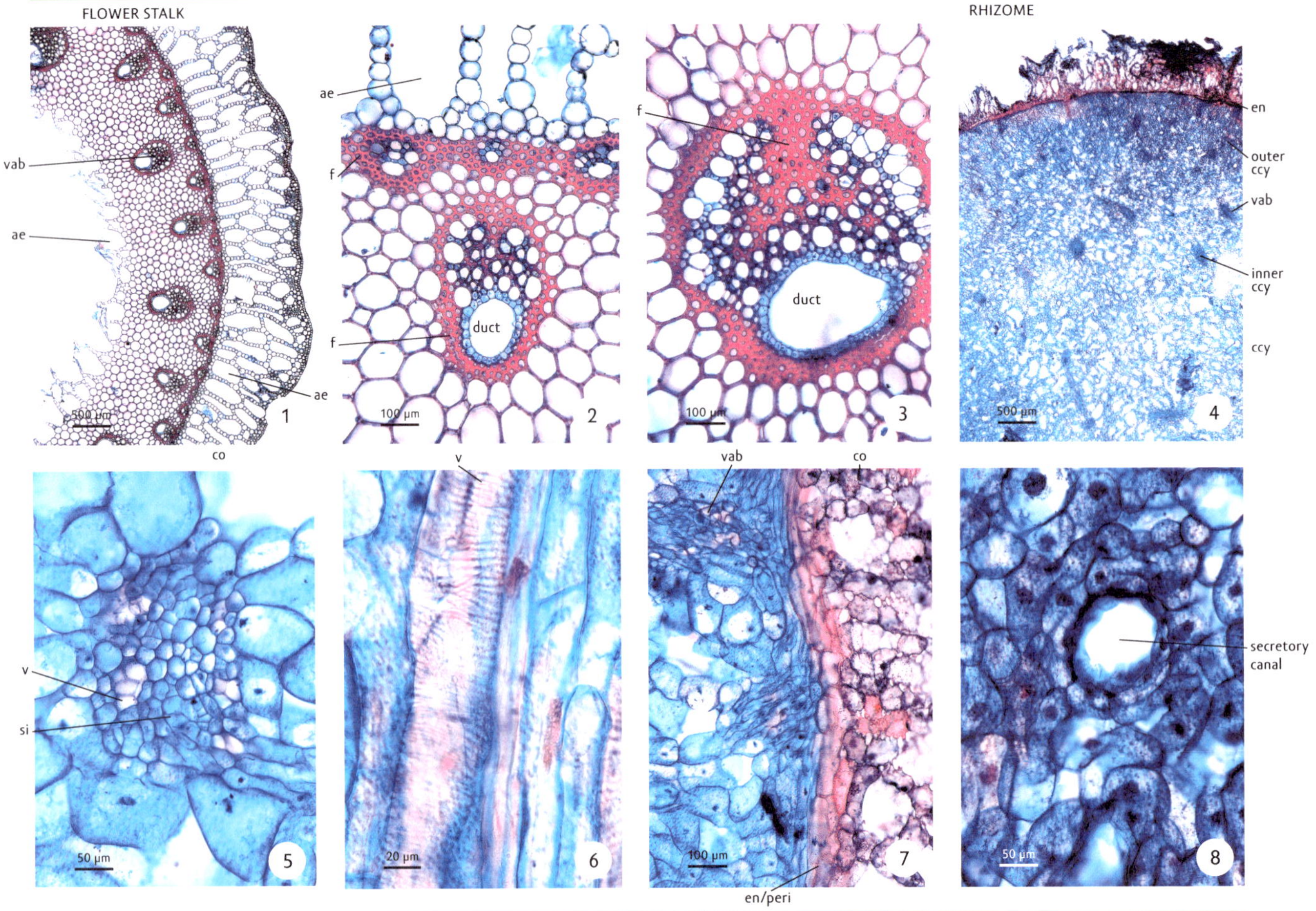

Caldesia parnassifolia (L.) Parl.

Perennial submerged, natant or emergent, fully amphibious plant, annual shoots, overwintering in turions. Leaves from shortened stem base, young submerged leaves ribbon-shaped, adult leaves 2–5 cm long, heart-shaped, with 5–30 cm long petioles. Unbranched, fibrous roots 5–10 cm long. Erect flower stems 0.3–0.5 m high, inflorescence in whorls, flowers 12–14 mm across, fruits achenes, torpedo-shaped turions 1–5 cm long, produced on flower stems or special turion-bearing stems.

Distribution and ecology

Extremely rare distribution: native to Europe, eastern Africa, Madagascar, and from China to Australia. In CE countries, only one site known in Germany and two in Hungary, extinct in AT, CH and PL. Shallow peat bogs or fen lakes (water depth 0–50 cm), peaty fishponds, strictly organic peaty bottom, dystrophic, acidic, soft water (pH 5.5–7), water level may fluctuate, in dry seasons, only seed bank present, seed germination very poor, plant ecology mostly unknown.

Anatomy

Circular flower stalk with a diameter of 4 mm (1).

Several closed, collateral vascular bundles occur within an aerenchymatic tissue (2).

A centripetal air duct dominates vascular bundles (2). The xylem of well-developed bundles consists of arc-like arranged thin-walled, polygonal vessels and parenchyma cells. The phloem is composed of a large sieve tubes and small companion cells (2). The xylem is absent in poorly developed vessels. The ground tissue is characterized by large net-like intercellulars (1). Special is the presence of secretory ducts (3) near the periphery.

The cortex consists of round and oval cells with large intercellulars (4). A few vascular bundles occur in the cortex (4). Crystals are absent.

Rhizome with a diameter of 6–7 mm (5).

Many non-axially directed concentric vascular bundles form the inner part (5). Groups of lignified vessels and unlignified parenchyma cells surround groups of sieve tubes (6). Vessels walls have scalariform intervessel pits or ring-like structures (7). Air ducts are absent. A poorly differentiated endodermis/pericycle zone separates the cortex from the central part (8).

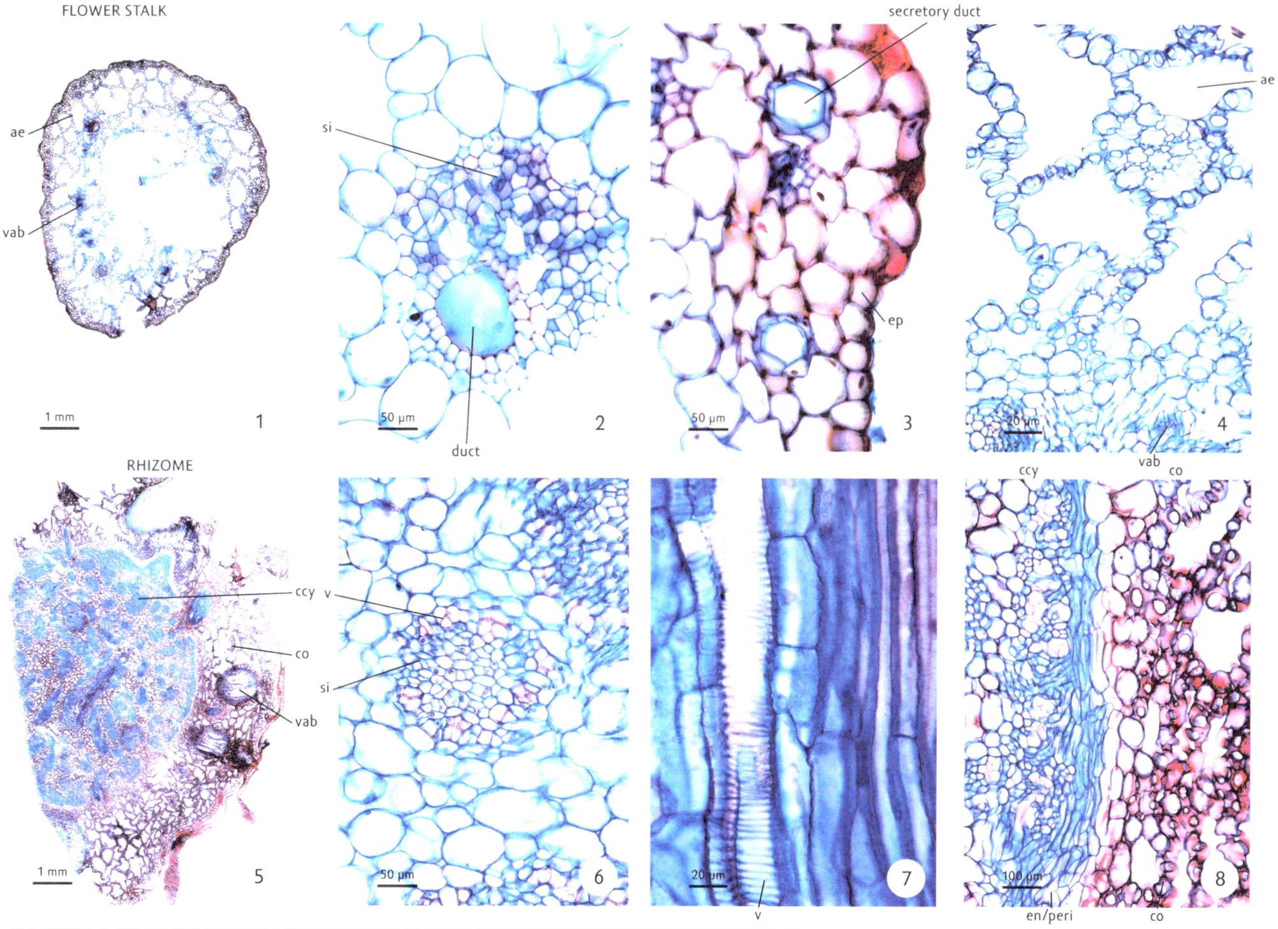

Sagittaria sagittifolia **Arrowhead** Alismataceae

Perennial emergent, amphibious or submerged plant, stoloniferous. Several leaf forms occur according to the developmental stage and environmental conditions. The first leaves at the stem base are ribbon-shaped; the next are spathulate; further leaves have heart-shaped bases and, finally, typical arrow-shaped leaves are formed. Inflorescence is a raceme usually with three unisexual flowers at each node. The bottom whorls of inflorescence consist mostly of female flowers, the upper whorls of male flowers. Tripetalous flowers have a white colour with purple bases of the petals. Fruit is an achene producing typical sphaerical syncarps.

Distribution and ecology

Euro-Siberian boreo-temperate element. Native to all CE countries. Occurs in oxbows, pools, fishponds, canals and drainage ditches, also in slowly flowing rivers from lowlands to uplands. Submerged plants possess only ribbon-shaped leaves, long linear floating leaves occur in rapidly flowing rivers. Terrestrial form known on drained bottoms. Prefers meso- to eutrophic waters with muddy bottoms. Hydrochorous, seeds germinate after a long period of dormancy. Overwintering and vegetative reproduction by tubers produced on stolons in the leaf axils.

Anatomy

Obtusely triangular culm with a diameter of 4 mm (1).

The epidermis consists of roundish, thin-walled cells (1, 2). The chlorenchyma is composed of several layers of round unlignified parenchyma cells (2). The cortex is formed of large intercellulars and vascular bundles (2). Vascular bundles consist of a xylem with a few large lateral thin walled vessels, a large intercellular and a phloem with sieve tubes and companion cells (3). Vascular bundles are surrounded by fibers (4).

Rhizome not available.

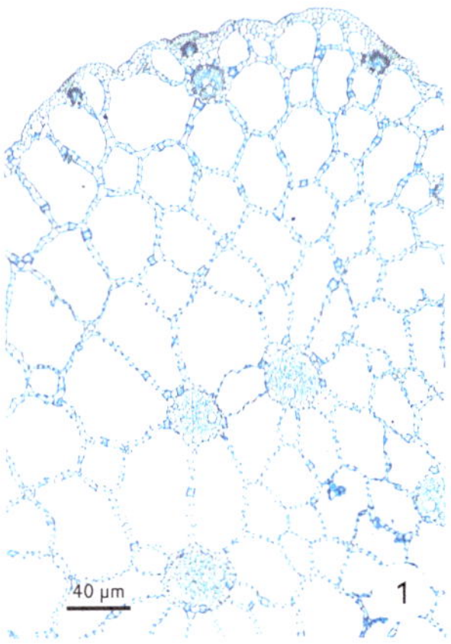

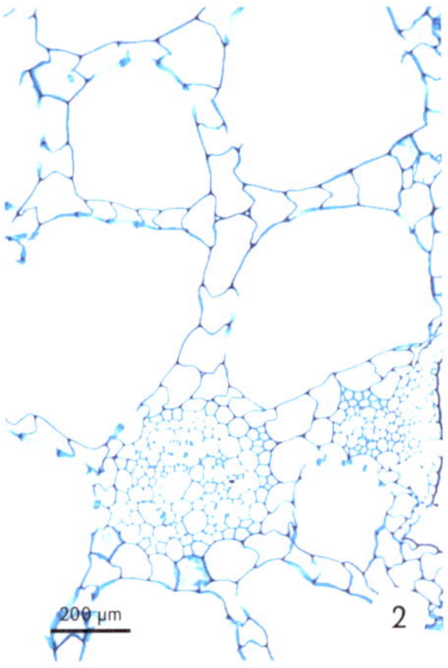

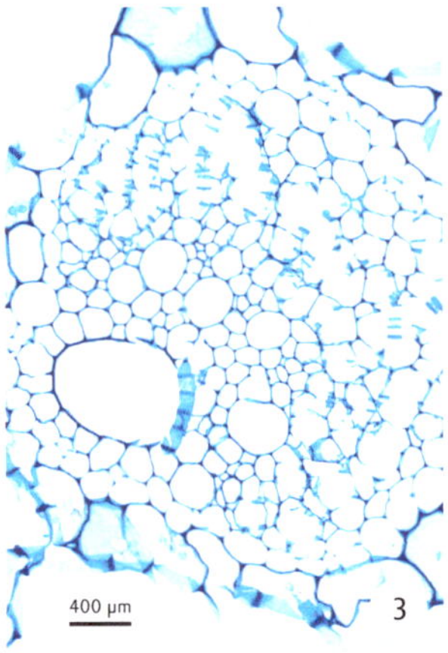

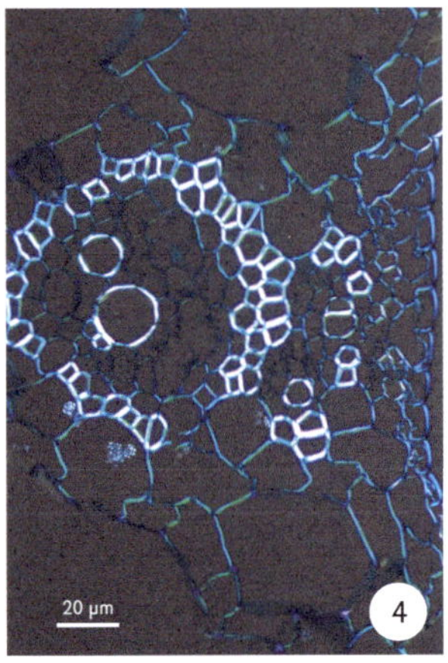

Perennial wetland bulbous herb forming clumbs, stems 30–60 cm high, compressed, narrowly double-winged. Umbel with 2–7 white flowers each with green patch near the apex.

Distribution and ecology

European southern-temperate element. On the northern spread, it reaches CZ and SK, native also in CH, AT, HU, introduced in DE, absent in PL. Native mostly in winter-flooded riverside *Alnus or Salix* carrs on alluvial soils with high nitrogen levels. Pollen limitation of reproduction in low-density populations. Optimal germination temperature 20–25 °C. Endangered due to destruction or alteration of its habitats and the harvesting of its bulbs for medical purposes (alkaloids).

Anatomy

Biangular oval flower stalk with a diameter of 2–3 mm (1).

The flower stalk consists of a parenchymatic tissue with angular intercellulars and some vascular bundles (1). Below the epidermis, a layer of palisade-like cells occurs (4).

Vascular bundles contain a few large air ducts, a few vessels (2, 3) and a phloem with a distinct separation of sieve tubes and companion cells (2). The external wall of the epidermis is unlignified and thick-walled (4). The parenchymatic tissue contains a few very long styloids (5).

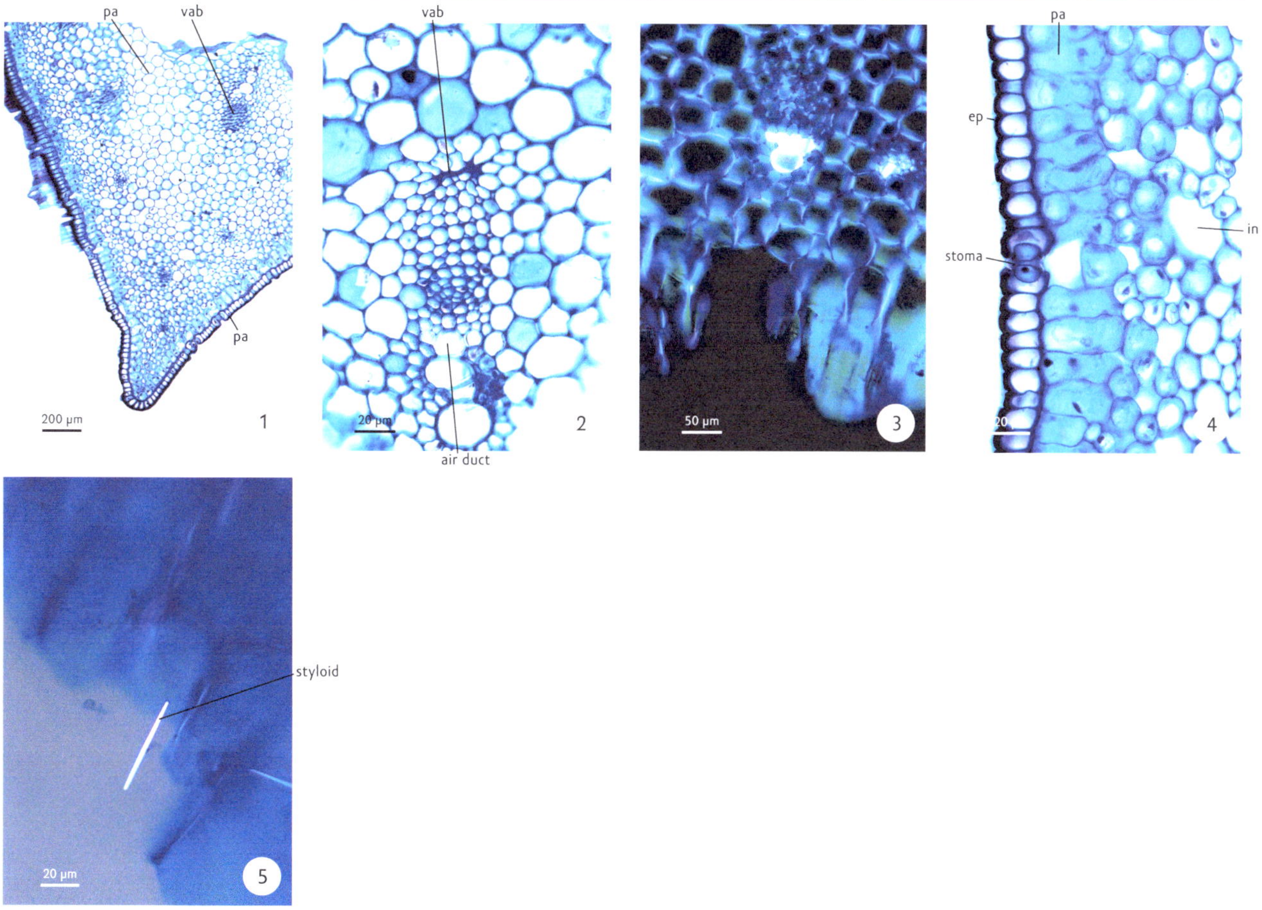

F. H. Schweingruber et al., *Anatomic Atlas of Aquatic and Wetland Plant Stems*, https://doi.org/10.1007/978-3-030-33420-8_58

Calla palustris L. **Bog Arum** Araceae

Perennial emergent herb, rhizomatous, leaves heart-shaped, with long petiole, shinning. Flowers mostly bisexual, inflorescence spadix, spathe white or greenish-white. Fruit is a red berry with several seeds.

Distribution and ecology

Circumpolar boreo-montane distribution, absent as a native from the Oceanic zone in Europe. In all CE countries, usually as floating mats at eutrophic margins of lakes and fishponds, or rooted in marshes and wet *Alnus* swamps. Reproduction vegetative by rhizome fragments, seeds eaten by ducks. Often in shade, also grown as ornamental plant.

Anatomy

Circular flower stalk with a diameter of 6–8 mm (1) and rhizome with 10–15 mm in diameter (2).

Both parts consist of an aerenchymatic tissue with vascular bundles. A cortex is absent (1, 2). Vascular bundles are distributed regularly on cross-sections (2). An arc-like xylem surrounds partially or totally the phloem, which consists of sieve tubes and companion cells. Air ducts within vascular bundles are absent. The ground tissue consists of a reticulate aerenchymatic tissue (1–3). Epidermis cells are covered with a thick cuticle (4). Note the presence of slime producing secretory cells (5) and raphids (6).

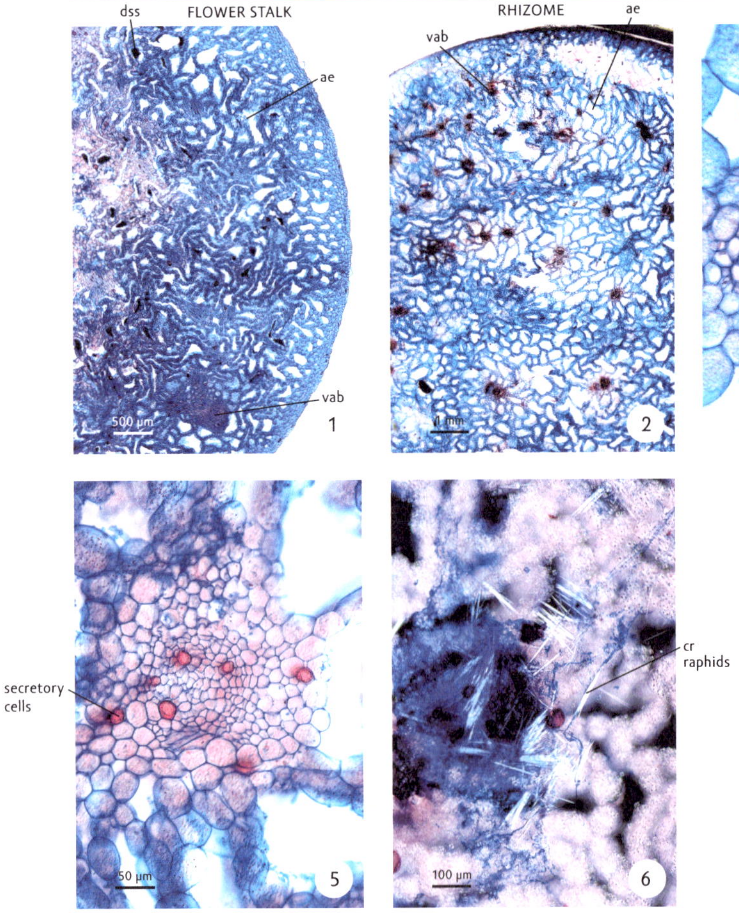

F. H. Schweingruber et al., *Anatomic Atlas of Aquatic and Wetland Plant Stems*, https://doi.org/10.1007/978-3-030-33420-8_59

Floating-leaved, roundish or obovoidal fronds usually swollen on lowerside (not in autumn-produced fronds, nor in starved plants) with larger air-spaces visible as a reticulum on frond lowerside. One unbranched root. Only rarely flowers in nature.

Distribution and ecology

European southern-temperate element; also in C Asia and N America. In all CE countries, usually in the zone of floating vegetation of stagnant waters in fishponds, ditches and canals, prefers eutrophic waters, tolerates brackish waters. It reproduces by vegetative budding, probably it flowers and sets seeds under unfavourable conditions, such as increasing salt concentrations and/or drying up of a waterbody. Widely used in ecotoxicological studies.

Anatomy

The poorly differentiated vegetative body consists primarily of unlignified parenchymatic cells. Characteristic are the large air pockets at the ventral side (1). The flat top-layer (epidermis and hypodermis) consists of extremely thin-walled irregularly formed cells (2). The air lacunes have basically the same structure but contain large intercellular spaces (3). Vascular bundles are absent (1–3) or extremely rare. Raphids are embedded in large parenchyma cells (4).

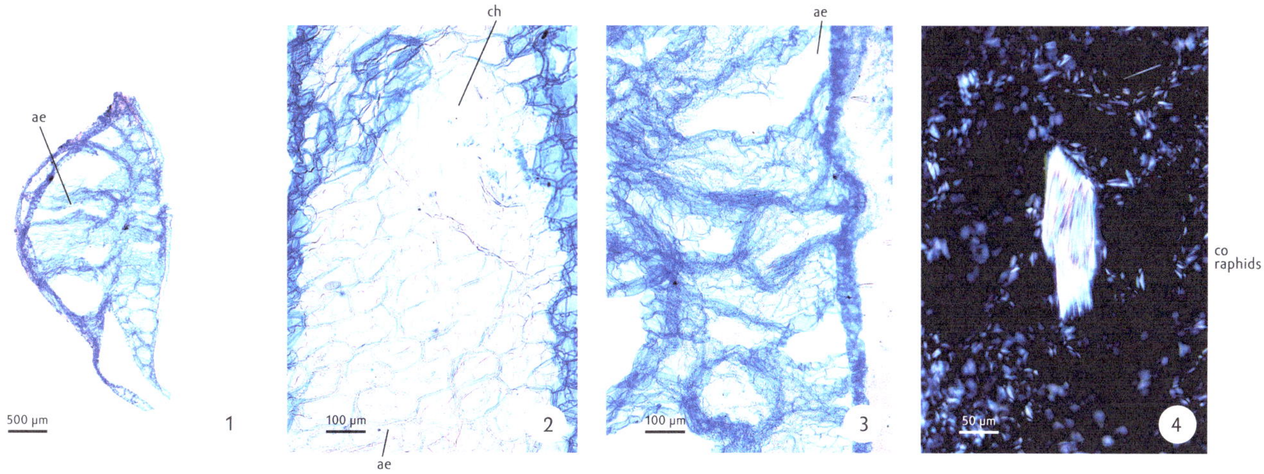

Perennial free-floating plant, very reduced body, leaves and stems modified in fronds, light-green fronds 1.5–4 mm large, widely ovate, flattened on both sides, air spaces only <0.3 mm across, no or one unbranched root 5–80 mm long, no morphologically distinct turions, only physiologically adapted winter fronds, terrestrial form known.

Distribution and ecology

Circumpolar southern-temperate element; widely naturalized outside its native range. Very common in all CE countries, from lowlands to mountains. Usually margins of fishponds, lakes, backwater pools and oxbows, fen lakes, reed stands, swamps, slowly flowing rivers, canals, ditches. Tolerates meso- to hypertrophic, slightly acidic to alkaline (pH ca. 6–9) and dystrophic waters, under favourable conditions very rapid propagation, winter fronds survive the inclusion in ice.

Anatomy

The poorly differentiated vegetative body consists primarily of unlignified parenchyma cells.

Strips of parenchyma cells surround large intercellular spaces (1). Denser parts contain very few small rudimentary vascular bundles (2). Their central small cells are probably sieve tubes and the unlignified, thin-walled larger surrounding cells probably vessels (2). The top layer of the body is covered by small, polymorph epidermis cells (3). Raphides are embedded in large cells (4).

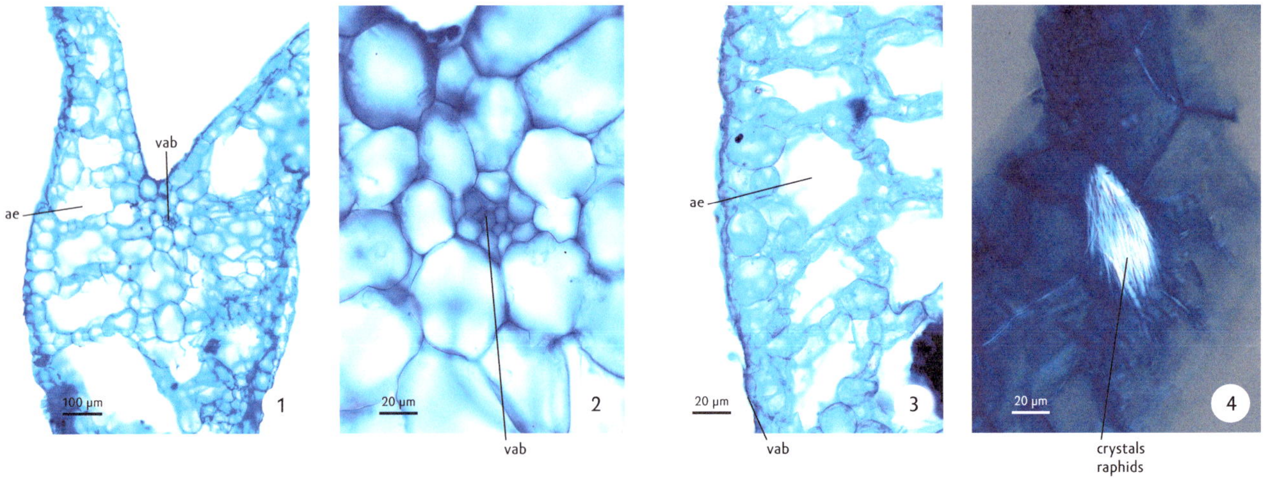

Lemna minuta Kunth

Least Duckweed

Araceae

Perennial free-floating tiny herb. Very reduced body, leaves and stems modified in fronds. Very similar to native L. minor, but fronds are characterized by 0–1 central vein and length <2 mm. Frond shape depends on growth conditions – a single green translucent oval body is formed in the shade but it generally grows in pairs in full sunlight. It produces an ephemeral membrane-bound flower.

Distribution and ecology

Native to temperate and subtropical N and S America; now widely naturalised in Europe. A neophyte already reported from some CE countries (DE, PL, HU) from lakes, fishponds, slowly flowing rivers, canals and ditches. Partly shade-tolerant. Reproduction is by vegetative budding – it belongs to the smallest and fastest growing flowering plants, often forming mats on the water surface.

Anatomy

The vegetative body is almost unstructured (1). The epidermis consists of small unlignified cells and the whole inner part of the body remains unstained (1, 2). Cells and vascular bundles cannot be recognized. Raphids occur in bundles (2).

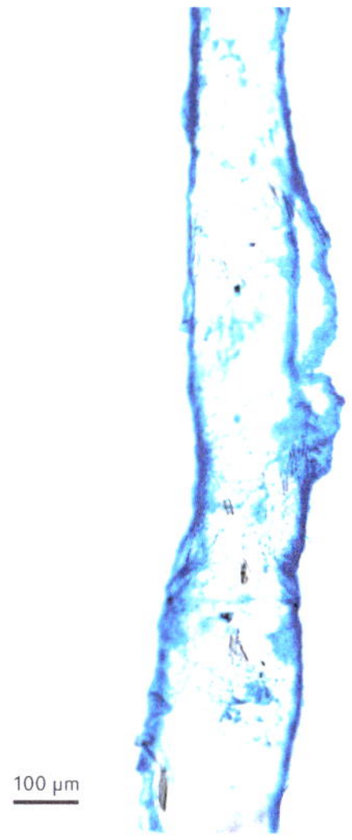

1

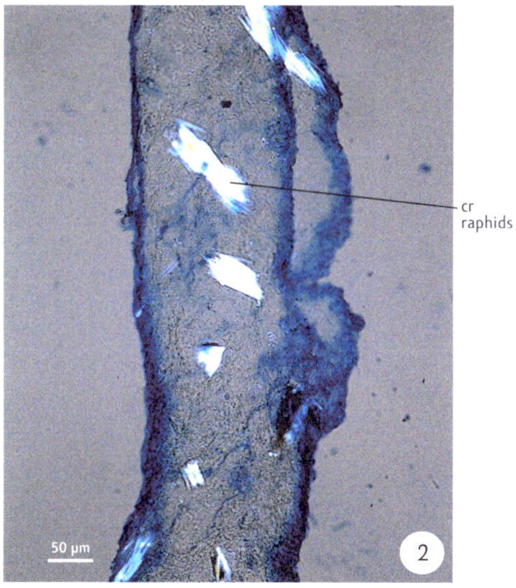

2

Perennial submerged or amphibious plant, reduced body, thin, translucent, flat petiolate fronds 5–10 mm large, no or one root. Connected fronds form larger colony, no turions, terrestrial form known.

Distribution and ecology

Circumpolar temperate element. Scattered irregularly but frequently in all CE countries, from lowlands to uplands. More frequent in fishpond and river floodplain areas, margins of fishponds and lakes, fen lakes, river floodplains, forest pools, soft and hard waters, usually dystrophic waters, pH ca. 5.5–8, facultative sciophyte, in hard waters, HCO_3^- use for photosynthesis.

Anatomy

The poorly differentiated vegetative body consists primarily on unlignified parenchyma cells, with large intercellular spaces (1, 2). Denser parts contain a very few small rudimentary vascular bundles (2). The top layer of the body is covered by small, polymorphic epidermis cells (2). Styloids are embedded in large cells (3).

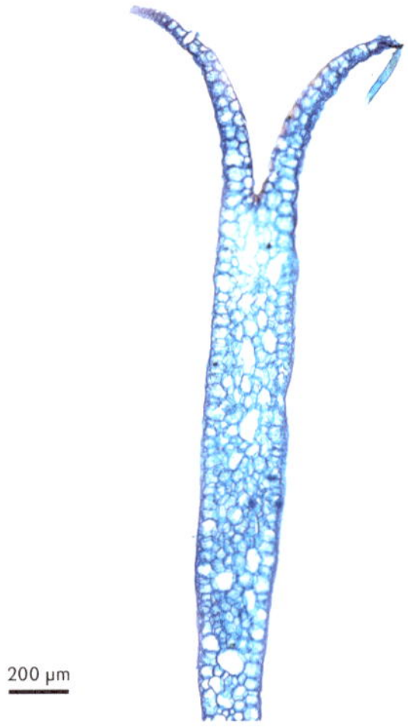

1

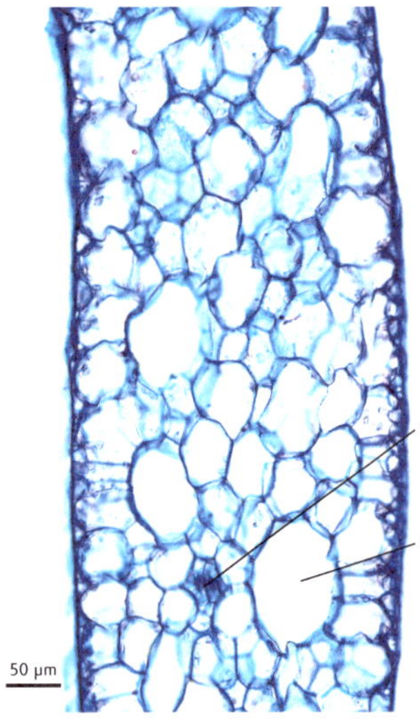

2

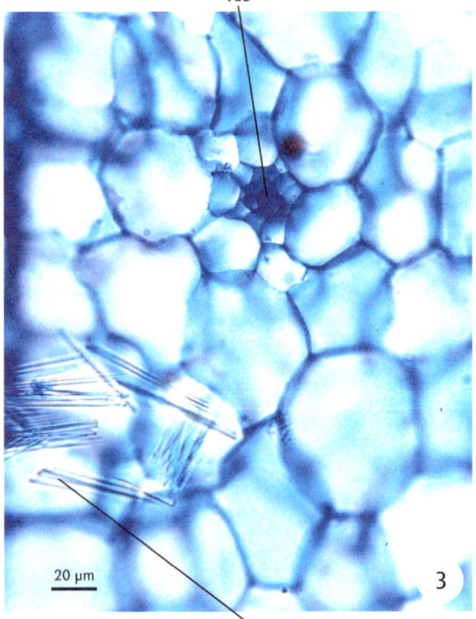

3

Spirodela polyrhiza (L.) Schleiden

Greater Duckweed Araceae

Perennial free-floating plant, reduced body, leaves and stems modified in light-green fronds, 4–10 mm large, nearly circular, flattened on both sides, dark red on lower side, several roots 5–30 mm long, circular flat turions 1.5–2 mm large, terrestrial form known.

Distribution and ecology

Circumpolar southern-temperate element. Scattered irregularly but frequently in all CE countries, from lowlands to uplands. More frequent in fishpond and river floodplain areas. Fishponds, lakes, backwater pools and oxbows, canals and ditches, tolerates slightly acidic, soft to alkaline, hard waters (pH ca. 6–9) and dystrophic, meso- to eutrophic (hypertrophic) waters. Very rapid propagation, facultative sciophyte, turions sink to the bottom during winter, float up in the spring pulled by air bubble.

Anatomy

The vegetative body consists primarily of a flat poorly differentiated parenchymatic top layer (1, 2) and a basal node of merging roots (1, 6). Large intercellular spaces characterize the upper part of the body. Vascular bundles do not exist but a few isolated vessels with ring-like stabilized cell walls have been observed (3, 7). The top layer of the body is covered by square epidermis cells (4). Crystal druses (8) and dark-staining substances in single cells (4, 5) as well as raphides are frequent (5).

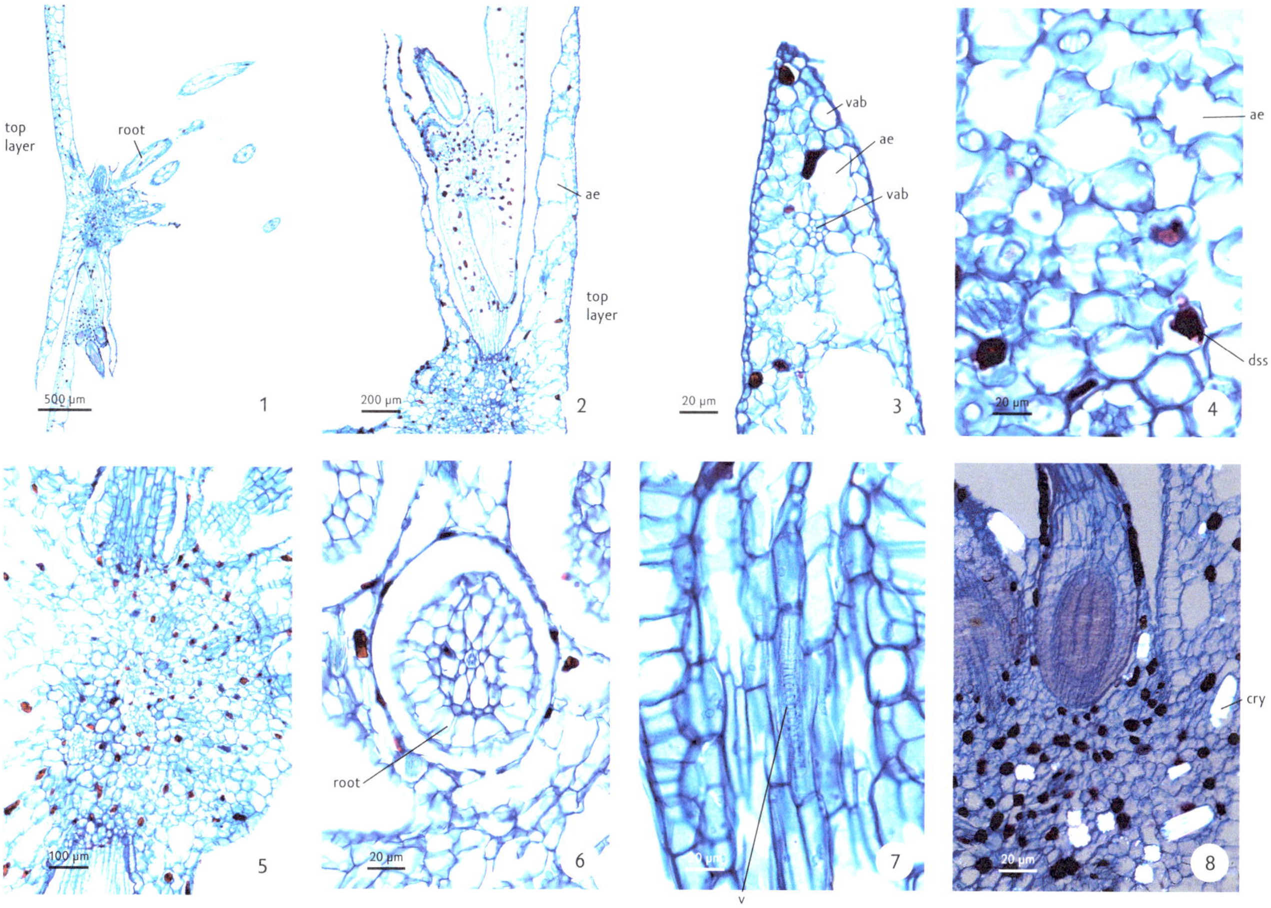

Butomus umbellatus L.

Flowering-rush Butomaceae

Perennial wetland, amphibious plant, erect flower stems 0.6–1.5 m tall, leaves usually slightly shorter. Linear erect leaves grow from rhizome in a basal rosette, leaves triangular in cross-section. Pink or purplish flowers in terminal umbel, submerged form known.

Distribution and ecology

Euro-Siberian temperate element; widely naturalized outside its native range. Scattered irregularly to frequently in all CE countries, lowlands to uplands. Shallow standing and slow-streaming rivers, fishponds, lakes, ditches, canals, backwater pools, sometimes introduced. Prefers shallow waters (10–20 cm), tolerates full submergence. Prefers nutrient-rich soils but not strongly eutrophic habitats. Facultative heliophyte.

Anatomy

Circular flower stalk with a diameter of 7 mm (1).

A distinct aerenchymatic cortex surrounds a large central cylinder. The central part of the flower stalk consists of a net-like aerenchyma containing few vascular bundles and a dense parenchymatic, peripheral part with many circular arranged vascular bundles (1, 4).

A centripetal air duct dominates vascular bundles (2). The xylem of well-developed bundles consists of arc-like arranged vessels and parenchyma cells and the phloem group is composed of a large sieve tubes and small companion cells (2). The xylem is absent in poorly developed vessels. The cortex is characterized by irregular small and large intercellulars (3). Crystals are absent.

Round rhizome with a diameter of 10–12 mm (5).

Loosely dispersed collateral vascular bundles form the inner part of the rhizome (5). Poorly lignified groups of vessels and unlignified parenchyma cells form the xylem (6). The phloem consists of groups of sieve tubes and companion cells (6). Vessel walls have scalariform intervessel pits or ring link structures. Air ducts are absent. The boundary between the cortex the central part is characterized by a layer of cells with prismatic crystals (7). The cortex consists of round and oval cells with tangentially large intercellulars (8). A few vascular bundles dles occur in the cortex (5). Crystals are absent.

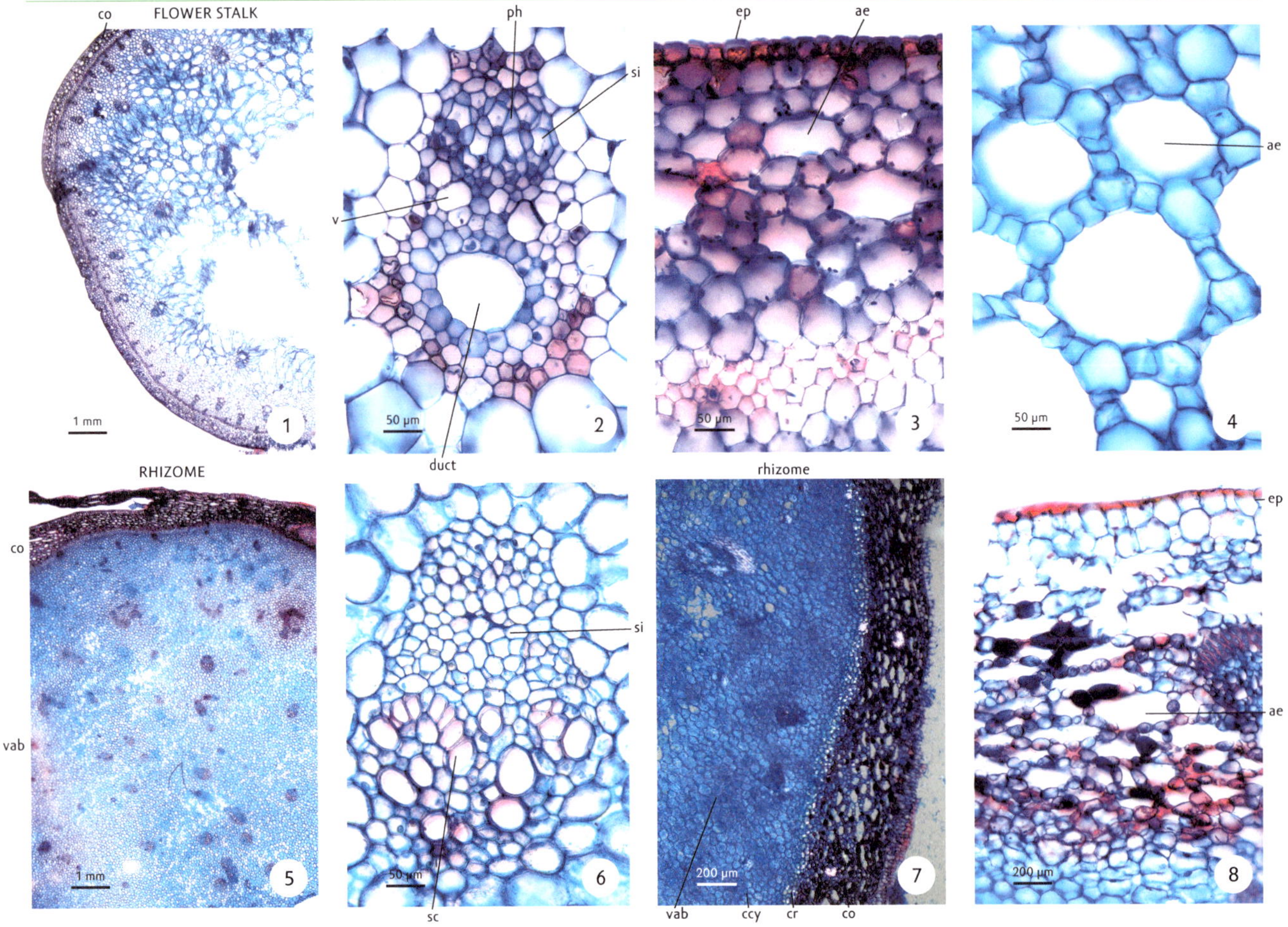

F. H. Schweingruber et al., *Anatomic Atlas of Aquatic and Wetland Plant Stems*, https://doi.org/10.1007/978-3-030-33420-8_60

Blysmus compressus (L.) Link

Flat-sedge Cyperaceae

Perennial wetland herb, stem erect, 5–40 cm high, single or tufted, roundish. Leaves flat to slightly keeled, grass-like. Inflorescence a flattened compact terminal spike, spikelets usually 10–20, reddish-brown, perianth bristles longer than nut.

Distribution and ecology

European temperate element, with a continental distribution in W Europe, also in C Asia. In all CE countries, on fen meadows, moist pastures, springheads and saline grounds. Usually in open areas, in short, sedge-rich grasslands which are subject to flooding. Declining due to drainage, eutrophication and the cessation of grazing.

Anatomy

Circular culm with a diameter of 1–2 mm.

The epidermis consists of roundish, externally lignified thick-walled cells (1–3). Guard cells of the stomata are thickened on the external and internal side (3). The chlorenchyma is composed of several layers of round unlignified parenchymatic cells (1, 2) and obtusely triangular groups of sclerenchyma (1–3). The sclerenchymatic sheath of a few vascular bundles is linked with the peripheral sclerenchymatic strands (1–3). The middle and inner parts of the culm contains round lignified parenchymatic cells, circular arranged vascular bundles and large oval intercellulars (1, 2). The large vascular bundles consist of a xylem with two slightly lateral vessels, an irregularly formed intercellular and a roundish phloem with sieve tubes and companion cells (2). Vascular bundles are surrounded by 1–3 layers of sclerenchymatic cells (1, 2). Secretory cells contain dark staining substances (2).

Rhizome with a diameter of 1–2 mm.

The cortex is composed of an epidermis, an external layer of lignified round parenchymatic cells and an inner part of thin-walled unlignified parenchymatic cells with large radial intercellulars (4). The central cylinder and the cortex are separated by a thick-walled lignified endodermis with U-shaped cells (5). The lignified parenchymatic ground tissue of the central cylinder contains amphivasal vascular bundles, with small sclerenchymatic sheaths (6).

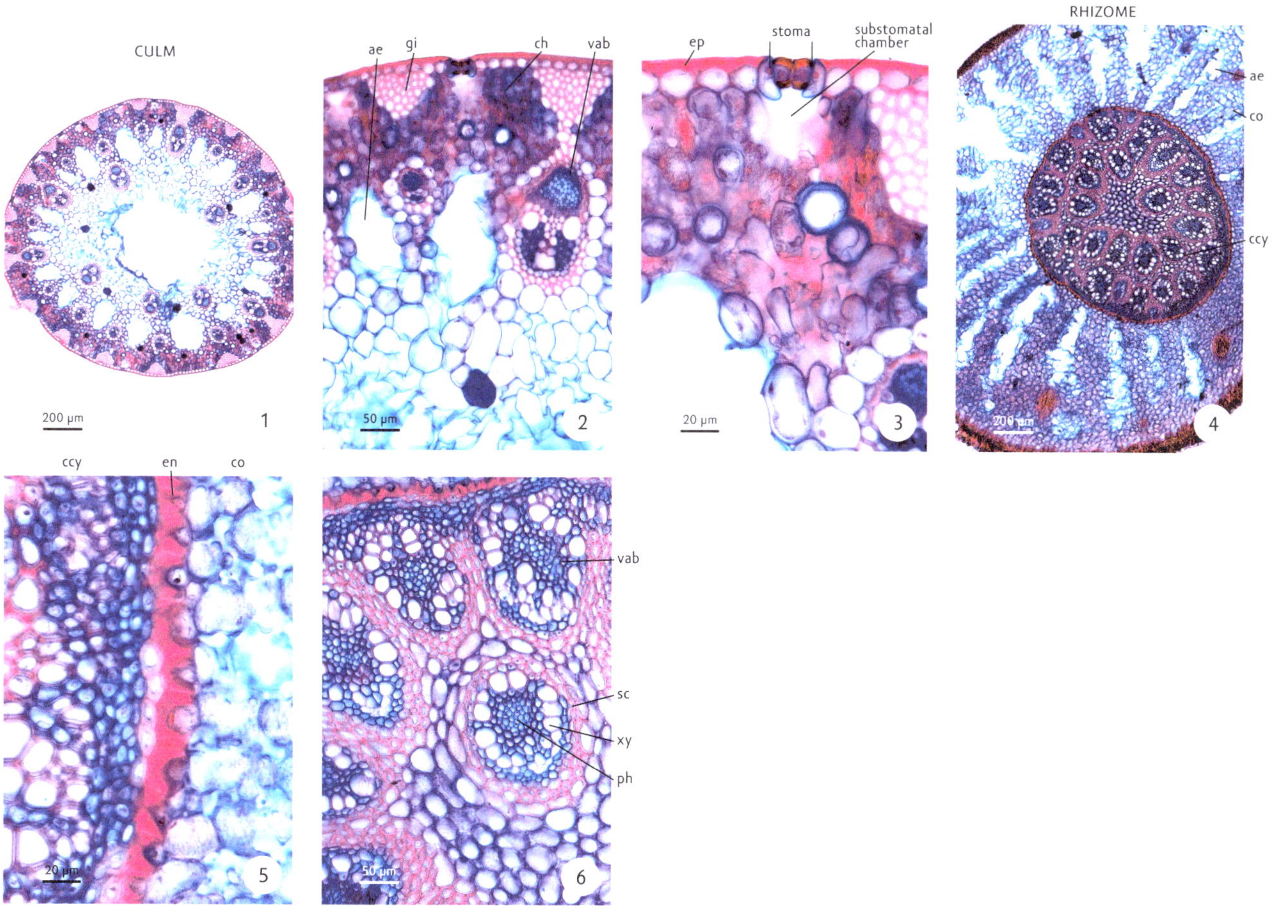

F. H. Schweingruber et al., *Anatomic Atlas of Aquatic and Wetland Plant Stems*, https://doi.org/10.1007/978-3-030-33420-8_61

Bolboschoenus maritimus (L.) Palla

(syn. *Scirpus maritimus*)

Sea Club-rush

Perennial wetland, rhizomatous herb with tubers. Stem sharply triangular, leafy, 0.7–1 m high. Inflorescence either head-like, consisting only of clustered, sessile spikelets, or formed by a central group of sessile spikelets and of 1–2 rays each bearing 1–4 spikelets. Flat-convex or biconvex fruit with exocarp ± 2 times thicker than sclerenchymatic mesocarp.

Distribution and ecology

Euro-Siberian southern-temperate element; widely naturalised outside its native range. In CE it is relatively frequent along sea coast and brackish habitats in river estuaries (DE, PL), in saline habitats around great lakes (AT, HU), rare in other inland saline wetlands (CZ, SK). Facultative halophyte. Fruits are buoyant. Air-filled exocarp cells enable plants to spread by floods.

Anatomy

Acutely triangular culm with a diameter of 2–3 mm.

The epidermis consists of roundish, externally lignified thick-walled cells (3). Guard cells of the stomata are thickened at the external and internal side (3). The chlorenchyma is composed of several layers of round unlignified parenchyma cells (1, 2) and round to obtusely triangular groups of sclerenchyma (1–3). The middle and inner part of the culm consist of round, thin-walled unlignified parenchyma cells, vascular bundles and large, square to irregularly formed intercellulars (1, 2). The large vascular bundles consist of a xylem with two large lateral vessels, an irregularly formed intercellular and a roundish phloem with sieve tubes and companion cells (2). Vascular bundles are surrounded by a radial bilateral sclerenchyma sheath with 1–4 layers of cells (1, 2). Secretory cells contain dark stained substances (2).

Rhizome with a diameter of 10–12 mm.

The slide of the cortex is missing. The central cylinder and the cortex are separated by a centripetal thick-walled layer of fibers (endodermis-like). The unlignified parenchymatic ground tissue of the central cylinder contains collateral vascular bundles, surrounded by sclerenchyma sheaths (4, 5). Vascular bundles are dispersed on the whole central cylinder. They consist of xylem with arc-like distributed vessels and a round group of phloem. Vascular bundles are surrounded by a 2–4 layered sheath of fibers (5).

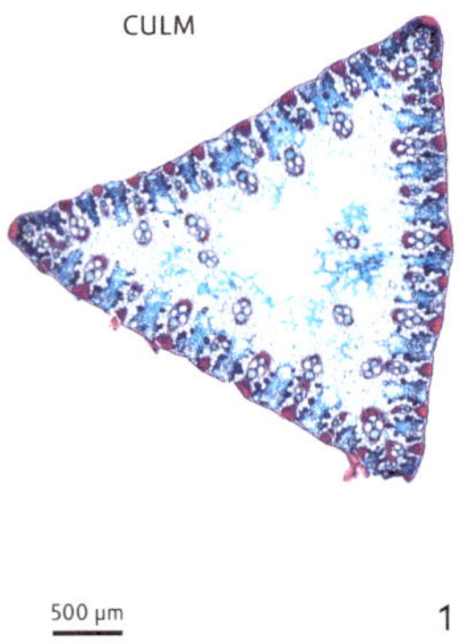

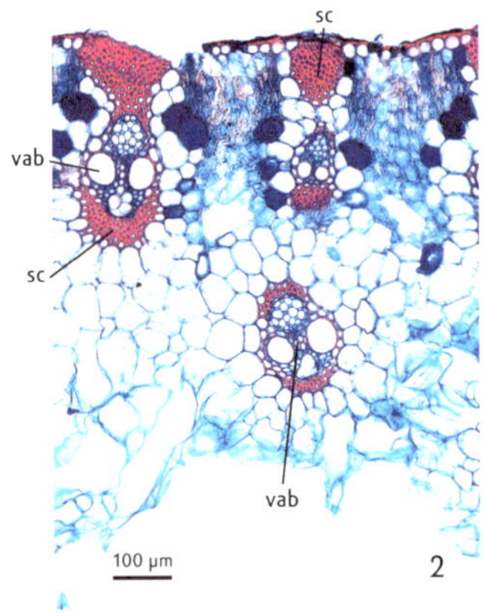

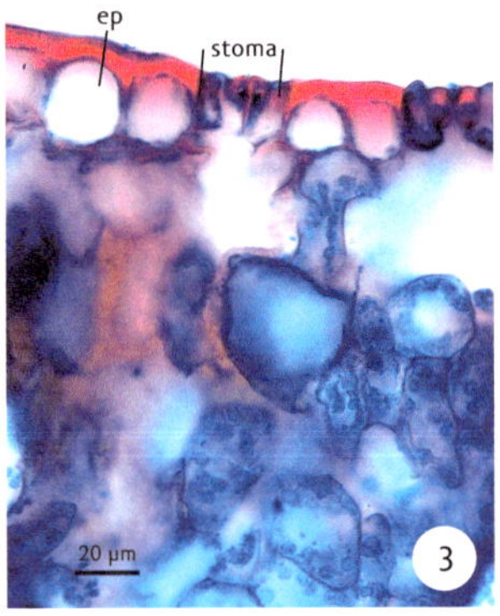

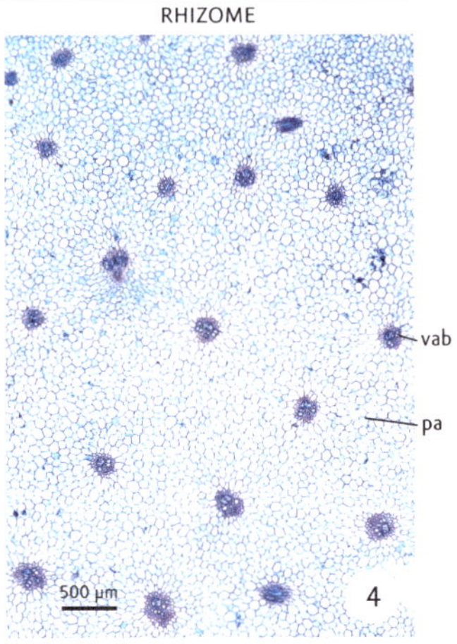

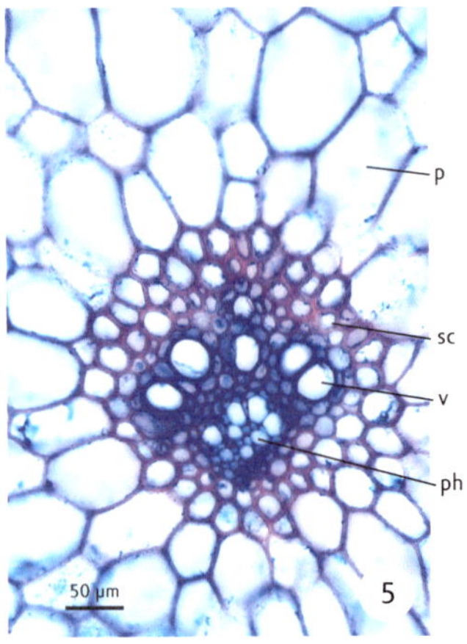

Bolboschoenus planiculmis

Perennial wetland, rhizomatous herb with tubers. Stem sharply triangular, leafy, 0.5–0.9 m high. Inflorescence head-like consisting of only sessile spikelets or poorly branched. Fruit is an achene, concave in cross section with exocarp thicker than mesocarp.

Distribution and ecology

Euro-Siberian southern-temperate element; widely naturalised outside its native range. In CE it is relatively frequent along sea coast and brackish habitats in river estuaries (DE, PL), in saline habitats around great lakes (AT, HU), rare in other inland saline wetlands (CZ, SK). Facultative halophyte. Fruits are buoyant. Air-filled exocarp cells enable plants to spread by floods.

Anatomy

Obtusely triangular culm with a diameter of 4–6 mm.

The epidermis consists of thin-walled roundish cells (1). The chlorenchyma is composed of a hypodermis with several layers of angular lignified parenchyma cells (4). The middle part consists of large square air spaces, which are loosely filled with stellate cells (2). The inner part consists of lignified, elongated cells around air spaces (2, 4). Vascular bundles consist of a xylem with 2–3 large lateral vessels and a roundish phloem with sieve tubes and companion cells and an air duct (2). Vascular bundles are surrounded by 1–3 layers of sclerenchyma cells (1, 2). The central part is characterized by very thin-walled, unlignified parenchyma cells and a few vascular bundles (1, 2).

Rhizome with a diameter of 8–12 mm.

The cortex is composed of an epidermis, an external layer of lignified parenchyma cells and an inner part of thin-walled, unlignified parenchyma cells without air spaces (5). It contains vascular bundles (5, 7). The central cylinder and the cortex are separated by a centripetal thick-walled layer of fibers (endodermis-like) (6). The lignified parenchymatic ground tissue of the central cylinder contains collateral vascular bundles, surrounded by sclerenchyma sheaths. Collateral vascular bundles are dispersed on the whole central cylinder. They consist of xylem with arc-like distributed vessels and a round group of phloem. Vascular bundles are surrounded by a 2–4 layered sheath of fibers. Secretory cells are frequent (7).

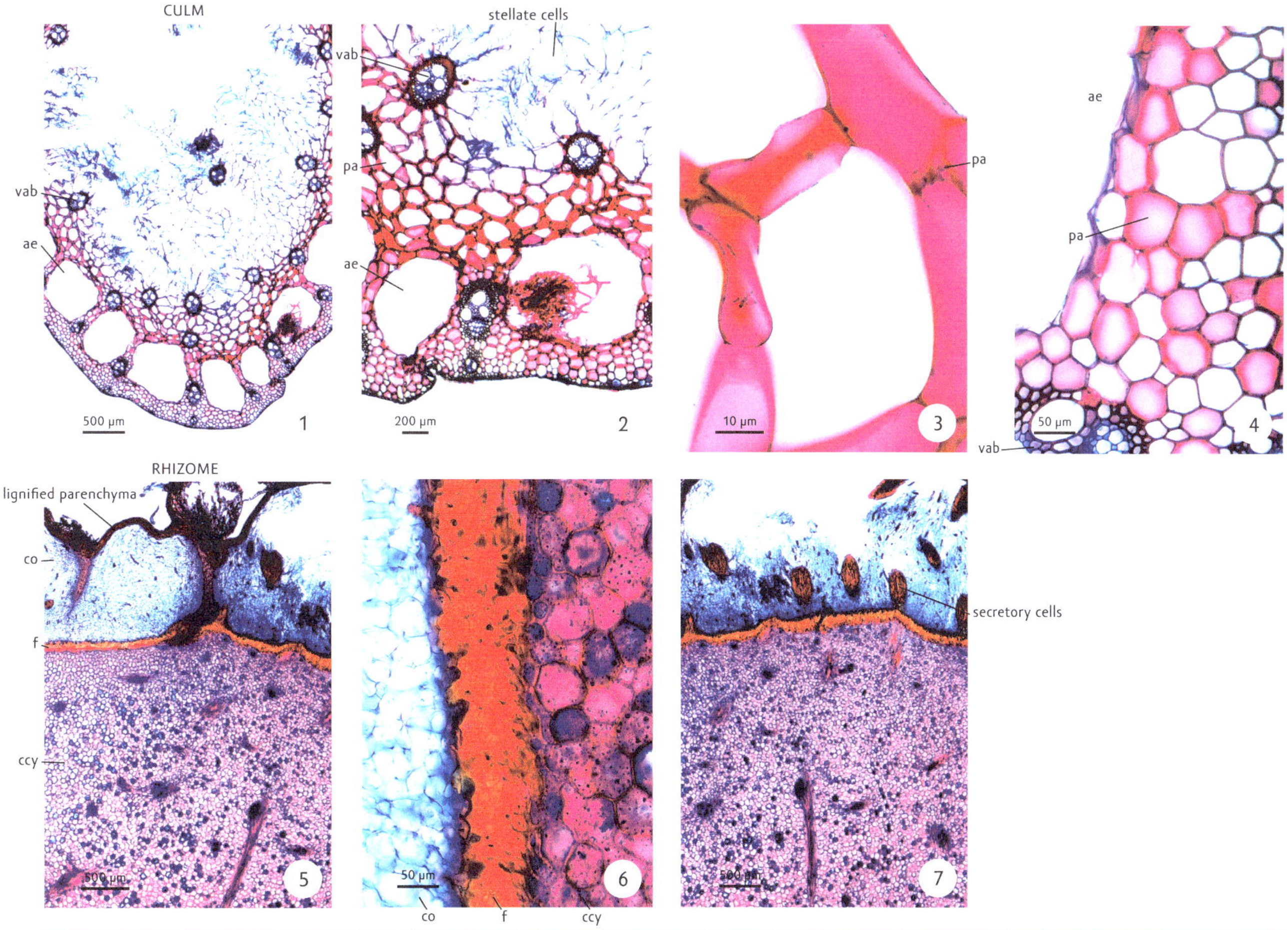

Bolboschoenus yagara

Perennial wetland, rhizomatous herb with tubers. Stem sharply triangular, leafy, 0.8–1.3 m high. Inflorescence branched, branches bearing usually 1-3 spikelets. Fruit is a narrow achene, trigonous in cross section with very thin exocarp.

Distribution and ecology

C Europe and E Asia. Probably in all CE countries, scattered irregularly in several fishpond basins (CZ, SK) from uplands to mountains, lacking in lowlands. It grows in fishpond littorals, only occasionally in other reservoirs, oxbows or secondary habitats. Prefers base-poor, acidic grounds with sandy, sand-clayey to peaty sediments, but occurs also in meso- to eutrophic fishponds, together with *B. laticarpus*. Vulnerable to eutrophication and an increasing intensity of fishpond management.

Anatomy

Obtusely triangular, slightly wavy culm with a diameter of 5–6 mm (1, 2).

The epidermis consists of thin-walled roundish cells (3). The chlorenchyma is composed of a hypodermis with a small belt of small thick-walled fibers (3), several layers of angular lignified parenchyma cells (2). The middle part consists of large air spaces, which are loosely filled with stellate cells (1, 2). The inner part consists of lignified, elongated cells around large square intercellulars (2). The vascular bundles consist of a xylem with 2 large lateral vessels, a roundish phloem with sieve tubes and companion cells and an air duct (2). Vascular bundles are surrounded by 1–3 layers of sclerenchyma cells (1, 2). The central part of the stem is characterized by very thin-walled, unlignified parenchyma cells and a few vascular bundles (1). Secretory cells are absent (1, 2).

Rhizome with a diameter of 8–12 mm.

The cortex is composed of a thin walled epidermis and of thin-walled unlignified parenchyma cells without air spaces (4). The central cylinder and the cortex are separated by 2–3 layers of centripetal thick-walled lignified endodermis-like cells (5). The lignified parenchymatic ground tissue of the central cylinder contains collateral vascular bundles, with sclerenchyma sheaths (6). Scalariform intervessel pits characterize vessel-walls (7).

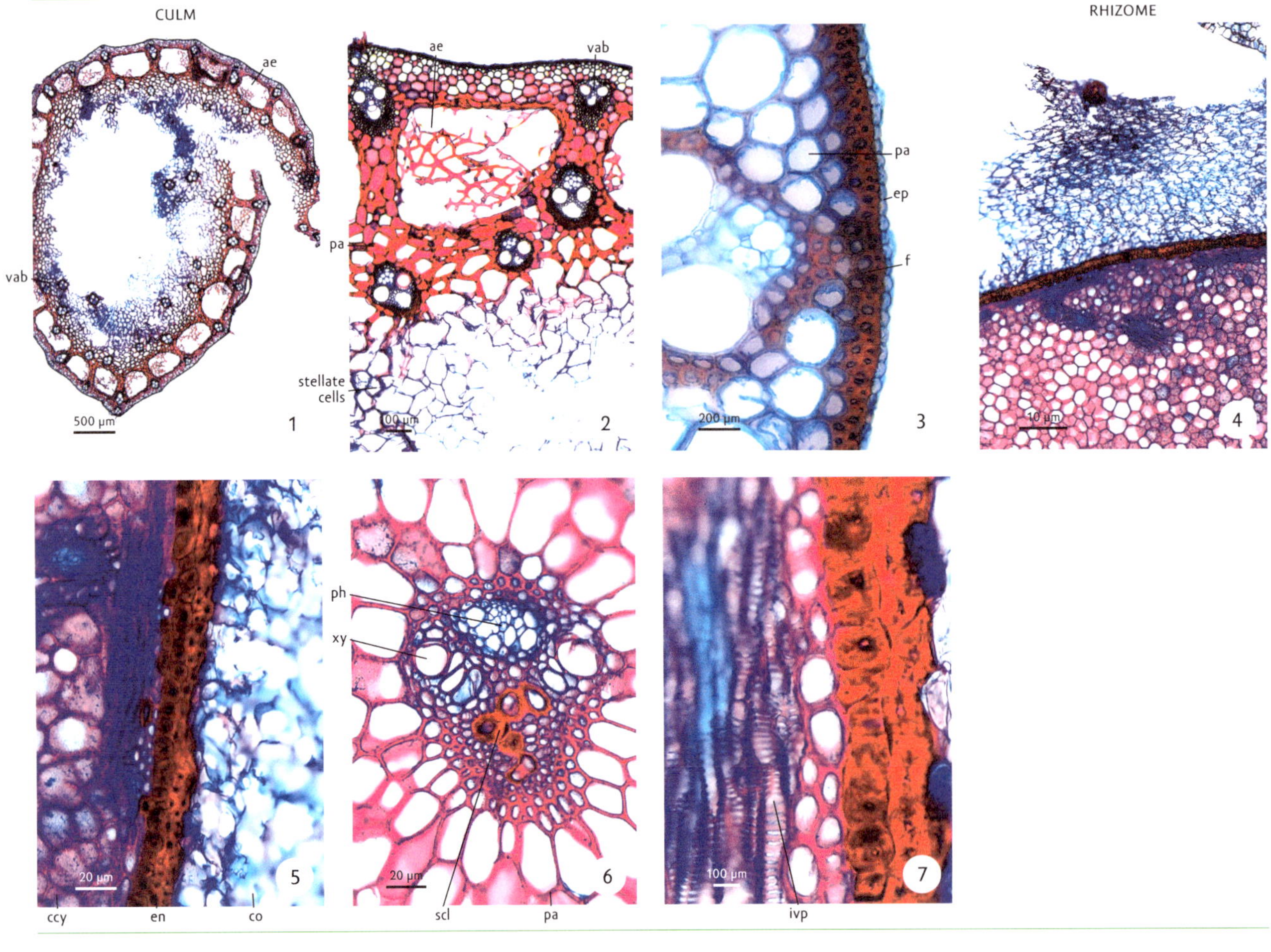

Perennial wetland, tufted, rhizomatous plant. Flower stems with distinctly acute angles, rough distally, 0.3–1.4 m tall, leaves dark green to grey-green.

Distribution and ecology

Euro-Siberian boreo-temperate element. Scattered commonly in all CE countries, from lowlands to uplands, frequently in peatland, fishpond or lake areas. Shorelines of lakes and fishponds, margins of peat bogs and fens, river banks, floodplains of rivers, wet meadows, canals, ditches, swamps, peaty, silty or sandy soils. Tolerates strongly acidic to neutral soils (pH ca. 4–7) and eutrophication, relatively eurytopic species.

Anatomy

Acutely triangular culm with a diameter of 2.5–3 mm (1).

The epidermis consists of bulliform, lignified cells (2). The chlorenchyma consists of external layer of unlignified parenchymatic cells. The middle part is dominated by large air spaces with spongy remains of cell walls (2) and collateral vascular bundles. Their fiber sheaths are externally linked with the epidermis with rectangular girders of lignified cells. The vascular bundles consist of a xylem with two large lateral vessels and a roundish phloem with sieve tubes and companion cells and an air duct (1, 2). Vascular bundles are centripetally surrounded by 3–4 layers of thin- to thick-walled sclerenchymatic cells (1, 2). The pith consists of unlignified thin-walled polygonal parenchymatic cells (1).

Rhizome with a diameter of 2.5–4 mm (3).

The cortex is composed of an external layer of parenchymatic cells with a few irregularly formed intercellulars. Few vascular bundles exist in the cortex (3). Both the central cylinder and the cortex are separated by a thick-walled endodermis with lignified round cells (4). The unlignified parenchymatic ground tissue of the central cylinder contains amphivasal vascular bundles with large sclerenchymatic sheaths (5).

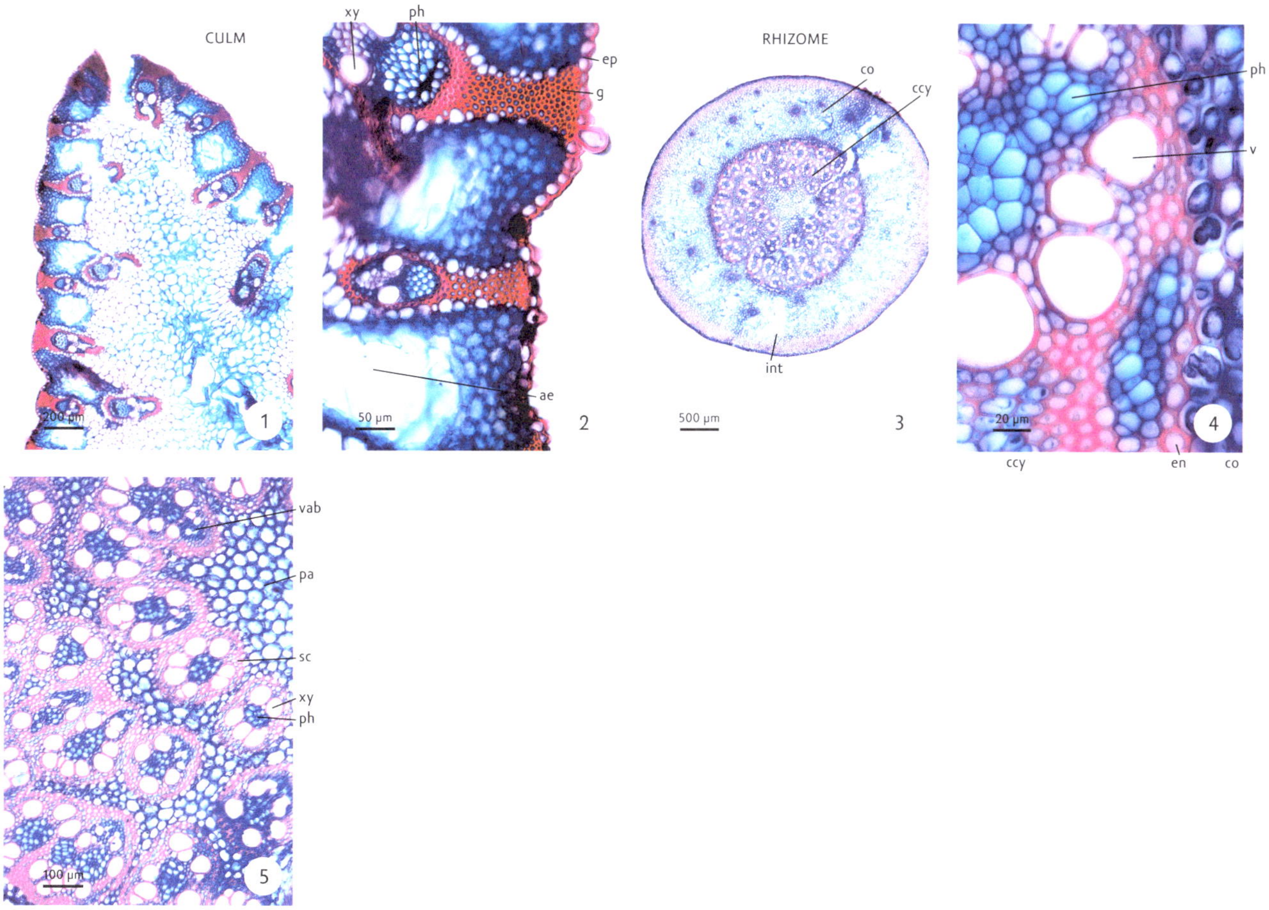

Carex acutiformis Ehrh.

Lesser Pond-sedge

Perennial wetland herb, robust, rhizomatous, long creeping tillers, forming extensive greyish stands. Stems erect, 40–140 cm high, sharply triangular, rough, basal sheaths scale-like, brown or reddish brown. Leaves 40–170 cm long, greyish-green or greyish, spikelets heteromorphic. Utricles oval-obovoid, greyish-green, apex gradually narrowed into short beak.

Distribution and ecology

Euro-Siberian temperate element, in all CE countries. In alder carrs, wet meadows and swamps, along ponds and streams. Mesotrophic to eutrophic sedge marshes, rare in highlands. Persistent seed bank, not germinating when buried at a soil depth of 7 cm.

Anatomy

Acutely triangular culm with a diameter of 3–4 mm (1).

The epidermis consists of bulliform, lignified cells (2, 3). The chlorenchyma consists of an external layer of unlignified parenchyma cells. The middle part is dominated by large air spaces with spongy remains of cell walls (2) and collateral vascular bundles. Their fiber sheaths are externally linked with the epidermis with girders of lignified cells. The vascular bundles consist of a xylem with two large lateral vessels, a roundish phloem with sieve tubes and companion cells and an air duct (3). Vascular bundles are surrounded by 1–3 layers of thin-to thick-walled sclerenchyma cells (3). The pith consists of unlignified thin-walled polygonal parenchyma cells (1).

Rhizome with a diameter of 5–6 mm (4).

The cortex is composed of an external layer parenchyma cells without intercellulars, subepidermal groups of sclerenchyma and an inner part of unlignified thin-walled parenchyma with many large, radial arranged intercellulars (4). The central cylinder and the cortex are separated by a thick-walled endodermis with lignified u-shaped cells (5). The unlignified parenchymateous ground tissue of the central cylinder contains amphivasal vascular bundles with large sclerenchyma sheaths (4, 6).

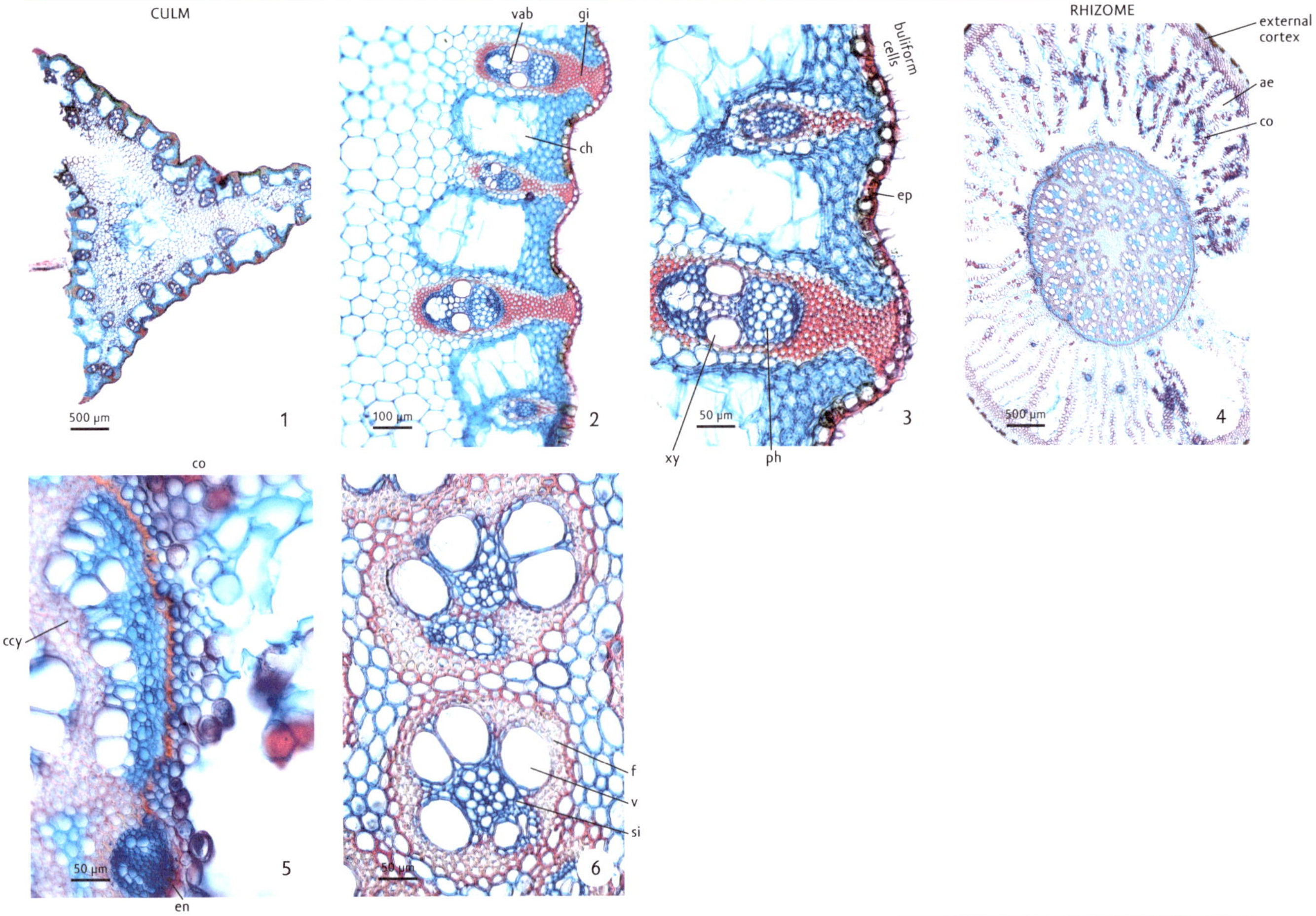

Carex appropinquata Schumacher **Fibrous Tussock-sedge** Cyperaceae

Perennial wetland herb, tufted, forming big tufts, stem rounded, basal part of stem with fibrous dark sheaths. Leaves 1–3 mm wide, old leaf sheaths becoming very fibrous. Inflorescence panicle, spikelets homomorphic, utricles greyish-brown, unwinged, markedly grooved, with beak.

Distribution and ecology

Euro-Siberian boreal-temperate element, with a continental distribution in W Europe. Scattered in all CE countries in areas with calcareous substrates, from lowlands to uplands. Tussock-forming perennial herb, mainly in fen meadows, peaty calcareous fens, alderwoods with springs or willow carrs. Persistent seed bank.

Anatomy

Obtusely triangular culm with a diameter of 1–2 mm (1).

The epidermis consists of lignified, round cells with externally thick walls (2, 3). Guard cells of the stomata are externally and internally thick-walled lignified (3). The chlorenchyma consists of unlignified parenchyma cells. The middle part is dominated by large air spaces with spongy remains of cell walls (2) and collateral vascular bundles. Their fiber sheaths are externally linked with the epidermis with girders of lignified cells. The vascular bundles consist of a xylem with two large lateral vessels, a roundish phloem with sieve tubes and companion cells and an air duct (2). Vascular bundles are surrounded by 1–3 layers of thin- to thick-walled sclerenchyma cells (3). The pith consists of unlignified thin-walled polygonal parenchyma cells (2).

Rhizome with a diameter of 2–3 mm (4).

The cortex is composed of an external layer of parenchyma cells and semicircular strands of lignified fibers and an inner part of unlignified thin-walled parenchyma with small intercellulars (4, 5). The central cylinder and the cortex are separated by a thick-walled endodermis with lignified U-shaped cells (5). The lignified parenchymatic ground tissue of the central cylinder contains amphivasal vascular bundles with small sclerenchyma sheaths (6).

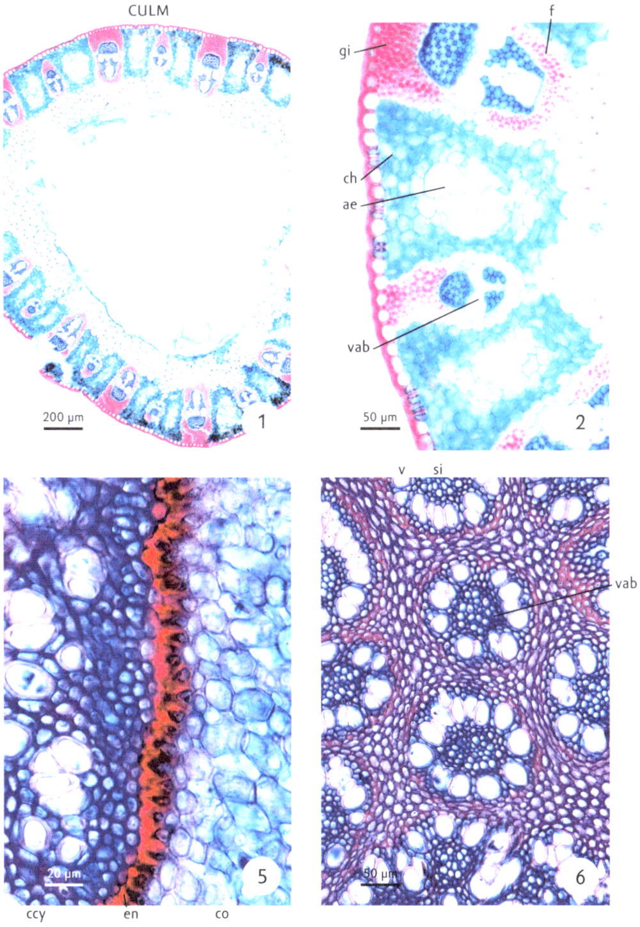

Carex atrata L.

Black Alpine-sedge Cyperaceae

Perennial wetland tufted herb. Stems 15–30 cm high, thin, smooth, upper part pendant. Leaves tough, grey-green. Spikelets heteromorphic, female spikes on long stalkes. Ovoid utricles yellow-green, purpur-black when matured.

Distribution and ecology

Circumpolar boreo-arctic montane element. In CE countries from subalpine to alpine zone (AT, CH, CZ, DE, SK) in deep wet soils, in alpine meadows, often on places with long-lying snow cover. Prefers calcareous soils.

Anatomy

Acutely triangular culm with a diameter of 1.5–2 mm (1).

The epidermis consists of bulliform, lignified cells (2). The chlorenchyma consists of external layer of unlignified parenchyma cells. The middle part is dominated by large air spaces with spongy remains of cell walls (2) and collateral vascular bundles. Their fiber sheaths are externally linked with the epidermis with rectangular girders of lignified cells. The vascular bundles consist of a xylem with two large lateral vessels and a roundish phloem with sieve tubes and companion cells and an air duct. Vascular bundles are surrounded by 3–4 layers of thin- to thick-walled sclerenchyma cells (1, 2). The pith is empty and contains remnants of unlignified cells (1).

Rhizome with a diameter of 2.5–3 mm (3).

The cortex is composed of an external layer of parenchyma cells without intercellulars. Few vascular bundles exist in the cortex (3). The central cylinder and the cortex are separated by a thick-walled endodermis with lignified U-shaped cells. The lignified parenchymatic ground tissue of the central cylinder contains amphivasal vascular bundles with large sclerenchyma sheaths (4).

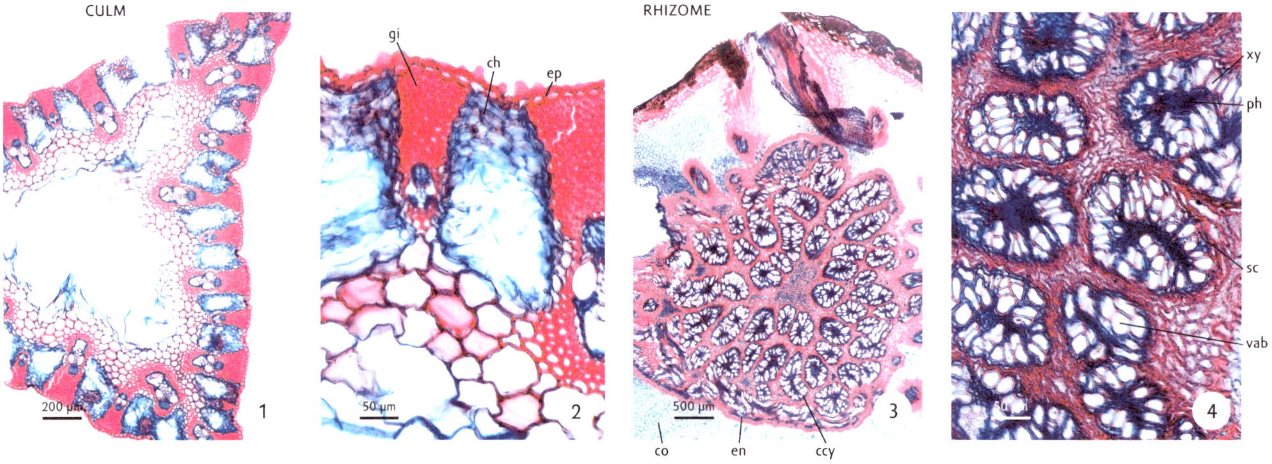

Perennial wetland herb, sparsely tufted with short runners. Triangular stems 5–20 cm high, feeble to procumbent. Basal leaves 1–2 mm wide, greyish to yellow green. Inflorescence umbelliform, from 2–5 sessile spikelets. Apical spike heteromorphic, other spikes only with female flowers. Utricles light green without beak.

Distribution and ecology

Arctic-alpine circumpolar distribution. In the Alps (AT, CH) in subalpine to alpine zone, on banks of small streams, wet soils in moraine and wet debris. Pioneer species on raw substrates without humus, also on human disturbed sites (along tracks).

Anatomy

Round culm with a diameter of <1 mm (1).

The epidermis consists of bulliform, lignified cells (2). The chlorenchyma consists of external layer of unlignified parenchyma cells. The middle part is dominated by air spaces with spongy remains of cell walls (2) and collateral vascular bundles. Their fiber sheaths are partially externally linked with the epidermis with girders of lignified cells. Vascular bundles are surrounded by 1–3 layers of thin-to thick-walled sclerenchyma cells. The pith consists of unlignified thin-walled parenchyma cells (2).

Rhizome with a diameter of 1 mm.

The cortex is composed of an external layer of parenchyma cells and semicircular strands of lignified fibers and an inner part of unlignified thin-walled parenchyma with large intercellulars and a few circular lignified strands or vascular bundles (3). The central cylinder and the cortex are separated by a thick-walled endodermis with lignified U-shaped cells (3). The partially lignified parenchymatic ground tissue of the central cylinder contains amphivasal vascular bundles with uniseriate sclerenchyma sheaths (4).

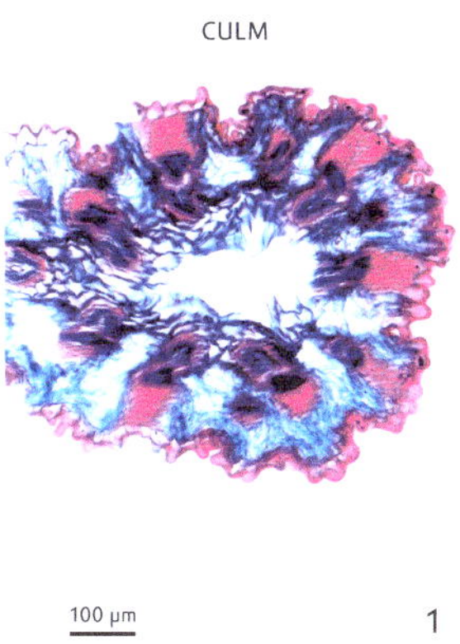

(section from dry material)

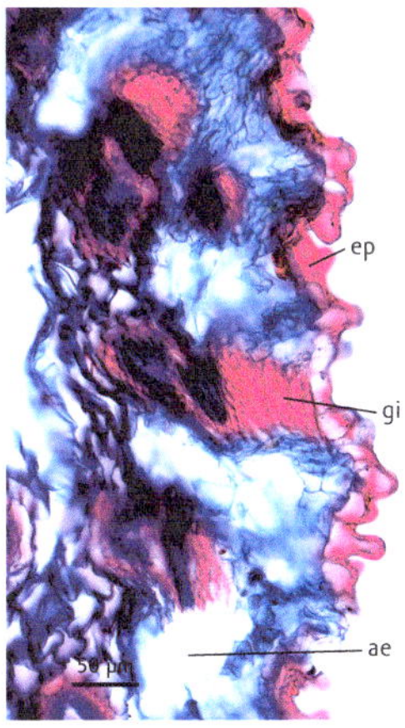

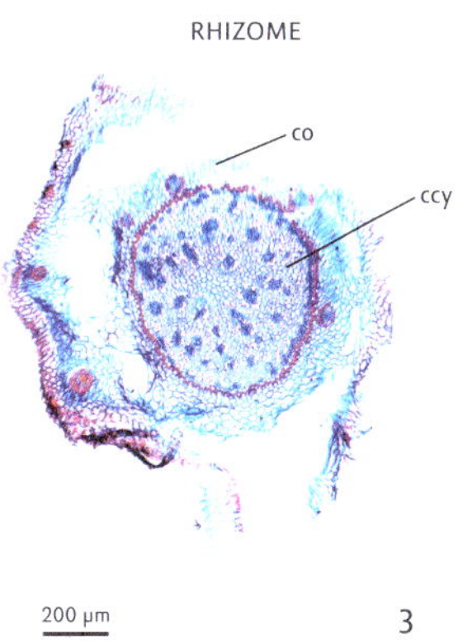

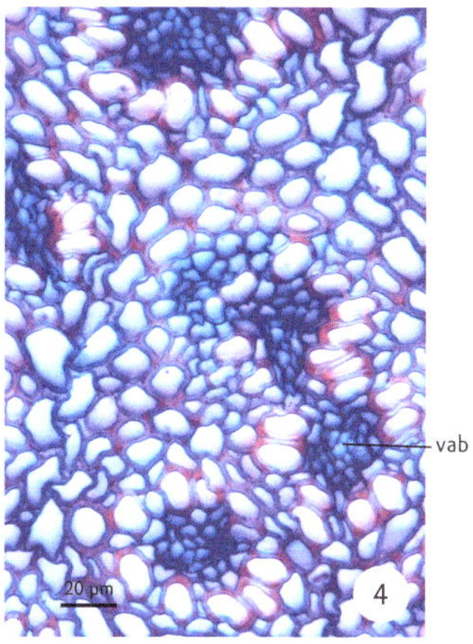

Annual to perennial wetland tufted plant. Fragile flower stem 10–30 cm tall, spikes in the inflorescence in a dense spherical cyme supported by 3 long bracts, leaves yellow-green.

Distribution and ecology

Euro-Siberian distribution, mainly in central Europe north of the Alps. Scattered irregularly in all CE countries, from lowlands to highlands, frequently in fishpond areas. Wet, temporarily denudated bottom of fishponds, margins of fishponds or lakes with oscillating water level, sandy, silty or muddy soils, usually at meso- to eutrophic sites, strictly photophilous.

Anatomy

Obtusely triangular culm with a diameter of 1–2 mm (1).

The epidermis consists of lignified, round cells with externally thick walls (2, 3). Guard cells of the stomata are externally and internally thick-walled lignified (3). The chlorenchyma consists of unlignified parenchymatic cells. The middle part is dominated by air spaces with spongy remains of cell walls (2) and collateral vascular bundles. Their fiber sheaths are partially externall linked with the epidermis with girders of lignified cells. The vascular bundles consist of a xylem with two large lateral vessels and a roundish phloem with sieve tubes and companion cells (2). Vascular bundles are surrounded by 1–3 layers of thin- to thick-walled cells (3). The pith consists of unlignified thin-walled polygonal parenchymatic cells (2).

Rhizome with a diameter of 2–3 mm (4).

The cortex is composed of an external layer of lignified parenchymatic cells and an inner part of unlignified thin-walled parenchyma with small intercellulars (4, 5). The central cylinder and the cortex are separated by a thick-walled endodermis with lignified U-shaped cells (5). The partially lignified parenchymatic ground tissue of the central cylinder contains amphivasal vascular bundles with uni- and biseriate sclerenchymatic sheaths (6).

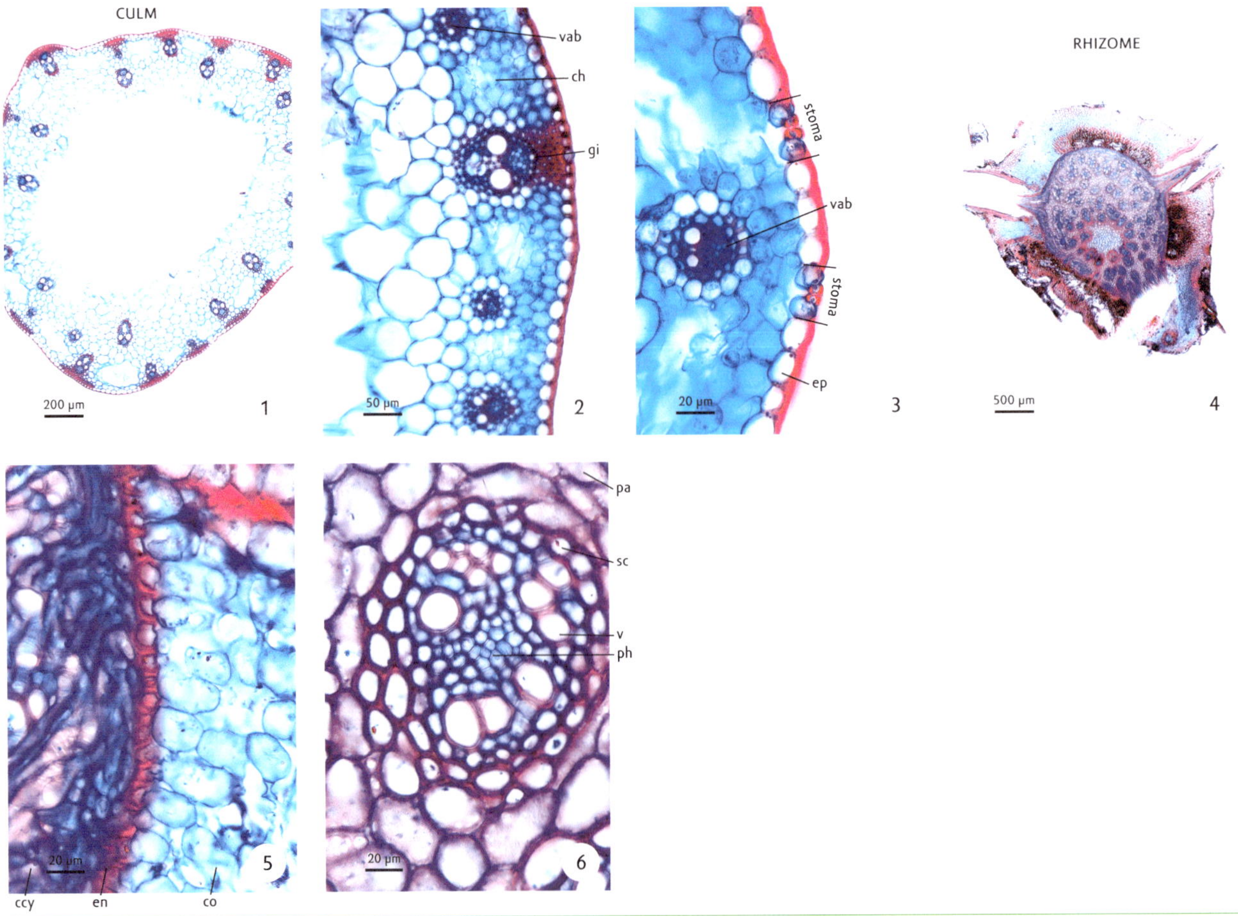

Carex brachystachys Schrank (syn. *Carex tenuis*)

Perennial wetland, dense tufted herb with not creeping, branched rhizomes. Stems thin, feeble, smooth, 10–40 cm high. Leaves shorter than stem, max. 1 mm wide, basal sheaths reddish to brown. Spikelets heteromorphic, 2–3 female spikes on smooth stalkes. Utricles light green to brown with long beak.

Distribution and ecology

European distribution. In CE countries in the Alps (AT, DE, CH) and Carpathians (SK, PL), usually in wet fissures and on wet rocks, also close to springs in wet gravel soils from submontane to alpine zone. Prefers wet, calcareous to neutral stony soils.

Anatomy

Round culm with a diameter of 0.3–0.5 mm (1).

The epidermis consists of lignified, round cells with externally thick walls (2). The chlorenchyma consists of external layer of unlignified parenchyma cells. The middle part contains collateral vascular bundles. Their fiber sheaths are externally linked with the epidermis with rectangular girders of lignified cells. Air spaces are represented by large unlignified cells (2). The vascular bundles consist of a xylem with two large lateral vessels and a roundish phloem with sieve tubes and companion cells. Air ducts are absent (2). Vascular bundles are surrounded centripetally by 3-4 layers of thin- to thick-walled sclerenchyma cells (1, 2). The pith is empty and contains remnants of unlignified cells (1).

Rhizome with a diameter of 1.5–2 mm (3).

The cortex is composed of an external layer of parenchyma cells without intercellulars. Few vascular bundles exist in the cortex. (3). Both central cylinder and cortex are separated by a thick-walled endodermis with lignified round cells. The lignified parenchymatic ground tissue of the central cylinder contains amphivasal vascular bundles with small sclerenchyma sheaths (4).

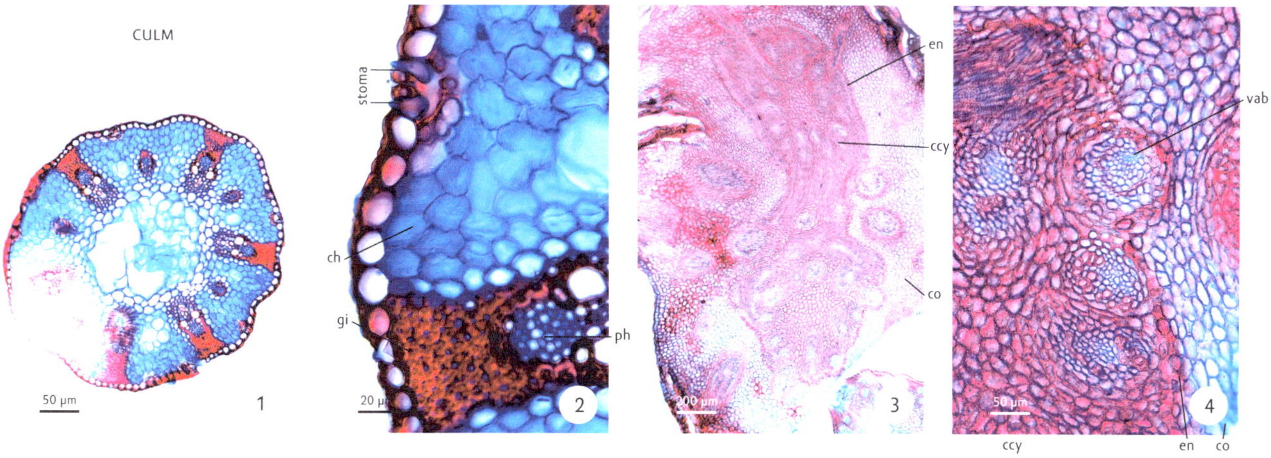

Perennial wetland herb, rhizomatous, usually forming dense stands, 30–70 cm high, procumbent later on. Basal sheaths pale reddish-brown, leaves flat, rough. Spikelets homomorphic, spikes short. Utricles thin, lanceolate, greenish-brown, gradually narrowed into beak.

Distribution and ecology

European temperate element. Scattered frequently to commonly in all CE countries, from lowlands to uplands. Mesophilous to wet meadows, along streams, in alderwood, wet deciduous and coniferous forests, on gleyed nutrient-poor soils, in lowlands in (partial) shade, in highlands in open places, invasive species in forests affected by human management.

Anatomy

Pentagonal hollow culm with a diameter of 0.6–1 mm (1).

The epidermis consists of lignified, round cells with externally thick walls (2). The chlorenchyma consists of external layer of unlignified parenchyma cells. The middle part is dominated by large air spaces with spongy remains of cell walls (1, 2) and collateral vascular bundles. Their fiber sheaths are externally linked with the epidermis with triangular girders of lignified cells. The vascular bundles consist of a xylem with two large lateral vessels, a roundish phloem with sieve tubes and companion cells and an air duct (2). Vascular bundles are surrounded centripetally by 3–4 layers of thin- to thick-walled sclerenchyma cells (1, 2). The center contains partially remnants of unlignified pith cells (1).

Rhizome with a diameter of 0.8–1.2 mm (3).

The cortex is composed of an external layer of thin-walled parenchyma cells with a few irregularly formed intercellulars (3, 4). The central cylinder and the cortex are separated by a rather thin-walled endodermis with lignified round cells. The outer part of the ground tissue of the central cylinder consists of lignified and the inner part of unlignified parenchymatic tissue. The amphivasal vascular bundles are surrounded by large sclerenchyma sheaths (3, 4).

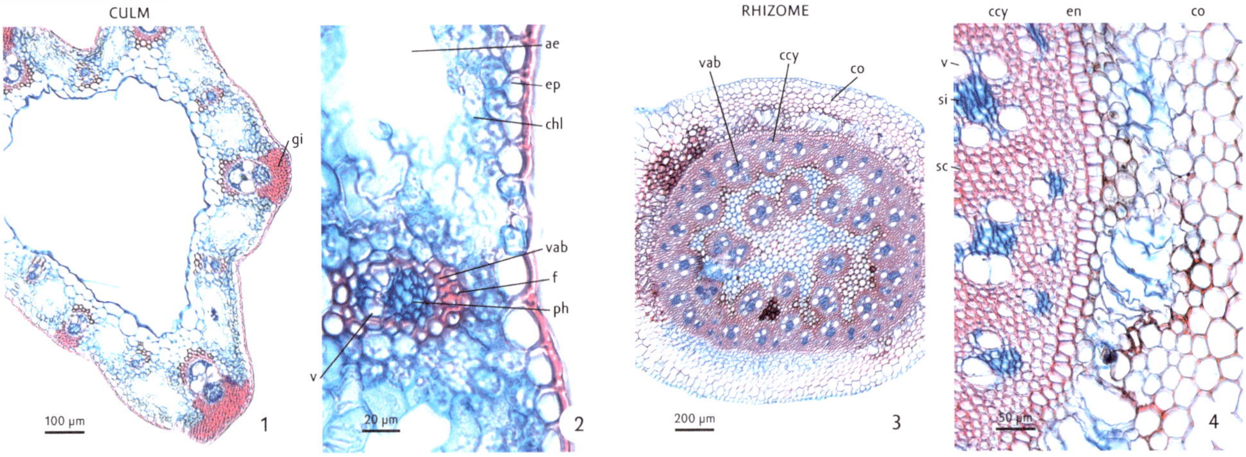

Carex buxbaumii Wahlenb.

Club Sedge

Perennial wetland herb, shortly rhizomatous, forming sparse stands, creeping tillers. Stem erect, 15–80 cm high, sharply triangular. Basal sheaths scale-like, stiff, purpur or dark brown, with a net on the margin. Female flowers in the upper half of the round terminal spike, on the base with collar of glumes of male flowers, 1–3 lateral female spikes. Utricle ellipsoid, papillose, with very short beak. Very similar to C. hartmanii.

Distribution and ecology

Euro-Siberian boreo-montane element with a continental distribution in W Europe; also in E Asia and N America, rare in all CE countries. Mesotrophic fens or near the margins of lakes, often bordering outflow of streams, always in periodically inundated areas, fen meadows, alder woods and mountain springs. It spreads by long rhizomes.

Anatomy

Acutely triangular culm with a diameter of 1 mm (1).

The epidermis consists of bulliform, lignified cells (2). The chlorenchyma consists of external layer of unlignified parenchymatic cells. The middle part is dominated by large air spaces with spongy remains of cell walls (2) and collateral vascular bundles. Their fiber sheaths are externally linked with the epidermis and girders. The vascular bundles consist of a xylem with two large lateral vessels and a roundish phloem with sieve tubes and companion cells. Vascular bundles are surrounded by 1–2 layers of thin- to thick-walled sclerenchymatic cells (2). The pith consists of unlignified thin-walled parenchymatic cells (1).

Rhizome with a diameter of 1.5–2 mm (3).

The cortex is composed of an external layer of lignified parenchymatic cells and an inner part of thin-walled, unlignified thin-walled parenchyma with a few large and many small intercellulars and a few circular lignified strands (3, 4). The central cylinder and the cortex are separated by a thick-walled endodermis with lignified U-shaped cells (4). The lignified parenchymatic ground tissue of the central cylinder contains amphivasal vascular bundles with 2–3-seriate sclerenchymatic sheaths (5).

CULM

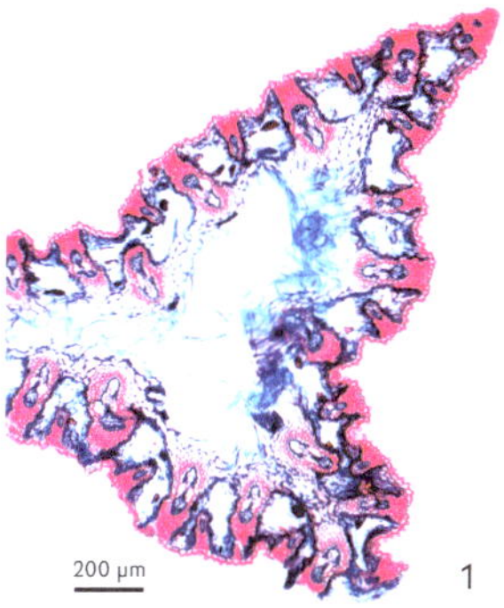

1

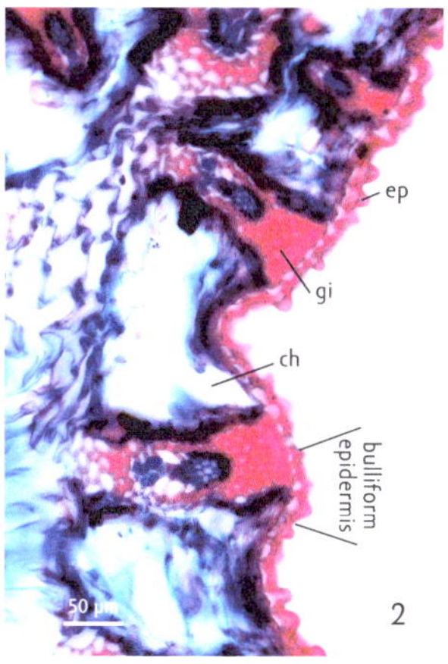

2

RHIZOME

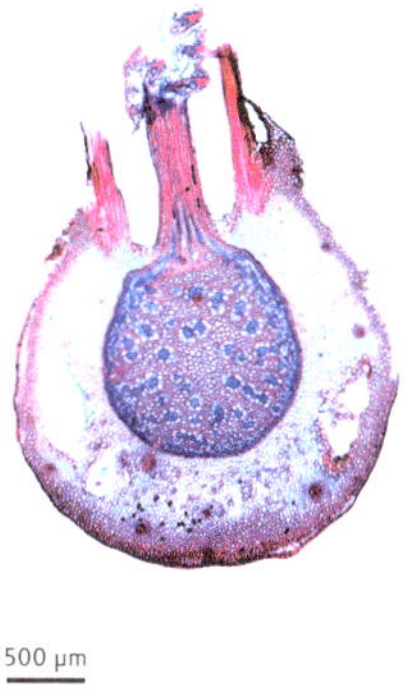

3

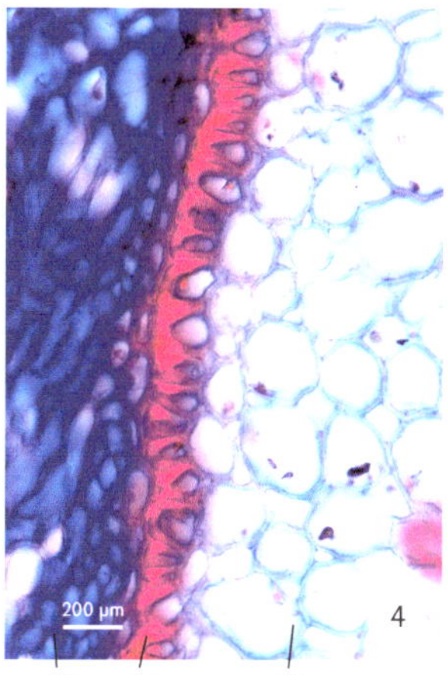

4

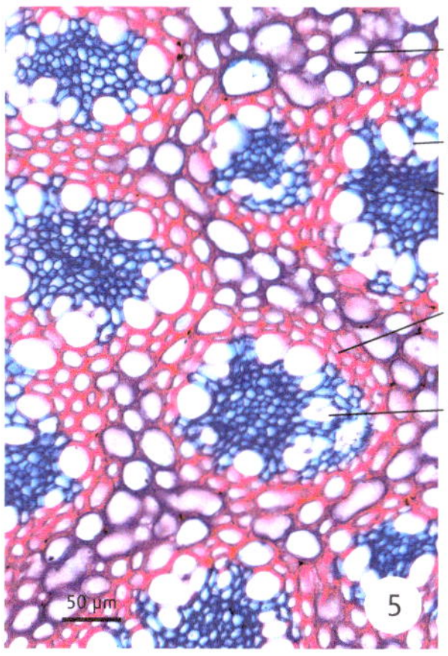

5

Perennial wetland, shortly rhizomatous, loosely tufted plant. Flower stems 20–50 cm tall, acute angles, upper part rough, leaves greyish to grey-green.

Distribution and ecology

Circumpolar boreo-montane element. Scattered irregularly and frequently in all CE countries, from highlands to mountains. More frequent in fishpond and peatland areas. Margins of oligotrophic peat bogs and acidic fens, shorelines of fishponds, wet meadows, wet peaty forests, forest paths, always acidic peaty or sandy soils.

Anatomy

Acutely triangular culm with a diameter of 1 mm (1).

The epidermis consists of bulliform, lignified cells (2). The chlorenchyma consists of unlignified parenchymatic cells. The middle part is dominated by air spaces with spongy remains of cell walls (2) and collateral vascular bundles. Their fiber sheaths are partially externally linked with the epidermis with girders of lignified cells. The vascular bundles consist of a xylem with two large lateral vessels and a roundish phloem with sieve tubes and companion cells (2). Vascular bundles are surrounded by 1–3 layers of thin- to thick-walled sclerenchymatic cells (2). The external part of the hollow pith consists of unlignified thin-walled, polygonal parenchymatic cells (1, 2).

Rhizome with a diameter of 2 mm (3).

The cortex is composed of an external layer of partially lignified parenchymatic cells and an inner part of thin-walled, unlignified parenchyma with many small intercellulars and a few vascular bundles (3, 4). The central cylinder and the cortex are separated by a thick-walled endodermis with lignified U-shaped cells (4). The unlignified parenchymatic ground tissue of the central cylinder contains amphivasal vascular bundles with uni- to biseriate sclerenchymatic sheaths (5).

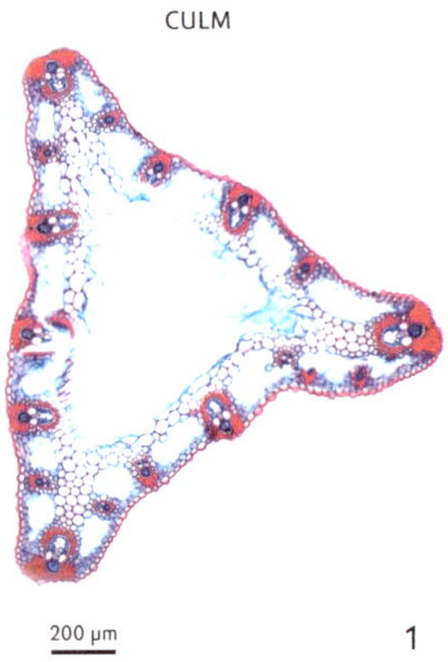

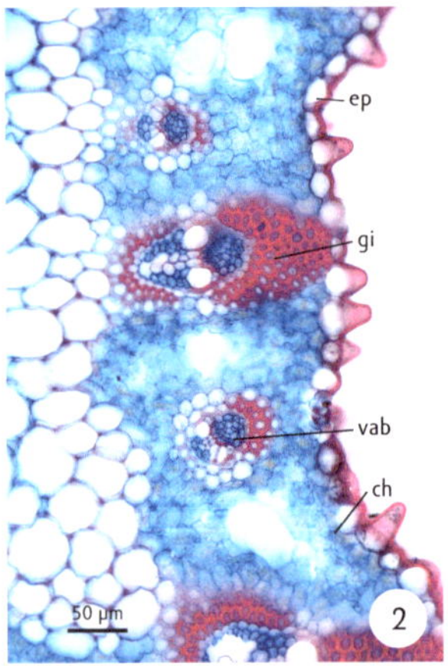

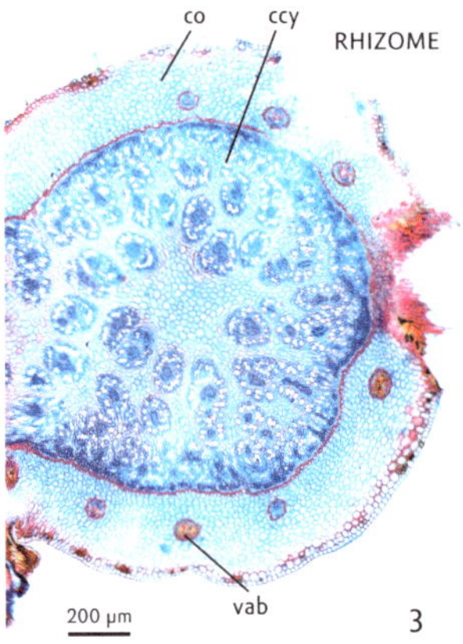

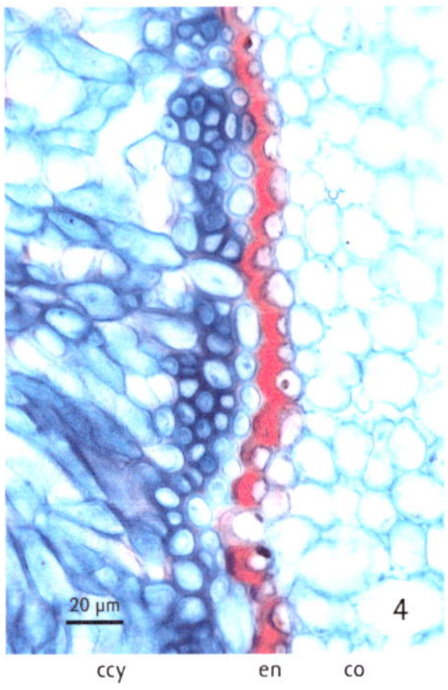

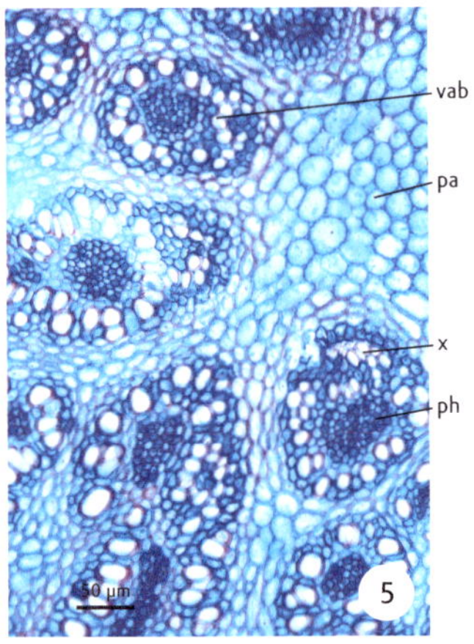

Perennial wetland, densely tufted herb. Stems erect, thin, usually 5–25 cm high, smooth. Leaves 0.5–2 mm wide, rough. Basal sheaths light brown to purpur brown, decaying. Heteromorphic spikelets, apical male spike and 2–4 female spikes on thin stalkes. Utricles with short beak.

Distribution and ecology

Circumpolar boreo-arctic montane element. In CE mainly in the Alps (AT, DE, CH), rarely in CZ and SK. Occurs mainly in subalpine to alpine meadows, particularly those flushed by calcareous springs, on wet, deep soils rich in nutrients. Prefers calcareous soils or base-rich fens. Often in species-rich communities.

Anatomy

Pentagonal culm with a diameter of 0.5–0.7 mm (1).

The epidermis consists of lignified, round cells with externally thick walls (2). The chlorenchyma consists of external layer of unlignified parenchyma cells. The middle part contains air spaces with spongy remains of cell walls (1, 2) and collateral vascular bundles. Their fiber sheaths are externally linked with the epidermis with triangular girders of lignified cells. The vascular bundles consist of a xylem with two large lateral vessels and a roundish phloem with sieve tubes and companion cells. Air ducts are absent (1, 2). Vascular bundles are surrounded centripetally by 3–4 layers of thin- to thick-walled sclerenchyma cells (1, 2). The center contains remnants of unlignified pith cells (1, 2).

Rhizome with a diameter of 0.7–0.9 mm (3).

The peripheral part of cortex is composed of a layer of thin- to thick-walled lignified and a layer with thin-walled unlignified parenchyma cells. The inner part consists of irregular formed intercellulars (3, 4). The central cylinder and the cortex are separated by an endodermis with lignified round cells. The outer part of the ground tissue the central cylinder consists of unlignified and the inner part of lignified parenchymatic tissue. The amphivasal vascular bundles are surrounded by small sclerenchyma sheaths (3).

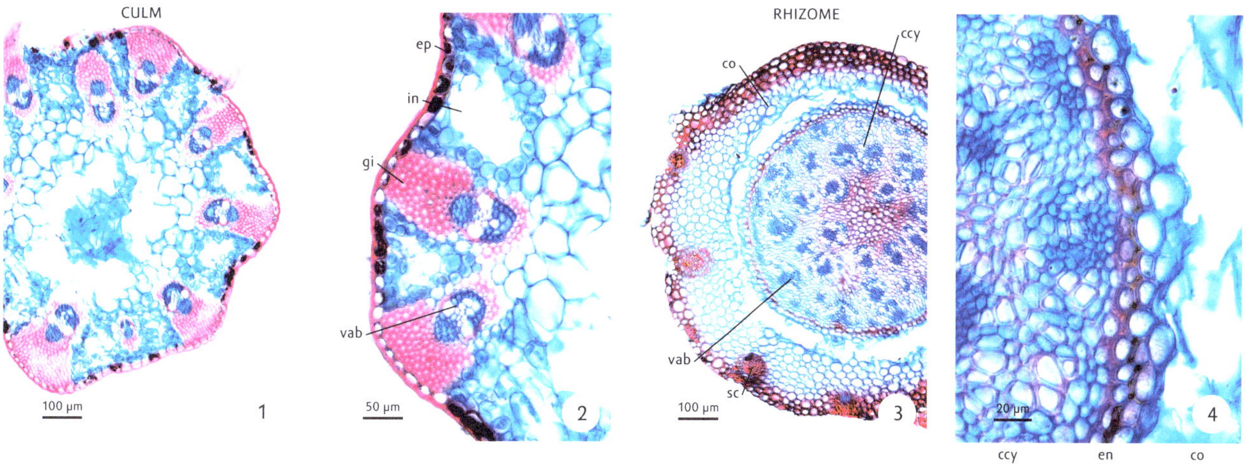

Carex cespitosa L.

Perennial wetland, dense tufted herb, often forming big tussocks. Stems 20–50 cm high, triangular, slender on the basis. Leaves are 1–4 mm wide, flat, green. Very similar to C. elata but has entire scale-like reddish-purplish basal sheaths. The lowermost bract is shorter than the inflorescence. Spikelets are heteromorphic, 1 apical male spike and usually 2 female spikes. Fruit is a nerveless utricle.

Distribution and ecology

Palearctic distribution. In CE common in DE and PL, scattered in CZ and SK and rare in AT, CH and HU. Occurs from lowlands to the mountains, usually in wet meadows and peat bogs, but can also be found along river banks. Prefers wet and waterlogged soils. High ecological plasticity, able to survive grazing or forest colonization.

Anatomy

Acutely to obtusely triangular culm with a diameter of 1.3–2 mm (1).

The epidermis consists of lignified, round cells with externally slightly thickened walls (1). The chlorenchyma consists of external layer of unlignified parenchyma cells and triangular spots of sclerenchyma. The middle part is dominated by large air spaces with spongy remains of cell walls (2) and collateral vascular bundles. Their fiber sheaths are externally partially linked with the epidermis with rectangular girders of lignified cells. The vascular bundles consist of a xylem with two large lateral vessels and a roundish phloem with sieve tubes and companion cells and an air duct (2). Vascular bundles are surrounded by 3–6 layers of thin- to thick and thick-walled sclerenchyma cells (1, 2). The pith consists of unlignified thin-walled polygonal parenchyma cells (1).

Rhizome with a diameter of 2.5–3 mm (3).

The cortex is composed of an external layer of parenchyma cells without intercellulars. Few vascular bundles exist in the cortex (3). Both central cylinder and the cortex are separated by a thick-walled endodermis with lignified U-shaped cells (4). The lignified parenchymatic ground tissue of the central cylinder contains amphivasal vascular bundles with small sclerenchyma sheaths (5).

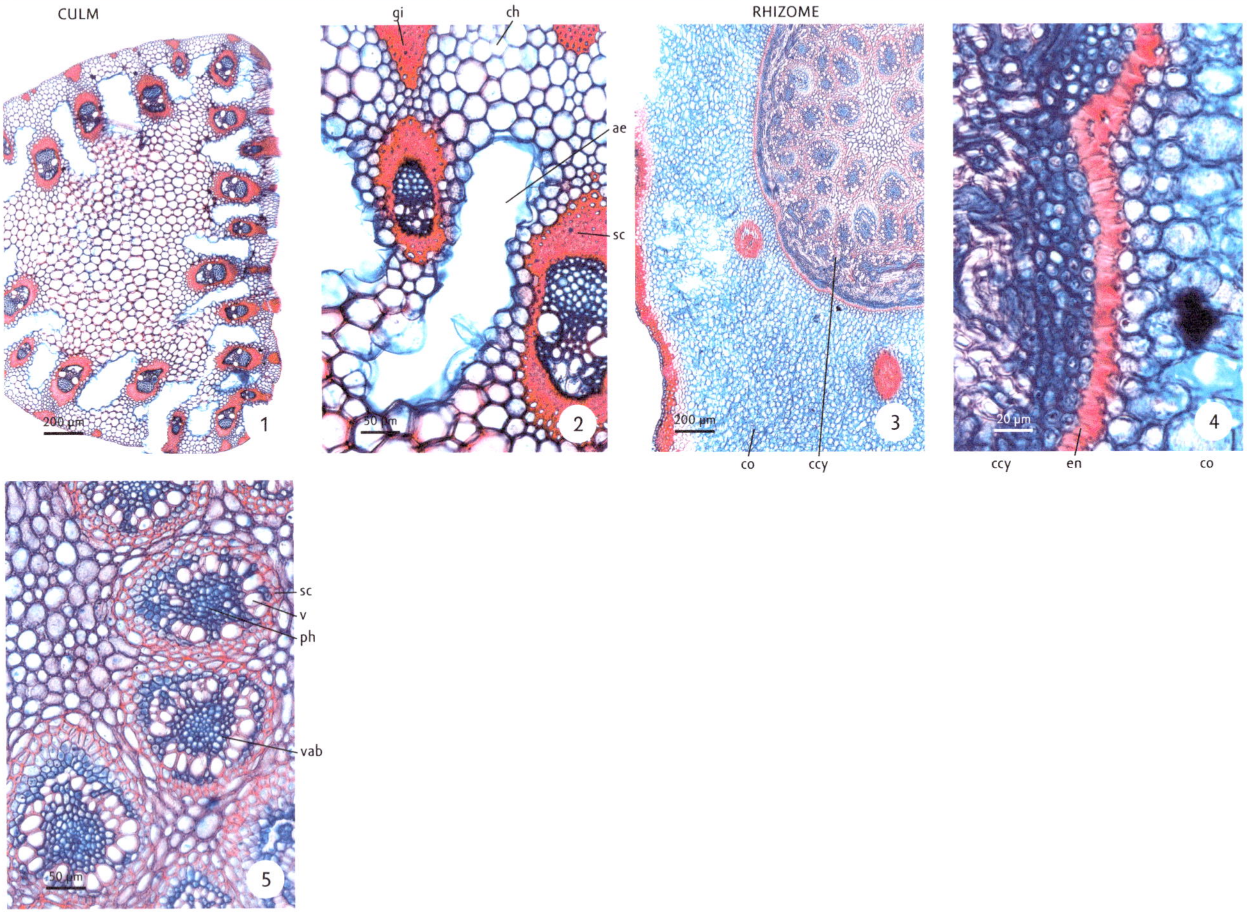

Carex chordorrhiza L. f.

String Sedge

Cyperaceae

Perennial wetland herb, stems 10–30 cm high, often creeping underground stems, leafy at base. Leaves narrow, flat, pale green to grassy green. Spikelets homomorphic, dense and short. Utricles elliptic, multiveinous, yellowish to partly brown.

Distribution and ecology

Circumpolar boreo-arctic montane element, with a continental distribution in W Europe. In all CE countries except HU. Transitional peat bogs and margins of raised bogs. Heliophilous species, very wet sites in the moss carpets. Propagation mainly vegetative by long trailing runners. Able to flow on the water surface. Competitive weak species, sensitive to drainage.

Anatomy

Obtusely triangular culm with a diameter of 1 mm (1).

The epidermis consists of lignified, round cells with externally thick walls (2). The chlorenchyma consists of unlignified parenchyma cells. The middle part is dominated by air spaces with spongy remains of cell walls (2) and collateral vascular bundles. Their fiber sheaths are externally linked with the epidermis with girders of lignified cells (1, 2). The vascular bundles consist of a xylem with two large lateral vessels and a roundish phloem with sieve tubes and companion cells (2). Vascular bundles are surrounded by 1–3 layers of thin- to thick-walled sclerenchyma cells (2). The hollow pith consists of unlignified thin-walled parenchyma cells (1, 2).

Rhizome with a diameter of 3 mm (3).

The cortex is composed of unlignified parenchyma cells (3). The central cylinder and the cortex are separated by a thick-walled endodermis with lignified U-shaped cells (4). The unlignified parenchymatic ground tissue of the central cylinder contains amphivasal vascular bundles with uni- to biseriate sclerenchyma sheaths (3, 4).

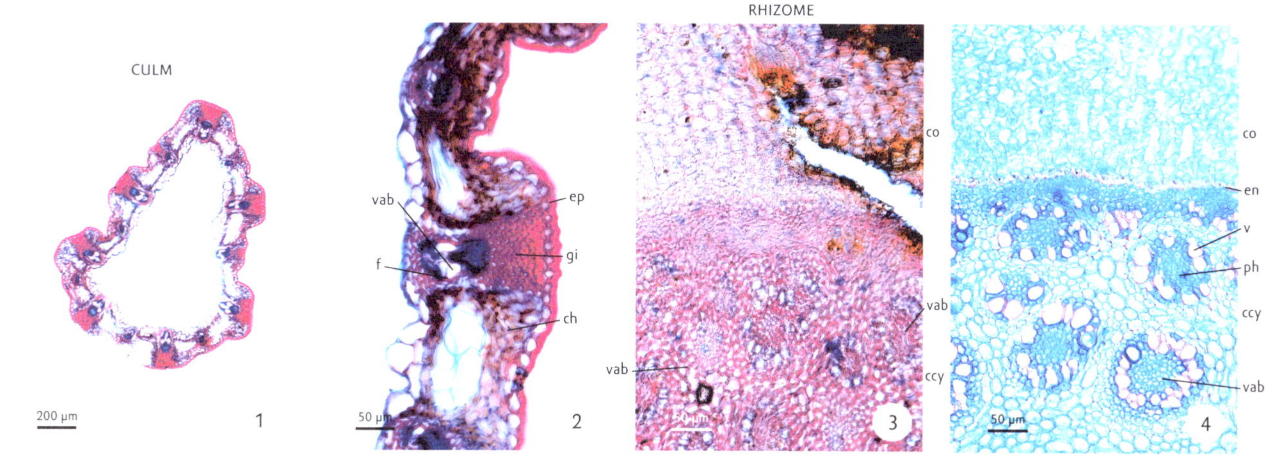

Carex davalliana Sm.

Davall's Sedge Cyperaceae

Perennial wetland herb, densely tufted, basal part with dark-brown fibres. Stems up to 40 cm long, with rounded angles, leaves bristled. Usually dioecious, only one terminal spike, utricles oriented horizontally, brown, with long beak.

Distribution and ecology

European temperate element with a continental distribution in W Europe. In all CE countries, common in CH. Wet fen and peaty meadows, minerotrophic fens, fen springs, prefers base-rich soils. Intolerant of water table fluctuation, heliophilous species, retreats due to partial abandonment and/or eutrophication. Persistent seed bank.

Anatomy

Circular culm with a diameter of <1mm (1).

The epidermis consists of lignified, round-shaped cells with thick walls. (2). The chlorenchyma consists of unlignified parenchymatic cells. The middle part is dominated by large air spaces with spongy remains of cell walls (2) and collateral vascular bundles. Their fiber sheaths are partially externally linked with the epidermis and girders of lignified cells (1, 2). The vascular bundles consist of a xylem with two large lateral vessels and a roundish phloem with sieve tubes and companion cells (2). Vascular bundles are surrounded by 1–3 layers of thin- to thick-walled sclerenchymatic cells (2). The hollow pith consists of unlignified thin-walled parenchymatic cells (1, 2).

Rhizome with a diameter of 1–2 mm (3).

The cortex is composed of an external layer of lignified parenchymatic cells and an inner part of thin-walled, unlignified, thin-walled parenchyma with many small intercellulars (3, 4). The central cylinder and the cortex are separated by a thick-walled endodermis with lignified U-shaped cells (4). The unlignified parenchymatic ground tissue of the central cylinder contains amphivasal vascular bundles with uniseriate sclerenchymatic sheaths (5).

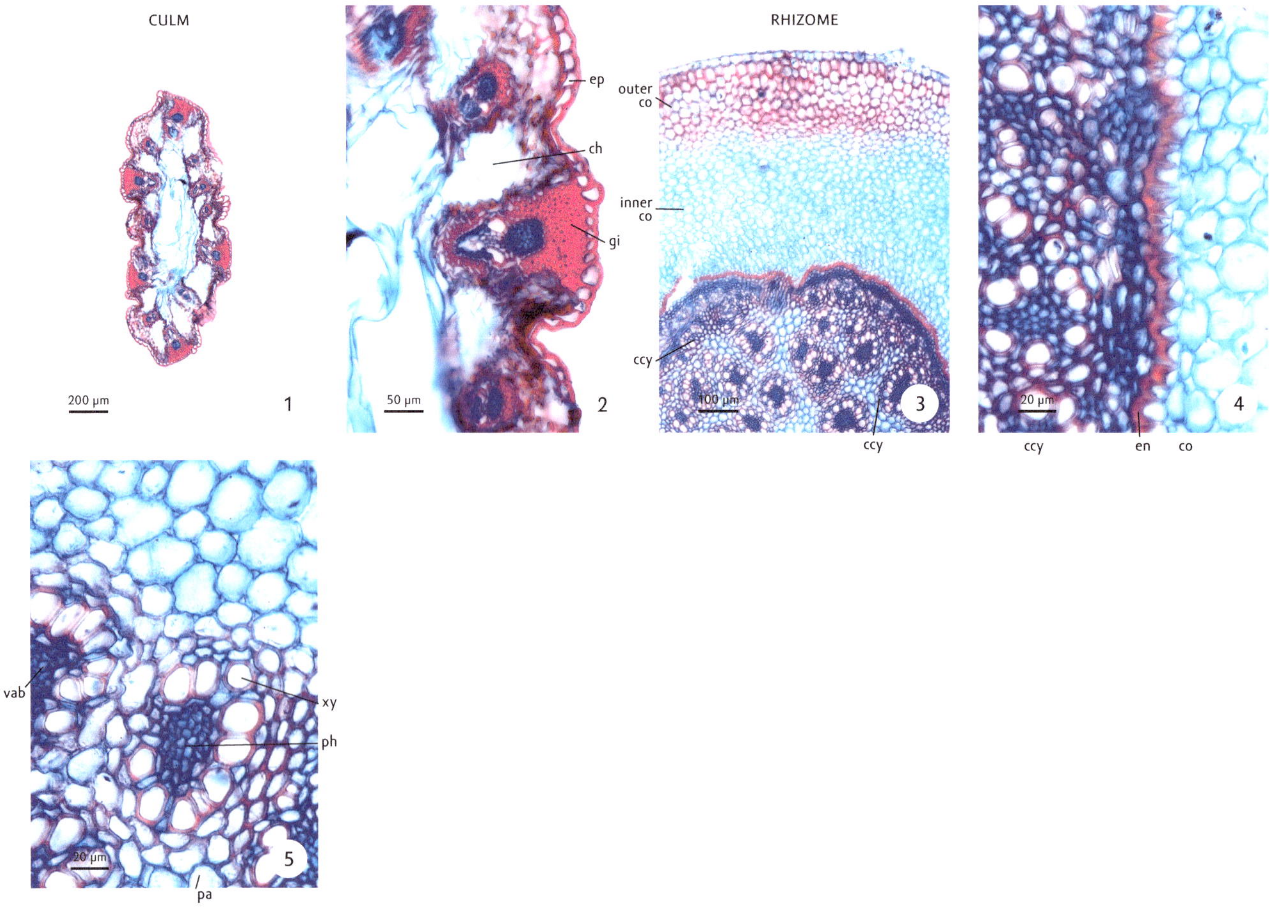

Perennial wetland herb, shortly rhizomatous, creeping, similar to Carex davalliana but underground tiller 10 cm long, thin. Basal sheaths pale brown, not decaying. Stems erect, 5–40 cm high, short-ascendent, rough under spike. Usually dioecious, sometimes both sexes in one spike. Utricles broad ovate, compressed, apex gradually narrowed into dentate beak.

Distribution and ecology

Circumpolar boreo-arctic montane element, in all CE countries, except HU. Very wet, neutral to base-rich mires. It grows particularly well in silty, calcareous mud and along the edges of lime-rich flushes and springs. Very sensitive to drainage and eutrophication. Leaves often infected by specific rust – spottiness on old ones.

Anatomy

Circular culm with a diameter of <1mm (1).

The epidermis consists of lignified, round cells with externally thick walls (2). The chlorenchyma consists of unlignified parenchyma cells. The middle part is dominated by air spaces with spongy remains of cell walls (2) and collateral vascular bundles. Their fiber sheaths are partially externally linked with the epidermis with girders of lignified cells (1, 2). The vascular bundles consist of a xylem with two large lateral vessels and a roundish phloem with sieve tubes and companion cells (2). Vascular bundles are surrounded by 1–3 layers of thin- to thick-walled sclerenchyma cells (2). The hollow pith consists of unlignified thin-walled parenchyma cells (1, 2).

Rhizome with a diameter of 1 mm (3).

The central cylinder and the cortex are separated by an endodermis consisting of U-shaped cell walls (3, 4).

CULM

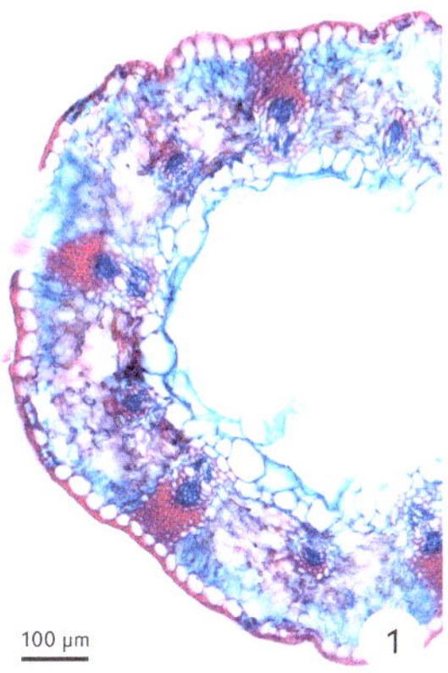

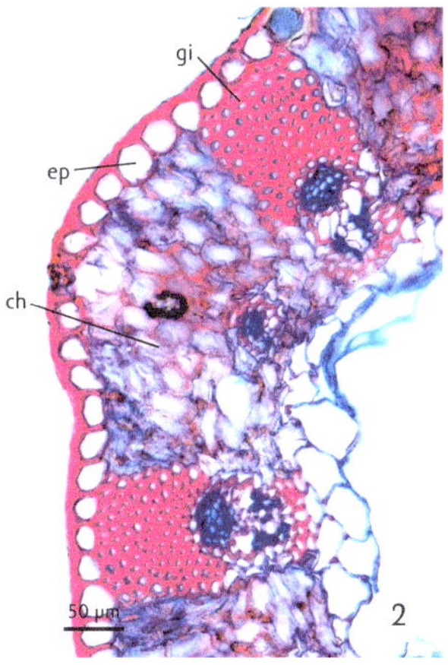

RHIZOME

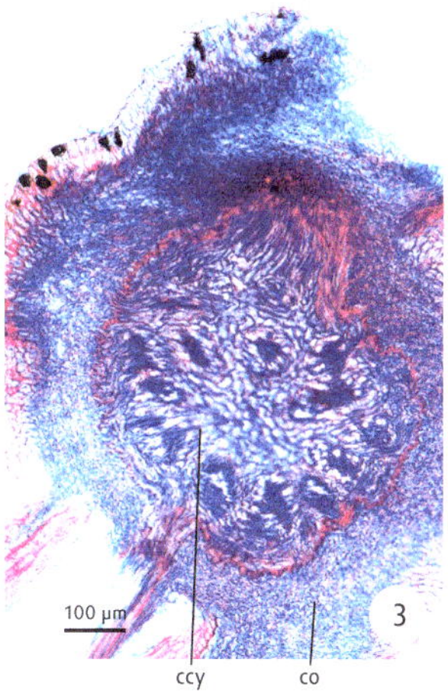

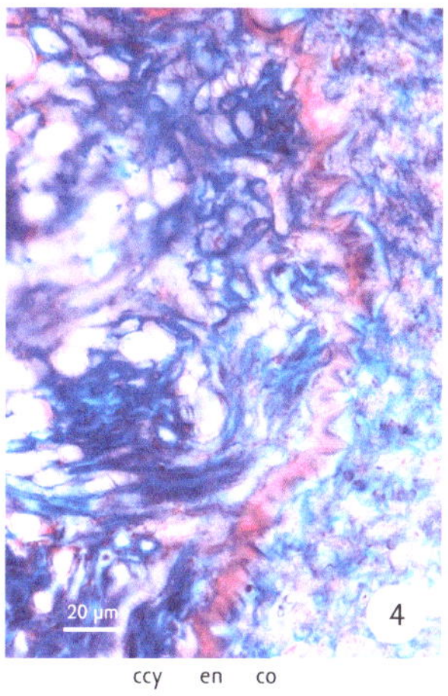

Perennial wetland herb, densely tufted, with ascending extravaginal tillers. Stems erect, 20–60 cm high, bluntly triangular, glabrous. Basal sheaths scale-like, reddish-brown or orange-brown. Leaves wintergreen, flat. Spikelets heteromorphic. Utricles obovate or ellipsoid, veined, yellowish-green, apex narrowed into beak. Very similar to C. hostiana.

Distribution and ecology

European southern-temperate distribution, in all CE countries. On sea-cliffs, rocky shores, in coastal grasslands and the uppermost parts of salt marshes. Inland, salty and fen meadows, springs in base-rich substrates, prefers heavy clayish, base-rich soils. Halophilous species with high exclusion capacity for soil ballast ions such as Na^+, Ca^{2+} and heavy metals and an effective overall regulation of ion absorption.

Anatomy

Obtusely triangular culm with a diameter of 1–1.5 mm (1).

The epidermis consists of lignified, round-shaped cells with externally thick walls. (2). The chlorenchyma consists of unlignified parenchymatic cells. The middle part is dominated by air spaces partially with spongy remains of cell walls (2) and collateral vascular bundles. Their fiber sheaths are partially externally linked with the epidermis and girders (1, 2). The vascular bundles consist of a xylem with two large lateral vessels and a roundish phloem with sieve tubes and companion cells (2). Vascular bundles are surrounded by 1–3 layers of thin- to thick-walled sclerenchymatic cells (2). The hollow pith consists of unlignified pentagonal, thin-walled parenchymatic cells (1, 2).

Rhizome with a diameter of 2–4 mm (3).

The cortex is composed of an external layer of partially lignified parenchymatic cells and an inner part of thin-walled, unlignified thin-walled parenchyma with many small intercellulars and a few vascular bundles (3, 4). The central cylinder and the cortex are separated by a thick-walled endodermis with lignified U-shaped cells (4). The lignified parenchymatic ground tissue of the central cylinder contains amphivasal vascular bundles with uni- to biseriate sclerenchymatic sheaths (5).

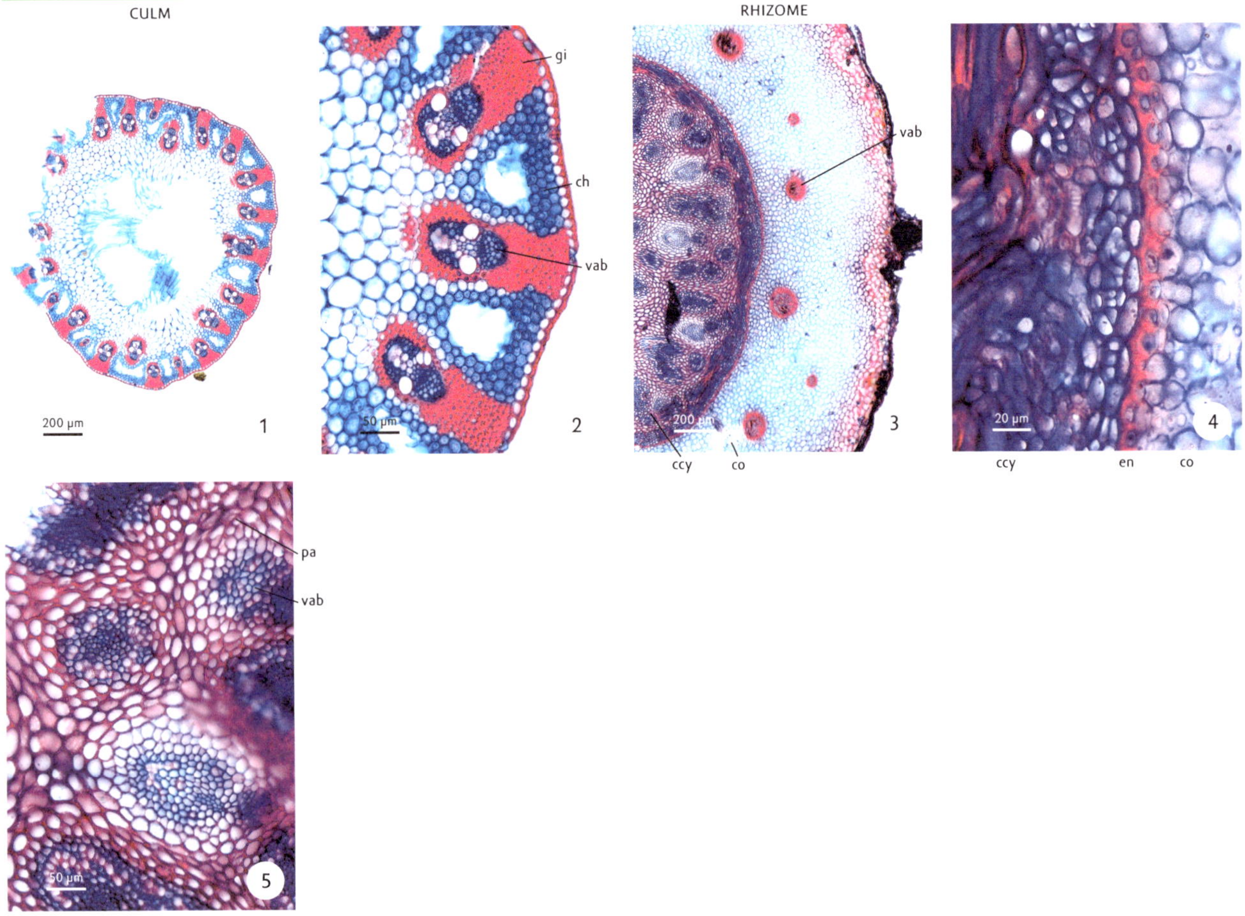

Carex disticha Huds.

Brown Sedge | Cyperaceae

Perennial wetland herb, rhizomatous, forming dense or sparse stands, rhizoms horizontal. Stem erect, 30–80 cm high, slender, steril stems leafed up to the top. Spikelets homomorphic. Utricles ovate, brown, upper part gradually narrowed into beak.

Distribution and ecology

Euro-Siberian temperate element, in all CE countries. Wet meadows and marshes, fens, along ditches and streams, margins of lakes and ponds, occasionally in dune-slacks. Heliophilous, particularly on heavy soils rich in mineral nutrients, areas with a fluctuating water table.

Anatomy

Acutely triangular culm with a diameter of 1.5–2 mm (1).

The epidermis consists of lignified, round cells with externally thick walls. Few epidermis cells are bulbiform (2). The chlorenchyma consists of external layer of unlignified parenchyma cells. The middle part contains collateral vascular bundles. Their fiber sheaths are externally linked with the epidermis with rectangular girders of lignified cells (1, 3). The vascular bundles consist of a xylem with two large lateral vessels and a roundish phloem with sieve tubes and companion cells. Air ducts are absent (1, 3). Vascular bundles are surrounded by 3-4 layers of thin- to thick-walled sclerenchyma cells (1, 3). The pith contains remnants of unlignified cells (1, 3).

Rhizome with a diameter of 2.5–3 mm (3).

The cortex is composed of an external layer of parenchyma cells with a net-like aerenchyma (4). The central cylinder and the cortex are separated by a thick-walled endodermis with lignified U-shaped cells (5). The unlignified parenchymatic ground tissue of the central cylinder contains amphivasal vascular bundles with large sclerenchyma sheaths (6).

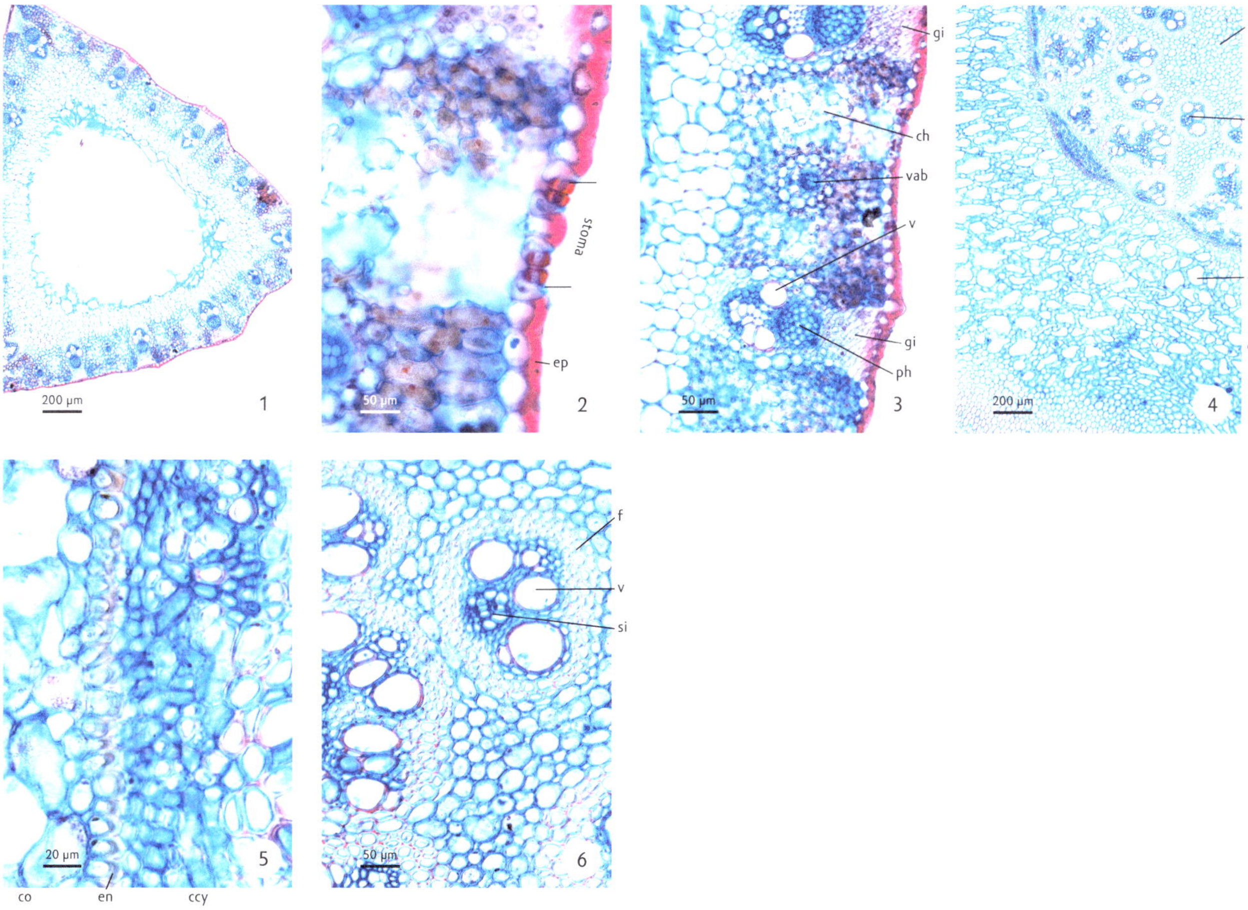

Carex echinata Murray

Star Sedge Cyperaceae

Perennial wetland herb, tufted. Stem erect, 15–40 cm high, askew protruding from the tuft. Leaves thin, green to dark green. Spikelets homomorphic, short spikes markedly bristly built. Utricles ovate, smooth, greenish to dark-brown, gradually narrowed into dentate beak.

Distribution and ecology

European boreo-temperate element; also in N America, in all CE countries. Peaty mesotrophic meadows, transitional mires, alderwoods, waterlogged spruce forests. Wet and moist acidic soils with low nutrient content. Prefers subatlantic climate.

Anatomy

Obtusely triangular culm with a diameter of 1–2 mm.

The epidermis consists of lignified, round cells with externally thick walls (2). The chlorenchyma consists of unlignified parenchyma cells. The middle part is dominated by indistinct or absent air spaces (2) and collateral vascular bundles. Their fiber sheaths are partially externally linked with the epidermis with girders of lignified cells (1, 2). The vascular bundles consist of a xylem with two large lateral vessels and a roundish phloem with sieve tubes and companion cells (2). Vascular bundles are surrounded by 1–2 layers of thin-to thick-walled sclerenchyma cells (2). The hollow pith consists of unlignified pentagonal thin-walled parenchyma cells (1, 2).

Rhizome with a diameter of 1–1.5 mm (3).

The cortex is composed of an external layer of parenchyma cells without intercellulars and an inner part of unlignified thin-walled parenchyma with many very large irregularly formed intercellulars (3). The central cylinder and the cortex are separated by a thick-walled endodermis with lignified U-shaped cells (4). The partially lignified parenchymatic ground tissue of the central cylinder contains amphivasal vascular bundles with 1–3-seriate sclerenchyma sheaths (4, 5).

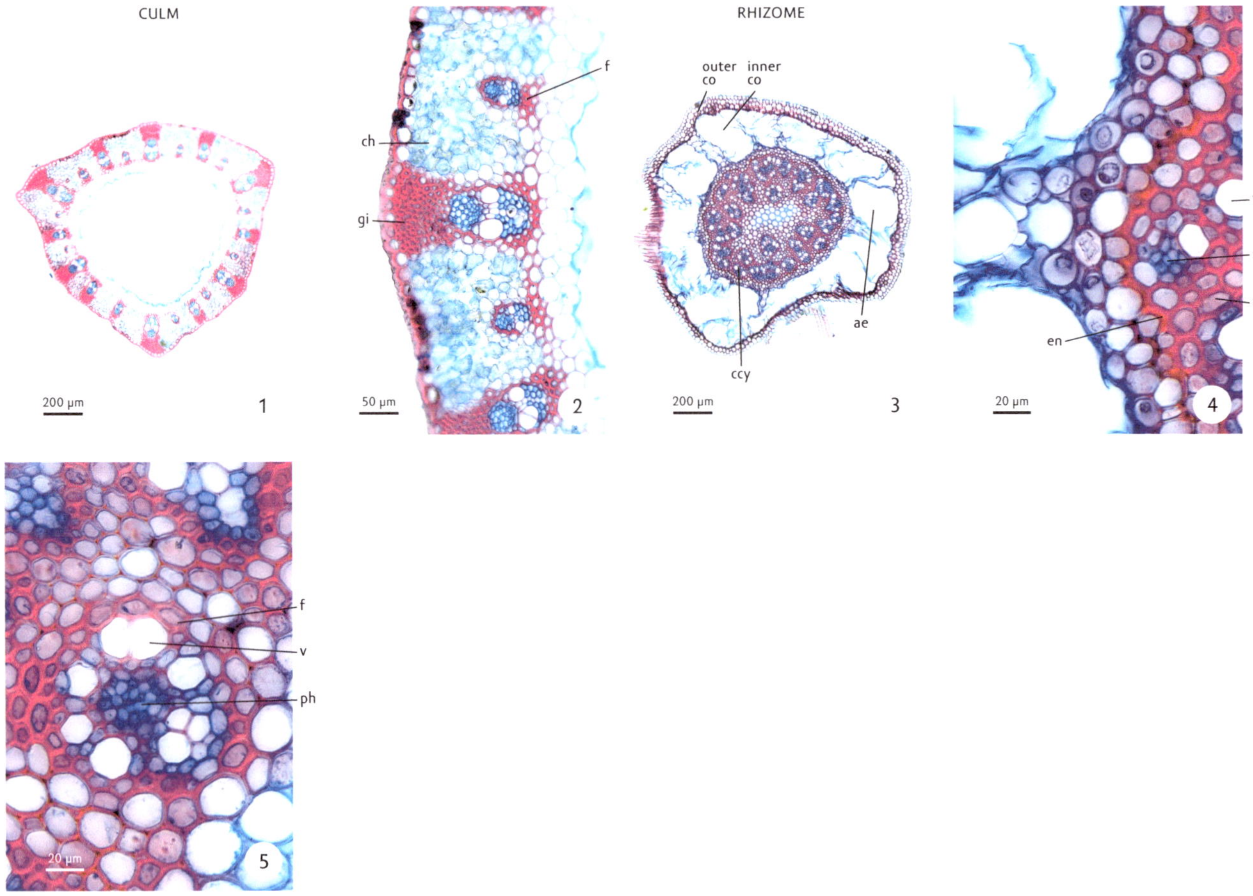

Perennial wetland tufted plant often forming high (0.3–1 m) tussocks. Flower stems 0.4–1.2 m all, acute angles, grey-green leaves 3–5 mm wide.

Distribution and ecology

Eurasian temperate element. Scattered irregularly in all CE countries, from lowlands to uplands, more frequent in fishpond areas. Fen swamps, shorelines of fishponds, river banks, wet meadows, reed swamps, alder carrs, always organic, meso- to eutrophic soils.

Anatomy

Acutely triangular culm with a diameter of 2 mm (1).

The epidermis consists of bulliform, lignified cells (3). The chlorenchyma consists of unlignified parenchymatic cells. The middle part is dominated by air spaces with spongy remains of cell walls (2) and collateral vascular bundles. Their fiber sheaths are partially externally linked with the epidermis with girders of lignified cells. The vascular bundles consist of a xylem with two large lateral vessels and a roundish phloem with sieve tubes and companion cells (2). Vascular bundles are surrounded by 1–3 layers of thin- to thick-walled sclerenchymatic cells (2). The external part of the partially hollow pith consists of unlignified thin-walled, polygonal parenchymatic cells (1, 2).

Rhizome with a diameter of 2.5–4 mm (4).

The cortex is composed of an external layer of parenchymatic cells without intercellulars, and an inner part of thin-walled, unlignified parenchyma with many large radial elongated intercellulars (4). The central cylinder and the cortex are separated by a thick-walled endodermis with lignified U-shaped cells (5). The unlignified parenchymatic ground tissue of the central cylinder contains amphivasal vascular bundles with 2–4-seriate thick-walled sclerenchymatic sheaths (5, 6).

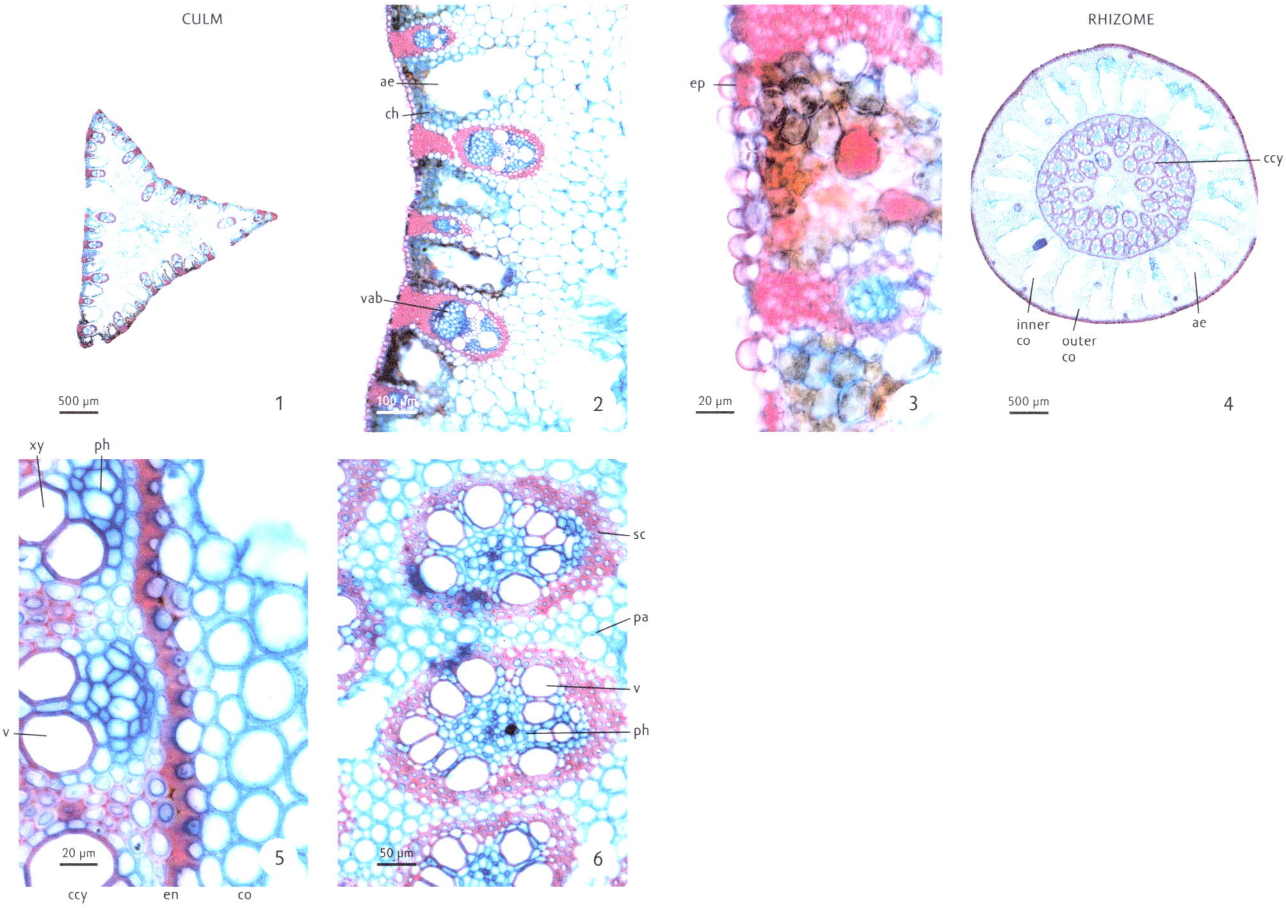

Carex ferruginea Scop.

Perennial wetland, sparse tufted herb. Stems smooth, rounded, 30–50 cm high. Leaves dark green, rough, 1–2 mm wide. Basal sheaths purpur to dark brown, slightly decaying. Spikelets heteromorphic, usually one apical male and 2–4 female spikes on up to 10 cm long stalkes, pendant. Utricles green, later black, glabrous, with short beak.

Distribution and ecology

European distribution. In CE in the Alps (AT, DE, CH) from uplands to the alpine zone. Occurs in springs, base-rich fens and alpine meadows on wet soils. Prefers base-rich soils. Seeds are conditionally dormant, require light for germination.

Anatomy

Circular wavy culm with a diameter of 0.5–0.7 mm (1).

The epidermis consists of lignified, round cells with externally thick walls (2). The chlorenchyma consists of external layer of unlignified parenchyma cells. The middle part contains air spaces with spongy remains of cell walls (1, 2) and collateral vascular bundles. Their fiber sheaths are externally linked with the epidermis with triangular girders of lignified cells. The vascular bundles consist of a xylem with two large lateral vessels and a roundish phloem with sieve tubes and companion cells (1, 2). Vascular bundles are surrounded centripetally by 3–4 layers of thin- to thick-walled sclerenchyma cells (1, 2). The center is empty and contains remnants of unlignified pith cells (1, 2).

Rhizome with a diameter of 0.7–0.9 mm (3).

The cortex is composed of an external layer of parenchyma cells without intercellulars (3). The central cylinder and the cortex are separated by a thick-walled endodermis with thick-walled, U-shaped lignified cells (4). The lignified parenchymatic ground tissue of the central cylinder contains amphivasal vascular bundles with large sclerenchyma sheaths (5).

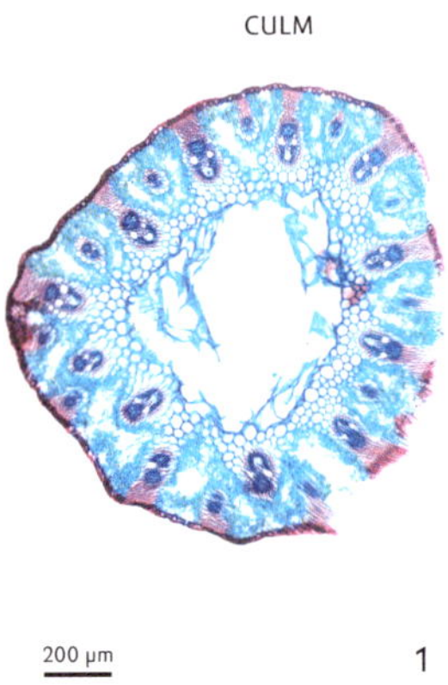

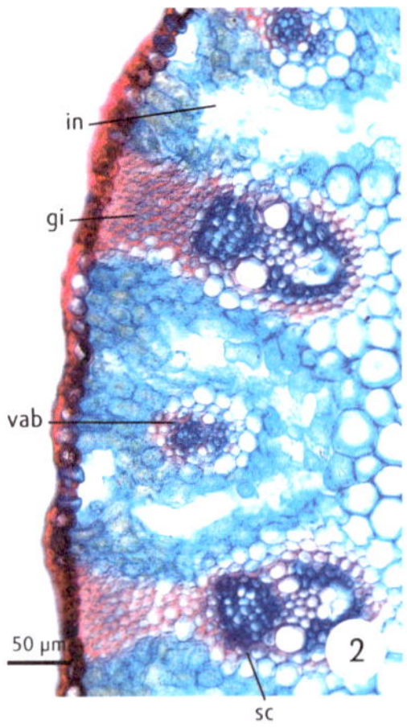

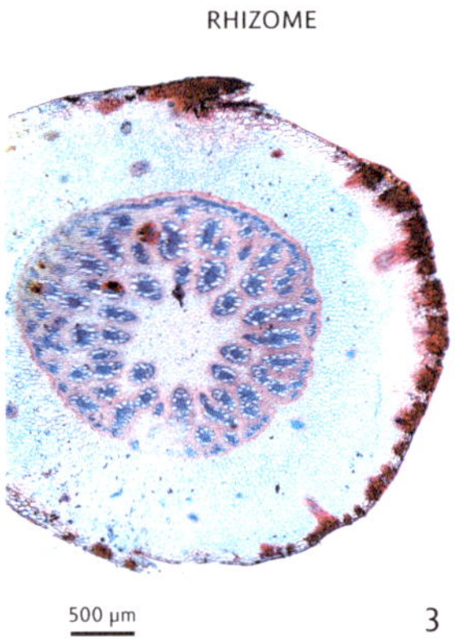

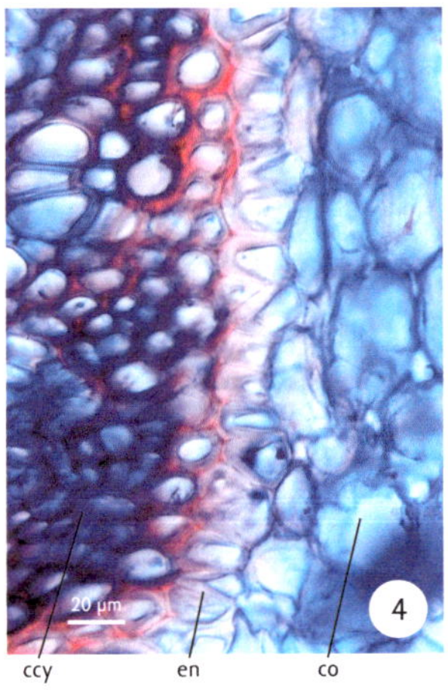

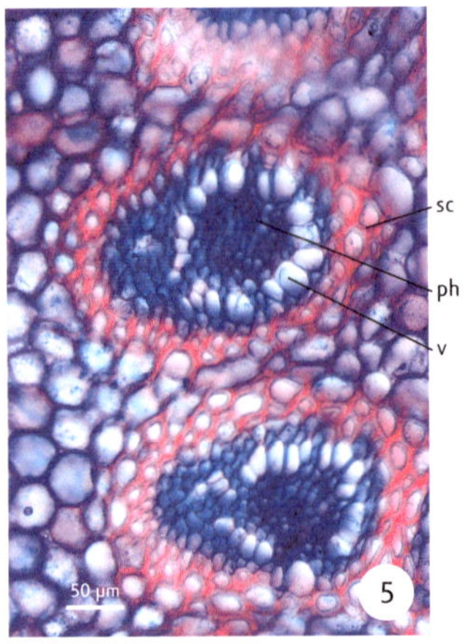

Carex flacca Schreber

Glaucous Sedge

Cyperaceae

Perennial wetland herb, rhizomatous, creeping rhizoms, tillers 2–15 cm long, leafy below. Stems erect, 20–60 cm high, stiff, bluntly triangular, ± glabrous. Youngest sheaths red or reddish. Leaves greyish-green or greyish underside. Spikelets heteromorphic. Female spikes with stalks, ultimately pendulous. Utricle oval or ovoid, shallow triangular, rounded, surface papillate, blackish-brown.

Distribution and ecology

European southern-temperate element; widely naturalized outside its native range. In all CE countries. Wet and alternating wet grasslands with heavy soils and base-rich fens but not waterlogged in summer, also in semi-xeroterm grasslands and light oak woods. Prefers base-rich soils. Very resistant to drought and grazing. Utricles buoyant, floating on water for 2–3 weeks. Persistent seed bank, seeds buried at a soil depth of 7 cm do not germinate.

Anatomy

Round culm with a diameter of 1.5–2 mm (1).

The epidermis consists of lignified thick-walled cells (2). The chlorenchyma consists of external layer of unlignified parenchyma cells. The middle part contains collateral vascular bundles and irregularly formed air spaces (1, 2). Fiber sheaths of vascular bundles are externally linked with the epidermis with rectangular girders of lignified cells (1, 2). The vascular bundles consist of a xylem with two large lateral vessels, a roundish phloem with sieve tubes, companion cells and air ducts (1, 2). Vascular bundles are surrounded by 1–2 layers of thin- to thick-walled sclerenchyma cells (1, 2). The pith is empty and contains remnants of unlignified cells (1).

Rhizome with a diameter of 2–2.5 mm (3).

The cortex is composed of an external layer of parenchyma cells without intercellulars (3). The central cylinder and the cortex are separated by an endodermis with thick-walled, U-shaped lignified cells (4). The lignified parenchymatic ground tissue of the central cylinder contains amphivasal vascular bundles with large sclerenchyma sheaths (5).

CULM RHIZOME

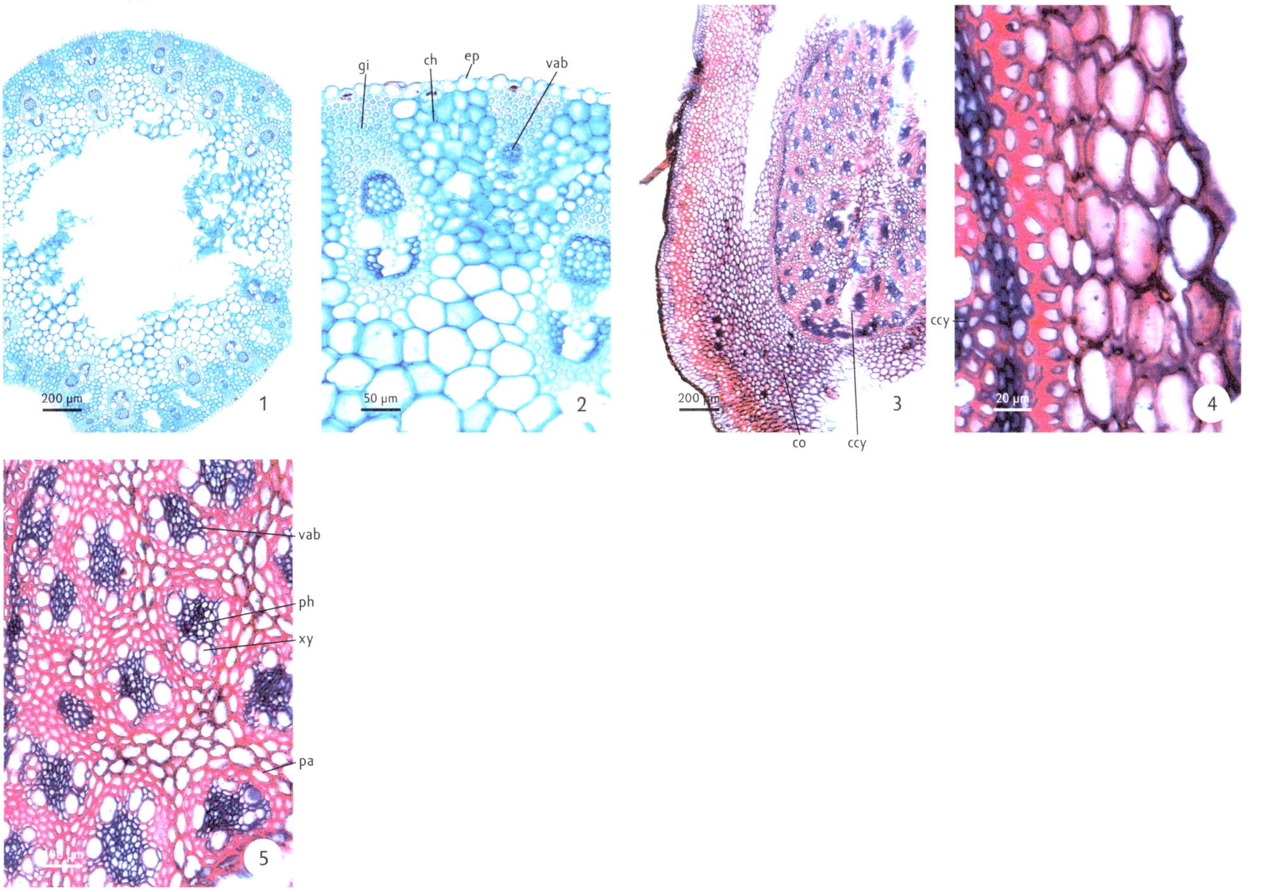

Carex flava L. s. str.

Large Yellow-sedge

Perennial wetland herb, tufted, stems erect, 20–80 cm high, glabrous, yellowish-green. Leaves pale green, of the same length as stem or slightly shorter. Spikelets heteromorphic. Only one male spike, very short. Utricles broad ellipsoid, inflated, curved, yellow, apex narrowed into 2 mm long beak, down directed. Part of the C. flava agg.

Distribution and ecology

European boreo-temperate element, also in C Asia and N America. In all CE countries. Fens and mesotrophic peaty meadows, springs on base-rich substrates. An open canopy or slight shade is preferred.

Anatomy

Obtusely triangular culm with a diameter of 1–1.5 mm (1).

The epidermis consists of lignified, round cells with externally thick walls (3). Guard cells of the stomata are externally and internally thick-walled lignified (3). The chlorenchyma consists of unlignified parenchymatic cells. The middle part is dominated by large air spaces with spongy remains of cell walls and collateral vascular bundles (2). Their fiber sheaths are mostly externally linked with the epidermis with girders of thick-walled, lignified cells (1–3). The vascular bundles consist of a xylem with two large lateral vessels and a roundish phloem with sieve tubes and companion cells (2). Vascular bundles are surrounded by many layers of thick-walled sclerenchymatic cells (2). The hollow pith contains vascular bundles and consists of lignified and unlignified pentagonal, thin-walled parenchymatic cells (1, 2).

Rhizome with a diameter of 2–3 mm (4).

The cortex is composed of an external layer of lignified parenchymatic cells and an inner part of thin-walled, unlignified thin-walled parenchyma with irregularly formed intercellulars and a few vascular bundles (4). The central cylinder and the cortex are separated by a thick-walled endodermis with lignified U-shaped cells (4). The lignified parenchymatic ground tissue of the central cylinder contains amphivasal vascular bundles with 1–3-seriate sclerenchymatic sheaths (4).

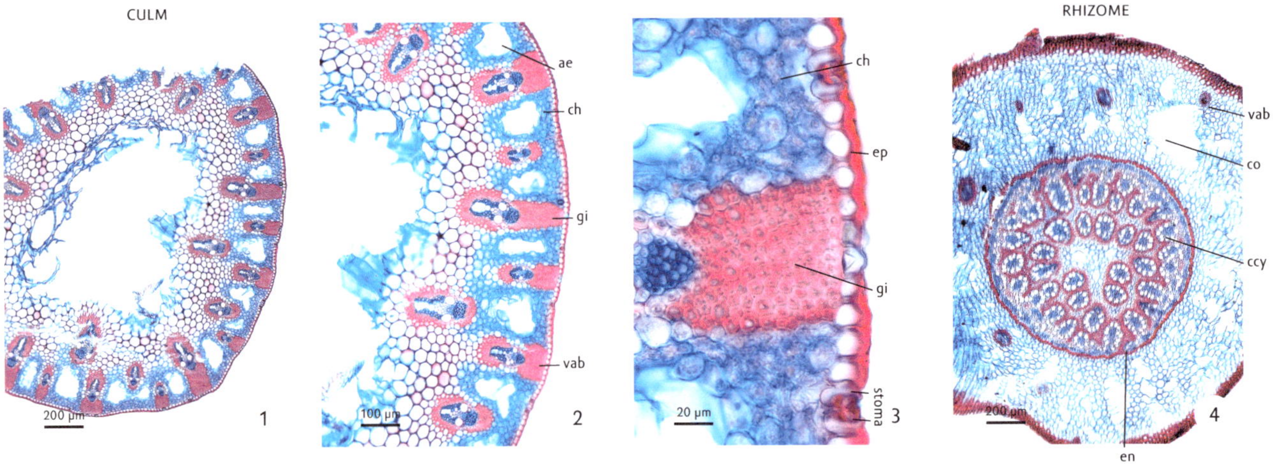

Carex hartmanii Cajander

Hartman's Sedge

Cyperaceae

Perennial wetland herb, creeping tillers 5–25 cm long. Stems erect, 25–80 cm high, triangular. Basal sheaths scale-like, brownish-red, in spring with net. All spikes similar, sessile. Utricles oval, greyish-brown or pale brown, apex with short beak. This species was not separated from the similar species Carex buxbaumii before Cajander (1935).

Distribution and ecology

Eurasian boreal element, in all CE countries. Mesotrophic peaty meadows, margins of minerotrophic peatbogs, alderwoods, moist oak woods on sandy soils.

Anatomy

Acutely triangular culm with a diameter of 1–1.5 mm (1).

The epidermis consists of bulliform, lignified cells (2). The chlorenchyma consists of unlignified parenchyma cells. The middle part is dominated by collateral vascular bundles. Air spaces are absent. Their fiber sheaths are mostly externally linked with the epidermis with girders of lignified cells. The vascular bundles consist of a xylem with two large lateral vessels and a roundish phloem with sieve tubes and companion cells (2). Vascular bundles are surrounded by 1–3 layers of thin- to thick-walled sclerenchyma cells (2). The external part of the partially hollow pith consists of unlignified thin-walled polygonal parenchyma cells (1, 2).

Rhizome with a diameter of 3–4 mm (3).

The cortex is composed of an external layer of partially lignified parenchyma cells and an inner part of unlignified thin-walled parenchyma with many small intercellulars and a few vascular bundles (3). The central cylinder and the cortex are separated by a thick-walled endodermis with lignified U-shaped cells (4). The lignified parenchymatic ground tissue of the central cylinder contains amphivasal vascular bundles with 2–3-seriate sclerenchyma sheaths (3).

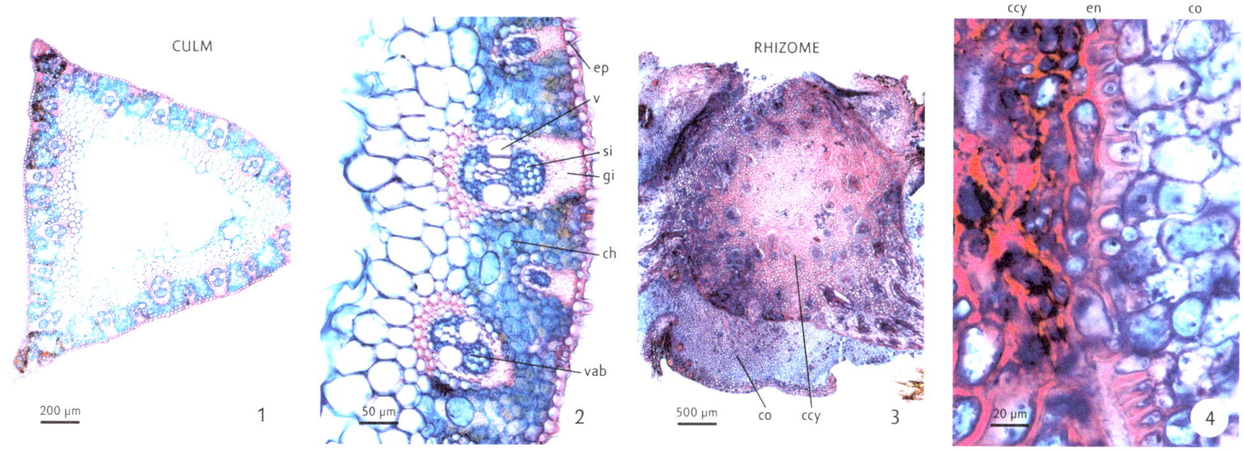

Carex hordeistichos Vill.

Perennial wetland herb, tufted, forming larger, dense tufts. Basal sheaths brown, matt. Leaves tough, dark green, upside shiny, with dark grooves on the underside. Spikelets heteromorphic. Utricles large, lanceolate, pale brown or yellow-brown, shiny, with teeth on the ridge.

Distribution and ecology

European submediterranean-continental element. In all CE countries except CH and PL. Inland salt marshes, salty margins of field and pastures, springs. It prefers heavy clayish soils.

Anatomy

Obtusely triangular culm with a diameter of 2 mm (1).

The epidermis consists of lignified, round cells with externally thick walls (2). The chlorenchyma consists of unlignified parenchyma cells. The middle part is dominated by large air spaces with spongy remains of cell walls and collateral vascular bundles (2). Their fiber sheaths are mostly externally linked with the epidermis with girders of thin-to thick-walled, lignified cells (1–2). The vascular bundles consist of a xylem with two large lateral vessels and a roundish phloem with sieve tubes and companion cells (2). Vascular bundles are surrounded by 1–4 layers of thick-walled sclerenchyma cells (2). The pith is hollow (1).

Rhizome with a diameter of 2–3 mm (3).

The cortex is composed of unlignified parenchyma cells with many small intercellulars (4). The central cylinder and the cortex are separated by a thick-walled endodermis with lignified U-shaped cells (4). The lignified parenchymateous ground tissue of the central cylinder contains amphivasal vascular bundles without a sclerenchyma sheath (4).

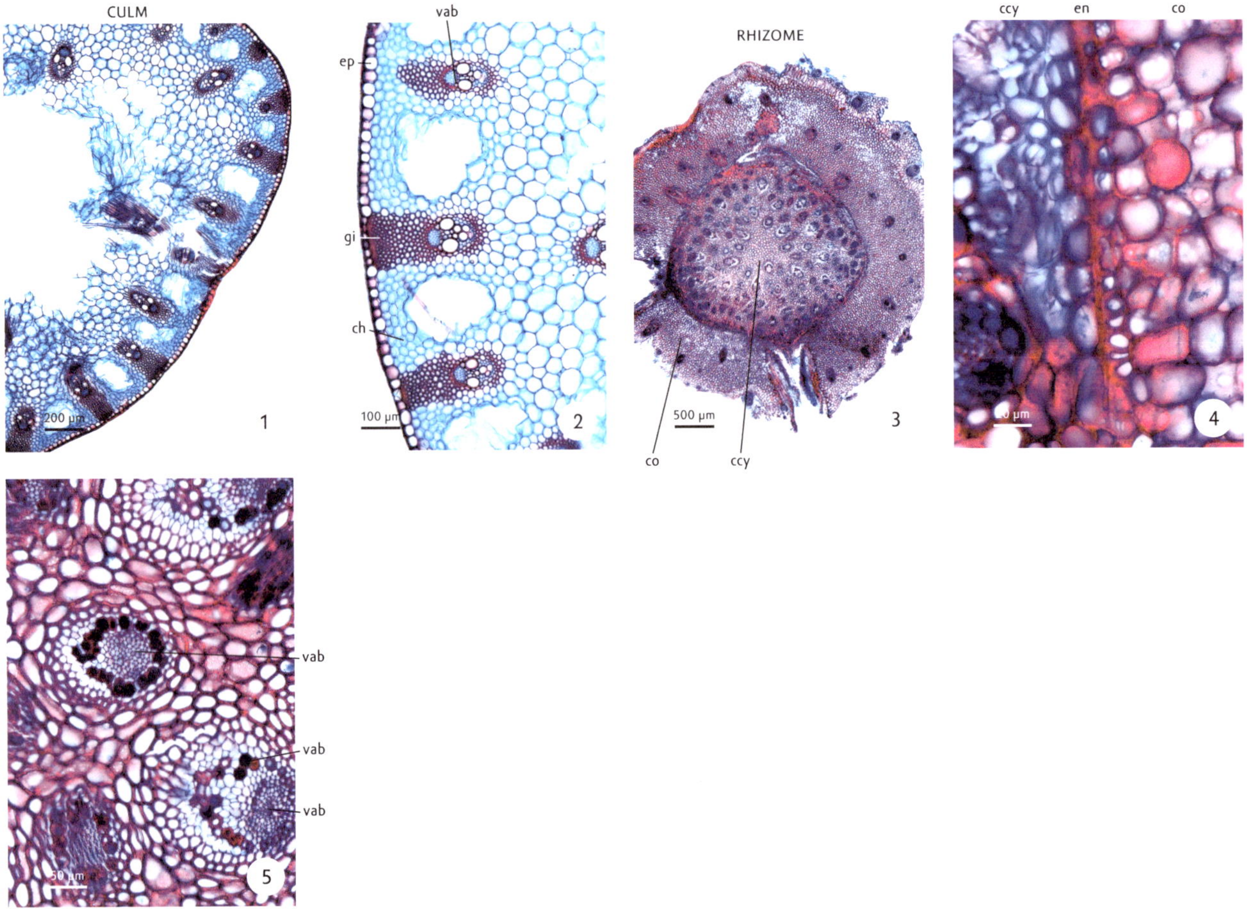

Carex lasiocarpa Ehrh.

Slender Sedge

Perennial wetland rhizomatous plant. Flower stem 0.4 to 1.0 m, with round angles in section, leaves 25–50 cm long, very narrow.

Distribution and ecology

Circumpolar boreo-montane element. Scattered irregularly to rarely in all CE countries, from lowlands to uplands, frequently in peatland, fishpond or lake areas. Mesotrophic peat bogs, fens, shorelines of lakes and fishponds, peaty meadows, always at very wet sites usually 5–30 cm deep, sometimes in nearly monospecific stands. Prefers acidic soils (pH ca. 5–6), weakly competitive species, does not tolerate eutrophication.

Anatomy

Roundish to obtusely triangular culm with a diameter of 1.5–2 mm (1).

The epidermis consists of lignified, round cells with externally thick walls (2). The chlorenchyma consists of external layer of unlignified parenchymatic cells. The middle part contains irregular air spaces with spongy remains of cell walls (2) and collateral vascular bundles. Their fiber sheaths are externally linked with the epidermis with rectangular to slightly triangular girders of lignified cells. Vascular bundles consist of a xylem with two large lateral vessels and a roundish phloem with sieve tubes and companion cells and air ducts (2). Vascular bundles are centripetally surrounded by 1–3 layers of thin- to thick-walled sclerenchymatic cells (2). The pith consists of unlignified thin-walled polygonal parenchymatic cells (2).

Rhizome with a diameter of 4–5 mm (3).

The cortex is composed of an external layer of parenchymatic cells with large, radially expanded intercellulars (3). Both central cylinder and the cortex are separated by a thin thick-walled endodermis with lignified round cells (4). The unlignified parenchymatic ground tissue of the central cylinder contains amphivasal vascular bundles with small sclerenchymatic sheaths (3).

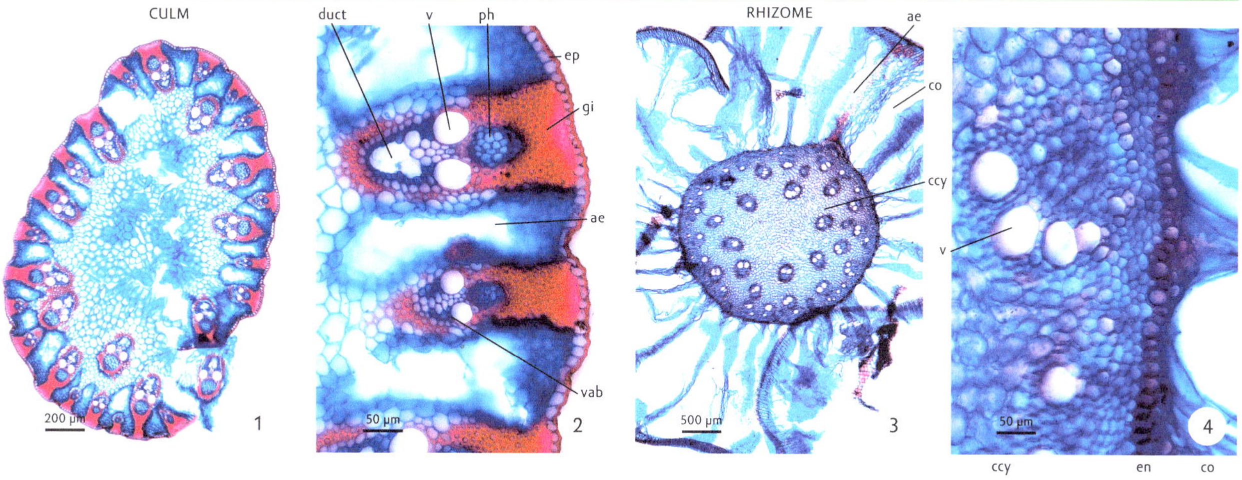

Carex lepidocarpa Tausch

Yellow Sedge

Cyperaceae

Perennial wetland herb, tufted, with extravaginal tillers. Stems erect, 20–80 cm high, faint scabrous bellow inflorescence. Leaves shorter than stems. Spikelets heteromorphic, one male spike with 10–30 mm long stalk, the lower female spike in the half of stem. Utricles obovate-ellipsoid, yellow, abruptly narrowed into long beak, directed down. Part of Carex flava agg.

Distribution and ecology

European boreal element. In all CE countries. Fen meadows, mesotrophic minerotrophic peat bogs, prefers base-rich substrates. Sensitive to changes in the water table and overgrowing due to cessation of traditional agricultural management (mowing and grazing).

Anatomy

Circular to obtusely trinangular wavy culm with a diameter of 1.5–2 mm (1).

The epidermis consists of lignified, round cells with externally thick walls (2). The chlorenchyma consists of external layer of unlignified parenchyma cells. The middle part contains air spaces with spongy remains of cell walls (1, 2) and collateral vascular bundles. Their fiber sheaths are externally linked with the epidermis with rectangular girders of lignified cells. The vascular bundles consist of a xylem with two large lateral vessels and a roundish phloem with sieve tubes and companion cells. Air ducts occur occasionally (1, 2). Vascular bundles are surrounded centripetally by 3–4 layers of thin- to thick-walled sclerenchyma cells (1, 2). The hollow center contains remnants of unlignified pith cells (2).

Rhizome with a diameter of 2–3 mm (3).

The cortex is composed of an external layer of parenchyma cells without intercellulars (3) but with a few vascular bundles. The central cylinder and the cortex are separated by an endodermis with round, thin- to thick-walled, lignified cells (3). The lignified parenchymatic ground tissue of the central cylinder contains amphivasal vascular bundles with small sclerenchyma sheaths (4).

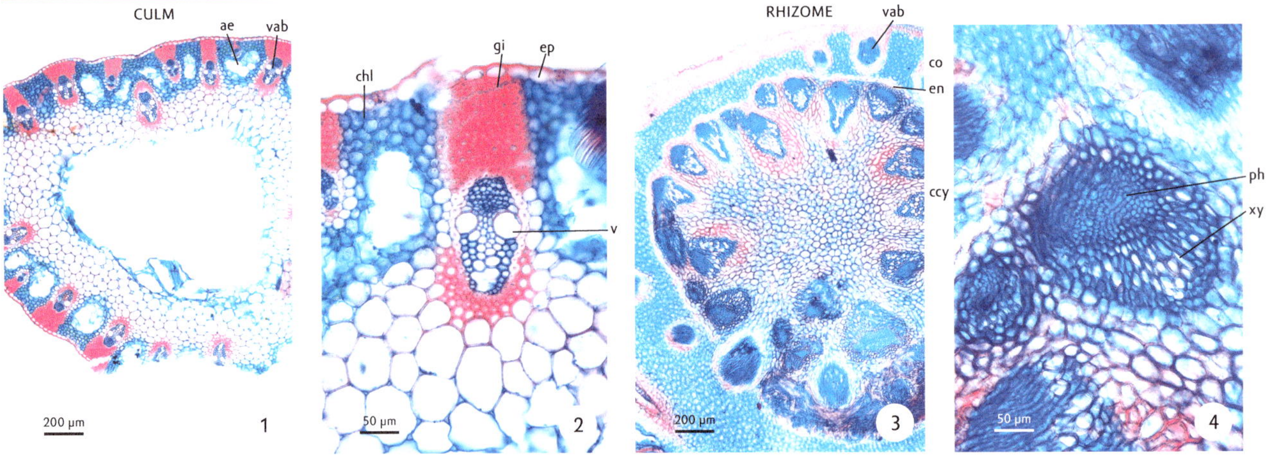

Perennial wetland, densely tufted plant. Stems leafy, with acute angles, up to 60 cm high. Leaves green, alternating. Homomorphic spikelets with female flowers in apical part. Utricles ovoidal with winged beak in upper half.

Distribution and ecology

Euro-Siberian boreo-temperate element; widely naturalised outside its native range. In all CE countries quite common, especially on acidic wet grasslands, frequent in upland pastures. It occurs also on damp meadows, woodland edges and in ruderal habitats.

Anatomy

Acutely triangular hollow culm with a diameter of 1.2–1.5 mm (1).

The epidermis consists of lignified, round cells with externally thick walls (2). The chlorenchyma consists of external layer of unlignified parenchyma cells. The middle part contains collateral vascular bundles. Their fiber sheaths are externally linked with the epidermis with rectangular girders of lignified cells (1, 2). The vascular bundles consist of a xylem with two large lateral vessels and a roundish phloem with sieve tubes and companion cells. Air ducts are partially present (1, 3). Vascular bundles are surrounded by 3–4 layers of thin- to thick-walled sclerenchyma cells (1, 2). The pith is empty and contains remnants of unlignified cells (1, 3).

Rhizome with a diameter of 1.5–1.8 mm (3).

The cortex is composed of an external layer of parenchyma cells without intercellulars (3). The central cylinder and the cortex are separated by a thick-walled endodermis with lignified U-shaped cells (4). The lignified parenchymatic ground tissue of the central cylinder contains amphivasal vascular bundles with large sclerenchyma sheaths (4).

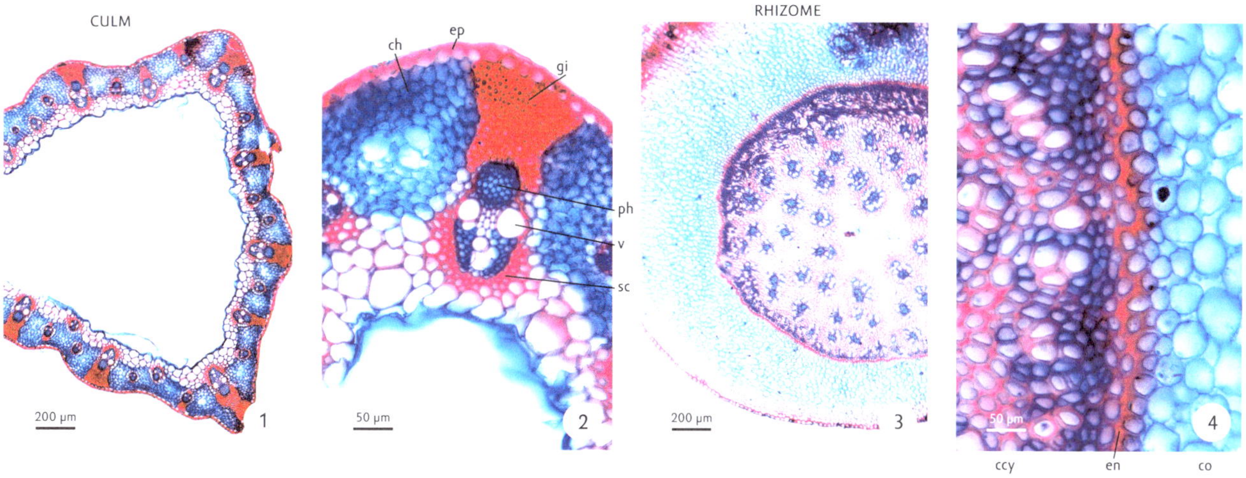

Perennial wetland rhizomatous plant, stems not or loosely tufted, distinctly stoloniferous (10–50 cm). Flower stems far extending the leaves, 10–40 cm long, flower spikes pendulate, narrow leaves 1–1.5 mm wide distinctly grey-green.

Distribution and ecology

Circumpolar boreo-montane element. Scattered irregularly and rarely in all CE countries, from lowlands to mountains, more frequently in uplands to mountains. Very wet habitats in peat bogs and fens (blanket- or valley bogs), wet peaty meadows, typically on the margin of bog lakes, sometimes as an amphibious plant, always very acidic, base-poor soils and waters. It does not tolerate eutrophication, competitively very weak species.

Anatomy

Obtusely triangular culm with a diameter of 1–2 mm (1).

The epidermis consists of bulliform, lignified cells (2). The chlorenchyma consists of unlignified parenchymatic cells. The middle part is dominated by air spaces and collateral vascular bundles (2). Their fiber sheaths are mostly externally linked with the epidermis with girders of lignified cells. The vascular bundles consist of a xylem with two large lateral vessels and a roundish phloem with sieve tubes and companion cells (2). Vascular bundles are surrounded by 1–3 layers of thin- to thick-walled sclerenchymatic cells (2). The external part of the partially hollow pith consists of unlignified thin-walled, polygonal parenchymatic cells (1, 2).

Rhizome with a diameter of 3–4 mm (3).

The cortex is composed of an external layer of lignified parenchymatic cells and an inner part of thin-walled, unlignified parenchyma with irregularly formed radially elongated intercellulars and a few vascular bundles (3). The central cylinder and the cortex are separated by a thick-walled endodermis with lignified U-shaped cells (3). The lignified parenchymatic ground tissue of the central cylinder contains amphivasal vascular bundles with uniseriate sclerenchymatic sheaths (3).

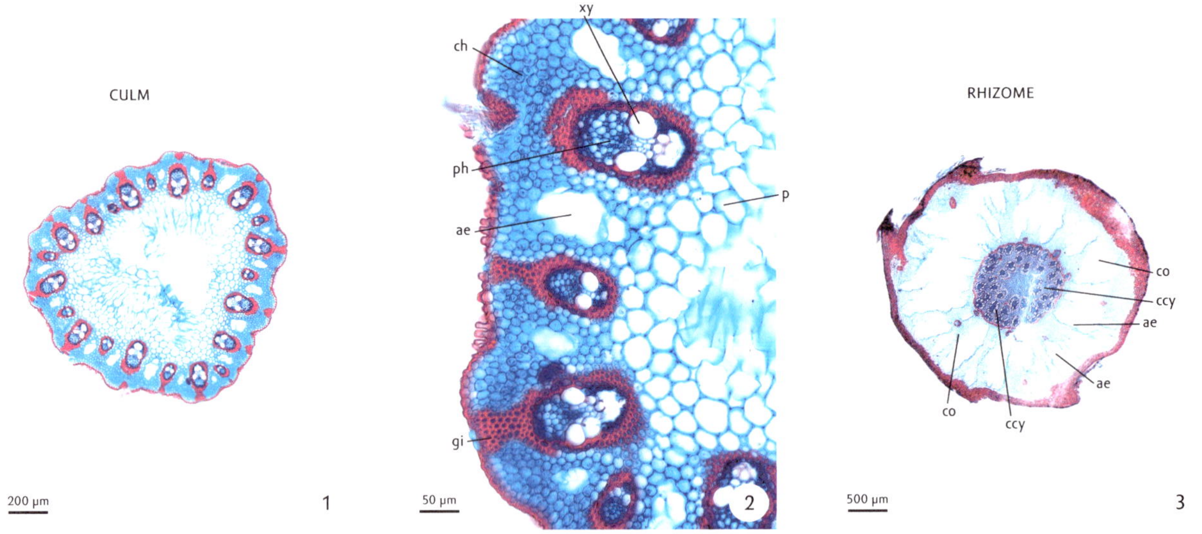

Carex magellanica subsp. *irrigua* (Wahlenb.) Hiitonen

Cyperaceae

(syn. *Carex paupercula*)

Tall Bog-sedge

Perennial wetland herb, short undergroung tillers forming sparse tufts. Stems erect, 10–30 cm high, slender, little inclined above, rounded. Basal sheaths scale-like, purpur or brown. Leaves 2–4 mm wide, green or grey-green. Spikelets heteromorphic, only 1 male spike, female spikes on long stalkes, later pendulous. Utricles broad ovate, glabrous, greyish or bluish-brown, apex with no or small beak.

Distribution and ecology

Circumpolar boreo-montane element. In all CE countries except HU. Oligotrophic and mesotrophic peaty meadows, raised bogs. Heliophilous, competitive weak species, sensitive to drainage.

Anatomy

Obtuse triangular hollow culm with a diameter of 1–1.5 mm (1).

The epidermis consists of lignified, round cells with externally thick walls (2). The chlorenchyma consists of external layer of unlignified parenchyma cells. The middle part is dominated by air spaces with spongy remains of cell walls (1–2) and collateral vascular bundles. Their fiber sheaths are externally linked with the epidermis with rectangular girders of lignified cells (1, 2). The vascular bundles consist of a xylem with two large lateral vessels and a roundish phloem with sieve tubes and companion cells. Air ducts are partially present (1, 2). Vascular bundles are surrounded by 1–2 layers of thin- to thick-walled sclerenchyma cells (1, 2). The pith is almost empty and contains remnants of unlignified cells (1, 3).

Rhizome with a diameter of 2–2.5 mm (3).

The cortex is composed of an external layer of parenchyma cells without intercellulars (3). It contains a few vascular bundles. The central cylinder and the cortex are separated by a thick-walled endodermis with lignified U-shaped cells (4). The unlignified parenchymatic ground tissue of the central cylinder contains amphivasal vascular bundles with large sclerenchyma.

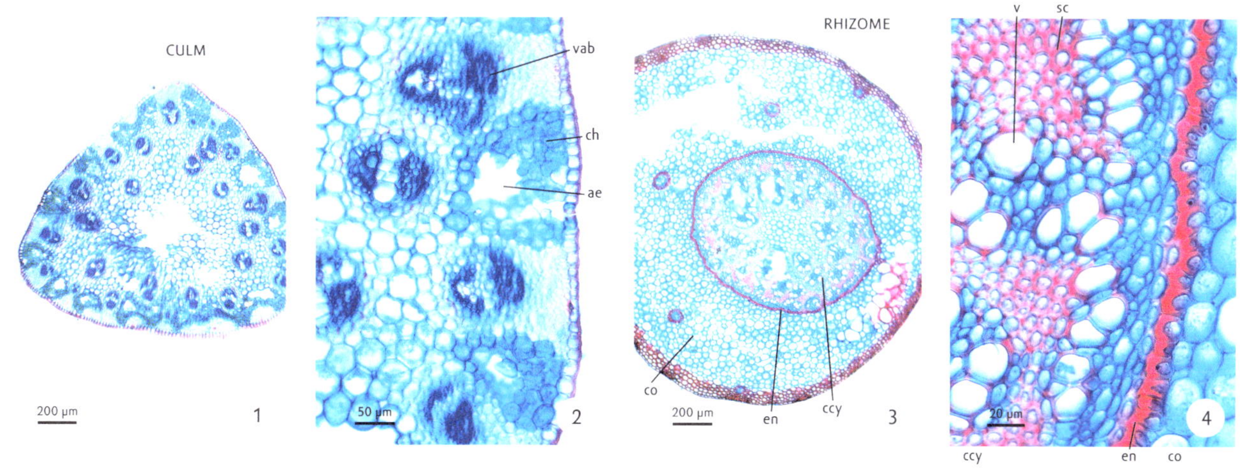

Carex melanostachya M. Bieb. ex Willd. **Great Plains Sedge** Cyperaceae

Perennial wetland herb, stems erect, 30–100 cm high, slender, bluntly triangular, glabrous, scabrous above. Spikelets heteromorphic. Leaves tough, longer than stems, dark green and shiny upside, greyish-green and matt underside. Utricles biconvex, obovate, markedly grooved, greyish-brown, beak short and broad.

Distribution and ecology

Eurasian subcontinental element. In all CE countries except CH. Floodplain lowland meadows, margins of inland salt marshes, rarely inside floodplain forests. It tolerates fluctuation of water level incl. short-term flooding.

Anatomy

Obtusely triangular culm with a diameter of 1–2 mm (1).

The epidermis consists of lignified, round cells with externally thick walls (2). The chlorenchyma consists of unlignified parenchyma cells. The middle part is dominated by large air spaces with spongy remains of cell walls and collateral vascular bundles (2). Their fiber sheaths are mostly externally linked with the epidermis with girders of thin- to thick-walled, lignified cells (1, 2). The vascular bundles consist of a xylem with two large lateral vessels and a roundish phloem with sieve tubes and companion cells (2). Vascular bundles are surrounded by 1–3 layers of thick-walled sclerenchyma cells (2). The pith consists of lignified and unlignified pentagonal thin-walled parenchyma cells (1, 2).

Rhizome with a diameter of 3–4 mm (3).

The cortex is composed of two external layers of unlignified and lignified parenchyma cells and an inner part of unlignified thin-walled parenchyma with many intercellulars formed by elongated parenchyma cells (3). The central cylinder and the cortex are separated by a thick-walled endodermis and a thick-walled pericycle (4). The unlignified parenchymatic ground tissue of the central cylinder contains amphivasal vascular bundles with 1–3-seriate sclerenchyma sheaths (5).

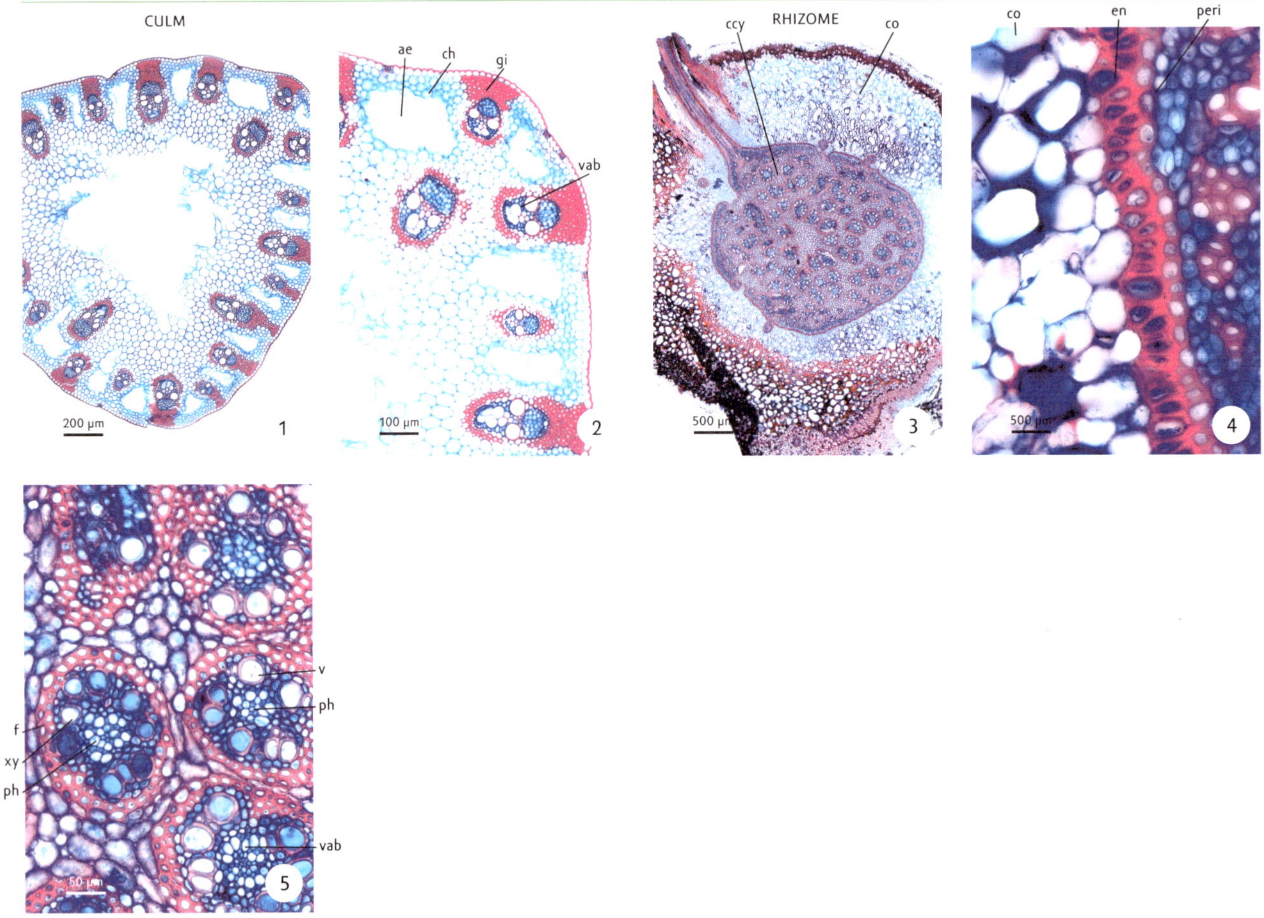

Perennial wetland shortly rhizomatous plant, single or tufted stems 10–70 cm tall, rounded angles, smooth.

Distribution and ecology

Euro-Siberian boreo-temperate element; also in N. America. Scattered irregularly and frequently in all CE countries, from lowlands to mountains. Wet and peaty meadows, margins of peat bogs and fens, subalpine meadows, shorelines of fishponds and lakes, wet paths, peaty and sandy soils, prefers acidic to neutral soils.

Anatomy

Acutely triangular culm with a diameter of 1–1.5 mm (1).

The epidermis consists of lignified, round cells with externally thick walls (2). The chlorenchyma consists of unlignified parenchymatic cells. The middle part is dominated by large air spaces filled with spongy remains of cell walls and collateral vascular bundles (1, 2). Their fiber sheaths are mostly externally linked with the epidermis with girders of thin- to thick-walled, lignified cells (1, 2). The vascular bundles consist of a xylem with two large lateral vessels and a roundish phloem with sieve tubes and companion cells (2). Vascular bundles are surrounded by 1–3 layers of thick-walled sclerenchymatic cells (2). The pith contains vascular bundles and consists of lignified and unlignified pentagonal, thin-walled parenchymatic cells (1, 2).

Rhizome with a diameter of 2–3 mm (3).

The cortex is composed of an external layer of lignified parenchymatic cells and an inner part of thin-walled, unlignified parenchyma with many radially elongated intercellulars and a few rudimentary vascular bundles (3). The central cylinder and the cortex are separated by a thick-walled endodermis and a thick-walled pericycle (3). The unlignified parenchymatic ground tissue of the central cylinder contains amphivasal vascular bundles with 1–3-seriate sclerenchymatic sheaths (3).

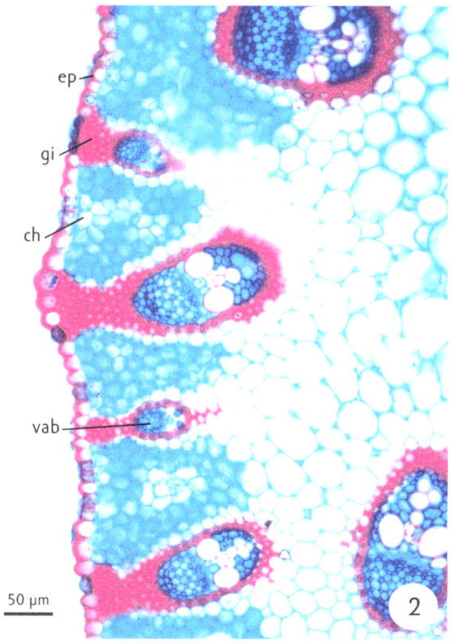

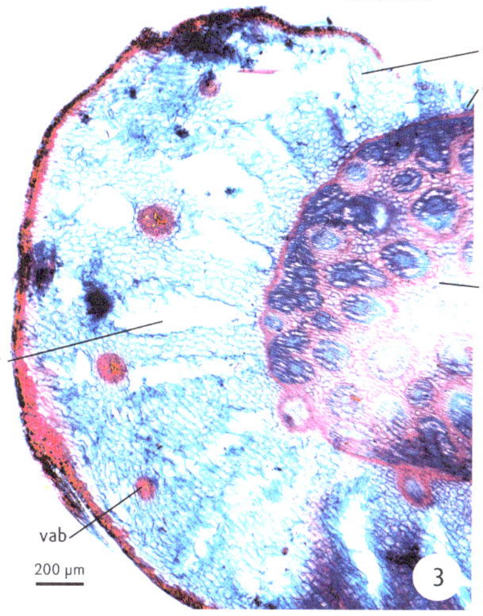

Carex oederi Retz. (syn. *C. serotina, C. viridula*) **Small-fruited yellow-sedge** Cyperaceae

Perennial wetland herb, small, tufted. Stems 5–25 cm high, erect, usually shorter than leaves. Leaves partly winter-green, tough, erect, pale green or yellowish-green. Spikelets heteromorphic, markedly clustered on the top of the stem, one male spike. Utricles obovate, greenish-yellow, sharply narrowed into long beak. Part of Carex flava agg.

Distribution and ecology

Eurasian boreo-temperate element. Scattered in all CE countries. Disturbed sites in fen meadows, in fen springs, exposed sand bottoms, partly halophilous.

Anatomy

Obtusely triangular culm with a diameter of 1 mm (1).

The epidermis consists of lignified, round cells with externally thick walls (2). The chlorenchyma consists of unlignified parenchyma cells. The middle part is dominated by air spaces, filled with spongy remains of cell walls and collateral vascular bundles (1, 2). Their fiber sheaths are externally mostly linked with the epidermis with large girders of thick-walled, lignified cells (2). The vascular bundles consist of a xylem with two large lateral vessels and a roundish phloem with sieve tubes and companion cells (2). The pith consists of unlignified thin-walled parenchyma cells (1, 2).

Rhizome with a diameter of 2 mm (3).

The cortex is composed of an external layer of partially lignified parenchyma cells and an inner part of unlignified thin-walled parenchyma with many very small intercellulars (3). The central cylinder and the cortex are separated by a thick-walled endodermis (3). The unlignified parenchymatic ground tissue of the central cylinder contains amphivasal vascular bundles with uni- to biseriate sclerenchyma sheaths (3).

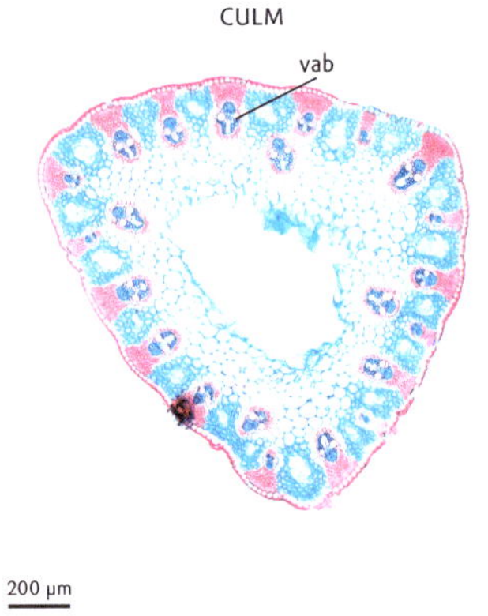

1

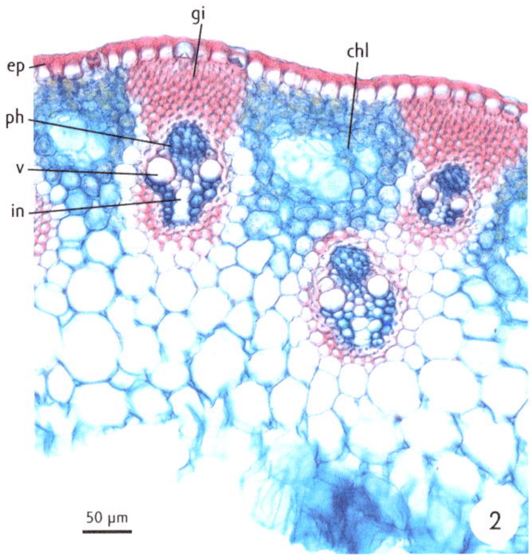

2

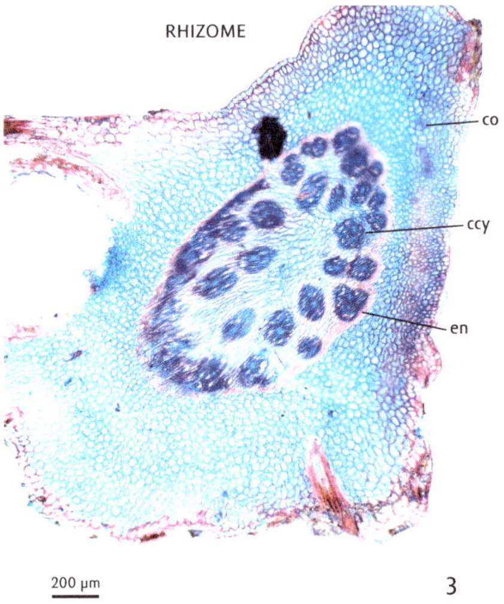

3

Carex otrubae Podpěra **False Fox-sedge** Cyperaceae

Perennial wetland herb, tufted, stems 30–105 cm high, triangular, askew oriented in tuft. Basal sheaths pale brown, not decaying. Leaves green to greyish-green, with 10–17 mm long pointed ligula. Spikelets homomorphic. Utricles broad ovate, pale or dark-brown, beak long. Similar to C. vulpina with distinctly winged stems and short rounded ligula.

Distribution and ecology

Euro-Siberian southern-temperate distribution. In all CE countries. Wet meadows on mineral rich sites, heavy clayish soils, with fluctuating water level, common also in ruderal stands, in nitrophilous vegetation, in ditches.

Anatomy

Acutely triangular culm with a diameter of 2–3 mm (1).

The epidermis consists of lignified, round-shaped cells with externally thick walls. (2). The chlorenchyma consists of unlignified parenchymatic cells. The middle part is dominated by air spaces filled with spongy remains of cell walls and collateral vascular bundles (1, 2). Their fiber sheaths are mostly externally linked with the epidermis and large girders of thick-walled, lignified cells (2). The vascular bundles consist of a xylem with two large lateral vessels and a roundish phloem with sieve tubes and companion cells (2). Vascular bundles are centripetally accompanied by a cup of thick-walled sclerenchymatic cells (2). The pith consists of lignified and unlignified pentagonal, thin-walled parenchymatic cells (1, 2).

Rhizome with a diameter of 3–4 mm (4).

The cortex is composed of an external layer of lignified parenchymatic cells and an inner part of unlignified thin-walled parenchyma with many very small intercellulars (4). The central cylinder and the cortex are separated by a thick-walled endodermis (4). The partially lignified parenchymatic ground tissue of the central cylinder contains amphivasal vascular bundles with uni- to biseriate sclerenchymatic sheaths (5).

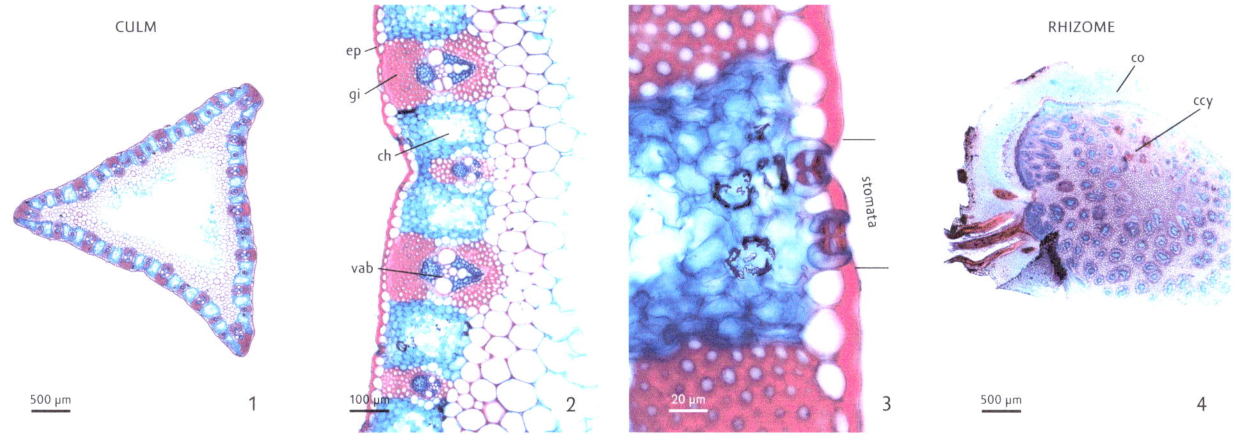

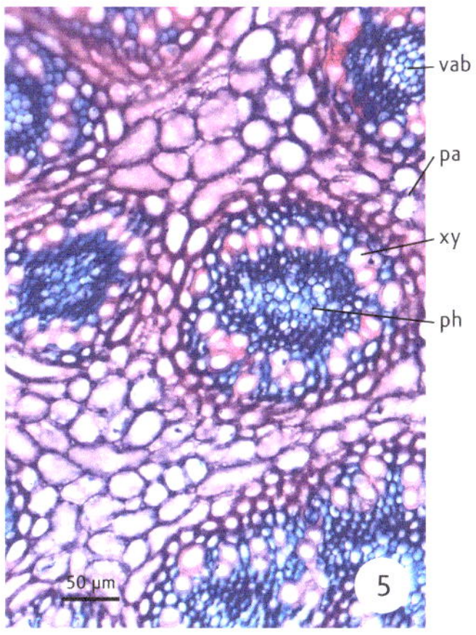

Perennial wetland herb, tufted, short extravaginal tillers. Basal sheaths scale-like, brown-red. Leaves grassy green, hairy on the upper side and along margins. Stems erect, 15–70 cm high, slender, sharply triangular, scabrous only above. Spikelets heteromorphic. Utricles prolonged obovate, sandy or reddish-sandy, bluntly triangular, nearly spherical without beak.

Distribution and ecology

Euro-Siberian boreo-temperate element; also in N America. In all CE countries.

Broad ecological amplitude, both forested and non-forested habitats, mesophilous, wet, peaty meadows and forests, disturbed sites in meadows and forests. Usually on moist, mildly acidic to neutral soils.

Anatomy

Acutely triangular hollow culm with a diameter of 1.5–1.8 mm (1).

The epidermis consists of lignified, round cells with externally thick walls (2). The middle part is dominated by air spaces with spongy remains of cell walls (1, 2) and collateral vascular bundles. The center contains collateral vascular bundles. Their fiber sheaths are externally linked with the epidermis with rectangular girders of lignified cells (1, 2). The vascular bundles consist of a xylem with two large lateral vessels, a roundish phloem with sieve tubes and companion cells and air ducts (1, 2). Vascular bundles are surrounded by 2–3 layers of thin- to thick-walled sclerenchyma cells (1, 2). The pith contains remnants of unlignified cells (1).

Rhizome with a diameter of 2.5–3.5 mm (3).

The cortex is composed of an external layer of parenchyma cells without air spaces (3). The central cylinder and the cortex are separated by a thick-walled endodermis (3) with lignified U-shaped cells. The unlignified parenchymatic ground tissue of the central cylinder contains amphivasal vascular bundles with large sclerenchyma sheaths (3).

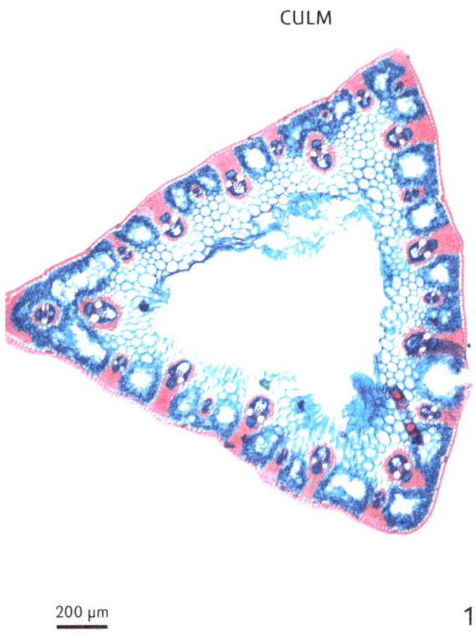

1

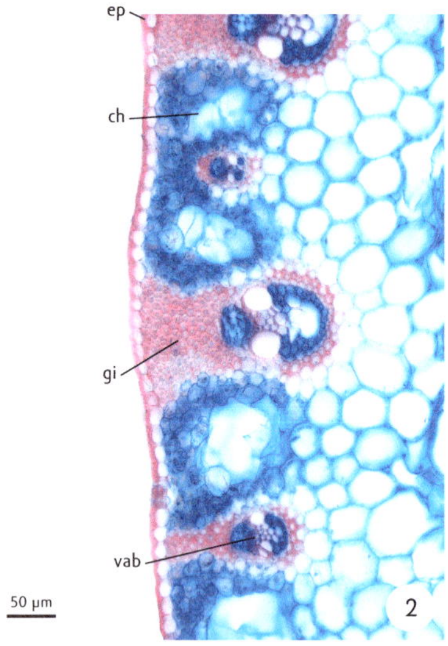

2

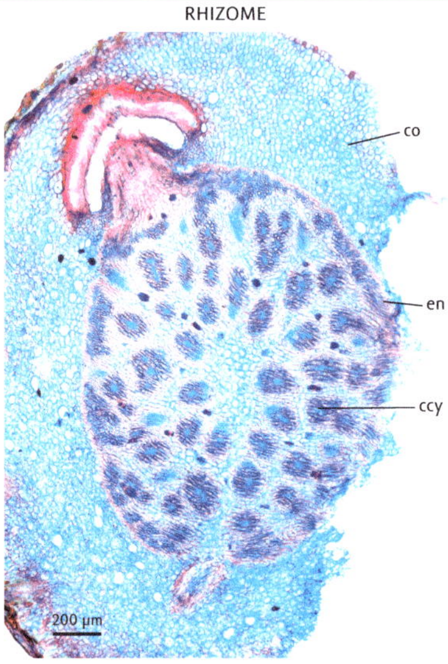

3

Carex panicea L.

Carnation Sedge

Cyperaceae

Perennial wetland herb, non-tufted, underground tillers 5–8 cm long. Stems erect, 10–50 cm high, slender, bluntly triangular, glabrous. Basal sheaths white or gray. Leaves same colour on both sides, grey-green or greyish, with keel underside. Spikelets heteromorphic. Utricles slightly asymmetric, elliptic, yellowish or reddish-brown, apex abruptly narrowed into short asymmetric beak.

Distribution and ecology

European boreo-temperate element, also in C Asia. In all CE countries. Wet and peaty grasslands, mesotrophic and minerotrophic fens, springs, alderwoods, on sandy and clayish soils. Growth probably limited more by N than P in boreal fen vegetation.

Anatomy

Round to obtuse triangular wavy or ribbed hollow culm with a diameter of 1.2–2 mm (1).

The epidermis consists of lignified, bulliform cells (3). The chlorenchyma consists of external layer of unlignified parenchyma cells. The middle part is dominated by air spaces with spongy remains of cell walls (1–3) and collateral vascular bundles. Their fiber sheaths are externally linked with the epidermis with rectangular girders of lignified cells (1, 2). The vascular bundles consist of a xylem with two large lateral vessels and a roundish phloem with sieve tubes and companion cells. Air ducts are partially present (1, 3). Vascular bundles are surrounded by 1–2 layers of thin- to thick-walled sclerenchyma cells (1, 2). The pith is empty and contains remnants of unlignified cells (1, 3).

Rhizome with a diameter of 2–2.5 mm (4).

The cortex is composed of an external layer of parenchyma cells without intercellulars (3). The central cylinder and the cortex are separated by a thick-walled endodermis with lignified U-shaped cells (4). The unlignified parenchymatic ground tissue of the central cylinder contains amphivasal vascular bundles with small sclerenchyma sheaths (4, 5).

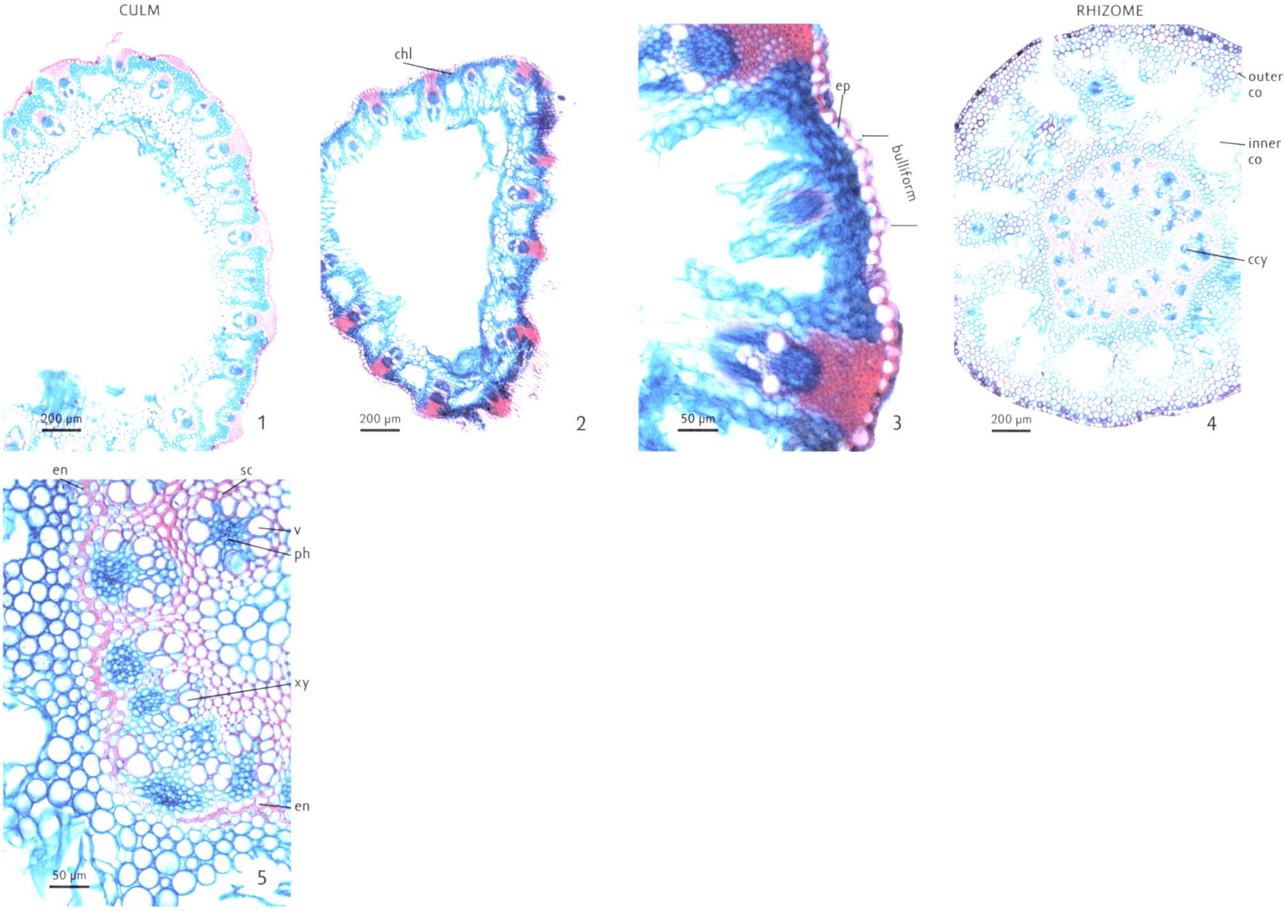

Carex pendula Hudson

Pendulous Sedge

Cyperaceae

Perennial wetland herb, densely tufted, huge greyish-green tufts, without tillers. Stems erect, 50–200 cm high, sharply triangular, rough, slender. Basal sheaths scale-like, dark purpur. Leaves dark-green, shine upside, matt underside. Spikelets heteromorphic, female spikes long, dense on long stalkes, pendulous. Utricles elliptic or ovate, inflated, pale green or brownish-green, glabrous, beak short and narrow.

Distribution and ecology

European southern-temperate element. In all CE countries. Banks of forest streams, forest springs, spring alderwoods. Germination in late spring and summer at disturbed, open sites (forest gaps) and the capability to form long-persistent seed bank. It accumulates considerable amounts of lead, particularly in root biomass.

Anatomy

Acutely triangular culm with a diameter of 5–8 mm (1).

The epidermis consists of bulliform, lignified cells (2). The chlorenchyma consists of external layer of unlignified parenchyma cells. The middle part is dominated by large air spaces with spongy remains of cell walls (2) and collateral vascular bundles. Their fiber sheaths are externally linked with the epidermis with triangular girders of lignified cells. The vascular bundles consist of a xylem with two large lateral vessels, a roundish phloem with sieve tubes and companion cells and an air duct (3). Vascular bundles are surrounded by 1–3 layers of thin- to thick-walled sclerenchyma cells (3). The pith consists of unlignified thin-walled polygonal parenchyma cells (1).

Rhizome with a diameter of 6–9 mm (4).

The cortex is composed of an external layer of parenchyma cells without intercellulars, but with many single large cells. Few vascular bundles exist in the cortex (4). The central cylinder and the cortex are separated by a thick-walled endodermis with lignified U-shaped cells (4). The unlignified parenchymatic ground tissue of the central cylinder contains amphivasal vascular bundles with large sclerenchyma sheaths (4,5).

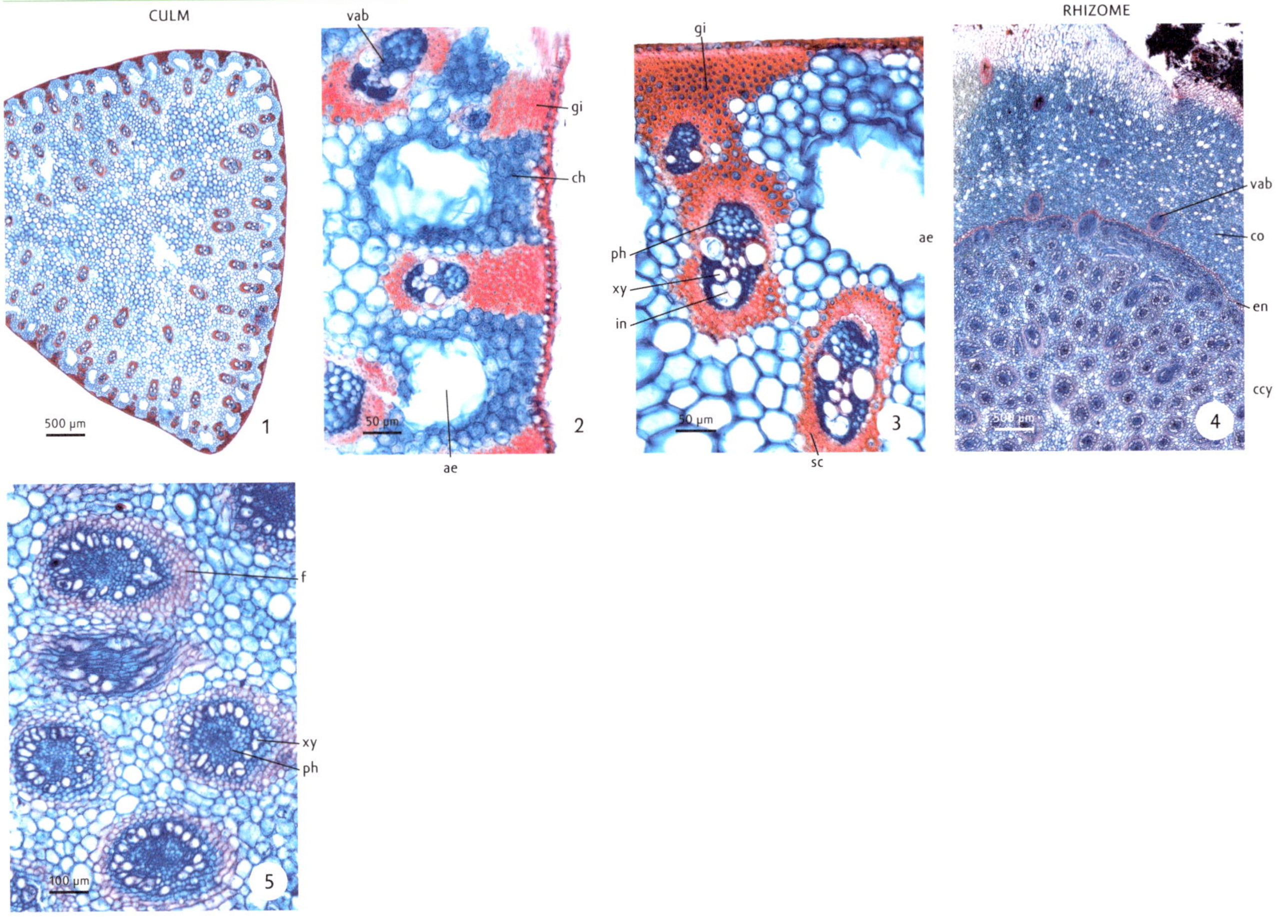

Carex pseudocyperus L.

Cyperus Sedge Cyperaceae

Perennial wetland rhizomatous plant, stems loosely tufted, 40–80 cm tall with acute angles, spikes distinctly pendulate, leaves light green to yellow-green.

Distribution and ecology

Euro-Siberian temperate element, also in E Asia and N America. Scattered irregularly in all CE countries, from lowlands to uplands, more frequently in fishpond or lake areas. Shorelines of lakes and fishponds, river banks, backwater pools and oxbows, canals, ditches, swamps, alder carrs, peaty, silty or sandy soils. Tolerates slightly acidic to neutral soils (pH ca. 5.5–7) and eutrophication. Photophilous species.

Anatomy

Acutely triangular culm with a diameter of 3–4 mm (1).

The epidermis consists of lignified, round cells with externally thickened walls (2). The small chlorenchyma consists of unlignified parenchymatic cells. The middle part is dominated by large, empty air spaces and collateral vascular bundles (1, 2). Their fiber sheaths are mostly externally linked with the epidermis with large girders of thick-walled, lignified cells (2). The vascular bundles consist of a xylem with two large lateral vessels and a roundish phloem with sieve tubes and companion cells (2). Some vascular bundles are centripetally accompanied by a cup of thick-walled sclerenchymatic cells (2). The pith consists of lignified and unlignified pentagonal, thin-walled parenchymatic cells.

Rhizome with a diameter of 4–5 mm (3).

The cortex is composed of an external layer of lignified parenchymatic cells and an inner part of thin-walled, unlignified parenchyma with many very small intercellulars (3). The central cylinder and the cortex are separated by a thick-walled endodermis. The lignified parenchymatic ground tissue of the central cylinder contains amphivasal vascular bundles with uni- to biseriate sclerenchymatic sheaths (4).

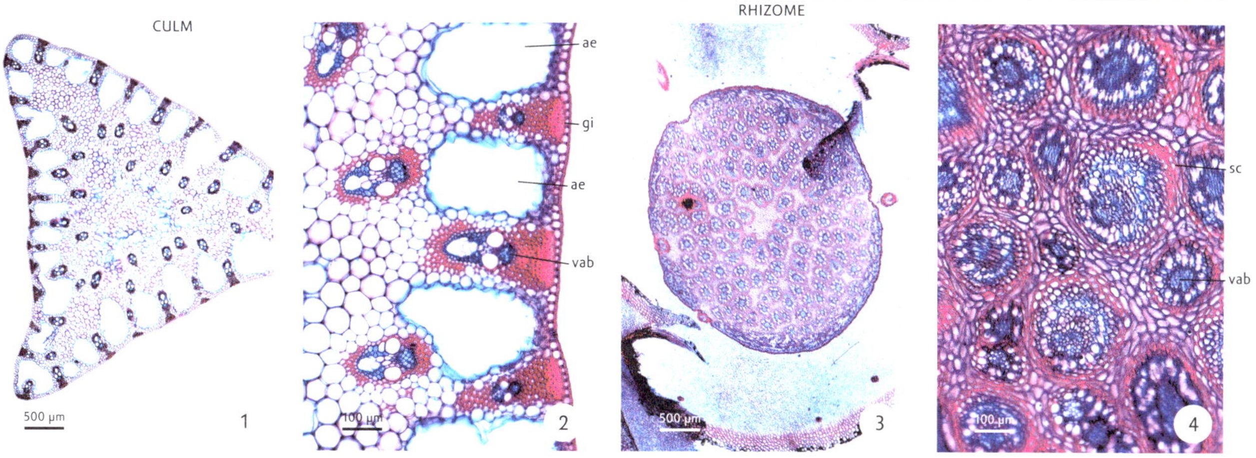

Carex pulicaris L.

Flea Sedge

Perennial wetland herb, soft dense tufts, markedly orange roots, short tillers. Stems erect, 5–40 cm high, thin, glabrous. Leaves tough, bristly, dark green. Spikelets homomorphic, one cylindrical spike. Utricles prolonged ellipsoid, acuminate to beak, smooth, dark brown to blackish-brown, later askew downwards oriented.

Distribution and ecology

Suboceanic temperate element. In all CE countries except HU. Peaty and fen meadows, peat bog margins, usually in stands with other small sedges and high cover of mosses. Prefers suboceanic climate. Sensitive to drainage and shade intolerant.

Anatomy

Obtusely pentagonal culm with a diameter of <1 mm (1).

The epidermis consists of lignified, round cells with externally thick walls (2). The chlorenchyma consists of unlignified parenchyma cells. The middle part is dominated by air spaces filled with spongy remains of cell walls and collateral vascular bundles (1, 2). Their fiber sheaths are externally linked with the epidermis with large triangular girders of thick-walled, lignified cells (1, 2). The vascular bundles consist of a xylem with two large lateral vessels and a roundish phloem with sieve tubes and companion cells (2). Vascular bundles are surrounded by a sheath of thick-walled sclerenchyma cells (2). The pith consists of unlignified thin-walled parenchyma cells (1).

Rhizome with a diameter of 1 mm (3).

The cortex is composed of unlignified parenchyma cells with many very small intercellulars (3, 4). The central cylinder and the cortex are separated by a thick-walled endodermis (4). The lignified parenchymatic ground tissue of the central cylinder contains amphivasal vascular bundles with uni- to biseriate sclerenchyma sheaths (3).

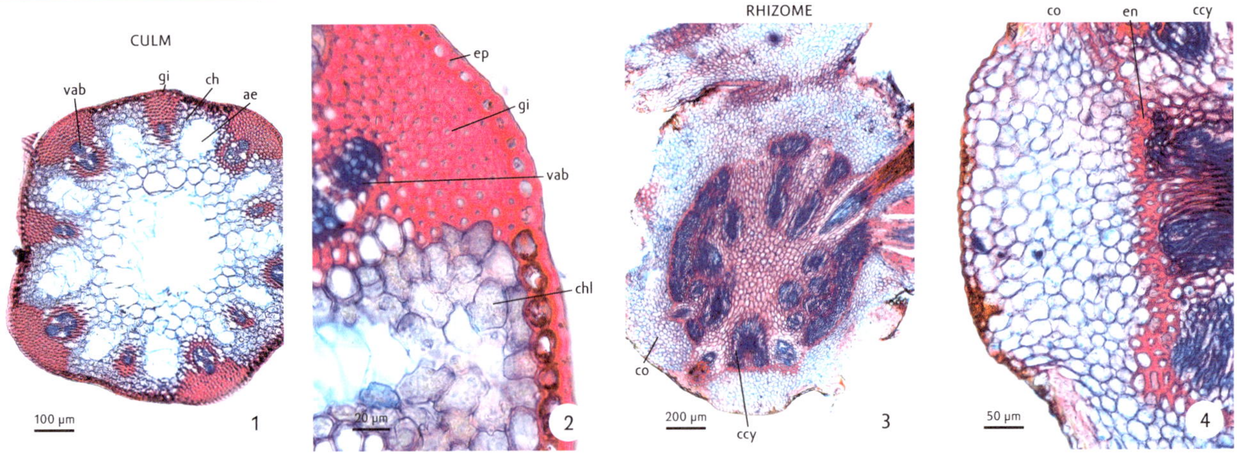

Carex remota L.

Remote Sedge

Cyperaceae

Perennial wetland herb, forming dense flat tufts, stems 30–60 cm high, thin, scabrous above, feeble to procumbent. Basal sheaths pale brown, decaying on thin fibres. Leaves very thin, green to yellowish-green. Spikelets homomorphic. Utricles ovoid-clavate, dorsal side convex, pale-brown, abruptly narrowed into beak.

Distribution and ecology

European temperate element, also in C and E Asia. In all CE countries. Stream banks, forest springs, floodplain forests. Tolerates soil compaction and shade conditions. The capability to form long-term, persistent seed banks. Tolerates high soil concentration of Fe and Mn. Flooding tolerant.

Anatomy

Obtusely pentagonal culm with a diameter of 1 mm (1).

The epidermis consists of lignified, round cells with externally thick walls (2). The small chlorenchyma consists of unlignified parenchyma cells. The middle part is dominated by air spaces, filled with spongy remains of cell walls and collateral vascular bundles (1, 2). Their fiber sheaths are rarely externally linked with the epidermis with large girders of thick-walled, lignified cells (2). The vascular bundles consist of a xylem with two large lateral vessels and a roundish phloem with sieve tubes and companion cells (2). The pith consists of unlignified thin-walled parenchyma cells (1).

Rhizome with a diameter of 2–2.5 mm (3).

The cortex is composed of an external layer of parenchyma cells without intercellulars (3). It contains a few vascular bundles. The central cylinder and the cortex are separated by a thick-walled endodermis with lignified U-shaped cells (4). The unlignified parenchymatic ground tissue of the central cylinder contains amphivasal vascular bundles with large sclerenchyma.

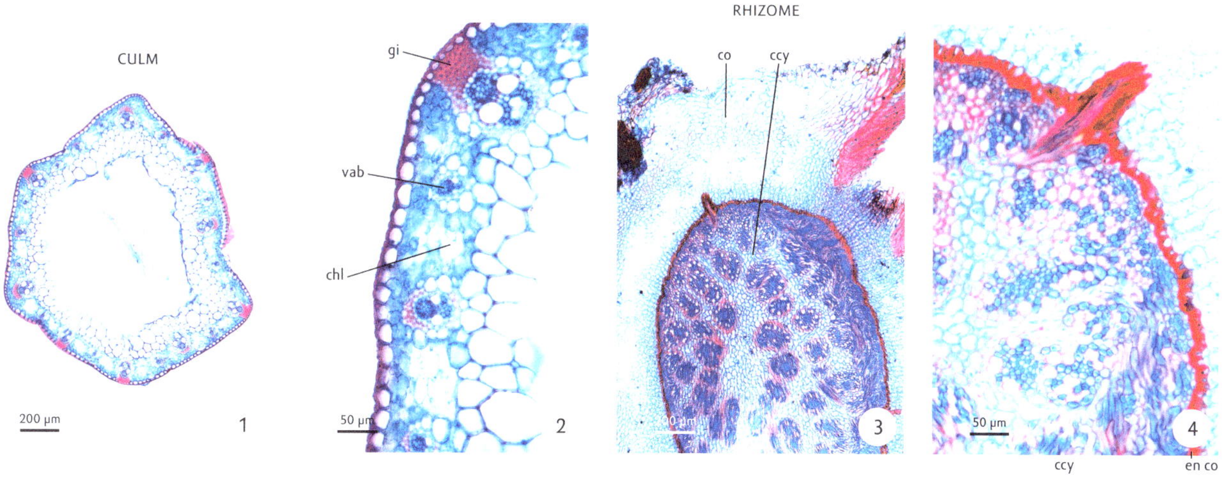

Perennial wetland tufted plant, long rhizomes. Flower stems 0.5–1.5 m tall, acute angles, leaves 10–25 mm wide.

Distribution and ecology

Euro-Siberian temperate element. Scattered frequently in all CE countries, from lowlands to uplands. Swamps, shorelines of fishponds, floodplains of rivers, wet meadows, wet forests, more common in lowlands and fishpond areas, meso- to eutrophic sites.

Anatomy

Acutely triangular culm with a diameter of 3–4 mm (1).

The epidermis consists of lignified, round cells with externally thick walls (2, 3). The chlorenchyma consists of unlignified parenchymatic cells. The middle part is dominated by air spaces, filled with spongy remains of cell walls and collateral vascular bundles (1, 2). Their fiber sheaths are externally linked with the epidermis with large girders of thick-walled, lignified cells (3). The vascular bundles consist of a xylem with two large lateral vessels and a roundish phloem with sieve tubes and companion cells (2). The pith contains vascular bundles and consists of unlignified thin-walled parenchymatic cells (1).

Rhizome with a diameter of 4–5 mm.

The cortex is composed of an external layer of parenchymatic cells without intercellulars and an inner layer with many large, radially arranged intercellulars (4). The central cylinder and the cortex are separated by a thin- to thick-walled endodermis (5). The unlignified parenchymatic ground tissue of the central cylinder contains closed collateral vascular bundles with large vessels and large sclerenchyma sheaths (6).

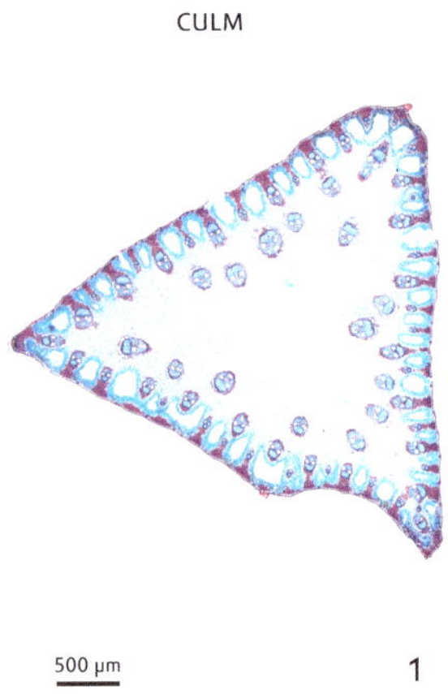

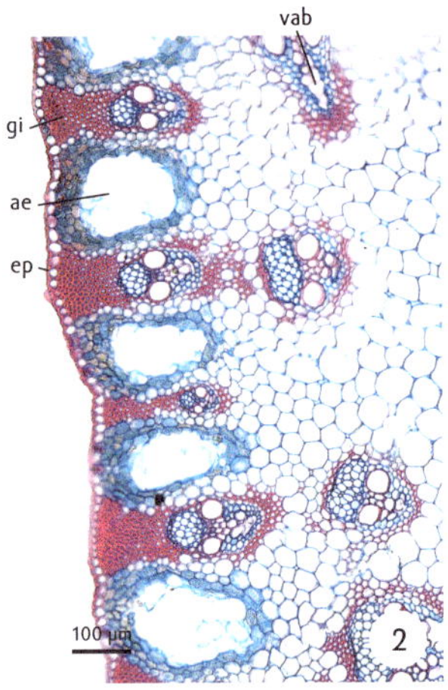

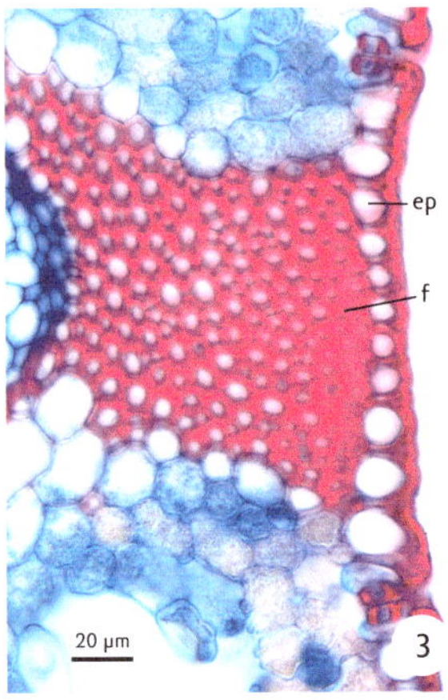

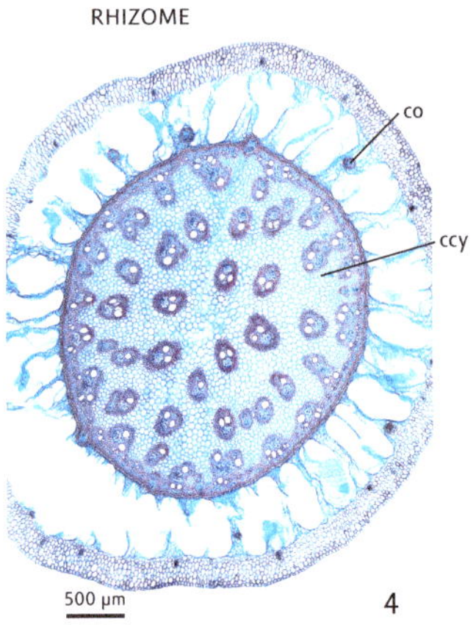

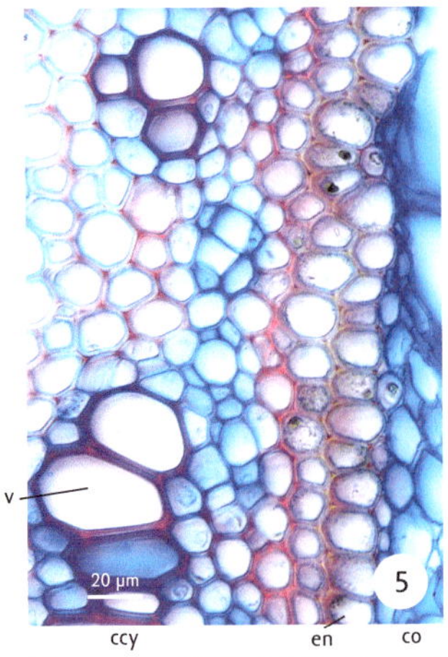

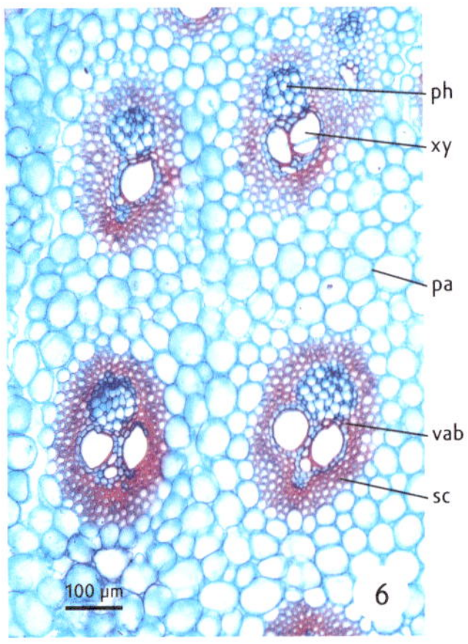

Perennial wetland plant, rhizomes 10–40 cm long. Flower stem 0.3–1.0 m high with rounded angles, distinctly grey-green leaves longer than the stem, leaf sheaths of lower leaves spongious to swollen.

Distribution and ecology

Circumpolar boreo-temperate element. Scattered commonly in all CE countries, from lowlands to mountains. Peat bogs, acidic fens, peaty pine forests, peaty meadows, fishpond and lake margins, alder carrs, usually very wet microhabitats, commonly 10–40 cm water depth. Oligo- to eutrophic sites, tolerates strongly acidic to neutral peaty soils and waters (pH ca. 3.5 to 7), typically base-poor soils, relatively eurytopic, usually lacking in lowlands, common in peatland areas in highlands and uplands.

Anatomy

Acutely triangular culm with a diameter of 1 mm (1).

The epidermis consists of lignified, round cells with externally thick walls (2). The chlorenchyma consists of unlignified parenchymatic cells. The middle part is dominated by indistinct air spaces, filled with spongy remains of cell walls and collateral vascular bundles (1, 2). Their fiber sheaths are externally mostly linked with the epidermis with large girders of thick-walled, lignified cells (2). The vascular bundles consist of a xylem with two large lateral vessels and a roundish phloem with sieve tubes and companion cells (2). The pith contains vascular bundles and consists of unlignified thin-walled parenchymatic cells (1).

Rhizome slide missing.

CULM

1

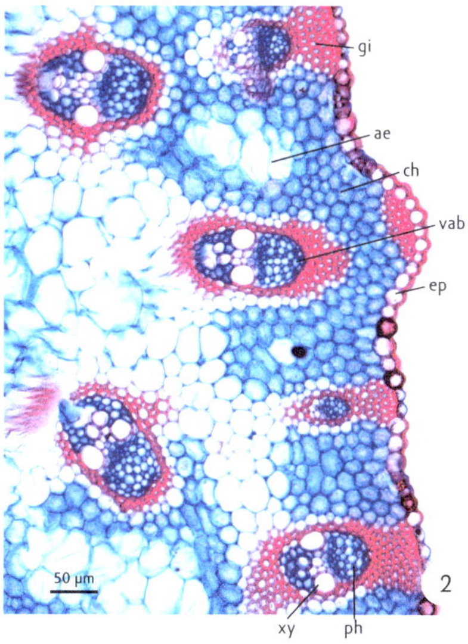

2

Carex secalina Willd. ex Wahlenberg

Perennial wetland herb, monoecious. Short-lived, densely tufted, tufts smaller, decaying. Basal sheaths chestnut-coloured or rust-brown. Stems erect or ascendent, 10–35 cm long. Leaves soft, pale green or green, shiny, with grooves. Spikelets heteromorphic. Utricles in spikes on the helix, pale brown, gradually narrowed into beak, with sharp teeth.

Distribution and ecology

Eurasian southern temperate to subcontinental element. In all CE countries except CH. Halophytic grassland communities, obligate halophyte. Salty field margins, wet ruderal and salty habitats, exposed banks of water reservoirs. On heavy clayish and salty soils. Utricles light, easily transportable by water on long distances.

Anatomy

Roundish culm with a diameter of 1–1.5 mm (1).

The epidermis consists of lignified, round cells with externally thick walls. The chlorenchyma consists of external layer of unlignified parenchyma cells. The middle part contains irregular air spaces (2) and collateral vascular bundles. Their fiber sheaths are externally linked with the epidermis with rectangular to triangular girders of lignified cells. Vascular bundles consist of a xylem with two large lateral vessels, a roundish phloem with sieve tubes and companion cells and sometimes with an air duct (2). Vascular bundles are centripetally surrounded by 1–3 layers of thin- to thick-walled sclerenchyma cells (3). The pith consists of unlignified thin-walled polygonal parenchyma cells (3).

Rhizome with a diameter of 1.5–2 mm (4).

The cortex is composed of an external layer of parenchyma cells without intercellulars. Few vascular bundles exist in the cortex (4). The central cylinder and the cortex are separated by a thick-walled endodermis with lignified U-shaped cells. The unlignified parenchymatic ground tissue of the central cylinder contains amphivasal vascular bundles with small sclerenchyma sheaths (5).

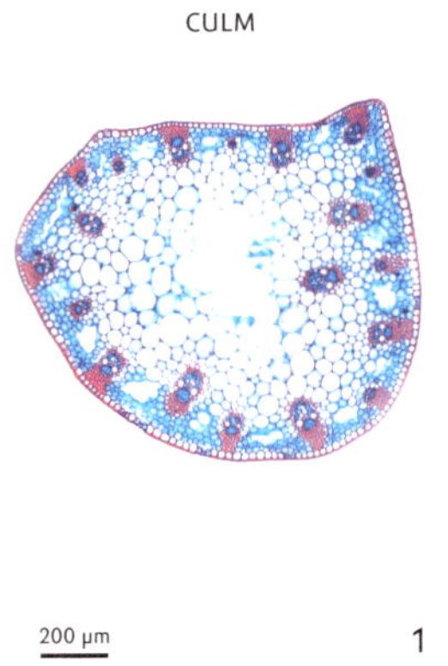

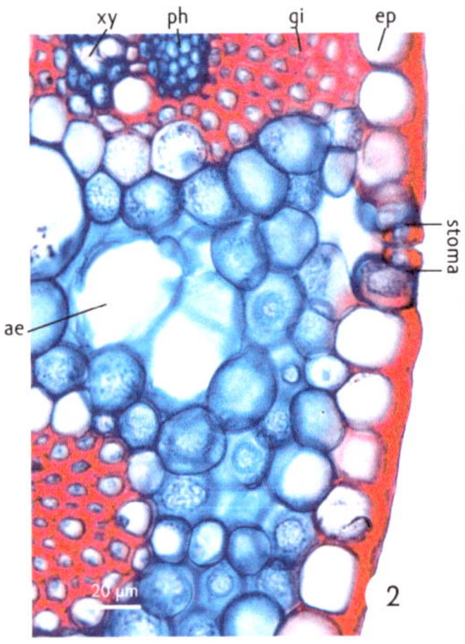

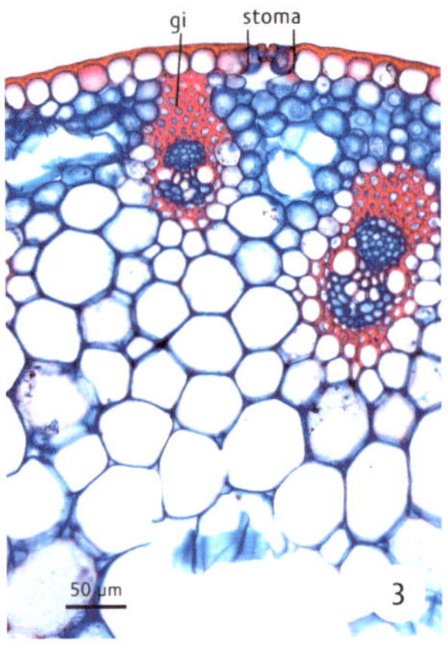

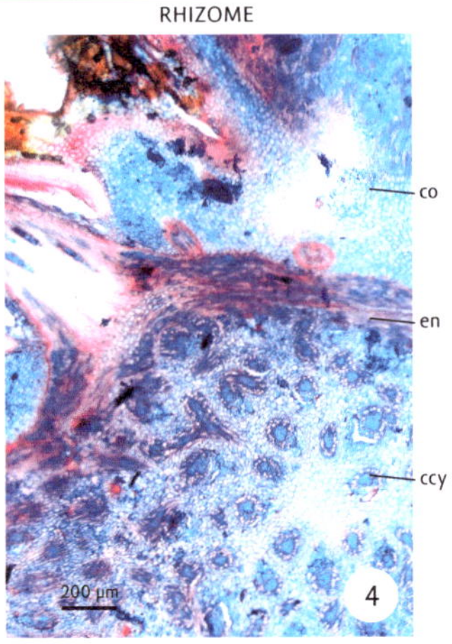

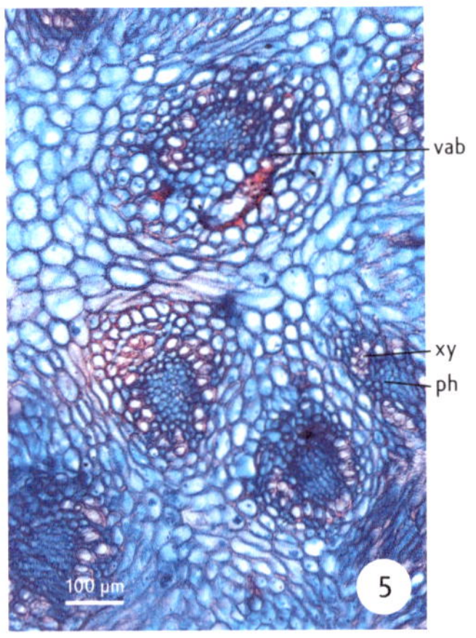

Carex vesicaria L.

Bladder-sedge

Perennial wetland, slightly tufted plant, very short rhizomes (3–6 cm), flower stem 0.3–1.2 m tall, obtusely triangular in section, of the same length as leaves, distinctly rough below the inflorescence.

Distribution and ecology

Circumpolar boreo-temperate element. Scattered commonly in all CE countries, from lowlands to mountains. Peat bogs, fens, wet meadows, fishpond and lake margins, usually very wet microhabitats, to 20 cm water depth. Meso- to eutrophic sites, tolerates acidic to neutral peaty soils and waters (pH ca. 4.5 to 7), base-poor to base-rich soils, relatively eurytopic, facultatively photophilous.

Anatomy

Acutely triangular culm with a diameter of 2 mm (1).

The epidermis consists of lignified, round-shaped cells with externally thick walls (2). The chlorenchyma consists of unlignified parenchymatic cells. The middle part is dominated by large air spaces, filled with spongy remains of cell walls, and collateral vascular bundles (1, 2). Their fiber sheaths are externally mostly linked with the epidermis with girders of thick-walled, lignified cells (2). The vascular bundles consist of a xylem with two large lateral vessels and a roundish phloem with sieve tubes and companion cells (2). The pith consists of unlignified thin-walled parenchymatic cells (1).

Rhizome with a diameter of 3–5 mm (3).

The cortex is composed of an external layer of partially lignified parenchymatic cells and an inner part of unlignified thin-walled parenchyma with many very small intercellulars (3). The central cylinder and the cortex are separated by a thick-walled endodermis with lignified u-shaped cells and a thick-walled pericycle (4). The lignified parenchymatic ground tissue of the central cylinder contains amphivasal vascular bundles with 2–3-seriate sclerenchymatic sheaths (5).

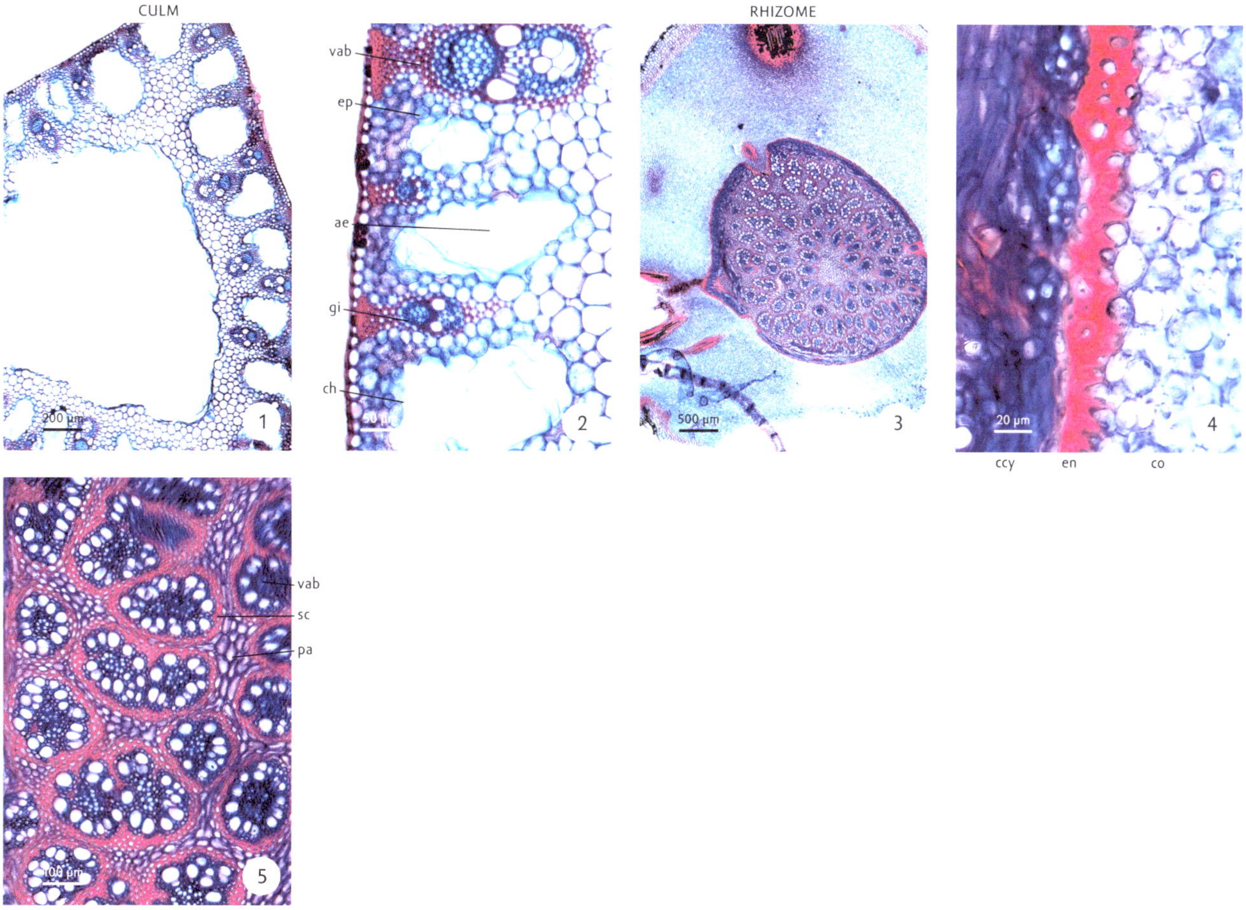

Carex vulpina L.

True Fox-sedge

Perennial wetland herb, forming big tufts. Basal sheaths grey, decaying into long brown fibres. Stems stiff, erect to inclined, 30–115 cm high, winged, considerably scabrous. Leaves green to dark green. Spikelets homomorphic. Utricles ovoid-ellipsoid, brown to dark-brown, gradually narrowed into dentate beak.

Distribution and ecology

Euro-Siberian temperate element, with a continental distribution in W Europe. In all CE countries. Moist and wet floodplain meadows, fishpond banks, eutrophic wetlands, floodplain forests. Usually on heavy clay soils that are flooded in winter and dry in summer, rarely sandy soils with high nutrient content.

Anatomy

Acute triangular culm with a diameter of 3–4 mm (1).

The epidermis consists of lignified, round cells with externally thick walls (2). Guard cells of the stomata are externally and internally thick-walled lignified (3). The chlorenchyma consists of unlignified parenchyma cells. The middle part is dominated by large air spaces, filled with spongy remains of cell walls and collateral vascular bundles (1, 2). Their fiber sheaths are externally mostly linked with the epidermis with girders of thick-walled, lignified cells (2). The vascular bundles consist of a xylem with two large lateral vessels and a roundish phloem with sieve tubes and companion cells (2). The pith consists of unlignified thin-walled parenchyma cells (1, 2).

Rhizome with a diameter of 3–5 mm (4).

The cortex is composed of an external layer of parenchyma cells without intercellulars, subepidermal semi-circular groups of sclerenchyma and an inner part of unlignified thin-walled parenchyma with many very small intercellulars (4). The central cylinder and the cortex are separated by a thick-walled endodermis with lignified U-shaped cells (4). The unlignified parenchymatic ground tissue of the central cylinder contains amphivasal vascular bundles with large sclerenchyma sheaths (4).

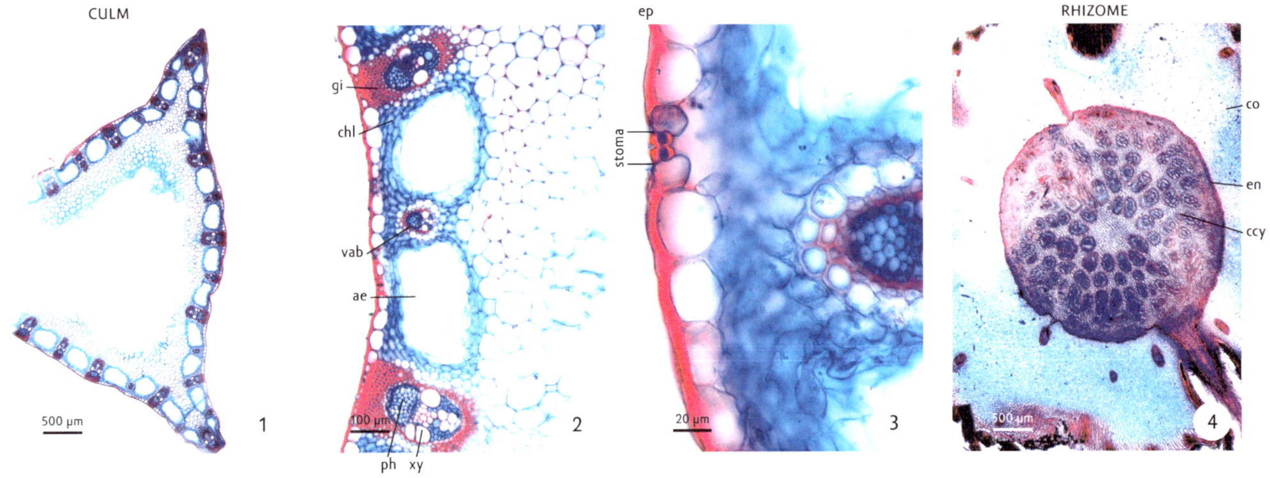

Cladium mariscus (L.) Pohl

Great Fen-sedge Cyperaceae

Perennial emergent herb, rhizomatous, stem erect up to 3 m with 3 rounded angles. Grey-green leaves up to 2 m high, channelled, with serrate edges and keel. Inflorescence branched with many rather compact heads.

Distribution and ecology

Euro-Siberian southern-temperate element, also in N America. In all CE countries, in base-rich areas in fens, in tall-herb fens and open fen carrs, along streams and fishponds, usually on peat. A relict species in fen mires in the middle course of the Labe (Elbe) River. Strong competitor forming compact and uniform stands. Inefficient dispersal of fruits and low survival ratio of seedlings.

Anatomy

Obtusely triangular culm with a diameter of 3–4 mm (1).

The epidermis consists of roundish, externally lignified thick-walled cells (2, 3). Guard cells of the stomata are thickened at the external and internal side (3). The chlorenchyma is composed of several layers of oval unlignified parenchymatic cells (2) and obtusely triangular to radially elongated groups of sclerenchyma (1, 2). The sclerenchymatic sheath of a few small vascular bundles is linked with the peripheral sclerenchymatic strands (1, 2). The middle part consists of large and small vascular bundles and radial-oval air spaces (1, 2). The large vascular bundles consist of a xylem with two large lateral vessels, an irregularly formed air duct and a roundish phloem with sieve tubes and companion cells (2). Vascular bundles are surrounded by 2–5 layers of sclerenchymatic cells (1, 2). Secretory cells are absent.

Circular rhizome with a diameter of 4–5 mm.

The cortex is composed of an epidermis, an external layer of lignified parenchymatic cells and an inner part of thin-walled unlignified parenchymatic cells with a net of air spaces (4). The central cylinder and the cortex are separated by a thin- to thick-walled, lignified endodermis/pericycle zone and U-shaped cells (4, 5). The unlignified parenchymatic ground tissue of the central cylinder contains irregularly formed intercellulars (6) and collateral vascular bundles, with sclerenchymatic sheaths of 1–2 cells at the periphery and 3–4 cells towards the center (6).

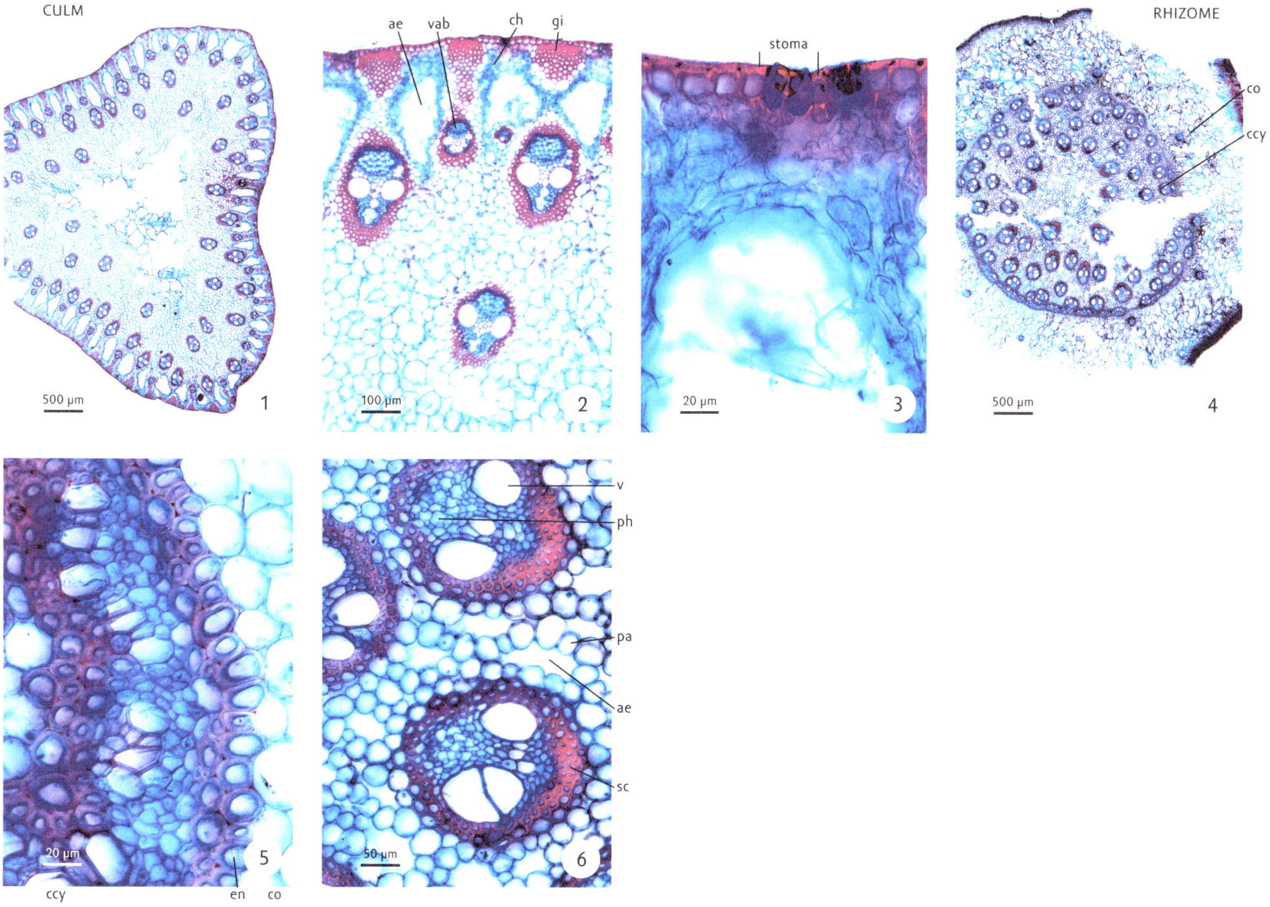

Cyperus eragrostis Lam.

Tall flatsedge

Perennial wetland, tufted, rhizomatous herb with several culms. Culms usually 40–60 cm high, obtusely trigonous, smooth. Rhizomes short and thick. Leaves dark green, flat or V-shaped, acute, linear, 4–10 mm wide. Leaf sheaths purple-brown. Inflorescences simple or compound with an umbel composed of 5–10 lateral branches. Spikelets densely crowded, compressed. Fruit is an obovoid, pale to dark brown achene.

Distribution and ecology

Native to tropical America. Neophyte already reported from several CE countries (CH, DE, HU, PL, temporarily from CZ). It is mostly found along the lower courses of rivers, in estuaries, on the banks of large rivers, roadsides, wet meadows and reedbeds.

Anatomy

Round to obtusely triangular culm with a diameter of 8–10 mm (1).

The epidermis consists of lignified externally thick-walled cells (1, 2). The chlorenchyma contains round parenchyma and collateral vascular bundles and square to rectangular intercellulars which are filled with a stellate-like tissue. Round strands of thick-walled fibers are located just below the epidermis (2). Vascular bundles are dispersed in the whole stem. Vascular bundles consist of a xylem with two large lateral vessels, an air duct and a round phloem (2). Vascular bundles are surrounded by a unilateral 3–5 layers thick-walled sclerenchyma sheath (3). Secretory cells are absent.

Rhizome with a diameter of 2 mm (4).

The cortex is composed of an external layer of small, unlignified parenchyma (4). The central cylinder and the cortex are separated by a lignified thick-walled endodermis (5). The lignified parenchymatic ground tissue of the central cylinder contains amphivasal vascular bundles with thin- to thick-walled sclerenchymatic sheaths (6). Secretory cells are frequent (6).

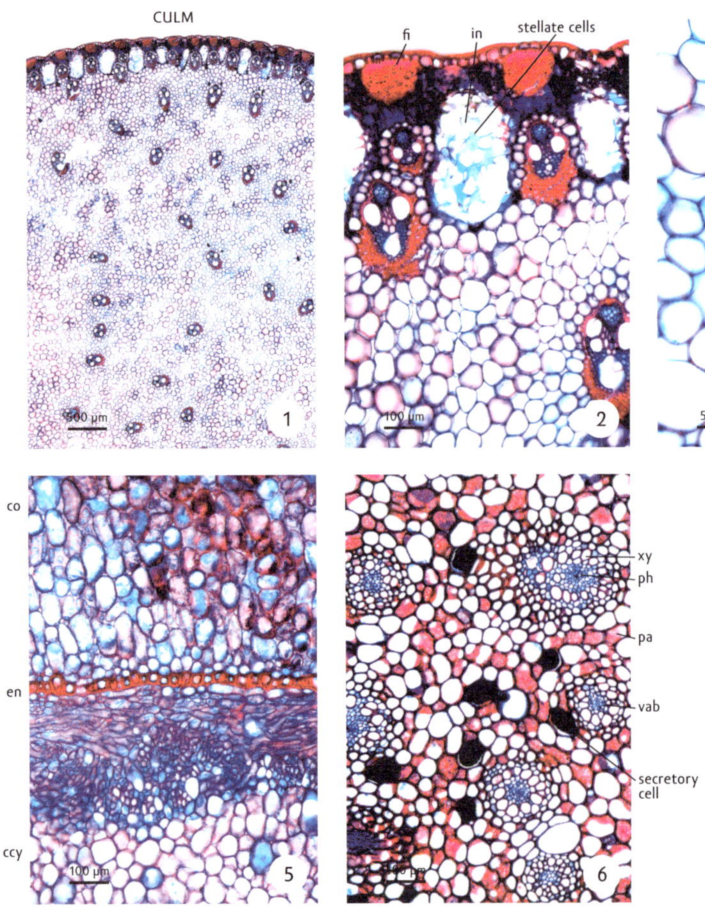

Annual wetland herb, densely tufted, stems ascendent, 5–50 cm high, thin, with two leaves below. Inflorescence simple umbel with grass-like many flowered spikelets.

Distribution and ecology

Native to Eurasia, secondary cosmopolitan, in all CE countries, in wet sandbanks, denudated fishpond bottoms, disturbed sites on fens, wet road sides. Requires open, disturbed places, markedly retreated in all Europe.

Anatomy

Acutely to obtusely triangular culm with a diameter of 1–2 mm (1).

The epidermis consists of roundish lignified, externally thick-walled cells (1, 2). The chlorenchyma contains round parenchyma, round intercellulars and collateral vascular bundles (2). Round strands of fibers are located just below the epidermis (2). Vascular bundles are absent in the parenchymatic central cylinder (1). Vascular bundles consist of a xylem with two large lateral and few small vessels arranged in an arc, an air duct and a roundish phloem with sieve tubes (2). Vascular bundles are surrounded by one layer of thick-walled sclerenchymatic cells (2). Secretory cells are absent.

Rhizome with a diameter of 2 mm.

The cortex is composed of an external layer of small, unlignified parenchymatic cells with large, irregularly formed intercellulars (3). The central cylinder and the cortex are separated by a lignified thin- to thick-walled endodermis (4). The unlignified parenchymatic ground tissue of the central cylinder contains amphivasal vascular bundles with unilateral small sclerenchymatic sheaths (4). Secretory cells are absent.

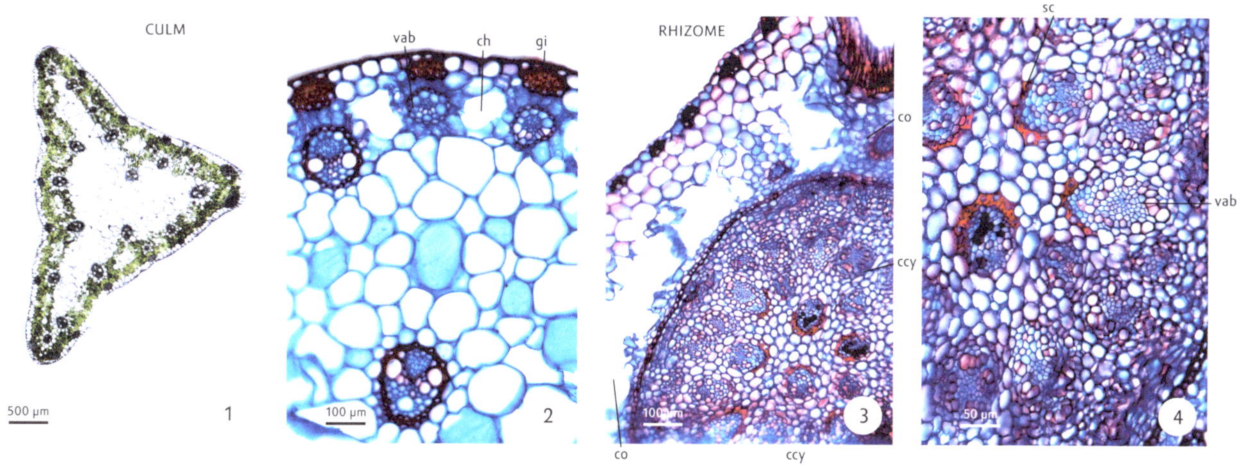

Cyperus fuscus L.

Brown Galingale

Annual wetland herb, tufted, stem erect up to 35 cm, leafy below. Inflorescence usually very compact umbel, with numerous dark brown spikelets.

Distribution and ecology

Euro-Siberian southern-temperate element, neophyte in N America. In all CE countries, in wet sand banks, denudated fishpond bottoms, fen meadows, springheads, wet road sides and banks of water ditches on humus-rich substrates. Prefers open disturbed ground, often subject to winter flooding. Seeds not set in cool summers.

Anatomy

Acutely triangular culm with a diameter of 1–2 mm (1).

The epidermis consists of roundish externally lignified, thick-walled cells (2). The chlorenchyma contains round parenchyma, large, irregularly formed intercellulars and collateral vascular bundles (2). Round strands of fibers are located just below the epidermis (2). Vascular bundles are absent in the parenchymatic central cylinder (1). Vascular bundles consist of a xylem with two large lateral vessels, an air duct and a roundish phloem with sieve tubes (2). Vascular bundles are surrounded by one layer of thick-walled sclerenchymatic cells (2). Secretory cells are absent.

Rhizome slide missing.

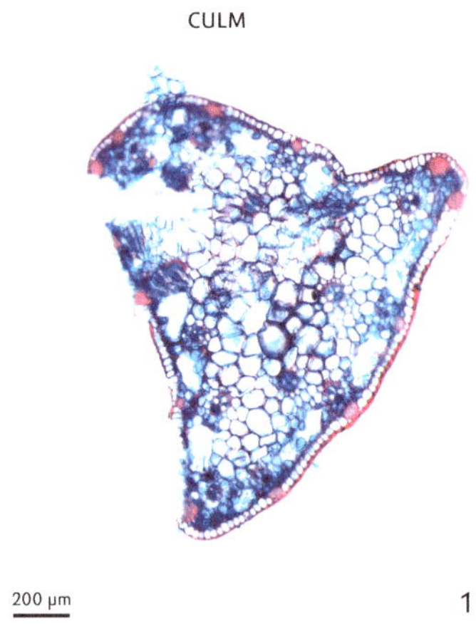

1

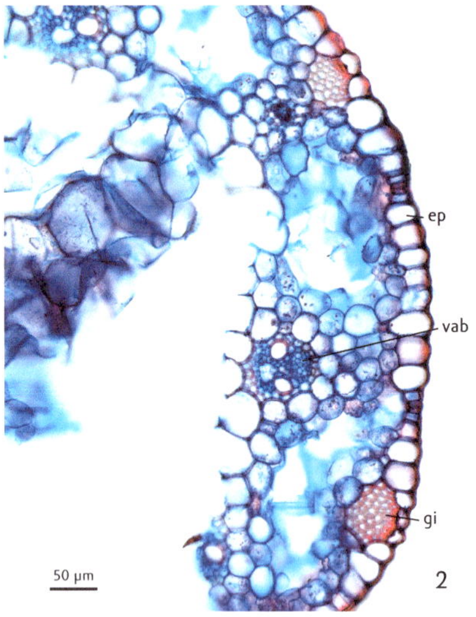

2

Cyperus michelianus L. Delile (syn. *Dichostylis micheliana*)

Annual wetland herb, tufted, stem erect or ascendent, up to 30 cm high, leafy on the base, leaves greyish-green. Inflorescence densely compacted spikes, with 3–6 cm long bracts resembling leaves undersides.

Distribution and ecology

Native to Eurasia, very rare in all CE countries, extinct in CH. Overflowed sand banks of fishponds, perennial pools in river floodplains, usually mesotrophic sandy or loamy soil. Prefers open disturbed ground, often subject to winter flooding. Long-term survival in seedbank.

Anatomy

Obtusely triangular culm with a diameter of 1 mm.

The epidermis consists of roundish lignified cells (1, 2). The chlorenchyma contains round parenchyma, round intercellulars and collateral vascular bundles. Small, irregularly formed intercellulars occur in the whole culm (1). Round strands of fibers are located just below the epidermis (2). Vascular bundles are dispersed in the whole stem. Vascular bundles consist of a xylem with two large lateral and a few small vessels arranged in an arc, an air duct and a roundish phloem with sieve tubes (2). Vascular bundles are surrounded by 1–3 layers of thick-walled sclerenchymatic cells (2). Secretory cells are absent. Thin-walled, unlignified cells and small intercellulars form the inner part of the stem (2).

Rhizome with a diameter of 2 mm.

The cortex is composed of an external layer of small, unlignified parenchymatic cells without large intercellulars (3). The central cylinder and the cortex are separated by a lignified thin- to thick-walled endodermis (3, 4). The lignified parenchymatic ground tissue of the central cylinder contains amphivasal vascular bundles without sclerenchymatic sheaths (4). Secretory cells are absent.

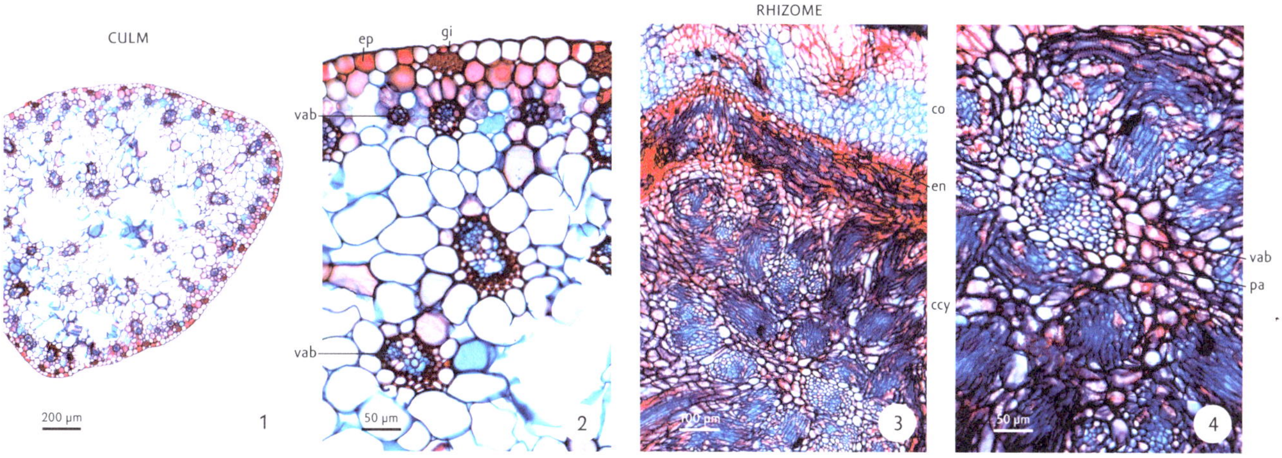

Eleocharis ovata (Roth) R. et Sch.

Annual wetland tufted plant, stems 5–30 cm tall, terminal flower spikes elliptic.

Distribution and ecology

Subatlantic-circumpolar element. Scattered irregularly and locally frequently in all CE countries, from lowlands to uplands. Denudated bottoms of fishponds and lakes, organic, clayish or sandy soils, tolerates slightly acidic to neutral, base-poor to base-rich soils, also eutrophic sites, strictly photophilous.

Anatomy

Circular culm with a diameter of <1 mm (1).

The epidermis consists mainly on roundish, externally lignified thick-walled cells (1, 2) and very small square and round groups of sclerenchyma cells (1, 2). Epidermis cells are not lignified outside of the sclerenchyma groups (2). The chlorenchyma is composed of several layers of oval to palisade-like unlignified parenchyma cells (1, 2). The middle part and inner part consist of round unlignified parenchyma cells, vascular bundles and large irregularly formed intercellulars (1). Most vascular bundles consist of a xylem with two large lateral vessels and an irregular intercellular (2). Vascular bundles are surrounded by one layer of thin- to thick-walled sclerenchyma cells (2). Secretory cells are absent.

Rhizome with a diameter of 1–2 mm.

The cortex is composed of parenchyma cells without intercellulars (3). The central cylinder and the cortex are separated by a thick-walled, lignified endodermis with U-shaped cells (4). The parenchymatic lignified ground tissue of the central cylinder contains amphivasal vascular bundles without sclerenchyma sheaths (5).

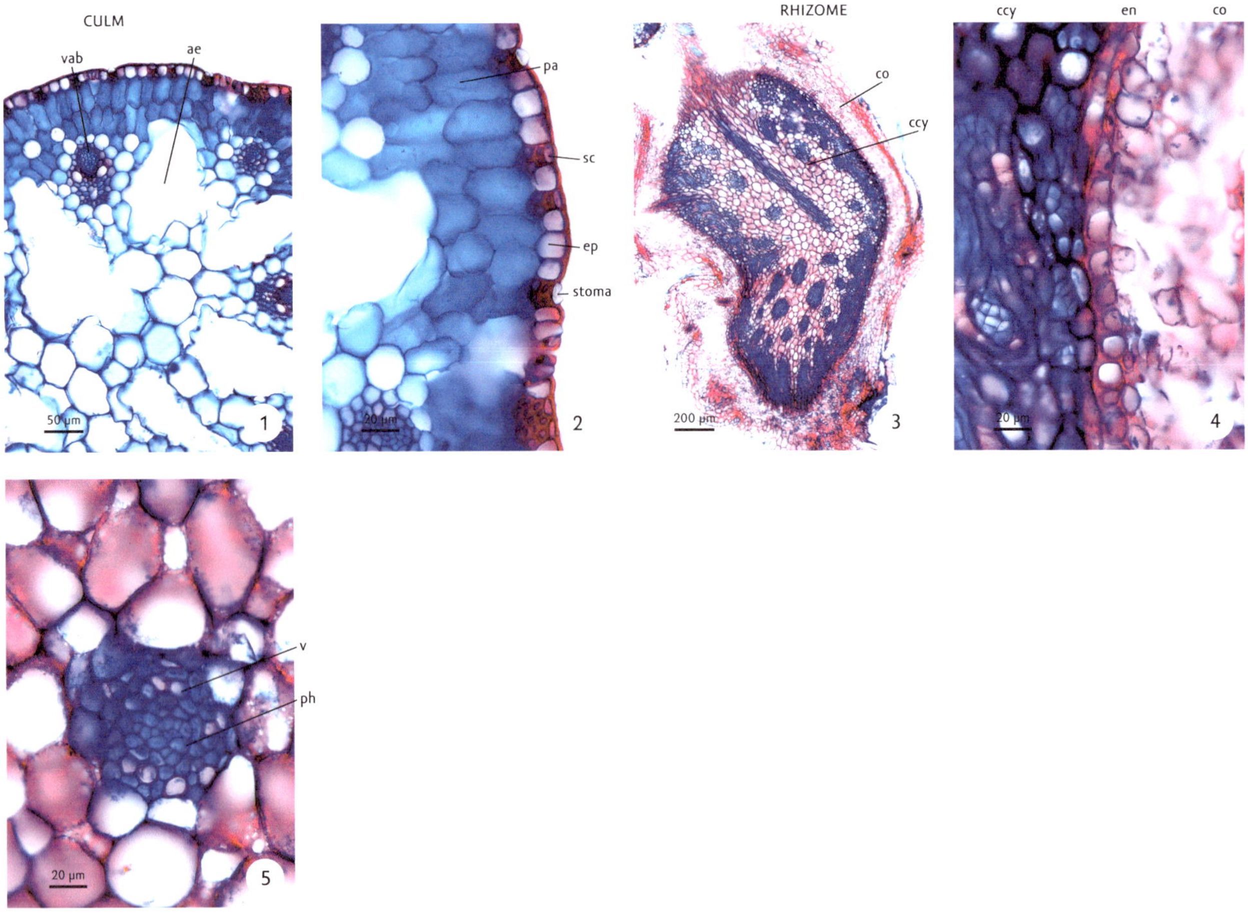

Perennial emergent or amphibious plant. Rhizomatous erect stems 20–60 cm high, loosely to densely tufted, needle-shaped leaves 1.5–2 mm wide.

Distribution and ecology

Eurasian wide-temperate element. Scattered irregularly and frequently in all CE countries, from lowlands to uplands. Shallow littorals of fishponds and lakes, riversides, ditches, alluvial meadows, water depth of 0–40 cm. Tolerates organic, silty or sandy soils, prefers slightly acidic to neutral, base-poor to base-rich soils.

Anatomy

Circular culm with a diameter of <1 mm (1).

The epidermis consists mainly of roundish, externally lignified thick-walled cells (1, 2) and very small groups of sclerenchyma cells (2). The chlorenchyma is composed of several layers of round unlignified parenchyma cells and small intercellulars (1, 2). The middle and inner parts consist of round unlignified parenchyma cells, vascular bundles and very large air spaces (1, 2). Vascular bundles consist of a xylem with two large lateral vessels and a roundish phloem with sieve tubes, companion cells and an irregularly formed intercellular (2). Vascular bundles are partially surrounded by one layer of sclerenchyma cells (2). Secretory cells are absent.

Rhizome with a diameter of 1–2 mm.

The cortex is composed of an epidermis, an external layer of lignified parenchyma cells and an inner part of thin-walled unlignified parenchyma cells with a net-like aerenchyma (3). The central cylinder and the cortex are separated by a thin-to thick-walled, lignified and an unlignified endodermis/pericycle zone (4). The unlignified parenchymatic ground tissue of the central cylinder (5) contains amphivasal vascular bundles, with or without sclerenchyma sheaths (5). Intercellulars are small (5).

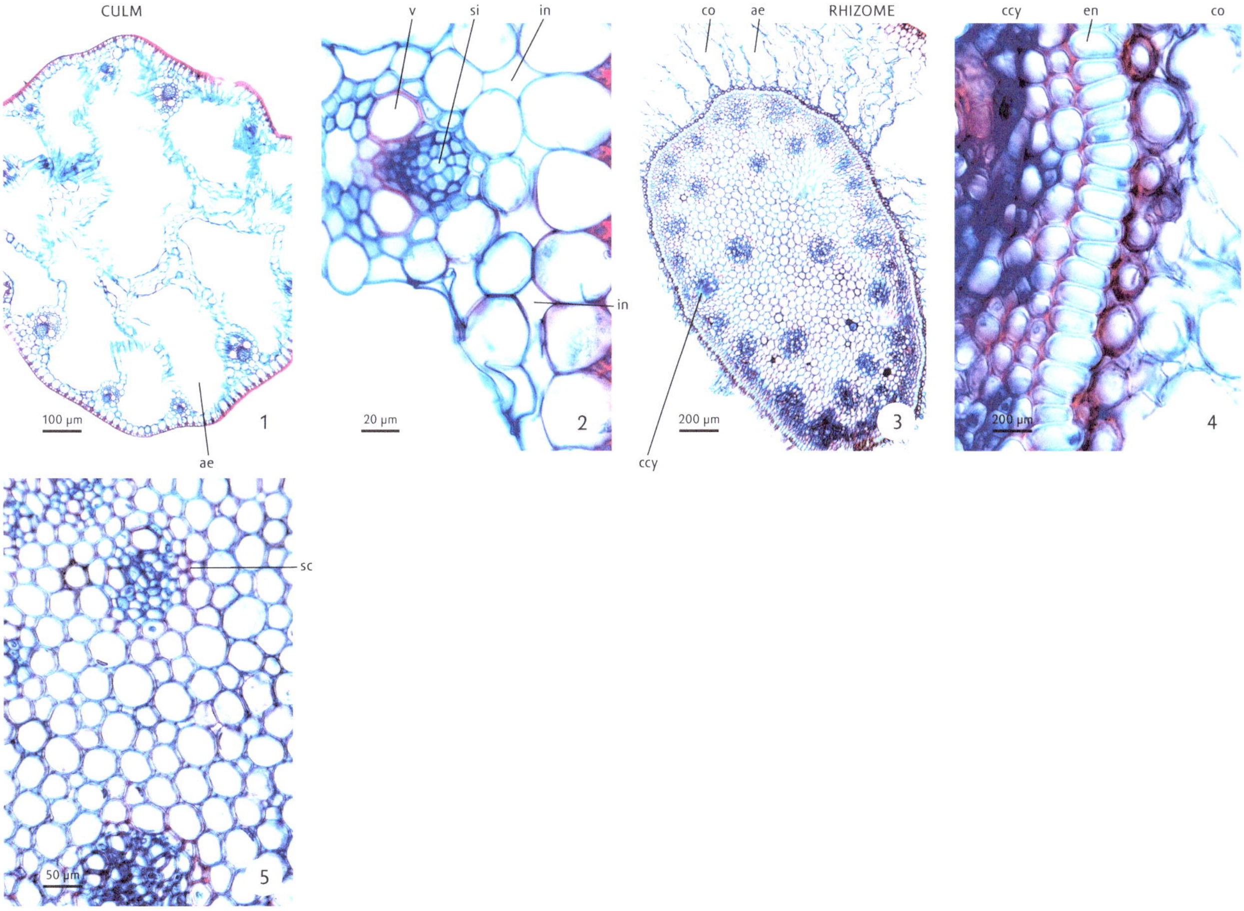

Eleocharis uniglumis (Link) Schult. **Slender Spike-rush** Cyperaceae

Perennial wetland herb, stem up to 60 cm high, firm but slender, smooth, often very shiny, with blade-less leaf-sheaths. Inflorescence of one terminal spikelet.

Distribution and ecology

Circumpolar temperate element. In all CE countries on saline or base-rich marshes, also coastal habitats (damp dune slacks, salt marshes).

Anatomy

Circular culm with ridges with a diameter of <1 mm (1).

The epidermis consists mainly of roundish, externally lignified thick-walled cells (1, 2) and very small groups of sclerenchymatic cells (2). The chlorenchyma is composed of several layers of round to palisade-like unlignified parenchymatic cells (1, 3). The middle part and inner part is composed of round unlignified parenchymatic cells, vascular bundles and large, irregularly formed intercellulars (1). Vascular bundles consist of a xylem with two lateral vessel cells, irregularly formed intercellulars and a roundish phloem with sieve tubes and companion cells (3). Vascular bundles are partially surrounded by one layer of thin-walled cells (3). Secretory cells are absent.

Rhizome with a diameter of 1–2 mm.

The cortex is composed of an epidermis, an external layer of lignified parenchymatic cells and an inner part of unlignified thin-walled parenchymatic cells with a reticulate aerenchyma. The central cylinder and the cortex are separated by a thin- to thick-walled, lignified and unlignified endodermis/pericycle zone (4, 5). The unlignified parenchymatic ground tissue of the central cylinder contains small intercellulars and amphivasal vascular bundles, without sclerenchymatic sheaths (5).

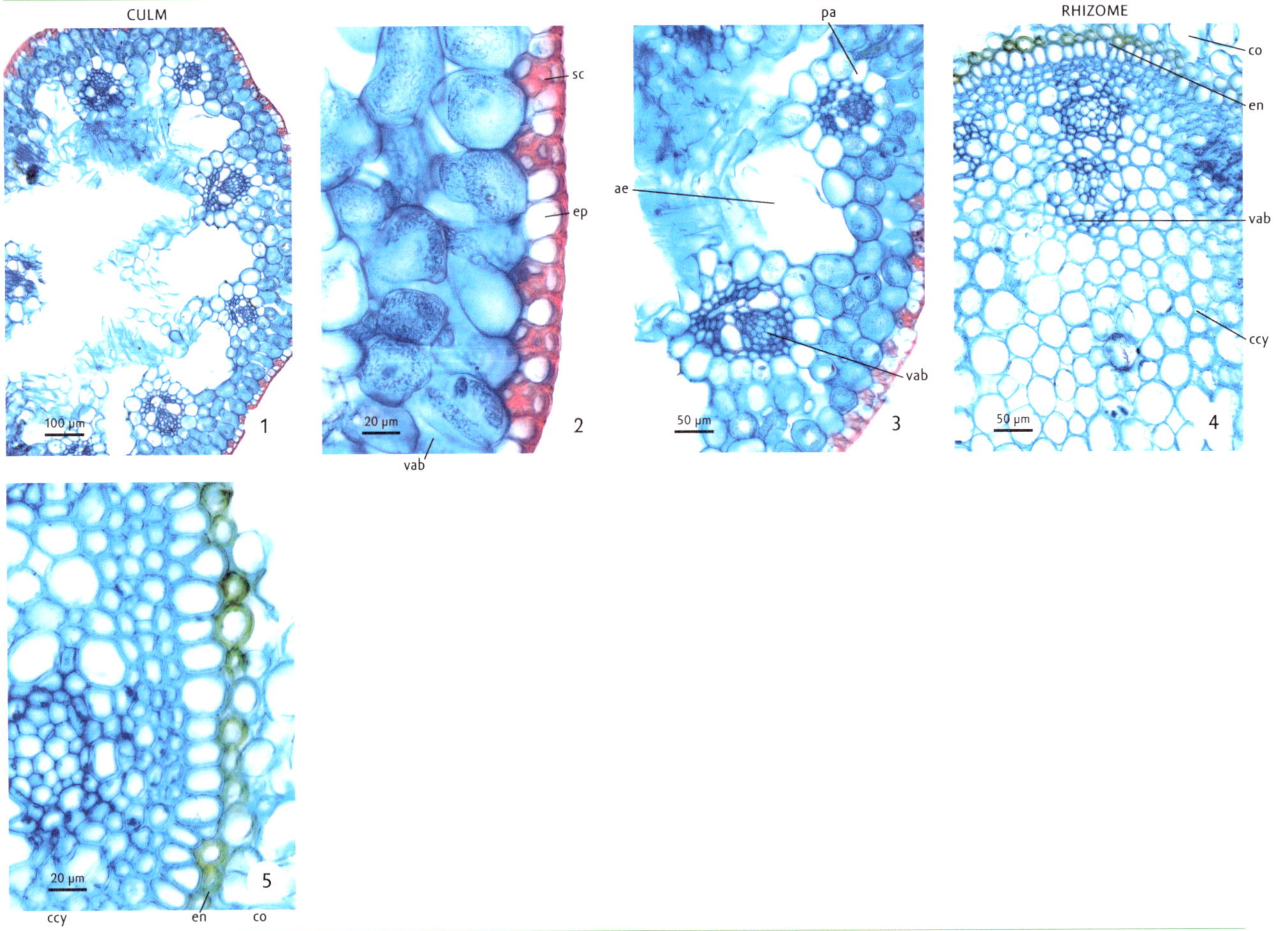

Eriophorum angustifolium Honck.

Common Cottongrass — Cyperaceae

Perennial wetland plant, erect flower stem 20–60 cm tall, stems scattered, creeping rhizomes. Leaves 2–6 mm wide, V-shaped in section, rust-coloured at tips, inflorescence of 3–7 pendulous spikelets, distinctive whitish perianth bristles 3–5 cm long.

Distribution and ecology

Circumpolar wide-boreal element. Scattered irregularly and frequently in all CE countries, usually from highlands to mountains, lacking in lowlands. Common in peatland areas, peat bogs, fens, peaty pine forests, peaty meadows, usually very wet microhabitats, commonly in 10–25 cm water depth as amphibious plant. Oligo- to mesotrophic sites, tolerates strongly acidic to neutral peaty soils (pH ca. 4–7), base-poor as well as base-rich soils, facultative photophilous, relatively eurytopic.

Anatomy

Circular culm with a diameter of 1.5–2 mm (1).

The epidermis consists of roundish, externally lignified thick-walled cells (1–3). Guard cells of the stomata are thickened at the external and internal side (3). The chlorenchyma is composed of several layers of angular unlignified parenchyma cells (1, 2). Girders are triangular (1, 2). The middle and inner part of the culm consist of round unlignified parenchyma cells and vascular bundles and large roundish intercellulars (1, 2). Large vascular bundles consist of a xylem with two large lateral vessels, and a roundish phloem with sieve tubes and companion cells (2). Vascular bundles are partially surrounded by 1–2 layers of sclerenchyma cells (1, 2). Secretory cells are absent (2).

Rhizome with a diameter of 2 mm.

The cortex is composed of a thin-walled epidermis and a thick-walled lignified hypodermis. The large inner part consists of thin-walled unlignified parenchyma cells with radially oriented intercellulars (4). The central cylinder and the cortex are separated by an internally thick-walled, lignified endodermis with U-shaped cells (5). The unlignified parenchymatic ground tissue of the central cylinder contains intercellulars and partially amphivasal vascular bundles, with partial 1–2 layered sclerenchyma sheaths (6).

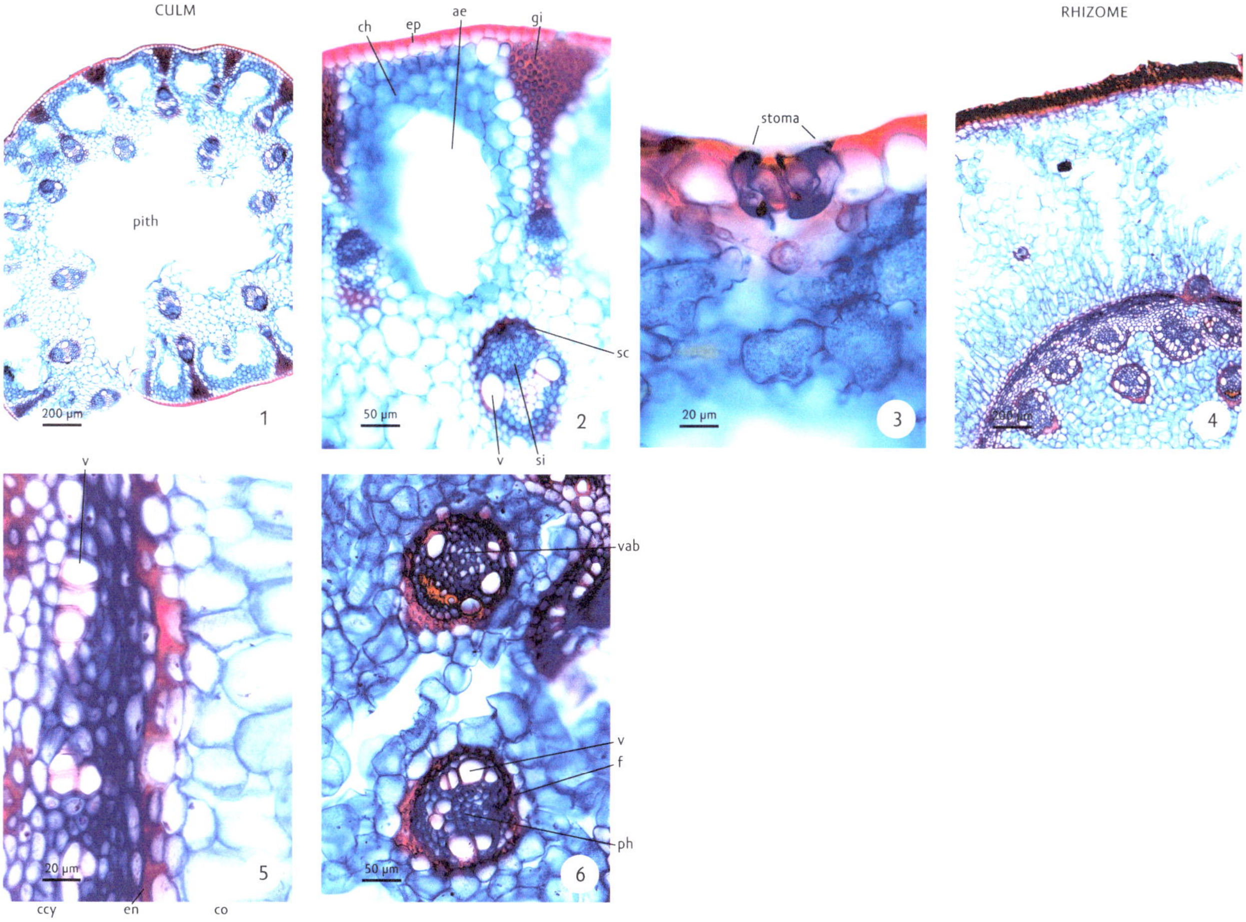

Eriophorum gracile W. D. J. Koch **Slender Cottongrass** Cyperaceae

Perennial wetland, loosely tufted plant, long creeping rhizomes. Erect flower stems 20–50 cm tall, obtusely triangular in section, leaves 0.5–2 mm wide, triangular in section, inflorescence of 3–5 spikelets on slightly rough peduncles, white perianth bristles 1–2.5 cm long.

Distribution and ecology

Circumpolar boreo-temperate element, rare and declining in Europe. Scattered irregularly and rarely in all CE countries, from lowlands to mountains. Fens, mesotrophic peat bogs, fen meadows, springs, acidic to neutral, base-poor to base-rich soils (pH ca. 5–6.5). Competitively very weak, declining especially at lowland eutrophicated sites.

Anatomy

Circular culm with a diameter of 1 mm (1).

The epidermis consists of roundish, externally lignified thick-walled cells (1–3). Guard cells of the stomata are thickened at three sides and have small lips (3). The chlorenchyma is composed of several layers of angular unlignified parenchyma cells (1, 2). Girders are triangular (1, 2). The sclerenchyma sheath of small vascular bundles is linked with the peripheral sclerenchyma strands (2). The middle and inner parts of the culm consist of round unlignified parenchyma cells, vascular bundles and irregularly formed intercellulars (1, 2). The large vascular bundles consist of a xylem with two large lateral vessels, an irregularly formed intercellular and a roundish phloem with sieve tubes and companion cells (2). Vascular bundles are partially surrounded by 1–2 layers of sclerenchyma cells (1, 2). Secretory cells contain dark staining substances (1, 2).

Rhizome with a diameter of 2 mm.

The cortex is composed of an epidermis, a hypodermis of thin- to thick-walled lignified cells and an inner part of unlignified thin-walled parenchyma cells with irregularly formed intercellulars (3, 4). The central cylinder and the cortex are separated by 1–2 layers of different cell sizes. An endodermis is absent (6). The unlignified parenchymatic ground tissue of the central cylinder contains intercellulars and collateral vascular bundles, with partial sclerenchyma sheaths (7). Secretory cells contain dark staining substances (5).

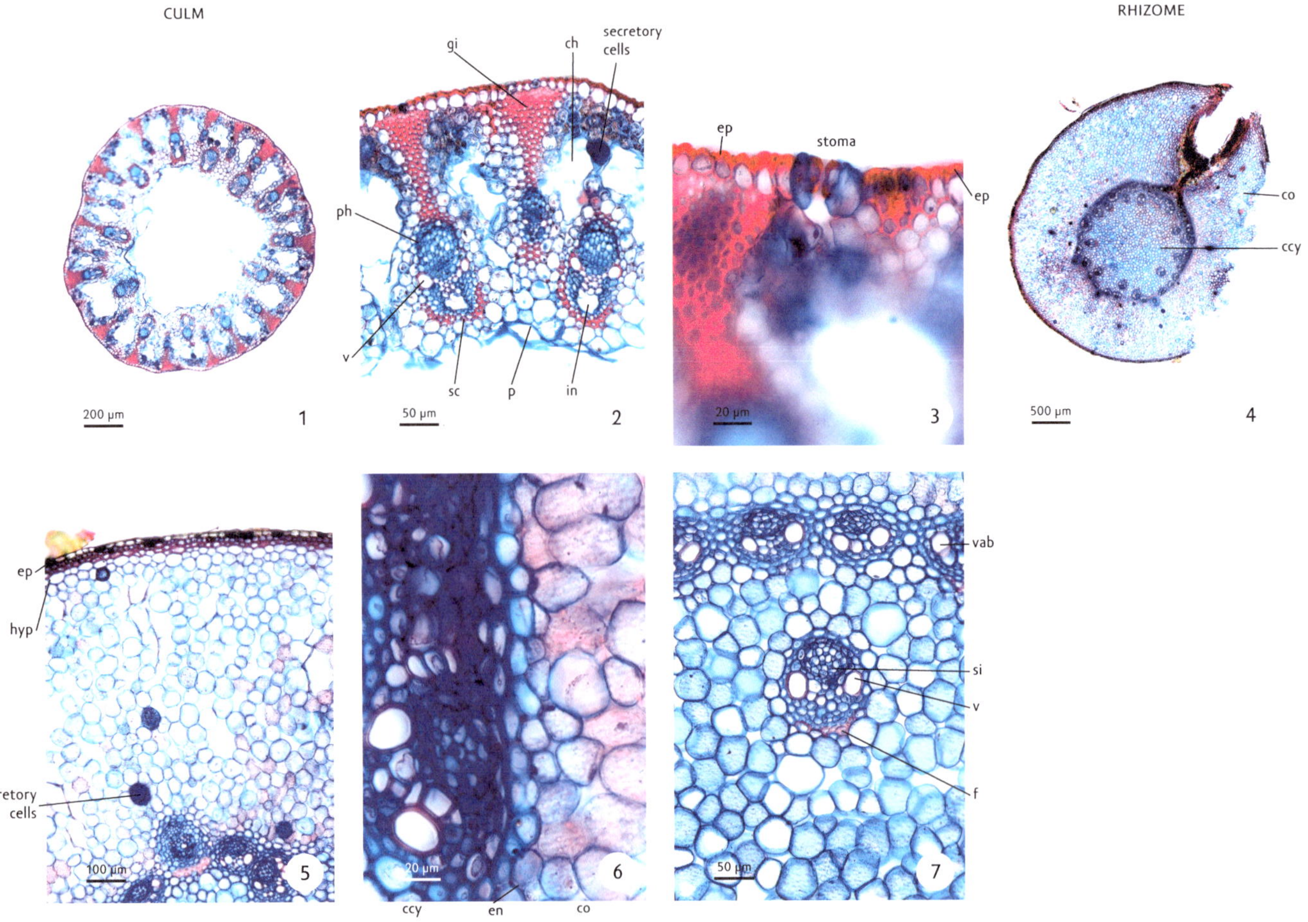

Eriophorum latifolium Hoppe **Broad-leaved Cottongrass** Cyperaceae

Perennial wetland plant, loosely tufted stems 20–60 cm tall, leaves flat, 3–8 mm wide. Inflorescence of 2–12 pendulous spikelets on rough peduncles, whitish perianth bristles 1.5–3 cm long.

Distribution and ecology

European boreo-temperate element. Scattered irregularly and rarely in all CE countries, from lowlands to mountains. Fens, neutral mesotrophic peat bogs, fen meadows, springs, always in neutral, base-rich soils (pH >7), competitively weak at lowland eutrophic sites.

Anatomy

Circular to slightly obtusely triangular culm with a diameter of 1–2.5 mm (1).

The epidermis consists of roundish, externally lignified thick-walled cells (1–3). Guard cells of the stomata are thickened at three sides and have small lips (3). The chlorenchyma is composed of several layers of angular unlignified parenchyma cells (1, 2). Girders are radially elongated (1, 2). The middle and inner parts of the culm consist of round unlignified parenchyma cells, vascular bundles and large, irregularly formed intercellulars (1, 2). The large vascular bundles consist of a xylem with two slightly large lateral vessels, an irregularly formed intercellular and a round phloem, consisting of sieve tubes and companion cells. Vascular bundles are surrounded by 1–2 layers of sclerenchyma cells (1, 2). Secretory cells contain dark staining substances (1, 2).

Rhizome with a diameter of 3–4 mm.

The cortex is composed of an epidermis, an external layer of lignified parenchyma cells and an inner part of thin-walled unlignified parenchyma cells without large intercellulars (4). The central cylinder and the cortex are separated by an unlignified endodermis-like layer of cells (5). The unlignified parenchymatic ground tissue of the central cylinder contains small intercellulars and amphivasal vascular bundles, with partial sclerenchyma sheaths (6). Secretory cells contain dark staining substances (4).

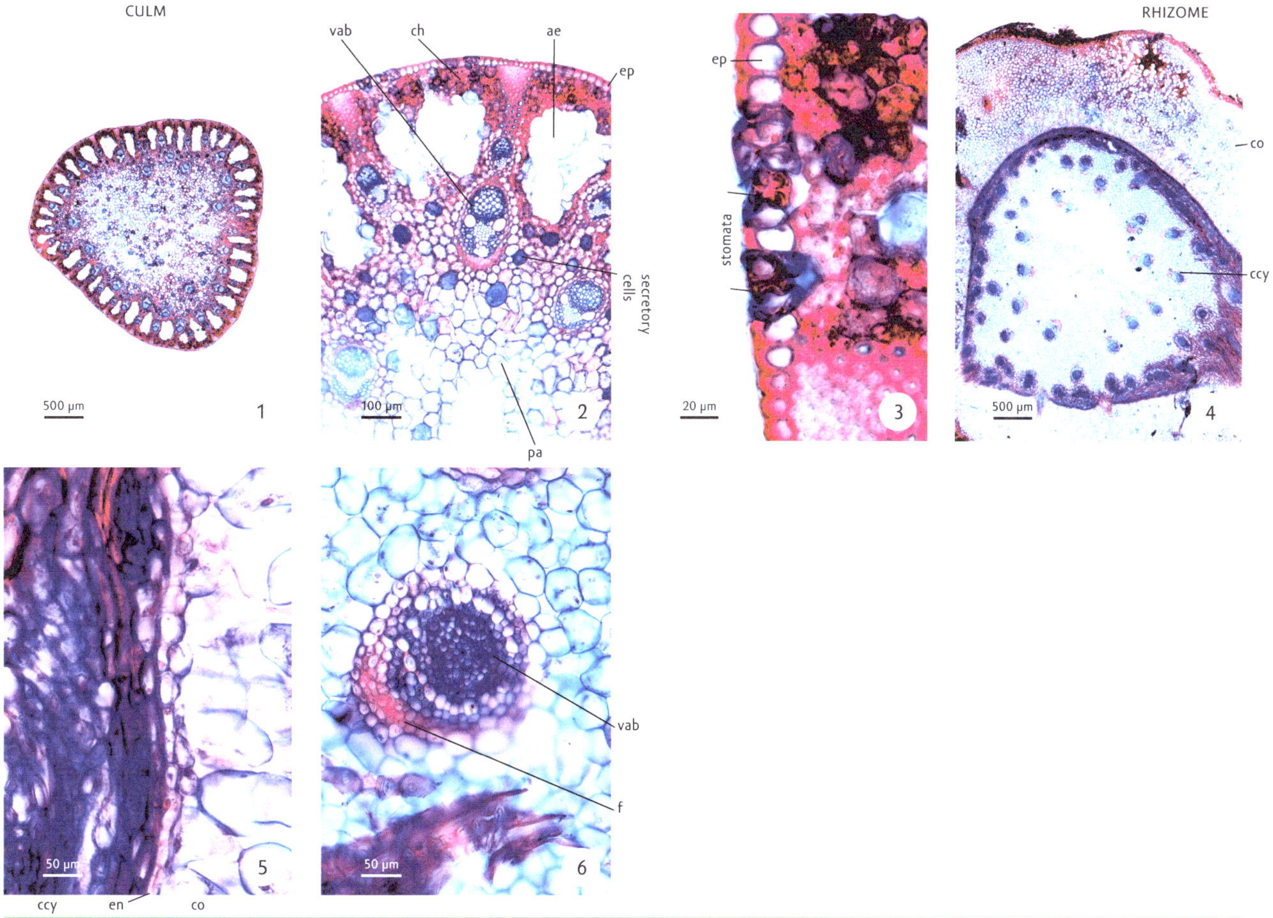

Eriophorum scheuchzeri Hoppe

Perennial wetland herb, stem single, erect, 10–35 cm high, roundish, leafy only at base. Inflorescence of one erect spikelet in a terminal umbel, reddish, cottony head in fruit.

Distribution and ecology

Native to Eurasia and N America, arcto-alpine element, with disjunct distribution – widespread in the Arctic, in C Europe only in high mountains above treeline (Alps, Carpathians), in AT, CH, DE and PL. Usually on peatbogs, springheads, overgrown water ditches, margins of shallow lakes.

Anatomy

Circular culm with a diameter of 1–1.5 mm (1).

The epidermis consists of roundish, externally lignified, very thick-walled cells (1–3). Guard cells of the stomata are thickened at three sides and have small lips (3). The chlorenchyma is composed of several layers of angular unlignified parenchymatic cells (1, 2) and radial girders of sclerenchyma (1, 2).

The middle and inner part of the culm consists of round unlignified parenchymatic cells, vascular bundles and radially oriented, irregularly formed intercellulars (1, 2). The large vascular bundles are composed of a xylem with two large lateral vessels, an irregularly formed intercellular and a roundish phloem with sieve tubes and companion cells (2). Vascular bundles are surrounded by 1–2 layers of sclerenchymatic cells, often endodermis-like with U-shaped cells (2, 4). Secretory cells contain dark staining substances (1, 2).

Rhizome with a diameter of 2–3 mm (4).

The cortex is composed of an epidermis, an external layer of lignified parenchymatic cells and an inner part of thin-walled unlignified parenchymatic cells with irregularly formed intercellulars (5). The central cylinder and the cortex are separated by an internally thick-walled, lignified endodermis (6). The parenchymatic ground tissue of the central cylinder contains intercellulars and amphivasal vascular bundles, with partial sclerenchymatic sheaths and large round groups of sclerenchyma (5).

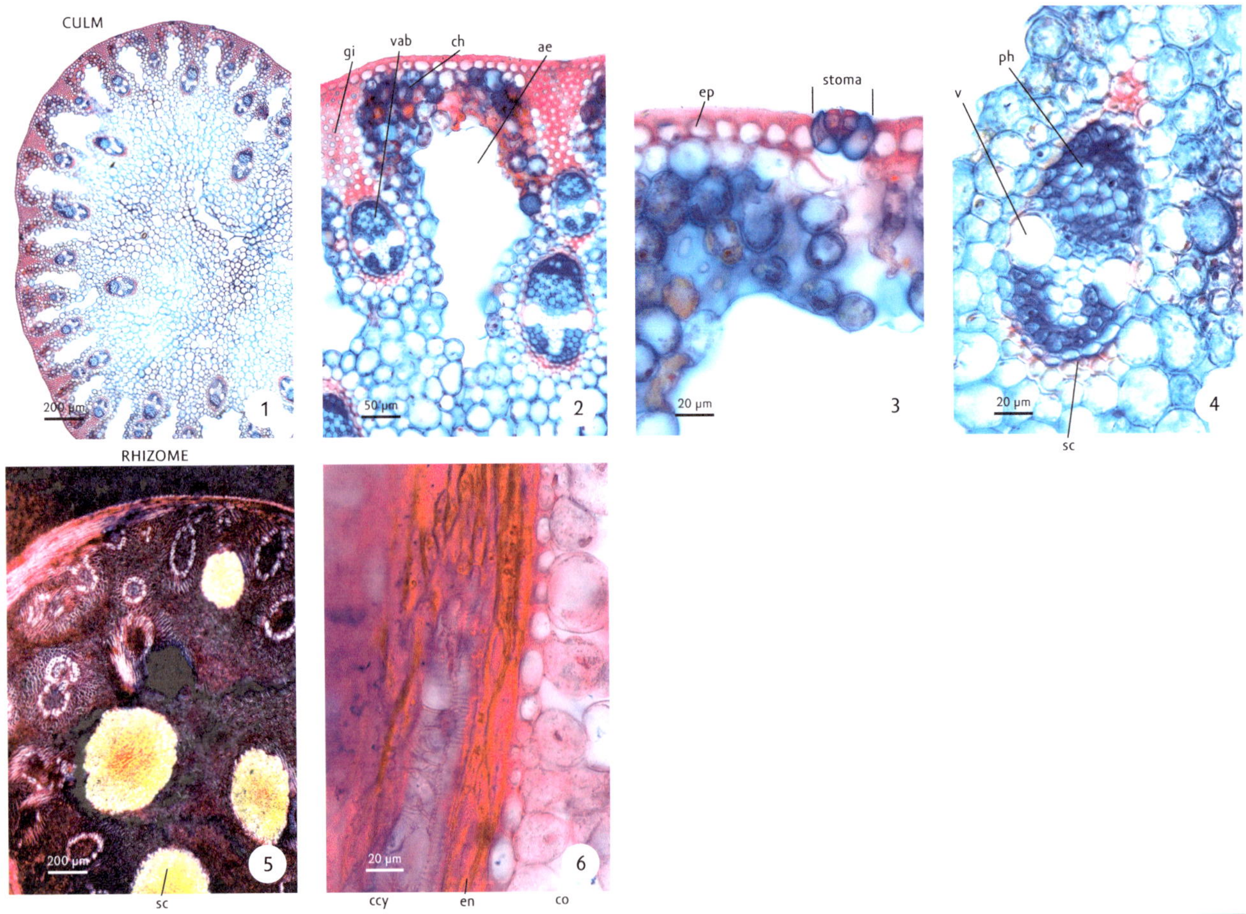

Perennial wetland herb, solitary flower spikelet on erect flower stem 30–60 cm high. Rhizomes very short, stems densely tufted, often tussock-forming, narrow leaf blades (to 1 mm) are triangular in section, distinctive white perianth bristles 2–3 cm long.

Distribution and ecology

Circumpolar boreo-arctic montane element. Scattered irregularly in all CE countries, usually from highlands to mountains. Peat bogs, acidic fens, peaty pine forests, usually very wet microhabitats. Oligotrophic, strongly acidic peaty soils (pH ca. 3.5–5), facultatively psychrophilous, more common in mountains.

Anatomy

Circular culm with a diameter of 2 mm (1).

The epidermis consists of roundish, externally lignified thin- to thick-walled cells (1, 2). The periphery contains small triangular girders of sclerenchyma (1, 2).

The middle and inner parts of the culm consist of round-shaped unlignified parenchymatic cells, vascular bundles and radially oriented, irregularly formed intercellulars (1, 2). The large vascular bundles are composed of a xylem with two large lateral and a few central angular vessels, an irregularly formed intercellular and a roundish phloem with sieve tubes and companion cells (2). Vascular bundles are surrounded by 1–2 layers of sclerenchymatic cells, and a U-shaped centripetal group of thick-walled fibers (1, 2).

Rhizome with a diameter of 2–3 mm (4).

The cortex is composed of an epidermis, an external layer of lignified parenchymatic cells and an inner part of thin-walled unlignified parenchymatic cells with irregularly formed intercellulars (3). The central cylinder and the cortex are separated by an internally thick-walled, lignified endodermis (3). The parenchymatic ground tissue of the central cylinder contains a few intercellulars and amphivasal vascular bundles, with partial sclerenchymatic sheaths and large round groups of sclerenchyma (3–5).

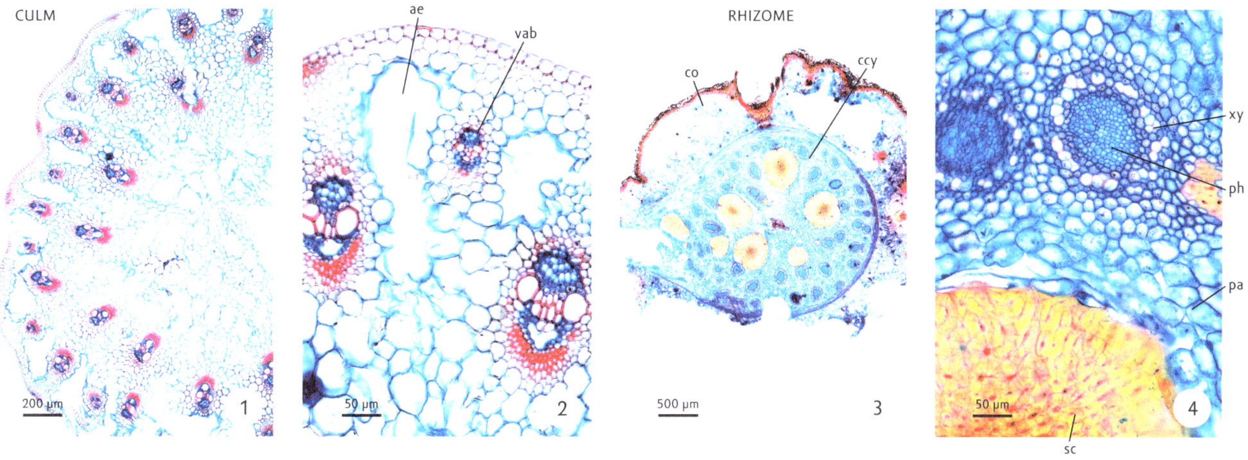

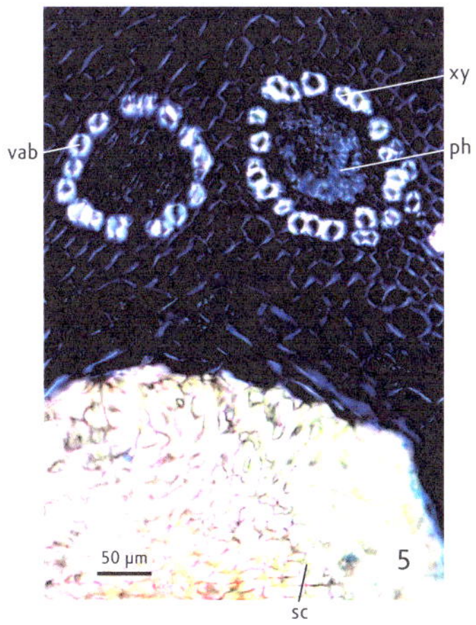

Isolepis cernua (Vahl) Roem. & Schult. **Slender Club-rush** Cyperaceae

(syn. *Scirpus cernuus*)

Perennial wetland, densely tufted herb. Stems erect to ascending, usually <1 cm high. Leaves crescent-shaped in section. Inflorescence of 1–4 sessile spikelets, lowest bract usually stem-like, slightly longer than inflorescence. Fruit is a matt nut.

Distribution and ecology

Subcosmopolitan distribution. Common in Western Europe, occurs in DE, it is usually found in wet, coastal grassland, in bare or open sites over damp sand, peat and mud, sometimes in trickles on rocky cliffs.

Anatomy

Circular culm with a diameter of <1 mm (1).

The epidermis consists of roundish, externally lignified thick-walled cells (2, 3). Guard cells of the stomata are thickened on three sides (3). The chlorenchyma is composed of several layers of round unlignified parenchyma cells (1, 2), small intercellulars and round groups of sclerenchyma (1–3). Round groups of sclerenchyma are located below the epidermis. The middle and inner part of the culm consist of round parenchyma cells and vascular bundles. Large irregularly formed intercellulars are inside of the chlorenchyma (1, 2). Vascular bundles consist of a xylem with two large lateral and a few arc-like arranged small vessels and a roundish phloem with sieve tubes and companion cells (2). Vascular bundles are surrounded by one layer of a uniseriate endodermis-like sheath with U-shaped sclerenchyma cells (4). Secretory cells are absent (2).

Circular rhizome with a diameter of 1 mm (5).

The cortex is composed of a thin-walled epidermis, an external layer of large lignified parenchyma cells and an inner part of small unlignified thin-walled parenchyma cells. Irregularly formed intercellulars occur partially (5). The central cylinder and the cortex are separated by two layers of thick-walled lignified endodermis-like cells (6). The unlignified parenchymatic ground tissue of the central cylinder contains intercellulars and mainly amphivasal vascular bundles, without sclerenchyma sheaths (6). Secretory cells are frequent (5, 6).

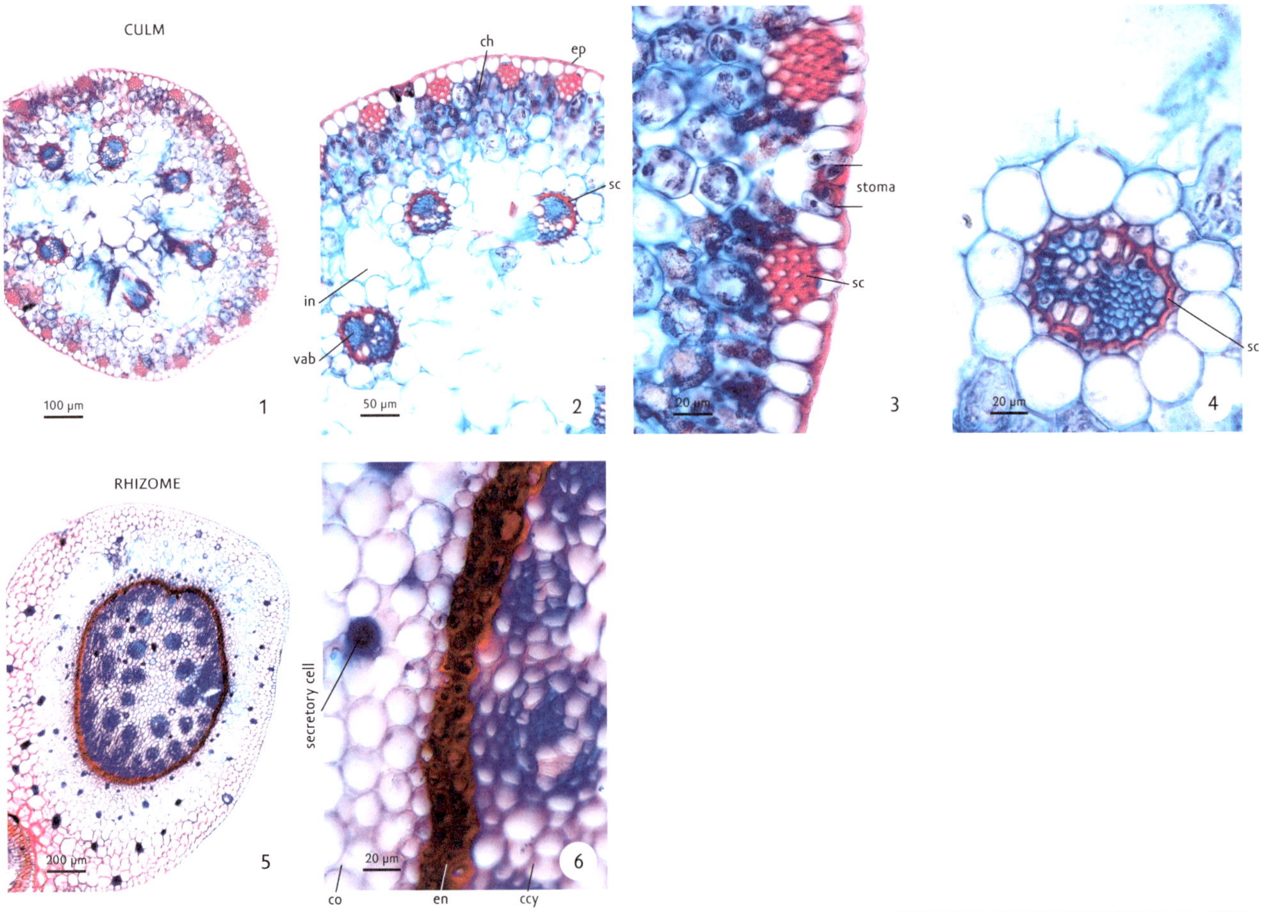

Rhynchospora fusca L. W. T. Aiton **Brown Beak-sedge** Cyperaceae

Perennial wetland plant, erect flower stems 10–30 cm long, thin rhizome 3–8 cm long gives rise to a new plant, neighbouring plants interconnected by rhizomes. Stems are biennial, inflorescence brown at flowering.

Distribution and ecology

Suboceanic boreo-temperate element; also in N. America. Scattered irregularly and very rarely in all CE countries (except for SK and HU) at a few sites, in lowlands to highlands, only in peatland areas. Prefers bare peat on peat bogs, acidic fens, usually very wet microhabitats, frequently to 5 cm water depth. Oligotrophic, strongly acidic peaty soils (pH ca. 4–5), usually on acidic sandy bed. Strictly photophilous, competitively very weak.

Anatomy

Obtusely triangular culm with a diameter of <1 mm (1).

The epidermis consists of roundish, lignified thick-walled cells (2, 3). Guard cells of the stomata are thickened at three sides (3). The chlorenchyma is composed of several layers of round unlignified parenchyma cells (2), irregularly formed large intercellulars and small groups of sclerenchyma (2, 3). The middle and inner parts of the culm consist of round unlignified parenchyma cells and vascular bundles. Vascular bundles consist of a xylem with two large lateral vessels and indistinct irregularly formed intercellular and a roundish phloem with sieve tubes and companion cells (2, 3). Vascular bundles are surrounded by 4–6 layers of thick-walled sclerenchyma cells, which touch the epidermis (2, 3). Secretory cells are absent (2).

Circular rhizome with a diameter of 1–2 mm (4).

The cortex is composed of a thin-walled epidermis, a small external layer of large unlignified parenchyma cells and an inner part of large unlignified thin-walled parenchyma cells with large irregularly formed intercellulars (5). The central cylinder and the cortex are separated by the presence of unlignified (cortex) and lignified (central cylinder) parenchyma cells. An endodermis is absent (5). The lignified parenchymatic ground tissue of the central cylinder contains collateral vascular bundles, with a 1–2 layered sclerenchyma sheath (5). Secretory cells are absent (4).

CULM

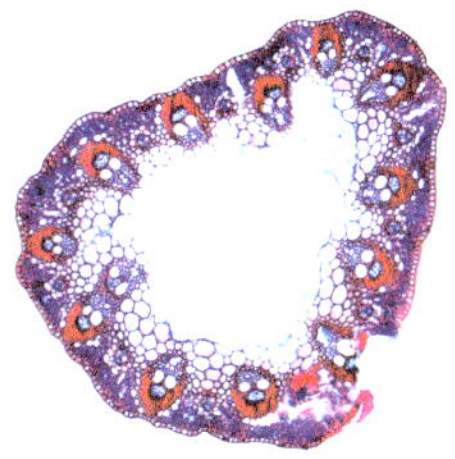

1

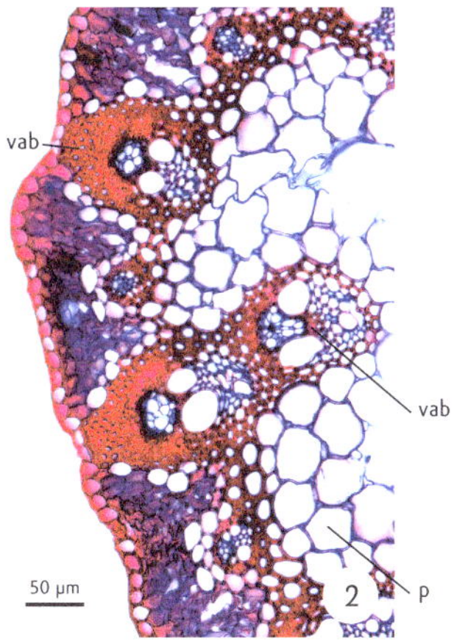

2

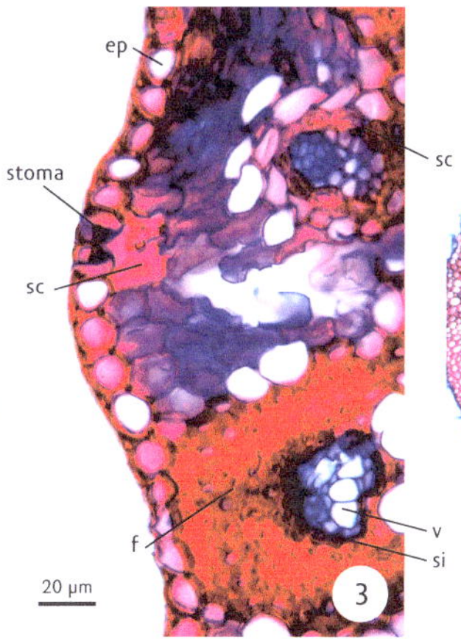

3

RHIZOME

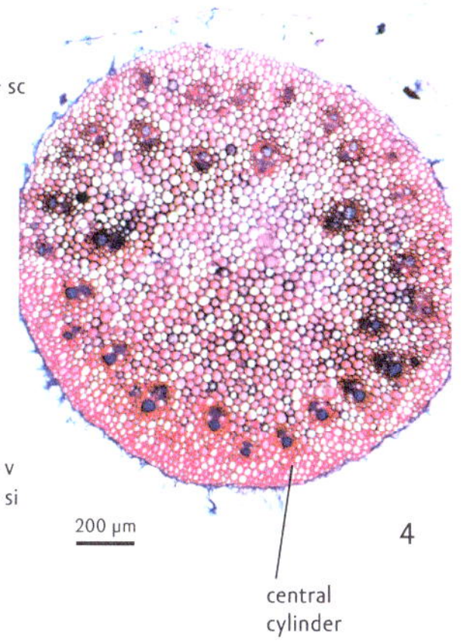

4

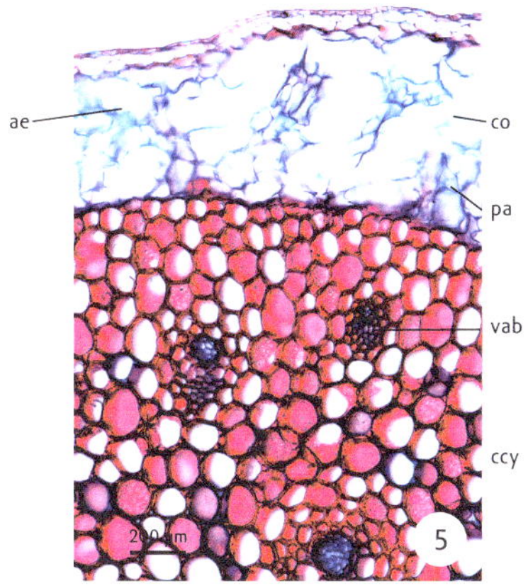

5

Schoenoplectus lacustris (L.) Palla

Common Club-rush

Perennial emergent herb, rhizomatous, stem rough, oval, 50–300 cm high, deeply green, ramified above, submerged form with ribbon-like leaves known. Leaf-sheaths mostly bladeless. Inflorescence of many variously stalked ovoid spikelets, lowest bract stem-like, making inflorescence appears lateral.

Distribution and ecology

Euro-Siberian wide-temperate element. In all CE countries, from lowlands to highlands. Stagnant or slowly flowing waters, from eutrophic and base-rich to oligotrophic and base-poor conditions. Especially sandy or gravel banks, usually in 0.3–1.5 m deep water, lack of pressurized ventilation, may establish on submersed substrates. Competition with *Phragmites australis* in eutrophic habitats.

Anatomy

Circular culm with a diameter of 5–8 mm (1).

The culm is divided into a thin layer of chlorenchyma and a large central part (1, 2).

The epidermis consists of roundish, lignified thin-walled cells (1–2) and roundish groups of thick-walled sclerenchymatic cells (2). Guard cells of the stomata are thickened on three sides. The chlorenchyma is composed of a few layers of palisade-like, unlignified thin-walled parenchymatic cells (2). Intercellulars are absent (2). The middle and inner part consist of round groups of fibers and vascular bundles (2). Vascular bundles consist of a xylem with two large lateral vessels and a roundish phloem with sieve tubes (2). At the central part of the vascular bundles, groups of thin- to thick-walled sclerenchymatic cells occur (1, 2). The large central part is composed of a net of polygonally arranged, small reticulate parenchyma and vascular bundles. Within the polygons with air spaces, unlignified thin-walled cells occur (1). Secretory cells are absent (2).

Rhizome slide missing.

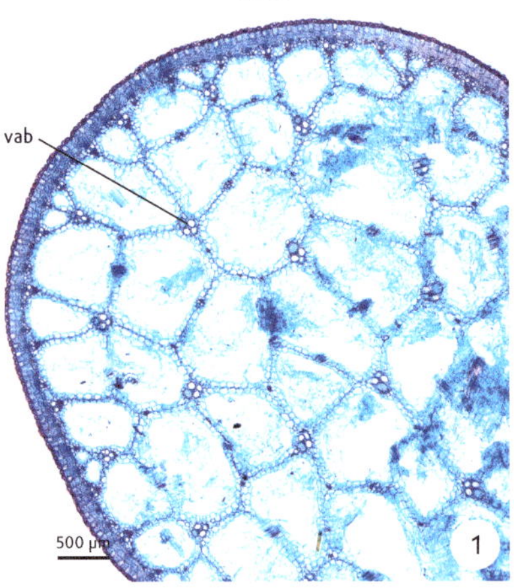

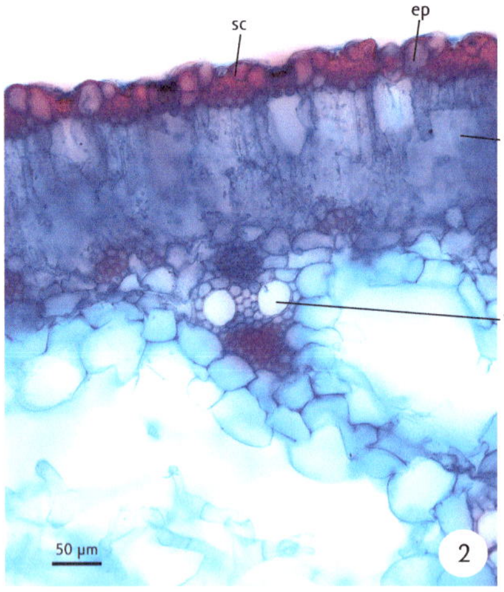

Schoenoplectus tabernaemontani (C. C. Gmel.) Palla

Grey Club-rush

Perennial emergent herb, rhizomatous, stem oval, up to 150 cm high, glaucous, grey or greyish-green, pale-brown below. Leaf-sheaths mostly bladeless. Inflorescence of many variously stalked ovoid spikelets, lowest bract stem-like, making inflorescence appear lateral.

Distribution and ecology

Eurasian southern-temperate element. In all CE countries. Reed banks of still waters, wet saline grounds, dune slacks, wet peaty places, mostly in coastal sites. Tolerant of partially soil drying. Indicator of salty soils or salty water.

Anatomy

Circular culm with a diameter of 3–4 mm (1).

The culm is divided into a thin layer of chlorenchyma and a large central part (1, 2).

The epidermis consists of roundish lignified thin-walled cells (1). Within the epidermis, triangular groups of thick-walled sclerenchyma cells occur (2). Guard cells of the stomata are thickened on three sides (2). The chlorenchyma is composed of a few layers of palisade-like unlignified parenchymatic cells (1, 2) Intercellulars are absent. The middle and inner part of the culm contain vascular bundles. The vascular bundles contain a xylem with two large lateral vessels, an irregularly formed intercellular and a roundish phloem with sieve tubes (2). At the central part of the vascular bundles, groups of thin- to thick-walled sclerenchymatic cells occur (1, 2). The large central part is composed of a net of polygonally arranged, small reticulate parenchyma and vascular bundles. Within the aerenchymatic polygons, unlignified thin-walled cells occur (1). Secretory cells are frequent (1, 2).

Round rhizome with a diameter of 6–8 mm (3).

The cortex is composed of an external layer of small unlignified parenchymatic cells without intercellulars and an inner part of thin-walled unlignified parenchymatic cells with many intercellulars (3). The central cylinder and the cortex are separated by unlignified layers of square, endodermis-like cells (4). The unlignified aerenchymatic, parenchymatic ground tissue of the central cylinder contains collateral vascular bundles with large oval partial sclerenchymatic sheaths (5). Secretory cells are frequent (5).

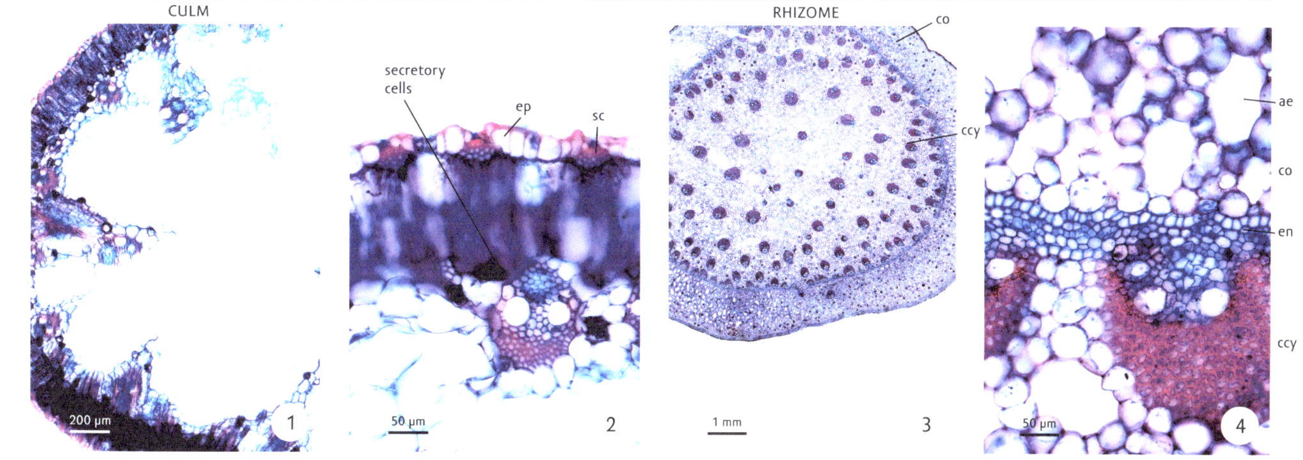

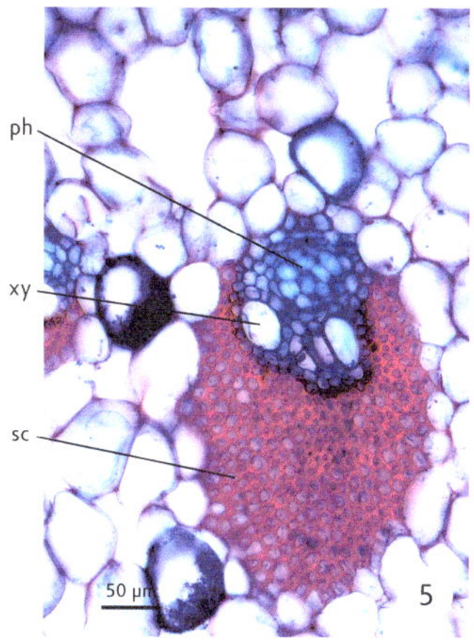

Schoenoplectus triqueter (L.) Palla

Triangular Club-rush Cyperaceae

Perennial emergent herb, rhizomatous, non-tufted, stem oval, deeply green, rigid, erect, 50–150 cm high, sharply triangular. Leaf sheaths dark-brown, only the highest with green blade. Inflorescence of many variously stalked ovoid spikelets, lowest bract stem-like, making inflorescence appears lateral.

Distribution and ecology

Eurasian temperate element. Endangered in all CE countries, extinct in CZ. Only in lowlands. Muddy banks of larger rivers and ditches, prefers warm and nutrient-rich water, tolerant of moderate salinity and inundation by the tides.

Anatomy

Obtusely triangular culm with a diameter of 3–5 mm.

The culm is distinctly divided into a thin layer of chlorenchyma and a large central part (1, 2). The epidermis consists of roundish lignified thin-walled cells (1–2). Guard cells of the stomata are thickened on three sides. The chlorenchyma is composed of a few layers of palisade-like, unlignified parenchymatic cells (2) and adjacent within the epidermis roundish groups of thick-walled sclerenchymatic cells (2). Intercellulars are absent. The middle and inner part of the culm contain a few small groups of fibers and vascular bundles (2). Vascular bundles consist of a xylem with two large lateral vessels and a roundish phloem with sieve tubes and companion cells (2). At the central part of the vascular bundles, groups of thick-walled sclerenchymatic cells occur (1, 2). The large central part is composed of a net of polygonally arranged parenchyma and vascular bundles. Within the aerenchymatic polygons, unlignified thin-walled cells occur. Secretory cells are absent (2).

Round rhizome with a diameter of 6–8 mm.

The cortex is composed of an external layer of small unlignified parenchymatic cells without intercellulars and an inner part of thin-walled unlignified parenchymatic cells with many intercellulars (3). The central cylinder and the cortex are separated by unlignified layers of square, endodermis-like cells. The unlignified ground tissue of the central cylinder contains collateral vascular bundles with large oval partial sclerenchymatic sheaths (3). Secretory cells are rare (3).

CULM

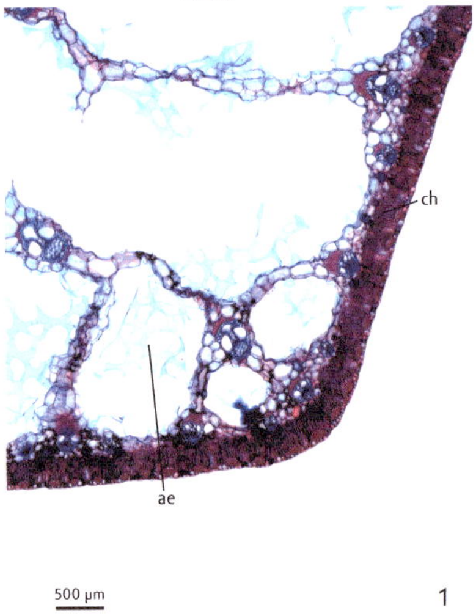

1

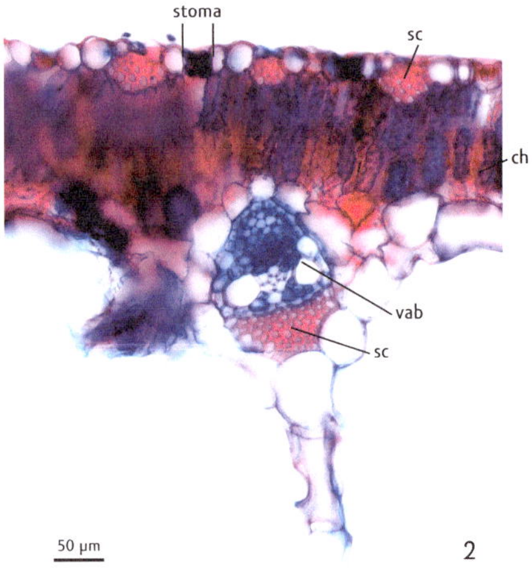

2

RHIZOME

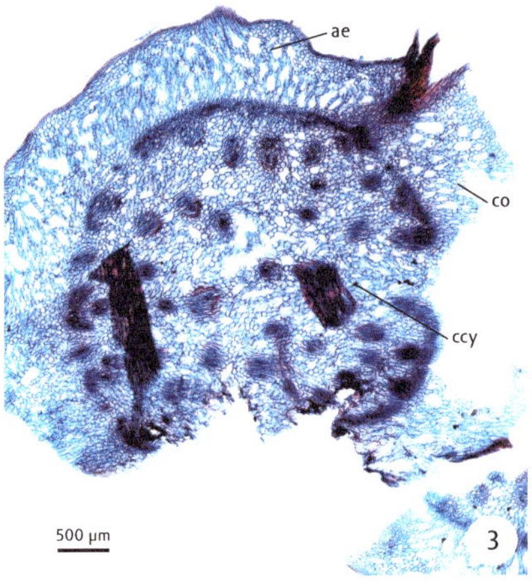

3

Schoenus ferrugineus L.

Brown Bog-rush Cyperaceae

Perennial wetland herb, densely tufted, stem erect, 10–30 cm high, leaf-sheaths only at or near base. Leaves shorter than half of the stem. Inflorescence of 1–3 spikelets in a compact head.

Distribution and ecology

European boreo-montane element. In all CE countries, especially in alpine foothills where it replaces *S. nigricans* at higher altitudes. Fenlands, dried fens, wet meadows, in the surroundings of mineral springs, requires high concentrations of carbonates and high, non-fluctuating water level. Low survival of seeds in upper soil layers. Seed is retained in the inflorescence for up to a year.

Anatomy

Circular culm with a diameter of <1 mm (1).

The epidermis consists of roundish peripheral, lignified thick-walled cells (1–2). The chlorenchyma is composed of several layers of oval unlignified parenchymatic cells (2), roundish groups of very thick-walled sclerenchyma and a few secretory cells (1, 2). Intercellulars are absent. The middle and inner part of the culm consist of round unlignified, thin-walled parenchymatic cells and vascular bundles. Vascular bundles consist of a xylem with two large lateral vessels, irregularly formed intercellulars and a roundish phloem with sieve tubes and companion cells (2). Vascular bundles are surrounded by 1–2 layers of thin- to thick-walled sclerenchymatic cells (1, 2).

Rhizome with a diameter of 1–2 mm.

The cortex is composed of a thin-walled epidermis, an external layer of lignified parenchymatic cells without intercellulars and an inner part of thin-walled unlignified parenchymatic cells with irregularly formed intercellulars (3). The central cylinder and the cortex are separated by a thin- to thick-walled, slightly lignified endodermis (4). The unlignified parenchymatic ground tissue of the central cylinder contains amphivasal vascular bundles with uniseriate, thin- to thick-walled sclerenchymatic sheath (5) and a central star-like sclerenchyma (3).

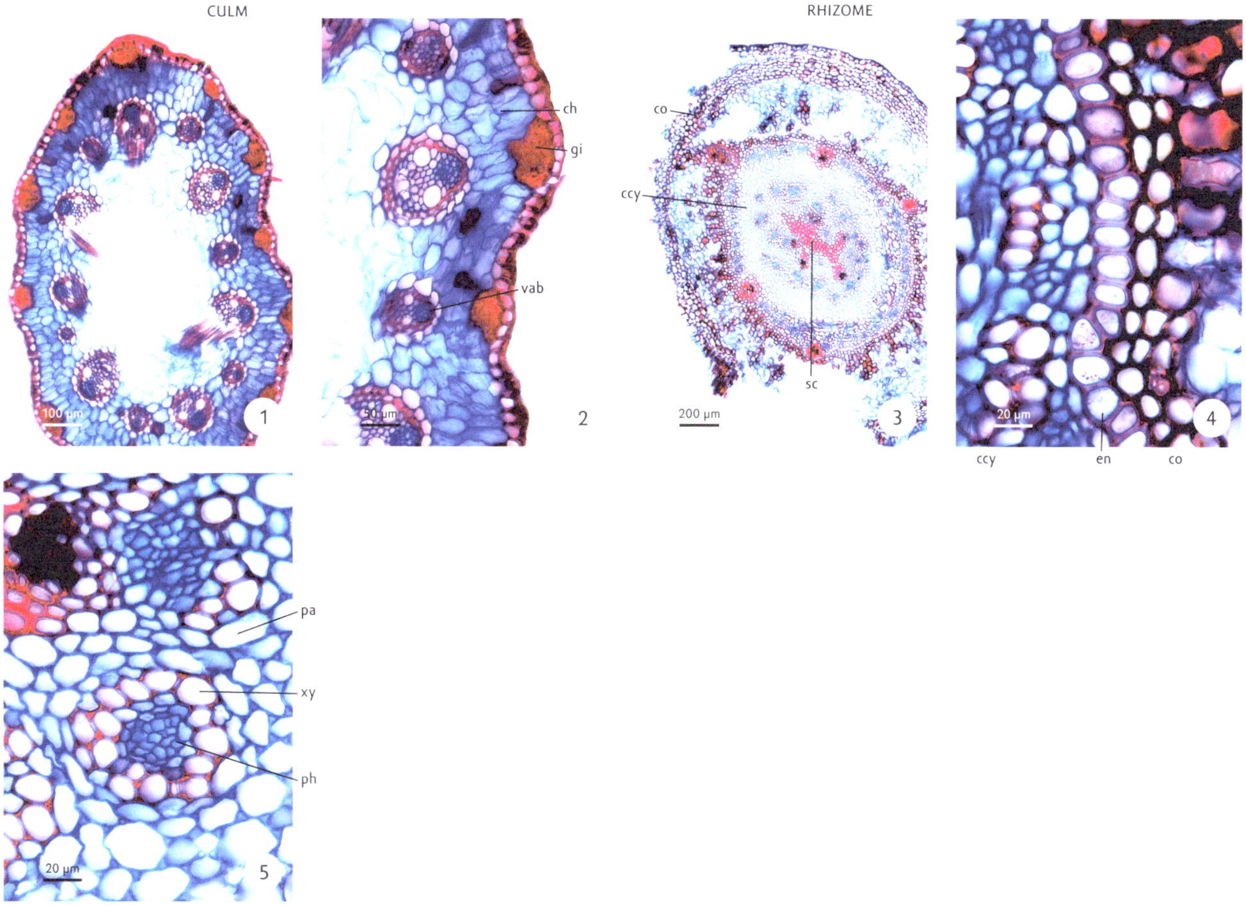

Schoenus nigricans L. **Black Bog-rush**

Perennial wetland herb, densely tufted, stem 15–60 cm high, grey-green. Leaves thin, rough, shorter than stem. Inflorescence with 5–10 spikelets, lowest bract usually conspicuously exceeding it.

Distribution and ecology

Euro-Siberian southern-temperate element; also in N America. In all CE countries, very rare in CZ, SK, from lowlands to highlands. Base-rich fens and wet meadows, waterlogged calcareous soils, salt marshes, serpentine heathlands, in the surroundings of the mineral springs. Requires mineral-rich calcareous soils, however poor in P. Forms persistent seed bank.

Anatomy

Circular culm with a diameter of <1 mm (1).

The epidermis consists of roundish lignified thin- to thick-walled cells (1–2). Guard cells of the stomata are thickened on three sides. The chlorenchyma is composed of several layers of oval to palisade-like unlignified parenchymatic cells (2) and roundish groups of very thick-walled sclerenchymatic cells (2). Intercellulars are absent. The middle and inner part of the culm consists of round unlignified parenchymatic cells. A circle of fibers is connected with vascular bundles laterally by lignified fibers. Vascular bundles consist of a xylem with two large lateral vessels and a roundish phloem with sieve tubes (2). Vascular bundles are surrounded by 1–2 layers of thin- to thick-walled sclerenchymatic cells (1, 2). Secretory cells are absent (2).

Circular rhizome with a diameter of 1–2 mm (3).

The cortex is composed of an external layer of unlignified parenchymatic cells with intercellulars and an inner part of thin-walled lignified parenchymatic cells without intercellulars (3). The central cylinder and the cortex are separated by two layers of a thick-walled, lignified endodermis-like sheath (4). The lignified parenchymatic ground tissue of the central cylinder contains amphivasal vascular bundles with sclerenchymatic sheaths and a central sclerenchyma (3).

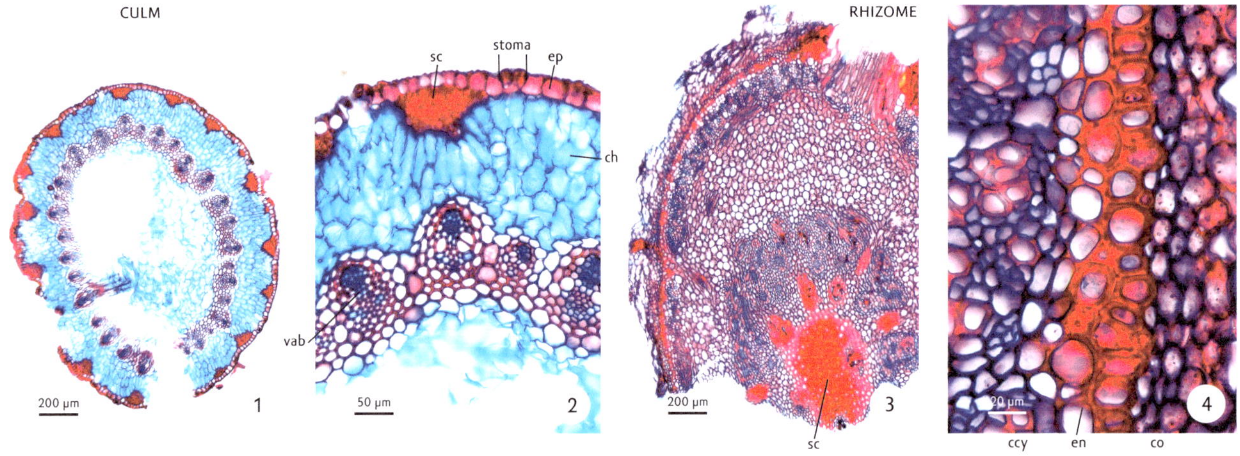

Scirpoides holoschoenus (L.) Soják **Round-headed Club-rush** Cyperaceae

Perennial wetland herb, rhizomatous, densely tufted, stem simple, erect, 30–150 cm high, roundish, grey-green. Leaves rather stem-like. Inflorescence of variously stalked or sessile globular brown heads. Lowest bract stem-like, making inflorescence appears lateral.

Distribution and ecology

Euro-Siberian southern-temperate element, also in N Africa. Rare and endangered in all CE countries. Moist sand banks, rare in still water banks and fen meadows. Also damp sandy places near sea (Great Britain). From lowlands to hilly areas.

Anatomy

Circular culm with a diameter of 1–2 mm (1).

The epidermis consists of roundish lignified, externally thick-walled cells (1, 2). Guard cells of the stomata are thickened on three sides. The chlorenchyma is composed of several layers of round lignified parenchymatic cells (2) and obtusely triangular groups of thick-walled sclerenchymatic cells (2). The middle and inner parts of the culm consist of round vascular bundles and air spaces. The vascular bundles consist of a xylem with two large lateral vessels, an irregularly formed intercellular and a roundish phloem with sieve tubes and companion cells (2, 3). Vascular bundles are surrounded by an endodermis-like layer of thin-walled cells and by 1–2 discontinuous layers of thin- to thick-walled sclerenchymatic cells (3). Secretory cells are absent (1, 2).

Rhizome with a diameter of 5–8 mm.

The cortex is composed of an external layer of small zone of sclerenchyma without intercellulars and an inner part of thin-walled unlignified parenchymatic cells with small intercellulars (4, 5). The central cylinder and the cortex are separated by a lignified thick-walled endodermis with U-shaped cells (5). The unlignified parenchymatic ground tissue of the central cylinder contains amphivasal vascular bundles with 3–5-layered sclerenchymatic sheaths (6). Secretory cells are absent.

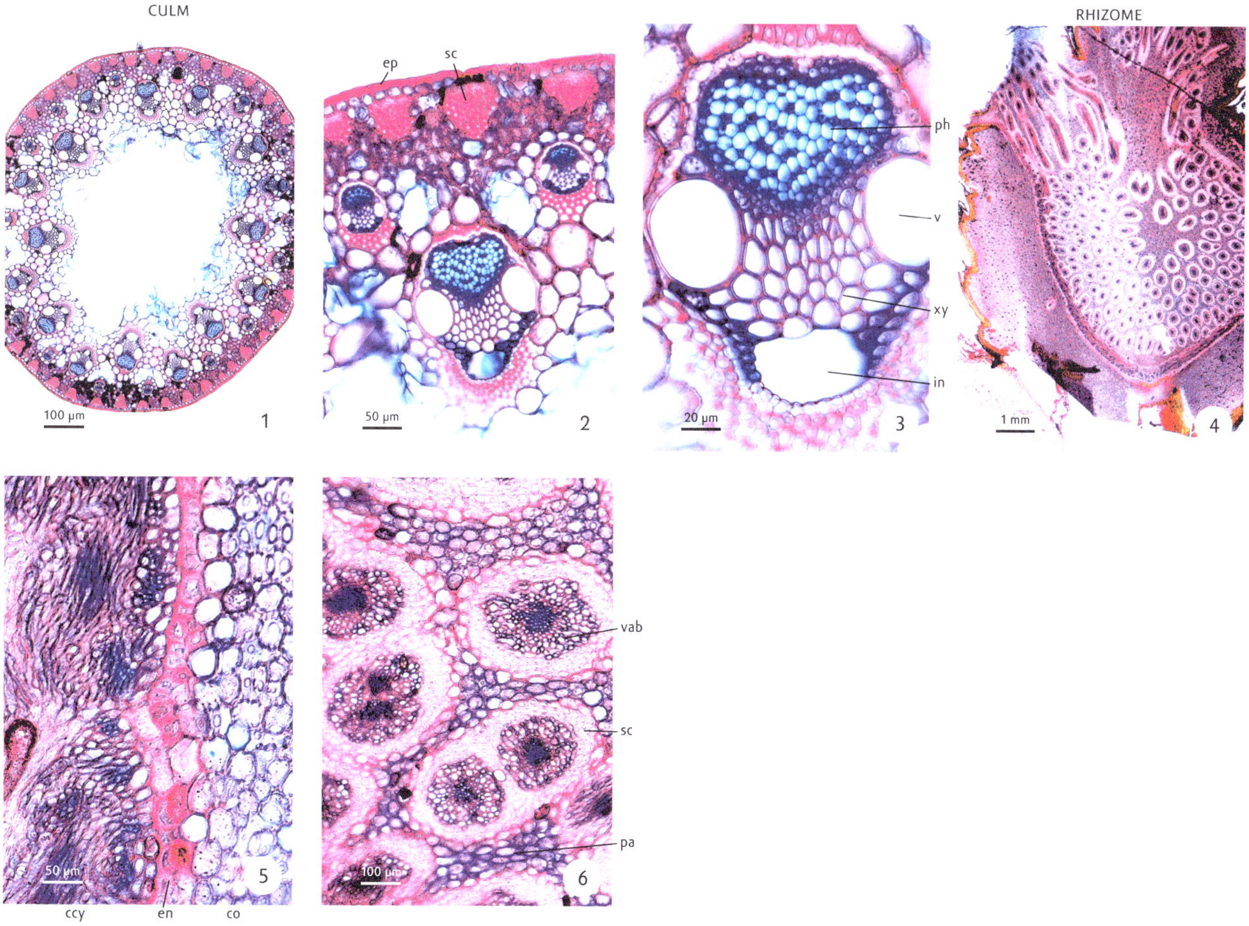

Scirpus radicans Schkuhr

Wood Bulrush

Perennial emergent herb, rhizomatous, densely tufted, stem sharply triangular, 40–100 cm high. Inflorescence of numerous spikelets on ends of diffusely branching panicle, usually one spikelet per branch.

Distribution and ecology

Native to Eurasia, in all CE countries except CH. Banks of larger rivers, water ditches and fishponds, denudated bottoms of fishponds, reed stands. Wet and nutrient-rich sediments, sometimes secondary habitats. From lowlands to highlands. Aboveground sprouts able to root.

Anatomy

Circular culm with a diameter of 6–8 mm (1).

The epidermis consists of roundish, externally lignified thick-walled cells (1–3). Guard cells of the stomata are thickened on three sides. The chlorenchyma is dominated by large square air spaces, which are loosely filled with thin-walled unlignified parenchymatic cells (1). Between the air spaces, round parenchymatic cells, small groups of fibers and small vascular bundles exist (2, 3). Centripetal to the epidermis, tangentially oval groups of fibers occur (3). The middle and inner part of the culm contain round vascular bundles. Vascular bundles consist of a xylem with two large lateral vessels, a small intercellular and a roundish phloem with sieve tubes and companion cells (2). Vascular bundles are surrounded by 1–3 layers of thick-walled sclerenchymatic cells (2). Secretory cells occur in the inner part of the chlorenchyma (2). The pith is filled with a net of very small parenchymatic cells (1, 4).

Circular rhizome with a diameter of 8–10 mm.

The cortex is composed of an external layer of small, lignified parenchymatic cells without intercellulars, and an inner part of thin-walled unlignified parenchymatic cells with many intercellulars (6). The central cylinder and the cortex are separated by a lignified endodermis-like layer of round cells (7). The unlignified parenchymatic ground tissue of the central cylinder contains amphivasal vascular bundles with 1–2 cell-layered sclerenchymatic sheaths (5, 7). Secretory cells are quite frequent.

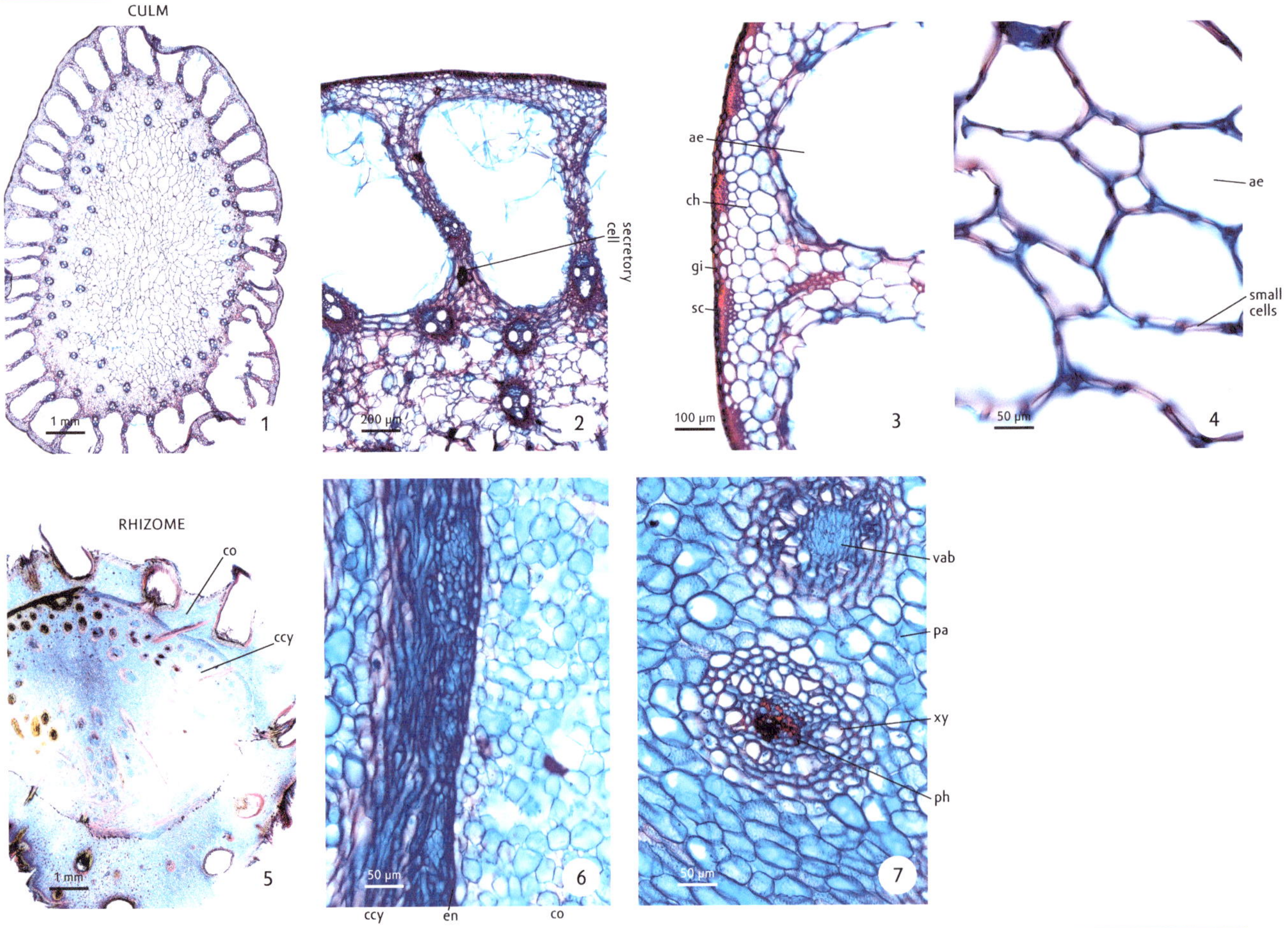

Perennial wetland herb, rhizomatous, stem strong, erect, up to 120 cm high, single, bluntly triangular, with a hollow inside. Leaves long, alternate. Inflorescence of numerous spikelets at the ends of diffusely branching panicle, usually more spikelets per branch.

Distribution and ecology

Euro-Siberian temperate element. In all CE countries, on muddy banks of still waters, fenlands, wet meadows and pastures, lowland woods, spring areas. It may form extensive stands, tolerates partly shade, typically grows on rather eutrophic silt, often iron-enriched.

Anatomy

Obtusely triangular culm with a diameter of 3–4 mm (1).

The epidermis consists of roundish lignified, externally thick-walled cells (2). Guard cells of the stomata are thickened on three sides. The chlorenchyma is dominated by large air spaces. Among them round parenchymatic cells and oval vascular bundles occur (2). Vascular bundles consist of a xylem with two large lateral vessels, irregularly formed intercellulars and a roundish phloem with sieve tubes, companion cells and an air duct (2). Vascular bundles are surrounded by 1–3 layers of thin- to thick-walled sclerenchymatic cells (2). Secretory cells are absent.

Rhizome with a diameter of 5–6 mm.

The cortex is composed of an external layer of small, unlignified parenchymatic cells without intercellulars and an inner part of thin-walled unlignified parenchyma cells with reticulate intercellulars (3, 4). Central cylinder and cortex are separated by a lignified thin-walled, endodermis-like sheath (4). The unlignified parenchymatic ground tissue of the central cylinder contains amphivasal vascular bundles with large oval sclerenchymatic sheaths on the inner side (5). Intercellulars are small (5). Secretory cells are absent.

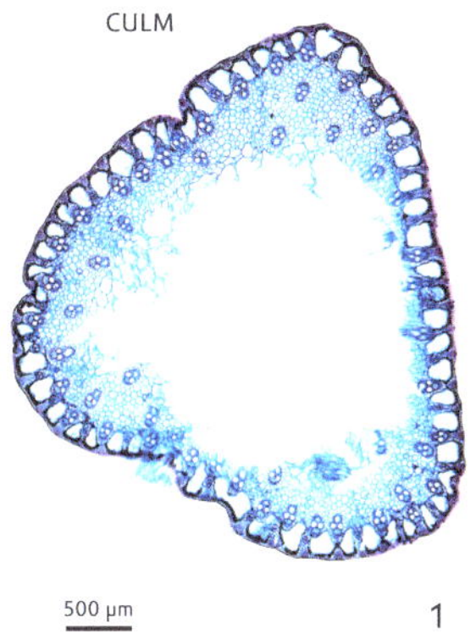

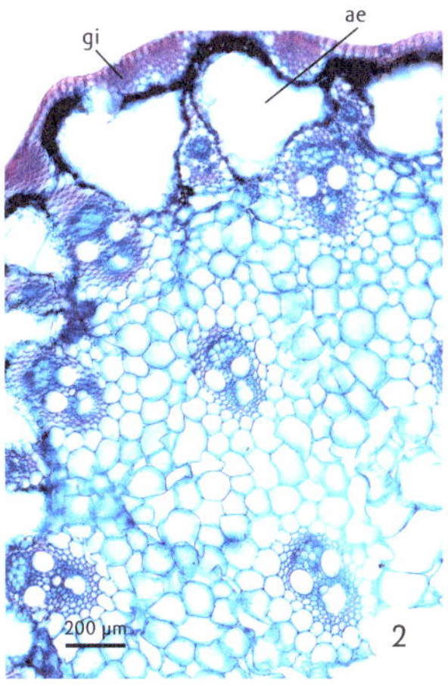

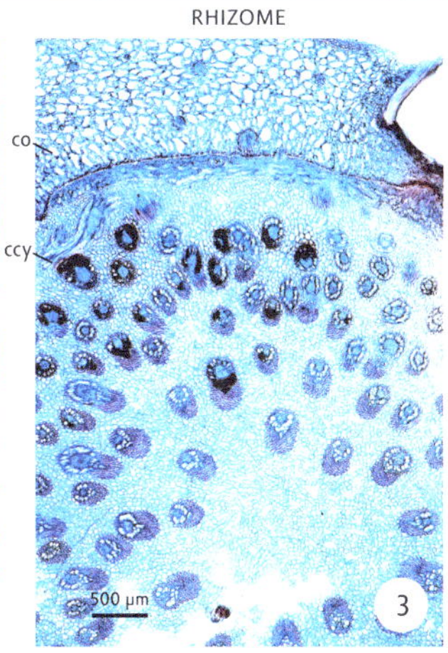

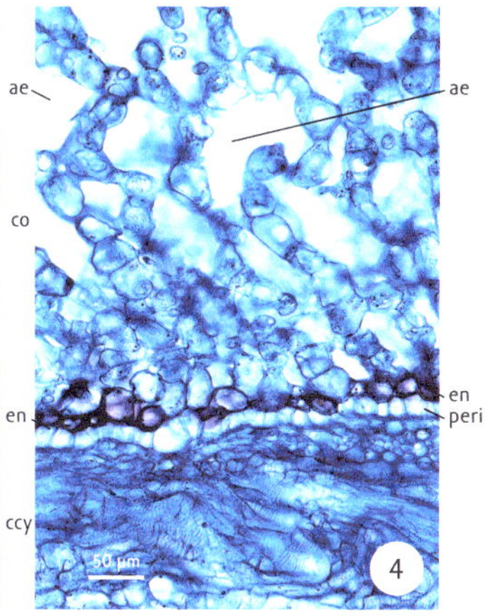

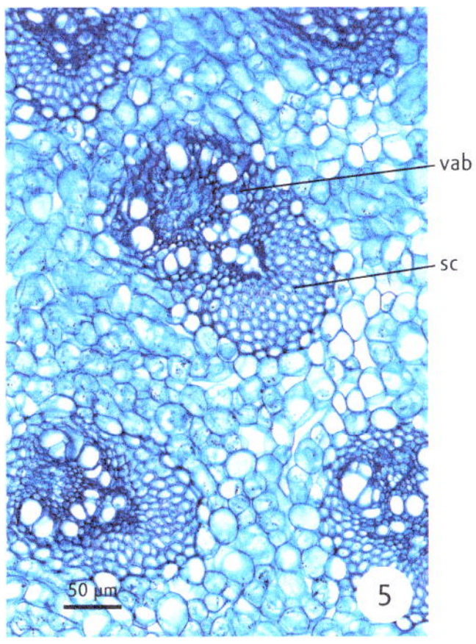

Trichophorum cespitosum (L.) Hartm.

(syn. *Scirpus cespitosus, Baeothryon caespitosum*)

Cyperaceae
Deergrass

Perennial wetland herb, densely tufted, with short rhizome, stem erect, 5–35 cm high, very slender, smooth, with numerous leaf-sheaths at base. Inflorescence of one terminal spikelet. Perianth-bristles remaining shorter than glumes, pale-brown.

Distribution and ecology

Circumpolar boreo-montane element, subatlantic. In all CE countries except HU, rare in SK, on peaty meadows and pastures, oligotrophic peatbogs, raised bogs, acidic soils, persisting even in burnt and heavily deer-grazed areas.

Anatomy

Circular culm with ridges with a diameter of <1 mm (1).

The epidermis consists mainly of roundish, externally lignified thick-walled cells (1–3). Guard cells of the stomata are thickened on three sides (3). Substomatal chambers are surrounded by sclerenchymatic cells (3). The chlorenchyma is composed of several layers of round and angular unlignified parenchymatic cells (1–3). The middle and inner part consists of round unlignified parenchymatic cells, radial rectangular girders of sclerenchyma and vascular bundles. Air spaces are absent (1, 2). Vascular bundles consist of a xylem with two large lateral vessels, an irregularly formed intercellular and a roundish phloem with sieve tubes (2). Vascular bundles are surrounded by 1–3 layers of sclerenchymatic cells (2, 3). Secretory cells are absent. The chlorenchyma is distinctly separated in the center by very thin-walled, unlignified parenchymatic cells (1, 2).

Rhizome with a diameter of 1–2 mm.

The cortex is composed of an external layer of small, lignified parenchymatic cells without intercellulars and an inner part of thin-walled unlignified parenchymatic cells with small intercellulars (4, 5). The central cylinder and the cortex are separated by a lignified thin- to thick-walled endodermis (5). The lignified parenchymatic ground tissue of the central cylinder contains amphivasal vascular bundles without distinct sclerenchymatic sheaths (6). Secretory cells are absent.

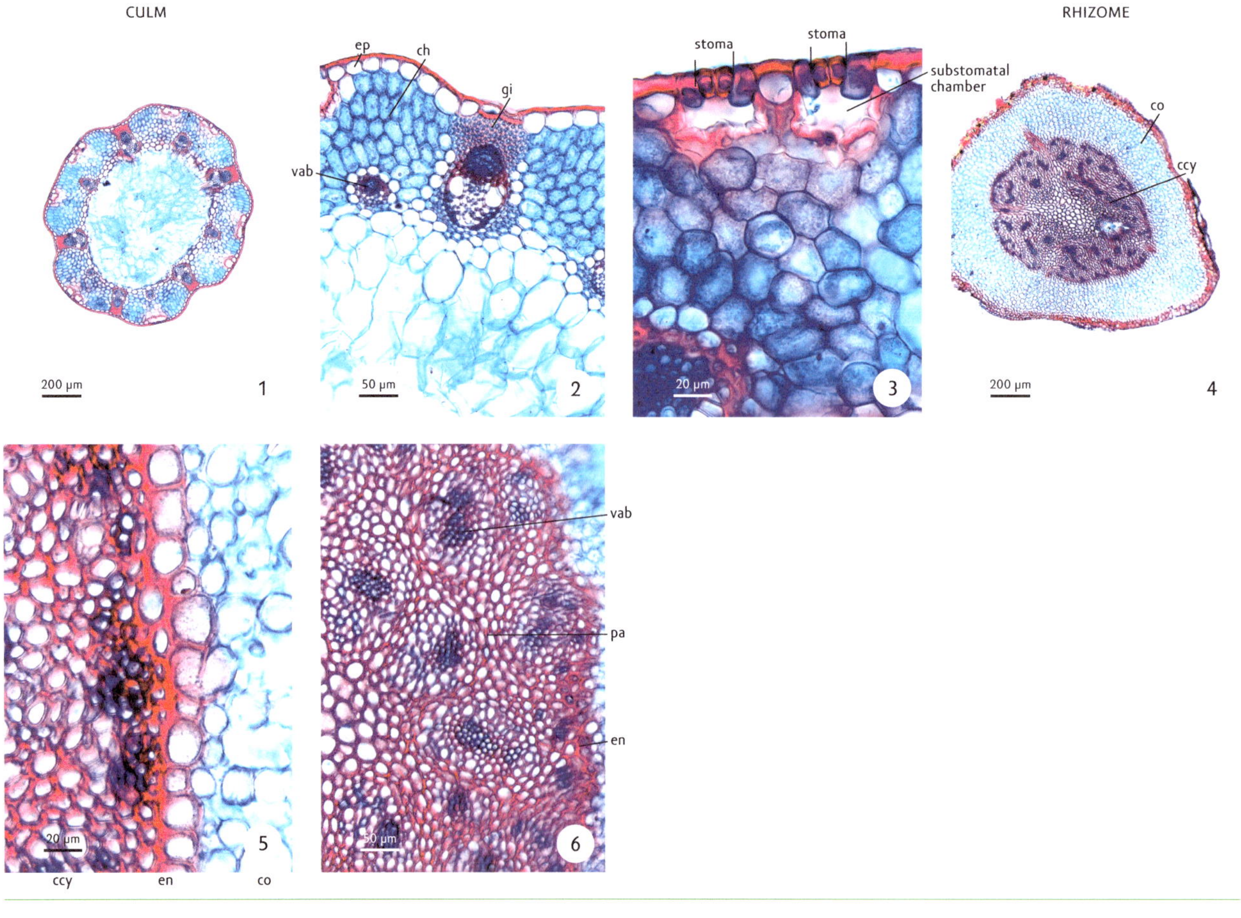

Trichophorum pumilum (Vahl) Schinz et Thell.

Perennial wetland herb, small, tufted, stem erect, similar to T. cespitosum, but only 3–15 cm high, with long rhizomes, greyish green, smooth, leafy only at base. Leaf-sheaths black-brown. Inflorescence of one terminal ovoid spikelet. Perianth bristles (3) remaining very short at fruiting.

Distribution and ecology

Native to Eurasia, disjunct distribution range, in Europe it currently occurs only in the Alps (CH), a small area in the Western Carpathians (SK), and Norway. Postglacial relict on calcareous fens in SK, usually those initiated during the late glacial and surrounded by extremely mineral-rich active travertine springs. Spring fen meadows, wet travertine rocks.

Anatomy

Circular culm with a diameter of <1 mm.

The epidermis consists mainly of roundish, externally lignified very thick-walled cells (1, 2) and very small groups of sclerenchymatic cells (2). Guard cells of the stomata are thickened on the external and internal side (2). The chlorenchyma is composed of several layers of round to oval angular unlignified parenchymatic cells (1, 2). The middle and inner parts consist of round unlignified parenchymatic cells, radial and rectangular girders of sclerenchyma and vascular bundles. Air spaces are absent (1, 2).

Vascular bundles consist of a xylem with two large lateral vessels, an irregularly formed intercellular and a roundish phloem with sieve tubes (2). Vascular bundles are surrounded by 1–4 layers of sclerenchymatic cells (2). Secretory cells are absent. The chlorenchyma is distinctly separated in the center by very thin-walled, unlignified parenchymatic cells (1, 2).

Circular rhizome with a diameter of <1 mm (3).

The cortex is composed of a thin-walled epidermis, an external layer of small lignified parenchymatic cells without intercellulars and an inner part of thin-walled unlignified parenchymatic cells with spongy intercellulars (3, 4). The central cylinder and the cortex are separated by a lignified, thick-walled endodermis with U-shaped cells (4). The lignified parenchymatic ground tissue of the central cylinder contains amphivasal vascular bundles without sclerenchymatic sheaths (5). Secretory cells are absent.

CULM

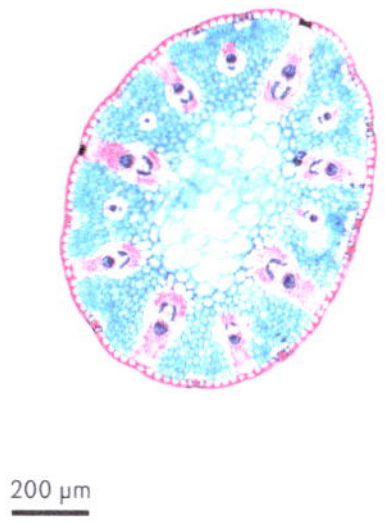

1

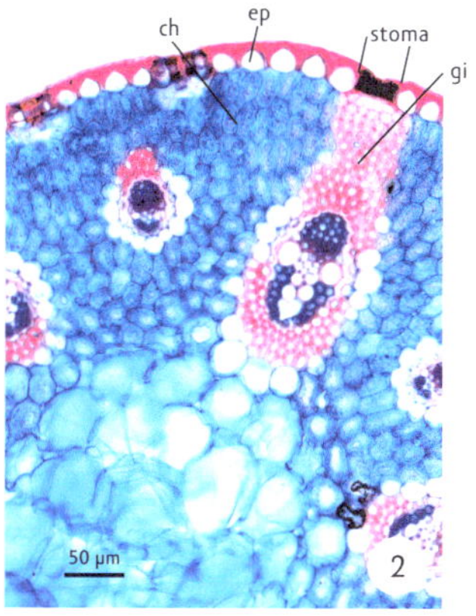

2

RHIZOME

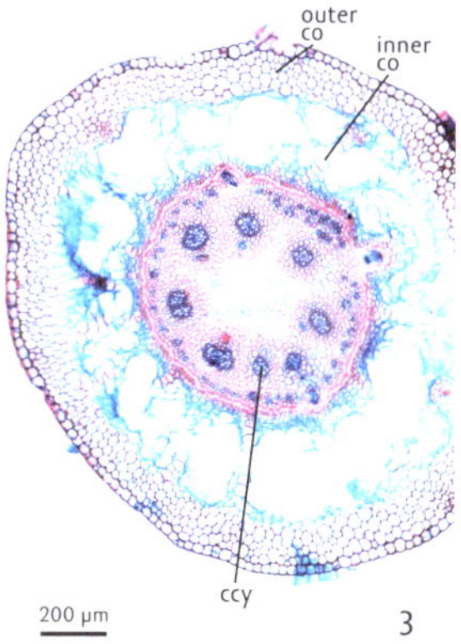

3

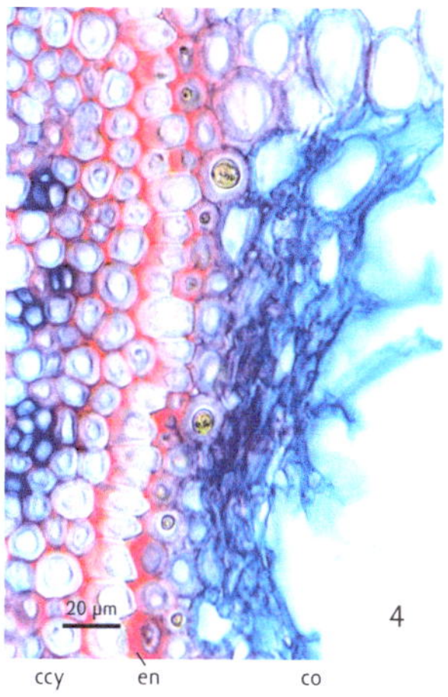

4

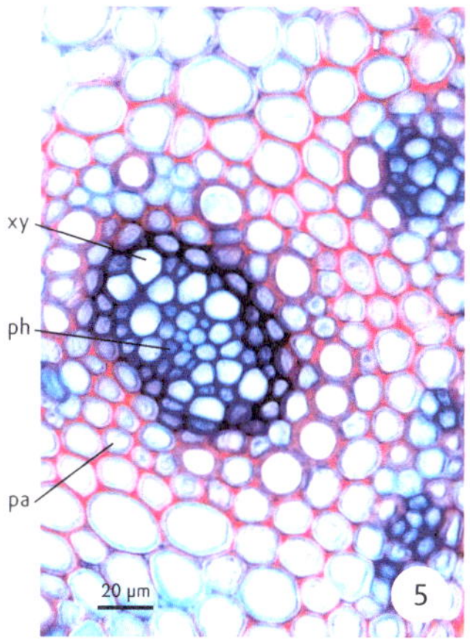

5

Elodea canadensis Michx.

Canadian waterweed

Hydrocharitaceae

Perennial submerged plant, thin, cylindrical, fragile, poorly branched stem, trinity of leaves in regular whorls, rooted in the sediment. Shoots 10–70 cm long, unbranched roots 5–30 cm long, adventitious roots at shoot bases, root hairs. Oblong leaves 5–16 mm long, leaf apices obtuse, leaves with two cell layers, dioecious plant: female solitery flowers float on the surface on peduncle (10–40 cm). Overwintering in apical dormant turions attached to the shoot.

Distribution and ecology

Native to temperate N America; widely naturalized in Europe, and Australasia. Common in all 7 CE countries mainly in lowlands by ca. 600 m a.s.l. Shallow standing waters, fishponds, lakes, peat bogs, fens, backwater pools and oxbows, canals and sand-pits, slow-streaming rivers and brooks, clear to mildly dystrophic waters, water depth 0.3–3 m, in oligo-meso- to mildly eutrophic waters, does not tolerate low water transparency <0.5 m, soft to hard waters (total alkalinity 0.15–4 meq/l), only vegetative propagation. "Polar leaf" photosynthesis, efficient HCO_3^- use, highly invasive species, very dense stands.

Anatomy

Round shoot with a diameter of 1–2 mm (1).

A large cortex and an epidermis surround a small central cylinder (1).

Central cylinder: An area of poorly differentiated unlignified cells surrounds a central air duct (2). Fibers are absent (2). Vessels can not be recognized. The endodermis is poorly developed (1, 2).

Cortex: Large parenchyma cells (70–90 µm) and round large air ducts form the cortex (3). A few small groups of cells, probably sieve tubes, are located near the periphery (3). Small intercellulars occur between the round cells (4).

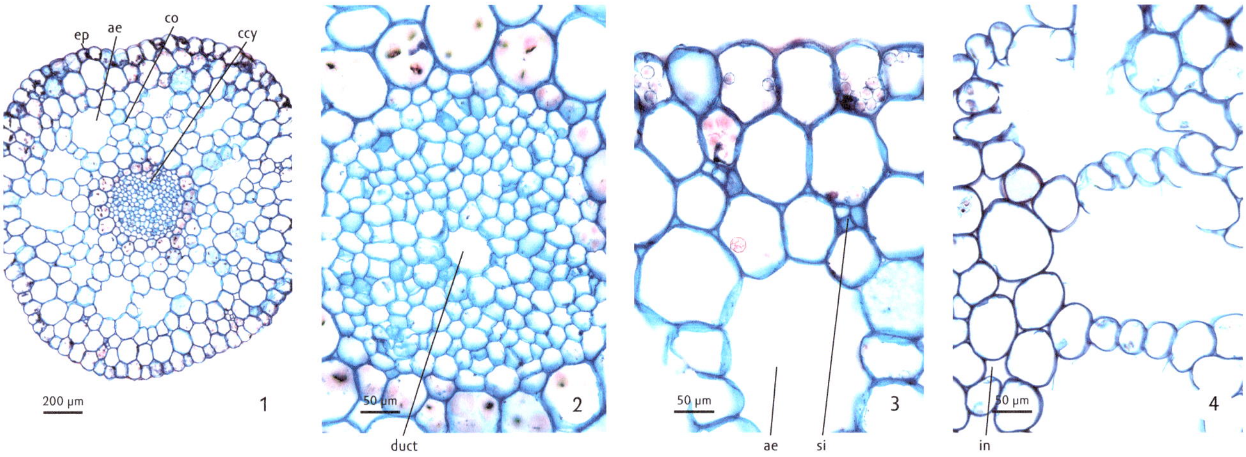

F. H. Schweingruber et al., *Anatomic Atlas of Aquatic and Wetland Plant Stems*, https://doi.org/10.1007/978-3-030-33420-8_62

Elodea nuttallii (Planch.) H. St. John **Nuttall's waterweed** Hydrocharitaceae

Perennial submerged plant, thin, cylindrical, fragile, poorly branched stem, trinity of leaves in regular whorls, rooted in the sediment. Shoot 10–70 cm long, unbranched roots 5–30 cm long, adventitious roots at shoot bases, root hairs. Oblong, very narrow leaves 0.8–1.5 mm wide and 6–20 mm long, leaf apices acute, leaves usually strongly recurved, only two cell layers, dioecious plant: female solitery flowers float on the surface on peduncle (10–30 cm), overwintering in non-dormant shoots.

Distribution and ecology

Native in temperate N. America; naturalized in Europe and Japan. Subatlantic distribution: distributed in all 7 CE countries, strong gradient between the West and East of CE: very common in CH, D, A, rarer in PL, SK, H and CZ. Shallow standing waters, fishponds, lakes, backwater pools and oxbows, canals, sand-pits, slow-streaming rivers and brooks, water depth 0.3–3 m, oligo-meso- to mildly eutrophic waters, does not tolerate low water transparency <0.5 m, soft to hard waters (total alkalinity 0.15–4 meq/l), only vegetative propagation. "Polar leaf" photosynthesis, efficient HCO_3- use, highly invasive species.

Anatomy

Round shoot with a diameter of <1 mm (1).

A large cortex and an epidermis surround a small central cylinder (1).

Central cylinder: An area of larger and smaller unlignified cells surrounds a center with an air duct (2). Polygonal cells represent sieve tubes, round cells and parenchyma cells (2). Fibers are absent. Vessels can be recognized only in longitudinal sections. Vessel walls have wide spaced ring-like thickenings (3). The endodermis is poorly developed (1, 2).

Cortex: Large parenchyma cells (70–90 µm) and round large air ducts form the cortex (4). A few sieve tubes are located near the periphery (4). Small intercellulars occur between the round cells. The slightly thickened epidermis-walls contain dark substances (4).

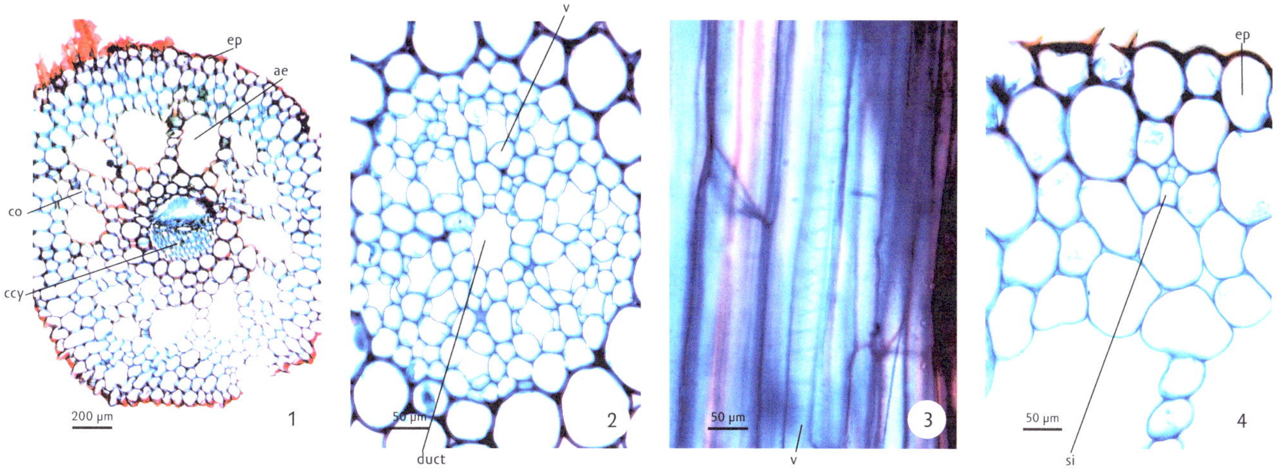

Hydrocharis morsus-ranae L. **Frogbit** Hydrocharitaceae

Perennial free-floating plant, rosette of circular-reniform, peltate floating leaves 2–5 cm large, tubular fragile petioles 4–15 cm long. Unbranched roots 15–40 cm long, from very shortened stem at petiole bases, dense and long root hairs, terrestrial form with shortened petioles known. Individual rosettes connected by stolons, solitary flowers 2–3 cm large, mostly dioecious, floating on the surface, seed set very rare. In autumn, 6–12 mm large, dark green elliptic scale-shaped turions formed on stolons, sink down after abscission, actively float up in spring.

Distribution and ecology

Euro-Siberian temperate element; widely naturalized outside its native range. Occurs in all 7 CE countries, more common in lowlands. Shallow standing waters. Prefers hardwaters, tolerates softwaters (total alkalinity ca. 0.2–4 meq/l), always organic, nutrient-rich bottom, meso- to slightly eutrophic waters, under optimal conditions, very rapid growth, able to cover the whole water surface.

Anatomy

Round submersed petiole with a diameter of 1–2 mm (1).

A large cortex and an epidermis surround a small central cylinder (1).

Central cylinder: The central cylinder is composed of indistinct vascular bundles, which are separated by round parenchyma cells. Some angular, unlignified, polygonal sieve tubes and small companion cells surround a central air duct of each vascular bundle. Fibers are absent (2). The endodermis is poorly developed (1, 2).

Cortex: Large parenchyma cells (10–150 µm) and large, radially oval intercellulars form the inner zone of cortex (1). The external zone consists of a belt of parenchyma cells which contain rudimentary, unlignified vascular bundles (3). The epidermis is thin-walled (3).

Round submersed rhizome with a diameter of 4–6 mm (4).

The central cylinder is composed of indistinct vascular bundles, which are separated by round parenchyma cells (5). The cell types within the bundles are poorly differentiated (6). Vessels appear bright in polarized light (7). Both fibers and intercellulars are absent.

The cortex consists of net-like intercellulars and rudimentary vascular bundles (4). Crystal sand is very frequent in the cortex (8).

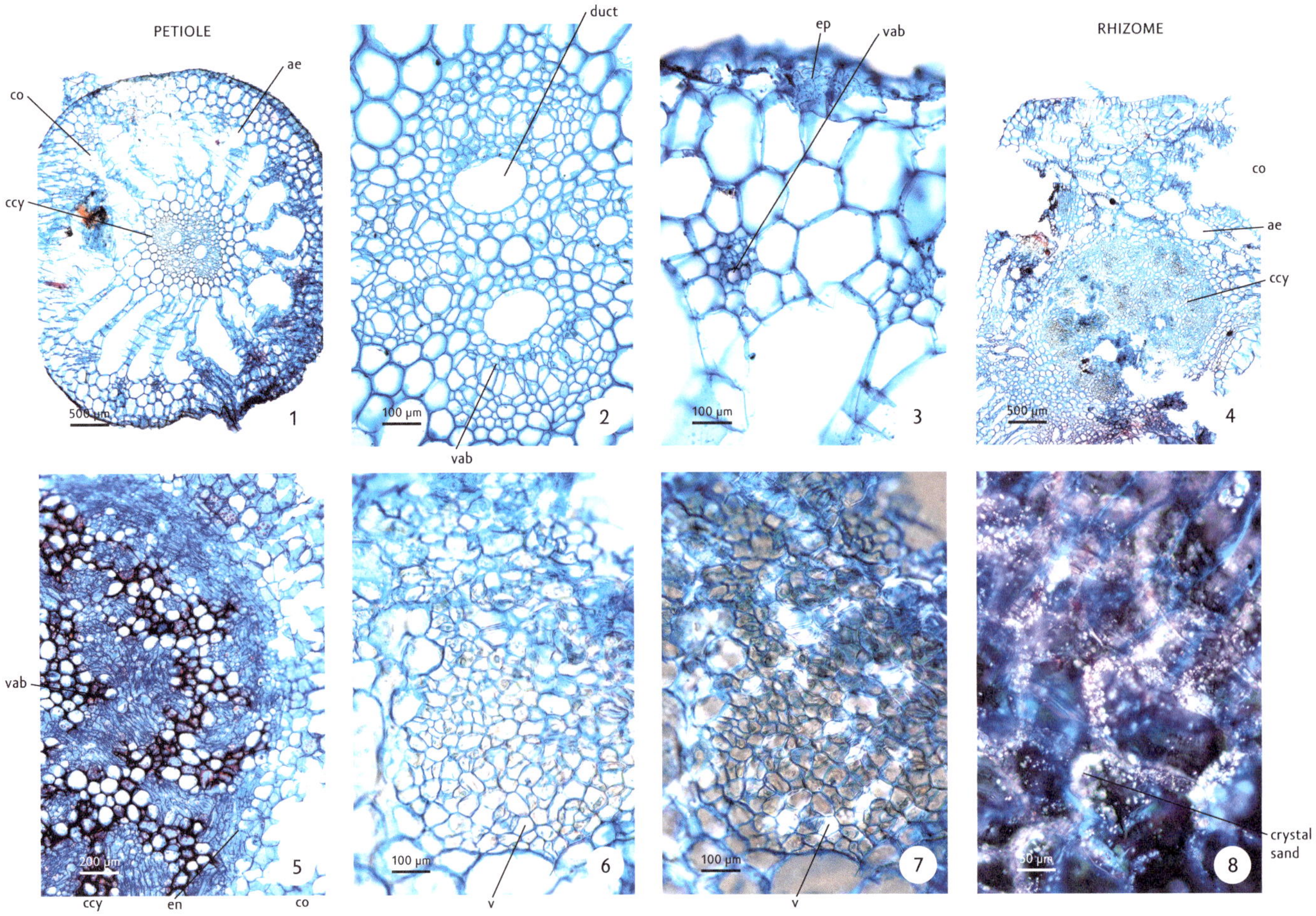

Najas marina L.

Holly-leaved naiad

Annual submerged herb rooted in the bottom. Flattened fleshy leaves dentate on margins, stems with spines, very fragile shoots 0.1–0.6 m long, unbranched roots 5–12 cm long with root hairs, also in leaf axils in shoot bases. Dioecious plant: solitary achlamydeous flowers formed in leaf axils, fruits achenes, ripe seeds launched by fruits to the surface where float for a limited time.

Distribution and ecology

Circumpolar southern-temperate element. Scattered in all 7 CE countries especially in lowlands or uplands in hard water regions. Standing waters in fishponds, lakes, dam reservoirs, backwater pools and oxbows of large rivers, also slow-streaming rivers and canals. Requires neutral to alkaline, medium to strongly hard water (pH 7–9, total alkalinity 1–5 meq/l), prefers meso- to eutrophic waters with organic, nutrient-rich sediments, water depth 0.5–3 m, distribution broadened as a result of fishpond liming and seed transfer by water birds (CZ, SK), does not tolerate hypertrophic fishponds, efficient HCO_3^- user.

Anatomy

Round shoot with a diameter of 1 mm (1).

A large cortex and an epidermis surround a small central cylinder (1).

Central cylinder: An area of poorly differentiated unlignified cells surrounds a central duct (2). Fibers and vessels are absent (2). Angular cells probably represent sieve tubes (2). The endodermis consists of unlignified thin-walled, round cells. (2).

Cortex: Large parenchyma cells (100–180 µm) form the cortex (3, 4). Some intercellulars are as big as parenchyma cells (3), some others are very small (4). The epidermis is thin-walled (3).

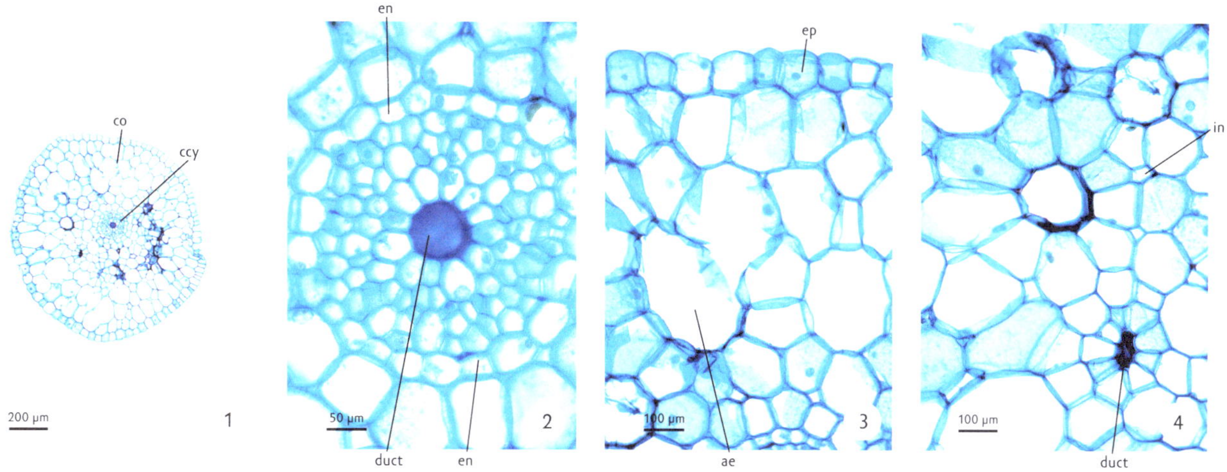

Najas minor All.

Minor naiad

Annual submerged herb rooted in the bottom. Narrow recurved leaves denticulate on margins, very fragile shoots 5–25 cm long. Unbranched roots with root hairs 3–8 cm long, also in leaf axils in shoot bases. Monoecious plant: solitary achlamydeous flowers formed in leaf axils, fruits achenes.

Distribution and ecology

Eurasian southern-temperate element, naturalized in N America. Scattered rarely in all 7 CE countries esp. in lowlands or uplands in hard water regions. Standing waters in fishponds, lakes, dam reservoirs, backwater pools and oxbows of large rivers, slow-streaming rivers and canals. Requires neutral to alkaline, medium to strongly hard water (pH 7–9, total alkalinity 0.8–5 meq/l), meso- to slightly eutrophic waters, organic, nutrient-rich sediments, also sand. Water depth 0.1–2 m, distribution broadened as a result of fishpond liming and seed transfer by water birds (CZ, SK), does not tolerate hypertrophic fishponds, competitively very weak, efficient HCO_3- user.

Anatomy

Round shoot with a diameter of <1 mm.

A large cortex and an epidermis surround a small central cylinder (1).

Central cylinder: An area of poorly differentiated unlignified cells surrounds a center with (2, 3) or without a central air duct (2). Fibers and vessels are absent (3). Angular cells probably represent sieve tubes (3). The endodermis consists of unlignified thin-walled round cells (3).

Cortex: Large parenchyma cells (80–150 μm) and radially oriented large intercellulars form the cortex (1, 3). The epidermis is thin-walled (1, 2).

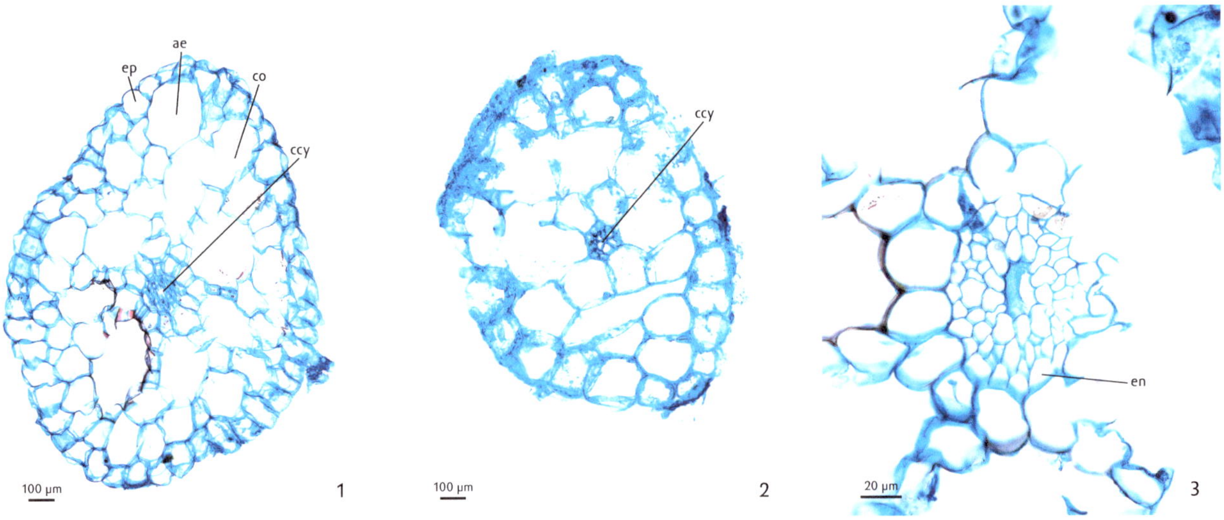

Stratiotes aloides L.

Water-soldier Hydrocharitaceae

Perennial floating, mostly submerged plant, rosette of sturdy amphibious leaves. Spinous-serrate, ribbon-like leaves 10–50 cm long, 12–20 mm wide, usually emerge by tips from water, terrestrial form known. Long unbranched fibrous roots (20–200 cm) usually root in the bottom, solitary white flowers 3–4 cm across on peduncles above the surface, flowers mostly dioecious, seed set very rare, overwintering under water in dormant turions.

Distribution and ecology

Euro-Siberian boreo-temperate element; widely naturalized outside its native range. Native to all CE countries, except CH (only as alien), very rare in CZ, more common esp. in PL and DE, commonly introduced by water gardeners. Shallow standing waters, fishponds, lakes, dam reservoirs, backwater pools and oxbows, irrigation or drainage canals, slow-streaming rivers, clear to strongly dystrophic waters, peat bogs, fen lakes, water depth 0.1–3 m, prefers lowland, absent from intensively agricultural areas, Ca-rich hardwaters, tolerates medium soft waters (total alkalinity ca. 1–5 meq/l), always organic, nutrient-rich bottom, meso- to slightly eutrophic waters. Submerged leaves without stomata with thin cuticle use HCO_3^-, emergent leaves with stomata strict CO_2 users, turions sensitive to high NH_4^+ concentration and anoxia.

Anatomy

Oval flower stalk with a diameter of 7 mm in the longer axis (1).

Several closed, collateral vascular bundles occur within an aerenchymatic tissue (1, 2). Xylem and phloem are difficult to distinguish. The xylem of well-developed bundles consists of a few angular, hardly lignified, large vessels and parenchyma cells and the phloem is composed of a poorly differentiated tissue (2). The xylem is absent in poorly developed vessels (3). Air ducts occur in a minority of vascular bundles (1, 2). An endodermis is absent.

Short basal part: Circular stem with a diameter of 10–12 mm (4). Many not axially directed vascular bundles form the inner part of the stem (4). Vessels walls have scalariform intervessel pits or ring-like structures (5). The cortex is indistinctly separated from the central part. The cortex consists of round cells with small intercellulars (6). Round large cells with blue staining substances are characteristic of basal part of the plant (7) just as small crystals with different forms (prismatic, tetragonal, irregular) (8).

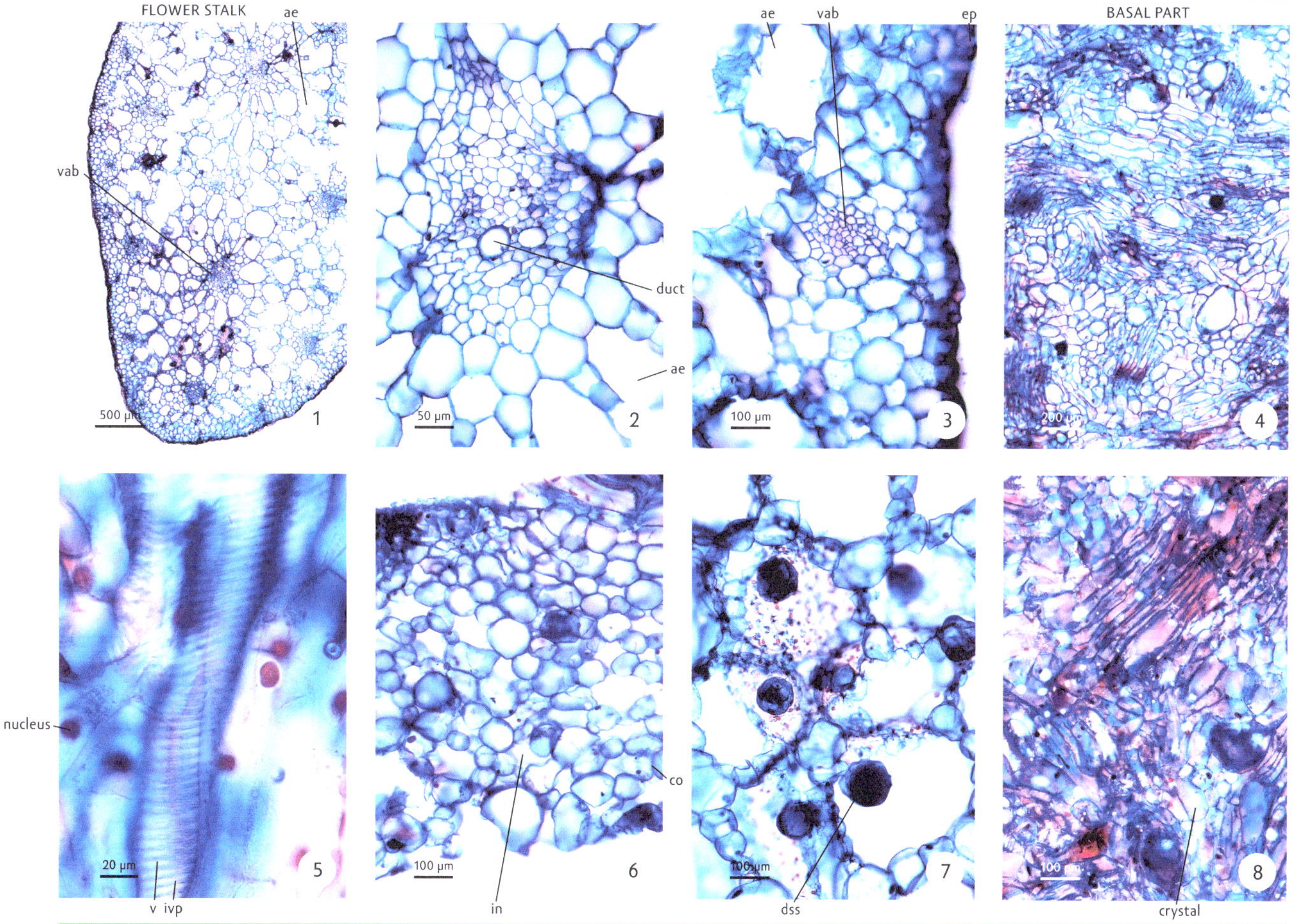

Perennial submerged plant, rooted in mud. All leaves in basal rosette, evergreen, linear, up to 1 m long, with pale-green to reddish colour. Female flowers on long stalks spiraling after flowering.

Distribution and ecology

Native to tropical and subtropical regions of S Europe, Africa, Asia and Australia. In CE reported only as a neophyte in AT, CH, DE, HU and PL where it grows in still or slowly flowing water in canals and rivers in lowland, usually where water is heated. Prefers nutrient-rich substrates. Reproduction by seed but more often it propagates by runners which can lead to dense stands. It is often grown by aquarists.

Anatomy

Very short, round transition zone between shoot and rhizome with a diameter of 3–5 mm.

Shoot: A distinct cortex surrounds a large central cylinder (1). The xylem of the closed collateral vascular bundles mainly consists of a few vessels with hardly lignified walls and surrounding parenchyma cells (2, 3). The phloem is composed of groups of sieve tubes and companion cells (2). Arcs of fibers limit the peripheral part of the bundles to the surrounding parenchyma (2). A ring of fibers separates the central cylinder and the cortex (1, 4). Small and large parenchyma cells and intercellulars of different form and size form the cortex (4). The epidermis is thin-walled (3).

Rhizome: A distinct cortex surrounds a large central cylinder (5). The cortex and the central cylinder are separated by a thin-walled endodermis (5). The cortex consists of radially oriented parenchyma cells and large intercellulars (5). Intercellulars in the pith are small. Mainly raphids and rarely crystal druses occur in the cortex. The xylem of the closed collateral vascular bundles mainly consists of an arc of lignified vessels and not lignified parenchyma cells (6). Vessel-wall pits are scalariform (7).

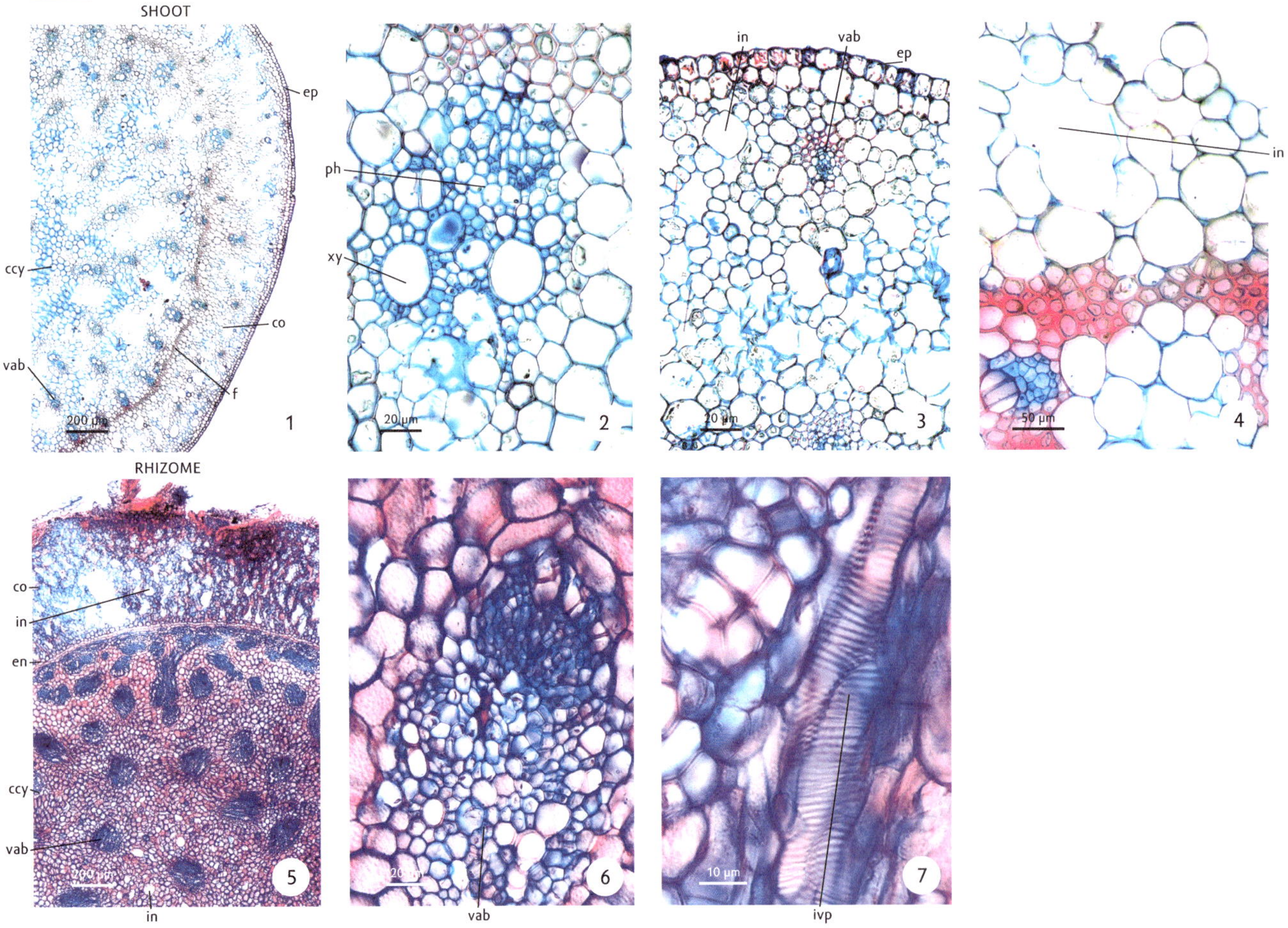

Perennial wetland herb, 30–80 cm high, erect, poorly ramified above. With underground tuber and sword-like leaves, the basal leaf obtuse. Tall stalks bear purple-red flowers collected in one-sided inflorescence, with strongly curved flower tube. Clonal plant.

Distribution and ecology

Native to E Europe, in all CE countries, very rare in AT and CH, from lowlands to uplands. Wet meadows, woods and fields of the montane zone, which dry during summer.

Anatomy

Circular flower stalk with a diameter of 2–3 mm (1).

An epidermis and a fiber belt surround a central thin-walled parenchymatic center (1).

The flower stalk consists of central unlignified parenchymatic cells, many collateral vascular bundles, a peripheral lignified belt of thick-walled fibers and an unlignified epidermis (1, 2). A sheath of lignified thin- to thick-walled fibers surrounds the triangular vascular bundles. The xylem of vascular bundles consists of an arc of vessels and a reniform part of phloem, consisting of large sieve tubes and small companion cells (3). Vessels of the secondary xylem have scalariform intervessel pits with large apertures. Vessels of the primary xylem have helical-like thickenings (4).

Circular rhizome with a diameter of 10–15 mm (5).

The center consists of parenchymatic cells and many vascular bundles. An unlignified endodermis-like belt separates the cortex from the central part (6). 2–3 layers of unlignified thin-walled vessels partially surround the sieve elements (7, 8). Vessel walls are unlignified (7). Vessels are surrounded by a uniseriate, endodermis-like sheath of unlignified cells (7).

The cortex consists of round parenchymatic cells and many small intercellulars and contains many vascular bundles (5, 6). Note the occurrence of cells with red-stained substances (7, 8). Crystals are absent.

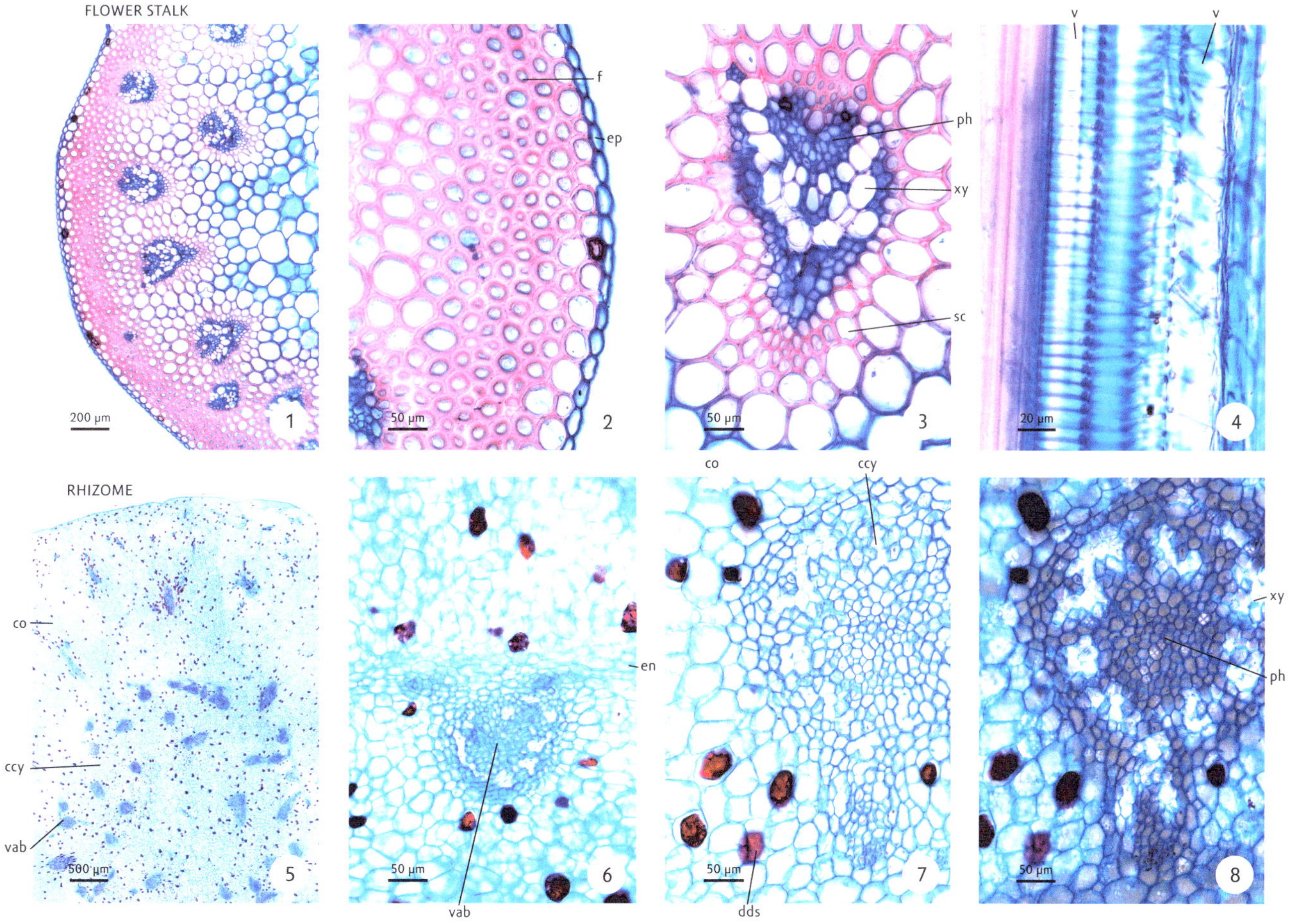

F. H. Schweingruber et al., *Anatomic Atlas of Aquatic and Wetland Plant Stems*, https://doi.org/10.1007/978-3-030-33420-8_63

Perennial wetland, rhizomatous herb. Stems 60–120 cm high, slightly compressed, bearing several flat leaves. Flowers 7–10 cm across, yellow with deeper yellow patch in outer tepals.

Distribution and ecology

European southern-temperate element; widely naturalized outside its native range. Common in all CE countries, in wet places near rivers, oxbows, water ditches, fishponds and reed swamps, common in warmer areas. Tolerates long-term anoxia.

Anatomy

Circular flower stalk with a diameter of 4–7 mm (1).

A small chlorenchyma surrounds a large central cylinder (1).

The central part of the flower stalk consists of unlignified parenchymatic cells, many collateral vascular bundles and a peripheral lignified belt of fibers (1). Note solitary excretion cells (2, 3). A sheath of lignified fibers surrounds the vascular bundles. The centripetal part of the vascular bundles consists of an arc of vessels and a round part of phloem, consisting of large sieve tubes and small companion cells (2). Vessel walls have scalariform intervessel pits (4). The chlorenchyma is mainly formed by large parenchymatic cells and small intercellulars (2).

Circular rhizome with a diameter of 10–20 mm (5).

The center consists of parenchymatic cells and many vascular bundles. A distinct lignified endodermis separates the cortex from the central part (6). 2–3 layers of thin- to thick-walled vessels partially surround the sieve elements. Vessel walls are lignified (6). The cortex consists of round parenchymatic cells and many small intercellulars (5, 6). Note the presence of very long crystals (7).

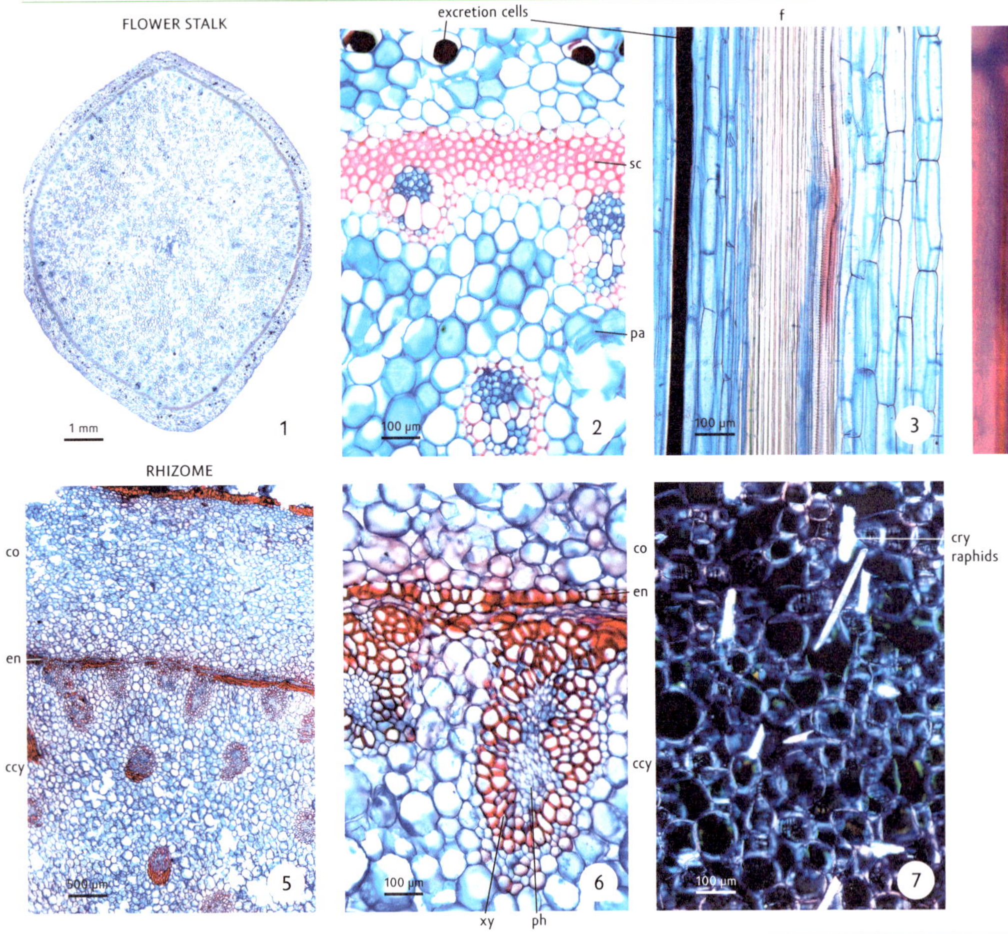

Perennial wetland, rhizomatous herb. Tufted, stems 50–120 cm high, erect, roundish or weakly compressed, ramified only above. Leaves flat. Flowers 6–7 cm across, blue or bluish-violet with white patch on outer tepals.

Distribution and ecology

Native to Euro-Siberia, scattered in all CE countries, from lowlands to uplands. Moist and peaty meadows, lowland woods, peaty forests and pastures on non-cultivated soils.

Anatomy

Circular flower stalk with a diameter of 4–5 mm (1).

A small, distinct aerenchymatic cortex surrounds a hollow, large central cylinder (1).

The central part of the flower stalk consists of unlignified parenchymatic part with many collateral vascular bundles and a peripheral lignified belt of fibers (1). Note solitary excretion cells. A sheath of lignified fibers surrounds the vascular bundles. The centripetal part of the vascular bundles consists of an arc of poorly lignified vessels and a round part of phloem, consisting of large sieve tubes and small companion cells (1).

The cortex is mainly formed by large parenchymatic cells and irregular intercellulars (1).

Circular rhizome with a diameter of 4–5 mm (2).

The center consists of parenchymatic cells and many vascular bundles. A distinct lignified endodermis separates the cortex from the central part (3). Thick-walled vessels surround partially the sieve elements (4). The thick-walled vessels are mostly unlignified (4). Vessel walls have scalariform intervessel pits or are stabilized by ring-like structures. The cortex consists of round parenchymatic cells and many small intercellulars (5). Note the presence of very long crystals (6, 7).

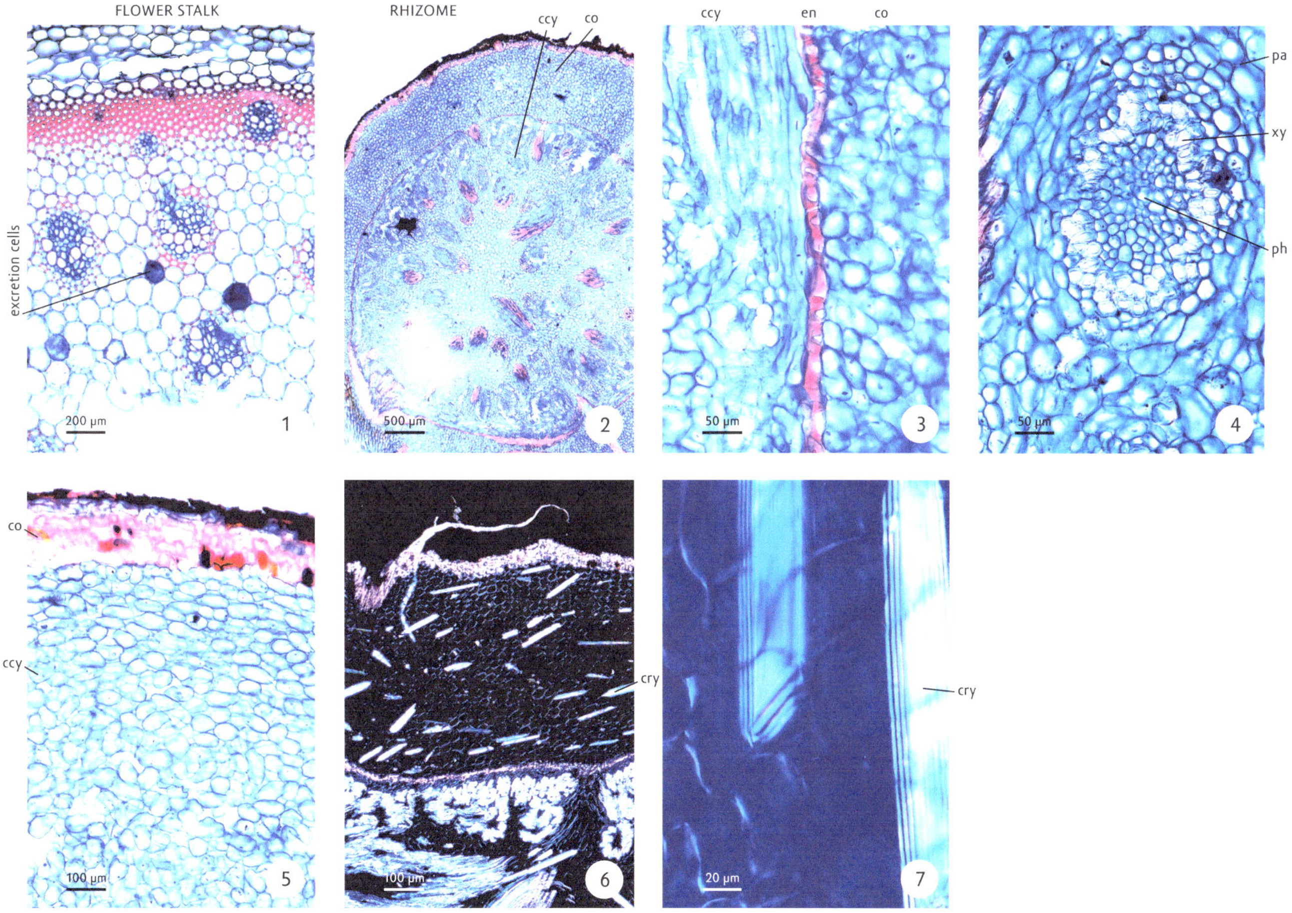

Perennial wetland herb, rhizomatous with longer rhizomes. Stems erect, up to 1.1 m high, leaves 2–4 mm wide, rounded. Inflorescence in two or more whorls, the tepals sharply pointed. Fruit is an elliptical, red-brown capsule, tapered to long points.

Distribution and ecology

European temperate element, also in N America. In all CE countries except HU, in marshes, peaty meadows, springs and wet forests. Wet or damp habitats on acidic soils. Iron deficiency chlorosis in soils with high nitrate levels.

Anatomy

Circular culm with a diameter of 2–3 mm.

The epidermis consists of roundish, externally lignified thick-walled cells (1–2). Guard cells of the stomata are peripherally lignified and have acuminate lips. The chlorenchyma is composed of 1–3 layers of round to oval unlignified parenchyma cells (1, 2). Epidermis and chlorenchyma cells are filled with dark stained substances (2). The culm consists of round parenchyma cells, small vascular bundles and air spaces (1, 2). The inner part is characterized by a continuous row of vascular bundles. The large vascular bundles consist of a xylem with two large lateral vessels, an air duct and a roundish phloem with sieve tubes and companion cells (2). The bundles are surrounded by a large sheath of fibers (1, 2). The pith consists of large thin walled parenchyma cells and a few vessels (1).

Rhizome with a diameter of 3–4 mm (3).

The cortex is composed of an epidermis, a peripheral layer of round thin-walled parenchyma cells and large round air spaces between radial zones of parenchyma cells (3). The central cylinder and the cortex are separated by a centripetal thick-walled endodermis and a thin-walled pericycle (4). The parenchymateous ground tissue of the central cylinder contains amphivasal vascular bundles with large vessels. The bundles are surrounded by a sclerenchyma sheath.

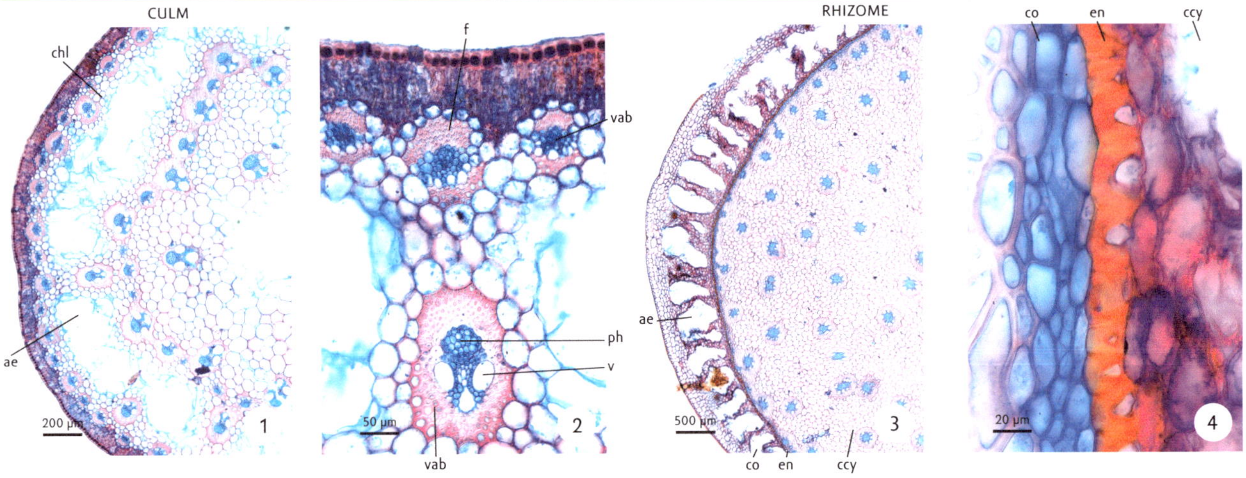

F. H. Schweingruber et al., *Anatomic Atlas of Aquatic and Wetland Plant Stems*, https://doi.org/10.1007/978-3-030-33420-8_64

Juncus alpinoarticulatus Chaix

Alpine Rush

Perennial wetland herb, rhizomatous. Stem erect to ascending, to 40 cm high, leaves with very distinct transverse septa. Inflorescence diffuse but often rather sparse, dark-brown tepals, the outer subacute to obtuse, the inner obtuse to rounded.

Distribution and ecology

Circumpolar boreo-montane element. In all CE countries, in wet sandy soils, peaty meadows, acidic fens and ditches.

Anatomy

Circular culm with a diameter of 1 mm.

The epidermis consists of roundish, externally slightly thickened cell walls (1, 2). Guard cells of the stomata are not lignified and have acuminate lips. The chlorenchyma is composed of 3–5 layers of round to oval, unlignified parenchyma cells (1). The middle and inner part of the culm consist of round unlignified parenchyma cells and vascular bundles (1). The large vascular bundles consist of a xylem with two slightly large lateral vessels and a roundish phloem with sieve tubes and companion cells (1, 2). Peripheral, moon-like groups of thick-walled fibers characterize the large bundles (1).

Rhizome with a diameter of 1–2 mm.

The cortex is composed of an epidermis, an external layer of thin-walled lignified parenchyma cells and an inner part of thin-walled unlignified parenchyma cells and air spaces (3, 4). The central cylinder and the cortex are separated by a centripetal thick-walled endodermis and partially an indistinct pericycle (3). The lignified parenchymatic ground tissue of the central cylinder contains amphivasal vascular bundles, without sclerenchyma sheaths.

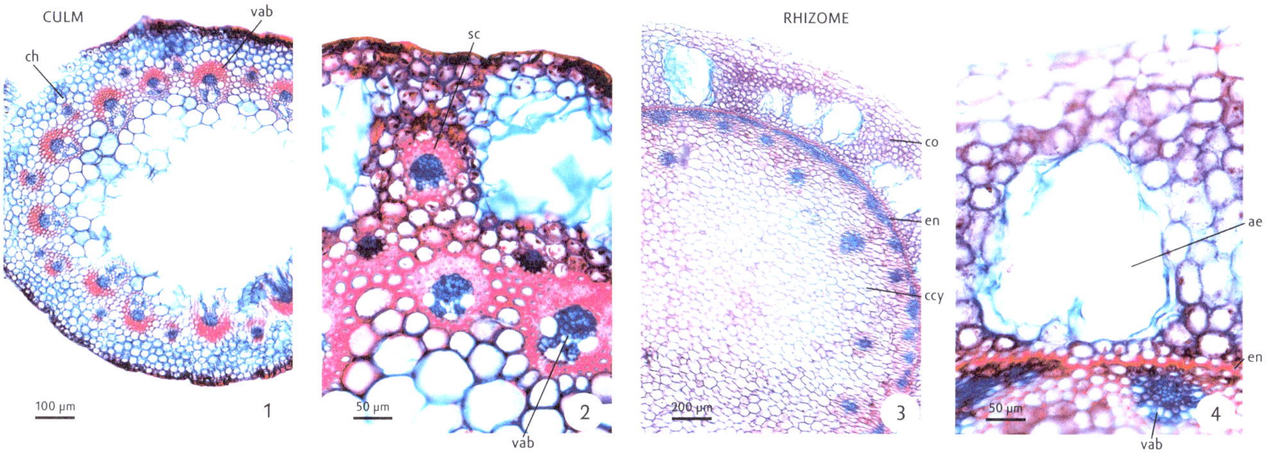

Perennial wetland, mat-forming herb with a horizontal system of rhizomes with 4–5 culms from current year in one row and numerous dead culms from previous years. Culms 30–40 cm, erect, terete, very finely ribbed, without leaves. Inflorescence a dense cyme, apparently lateral on the culm. Fruit is a dark reddish-brown to blackish capsule.

Distribution and ecology

Arctic and circumpolar boreal distribution. In CE in the Alps (AT, CH), especially in banks of rivers and lakes, on alluvial terraces and fens, in open vegetation with little competition. Occurs also on the arctic seashores, it is able to tolerate saline environment. Vegetative reproduction by extensive growth and fragmentation of rhizome.

Anatomy

Culm with a diameter of 1–2 mm.

The epidermis consists of roundish, lignified thick-walled cells (1, 2). Guard cells of the stomata are centrically lignified and have small lips (3). The chlorenchyma is composed of 3–4 layers of oval to palisade-like unlignified parenchyma cells (2, 3). The inner part of the hollow culm consists of round parenchyma cells, large and small vascular bundles and small air spaces (2). The large vascular bundles consist of a xylem with two large lateral vessels, rarely an air duct and a roundish phloem with sieve tubes and companion cells (2). Vascular bundles are surrounded by a large sheath of fibers (1, 2).

Rhizome with a diameter of 3–5 mm.

The cortex is composed of an epidermis, a peripheral layer of small round parenchyma cells, and large radial air spaces between radial layers of parenchyma cells (4). The central cylinder and the cortex are separated by a thick-walled endodermis and an indistinct pericycle (5). The parenchymatic ground tissue of the central cylinder contains amphivasal vascular bundles, which are surrounded by a sheath of sclerenchyma (6).

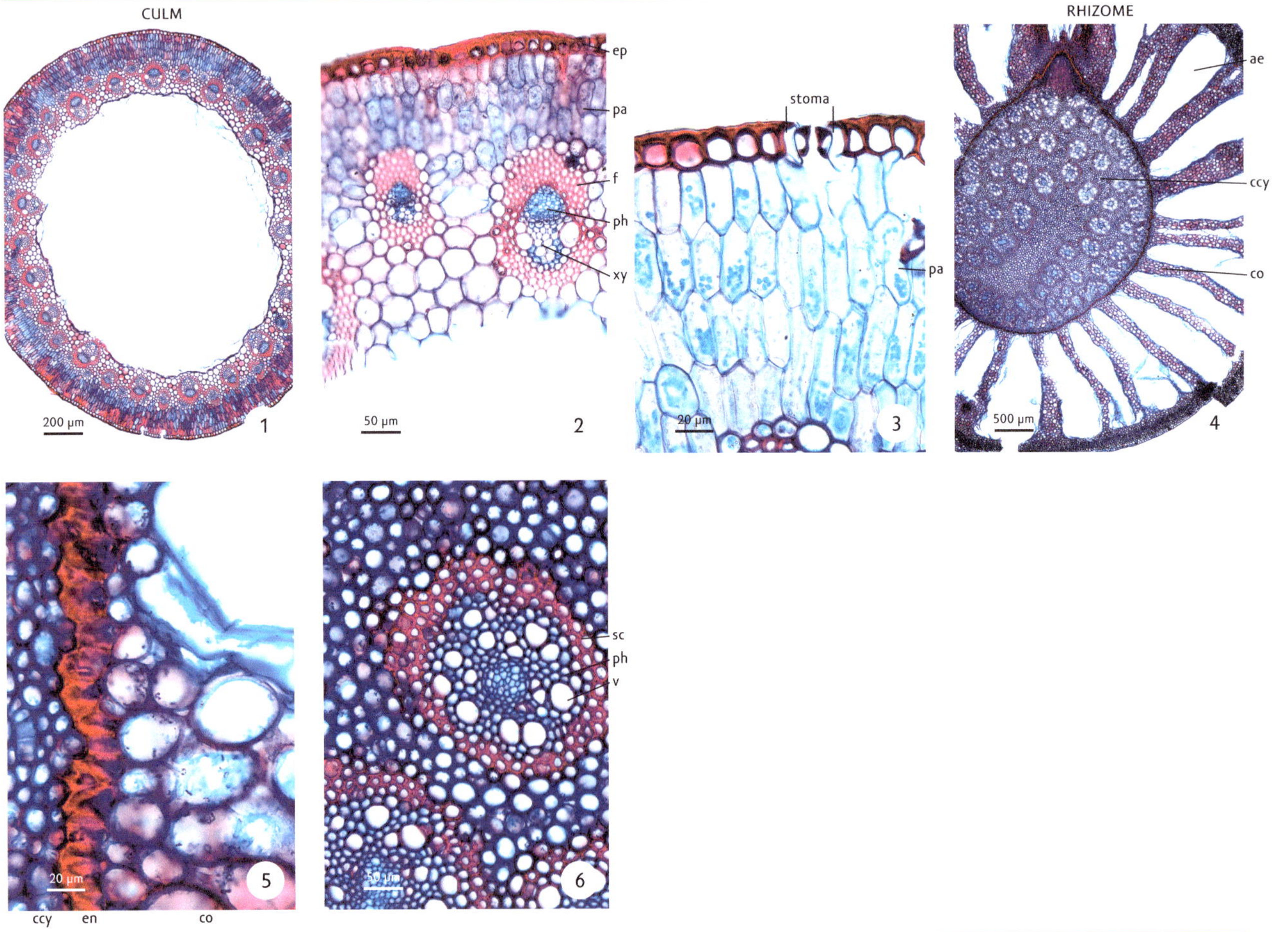

Perennial wetland herb, rhizomatous, very variable. Stems erect to decumbent, to 80 cm high, leaves with very distinct transverse septa. Inflorescence diffuse, dark brown tepals, the outer acute, the inner subacute to acute. Fruit is an ovoid, shiny dark brown capsule.

Distribution and ecology

Euro-Siberian southern-temperate element, but widely naturalized. In all CE countries, in a wide range of wet or damp habitats. Wet grasslands, ditches, margins of rivers and ponds, wet tracks. Roots are able to alkalize the rhizosphere.

Anatomy

Circular culm with a diameter of 1.5–2.5 mm.

The epidermis consists of roundish, externally lignified thick-walled cells (1, 2). Guard cells of the stomata are centrically lignified and have acuminate lips (3). The chlorenchyma is composed of 1–3 layers of round to oval unlignified parenchyma cells (1, 2, 3). The middle part of the culm consists of round parenchyma cells, small vascular bundles and air spaces (2). The inner part is characterized by a continuous belt of fibers and large vascular bundles. The large vascular bundles consist of a xylem with two large lateral vessels, an air duct and a roundish phloem with sieve tubes and companion cells (2). The bundles are surrounded by a sheath of fibers (1, 2). The pith consists of large thin walled parenchyma cells and a few vessels (1, 2).

Rhizome with a diameter of 3–4 mm.

The cortex is composed of an epidermis, a peripheral layer of round parenchyma cells, and large round air spaces between radial zones of parenchyma cells (4). The central cylinder and the cortex are separated by a centripetal thick-walled endodermis and a thin-walled pericycle (4). The parenchymatic ground tissue of the central cylinder contains amphivasal vascular bundles, which are surrounded by a sclerenchyma sheath (5).

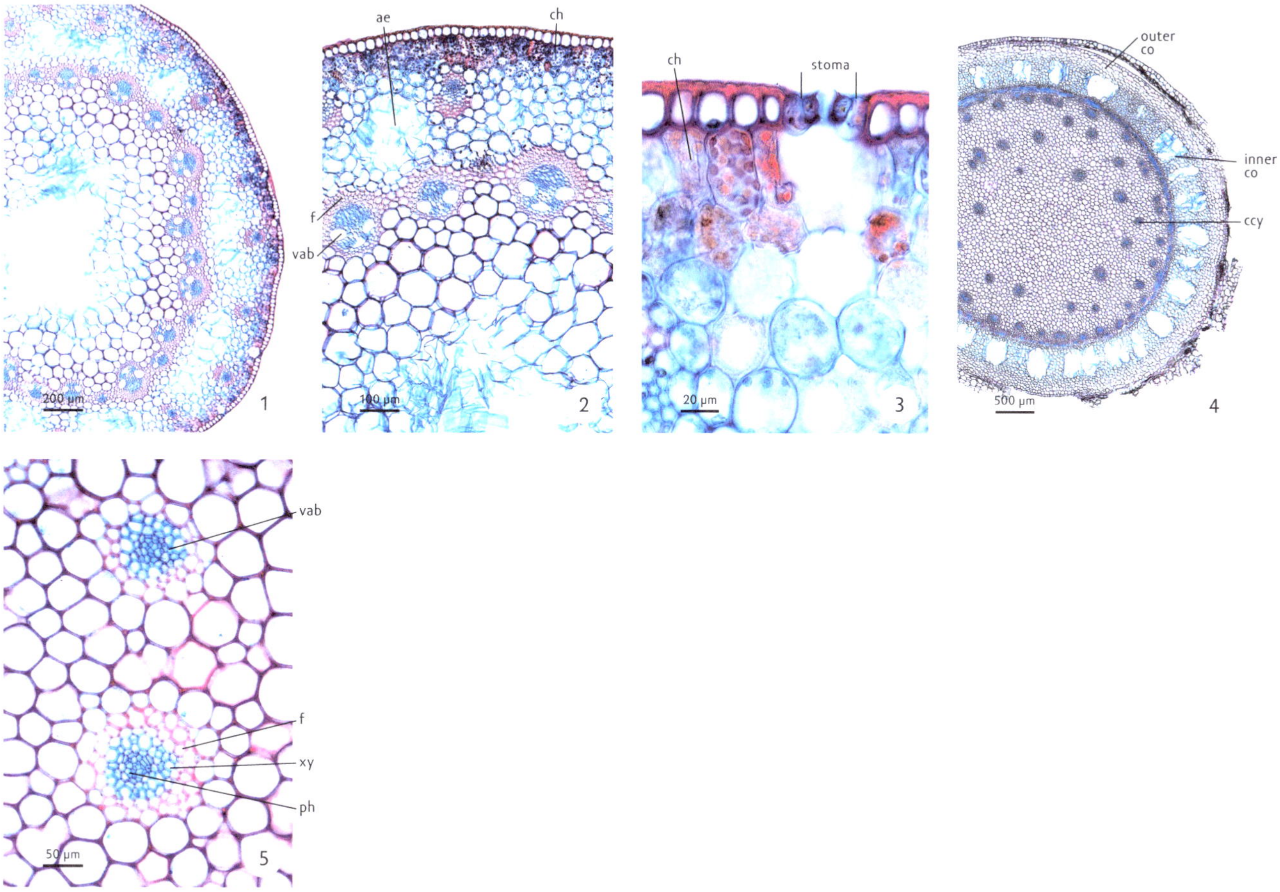

Perennial wetland herb, rather short-lived small tussocks from a short, branched rhizome. Stems erect, 3–15 cm high. All or almost all leaves basal. One inflorescence with flowers in a vertical row, usually exceeded by lowest bract. The entire inflorescence turns blackish with age. Retuse capsule.

Distribution and ecology

Circumpolar arctic-alpine element. It occurs only at one site in the Alps (AT). Well-watered damp rocky or gravelly places and flushes or marshes with short open vegetation. Prefers base-rich, but relatively competition-free habitats. Sexual reproduction by seeds, no vegetative reproduction. Adapted to wind pollination.

Anatomy

Circular culm with a diameter of <1 mm (1).

The epidermis consists of roundish, externally lignified thick-walled cells (1, 2). Guard cells of the stomata are centrically lignified and have acuminate lips (3). The chlorenchyma is composed of 2–4 layers of round unlignified parenchymatic cells (1, 2, 3). The middle part consists of round parenchyma cells and small vascular bundles (2). The inner part of the hollow culm is characterized by a continuous wavy belt of fibers and large vascular bundles. The large vascular bundles consist of a xylem with two slightly large lateral vessels, an air duct and a roundish phloem with sieve tubes and companion cells (2). The large bundles are embedded in the fiber belt (1, 2). The pith consists of large thin-walled, parenchymatic cells and a few vessels (1, 2).

Rhizome with a diameter of 1 mm (4).

The cortex is composed of an epidermis, a peripheral layer of round parenchymatic cells and air spaces between radial zones of parenchymatic cells (4). The central cylinder and the cortex are separated by a centripetal thin- to thick-walled endodermis and a thin-walled pericycle (5). The unlignified parenchymatic ground tissue of the central cylinder contains amphivasal vascular bundles, which are surrounded by an interrupted sclerenchymatic sheath (6).

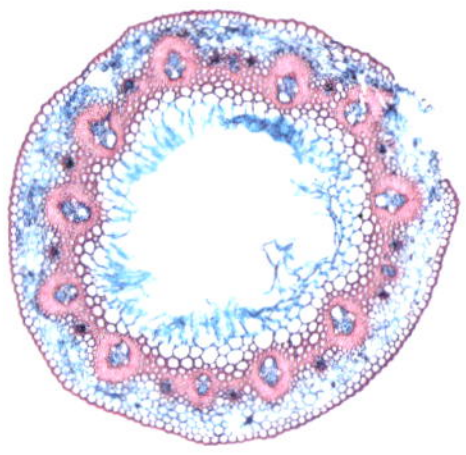

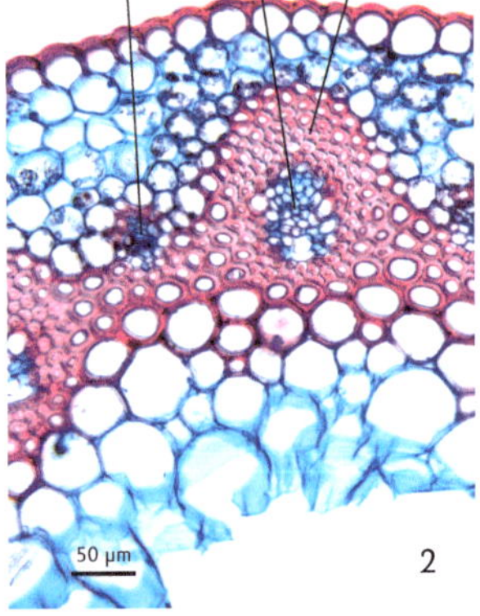

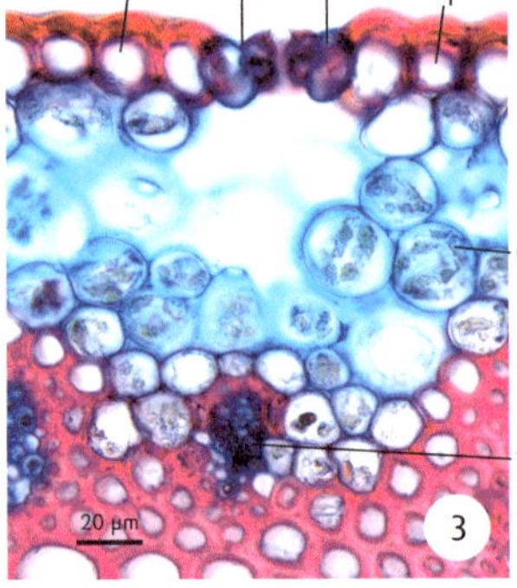

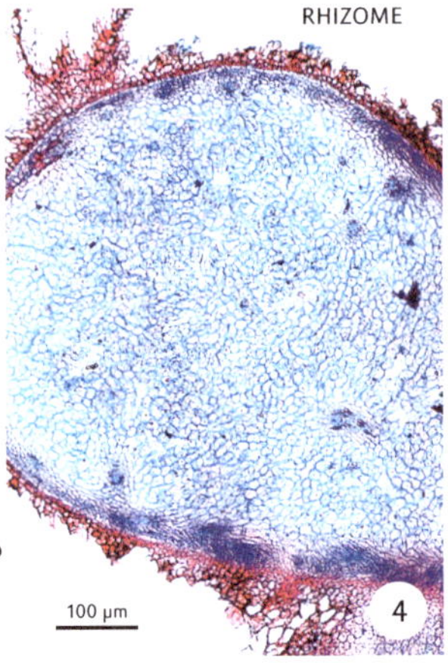

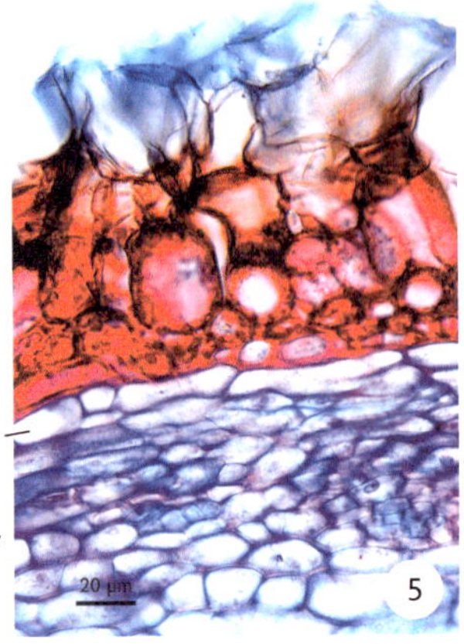

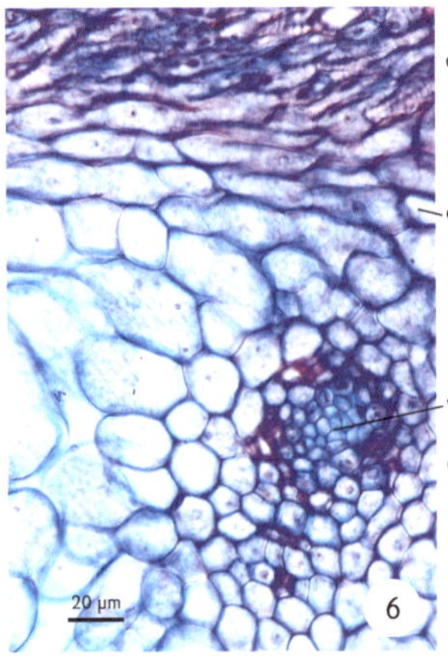

Juncus bufonius L.

Toad Rush

Juncaceae

Annual wetland herb, loosely tufted. Stems erect to procumbent, to 35 cm high, variable branching. Leaves 2 mm wide, grooved, at base and on stem. Flowers in inflorescence well spaced to tightly bunched. Fruit is an acute to subacute brown capsule, shorter than tepals.

Distribution and ecology

Circumpolar wide-temperate element; widely naturalized outside its native range. In all CE countries in habitats where the water table is at least seasonally high and with little competition. Exposed bottoms, margins of ponds, lakes, streams and rivers, marshes and dune-slacks. Often a weed of disturbed ground, including wet tracks and roadsides.

Anatomy

Circular culm with a diameter of 0.5–1 mm.

The epidermis consists of roundish, lignified, slightly thickened-walled cells (1). Guard cells of the stomata are not lignified (2). The chlorenchyma is composed of 3–5 layers of round to oval unlignified parenchyma cells. Air spaces are absent (1). The middle part consists of round parenchyma cells and small vascular bundles (1, 2). The inner part of the filled culm is characterized by a continuous wavy belt of fibers and large vascular bundles. The large vascular bundles consist of a xylem with a few lateral vessels and a roundish phloem with sieve tubes and companion cells (1). The large bundles are surrounded by 1–4 rows of thick-walled fibers (1).

Rhizome with a diameter of 1.5–2 mm (3).

The cortex is composed of an epidermis, an external layer of thin-walled lignified parenchyma cells and an inner part of thin-walled unlignified parenchyma cells without air spaces (3). The central cylinder and the cortex are separated by a centripetal thick-walled endodermis and a pericycle (4). The lignified parenchymatic ground tissue of the central cylinder contains amphivasal vascular bundles, with small sclerenchyma sheaths (3).

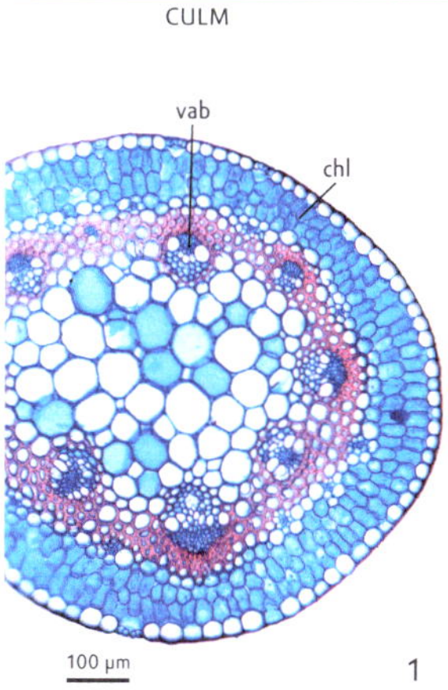

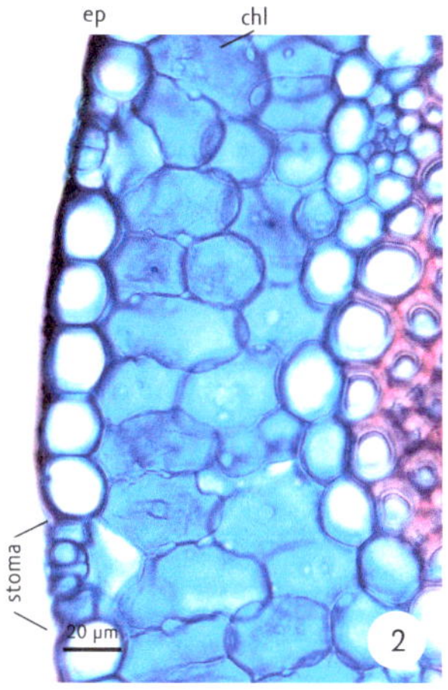

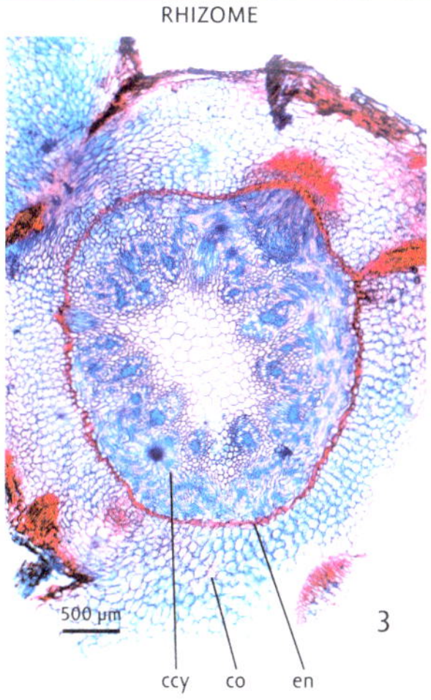

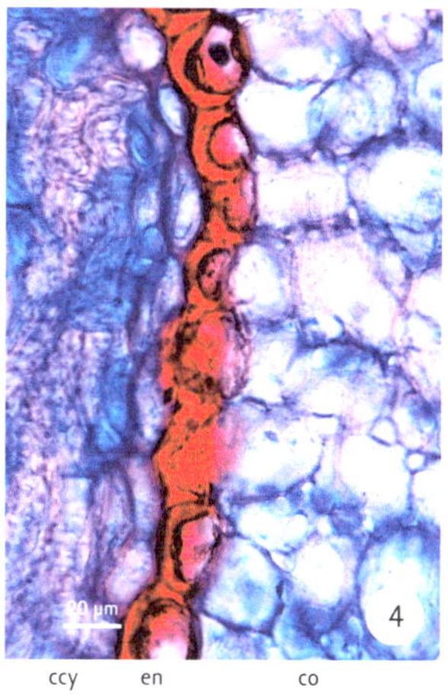

Juncus bulbosus L.

Bulbous rush

Perennial amphibious tufted plants without rhizomes, extremely variable in habit. Stems erect to procumbent, floating or submerged, 2–20 cm long, bulbous and red coloured at base, leaves semi-circular, filamentous or narrowly subulate, inflorescence often proliferating (viviparous).

Distribution and ecology

European boreo-temperate element; also in N America. Commonly in all CE countries, from lowlands to mountains. Shorelines of fishponds and lakes, margins of peat bogs and fens, banks of rivers and brooks, backwater pools and oxbows, wet meadows, mountainous lakes and streams, sand-pits. Extremely ecologically plastic and tolerant, strongly acidic to slightly alkaline waters (pH ca. 3–8), base-poor to base-rich waters, dystrophic or clear, oligo- to eutrophic waters, peaty, silty or sandy soils, facultative photophilous, competitively very strong.

Anatomy

Circular culm with a diameter of 0.8–1.2 mm (1).

The epidermis consists of roundish, externally lignified slightly thickened-walled cells (1, 2). Guard cells of the stomata are laterally lignified and have acuminate lips (2). The chlorenchyma is composed of 2–4 layers of round to oval unlignified parenchyma cells (1, 2). The middle part consists of round parenchyma cells and small vascular bundles (2). The inner part of the hollow culm is characterized by a continuous wavy belt of fibers and large vascular bundles. The large vascular bundles consist of a xylem with a few lateral vessels and a roundish phloem with sieve tubes and companion cells (2). The large bundles are internally surrounded by 1–3 rows of thin- to thick-walled and externally by thick-walled fibers (2).

Rhizome with a diameter of 1.5–2 mm (3).

The cortex is composed of an epidermis, an external layer of thin-walled unlignified parenchyma cells and an inner part of thin-walled unlignified parenchyma cells. Air spaces are irregularly formed (3). The central cylinder and the cortex are separated by a centripetal thin-to thick-walled endodermis (3). The lignified thin-walled parenchymatic ground tissue of the central cylinder contains amphivasal vascular bundles. They are surrounded with 1–2 layers of thin- to thick-walled sclerenchyma cells (4).

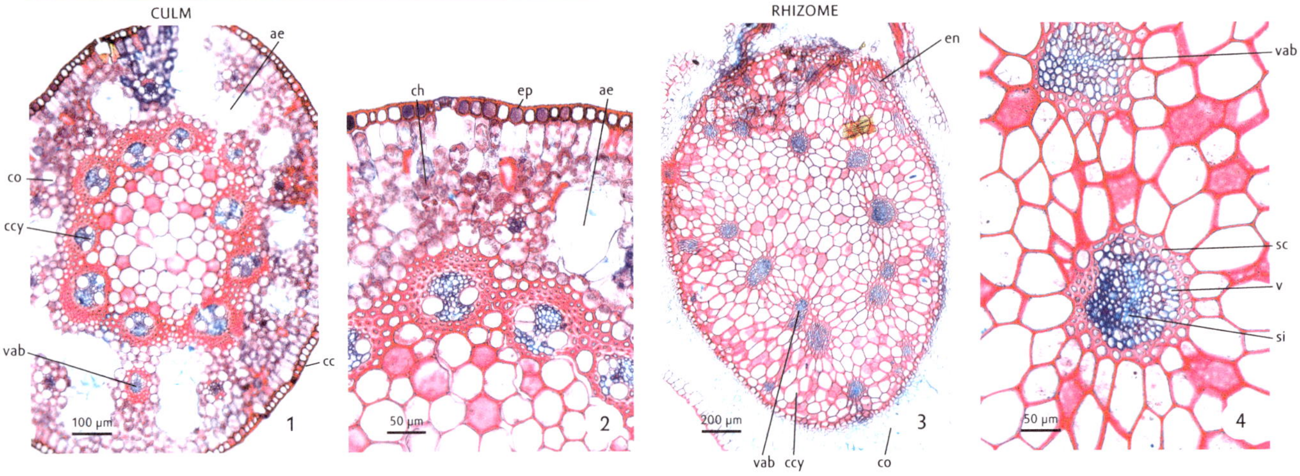

Juncus capitatus Weigel

Dwarf Rush

Annual wetland herb. Stems erect, leafless, 1–20 cm high. Basal leaves thin, <1 mm wide, wiry, grooved. Inflorescence terminal, a tight cluster of 1–4 heads. Fruit is an ovoid, pale brown capsule, markedly shorter than tepals.

Distribution and ecology

European southern-temperate element, widely naturalized outside its native range. In all CE countries. Diminutive, autumn-germinating annual of bare ground often kept open by standing water in winter and drained in summer. Wet sandy soils, exposed fishpond bottoms, around serpentine rock outcrops, in dune-slacks and sometimes in quarries.

Anatomy

Circular culm with a diameter of 0.5–1 mm.

The epidermis consists of roundish, unlignified thin-walled cells (1). Guard cells of the stomata are thin-walled, unlignified, without acuminate lips. The chlorenchyma is composed of 3–5 layers of round to oval unlignified parenchyma cells (1). The middle part consists of round parenchyma cells and small vascular bundles (1). The inner part of the filled culm is characterized by a continuous small belt of fibers and large vascular bundles. The large vascular bundles consist of a xylem with a few lateral vessels and a roundish phloem with sieve tubes and companion cells and an air tube (1). The large bundles are surrounded internally by 1–2 rows and externally by 5–6 layers of thick-walled fibers (1).

Rhizome with a diameter of 1–1.5 mm (2).

The cortex is composed of a thin-walled epidermis and a uniform layer of unlignified, thin-walled cells. Air spaces are absent (2). The central cylinder and the cortex are separated by a centripetal thin- to thick-walled endodermis and a pericycle (2). The lignified parenchymatic ground tissue of the central cylinder contains indistinct vascular bundles without sclerenchyma sheaths (2).

CULM

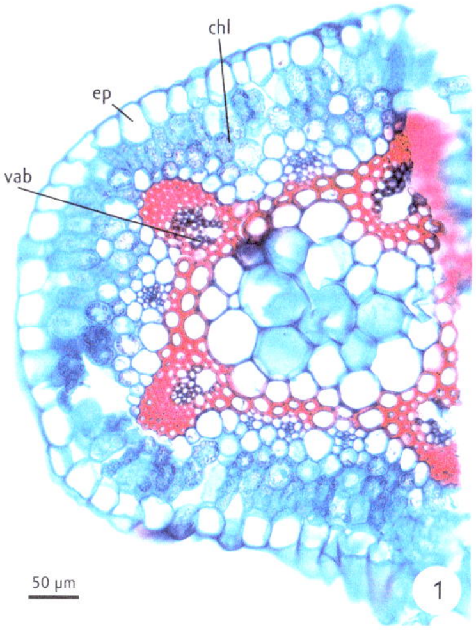

RHIZOME

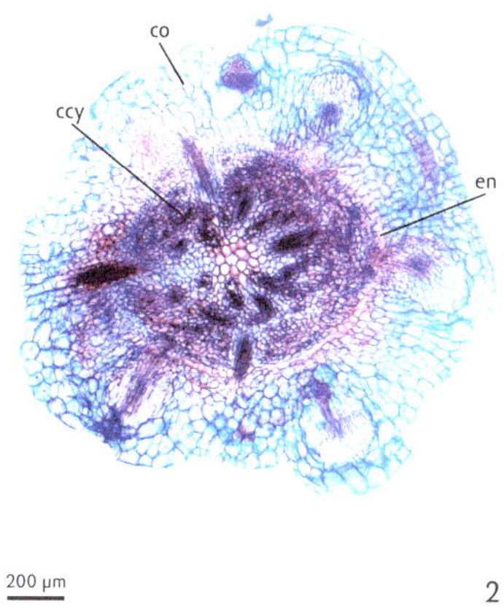

2

Juncus compressus Jacq.

Round-fruited Rush Juncaceae

Perennial wetland herb, loosely tufted to extensive patches, stems erect, to 50 cm high. Leaves 1 mm wide, mostly basal, flattened, often inrolled. Inflorescence is a loose forking cluster, the lowest bract usually exceeding inflorescence. Fruit is a dark brown capsule, as long as or longer than the tepals.

Distribution and ecology

Euro-Siberian temperate element, widely naturalized outside its native range. In all CE countries. Marshes, wet meadows and pastures, exposed wet sites, road sides, often near the sea and sometimes in brackish conditions.

Anatomy

Circular culm with a diameter of 2–3 mm (1).

The epidermis consists of roundish, externally lignified thick-walled cells (1–2). Guard cells of the stomata are lignified and have acuminate lips. The chlorenchyma is composed of 3–5 layers of round unlignified parenchyma cells, which are filled with dark stained substances (1–2). The middle and inner parts of the hollow culm consist of round unlignified parenchyma cells and vascular bundles (2). The large vascular bundles consist of a xylem with two lateral large vessels, an air duct and a roundish phloem with sieve tubes and companion cells (2). All vascular bundles are surrounded by a thick sheath of fibers (1, 2).

Rhizome with a diameter of 2–3 mm (3).

The cortex is composed of an epidermis and a large layer of thin-walled parenchyma cells with air spaces (3). The central cylinder and the cortex are separated by a centripetal thick-walled endodermis and an indistinct pericycle (3). The unlignified thin-walled parenchymatic ground tissue of the central cylinder contains amphivasal vascular bundles with sclerenchyma sheaths (4).

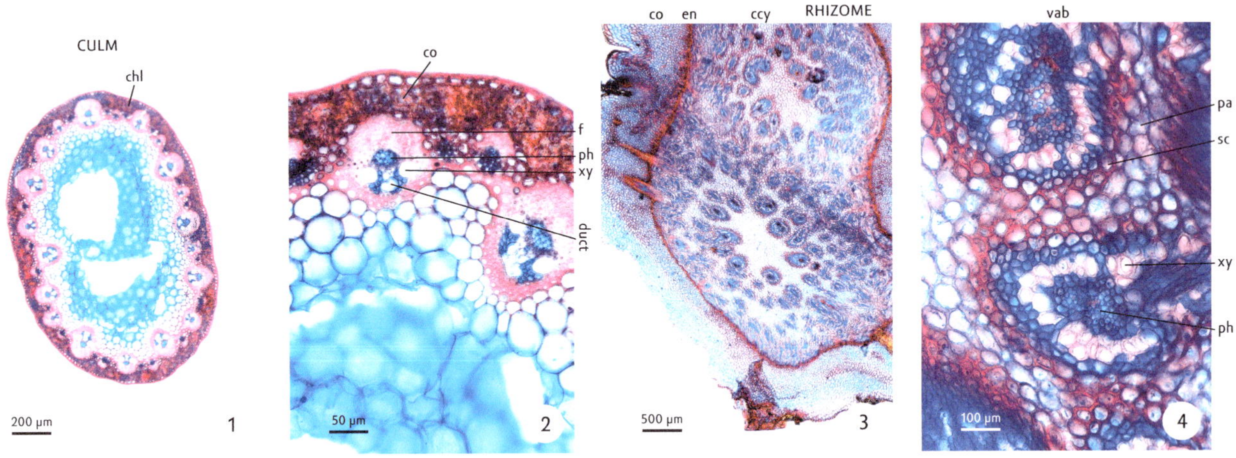

Juncus conglomeratus L.

Compact Rush

Perennial wetland herb, densely tufted. Stems erect, to 1 m high, distinctly ridged and dull, with continuous pith. Leaves reduced to brown sheaths at base of the stem. Inflorescence is a densely-branched, rounded, brown head, very compact. Fruit is an ovoid capsule, with a mucronate point in the tip.

Distribution and ecology

European temperate element, widely naturalized outside its native range. In all CE countries. Wet pastures, wet abandoned meadows, along rivers and fishponds, prefers disturbed sites, road sides, ditches, mostly on acidic soils, calcifuge. It is able to sustain metal-contaminated environment.

Anatomy

Ribbed, circular culm with a diameter of 2–3 mm (1).

The epidermis consists of roundish, externally lignified thick-walled cells (1, 2, 3). Guard cells of the stomata are centrically lignified and have slightly acuminated lips (4). The chlorenchyma is composed of 2–4 layers of round to oval unlignified parenchyma cells and groups of fibers (1, 2, 3). The middle part consists of round parenchyma cells, small vascular bundles and often of large air spaces (2). The inner part of the cortex is characterized by large vascular bundles. The large vascular bundles consist of a xylem with two large lateral vessels, 1–2 air ducts and a roundish phloem with sieve tubes and companion cells (2, 3). The large bundles are surrounded by a fiber belt (1, 2). The large pith consists of stellate parenchyma cells (1, 2).

Rhizome with a diameter of 4–5 mm (5).

The cortex is composed of an epidermis, an outer part of thin- to thick-walled parenchyma cells and strands of sclerenchyma cells (5) and an inner part of large air spaces (5). The central cylinder and the cortex are separated by a centripetal thick-walled endodermis and an indistinct pericycle (6). The lignified parenchymatic ground tissue of the central cylinder contains amphivasal vascular bundles, which are surrounded by a sclerenchyma sheath (7).

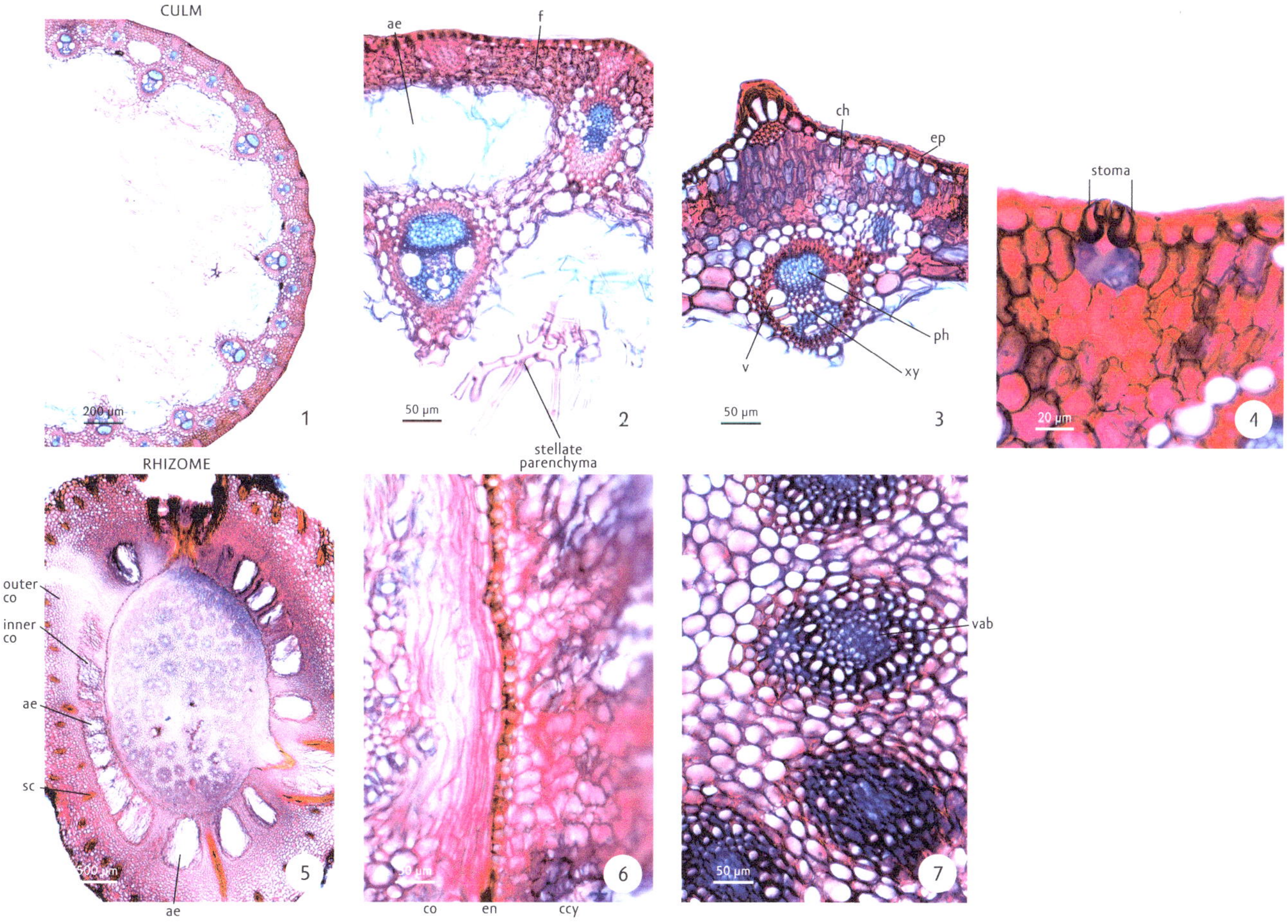

Perennial wetland herb, densely tufted, sometimes forming larger patches. Stems erect, to 1.2 m high, smooth and glossy, with continuous pith. Leaves reduced to brown sheaths at base of stem. Inflorescence lateral, lax with suberect to widely divergent branches, or compact. Tepals pale brown. Fruit is a brown capsule, without a mucronate point in the tip.

Distribution and ecology

European southern-temperate element; also disjunctly in C and E Asia and N America and widely naturalized outside its native range. In all CE countries, on wet pastures and abandoned meadows, bogs, along rivers and fishponds. It prefers disturbed sites, road sides, ditches, mostly on acidic sandy or peaty soils.

Anatomy

Ribbed, circular culm with a diameter of 2-3 mm.

The epidermis consists of roundish, externally lignified thick-walled cells (1, 2). Guard cells of the stomata are centrically lignified and have acuminate lips (3). The chlorenchyma is composed of 2–4 layers of oval unlignified parenchyma cells and triangular groups of fibers (1, 2). The middle part consists of round parenchyma cells and small vascular bundles (2). The inner part of the hollow culm is characterized by large vascular bundles. The large vascular bundles consist of a xylem with two slightly large lateral vessels, an air duct and a roundish phloem with sieve tubes and companion cells (2). The large bundles are surrounded by a sheath of fibers (1, 2).

Rhizome with a diameter of 4–5 mm (4).

The cortex is composed of an epidermis, an external layer of thin-walled lignified parenchyma cells and an inner part of thin-walled unlignified parenchyma cells and air spaces (4, 5). The central cylinder and the cortex are separated by a centripetal thick-walled endodermis and partially an indistinct pericycle (4, 5). The lignified parenchymatic ground tissue of the central cylinder contains amphivasal vascular bundles, with sclerenchyma sheaths.

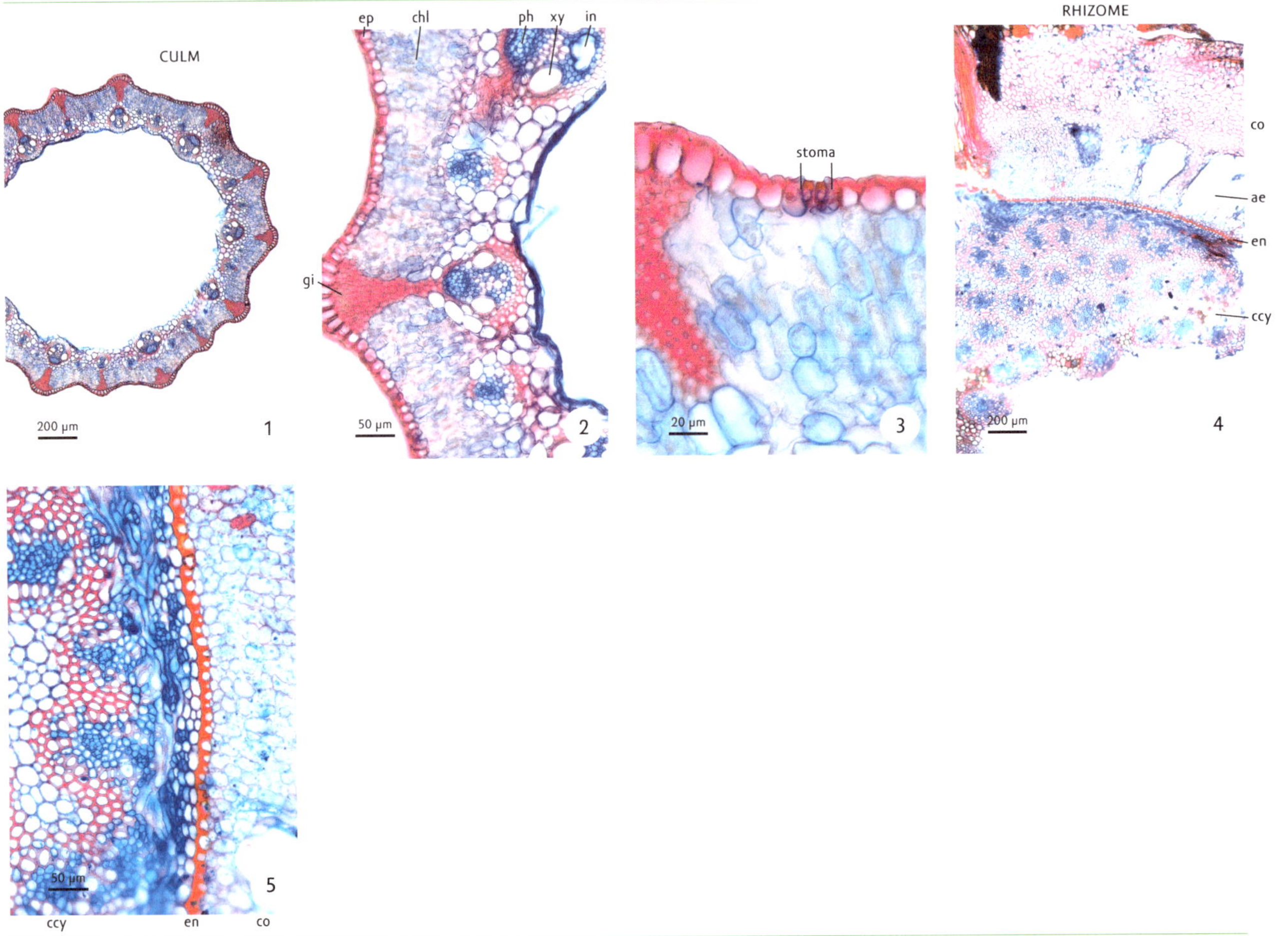

Juncus filiformis L.

Thread Rush

Juncaceae

Perennial wetland herb, rather weekly rhizomatous. Stems erect, to 45 cm high, thin, very slightly ridged with continuous pith. Leaves reduced to brown sheaths at base of stem. Inflorescence is a loose cluster, fewer greenish to yellowish flowers. Fruit is an almost globular capsule.

Distribution and ecology

Circumpolar boreo-montane element. In all CE countries except HU. Acidophilous, wet and peaty meadows, peat bogs, waterlogged spruce forests in highlands, also margins of lakes, fishponds or reservoirs. Long-term persistent seed bank.

Anatomy

Ribbed, circular culm with a diameter of 1–1.5 mm (1).

The epidermis consists of roundish, externally lignified thick-walled cells (2). Oval groups of sclerenchyma cells below the epidermis. Guard cells of the stomata are centrically lignified and have slightly acuminate lips. The chlorenchyma is composed of 4–6 layers of oval to palisade-like unlignified parenchyma cells and tangentially oval groups of fibers (1, 2). The middle part consists of round parenchyma cells, small vascular bundles and irregular air spaces (2). The inner part of the hollow culm is characterized by large vascular bundles (2). The large vascular bundles consist of a xylem with two large lateral vessels and a roundish phloem with sieve tubes and companion cells (2). The large vascular bundles are surrounded internally by one layer of thin- to thick-walled and externally by 2–3 layers of thick-walled fibers (1, 2).

Rhizome with a diameter of 4–6 mm (3).

The cortex is composed of an epidermis, an outer part of thin- to thin-walled parenchyma cells and an inner part of large air spaces (3). The central cylinder and the cortex are separated by a centripetal thick-walled endodermis (4). The unlignified parenchymatic ground tissue of the central cylinder contains amphivasal vascular bundles, which are surrounded by a small sclerenchyma sheath (5).

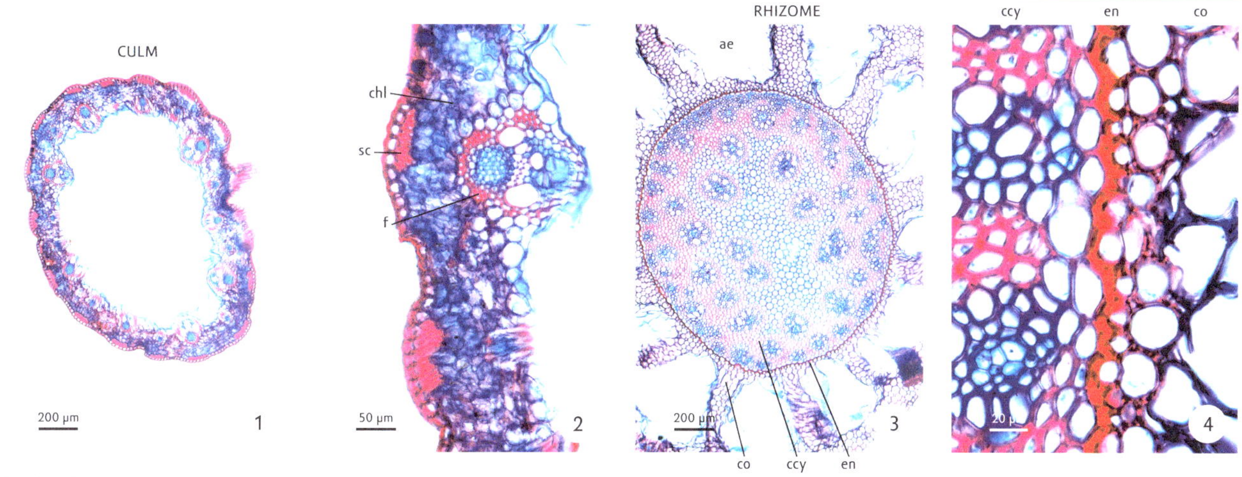

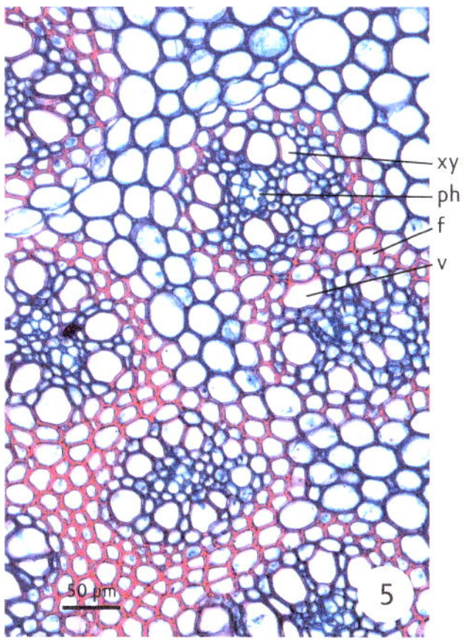

Juncus gerardii Loisel.

Saltmarsh Rush Juncaceae

Perennial wetland herb, extensively rhizomatous, stems erect, up to 50 cm high. Leaves mostly basal. Lowest bract usually shorter than inflorescence. Tepals usually dark brown, capsule scarcely exceeding tepals.

Distribution and ecology

Circumpolar wide-temperate element. In all CE countries except CH. Confined to saline habitats, mostly in the uppermost parts of coastal salt marshes and in inland saline areas and salty meadows.

Anatomy

Circular culm with a diameter of 1 mm.

The epidermis consists of roundish, externally lignified thick-walled cells (1, 2). Guard cells of the stomata are concentrically lignified and have rounded lips (3). The chlorenchyma is composed of 2–4 layers of round to oval unlignified parenchymatic cells (1, 2). The middle part consists of round parenchymatic cells and small vascular bundles (2). The inner part of the hollow culm is characterized by a continuous wavy belt of fibers and large vascular bundles. The large vascular bundles consist of a xylem with two large lateral vessels and a roundish phloem with sieve tubes and companion cells (2). The large bundles are embedded in the fiber belt (1, 2).

Rhizome with a diameter of 2–3 mm (4).

The cortex is composed of an epidermis, an outer part on a layer of thin- and a layer of thick-walled parenchymatic cells and an inner part with large air spaces (4). The central cylinder and the cortex are separated by a centripetal thick-walled endodermis and an indistinct pericycle (5). The lignified parenchymatic ground tissue of the central cylinder contains amphivasal vascular bundles surrounded by a sclerenchymatic sheath (6).

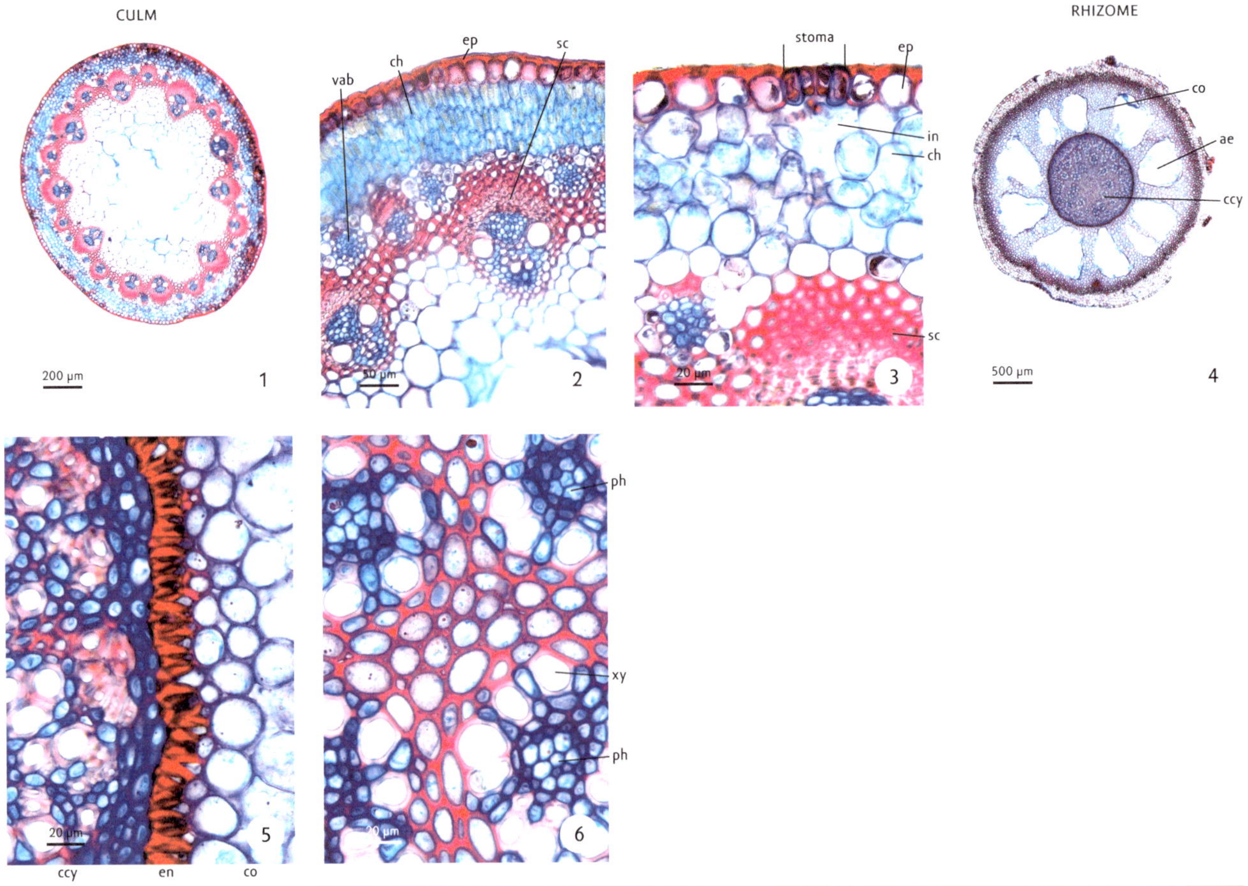

Juncus inflexus L. **Hard Rush** Juncaceae

Perennial wetland herb, densely tufted. Stems erect, to 1.2 m high, glaucous, greyish-green, very strongly ridged, with interrupted pith. Leaves reduced to red-brown sheaths at base of stem. Inflorescence is a loose cluster, rather lax, with suberected branches. Tepals dark brown. Fruit is an ovoid, dark brown capsule.

Distribution and ecology

Euro-Siberian southern-temperate element, widely naturalized outside its native range. In all CE countries. Wet meadows, meadow springs, usually on neutral or base-rich heavy soils. Prefers disturbed sites in wet grasslands, occasionally dune-slacks and fens and other salty habitats (road sides, ditches).

Anatomy

Ribbed, circular culm with a diameter of 1–1.5 mm (1).

The epidermis consists of roundish, externally lignified thick-walled cells (1–3). Guard cells of the stomata are centrically lignified and have slightly acuminate lips (3). The chlorenchyma is composed of 4–6 layers of oval to palisade-like unlignified parenchyma cells and triangular groups of girders (1, 2). The middle part consists of round parenchyma cells and small vascular bundles (2). The inner part of the hollow culm is characterized by large vascular bundles (2). The large vascular bundles consist of a xylem with two large lateral vessels and a roundish phloem with sieve tubes and companion cells (2). All vascular bundles are surrounded by a sheath of fibers (1, 2).

Rhizome with a diameter of 4–6 mm (4).

The cortex is composed of an epidermis, an outer part of thin- to thick-walled parenchyma cells and strands of sclerenchyma cells (4) and an inner part of large air spaces (4). The central cylinder and the cortex are separated by a centripetal thick-walled endodermis and an indistinct pericycle (5). The lignified parenchymatic ground tissue of the central cylinder contains amphivasal vascular bundles, which are surrounded by a large sclerenchyma sheath (6).

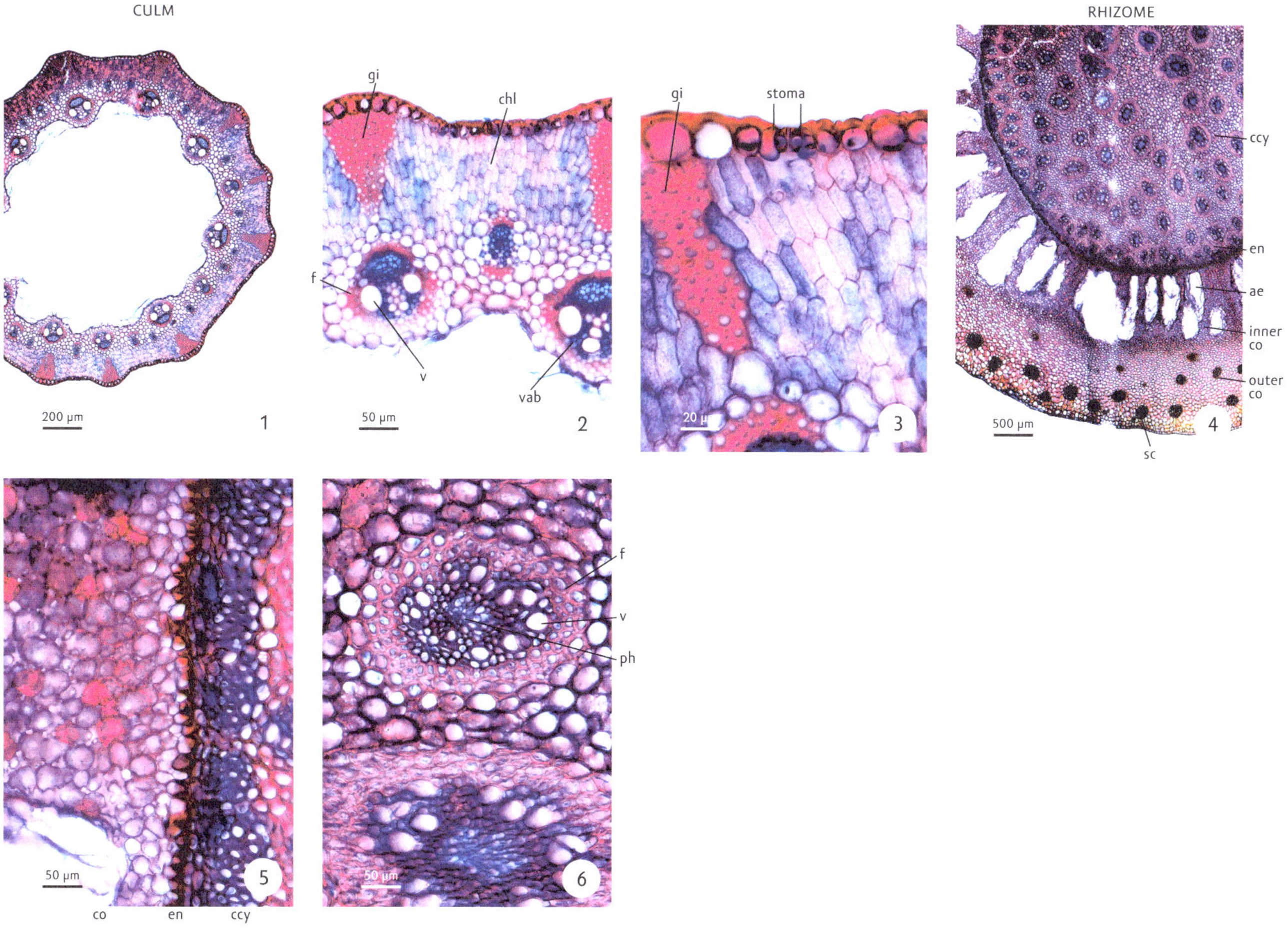

Perennial wetland, tufted herb with short runners. Stems erect, terete, very finely ribbed, 10–30 cm high. One leaf in the upper part of the stem, longer than the inflorescence. Inflorescence is a cluster of flowers, apparently lateral on the culm, tepals shiny, brown-blackish. Fruit is a shiny brown-blackish capsule.

Distribution and ecology

Native to central to southeastern Europe. In AT, DE and CH from subalpine to alpine zone usually on banks of small streams, wet alpine meadows and springs. Usually forms very dense stands.

Anatomy

Circular culm with a diameter of 1–1.5 mm (1).

The epidermis consists of roundish, externally lignified thick-walled cells (1, 2). Guard cells of the stomata are concentrically lignified and have acuminate lips (3). The chlorenchyma is composed of 2–4 layers of oval to palisade-like unlignified parenchyma cells and large radial groups of fibers (1, 2). The middle part consists of round parenchyma cells and small vascular bundles (1, 2). The inner part of the hollow culm is characterized by large vascular bundles. The large vascular bundles consist of a xylem with two large lateral vessels, an air duct and a roundish phloem with sieve tubes and companion cells (2). The large and small bundles are surrounded by a sheath of fibers (1, 2).

Rhizome with a diameter of 2 mm (4).

The cortex is composed of an epidermis, an outer part of a layer of thick-walled and a centripetal part of thin-walled parenchyma cells and an inner part of large air spaces (4). The central cylinder and the cortex are separated by a centripetal thick-walled endodermis (5). The unlignified parenchymatic ground tissue of the central cylinder contains amphivasal vascular bundles, without sclerenchyma sheaths (6).

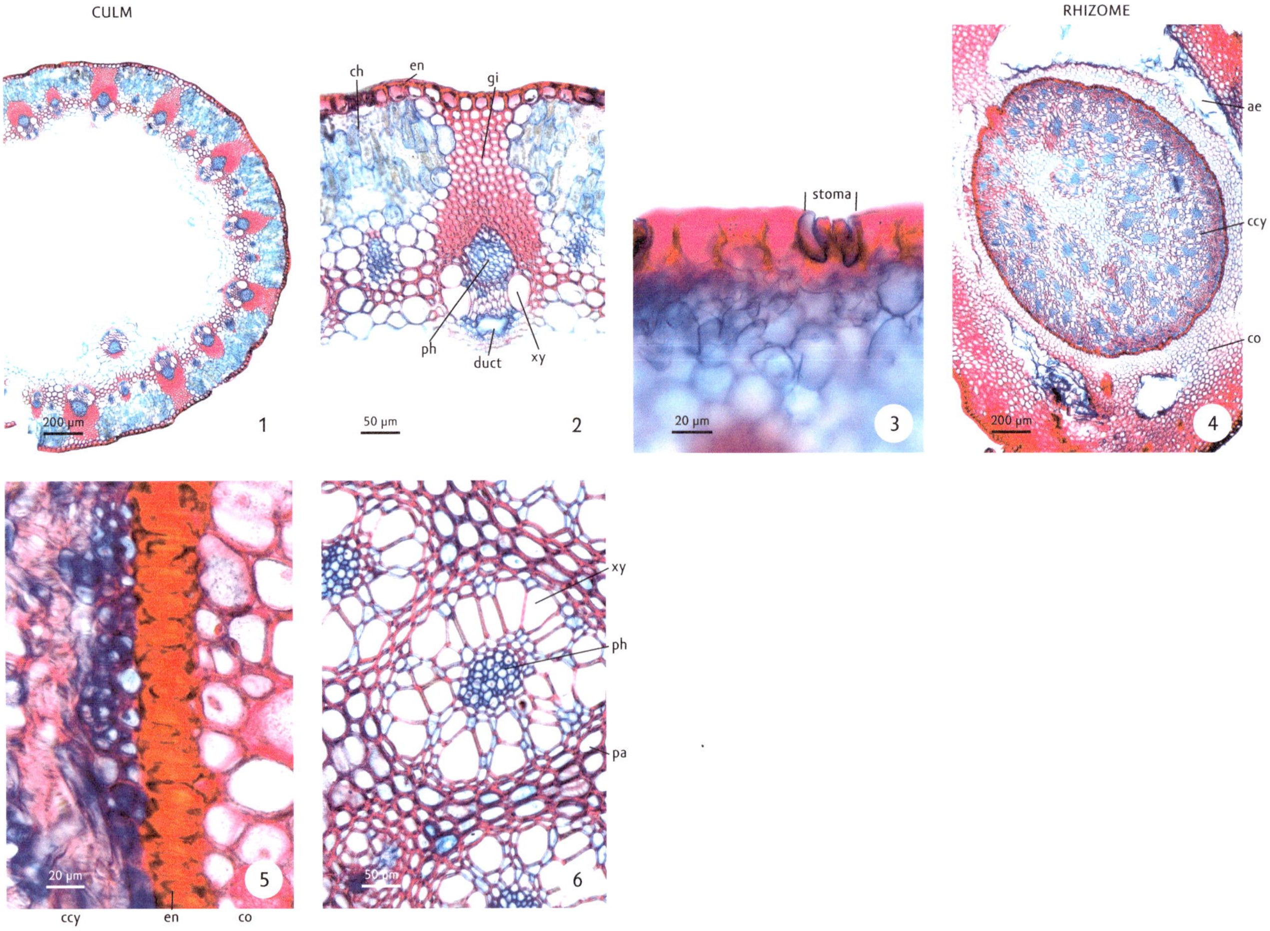

Juncus maritimus Lam.

Sea Rush Juncaceae

Perennial wetland herb, robust. Stems densely to scarcely tufted, erect, very stiff, to 1 m high. Leaves at base, 2 mm wide, rounded, sharply pointed. Inflorescence a loose forking cluster below a long, sharply pointed bract. Fruit is a brown capsule, as long as or slightly longer than the tepals.

Distribution and ecology

European southern-temperate element, widely naturalized outside its native range. Only in AT, DE and HU. Typical of saltmarshes and saline dune-slacks, tolerant of a wide range of salinities and soil moisture, occurring at all levels in salt marshes and in both silty and sandy substrates.

Anatomy

Circular culm with a diameter of 2 mm (1).

The epidermis consists of square, externally lignified thick-walled cells (1, 2). Guard cells of the stomata are centrically lignified and have acuminate lips. The chlorenchyma is composed of 1–3 layers of round to oval unlignified parenchyma cells (1, 2). The middle and inner part of the culm consists of round lignified parenchyma cells and vascular bundles (2). The large vascular bundles consist of a xylem with two large lateral vessels and a roundish phloem with sieve tubes and companion cells (2). The pith consists of large thin-walled parenchyma cells and a few vessels (1, 2). All vascular bundles are partially surrounded by a sheath of fibers (1, 2). The pith consists of thin-walled parenchyma cells, irregular intercellulars and vascular bundles (2).

Rhizome with a diameter of 4–6 mm (3).

The cortex is composed of an epidermis, an external part of thin- to thick-walled parenchyma cells and an internal part of thin-walled parenchyma cells containing a few strands of sclerenchyma cells (4) and an inner part with large air spaces (4). The central cylinder and the cortex are separated by a centripetal thick-walled endodermis and a distinct pericycle (4). The unlignified parenchymatic ground tissue of the central cylinder contains amphivasal vascular bundles, which are surrounded by a large sclerenchyma sheath (5).

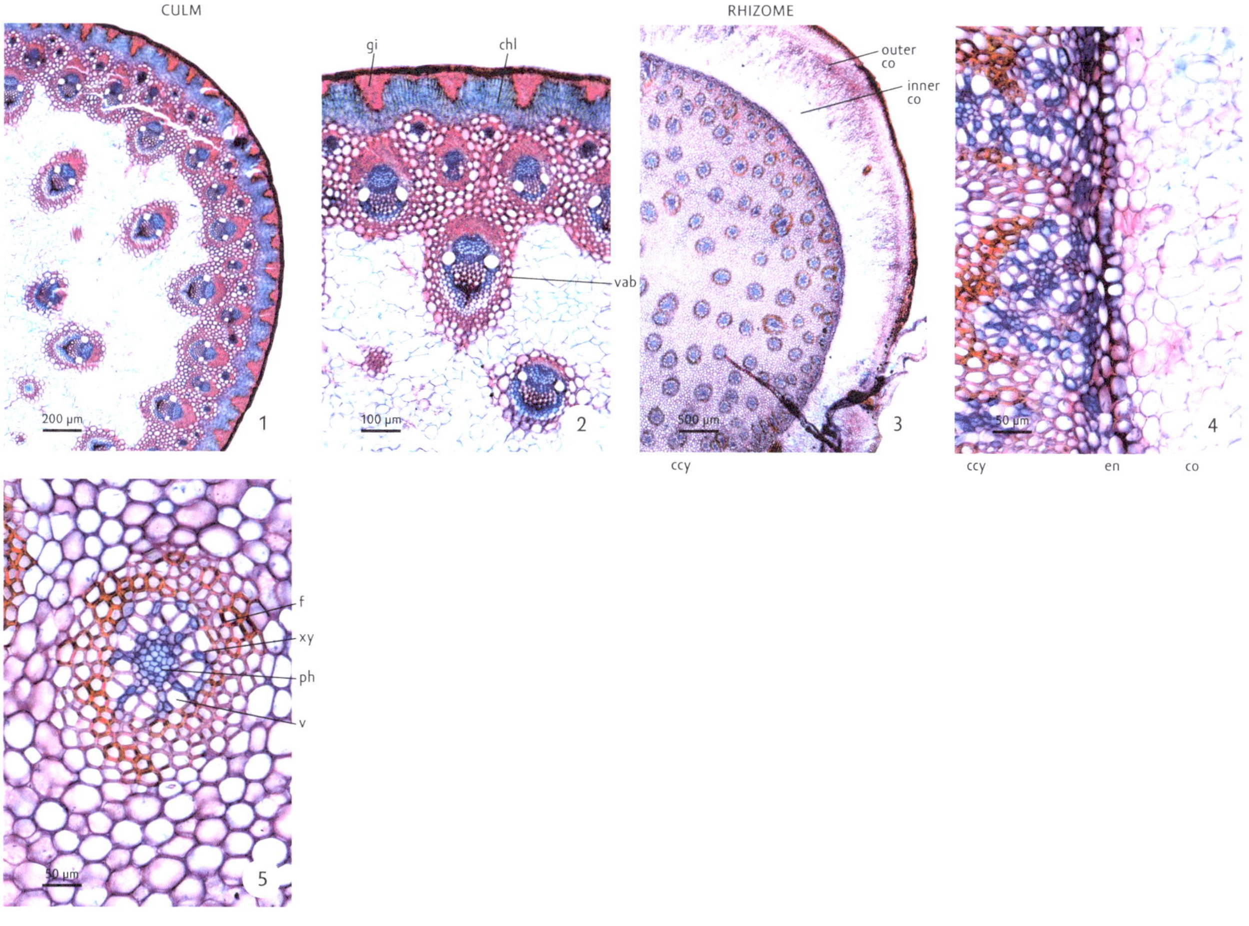

Annual wetland herb, stems feeble, procumbent, 5–30 cm high, ramified, with numerous basal leaves. Loose, branched inflorescence, flowers already in lower plant part. Fruit is an ovoid, light-brown capsule, much shorter than tepals.

Distribution and ecology

Native to circumpolar territories. In AT, CZ, DE and HU. River and fishpond banks, ditches, usually on sandy wet soils, competitively weak.

Anatomy

Circular culm with a diameter of 0.6–0.8 mm.

The epidermis consists of roundish; externally slightly lignified thin-walled cells (1). Guard cells of the stomata are not lignified (2). The chlorenchyma is composed of 2–3 layers of round unlignified parenchyma cells (1, 2). The middle part consists of round parenchyma cells and small vascular bundles (2). The inner part of the filled culm is characterized by a continuous small belt of fibers and large vascular bundles. The large vascular bundles consist of a xylem with a few lateral vessels and a roundish phloem with sieve tubes and companion cells and an air tube. The large bundles are surrounded internally by 1–2 rows and externally by 5–6 layers of thick-walled fibers (2).

Rhizome with a diameter of 1–1.5 mm (3).

The cortex is composed of a thin-walled epidermis and a uniform layer of unlignified, thin-walled cells. Air spaces are absent (3). Endodermis and a pericycle around the central cylinder are indistinct (3). The lignified parenchymatic ground tissue of the central cylinder contains indistinct vascular bundles without sclerenchyma sheaths (3).

CULM

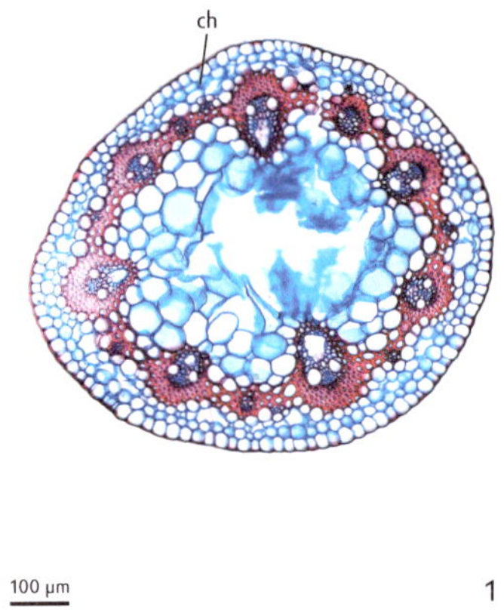

1

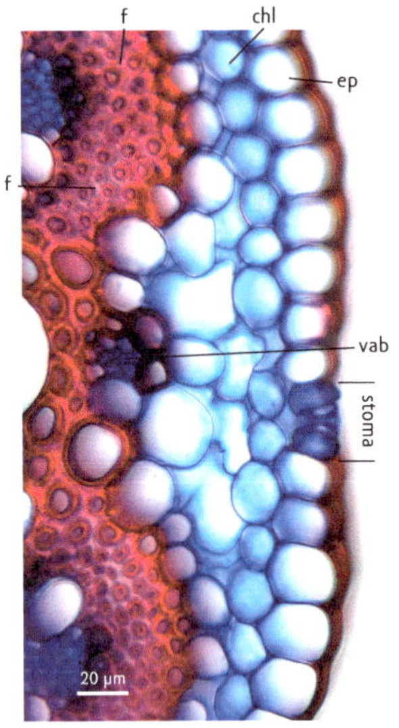

2

RHIZOME

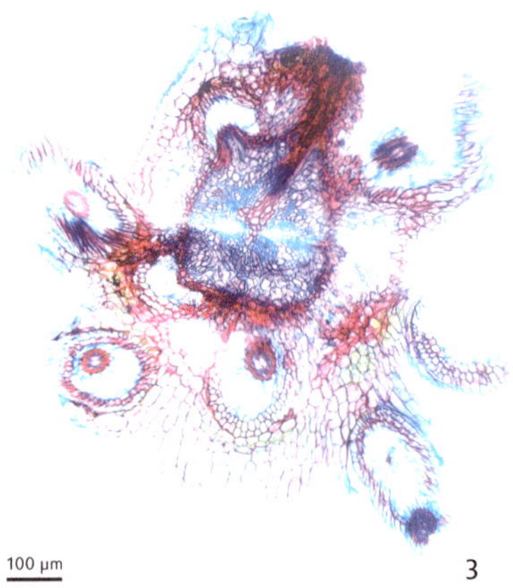

3

Perennial wetland herb, densely tufted. Stems erect, to 50 cm high. Leaves tough, rounded on lowerside, in a basal rosette. Inflorescence a cluster of tight heads, lowest bract much shorter than inflorescence. Fruit is an ovoid, brown capsule, much shorter than tepals.

Distribution and ecology

Suboceanic temperate element, also in Greenland. In all CE countries except HU. Wet abandoned meadows, margins of peatbogs, wet heaths, wet forest clearings, on acidic soils.

Anatomy

Circular culm with a diameter of 1 mm (1).

The epidermis consists of oval, lignified thin- to thick-walled cells (1, 2). Guard cells of the stomata are not lignified and have rounded lips (3). The chlorenchyma is composed of 3–5 layers of round to oval unlignified parenchyma cells, intercellulars and large air spaces (1, 2). The middle part consists of round parenchyma cells and small vascular bundles (2). The inner part of the culm is characterized by large vascular bundles (1, 2). The large vascular bundles consist of a xylem with two large lateral vessels, an air duct and a roundish phloem with sieve tubes and companion cells (2). All vascular bundles are partially surrounded by a sheath of fibers (1, 2). The pith consists of thin-walled parenchyma cells and irregular intercellulars (1).

Rhizome with a diameter of 2–3 mm (4).

The cortex is composed of an epidermis, an external layer of thin-walled parenchyma cells and an inner part with large air spaces (4). The central cylinder and the cortex are separated by a centripetal thick-walled endodermis and an indistinct pericycle (5). The lignified parenchymatic ground tissue of the central cylinder contains amphivasal vascular bundles, which are surrounded by a large sclerenchyma sheath (6).

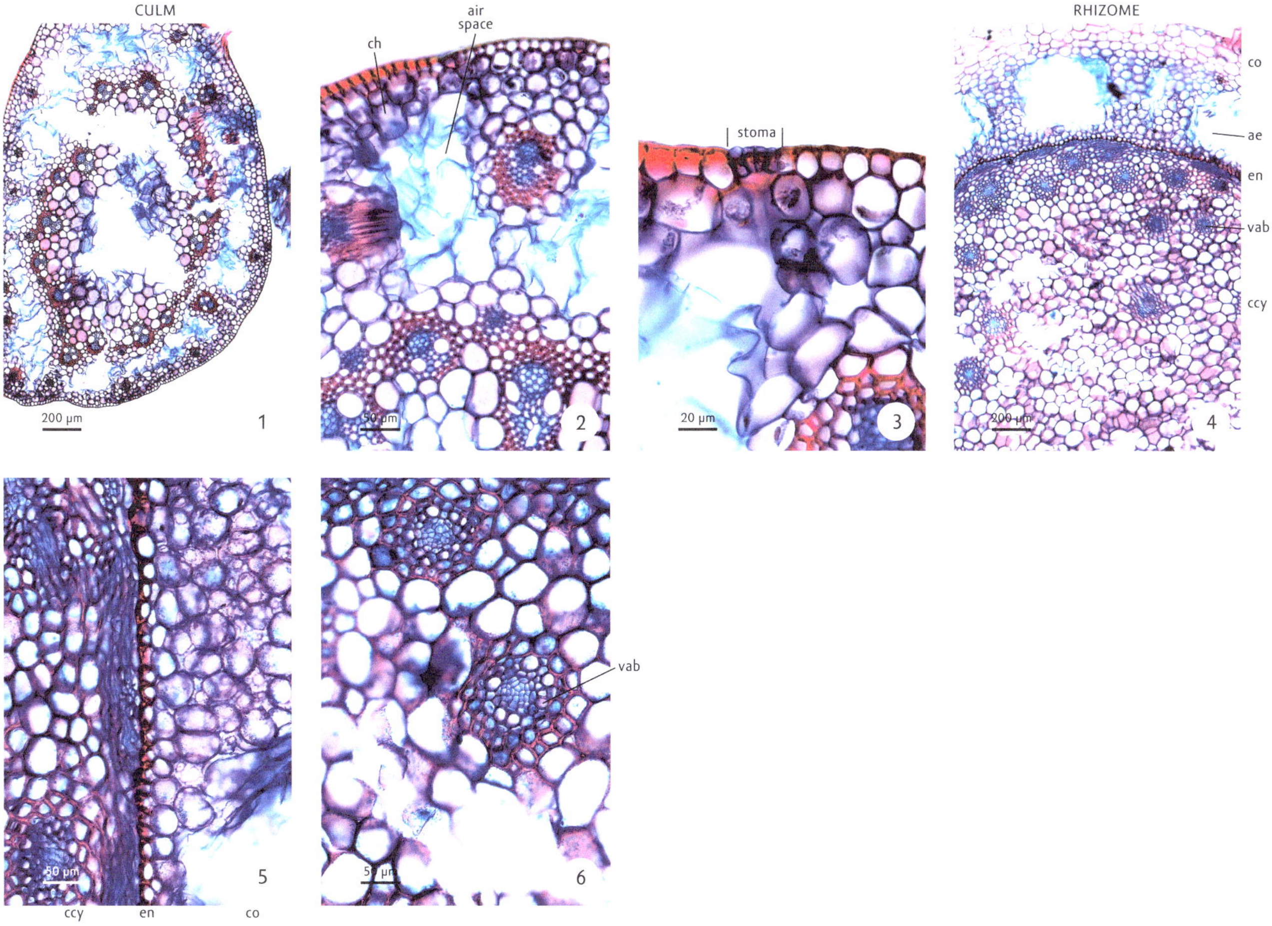

Juncus subnodulosus Schrank

Blunt-flowered Rush

Perennial wetland herb, rhizomatous. Stem erect, to 1.2 m high. Leaves rounded, with very distinct transverse septa. Inflorescence diffuse, with very widely spreading to reflexed branchlets. Fruit is a pale brown capsule.

Distribution and ecology

European southern-temperate element. In all CE countries. Dense stands in base-rich fens, marshes, wet meadows, ditches, dune-slacks on peaty base-rich soils. Often in spring-fed fens with high concentrations of iron marked by superficial precipitated iron ochre, also in brackish water.

Anatomy

Circular culm with a diameter of 2 mm.

The epidermis consists of roundish, externally lignified thick-walled cells (1–3). Guard cells of the stomata are not lignified and have acuminate lips (3). The chlorenchyma is composed of 2–4 layers of palisade-like unlignified parenchyma cells (1, 2). The middle and inner part of the hollow culm consists of round parenchyma cells and vascular bundles (2). The large vascular bundles consist of a xylem with two lateral, large vessels and a roundish phloem with sieve tubes and companion cells. Air ducts are rare (1, 2). All vascular bundles are surrounded by a sheath of fibers (1–3).

Rhizome.

Slide missing.

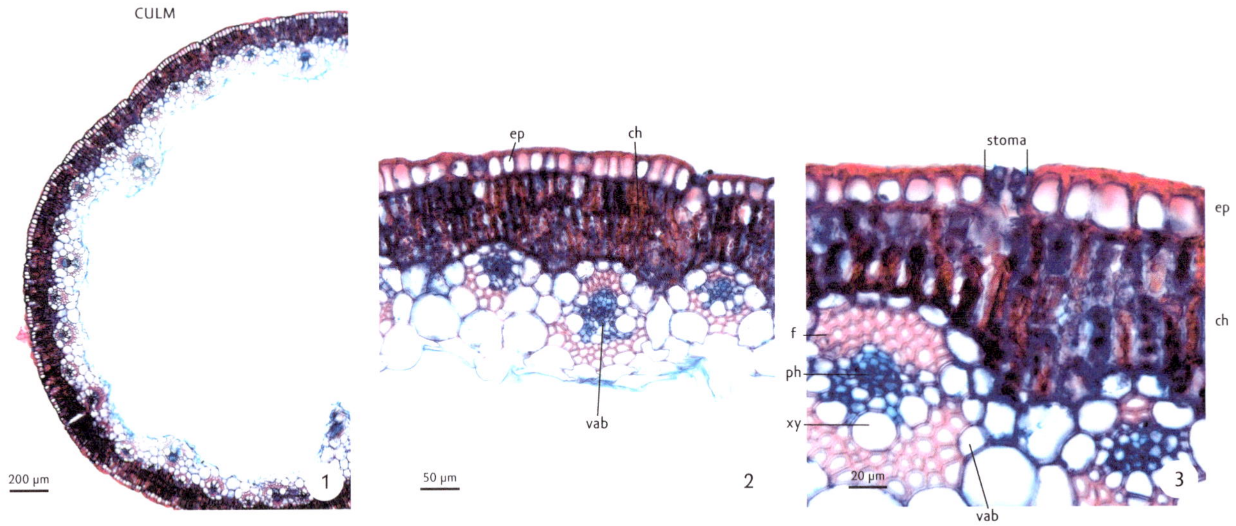

Annual wetland herb, tufted. Stems erect, 5–25 cm high, ramified. Leaves short, thread-like. Loose, branched inflorescence, starting in the upper third of plant. Fruit is a globular, dark-brown capsule, equalling tepals.

Distribution and ecology

European southern-temperate element. In all CE countries on moist, regularly disturbed sites, wet sandy soils, exposed bottoms of fishponds and rivers. Occurs on nitrogen-poor but relatively phosphorus-rich soils. Retreat in many European countries due to eutrophication and acidification, competitively weak.

Anatomy

Circular culm with a diameter of 0.5–0.6 mm.

The epidermis consists of roundish, externally slightly lignified thin-walled cells (1, 2). The chlorenchyma is composed of 3–4 layers of round unlignified parenchyma cells (1). Air spaces are small and irregulary formed. The middle part consists of round parenchyma cells and very small, rudimentary vascular bundles (1). The inner part of the filled culm is characterized by a continuous small belt of fibers and large vascular bundles. The large vascular bundles consist of a xylem with a few lateral vessels and a roundish phloem with sieve tubes and companion cells and an air tube (1). The large bundles are surrounded internally by 1 row and externally by 5–6 rows of thick-walled fibers (1, 2).

Rhizome slide missing.

CULM

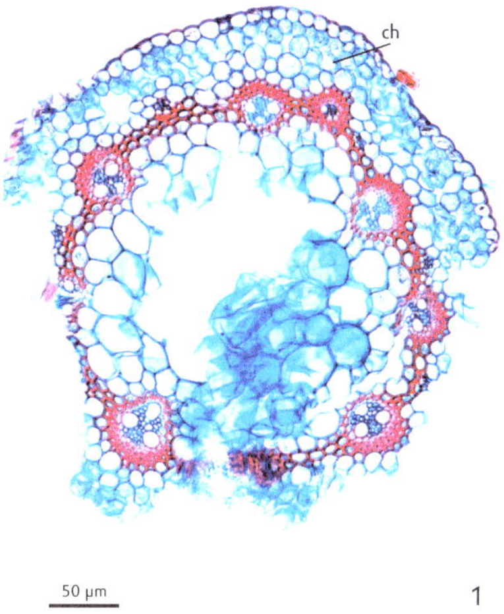

1

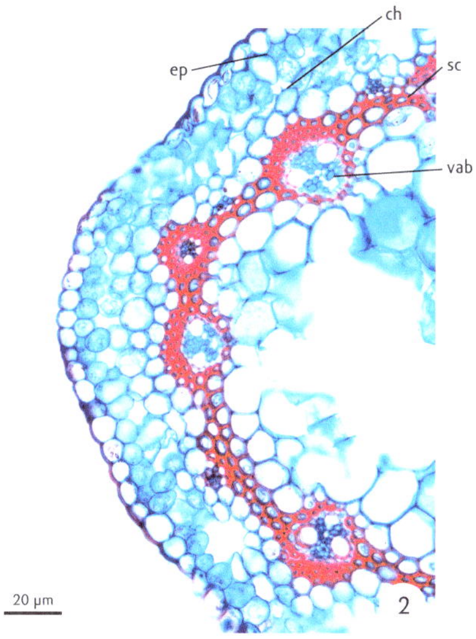

2

Juncus tenuis Willd. (syn. *J. macer*) **Slender Rush** Juncaceae

Perennial wetland herb, densely tufted. Stems erect, to 80 cm high. Leaves 1–2 mm wide, flat or grooved, nearly all at base of stem. Inflorescence a cluster of small heads, the two lowest bracts usually much longer than the inflorescence. Fruit is a grey-brown, ovoid capsule, shorter than tepals.

Distribution and ecology

Neophyte, native to N and S America, naturalized in Europe. In all CE countries in damp open ground along roads and lakes, on wet tracks and in woodland and forest rides.

Anatomy

Circular culm with a diameter of <1–1.5 mm (1).

The epidermis consists of roundish, externally lignified thick-walled cells (1–3). Guard cells of the stomata are not lignified and have acuminate lips (3). The chlorenchyma is composed of 3–5 layers of round unlignified parenchyma cells and a few large square groups of fibers (1, 2). The middle and inner part of the hollow culm consists of round lignified parenchyma cells and vascular bundles (2). The large vascular bundles consists of a xylem with two lateral large vessels, an air duct and a roundish phloem with sieve tubes and companion cells (1, 2). All vascular bundles are surrounded by a sheath of fibers (1–3).

Rhizome with a diameter of 2–3 mm (4).

The cortex is composed of an epidermis and a large layer of thin-walled parenchyma cells without air spaces (4). The central cylinder and the cortex are separated by a centripetal thick-walled endodermis and a distinct pericycle (5). The unlignified parenchymatic ground tissue of the central cylinder contains amphivasal vascular bundles without sclerenchyma sheaths (6).

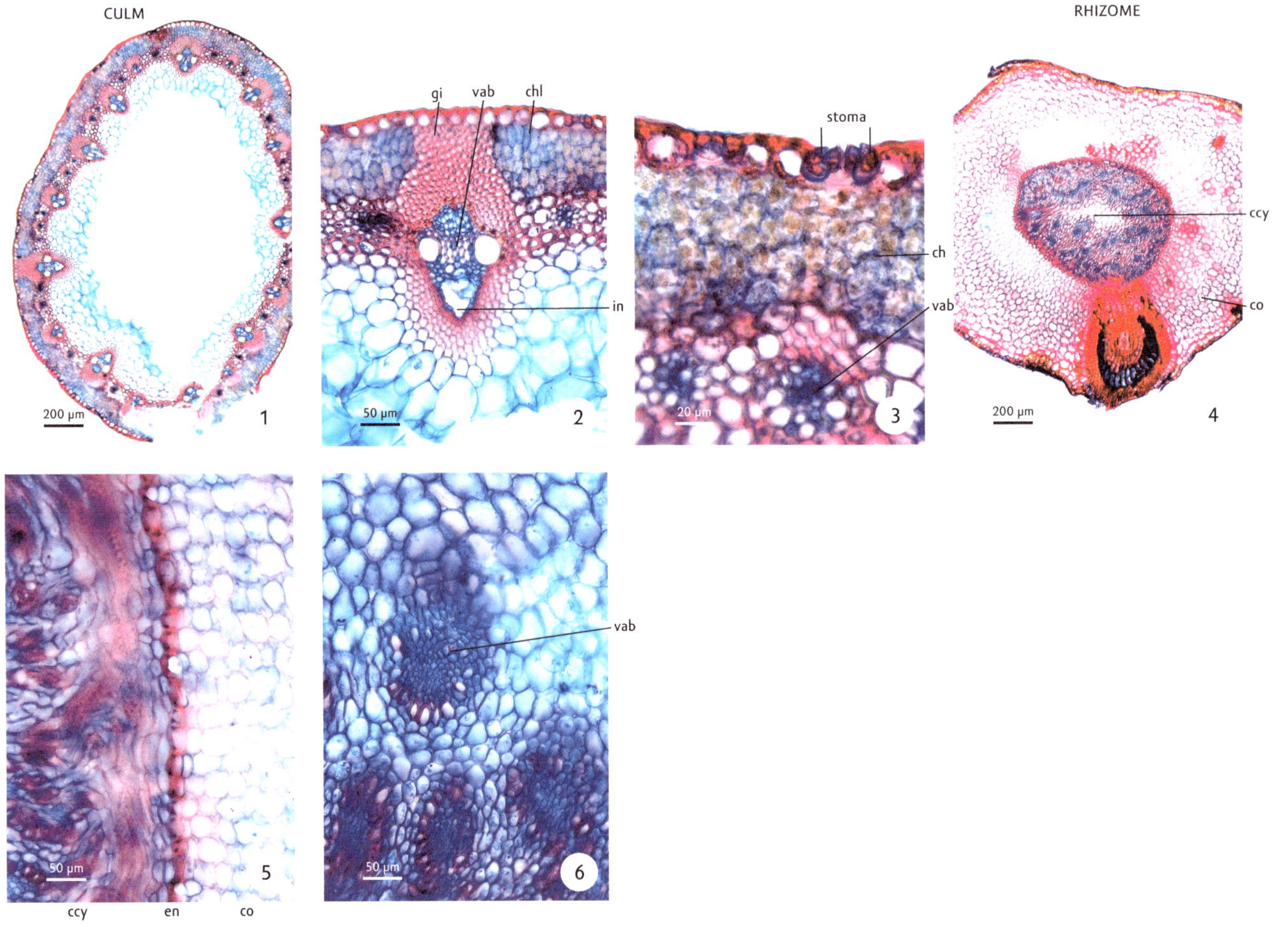

Perennial wetland, tufted herb. Stems erect, up to 20 cm high. Leaves basal. One inflorescence with flowers in a horizontal row, usually not exceeded by lowest bract. Obtuse to rounded capsule.

Distribution and ecology

Circumpolar arctic-montane element. In CE only in the Alps (AT, CH, DE) and Carpathians (SK). Base-rich, damp rocky or gravelly places, flushes and small marshes with short open vegetation in mountains.

Anatomy

Circular culm with a diameter of <1 mm.

The epidermis consists of roundish, externally lignified thick-walled cells (1–3). Guard cells of the stomata are unlignified and have acuminate lips (3). The chlorenchyma is composed of 3–5 layers of round to oval, unlignified parenchymatic cells (1, 2). The middle and inner part of the culm consists of round lignified parenchymatic cells and vascular bundles (2). The large vascular bundles consist of a xylem with two slightly lateral, large vessels and a roundish phloem with sieve tubes and companion cells (1, 2). Peripheral, half moon-like groups of thick-walled fibers characterize the large bundles (1, 2).

Rhizome with a diameter of 1–2 mm.

The cortex is composed of an epidermis, an external layer of thin-walled lignified parenchymatic cells and an inner part of thin-walled unlignified parenchymatic cells without air spaces (4). The central cylinder and the cortex are separated by a centripetal thick-walled endodermis and an indistinct pericycle (5). The lignified parenchymatic ground tissue of the central cylinder contains amphivasal vascular bundles, without sclerenchymatic sheaths (6).

CULM

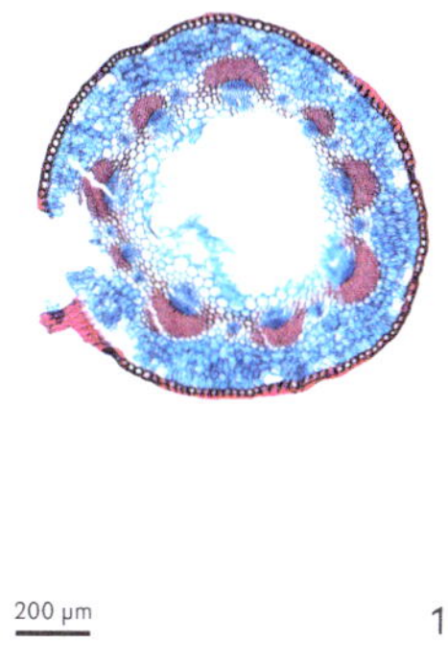

1

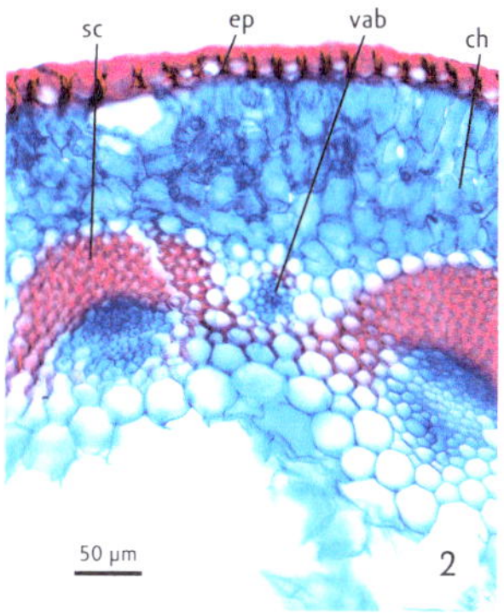

2

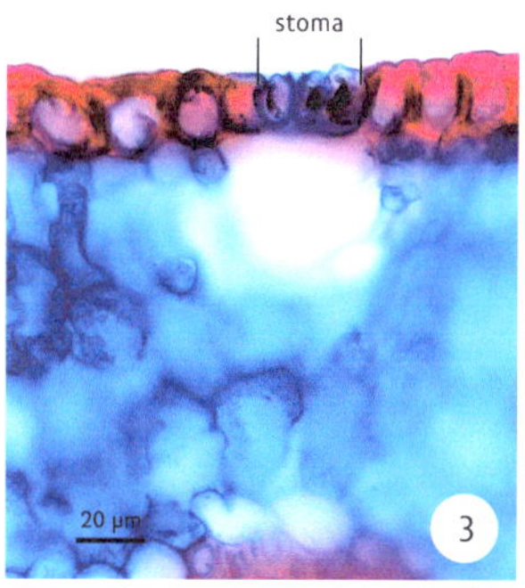

3

RHIZOME

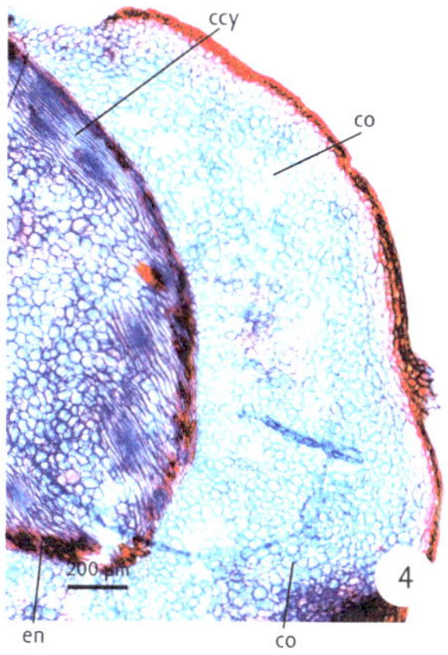

4

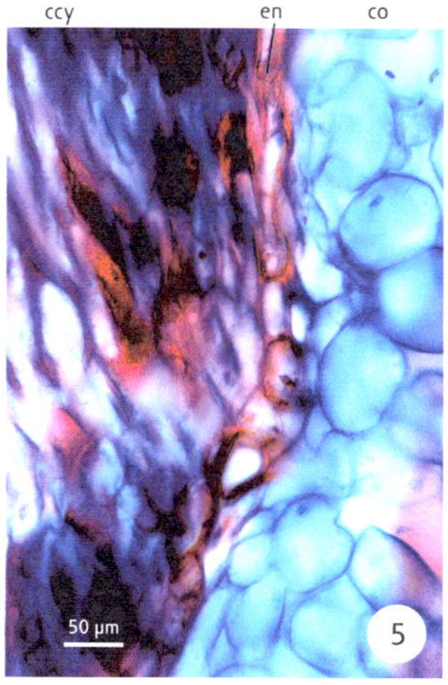

5

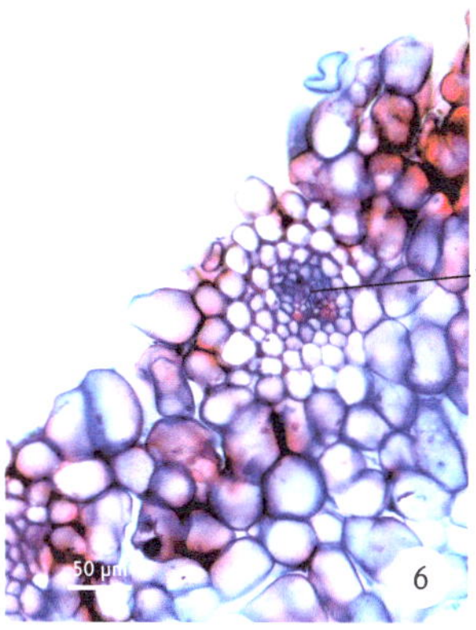

6

Perennial wetland herb, with rhizomes, flowering stem 10–60 cm high, erect, leaves usually flat on upperside near the base. Many flowers in a bractless simple terminal raceme.

Distribution and ecology

Circumpolar boreo-temperate element. In all CE countries except CH, in CZ probably extinct. In salt marshes and salt-sprayed grassland, round coast, rare in inland salty areas (close to mineral springs). Seeds tolerate high levels of salinity and germinated best at summer temperatures during periods of highest soil salinity.

Anatomy

Circular flower stalk with a diameter of 1–2 mm (1).

A small, distinct cortex surrounds a hollow, large central cylinder (1).

The central part of the flower stalk consists of large, poorly lignified parenchymatic cells and the outer part of lignified fibers. Collateral vascular bundles are circular arranged (1). A sheath of lignified fibers surrounds the vascular bundles (1). The centripetal part of the vascular bundles consists of an arc of vessels and a round part of phloem, consisting of large sieve tubes and small companion cells (1). The cortex is mainly formed by round parenchymatic cells and irregular intercellulars (1). Vascular bundles are absent in the cortex.

Circular rhizome with a diameter of 2–3 mm (2).

The center consists of parenchymatic cells, many vascular bundles and a group of sclereids (3). A distinct lignified endodermis separates the cortex from the central part (3, 4).

Central cylinder: The center consists of a group of thick-walled sclereids (5, 6). It is surrounded by a parenchymatic belt with concentric vascular bundles (6). The xylem consists of angular lignified vessels and a center with parenchymatic cells and sieve tubes (5, 6).

Cortex: Intercellulars are irregularly formed and small (4). The periphery consists of a thin-walled epidermis (7). The hypodermis is formed by a belt of thick-walled cells (7).

Crystals are absent in both parts.

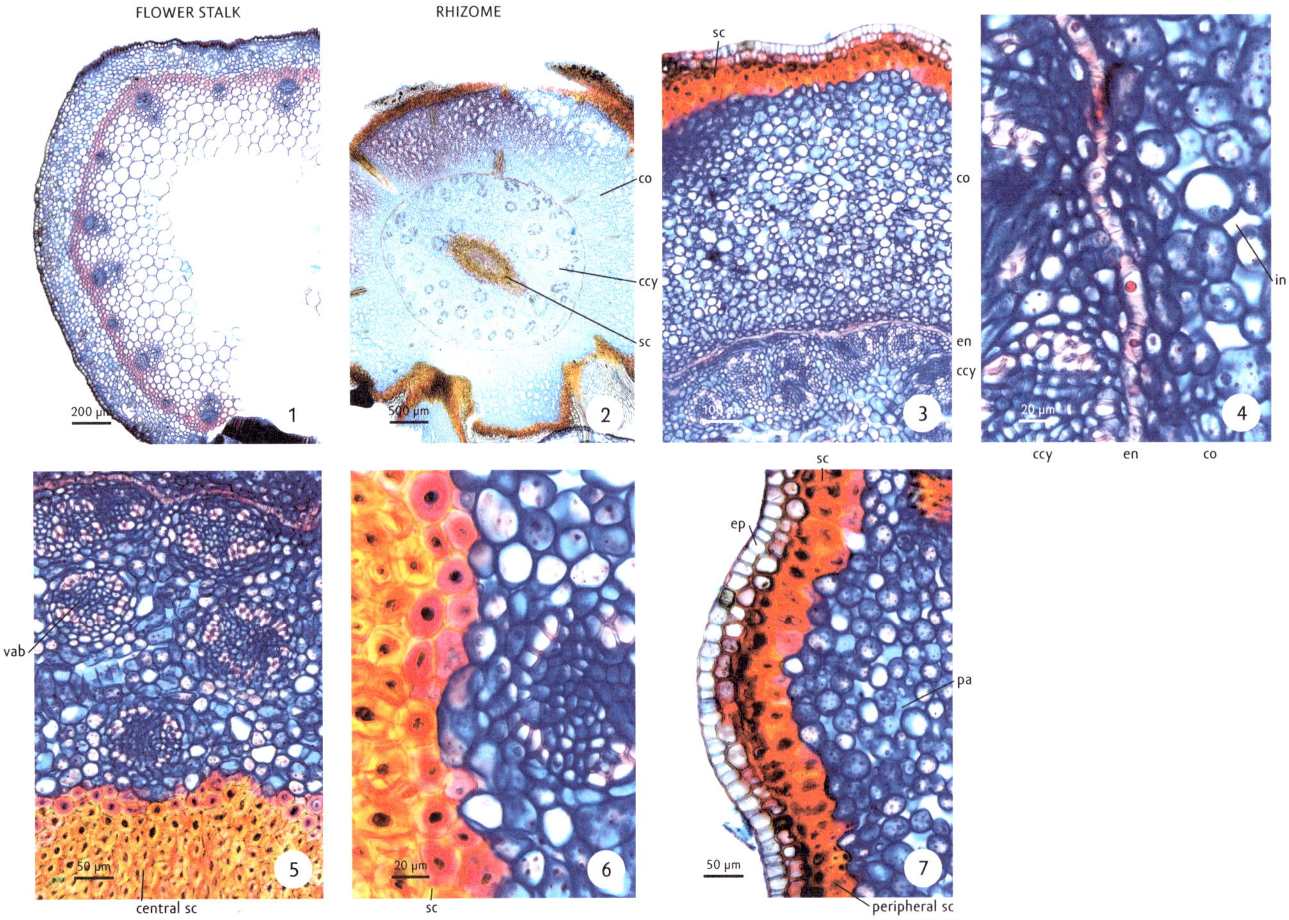

F. H. Schweingruber et al., *Anatomic Atlas of Aquatic and Wetland Plant Stems*, https://doi.org/10.1007/978-3-030-33420-8_65

Perennial wetland herb, with rhizomes, flowering stem 10–60 cm high, erect, leaves usually deeply furrowed on upperside near the base. Many flowers in bractless simple terminal raceme.

Distribution and ecology

Circumpolar boreo-temperate element. In all CE countries, in swampy meadows, drying up fens, springhead, from basic to acidic soils, more often on calcareous substrates. Short-term persistent seedbank.

Anatomy

Circular flower stalk with a diameter of 1–2 mm (1).

A distinct cortex surrounds a hollow, large central cylinder (1).

The central part of the flower stalk consists of large, poorly lignified parenchymatic cells and the outer part of lignified fibers. Collateral vascular bundles are circularly arranged (1). A sheath of lignified fibers surrounds the vascular bundles (2). The centripetal part of the vascular bundles consists of an arc of vessels and a round part of phloem, consisting of large sieve tubes and small companion cells (2). The cortex is mainly formed by round parenchymatic cells and irregular intercellulars (2). Vascular bundles are absent in the cortex.

Circular rhizome with a diameter of 2–3 mm (3).

The center consists of parenchymatic cells, many vascular bundles and a group of sclereids (3). A distinct lignified endodermis separates the cortex from the central part (4).

Central cylinder: The center consists of a group of sclereids (3, 5). It is surrounded by irregularly distributed concentric vascular bundles (3). The xylem consists of angular lignified vessels and a center with parenchymatic cells and sieve tubes (5). Vessels walls have scalariform intervessel pits or are stabilized by ring-like structures (6).

Cortex: Intercellulars are irregularly formed and small (4). The periphery consists of a thin-walled epidermis (3). The hypodermis is formed by a belt of thin-walled cells (3).

Crystals are absent in both parts.

The anatomical structure is almost identical with that of *Triglochin maritima*. Differences in the rhizome – the absence of the peripheral belt of thick-walled fibers (3).

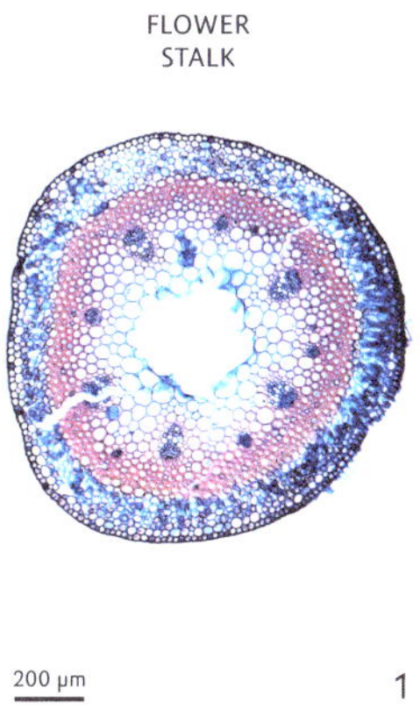

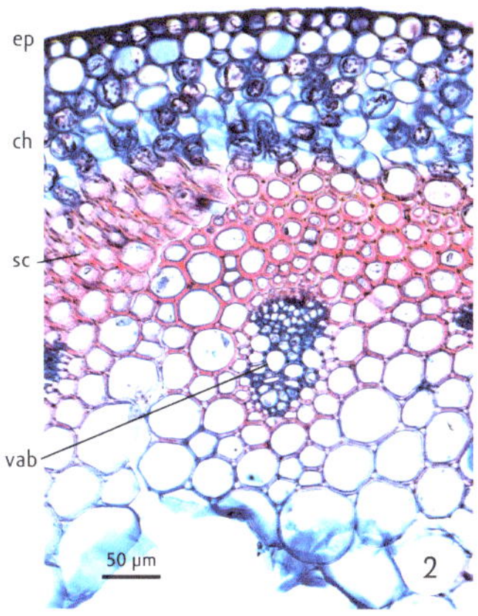

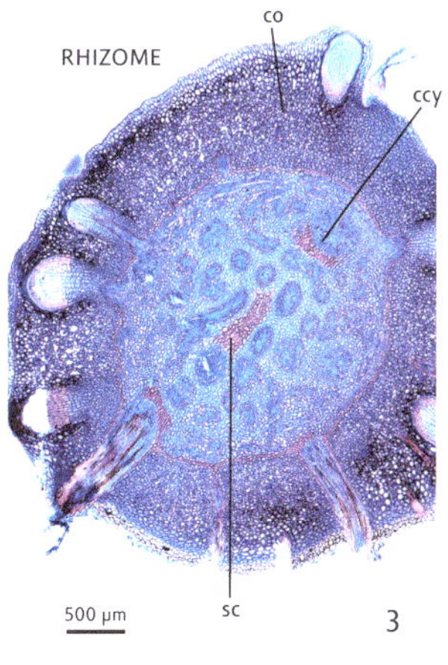

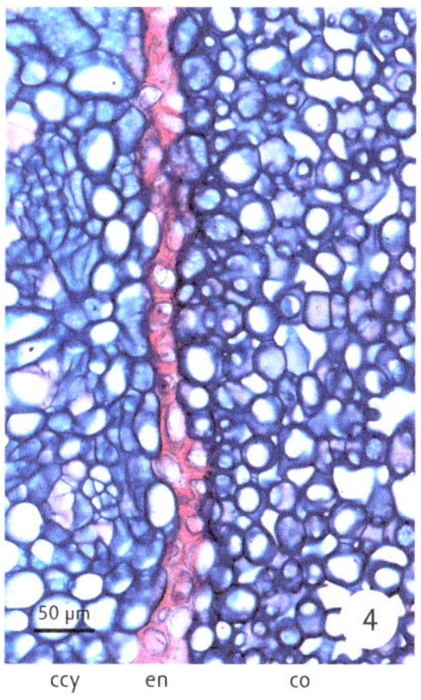

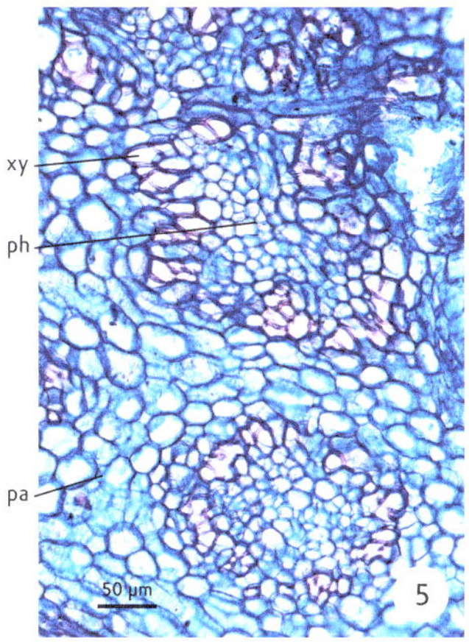

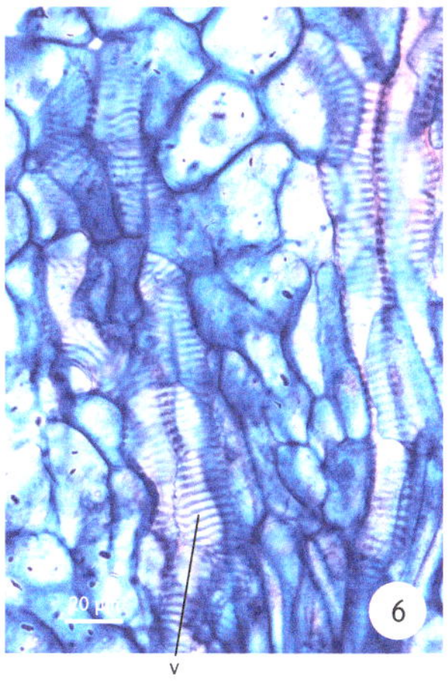

Dactylorhiza incarnata (L.) Soó

Early Marsh-orchid

Orchidaceae

Perennial wetland orchid, bulbs flattish, stem stout, hollow, erect, 30–60 cm high, slightly grooved, scaly at base, leafy. Leaves light-green, non-spotted, lanceolate. Flowers red, rarely white. Labellum obscurely lobed, the sides usually strongly reflexed.

Distribution and ecology

Euro-Siberian boreo-temperate element. In all CE countries, from lowlands to uplands. Wet meadows, marshes, peatbogs, stagnant water banks, calcareous and non-calcareous habitats, light-demanding. Decreasing due to land-drainage and management changes. Very variable.

Anatomy

Circular hollow flower stalk with a diameter of 5–7 mm (1).

The epidermis of unlignified square cells (2, 3). The stem consists of three principal layers (1, 2). The external part is composed of round, thin-walled unlignified parenchymatic cells and small and large irregularly formed intercellulars (3). A belt of round axially elongated, slightly lignified parenchymatic cells without intercellulars, forms the middle part (3). Vascular bundles are irregularly dispersed in all three layers (1). The xylem of vascular bundles contains rectangular, lignified vessels, which are surrounded by unlignified parenchymatic cells (4). The phloem consists of sieve tubes and small companion cells (4). A surrounding sheath of special cells is absent (4). Crystals are absent.

FLOWER STALK

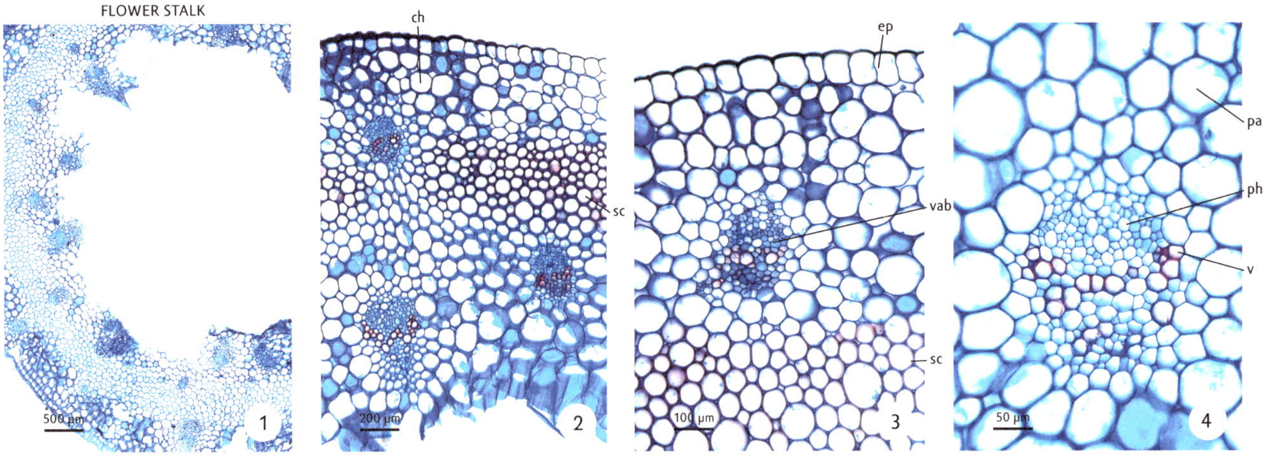

F. H. Schweingruber et al., *Anatomic Atlas of Aquatic and Wetland Plant Stems*, https://doi.org/10.1007/978-3-030-33420-8_66

Perennial wetland orchid, rhizomatous, stem single, erect, 10–70 cm high, grooved, scaly below, downy above. Leaves spiral. Flowers pedicellate, large brown or purplish-white.

Distribution and ecology

Euro-Siberian temperate element. In all CE countries, in calcareous fens and on peat, but also on sandy substrates overlying heavy clayish or loamy soils. Springheads, peaty meadows, wet dune slacks, also disused gravel-pits. It prefers calcareous but nutrient-poor soils, moist to wet substrates. It grows mainly in full sunlight, rarely under shade.

Anatomy

The circular flower stalk with a diameter of 4–5 mm is filled with unlignified cells, vascular bundles and large intercellulars (1).

The external part of the flower stalk is composed of round-shaped, thin-walled epidermal cells, a large (6–10-seriate) unlignified belt of parenchymatic cells with thin-walled, small and large irregularly formed intercellulars (1–3). A lignified belt of cells in the middle part is absent (1, 2). The inner part consists of large, round-shaped unlignified parenchymatic cells (2, 3). Vascular bundles are circularly arranged at the outside of the flower stalk and irregularly dispersed in the center (1). The xylem of vascular bundles contains angular, lignified thin-walled vessels (4). Lignified vessels are arranged in a tangential line. The phloem consists of round sieve tubes and small angular companion cells (4). A surrounding sheath consists of one layer of thin-walled fibers centripetally and 2–3 layers thin- to thick-walled fibers on the external side (4). Crystals are absent. Small intercellulars occur in all parts of the flower stalk.

FLOWER STALK

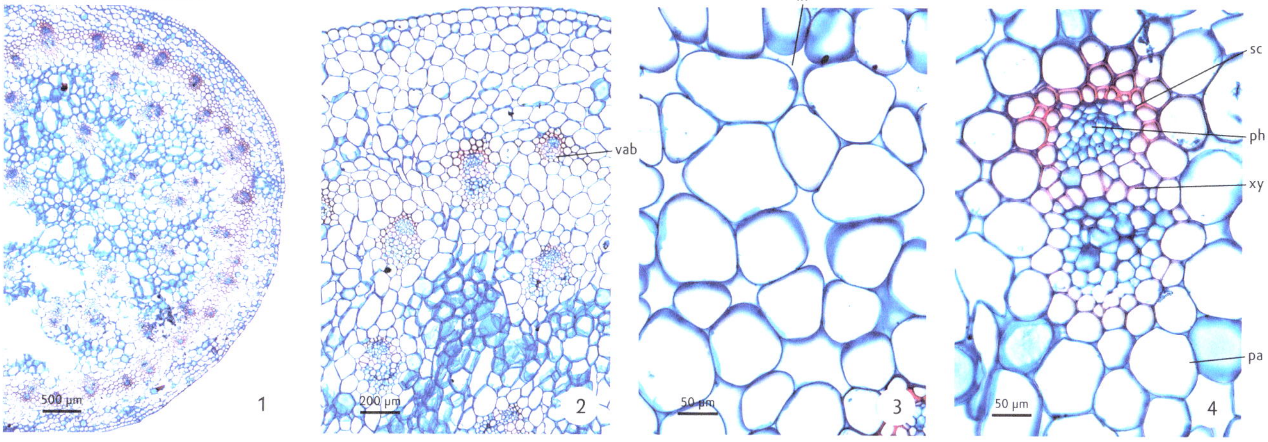

Perennial wetland herb. Stem erect, slender, up to 60 cm high, with looser, long spike of rather big flowers. Leaves 15–20 mm wide, unspotted, arranged in a spiral from basal rosette up to stem. Flowers usually purple, rarely pink or white. Lip spotted, midlobe of the lip narrow, tooth-like, equalling or longer than side-lobes. Fruit is a capsule.

Distribution and ecology

Euro-mediterranean distribution. Very sparsely found in all CE countries. It grows in base-rich fens, fen meadows or salt marshes, prefers wet and muddy soils. In Mediterranean sites it grows exclusively in salt marsh habitats, typically in retrodunal areas along the coast.

Anatomy

Circular hollow flower stalk with a diameter of 2–3 mm (1).

The external part of the flower stalk is composed of round, thin-walled epidermis cells and a small (1–5-seriate) unlignified belt of parenchyma cells with small intercellulars (1, 2). A belt of round, axially elongated, lignified parenchyma cells with very small intercellulars forms the middle part (2, 3). The inner part consists of large, angular unlignified parenchyma cells (2). Intercellulars are absent in inner parts and present in outer parts.

Vascular bundles occur in the most inner parts of the flower stalk (1). The xylem of vascular bundles contains angular, lignified vessels (3). Vessels are arranged in a tangential line. The phloem consists of round sieve tubes and small angular companion cells (3). A lignified small sheath exists mainly at the outside of the bundle (3). Crystals are absent.

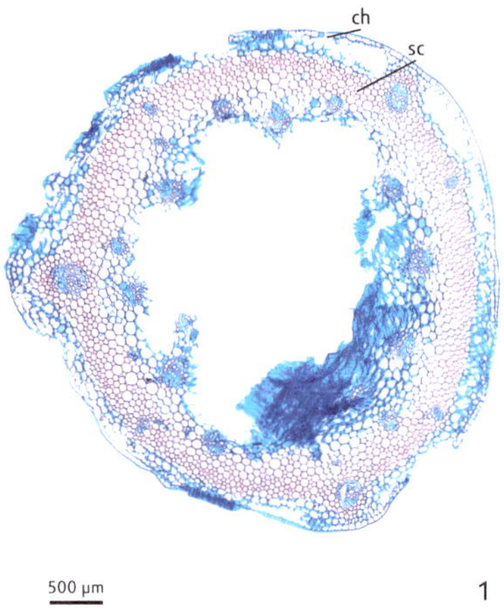

1

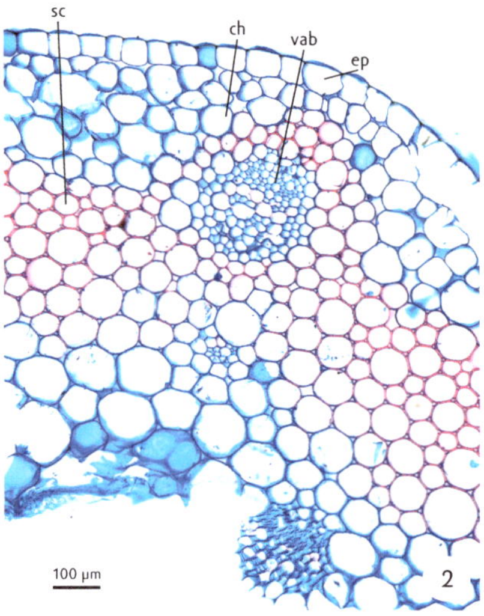

2

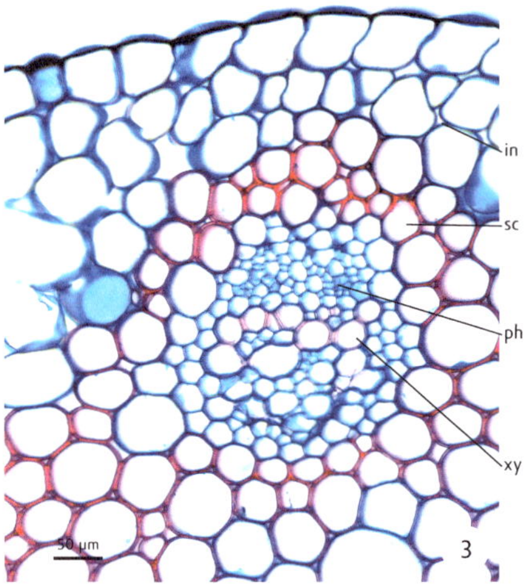

3

Perennial wetland herb, stoloniferous, without rhizomes, free-tufted. Culms numerous, erect, 10–70 cm high, lower nodules rooting down. Leaves narrow, 1–2 mm wide, ligules long, pointed. Panicles loosely contracted at fruiting.

Distribution and ecology

Circumpolar boreo-temperate element, with a disjunct distribution. In all CE countries, on wet meadows and pastures, pond banks, ditches, peatbogs and springs. It prefers acid infertile soils and peaty soils in mires. It may exploit vegetation gaps or act as a dominant. It spreads by stolon fragments.

Anatomy

Circular hollow culm with a diameter of 2 mm (1).

The epidermis consists of lignified, round thin- to thick walled cells (2). The chlorenchyma is composed of unlignified parenchymatic cell groups with tangential elliptical outlines. The sclerenchyma between the chlorenchyma is formed by small, square to tangential rectangular girders, consisting of thin- to thick-walled fibers (2). A 5–9-seriate sclerenchymatic belt inside the chlorenchyma is formed by thick-walled cells. The vascular bundles consist of a xylem with two large lateral vessels, an intercellular around the protoxylem and a roundish phloem with uniform cells (2). The bundles are surrounded by a row of thin- to thick-walled sheath of sclerenchyma (3). The center of the culm is surrounded by unlignified, large, thin-walled parenchyma cells (1).

CULM

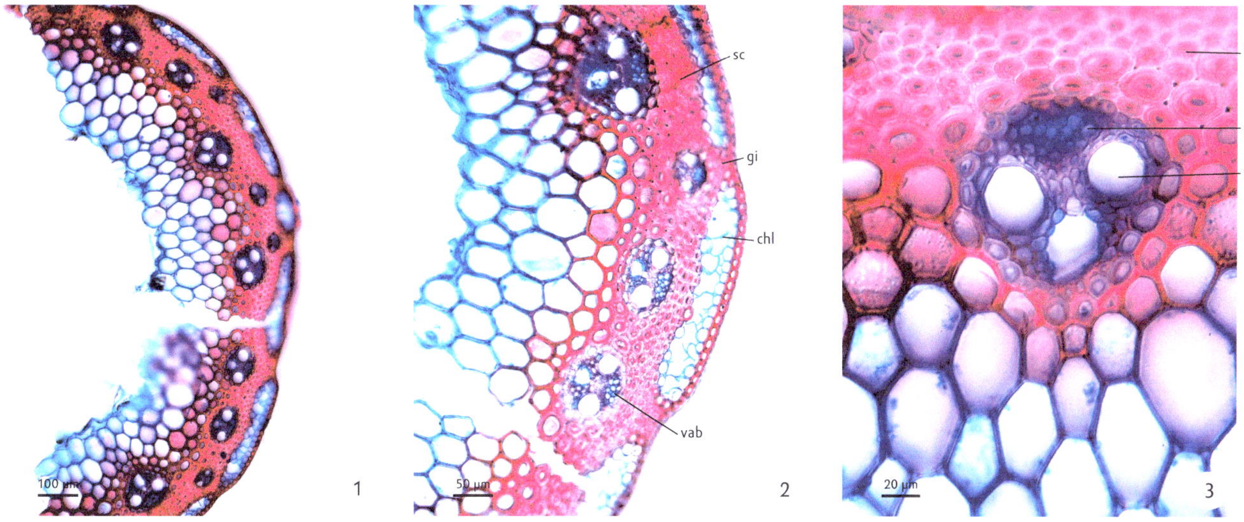

F. H. Schweingruber et al., *Anatomic Atlas of Aquatic and Wetland Plant Stems*, https://doi.org/10.1007/978-3-030-33420-8_67

Perennial wetland, rhizomatous herb, 0.4–1.3 m high, with thick, short underground rhizoms. Leaves flat, lanceolate. Ligules truncate or rounded, ≤ 6 mm long. Panicle diffuse at fruiting, usually wider than 5 cm, triangular, with spikelets in small clusters at branch tips, each containing only one flower. Fruit is a small, yellow caryopsis.

Distribution and ecology

Eurasian southern-temperate distribution, but it is naturalised in N. America. Scattered in all CE countries, mostly on disturbed wet soils, along ditches, river banks and roads, in wet meadows, also as a weed in arable fields.

Anatomy

Circular hollow culm with a diameter of 3 mm (1).

The epidermis consists of lignified, round thin- to thick-walled cells (2). The chlorenchyma forms almost a continuous periphery. The sclerenchymatic tissue inside the chlorenchyma is formed by a continuous small 3–4-seriate band of lignified thin- to thick-walled cells (1, 2). Large vascular bundles consist of a xylem with two large lateral vessels, a protoxylem vessel and an oval phloem with sieve tubes and companion cells (2). The bundles are surrounded by uniseriate sheath (2). The center of the culm is surrounded by unlignified, large thin-walled parenchyma cells (1, 2).

CULM

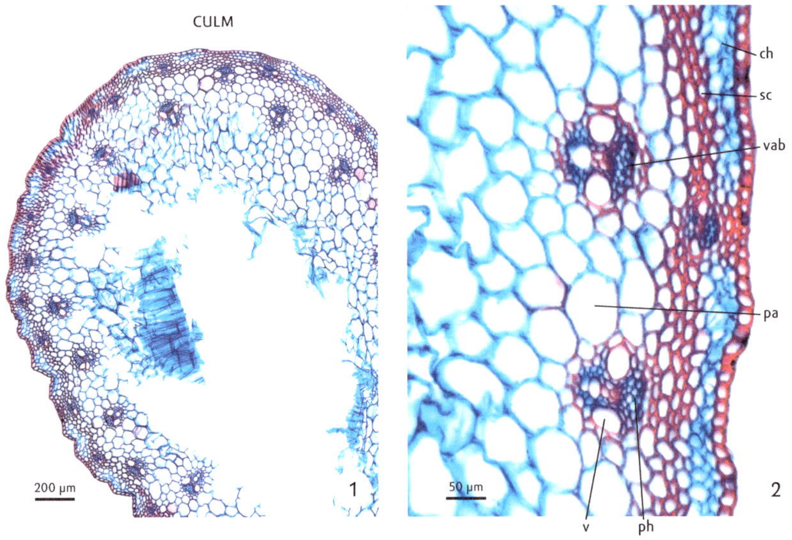

Perennial wetland or emergent herb, stoloniferous. Culm erect, 20–150 cm high, richly branched and procumbent at base. Leaves flat, ligules subacute to rounded, <5 mm long. Panicles contracted when fruiting. Very similar is A. gigantea with underground rhizomes.

Distribution and ecology

Circumpolar wide-temperate element, widely naturalized outside its native range. In all CE countries on moist meadows and pastures, ditches, marshes, along lakes, ponds and rivers, fields, also saline soils (dune slacks). A wide range of open and disturbed habitats.

Anatomy

Circular hollow culm with a diameter of 2 mm (1).

The epidermis consists of lignified, round thin- to thick-walled cells (2). The chlorenchyma is composed of unlignified parenchymatic cell groups with round to elliptical outlines (2). The sclerenchyma between the chlorenchyma is formed by square girders, consisting of thick-walled fibers (2). A sclerenchymatic belt inside the chlorenchyma is absent. The vascular bundles consist of a xylem with two large lateral vessels, an intercellular around the protoxylem and a roundish phloem with uniform cells (2). The bundles are surrounded by a row of thin- to thick-walled sheath of sclerenchyma (2). The center of the culm is surrounded by unlignified, large thin-walled parenchyma cells (1, 2).

CULM

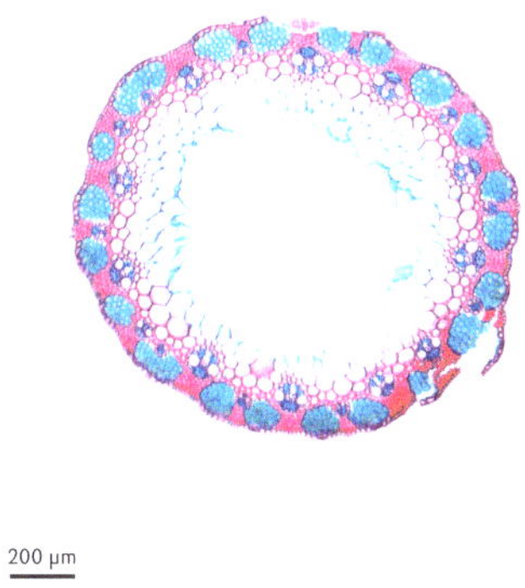

1

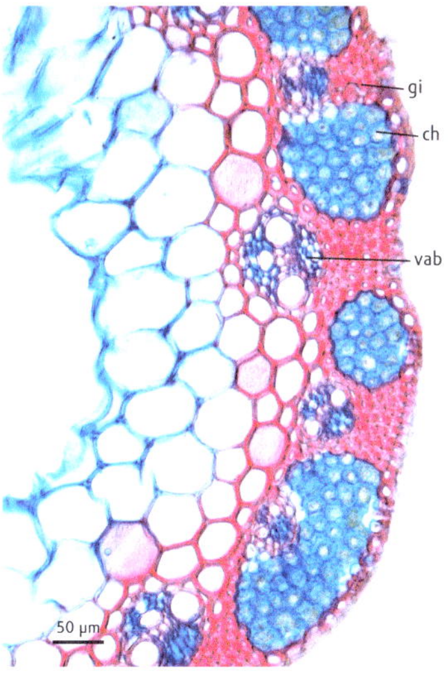

2

Annual to biennial (winter-annual), wetland herb. Culms numerous, procumbent or geniculately ascendent, central culm erect, to 40 cm high. Leaves flat, with pointed ligule. Very contracted spike-like panicle, glumes with hairs to 0.5 mm, anthers orange.

Distribution and ecology

Circumpolar boreal-temperate element. In all CE countries, on wet meadows, marshes, ditches, pond banks, exposed drained bottoms, denuded riverbanks or flooded fields. It prefers drying mud.

Anatomy

Circular hollow culm with a diameter of 1 mm (1).

The epidermis consists of lignified, round thin- to thick walled cells (2). The chlorenchyma is composed of unlignified parenchymatic cell groups with tangential elliptical outlines. The sclerenchymatic tissue between the chlorenchyma is formed by small, square to tangential rectangular girders, consisting of thin to thick-walled fibers, and vascular bundles (2). A sclerenchymatic belt inside the chlorenchyma is absent (2). The small vascular bundles consist of a xylem with two large lateral vessels, an intercellular around the protoxylem and a roundish phloem with uniform cells (2). The bundles are surrounded by a row of thin-walled sheath of sclerenchyma (3). The center of the culm is surrounded by unlignified, large, thin-walled parenchyma cells (1, 2).

CULM

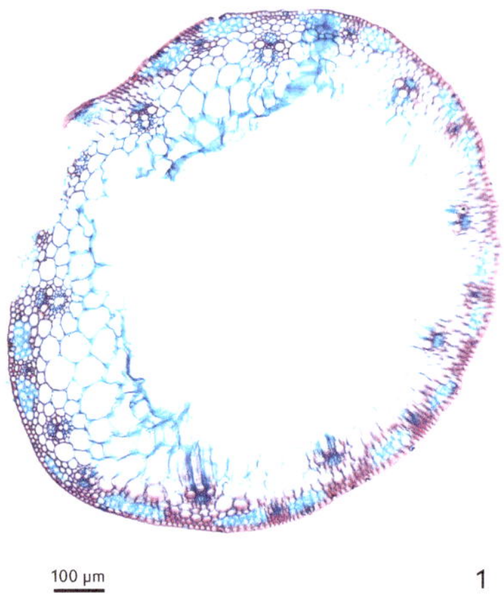

1

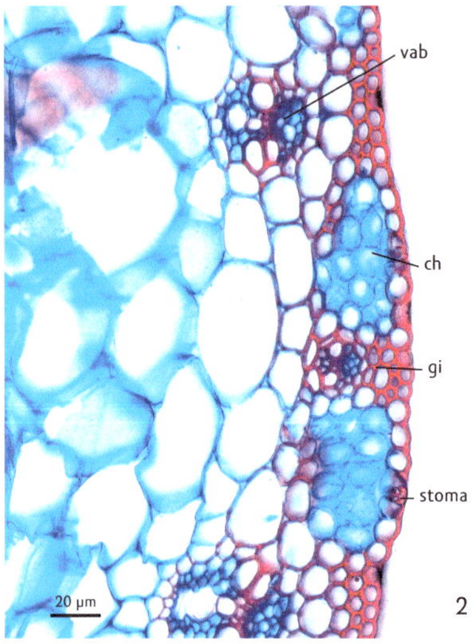

2

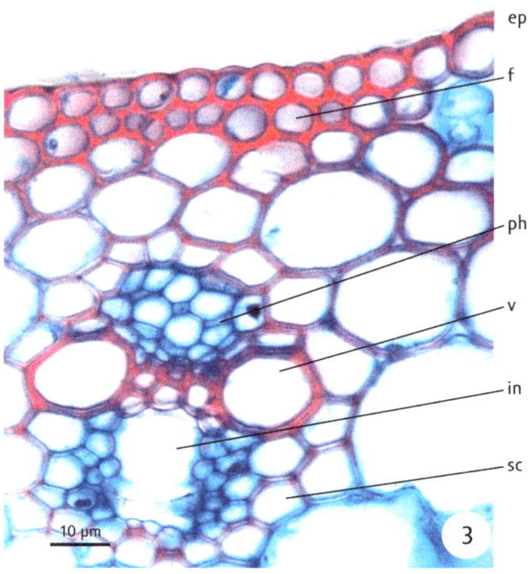

3

Alopecurus geniculatus L.

Marsh Foxtail

Poaceae

Annual or biennial, terrestrial wetland herb. Stems procumbent to ascendent, up to 40 cm long, rooting at the nodes. Leaves linear-lanceolate, rough above, with inflated sheaths and ligules 5 mm long. Inflorescence cylindrical, foxtail-like, blunt, 2–4 cm long. Densely packed spikelets, silky-hairy glumes, awns longer than spike. Anthers later rusty brown. Fruit is a caryopsis.

Distribution and ecology

Native to much of Eurasia and introduced to North and South America and Australia. In all CE countries on wet soils, usually flooded in winter, frequent on muddy ditch banks, pond margins, wet arable fields and meadows. It is able to colonise disturbed wet sites, but avoids both the most acidic and the most alkaline soils.

Anatomy

Circular hollow culm with a diameter of 1 mm (1).

The epidermis consists of lignified, round thin- to thick-walled cells (2). The chlorenchyma is composed of unlignified parenchymatic cells in columns with tangentially elliptical outlines. The sclerenchymatic tissue inside the chlorenchyma is formed by a continuous small band of lignified thin- to thick-walled cells (1, 2). Large vascular bundles consist of a xylem with two large lateral vessels, a protoxylem and a roundish phloem with sieve tubes and companion cells (2). The bundles are surrounded by uniseriate sheath of cells (2). The center of the culm is surrounded by unlignified, large thin-walled parenchyma cells (1, 2).

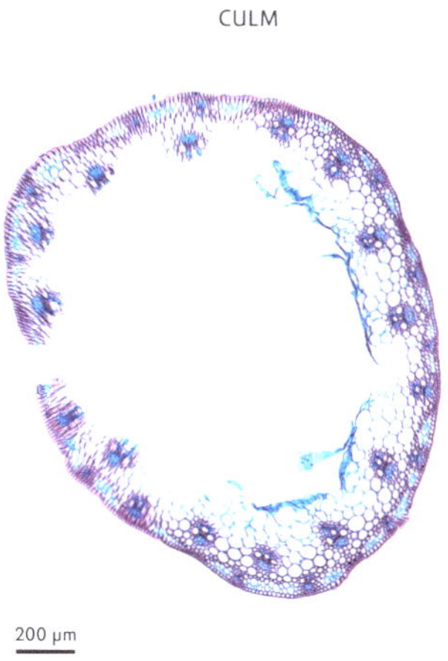

1

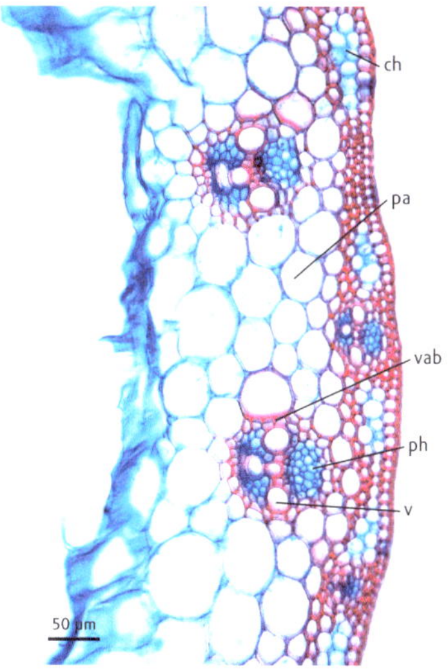

2

Beckmannia eruciformis (L.) Host

European Slough-grass Poaceae

Perennial wetland herb, culm erect, 40–150 cm high, onion-like roughened at base, with nodules rooting down. Leaves flat, 4–10 mm wide. Inflorescence is a long raceme of closely packed appressed spikes forming a long narrow inflorescence.

Distribution and ecology

Native to Eurasia. Only rarely in CZ, SK and PL, more often in HU. It occurs mainly in lowlands, on wet meadows or water banks. It prefers gleyish or muddy soils, base-rich or saline substrates, moist in spring, dry in summer.

Anatomy

Circular hollow culm filled with parenchyma cells and vascular bundles with a diameter of 1.5 mm (1).

The epidermis consists of lignified, round, laterally and externally thick-walled cells (2). The large chlorenchyma is composed of unlignified parenchymatic groups of thin-walled, unlignified cells with irregular tangential outlines and intercellulars (1, 2). The thin- to thick-walled sclerenchymatic tissue is formed by a few conic girders and a 3–6-seriate band inside the chlorenchyma (1, 2). Large vascular bundles consist of a xylem with two large lateral vessels, a protoxylem-vessel, an intercellular and a roundish phloem with sieve tubes and companion cells (2). The bundles are embedded in a uni- to biseriate sheath (2). The external part of the pith consists primarily of lignified thin- to thick-walled polygonal parenchyma cells (1, 2).

CULM

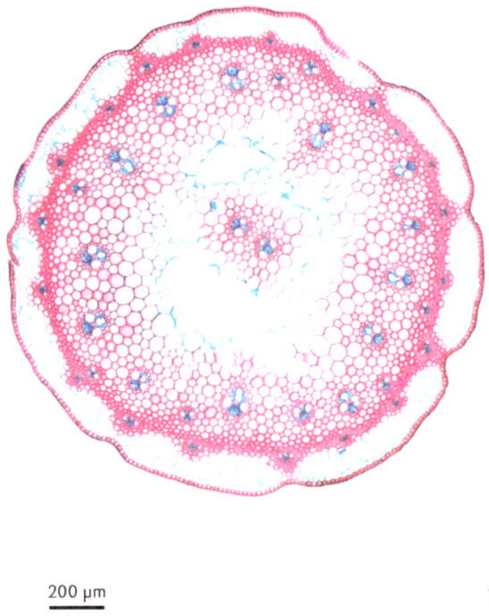

1

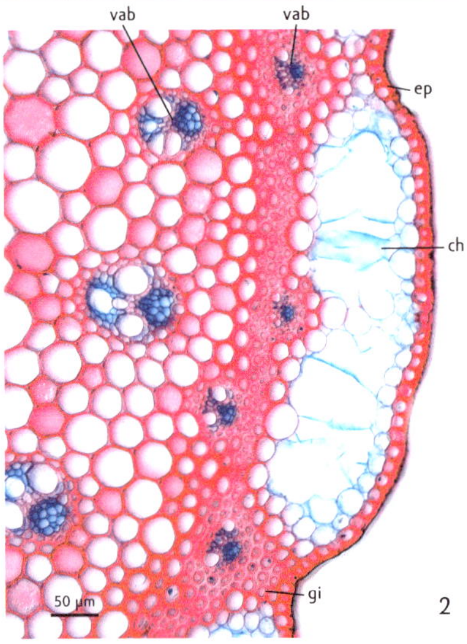

2

Calamagrostis canescens (F. H. Wigg.) Roth **Purple Small-reed** Poaceae

Perennial wetland, rhizomatous herb, slightly tufted. Culm erect, 50–150 cm high, thin, with 4–6 nodules, often branched. Leaves to 8 mm wide, shortly hairy above. Inflorescence is a diffuse panicle.

Distribution and ecology

Euro-Siberian boreal-temperate element, with a continental distribution in W Europe. In all CE countries, usually in water banks, wet and fen meadows, tall herb fens and *Alnus* or *Salix* carrs, on moister grassy forest clearings, sometimes forming extensive stands.

Anatomy

Circular hollow culm with a diameter of 2 mm (1).

The epidermis consists of lignified, round, externally very thick-walled cells (2). The chlorenchyma is composed of unlignified thin- to thick-walled parenchymatic cell groups with tangential outlines. The sclerenchyma between the chlorenchyma is formed by square to tangential to conic girders, consisting of thick-walled fibers and vascular bundles (2). Large vascular bundles consist of a xylem with two large lateral vessels, a protoxylem and a roundish phloem with uniform cells (3). The bundles are surrounded by a uni- to biseriate row of thin- to thick-walled sheath (2). The center of the culm is surrounded by a few unlignified, large, thin-walled parenchyma cells (1, 2).

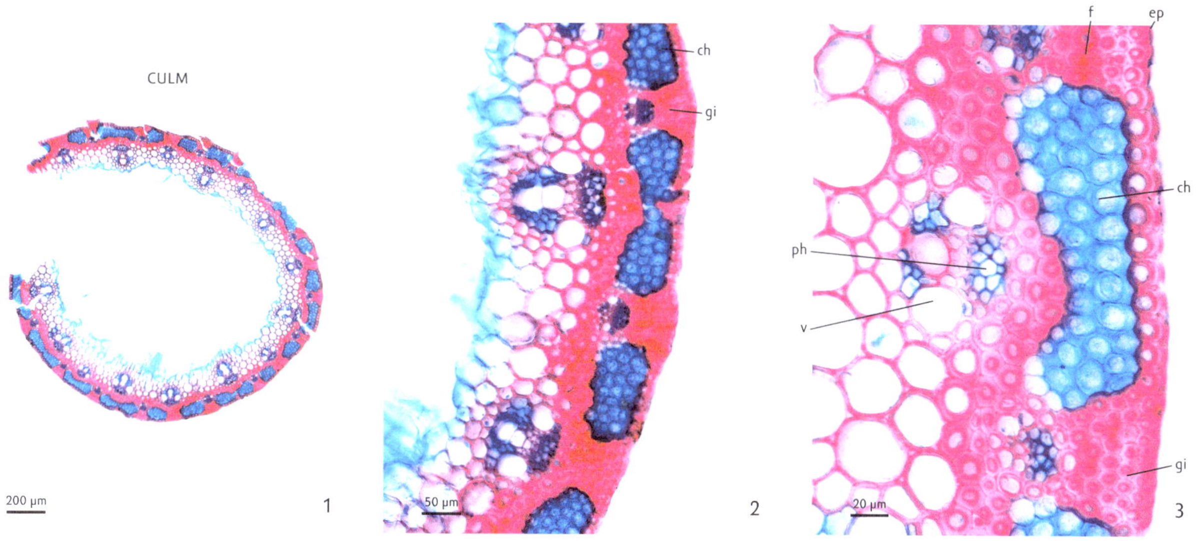

Calamagrostis pseudophragmites (Haller f.) Koeler

Poaceae

Coastal Small-reed

Perennial wetland herb, tufted with creeping rhizomes. Culm 50–150 cm high, with 4–6 nodules, smooth below panicle. Leaves grey-green, rough, veined, to 10 mm wide. Sparse panicle with thin branches, later pendant. Violet anthers on long filaments.

Distribution and ecology

Native to Eurasia, in all CE countries, only rarely in CZ. In brook gravel deposits and coastal shrubberies.

Anatomy

Circular hollow culm with a diameter of 2 mm (1).

The epidermis consists of lignified, round, externally thick-walled cells (2). The chlorenchyma is composed of an unlignified parenchymatic continuous band of cells or groups with tangential extended outlines (1, 2). The thin- to thick-walled sclerenchymatic band consists of rare conic girders between the chlorenchyma and a 2–4-seriate band inside the chlorenchyma (1, 2). Large vascular bundles consist of a xylem with two large lateral vessels, protoxylem vessels, an intercellular and a round to oval phloem with sieve tubes and companion cells (2). The bundles are surrounded by a uniseriate sheath (2). The center of the culm is surrounded by poorly lignified, large, thin-walled parenchyma cells (1, 2).

CULM

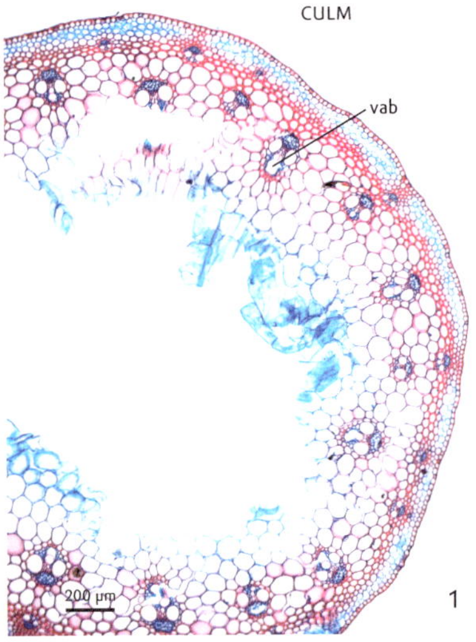

1

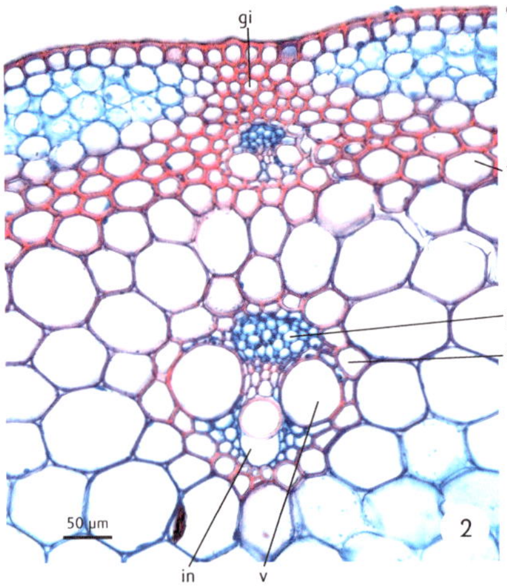

2

Perennial wetland herb, with stolons, without rhizomes. Culms erect to decumbent, 20–75 cm high. Leaves broad, soft, to 10 mm wide, sweet-tasting, ligules long. Inflorescence is a diffuse panicle, spikelets purplish.

Distribution and ecology

European boreo-temperate element; also in C Asia and N America. In all CE countries, on swampy and muddy places, close to stagnant or slowly flowing waters in lowlands and uplands. Often on muddy fishpond margins, cattle-poached ditches, canals and sluggish streams.

Anatomy

Circular hollow culm with a diameter of 1.5 mm (1).

The epidermis consists of lignified, round, thin- to thick-walled cells (2). The chlorenchyma is composed of unlignified parenchymatic cells with tangential extended outlines (1, 2). The thin- to thick-walled sclerenchymatic band consists of tangentially very extended girders between the chlorenchyma and a small uniseriate band inside the chlorenchyma (1, 2). Large vascular bundles consist of a xylem with two large lateral vessels, a few protoxylem vessels and a round homogeneous phloem (2). The bundles are surrounded by a uniseriate sheath (2). The center of the culm is surrounded by unlignified, large, thin-walled parenchyma cells (1, 2).

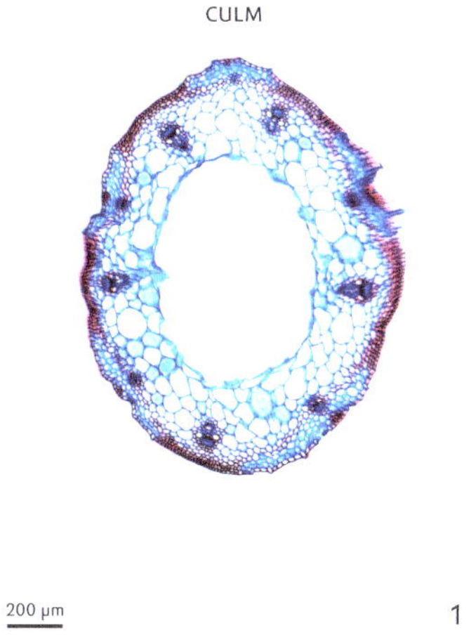

1

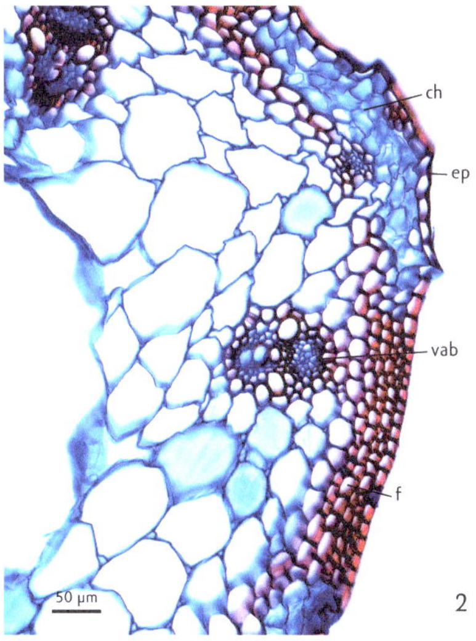

2

Coleanthus subtilis (Tratt.) Seidel (syn. *Schmidtia subtilis*)

Annual wetland ephemeral plant. Culms procumbent to erect, 3–8 cm long, leaf sheaths strongly inflated, the uppermost sheath broadened.

Distribution and ecology

Circumpolar temperate element, with a disjunct distribution. Probably extinct in AT, scattered irregularly and rare in CZ, in lowlands to uplands, usually in fishpond basins. Wet denudated shores of fishponds, periodically flooded depressions, muddy, organic soils. Strict heliophyte, competitively very week, pioneering species on summer-drained fishpond bottoms. The whole growth cycle can complete only during 3–4 weeks, 2–3 generations per season possible.

Anatomy

Circular culm with a diameter of 0.5 mm (1).

The epidermis consists of lignified, round thin- to thick walled cells (1). The chlorenchyma is composed of unlignified palisade-like cells. Sclerenchymatic cells are absent. Vascular bundles consist of a xylem with small vessels (5 µm), and a small undifferentiated phloem (1). The bundles are surrounded by a row of thin-walled cells (1).

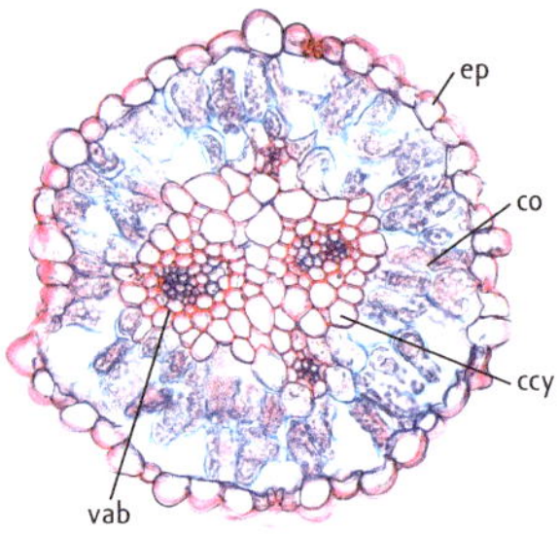

1

Deschampsia cespitosa (L.) P. Beauv.

Tufted Hair-grass

Perennial wetland herb, densely tufted, without rhizomes or stolons. Culms numerous, erect, 30–200 cm high. Leaves grooved, with very rough edges. Inflorescence is a very diffuse panicle with fine branches.

Distribution and ecology

Circumpolar wide-boreal element; widely naturalized outside its native range. Common in all CE countries. It occurs on poorly-drained, mildly acidic or neutral soils in woodlands, moist meadows and pastures, grass-heath, and a wide range of montane habitats. It can rapidly colonise bare ground and tolerates some disturbance. It tolerates high concentrations of heavy metals in soil. It often persists in overgrazed pasture because its unpalatable leaves have high silica content. The closely related *D. littoralis* (Gaudin) Reut [syn. *D. rhenana*] is adapted to long-lasting inundation along Lake Constance, including pseudo-vivipary.

Anatomy

Circular hollow culm with a diameter of 1.5 mm (1).

The epidermis consists of lignified, round, externally very thick-walled cells (2). The chlorenchyma is composed of unlignified parenchymatic cell groups with irregular outlines. The sclerenchymatic band between the chlorenchyma is formed by large girders, consisting of thick-walled fibers and small vascular bundles (1, 2). Large vascular bundles consist of a xylem with two large lateral vessels, a vessel of the protoxylem and an oval phloem with sieve tubes and companion cells (2). The bundles are surrounded by a uniseriate row of thin-walled sheath (2). The center of the culm is surrounded by lignified, large, thin-walled parenchyma cells (2).

CULM

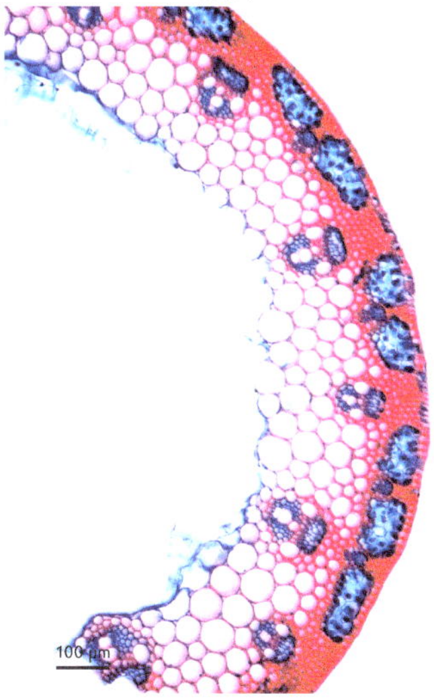

1

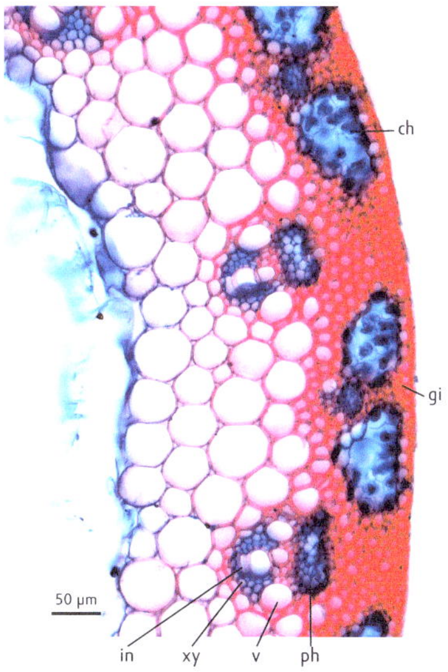

2

Annual wetland herb, free-tufted. Culm erect or ascendent, 30–100 cm high, nodules hirsute. Leaves flat, ligule absent. Inflorescence is a 5–10 cm long panicle of one-sided racemes, usually strongly purplish-tinged.

Distribution and ecology

Native of warm-temperate and tropical regions of Europe, Asia and N America, widely introduced as a fodder grass. Naturalized in all CE countries. Weed in fields, drained fishpond bottoms, muddy sediments. Strong competitor in rice and other crop fields, produces large number of seeds and also have the seed dormancy, which ensures its propagation as well as seed bank build up in the soil profile, C4 plant.

Anatomy

Circular hollow culm with a diameter of 4–6 mm (1).

The epidermis consists of lignified, round, thin- to thick-walled cells (2). The chlorenchyma is formed by a large, unlignified continuous parenchymatic band below the hypodermis (1, 2). The sclerenchyma is formed by a continuous band of thick-walled fibers and small vascular bundles (1, 2). Large vascular bundles centripetal of the sclerenchyma belt are arranged in two concentric rows and are embedded in a thin-walled, unlignified tissue (1). Large vascular bundles consist of a xylem with two lateral large vessels (2). Instead of the protoxylem vessel, there is often an intercellular canal. The phloem consists of sieve tubes and companion cells (2). The bundles are surrounded by a uni- to biseriate lignified sheath (2). The external part of the hollow pith consists of unlignified, thin- to thick-walled polygonal parenchyma cells with very small intercellulars (1, 2).

CULM

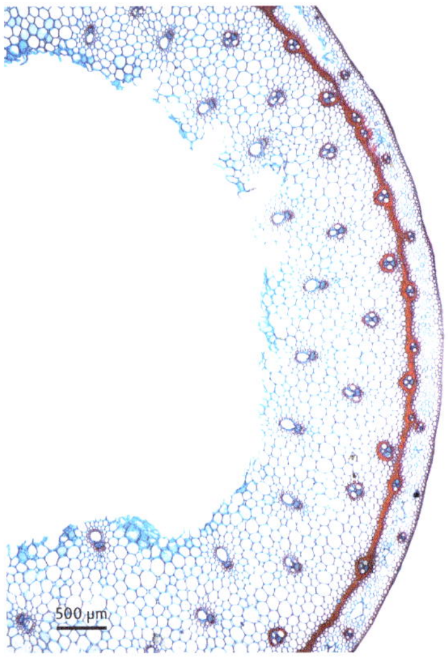

1

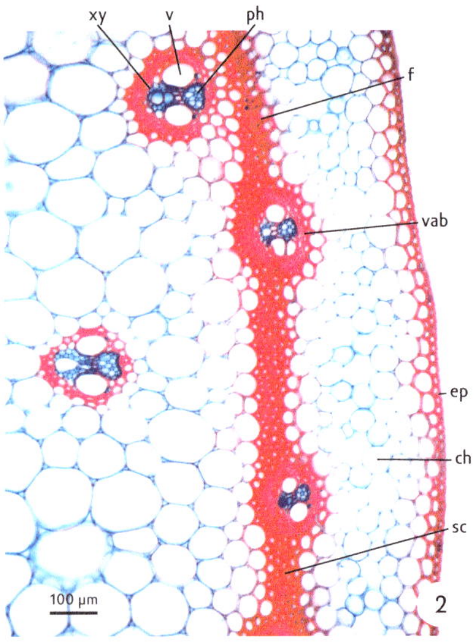

2

Festuca arundinacea Schreb.

Tall Fescue | Poaceae

Perennial wetland, tufted herb. Culms 50–150 cm high, leaves usually flat, upper side sulcate, rounded auricles usually minutely pubescent at margins. Panicle up to 20 cm long, erect or pendent. Fruit is a caryopsis.

Distribution and ecology

Euro-Siberian southern-temperate element; widely naturalised outside its native range. In all CE countries from lowland to the mountains. Occurs on wet meadows and pastures, in ditches, along stream banks and roadsides, often on disturbed sites. Prefers wet, heavy, neutral or basic soils, often with higher concentration of salts.

Anatomy

Circular hollow culm with a diameter of 3–4 mm (1).

The epidermis consists of lignified, round thick-walled cells (2). The chlorenchyma is composed of small unlignified groups of parenchymatic cells with tangential extended outlines (1, 2). The sclerenchymatic tissue is formed by rectangular girders and a 5–7 seriate band inside the chlorenchyma (1, 2). The vascular bundles consist of a xylem with two large lateral vessels, a few protoxylem vessels and a roundish phloem with sieve tubes and companion cells (2). The bundles are surrounded by 1–2-seriate layer of sheath cells (2). The pith is surrounded by thin-walled, lignified cells and a circle of vascular bundles (1, 2).

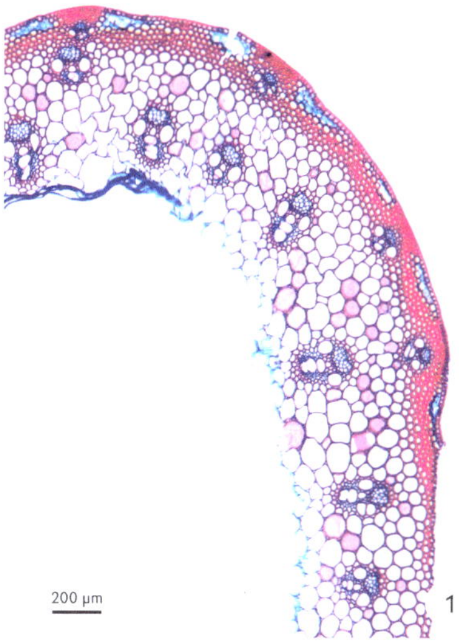

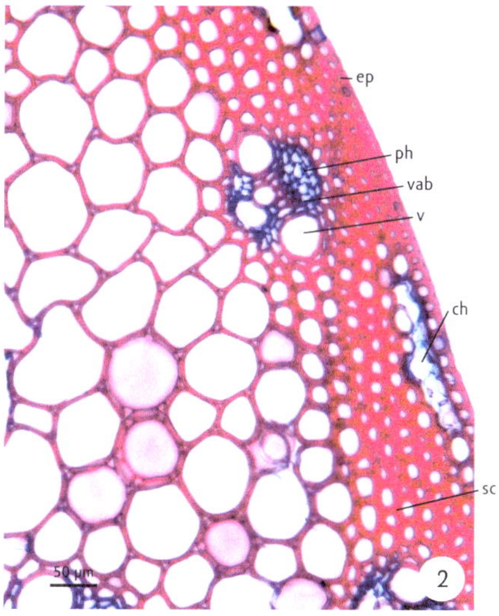

Glyceria fluitans (L.) R. Br.

Floating Sweet-grass

Perennial wetland, floating-leaved or emergent herb. Culm ascendent or crawling and rooting or swimming in water, 40–120 cm long, rhizomatous. Leaves flat, rough; ligule to 15 mm long, jagged. Panicles sparsely branched, with cigar-shaped grey-green spikelets.

Distribution and ecology

European temperate element, widely naturalized outside its native range. Common in all CE countries on mud or shallow water in ponds, rivers, canals, ditches and wet meadows, springheads, light lowland woods. It is able to form floating rafts in shallow water. It tolerates high levels of disturbance and nutrient enrichment.

Anatomy

Circular hollow culm with a diameter of 1 mm (1).

The epidermis consists of lignified, round, thick walled cells (3). The chlorenchyma is composed of thin-walled unlignified parenchymatic cell groups and large intercellulars (air spaces). The sclerenchymatic belt is indicated by a uniseriate row of thin- to thick-walled cells just below the epidermis (2, 3). The vascular small bundles consist of a xylem with two large lateral vessels, a vessel of the protoxylem and a round phloem with sieve tubes and companion cells (2). The bundles are surrounded by a 1-3-seriate sheath of thick-walled cells (2). The center of the culm is surrounded by large, thin-walled unlignified and lignified parenchyma cells (1).

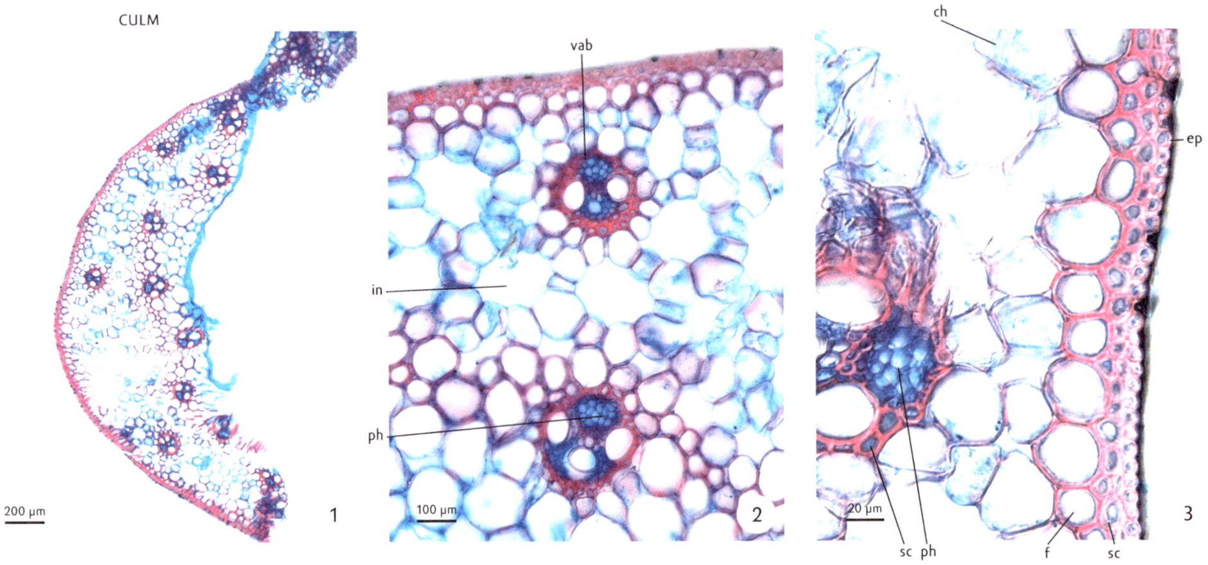

Glyceria maxima (Hartm.) Holmb.

Reed Sweet-grass

Poaceae

Perennial wetland, floating-leaved or emergent herb with rhizomes. Culm stiff, erect, 80–200 cm high, up to 10 mm in diameter, smooth. Leaves broad, to 18 mm wide, bright green, with a brown mark at the stem-junction. Inflorescence is a well-branched panicle.

Poaceae

Distribution and ecology

Circumpolar temperate element, widely naturalized outside its native range. Common in all CE countries on banks of stagnant and slow-flowing waters, swamps and reeds; also in seasonally-flooded grasslands. It can form floating rafts. It was formerly cultivated as a fodder crop. It prefers nutrient-rich water.

Anatomy

Circular hollow culm with a diameter of 4–6 mm (1).

The epidermis consists of lignified, round, thin- to thick-walled cells (2). The culm is dominated by large columns of chlorenchyma with tangentially elliptical outlines (1, 2). The sclerenchymatic tissue is formed by conic girders and a wavy 3–5-seriate band inside the chlorenchyma (1, 2). The vascular bundles consist of a xylem with two large lateral vessels, an air space inside o the protoxylem and a roundish phloem with sieve tubes and companion cells (2). The bundles are surrounded by uni- to biseriate layer of sheath cells (2). The pith is surrounded by thin-walled, unlignified cells and a circle of vascular bundles (1, 2).

CULM

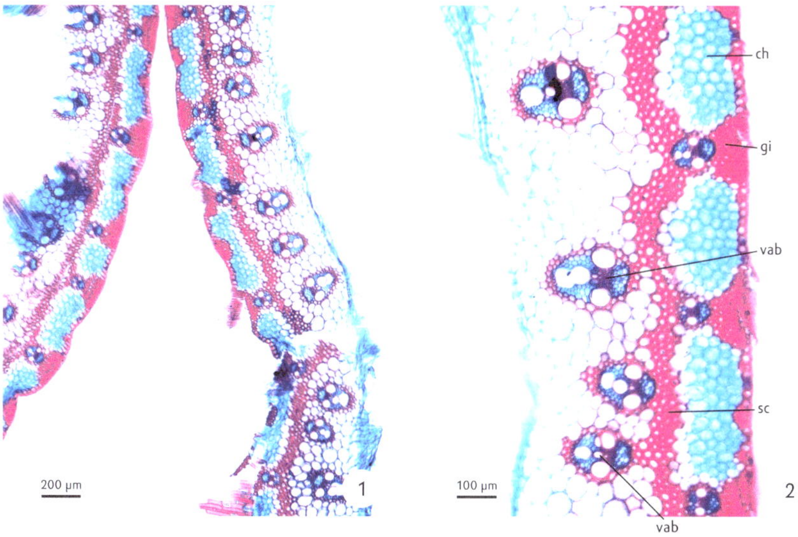

Hierochloë odorata (L.) P. Beauv.

Holy-grass

Perennial wetland, rhizomatous herb, with up to 30 cm long underground rhizoms. Stems 20–60 cm high, leaves 3–6 mm wide. Inflorescence is a panicle with patent to erecto-patent thin branches. Spikelets greenish-purple, later becoming golden-brown. Fruit is a caryopsis.

Distribution and ecology

Circumpolar boreo-montane element. Rarely in CZ and PL, absent in HU, scattered in other CE countries in open vegetation on wet sandy or peaty soils. In CE it occurs especially in fens, sedge swamps and wet meadows. Medicinal plant.

Anatomy

Circular hollow culm with a diameter of <1 mm (1).

The epidermis consists of lignified, round, laterally and externally thick-walled cells (2). The chlorenchyma dominates the culm and is composed of unlignified parenchymatic cells in large columns with irregular tangential outlines (1, 2). The thick-walled sclerenchymatic tissue is formed by conic girders below the epidermis and a wavy band inside the chlorenchyma (1, 2). Large vascular bundles consist of a xylem with 1–2 large lateral vessels, a vessel of the protoxylem and a roundish phloem with sieve tubes and companion cells (2). The bundles are externally embedded in the sclerenchymatic tissue (2). The hollow pith is surrounded by lignified thin- to thick walled polygonal parenchyma cells (1, 2).

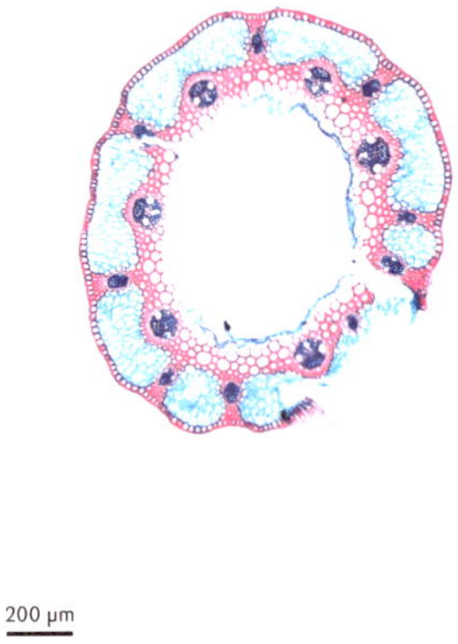

1

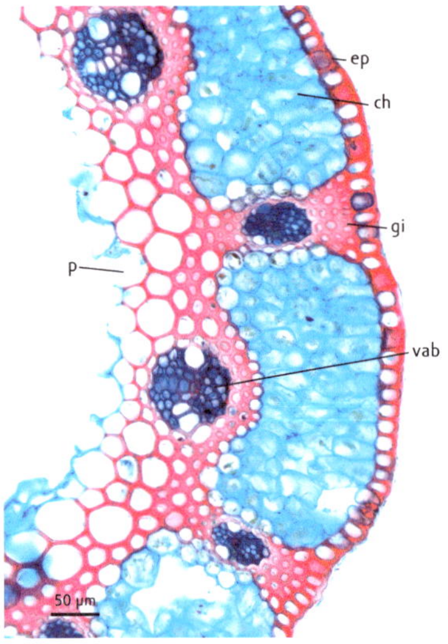

2

Perennial wetland herb, culms up to 1.2 m high. Leaves flat, pale yellowish-green, rough-edged. Sheaths inflated, often partly or even wholly enfolding the inflorescence. Inflorescence a panicle, spreading when freed from its sheath.

Distribution and ecology

European temperate element, also in C and E Asia and N America. In all CE countries on banks of stagnant and slow-flowing waters, dead arms and ditches, wet meadows. Prefers nutrient-rich muddy substrates, formerly also around the cattle-trampled margins of lakes and ponds.

Anatomy

Circular hollow culm with a diameter of 2–3 mm (1).

The epidermis consists of lignified, round, thin- to thick-walled cells (2). The chlorenchyma is composed of a belt of thin-walled unlignified parenchymatic cell groups and large intercellulars. The sclerenchymatic belt is reduced to a uniseriate hypodermis (2). The vascular bundles consist of a xylem with two large lateral vessels, a small protoxylem vessel and a round phloem with sieve tubes and companion cells (2). The bundles are surrounded by a uniseriate sheath of thin- to thick-walled cells (2). The center of the culm is surrounded by large, lignified, thin-walled parenchyma cells (1, 2).

CULM

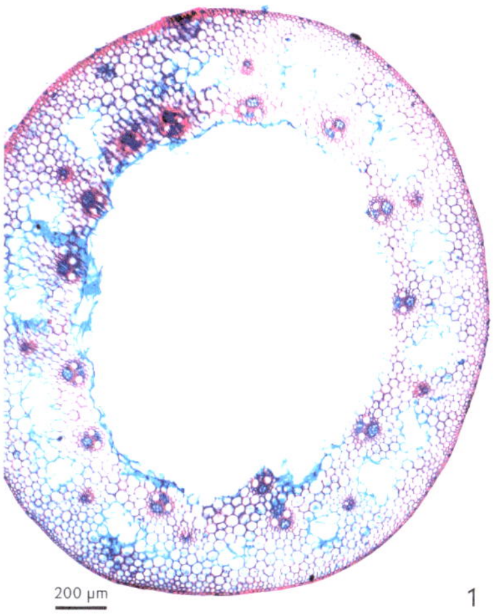

1

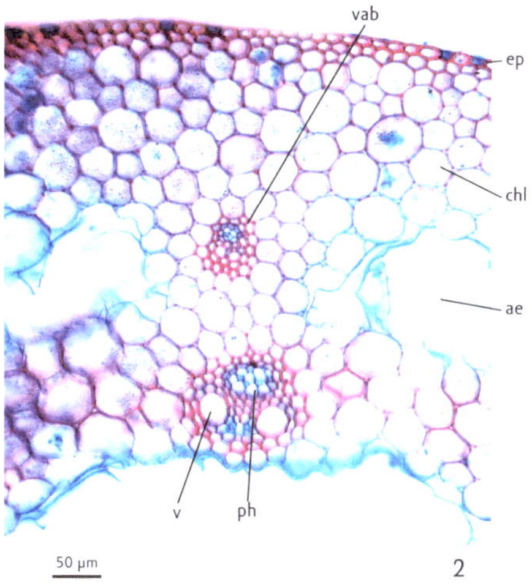

2

Leymus arenarius (L.) Hochst.

Lyme Grass

Perennial wetland herb with long rhizomes. Culm stiff, erect, 50–150 cm high, very glaucous. Leaves bluish-grey, flat, becoming inrolled, 8–20 mm wide. Inflorescence is a spike, the spikelets in pairs alternating up opposite sides of the axis.

Distribution and ecology

European boreo-arctic montane element; also in E Asia and N America. Native to the North and the Baltic Sea. It occurs in DE and PL on coastal sand dunes, sands of littoral and continental areas. An important species for the stabilisation of mobile dunes and widely planted as a sand binder.

Anatomy

Circular hollow culm with a diameter of 2 mm (1).

The epidermis consists of lignified, round, thick-walled cells (2). The chlorenchyma is almost absent (1, 2). The sclerenchymatic tissue is formed by a continuous band of thick-walled fibers (2). Large vascular bundles inside of the sclerenchyma belt consist of a xylem of two large lateral vessels, a protoxylem vessel and a roundish phloem with sieve tubes and companion cells (2). The bundles are surrounded by a thin- to thick-walled excentric sheath (2). The external part of the hollow pith consists of lignified thin-walled polygonal parenchyma cells (1, 2).

CULM

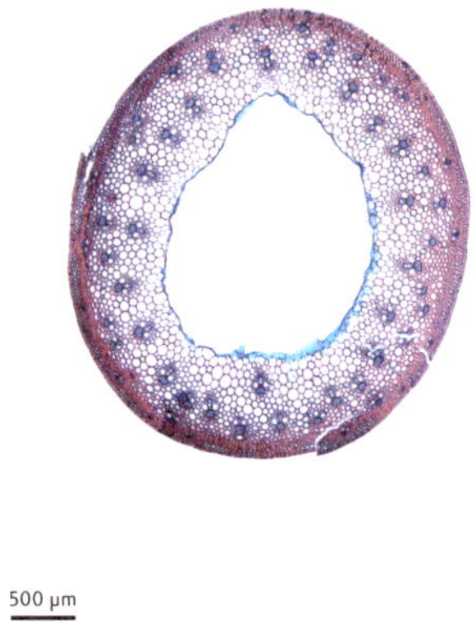

1

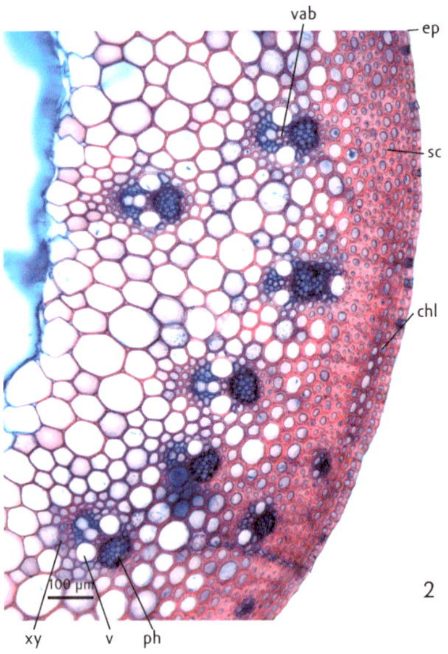

2

Molinia caerulea (L.) Moench

Purple Moor-grass

Perennial wetland herb, often forming tussocks. Culms up to 1.3 m high, erect, with one roundish node, basal article 5 cm long. Leaves flat, greyish, 3–6 mm wide; ligule a ring of hairs. Inflorescence a spike-like panicle, often purplish.

Distribution and ecology

Euro-Siberian boreo-temperate element. Common in all CE countries in wide range of habitats, especially on moist meadows, peat bogs, fens, pond-shores, swamps, always on seasonally wet ground. It is found on mildly basic to strongly acidic peats, mineral to barren soils which are permanently or seasonally wet. Dominant grass in low productive herbaceous vegetation in the absence of management.

Anatomy

Circular hollow culm with a diameter of 1 mm (1).

The epidermis consists of lignified, round, thin- to thick-walled cells (2). The chlorenchyma is composed of unlignified parenchymatic groups of cells with irregular tangential outlines (1, 2). Sclerenchymatic tissues dominate the culm. The thick-walled sclerenchymatic tissue is formed by conic girders and a 4–6-seriate band inside the chlorenchyma (1, 2). Large vascular bundles consist of a xylem with two large lateral vessels, an air space, remnants of protoxylem vessels and a roundish phloem with sieve tubes and companion cells (2). The bundles are embedded in a uniseriate sheath (2). The external part of the hollow pith consists primarily of lignified thin- to thick-walled polygonal parenchyma cells (1, 2).

CULM

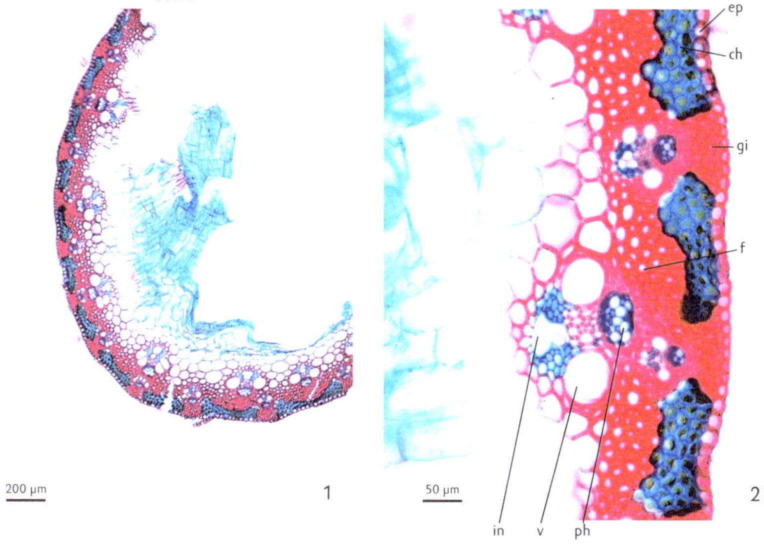

Phalaris arundinacea L.

Reed Canary-grass Poaceae

Perennial, emergent or terrestrial wetland plant. Culms up to 2 m high, with long rhizomes, sterile tillers in leaf axils. Leaves rough-edged, to 18 mm wide; ligules long. Inflorescence a compact lobed panicle, up to 30 cm long.

Distribution and ecology

Circumpolar boreo-temperate element; widely naturalized outside its native range, invasive in N America. Very common in all CE countries, from lowlands to mountains. It forms extensive monospecific stands along river banks, ditches, in wet meadows, floodplain forests, reed beds, willow carrs and at the margins of canals, reservoirs, lakes and fishponds, growing especially well where the water table fluctuates markedly. It is able to tolerate summer-dry conditions. It does not germinate in the dark.

Anatomy

Circular hollow culm with a diameter of 2–4 mm (1).

The epidermis consists of lignified, round, thin- to thick-walled cells (2). The chlorenchyma is composed of a small, unlignified continuous parenchymatic band below the hypodermis (1, 2). Sclerenchymatic tissues dominate the culm. The thin- to thick-walled sclerenchyma is formed by a large continuous peripheral zone, which contain vascular bundles (1, 2). Large vascular bundles consist of a xylem with a tangential row of vessels, a few protoxylem vessels and a roundish phloem with sieve tubes and companion cells (2). The bundles are surrounded by a 2–3-seriate lignified sheath (2). The external part of the hollow pith consists of lignified thin- to thick-walled polygonal parenchyma cells (1, 2).

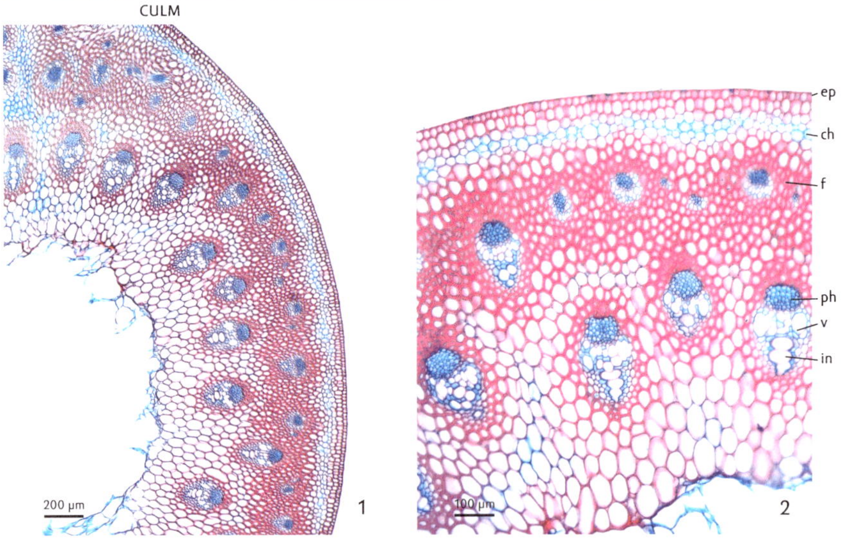

Perennial emergent herb, extensively rhizomatous, culms up to 5 m high, but also <1 m high. Not overwintering leaves, greyish-green, without ligula, only hairs present. Very large spreading panicle, usually purple, silky hairs very conspicuous in fruit.

Distribution and ecology

Circumpolar wide-temperate element. Very common in all CE countries. It forms extensive monospecific stands along water banks, in reed beds, in shallow water around lakes and ponds, also in marshes, fens, wet meadows and waterlogged woods. Pressurized ventilations of roots allow the growth in deep water. Local die-back of reed-beds noted in hyper-eutrophicated waters.

Anatomy

Circular hollow culm with a diameter of 5–10 mm (1).

The culm structure is variable along the axis. Here we describe a part in the middle of the culm but surrounded by a leaf sheath.

The epidermis consists of lignified, round, thin-to thick-walled cells (2). The chlorenchyma is consists of unlignified cells (1, 2). Large intercellulars occur between the chlorenchyma (2). The thin- to thick-walled sclerenchyma is formed by a small continuous zone centripetal of the intercellular zone (3). Large vascular bundles consist of a xylem with two large lateral vessels and a roundish phloem with sieve tubes and companion cells (3, 4). The bundles are surrounded by a uni- to biseriate lignified sheath (4). The external part of the hollow pith consists of lignified, thin- to thick-walled polygonal parenchyma cells (1, 3).

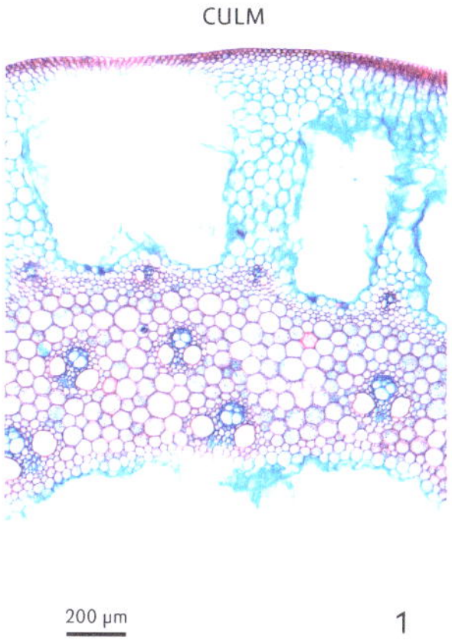

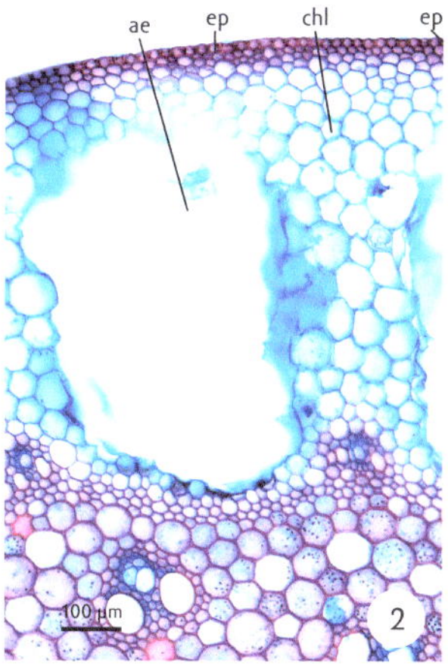

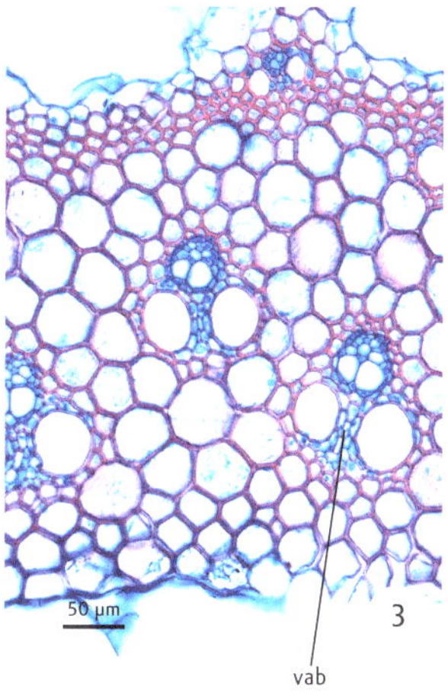

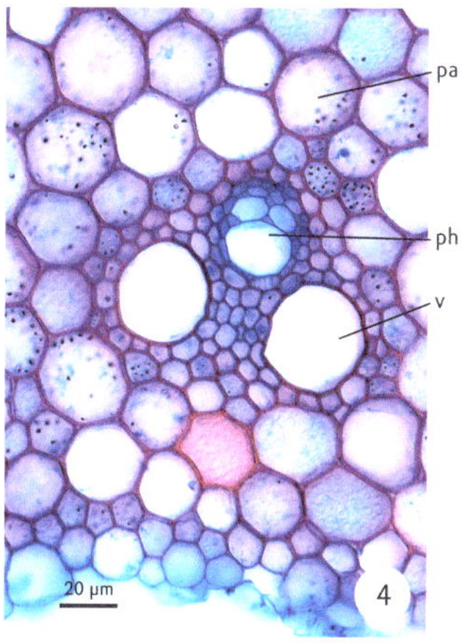

Perennial, emergent or terrestrial wetland herb, free-tufted, with short sterile tillers. Culms erect, to 1 m high, smooth. Leaves narrow, smooth, green to greyish-green; acute ligula 2–3 mm long. Inflorescence a narrowly pyramidal panicle, up to 10–25 cm long.

Distribution and ecology

A circumpolar boreo-temperate species, scattered to frequent in all CE countries. Short-lived grass growing in marshes, wet meadows and fens, at the banks of rivers, lakes and fishponds, also in ditches and in willow carrs.

Anatomy

Circular hollow culm with a diameter of 2.5 mm (1).

The epidermis consists of lignified, round, externally thick-walled cells (2). The chlorenchyma is composed of unlignified parenchymatic cells with tangential extended outlines (1, 2). Sclerenchymatic tissues dominate the culm. The thick-walled sclerenchyma is formed by square to rectangular girders and a continuous belt inside the chlorenchyma (1, 2). Large vascular bundles consist of a xylem with two large lateral vessels, a protoxylem vessel and a roundish phloem with sieve tubes and companion cells (2). The bundles are surrounded by a uniseriate sclerenchymatic sheath (2). The external part of the hollow pith consists of a large zone of unlignified polygonal parenchyma cells (1, 2).

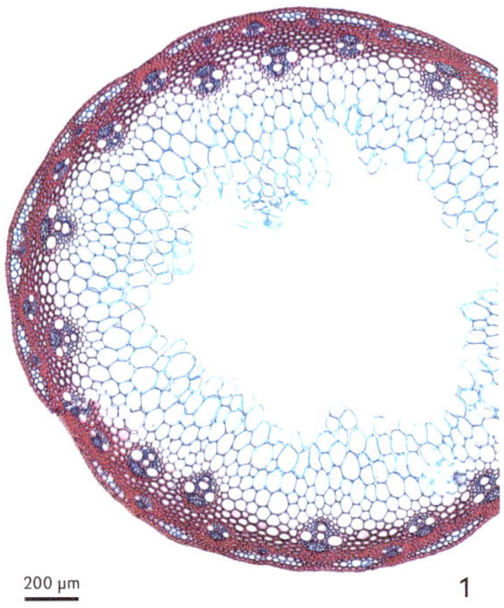

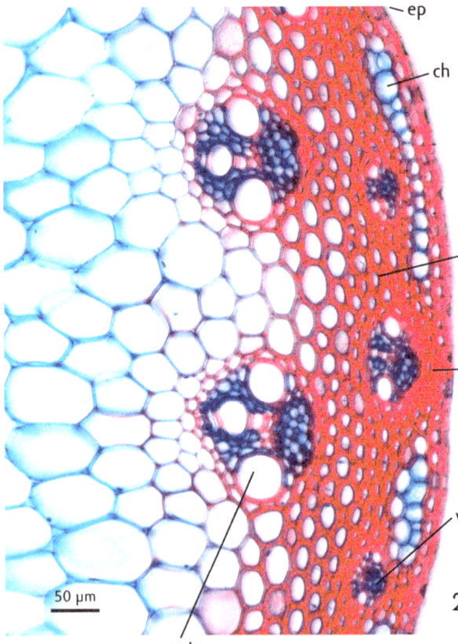

Perennial emergent rhizomatous herb. Stem erect, up to 1.2 m high. Leaves are heteromorphic, sessile linear leaves in the basis and petiolate lanceolate to cordate leaves emerge above water level. Inflorescence is a terminal raceme of purple flowers with two yellow markings. The fruit is a capsule.

Distribution and ecology

Native to both N and S America, but secondarily also in Australia and S Africa. In CE as a neophyte, rarely reported from DE, CH and CZ. It grows in variety of wetlands, especially on pond and lake margins, rich in nutrients. Seeds are dormant at the time of dispersal. It reproduces also by rhizomes, forming large clonal stands.

Anatomy

Round flower stalk with a diameter of 5–8 mm (1).

The flower stalk consists of a net-like aerenchymatic tissue with irregularly distributed vascular bundles. A cortex is absent (1). The vascular bundles consist of a centripetal air duct and a peripheral large group of angular sieve tubes with companion cells (2). Vessels are absent. Small rudimentary vascular bundles occur just inside the epidermis (1).

Round rhizome with a diameter of 10–15 mm (3).

The rhizome consists primarily of an aerenchymatic net-like tissue (3, 7). An endodermis separates the central part from the cortex (3, 4). Vascular bundles are distributed in the whole central part. They consist of a few lignified vessels, groups of sieve tubes, parenchyma cells and a central duct (5). Vessel walls have scalariform intervessel pits or ring-like structures (6). Raphids and dark staining substances occur in large cells (4, 7).

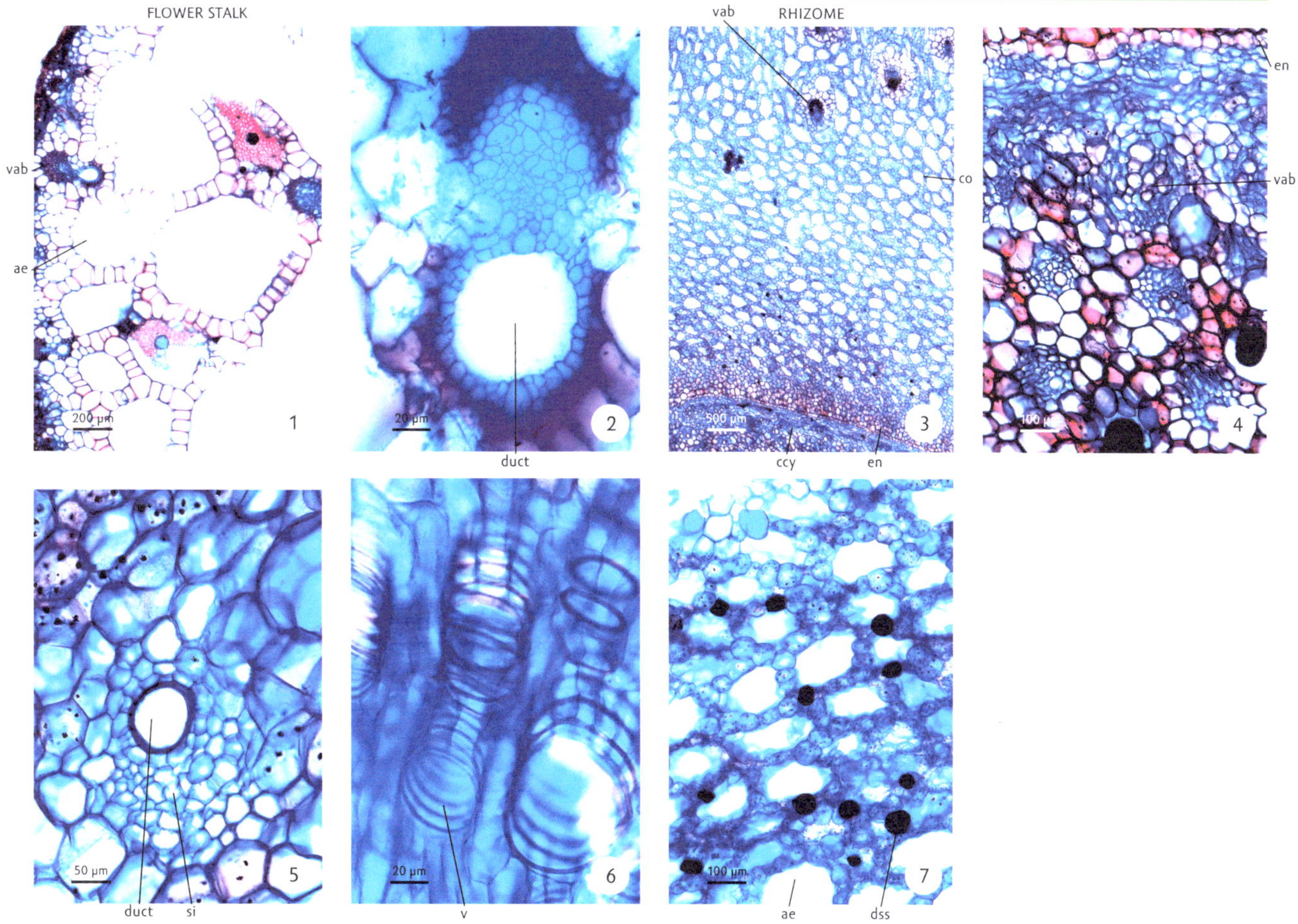

F. H. Schweingruber et al., *Anatomic Atlas of Aquatic and Wetland Plant Stems*, https://doi.org/10.1007/978-3-030-33420-8_68

Groenlandia densa (L.) Fourr. **Opposite-leaved Pondweed** Potamogetonaceae

(syn. *Potamogeton densus*)

Perennial strictly submerged plant rooted in the bottom. Shoot 0.1–0.7 m long, ovate to lanceolate, opposite leaves 8–35 mm long, unbranched roots 5–8 cm long, root hairs, growth from leaf axils. Very poor inflorescences open under water, pollen grains transferred from anthers to stigma on the surface of gas bubble evolved by stigma, fruits achenes.

Distribution and ecology

European temperate element. Subatlantic distribution, very rare spread in eastern part of CE (SK, HU, CZ, PL), more common in W part of CE (DE). Standing waters, fishponds, lakes, dam reservoirs, fast-flowing rivers and canals, usually lowlands, colder water. Stenotopic, requires transparent, alkaline hard water (pH 8–9, total alkalinity 1.5–5 meq/l), meso- to slightly eutrophic waters rich in nitrates and basic cations, organic or sandy, nutrient-rich sediments, efficient HCO_3^- user, "polar leaf" photosynthesis. Overwintering in perennial shoots and rhizomes.

Anatomy

Circular to oval stem with a diameter of 1.5–2 mm (1).

A large aerenchymatic cortex and an epidermis surround a small central cylinder (1).

Central cylinder: An area of poorly differentiated unlignified cells surrounds a central and two lateral air ducts (2). Fibers are absent (2). The endodermis consists of unlignified thin-walled round cells. (2).

Cortex: Uniseriate rows of parenchyma cells form a net-like aerenchyma. The epidermis is thin-walled (2). Vascular bundles are absent in the cortex (3).

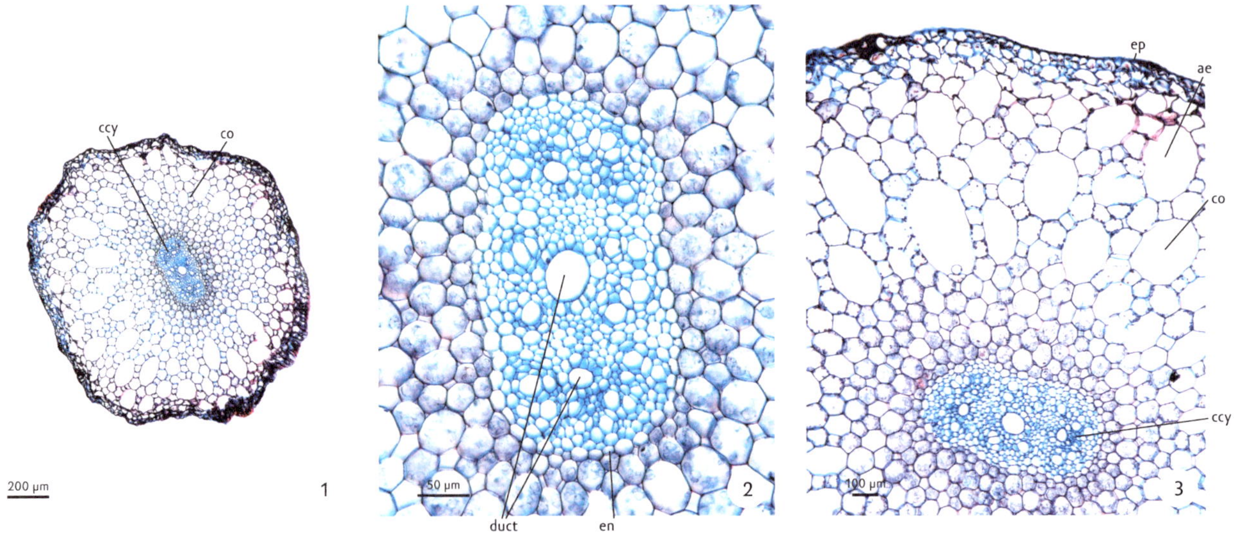

F. H. Schweingruber et al., *Anatomic Atlas of Aquatic and Wetland Plant Stems*, https://doi.org/10.1007/978-3-030-33420-8_69

Potamogeton acutifolius Link

Sharp-leaved Pondweed Potamogetonaceae

Perennial, submerged plant rooted in the bottom, overwintering only in turions. Annual shoots 0.2–1 m long, stems strongly compressed, dark green linear leaves 3.5–10 cm long, distinctly acuminate at apex, terrestrial form not known. Unbranched roots 3–6 cm long, root hairs, root growth from leaf axils. Small flower spikes open above water, round-shaped fruits achenes. Overwintering in green, flattened turions, formed at shoot apices.

Distribution and ecology

European temperate element. Scattered commonly to rarely in all 7 CE countries, from lowlands to uplands, more commonly in fishpond basins. Standing waters, fishponds, lakes, dam reservoirs, backwater pools and oxbows, fen lakes, slow-streaming rivers and canals, water depth 0.2–2 m. Tolerates slightly acidic to slightly alkaline, soft to hard water (pH 6.0–8.5, total alkalinity 0.1–5 meq/l), clear to dystrophic waters, organic or clayish bottom, prefers mesotrophic to slightly eutrophic waters, slightly acidic or alkaline soils, does not tolerate strong eutrophy, efficient HCO_3^- user, "polar leaf" photosynthesis.

Anatomy

Oval stem with a diameter of 1–1.5 mm (1).

A large aerenchymatic cortex and an epidermis surround a small central cylinder (1).

Central cylinder: An area with smaller and larger cells surrounds a central air duct (2). Larger polygonal cells represent sieve tubes (2). Fibers are absent. The endodermis consists of lignified, thick-walled U-shaped cells (2).

Cortex: Uniseriate rows of parenchyma cells form a net-like aerenchyma. Rudimentary vascular bundles occur in nodes in the middle of the net (1). These bundles consist of smaller and larger, lignified cells with secondary walls (2).

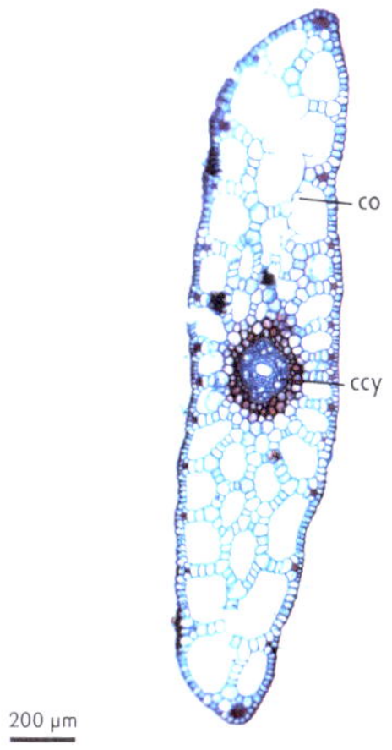

1

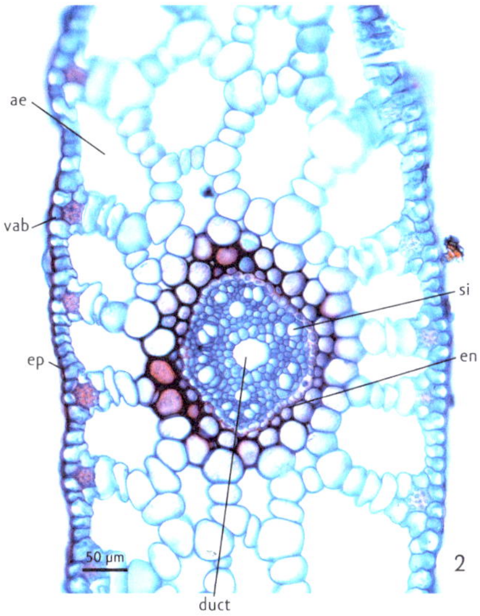

2

Potamogeton alpinus Balb. **Red Pondweed** Potamogetonaceae

Perennial floating-leaved or submerged plant rooted in the bottom. Shoot 0.2–1.3 m long, floating leathery leaves elliptic, 5–9 cm long, membraneous, translucent, elliptic-oblong submerged leaves, 8–22 cm long, often tinged reddish-brown in bright light, terrestrial form with small leathery leaves known. Unbranched roots 5–10 cm long, root hairs. Flower spikes above water, fruits achenes, very poor germination. Overwintering in perennial shoots and narrow pale turions attached to the end of rhizomes.

Distribution and ecology

Circumpolar boreo-montane element. Scattered irregularly and rarely in all 7 CE countries, rare in lowlands, more common in highlands. Standing waters, fishponds, lakes, dam reservoirs, peat bogs, fen lakes, backwater pools and oxbows, fast streaming rivers, canals, lowlands to highlands, water depth 0.1–2 m. Tolerates acidic, neutral to slightly alkaline soft to hard water (pH 5–8.5, total alkalinity 0.2–3 meq/l), clear to strongly dystrophic waters, organic, clayish or sandy bottom, does not tolerate eutrophic waters, facultative psychrophilic and sciophilous, HCO_3^- user, "polar leaf" photosynthesis.

Anatomy

Circular to oval stem with a diameter of 1–1.5 mm (1).

A large aerenchymatic cortex and an epidermis surround a small central cylinder (1).

The central cylinder is composed of several hardly differentialed vascular bundles (2). Vascular bundles consist of an air duct, lignified vessels and polygonal small sieve tubes. The endodermis consists of round cells without cell wall thickenings (2, 3).

Cortex: Uniseriate rows of parenchyma cells form a net-like aerenchyma. The epidermis is thin-walled (1). Vascular bundles are absent in the cortex (1, 4).

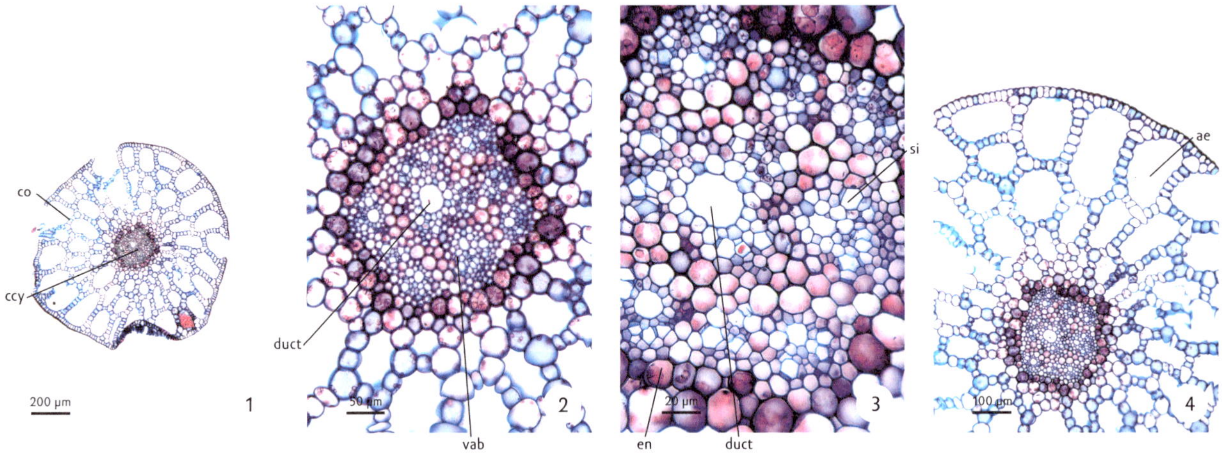

Potamogeton berchtoldii Fieber

Small Pondweed

Perennial, submerged plant rooted in the bottom, overwintering only in turions. Annual shoots 0.1–1 m long, stems very slightly compressed, narrow linear leaves 0.5–2.3 mm wide, 3–6 cm long, terrestrial form not known. Unbranched roots 3–6 cm long, root hairs, root growth from leaf axils. Small flower spikes open above water, fruits achenes. Green, flattened turions 13–24 mm long formed in shoot apices or axils.

Distribution and ecology

Circumpolar boreo-temperate element. Scattered commonly to rarely in all 7 CE countries, from lowlands to uplands, frequently in fishpond basins. Standing waters, fishponds, lakes, dam reservoirs, backwater pools and oxbows, fen lakes, slow-streaming rivers, brooks, canals, water depth 0.2–2 m. Tolerates slightly acidic to slightly alkaline, soft to hard water (pH 6.5–8.5, total alkalinity 0.3–3 meq/l), clear to slightly dystrophic waters, organic, sandy or clayish bottom. Prefers mesotrophic to slightly eutrophic waters, slightly acidic or neutral soils, does not tolerate strong eutrophy, facultative sciophyte.

Anatomy

Circular stem with a diameter of <1 mm (1).

A large aerenchymatic cortex and an epidermis surround a small central cylinder (1).

Central cylinder: A ring of poorly differentiated unlignified cells surrounds a central air duct (2). Fibers are absent. Polygonal cells represent sieve tubes (2). The endodermis consists of cells with a homogeneous cell-wall thickening (2, 3).

Cortex: Radial strips of parenchyma cells separate large air ducts. The epidermis is thin-walled (2). Vascular bundles in the cortex are absent.

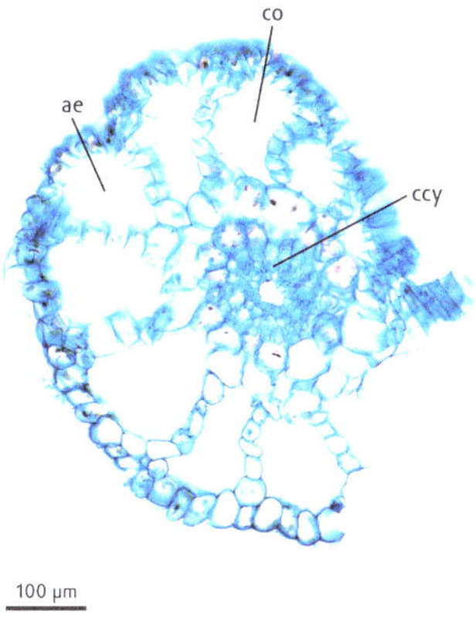

1

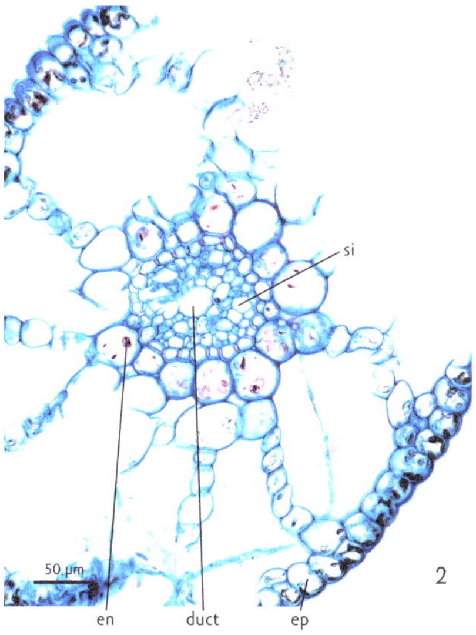

2

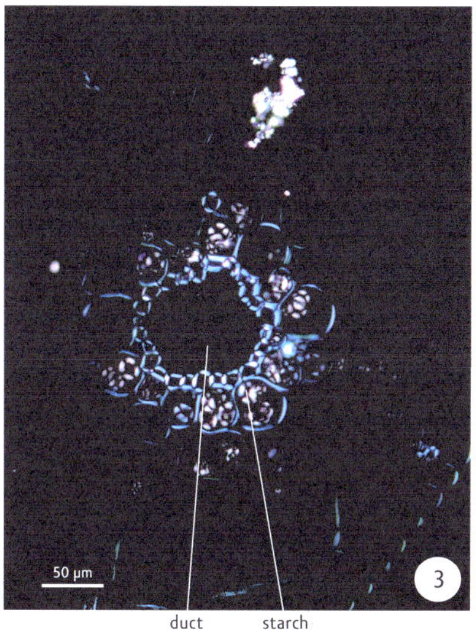

3

Potamogeton coloratus Hornem. **Fen Pondweed** Potamogetonaceae

Perennial, submerged plant with floating leaves rooted in the bottom. Shoot 0.1–0.7 m long, petiolate submerged membraneous leaves narrowly elliptic to oblanceolate, floating leaves leathery, ovate, 2–10 cm long, both leaves distinctly brownish to reddish due to anthocyanins, terrestrial form known. Unbranched roots 4–8 cm long, root hairs, root growth from leaf axils. Very thin (3–5 mm) flower spikes open above water, fruits achenes, common germination. Overwintering in shoots or rhizomes.

Distribution and ecology

European southern-temperate element. Scattered very rarely only in warm western parts of CE (DE, CH, AT, PL, CZ), only lowlands. Standing shallow dystrophic waters, fen lakes, fen pools, water depth 0.05–0.6 m. Strictly requires hard, neutral to slightly alkaline, base-rich dystrophic waters (pH 7–8.5, total alkalinity 2–5 meq/l), base-rich fen soils, prefers meso- to slightly eutrophic waters, facultative heliophyte, competitively very weak, terrestrial form very sensitive to frosts: plants lack any overwintering organ, efficient HCO_3^- user, "polar leaf" photosynthesis.

Anatomy

Circular to oval stem with a diameter of 1–1.5 mm (1).

A large aerenchymatic cortex and an epidermis surround a small central cylinder (1).

Central cylinder: Several vascular bundles are concentrically grouped (2). Each bundle is surrounded by one or two rows of fibers (2, 3) and it consists of a large centripetal air duct, a tangential row of polygonal, thin-walled, unlignified sieve tubes and a small zone of phloem (3). The endodermis surrounds the whole central cylinder (2) and consists of U-shaped, thick-walled cells (3).

Cortex: Uniseriate rows of parenchyma cells form a net-like aerenchyma. Vascular bundles are absent in the cortex (1). The epidermis is thin-walled (1).

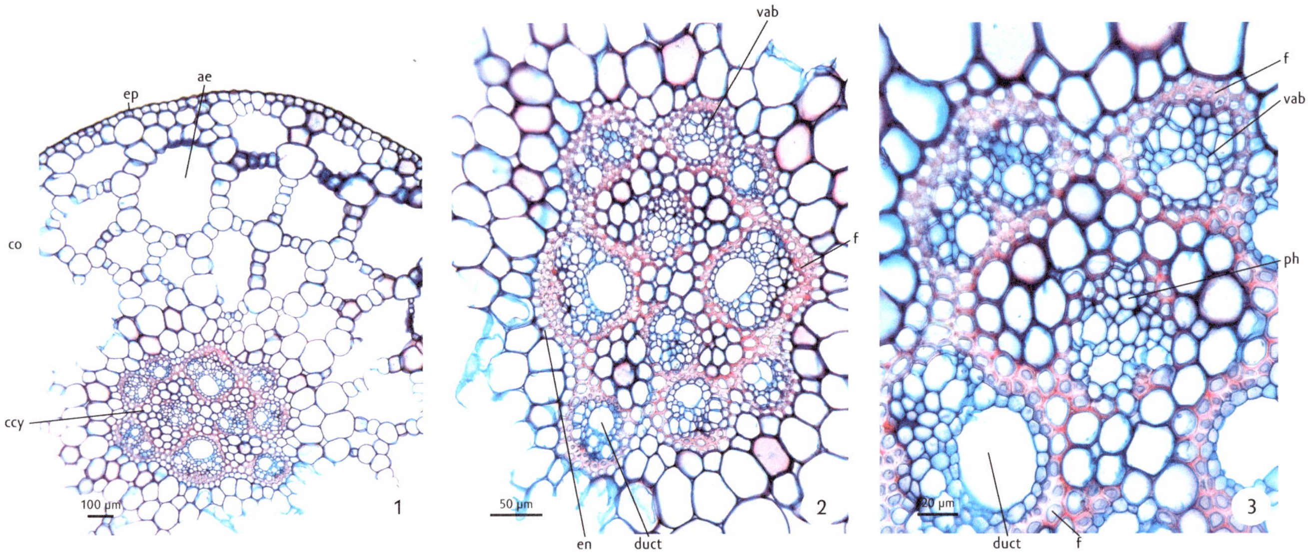

Perennial, strictly submerged plant rooted in the bottom. Annual shoot 0.2–1.5 m long, stem compressed, sessile leaves narrowly oblong-linear, minutely serrate, usually undulate, 3–10 cm long, unbranched roots 5–10 cm long, root hairs, root growth from leaf axils. Flower spikes open above water, fruits achenes, very poor germination. Overwintering or aestivation in propeller-shaped fleshy turions formed in shoot apices or leaf axils, rest of shoots decays.

Distribution and ecology

Eurasian southern-temperate element, but naturalized in N America. Scattered frequently in all 7 CE countries, from lowlands to highlands, mainly base-rich soils. Standing waters, fishponds, lakes, dam reservoirs, fen lakes, backwater pools and oxbows, sand-pits, fast streaming rivers and canals, water depth 0.2–3 m. Prefers slightly acidic to alkaline, soft to hard water (pH 6–9, total alkalinity 0.5–5 meq/l), clear to slightly dystrophic water, organic, clayish or sandy bottom, rich in nutrients, rather eurytopic species, tolerates turbid hypertrophic waters, efficient HCO_3^- user, "polar leaf" photosynthesis.

Anatomy

Circular to oval stem with a diameter of 1 mm (1).

A large aerenchymatic cortex and an epidermis surround a small central cylinder (1).

Central cylinder: An area of poorly differentiated unlignified cells surrounds a central and two lateral air ducts (2). Vessel-walls seem to be homogeneous (3). Fibers are absent (2). The endodermis consists of unlignified thin-walled round cells (2).

Cortex: Uniseriate rows of parenchyma cells form a net-like aerenchyma. The epidermis is thin-walled (2). Vascular bundles are absent in the cortex (3).

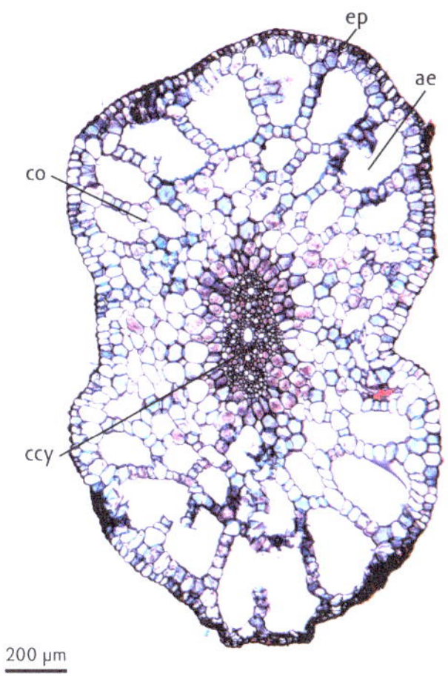

1

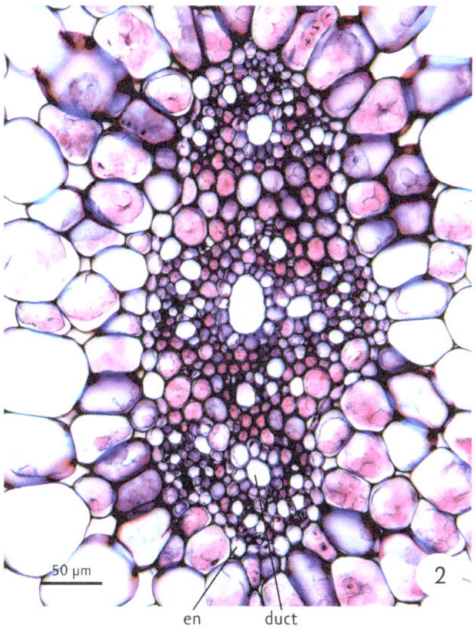

2

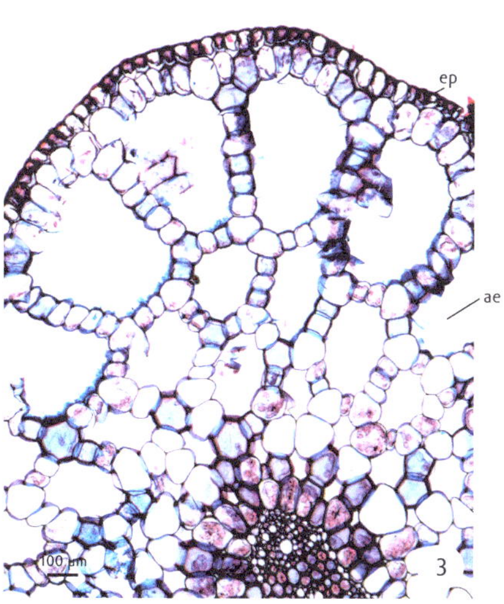

3

Perennial, submerged plant with floating leaves rooted in the bottom. Shoot 0.2–1.2 m long, submerged sessile leaves narrowly elliptic to oblanceolate, floating leaves leathery, elliptic, 2–8 cm long, terrestrial form known. Unbranched roots 4–10 cm long, root hairs, root growth from leaf axils. Flower spikes open above water, fruits achenes, very poor germination. Overwintering in shoots or pale turions, formed at apices of rhizomes.

Distribution and ecology

Circumpolar boreo-temperate element. Scattered rarely in all 7 CE countries, from lowlands to uplands. Standing waters, fishponds, lakes, dam reservoirs, backwater pools and oxbows, slow streaming rivers and canals, water depth 0.1–1 m. Prefers slightly acidic to neutral, soft to hard water (pH 6.5–8, total alkalinity 0.5–4 meq/l), organic (peaty) or clayish bottom, mesotrophic to slightly eutrophic waters, slightly acidic or neutral soils, does not tolerate eutrophic waters, efficient HCO_3- user, "polar leaf" photosynthesis.

Anatomy

Circular stem with a diameter of <1 mm.

A large aerenchymatic cortex and an epidermis surround a small central cylinder (1).

Central cylinder: An area with smaller and larger cells surrounds a central air duct (2). Larger polygonal cells represent sieve tubes. Fibers are absent (3). The endodermis consists of lignified, thick-walled U-shaped cells (2, 3).

Cortex: Uniseriate rows of parenchyma cells form a net-like aerenchyma. Rudimentary vascular bundles occur in nodes of the net near the periphery (4, 5). These bundles consist of smaller and larger, unlignified cells with secondary walls (blue light in polarized light) (4).

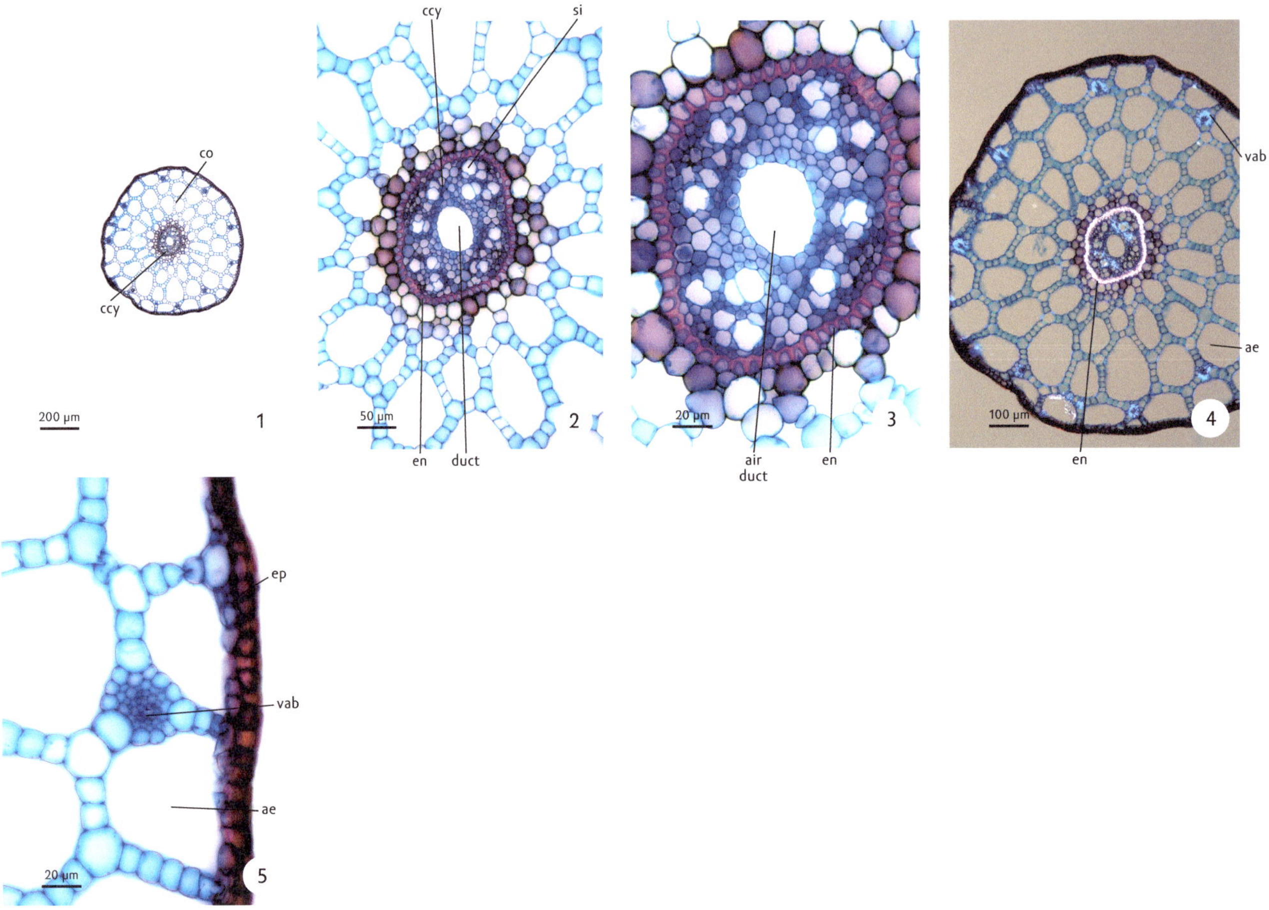

Potamogeton lucens L.

Shining Pondweed

Potamogetonaceae

Perennial, strictly submerged plant rooted in the bottom. Shoot 0.4–2.5 m long, leaves narrowly elliptic, cuspidate, 4–20 cm long, unbranched roots 5–12 cm long, root hairs, root growth from leaf axils. Flower spikes open above water, fruits achenes, very poor germination. Overwintering in perennial shoots and narrow pale turions attached to the end of rhizomes.

Distribution and ecology

Euro-Siberian temperate element. Scattered commonly in all 7 CE countries, from lowlands to highlands, mainly on base-rich soils. Standing waters, fishponds, lakes, dam reservoirs, fen lakes, backwater pools and oxbows, sand-pits, slow streaming big rivers and canals, lowlands to highlands, water depth 0.5–3 m. Prefers neutral to alkaline, medium hard to hard water (pH 6.5–9, total alkalinity 0.8–5 meq/l), clear to slightly dystrophic waters, organic, clayish or sandy bottom rich in bases, does not tolerate hypertrophic waters, efficient HCO_3- user, "polar leaf" photosynthesis.

Anatomy

Circular to oval stem with a diameter of 1–2 mm (1).

A large aerenchymatic cortex and an epidermis surround a small central cylinder (1).

Central cylinder: Vascular bundles are not separated. An area of larger and smaller cells surrounds a central and two lateral air ducts (2). Larger polygonal cells represent probably sieve tubes (3). Fibers are absent (3). The endodermis consists of lignified, thick-walled U-shaped cells (3).

Cortex: Uniseriate rows of parenchyma cells form a net-like aerenchyma. Rudimentary vascular bundles occur in nodes of the whole net (1). Fully developed vascular bundles are absent in the cortex.

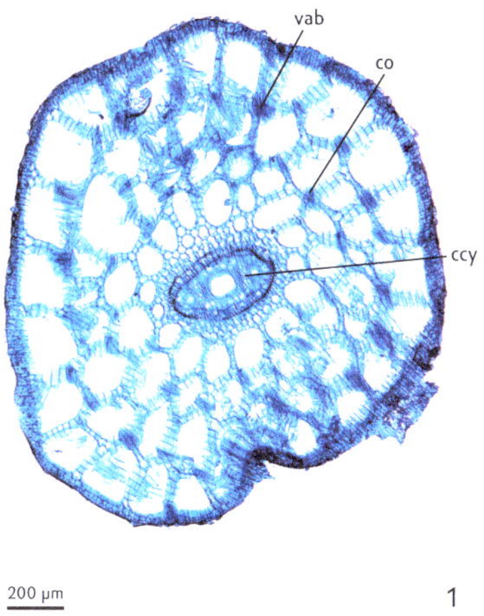

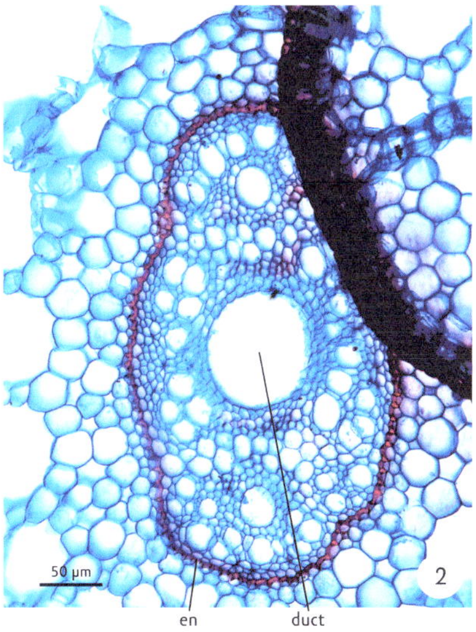

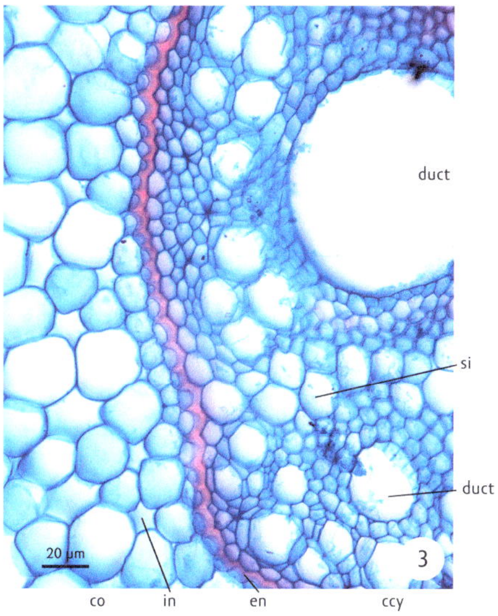

Potamogeton natans L.

Broad-leaved Pondweed

Potamogetonaceae

Perennial floating-leaved plant rooted in the bottom. Shoot 0.2–1.8 m long, floating leaves elliptic to ovate 3–12 cm long, 1.8–5 cm wide, subulate, linear submerged leaves 6–15 cm long, only 2–3.5 mm wide, terrestrial form with small leathery leaves known. Unbranched roots 5–12 cm long, root hairs, root growth from leaf axils. Flower spikes open above water, fruits achenes, poor germination. Overwintering in perennial shoots and rhizomes.

Distribution and ecology

Circumpolar boreo-temperate element. Very common in all 7 CE countries. Standing waters, fishponds, lakes, dam reservoirs, sand-pits, backwater pools and oxbows, slow-streaming rivers, canals, lowlands to highlands, water depth 0.2–2 m. Very eurytopic, tolerates strongly acidic, neutral to alkaline hard water (pH 3.8–9, total alkalinity 0–5 meq/l), clear to strongly dystrophic water, oligo- to eutrophic waters, organic, clayish or sandy bottom, strict CO_2 user.

Anatomy

Circular to oval stem with a diameter of 2–3 mm (1).

A large aerenchymatic cortex and an epidermis surround a small central cylinder (1).

Central cylinder: Several vascular bundles are concentrically grouped (2). Each bundle is surrounded by a row of parenchyma cells and consists of a large centripetal air duct, a tangential row of large, polygonal, thin-walled, unlignified sieve tubes and small vessels. Vessel walls contain ring-like thickenings (4). Small groups of fibers occur at the periphery of the vascular bundles (3). The endodermis surrounds the whole central cylinder (2) and consists of U-shaped, thick-walled cells (2, 5).

Cortex: Uniseriate rows of parenchyma cells form a net-like aerenchyma. Rudimentary vascular bundles (7) occur in nodes of the whole net (6). The epidermis is thin-walled (1).

Petiole: The outline of petioles is boat-like (8). A few collateral vascular bundles are arranged in a line (8).

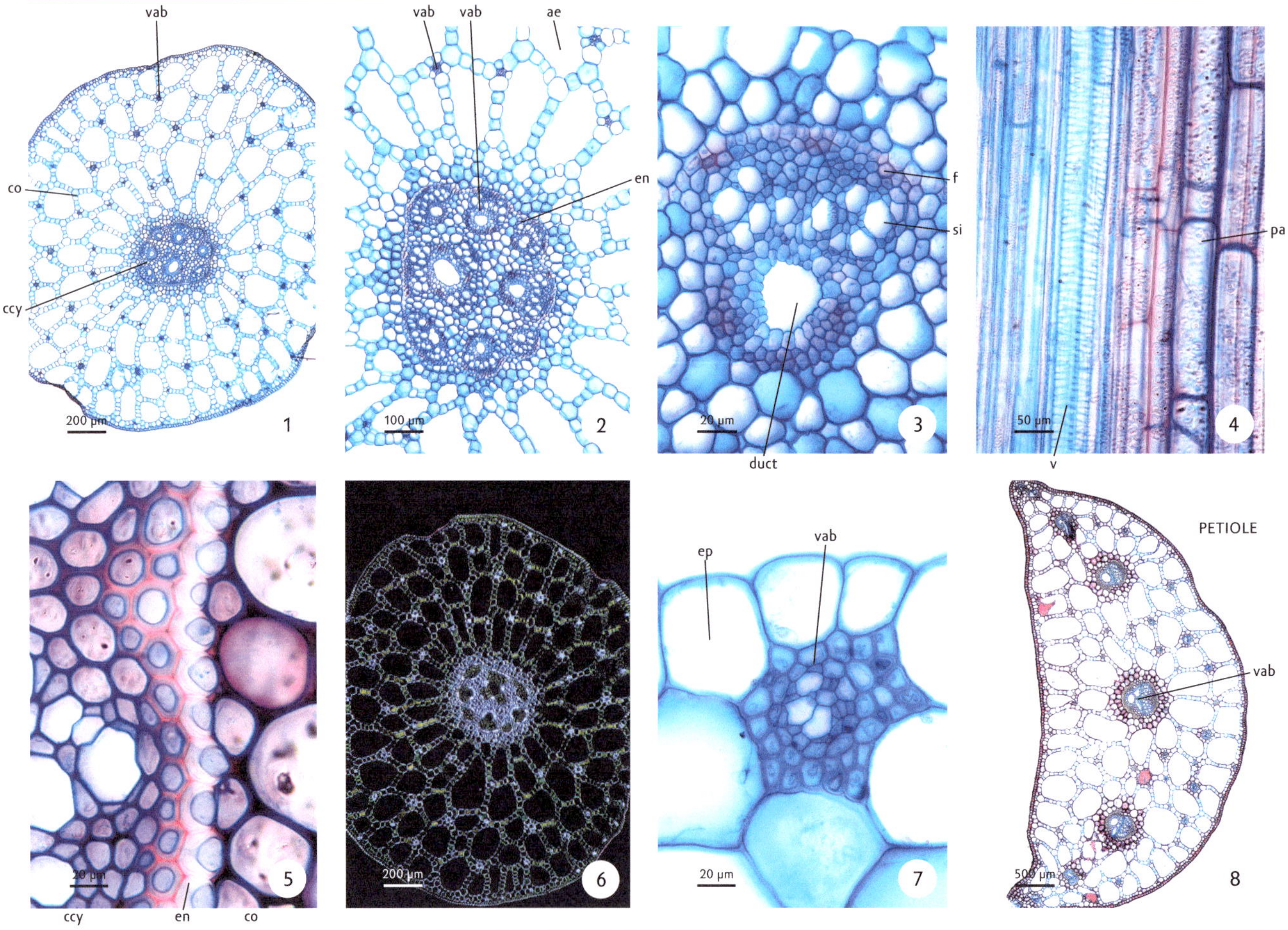

Potamogeton nodosus Poir.

Loddon Pondweed

Potamogetonaceae

Perennial, submerged plant rooted in the bottom. Perennial shoot 0.2–2 m long, petiolate submerged membraneous leaves narrowly elliptic to oblanceolate, floating leaves leathery, elliptic to narrowly so, 6–15 cm long. Unbranched roots 5–10 cm long, root hairs, root growth from leaf axils. Flower spikes open above water, fruits achenes, very poor germination. Overwintering in shoots or pale turions formed at apices of rhizomes.

Distribution

Circumpolar southern-temperate element. Scattered rarely in all 7 CE countries, from lowlands to uplands. Standing waters, lakes, dam reservoirs, backwater pools and oxbows, sand-pits, slow-streaming rivers and canals, lowlands to highlands, water depth 0.4–3 m. Prefers neutral to slightly alkaline, soft to hard water (pH 7–8.5, total alkalinity 0.5–4 meq/l), clear, transparent water, organic, clayish or sandy bottom, prefers mesotrophic water, does not tolerate eutrophic water, efficient HCO_3^- user, "polar leaf" photosynthesis.

Anatomy

Circular to oval stem with a diameter of 1.5–2 mm (1).

A large aerenchymatic cortex and an epidermis surround a small central cylinder (1).

Central cylinder: A few vascular bundles are concentrically grouped (2). Each bundle is surrounded by a few continuous rows of parenchyma discontinuous rows of fibers (2). The bundle consists of a large centripetal air duct, a tangential row of polygonal, thin-walled, unlignified sieve tubes. The endodermis surrounds the whole central cylinder (2) and consists of U-shaped, thick-walled cells (3).

Cortex: Uniseriate rows of parenchyma cells form a net-like aerenchyma. Rudimentary vascular bundles (4) occur in nodes of the whole net (4).

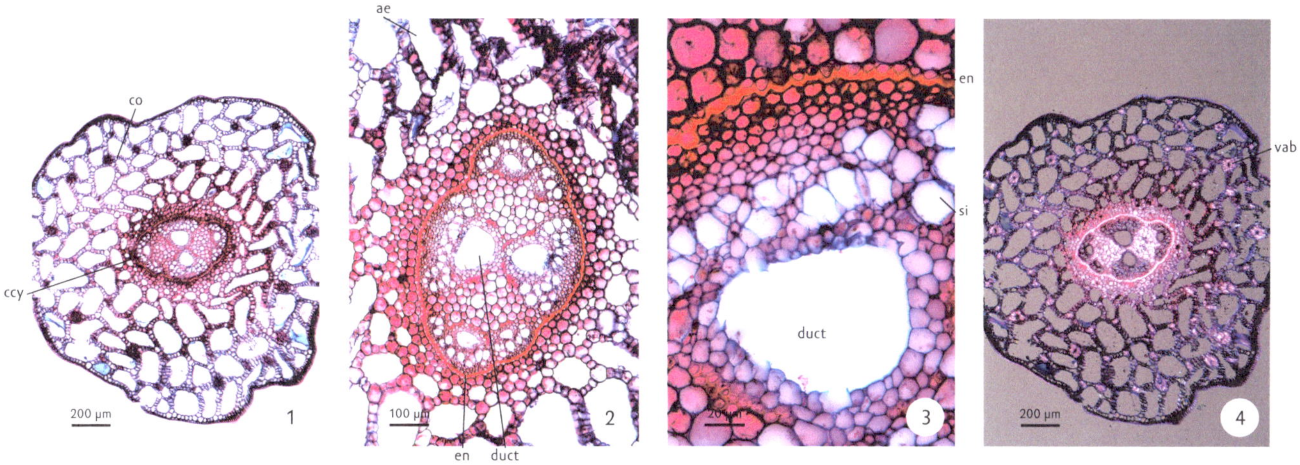

Potamogeton obtusifolius Mert. & W. D. J. Koch

Potamogetonaceae

Blunt-leaved pondweed

Perennial, submerged plant rooted in the bottom, overwintering only in turions. Annual shoots 0.1–0.8 m long, stems slightly compressed, linear leaves 3–9 cm long, obtuse, apiculate at apex, waxy, brown-reddish tinge, terrestrial form not known. Unbranched roots 3–6 cm long, root hairs, root growth from leaf axils. Small flower spikes open above water, fruits achenes. Overwintering in green, flattened turions formed at shoot apices.

Distribution and ecology

Circumpolar boreo-temperate element, with a disjunct distribution. Scattered commonly to rarely in all 7 CE countries, from lowlands to uplands, more commonly in fishpond basins. Standing waters, fishponds, lakes, backwater pools and oxbows, bog pools, fen lakes, slow-streaming rivers and canals, water depth 0.2–1 m. Tolerates slightly acidic to slightly alkaline, soft to hard water (pH 6.0–8.5, total alkalinity 0.1–4 meq/l), clear to dystrophic waters, organic (peaty) or clayish bottom, prefers mesotrophic to slightly eutrophic waters, slightly acidic or alkaline soils, does not tolerate strong eutrophy, efficient HCO_3^- user, "polar leaf" photosynthesis.

Anatomy

Oval stem with a diameter of 1.5–2 mm (1).

Central cylinder: Vascular bundles are anatomically not separated. Around a central and two lateral air ducts (2) are larger polygonal thin-walled sieve-tubes. Vessels are anatomically not recognizable (2). Fibers are absent (2). The endodermis consists of lignified, thick-walled cells (2).

Cortex: Uniseriate rows of parenchyma cells form a net-like aerenchyma (3). Rudimentary vascular bundles or just a few fibers occur in the hypodermis (1, 4). The epidermis is thin-walled (4).

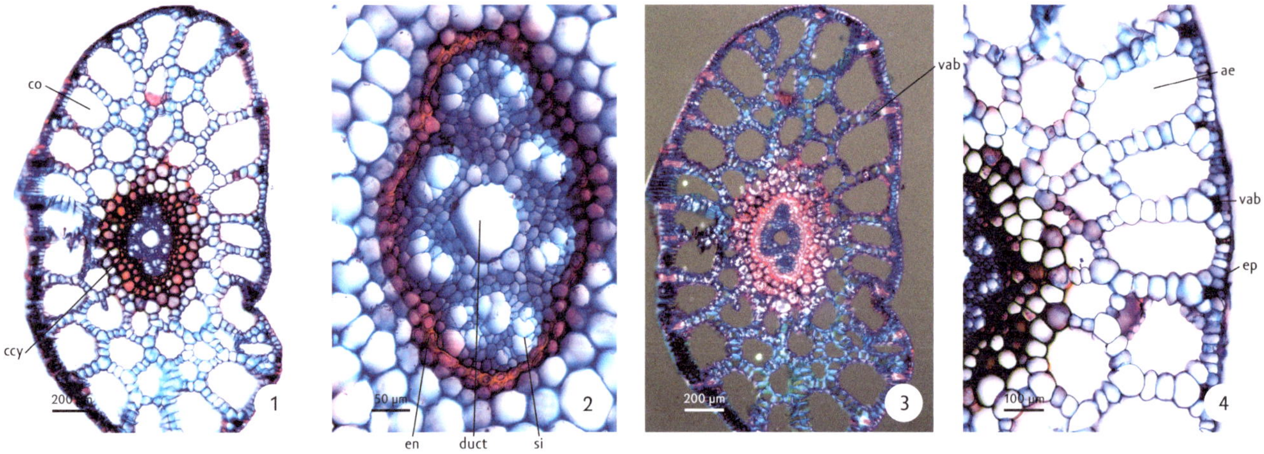

Perennial, strictly submerged plant rooted in the bottom. Perennial shoot 0.3–3 m long, opposite leaves with distinct parallel ribs, narrowly oblong-ovate or cordate, stem-clasping, obtuse, rarely undulate, 2–8 cm long. Unbranched roots 5–10 cm long, root hairs, root growth from leaf axils. Flower spikes open above water, fruits achenes, very poor germination. Overwintering in shoots or pale turions formed at apices of rhizomes

Distribution and ecology

Circumpolar boreo-temperate element, with a disjunct distribution. Scattered rarely to commonly in all 7 CE countries, from lowlands to uplands. Standing waters, fishponds, lakes, dam reservoirs, backwater pools and oxbows, sand-pits, slow-streaming rivers and canals, lowlands to highlands, water depth 0.4–3 m. Prefers neutral to slightly alkaline, soft to hard water (pH 7–9, total alkalinity 0.5–4 meq/l), clear, transparent water, organic, clayish or sandy bottom, prefers mesotrophic, does not tolerate eutrophic waters, efficient HCO_3^- user, "polar leaf" photosynthesis.

Anatomy

Circular stem with a diameter of 2–2.5 mm (1).

A large aerenchymatic cortex and an epidermis surround a small central cylinder (1).

Central cylinder: An area with smaller and larger cells surrounds a central air duct (2). Larger polygonal cells represents sieve tubes and round large cells parenchyma cells. Fibers are absent (3). Vessel-walls are not structured (4). The endodermis consists of round cells with lateral lignified thick walls (3).

Cortex: Uniseriate rows of parenchyma cells form a net-like aerenchyma. Rudimentary vascular bundles occur in nodes of the net near the periphery. The external wall of the epidermis is thickened (1).

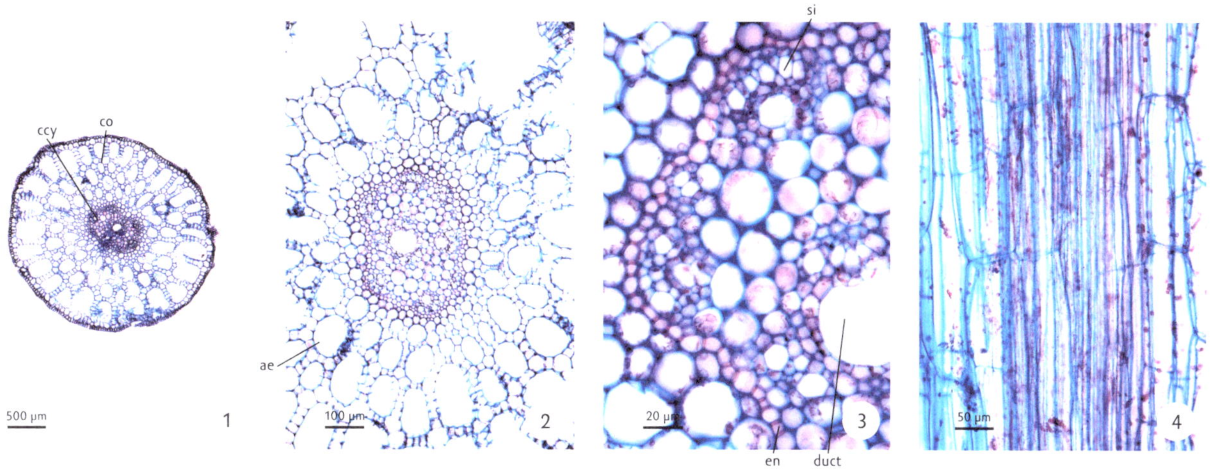

Potamogeton polygonifolius Pourr. **Bog Pondweed** Potamogetonaceae

Perennial, submerged plant with floating leaves rooted in the bottom. Perennial shoots 0.1–0.7 m long, petiolate submerged membraneous leaves narrowly elliptic to oblanceolate, floating leaves leathery, elliptic, 2–8 cm long, terrestrial form known. Unbranched roots 4–8 cm long, root hairs, root growth from leaf axils. Flower spikes open above water, fruits achenes, very poor germination. Overwintering in shoots or pale turions formed at apices of rhizomes.

Distribution and ecology

Suboceanic temperate element; also in N. America. Scattered very rarely only in western part of CE (DE, CH, AT, PL, CZ), from lowlands to uplands. Standing waters, peaty lakes or fishponds, bog or fen pools, dystrophic brooks, water depth 0.1–0.6 m, requires acidic, soft dystrophic waters (pH 5–7, total alkalinity 0–0.5 meq/l), acidic or peaty soils, prefers oligo-mesotrophic, does not tolerate eutrophic waters, probably a strict CO_2 user.

Anatomy

Circular to oval stem with a diameter of 1–1.5 mm (1).

A large aerenchymatic cortex and an epidermis surround a small central cylinder (1).

Central cylinder: Several vascular bundles are concentrically grouped (2). Each bundle is partially surrounded by a row of fibers (2, 3) and it consists of a large centripetal air duct, a tangential row of polygonal, thin-walled, unlignified sieve tubes and a small zone of phloem (3). The endodermis surrounds the whole central cylinder (2) and consists of U-shaped, thick-walled cells (3).

Cortex: Uniseriate rows of parenchyma cells form a net-like aerenchyma. Rudimentary vascular bundles occur in nodes in the hypodermis (5). The epidermis is thin-walled (4).

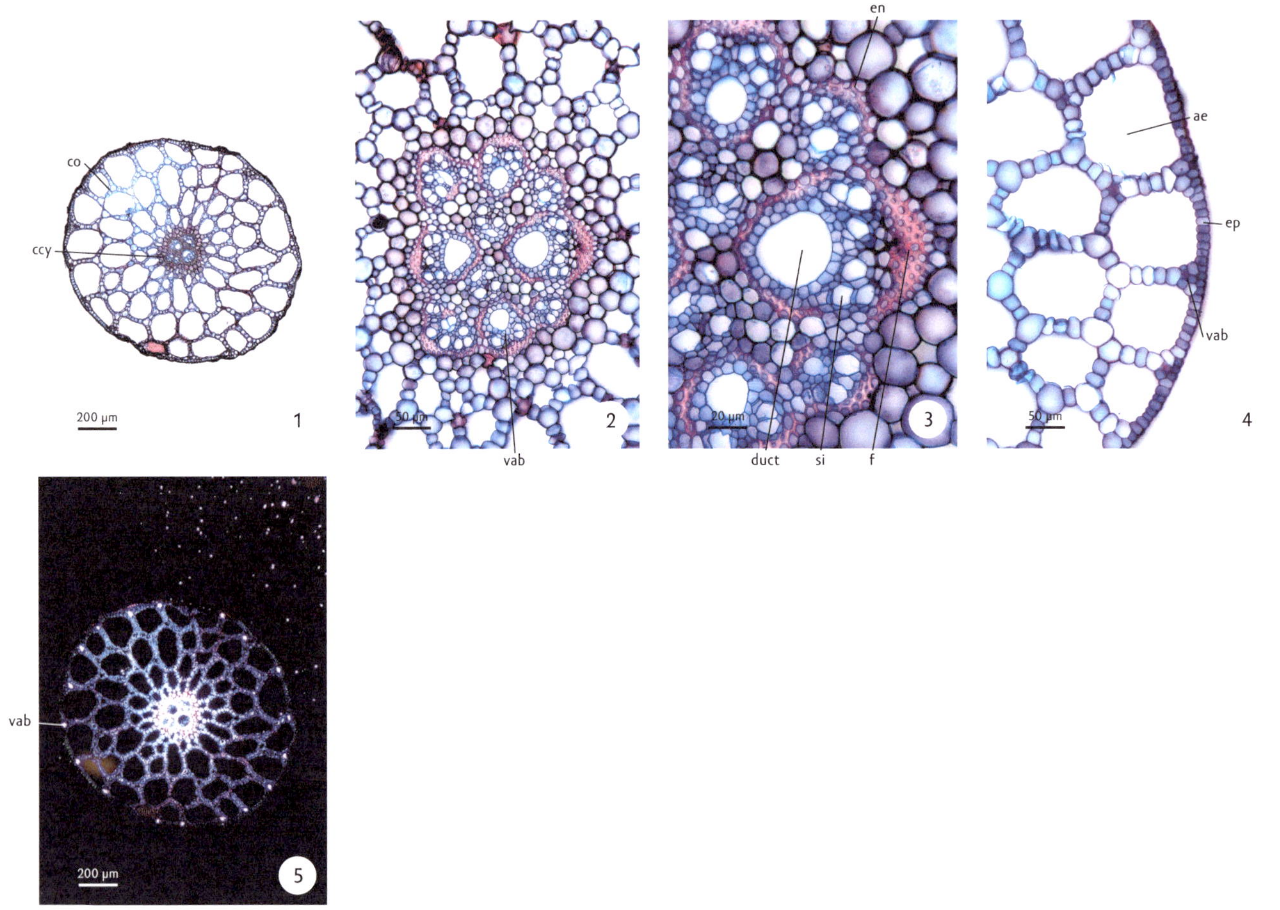

Potamogeton praelongus Wulfen **Long-stalked Pondweed** Potamogetonaceae

Perennial strictly submerged plant rooted in the bottom. Shoot 0.3–2 m long, stems usually curved in zig-zag nodes. Oblong-ovate, dark green leaves 5–20 cm long, distinct parallel ribs, unbranched roots 5–12 cm long, root hairs, root growth from leaf axils. Flower spikes open above water, fruits achenes, germinate very poorly. Overwintering in perennial shoots and narrow, banana-shaped pale turions, formed at the apex of rhizomes.

Distribution and ecology

Circumpolar boreal-montane element. Scattered very rarely in W part of CE but lacking in eastern areas (SK, HU). Standing waters, fishponds, lakes, dam reservoirs, backwater pools and oxbows, rivers, streams and canals, in lowlands and uplands, always in deep colder water, prefers partly shaded habitats. Very stenotopic, requires deep, transparent, neutral to alkaline hard water (pH 7–9, total alkalinity 0.8–4 meq/l) rich in basic cations, meso- to slightly eutrophic waters, organic, clayish or sandy, nutrient-rich sediments, efficient HCO_3^- user, "polar leaf" photosynthesis.

Anatomy

Circular to oval stem with a diameter of 2–2.5 mm (1).

A large aerenchymatic cortex and an epidermis surround a small central cylinder (1).

Central cylinder: Several vascular bundles are concentrically grouped (2). Each bundle is partially surrounded by a row of thin-walled slightly lignified vessels (2, 3) and polygonal, thin-walled, unlignified sieve tubes. The endodermis surrounds the whole central cylinder (2) and consists of U-shaped, not very thick-walled cells (3).

Cortex: Uniseriate rows of parenchyma cells form a net-like aerenchyma (1). Rudimentary vascular bundles occur in most nodes of the net (4, 5). The epidermis is thin-walled (1).

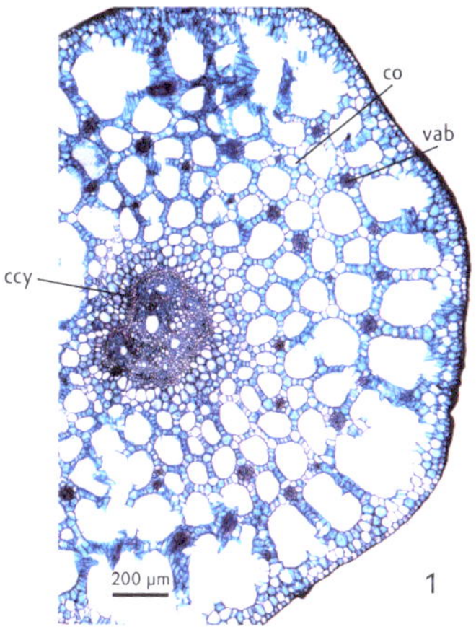

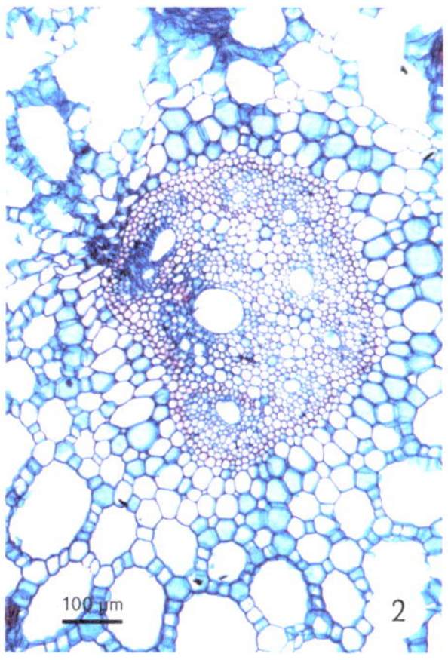

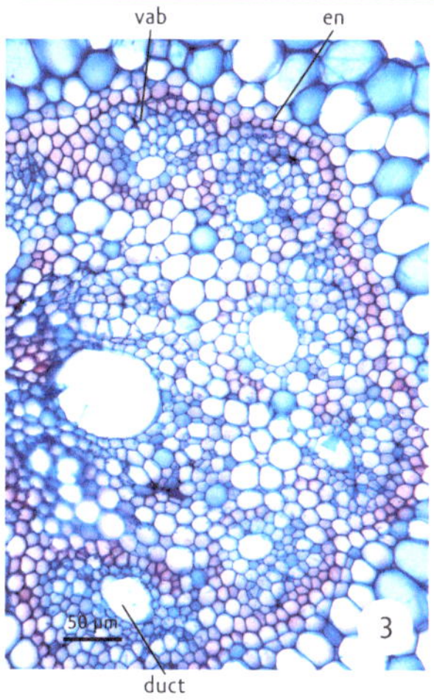

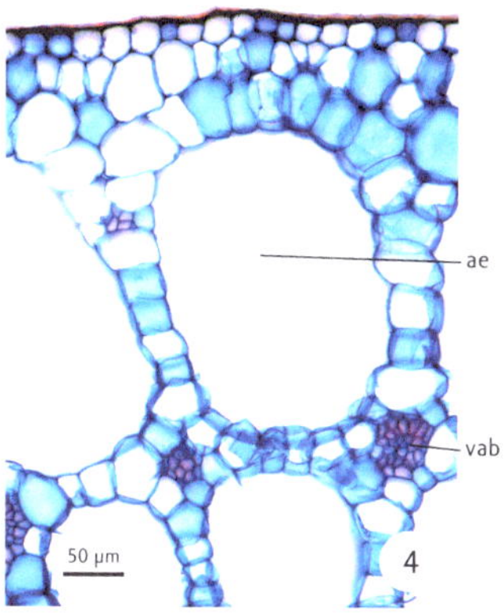

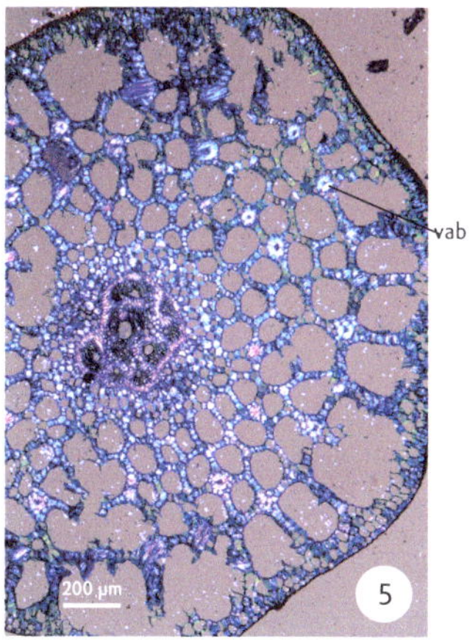

Potamogeton pusillus L.

Lesser Pondweed

Perennial, submerged plant rooted in the bottom, overwintering only in turions. Annual shoots 0.1–1 m long, stems slightly compressed, narrow linear leaves 2–6 cm long, terrestrial form not known. Unbranched roots 3–6 cm long, root hairs, root growth from leaf axils. Small flower spikes open above water, fruits achenes. Green, flattened turions 9–15 mm long formed in shoot axils.

Distribution and ecology

Circumpolar southern-temperate element. Scattered commonly to frequently in all 7 CE countries, from lowlands to uplands, frequently in fishpond basins in base-rich soils. Standing waters, fishponds, lakes, dam reservoirs, sand-pits, backwater pools and oxbows, fen lakes, slow-streaming rivers and canals, water depth 0.2–2 m. Tolerates slightly acidic to strongly alkaline, soft to hard water (pH 6.5–10, total alkalinity 0.5–5 meq/l), clear to slightly dystrophic waters, organic, sandy or clayish bottom, prefers mesotrophic to eutrophic waters, slightly acidic or alkaline soils, tolerates strongly eutrophic waters, efficient HCO_3^- user, "polar leaf" photosynthesis.

Anatomy

Circular stem with a diameter of 0.5–1 mm (1).

A large aerenchymatic cortex and an epidermis surround a small central cylinder (1).

Central cylinder: An area with smaller and larger cells surrounds a central air duct (2). Larger polygonal cells represent sieve tubes. Vascular bundles and fibers are absent (2, 3). The endodermis consists of lignified, thick-walled U-shaped cells (3).

Cortex: Uniseriate rows of parenchyma cells form a net-like aerenchyma. The epidermis is thin-walled. Rudimentary vascular bundles occur in nodes in the hypodermis.

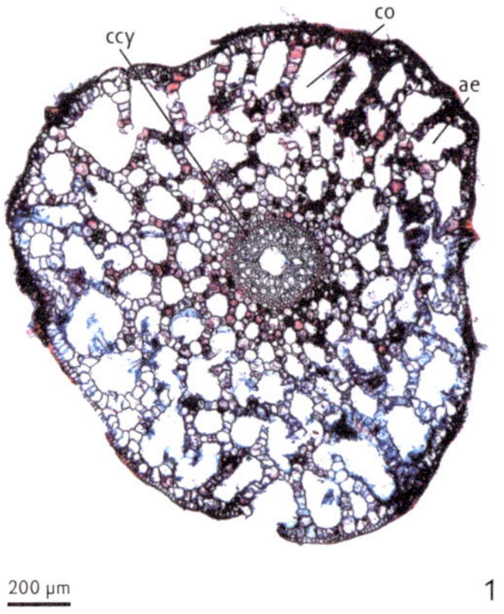

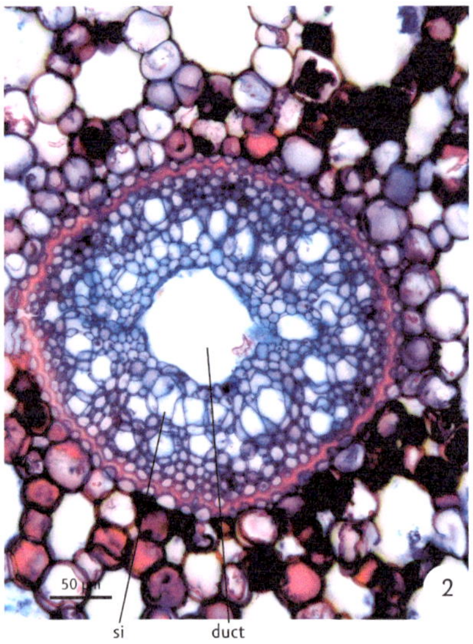

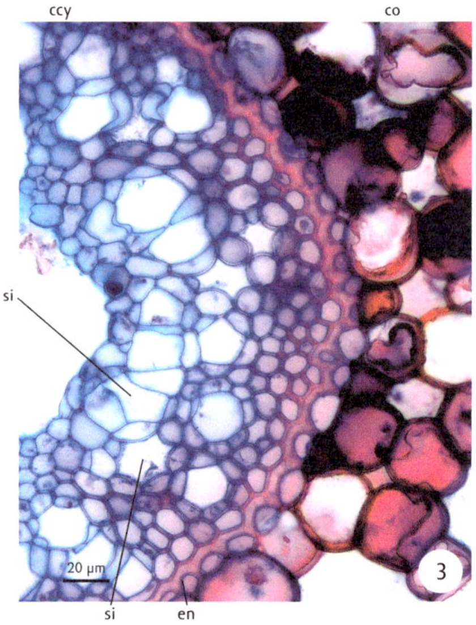

Potamogeton trichoides Cham. et Schltdl. **Hairlike Pondweed** Potamogetonaceae

Perennial, submerged plant rooted in the bottom, overwintering only in turions. Annual shoots 0.2–1 m long, stems not compressed, narrow linear leaves 2–6 cm long, terrestrial form not known. Unbranched roots 3–6 cm long, root hairs, root growth from leaf axils. Small flower spikes open above water, fruits achenes. Green, flattened turions formed at shoot apex.

Distribution and ecology

Euro-Siberian southern-temperate element. Scattered commonly in all 7 CE countries, from lowlands to highlands, more commonly in fishpond basins. Standing waters, fishponds, lakes, dam reservoirs, backwater pools and oxbows, fen lakes, slow-streaming rivers and canals, water depth 0.2–2 m. Tolerates slightly acidic to strongly alkaline, soft to hard water (pH 6.5–10, total alkalinity 0.5–5 meq/l), clear to slightly dystrophic waters, organic, sandy or clayish bottom, prefers mesotrophic to eutrophic waters, slightly acidic or alkaline soils, tolerates strongly eutrophic waters, efficient HCO_3- user, "polar leaf" photosynthesis.

Anatomy

Circular stem with a diameter of <1 mm (1).

A large aerenchymatic cortex and an epidermis surround a small central cylinder (1).

Central cylinder: An area with smaller and larger cells surrounds a central air duct (2). Larger polygonal cells represent sieve tubes (3). Fibers are absent (3). The endodermis consists of lignified, thick-walled cells (2).

Cortex: Uniseriate rows of parenchyma cells form two circles of a net-like aerenchyma (1). Rudimentary vascular bundles occur in the hypodermis (1). These bundles consist of smaller and larger, unlignified cells with secondary walls.

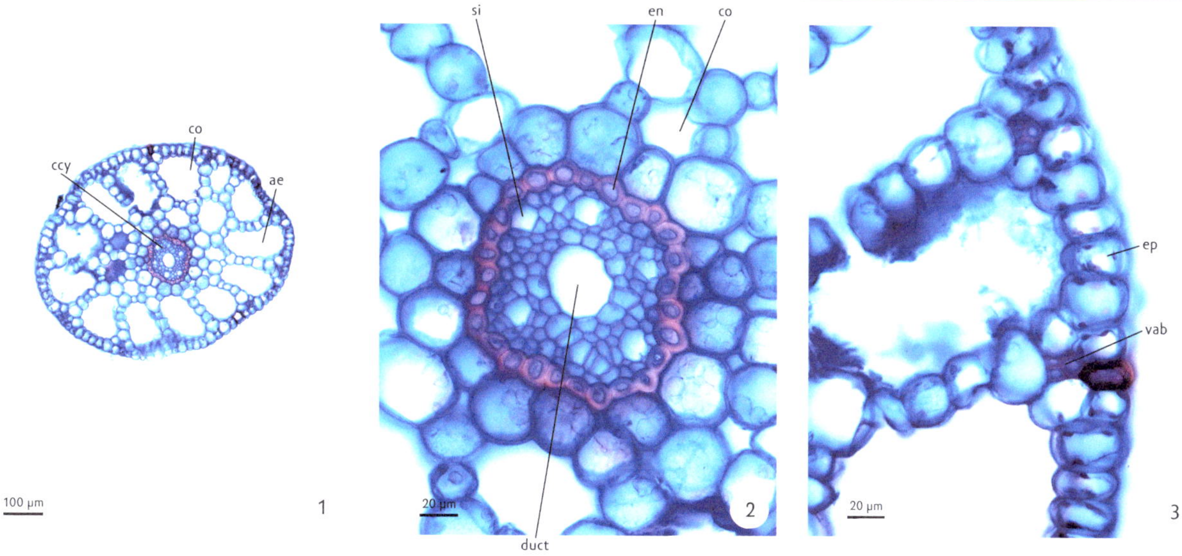

Stuckenia pectinata (L.) Börner

(syn. *Potamogeton pectinatus*)

Fennel pondweed Potamogetonaceae

Perennial, submerged plant rooted in the bottom, overwintering only in small belowground, starchy rhizome tubers. Annual shoots 0.2–2 m long, stems not compressed, younger stems zig-zag curved in nodes. Terrestrial form not known. Narrowly linear to filamentous sessile leaves 4–12 cm long, elliptic in cross-section, with distinct leaf sheath, stems and leaves often with reddish tinge. Unbranched roots 3–6 cm long, root hairs, root growth from leaf axils. Small flower spikes open, pollinate and ripen under water, fruits achenes.

Distribution and ecology

Circumpolar wide-temperate element. Scattered commonly to frequently in all 7 CE countries, from lowlands to uplands, frequently in fishpond basins in base-rich soils. Standing waters, fishponds, lakes, dam reservoirs, sand-pits, backwater pools and oxbows, fen lakes, rivers, canals, water depth 0.2–3 m. Tolerates slightly acidic to strongly alkaline, soft to hard water (pH 6.5–10, total alkalinity 0.5–5 meq/l), clear to slightly dystrophic waters, organic, sandy or clayish bottom, prefers mesotrophic to eutrophic waters, neutral or alkaline soils, tolerates strongly eutrophic, slightly brackish waters, efficient HCO_3- user.

Anatomy

Anatomy of the annual shoot

Circular to oval stem with a diameter of 1 mm.

A large aerenchymatic cortex and an epidermis surround a small central cylinder (1).

Central cylinder: An area of poorly differentiated unlignified cells surrounds a central air duct (2). Larger polygonal cells represent sieve tubes. Fibers are absent (2). The endodermis consists of lignified, thick-walled U-shaped cells (2, 3).

Cortex: Uniseriate rows of parenchyma cells form a net-like aerenchyma. The epidermis is thin-walled (1). Vascular bundles are absent in the cortex (4).

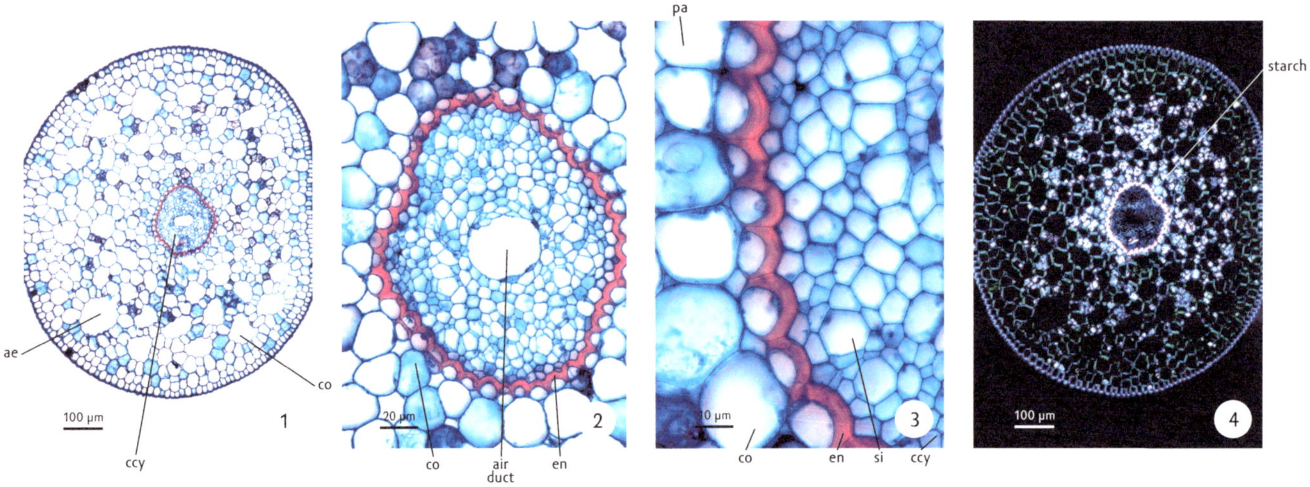

Zannichellia palustris L. **Horned Pondweed** Potamogetonaceae

Annual submerged plant rooted in the bottom. Leaves alternate along vegetative shoots, seemingly opposite along fertile shoots, leaves very narrow, 2–10 cm long. Thin erect shoots 5–40 cm high, unbranched, weak roots with root hairs 3–6 cm long, also in leaf axils in shoot bases. Monoecious plant: solitary achlamydeous flowers formed in leaf axils, fruits banana-shaped achenes 4–6 mm long.

Distribution and ecology

Circumpolar southern-temperate element. Scattered commonly in all 7 CE countries in lowlands or uplands in freshwaters and slightly brackish waters. Standing waters in fishponds, lakes, dam reservoirs, backwater pools and oxbows of large rivers, slow-streaming rivers and canals. Tolerates neutral to alkaline, medium to strongly hard or slightly brackish water (pH 7–9, total alkalinity 0.8–5 meq/l), prefers harder waters, meso- to hypertrophic waters, organic, nutrient-rich sediments, also sand. Water depth 0.05–2 m, distribution broadened as a result of fishpond liming and seed transfer by water birds (CZ, SK), efficient HCO_3^- user, seasonal life cycle very short, fruits without dormancy.

Anatomy

Round shoot with a diameter of <1 mm.

A large cortex and an epidermis surround a small central cylinder (1).

Central cylinder: An area of poorly differentiated unlignified cells surrounds a central air duct (2). Vessels and fibers are absent. Sieve tubes are represented by unlignified angular cells and small neighboring companion cells. The central cylinder and the cortex are separated by a poorly developed endodermis (1–3).

Cortex: Parenchyma cells and large intercellulars form the cortex (1, 2). The epidermis is thin-walled. Crystals are absent.

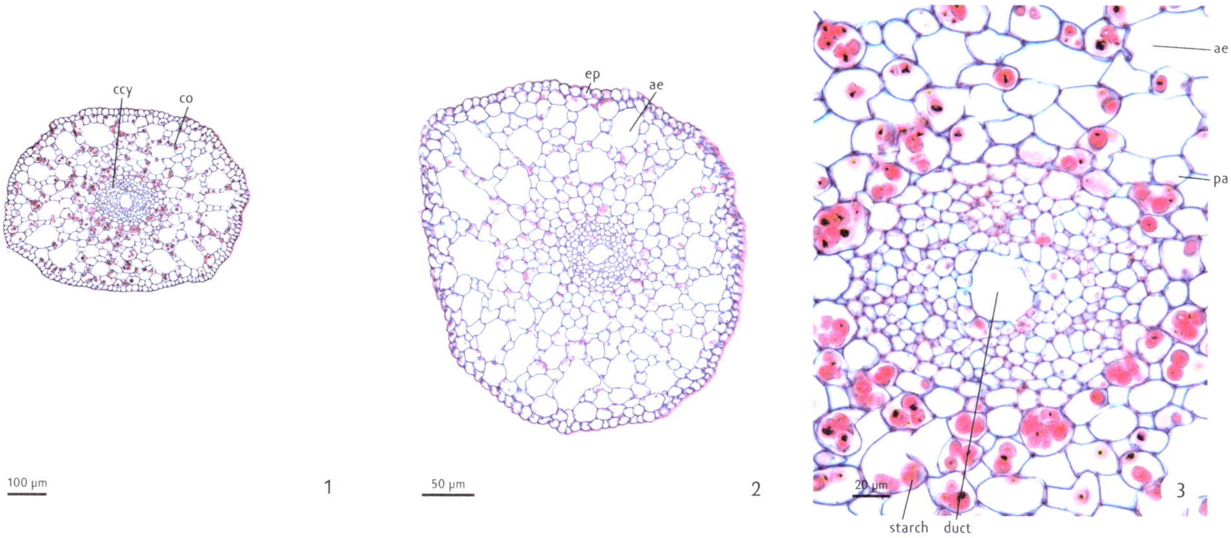

Scheuchzeria palustris L.

Rannoch rush

Scheuchzeriaceae

Perennial wetland to amphibious plant. Erect stems 10–40 cm tall, circular in section, arising from a deep creeping rhizome. Narrow linear alternating leaves usually superior to the inflorescence, form basal sheath, flowers monoecious, fruits follicules.

Distribution and ecology

Circumpolar boreal-montane element. Scattered irregularly and rarely in all CE countries except HU, mainly in uplands to mountains comprising peatlands. Very wet habitats in peat bogs and raised bogs (blanket bogs), typically margins of bog lakes, sometimes as amphibious, always very acidic, base-poor soils and waters (pH ca. 3.5–5). Does not tolerate eutrophication, facultative psychrophilic and stenotopic.

Anatomy

Circular flower stalk with a diameter of 2 mm (1).

A small, compact peripheral zone surrounds a central net-like aerenchymatic tissue (1). Vascular bundles occur in both parts (1).

Numerous collateral vascular bundles are circularly arranged at the periphery and a few are irregularly distributed in the center (1). A sheath of lignified fibers surrounds the vascular bundles (2, 3). The centrifugal part of the vascular bundles consists of an arc of vessels and a round to oval part of phloem consisting on large sieve tubes and small companion cells. A few large cells at the periphery contain dark stained substances (3). Crystals are absent.

Circular rhizome with a diameter of 3–4 mm (4).

The central part is surrounded by an unlignified endodermis (5) and a large zone of a net-like aerenchyma, which contains vascular bundles (4, 5, 6). The center consists of an aerenchymatic tissue and circularly arranged circular concentric vascular bundles (4, 5). A sheath of hardly lignified fibers exists mainly in the centripetal part of the vascular bundles (5). Unlignified vessels surround totally or partially a round phloem (7). The centripetal lignification of the epidermis cells is something special (7). Crystals are absent (8).

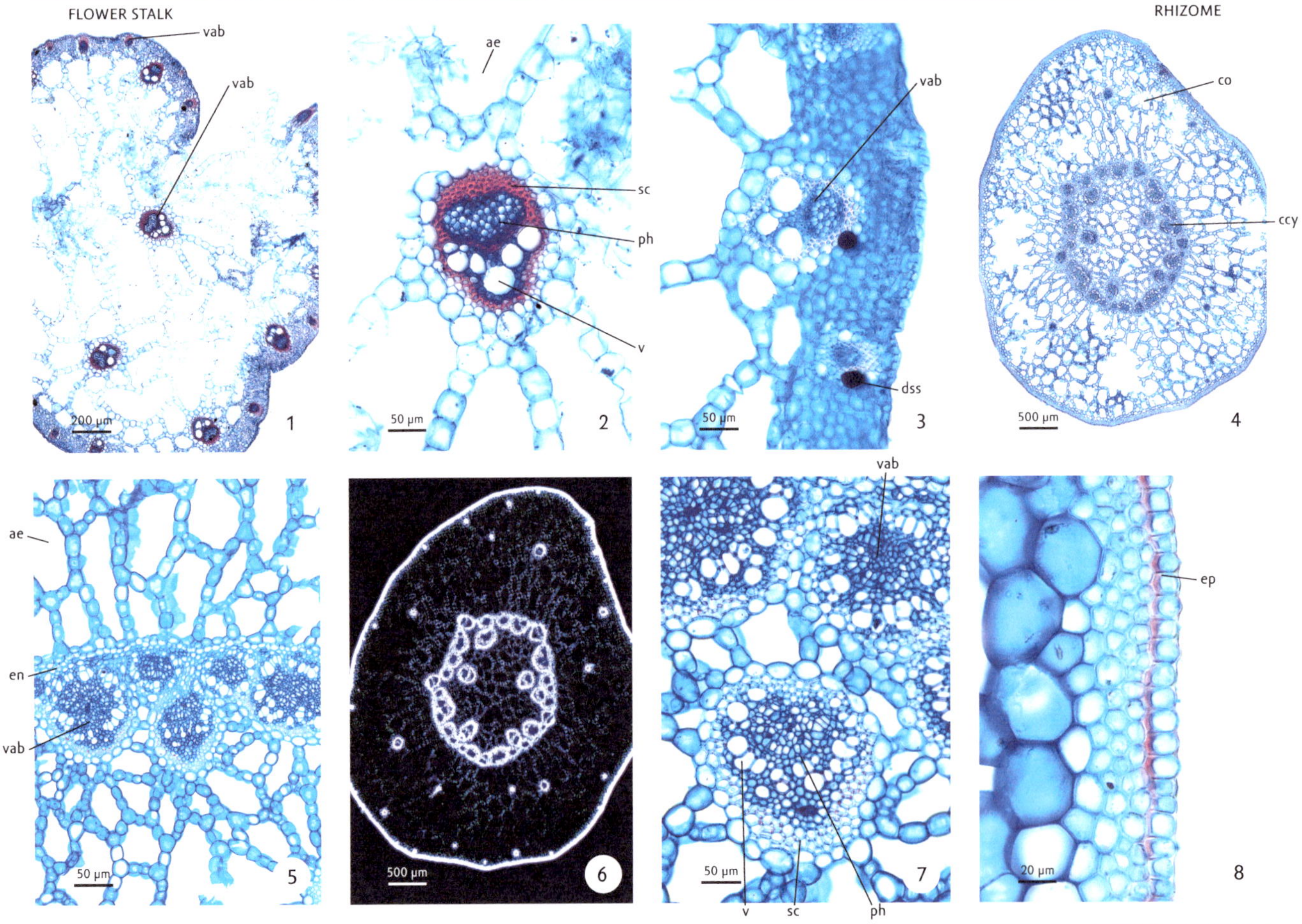

F. H. Schweingruber et al., *Anatomic Atlas of Aquatic and Wetland Plant Stems*, https://doi.org/10.1007/978-3-030-33420-8_70

Tofieldia calyculata (L.) Wahlenb. **Mountain Scottish Asphodel** Tofieldiaceae

Perennial wetland, erect, hairless herb, 0.1–0.45 m high, with short rhizomes. Leaves mostly in a basal rosette. Inflorescense is an apical raceme, flowers with short stalks and oblong yellow perianth. Fruit is a capsule.

Distribution and ecology

European distribution. Very scattered in all CE countries except HU, from lowlands to alpine zone, very rare in CZ. It grows especially in fen meadows, in springs, wet rocks and mountain peat bogs usually on very wet, humic or clayish soils. Calciphilous plant.

Anatomy

Circular flower stalk with a diameter of 1–2 mm (1).

A small, distinct unlignified cortex surround a large lignified central cylinder (1, 2).

The central part of the flower stalk consists of large, poorly lignified parenchyma cells and the outer part of small, lignified fibers (2). Collateral vascular bundles are circularly arranged in two rows (1). Groups of lignified fibers occur at the central and peripheral zone of the vascular bundles (2, 3). The central part of the vascular bundles consists of an arc of lignified angular vessels. On the centripetal and the centrifugal part of the vessels occur unlignified, poorly differentiated cells (3). The cortex is mainly formed by round parenchyma cells and small irregular intercellulars (2). Vascular bundles are absent in the cortex (1, 2).

Circular rhizome with a diameter of 2–3 mm (4).

A lignified extremely thick-walled endodermis separates the cortex from the central part (5). The center consists of densely packed vascular bundles (6). The xylem of the concentric vascular bundles consists of angular lignified vessels, which surround an irregularly, unlignified phloem (6). Vessels walls have scalariform intervessel pits with ring-like structures (7). The cortex consists of round dark stained parenchyma cells without intercellulars (8). The epidermis is thin-walled (4). Intercellulars and crystals are absent in both parts (8).

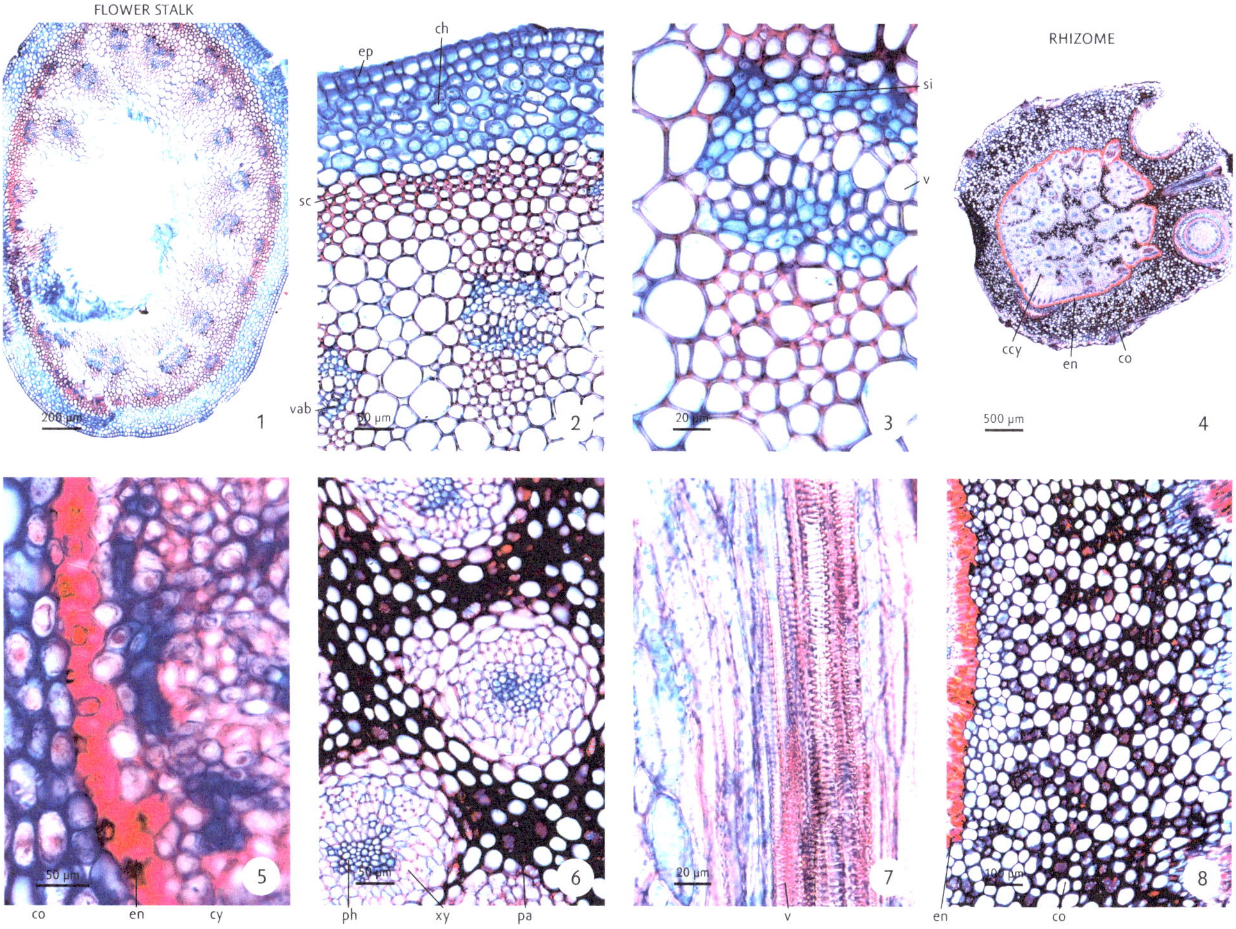

F. H. Schweingruber et al., *Anatomic Atlas of Aquatic and Wetland Plant Stems*, https://doi.org/10.1007/978-3-030-33420-8_71

Tofieldia pusilla (Michx.) Pers.

Scottish Asphodel Tofieldiaceae

Perennial wetland, rhizomatous, erect, evergreen hairless herb, 5–12 cm high. Leaves mostly in basal rosette, all flattened in one plane. Inflorescence is a short, dense raceme, flowers are on short stalks. Oblong greenish-cream perianth. Fruit is a short oval capsule.

Distribution and ecology

Circumpolar arctic-montane element. Rarely scaterred in the Alps (DE, AT, CH) in subalpine to alpine zone, very rare in SK, absent in CZ, HU and PL. It grows usually close to streams, in calcareous springs or in snow fields. Requires constant moisture but not waterlogged conditions. Distinctly calcicole plant on both mineral and organic soils.

Anatomy

Circular flower stalk with a diameter of <1mm (1).

A small, distinct unlignified cortex surrounds a large lignified central cylinder (1, 2).

The central part of the flower stalk consists of large, poorly lignified parenchyma cells and the outer part of small, poorly lignified fibers (2). The central part of the flower stalk consists of large, unlignified parenchyma cells (2). Collateral vascular bundles are circularly arranged in two rows (1, 2). Vascular bundles consist of an arc of poorly lignified vessels and undifferentiated unlignified round cells (3). Vascular bundles are absent in the chlorenchyma.

Circular rhizome with a diameter of 1–2 mm (4).

A lignified extremely thick-walled endodermis separates the cortex from the central part (4, 5). The center consists of densely packed vascular bundles (4). The xylem of the concentric vascular bundles consists of angular lignified vessels, which surround an irregularly, unlignified phloem (6). The cortex consists of round, unlignified parenchyma cells without intercellulars (6). Crystals are absent.

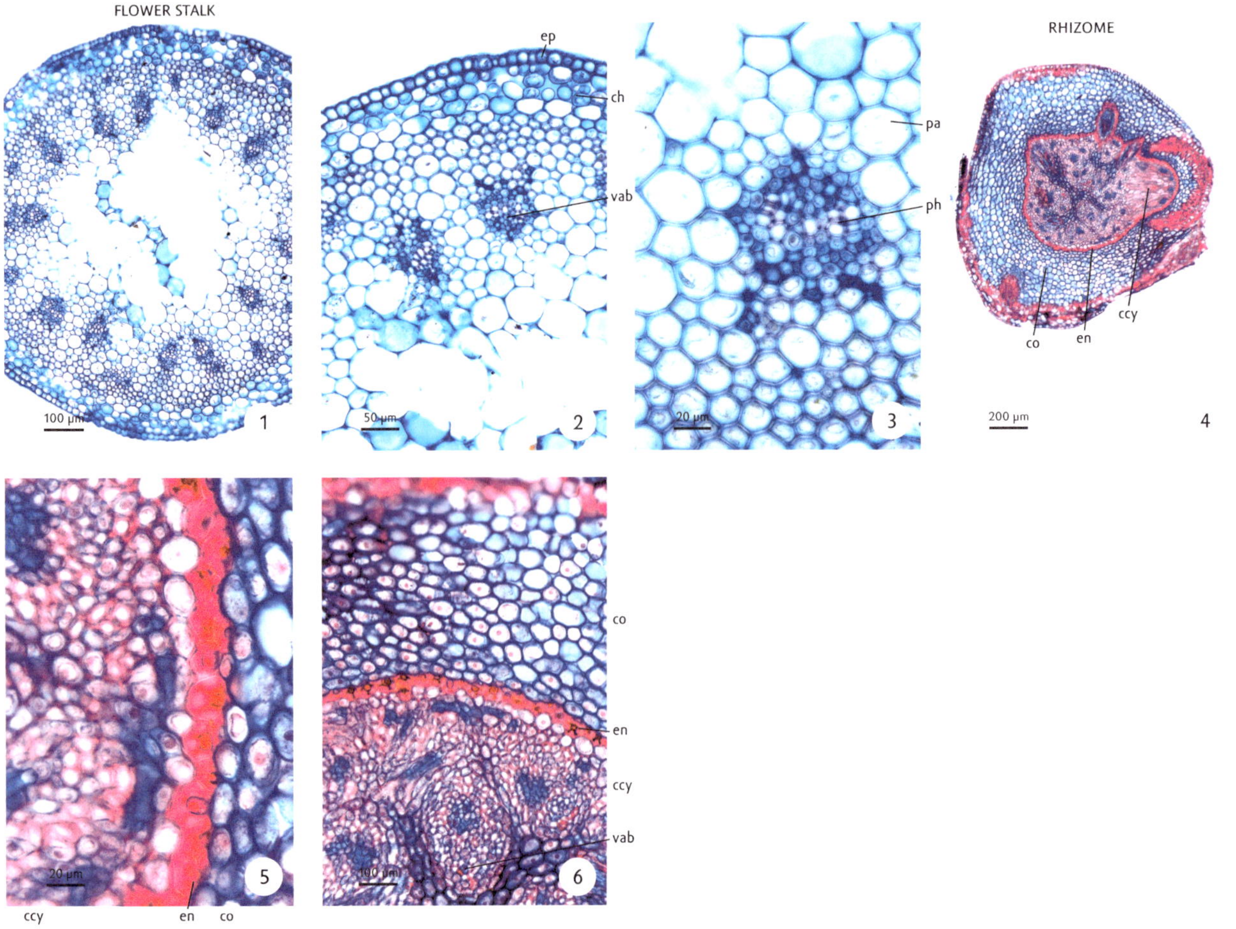

Sparganium angustifolium Michx. **Floating Bur-reed** Typhaceae

Perennial floating-leaved herb rooted in the bottom with stems 0.1–1 m long. All simple flat leaves alternating, with sheathing inflated bases. Unbranched inflorescence has usually 2 male and 2–4 female heads. The flowers are small, sessile. A drupe-like fruit has a persistent style.

Distribution and ecology

European boreo-montane element; also in E. Asia and N. America. Rarely in all CE countries, except HU. It grows in shallow banks of upland lakes. Prefers softwater lakes with clear, oligotrophic water, only rarely occurs in mesotrophic conditions. Most likely it can partly take up CO_2 by the roots.

Anatomy

Circular flower stalk with a diameter of 2–3 mm (1).

A distinct aerenchymatic chlorenchyma surrounds a large central part (1, 3). The xylem of the closed collateral vascular bundles mainly consists of few vessels with unlignified walls and surrounding parenchyma cells (2). Some vascular bundles contain also air ducts (2). The phloem is composed of groups of sieve tubes and companion cells (2). A few vascular bundles exist in the chlorenchyma (3). A net-like aerenchyma forms the chlorenchyma (3) The epidermis it thin-walled. Bundles of styloids occur in the cortex (4).

Rhizome with a diameter of 4–6 mm (5).

A distinct cortex surrounds a large central cylinder (7). The xylem of the concentric vascular bundles mainly consists of an arc of poorly lignified vessels and unlignified parenchyma cells (6, 8). Vessel-wall pits are scalariform. The cortex consists of radially oriented parenchyma cells and large intercellulars (5). Many crystal druses and styloids occur in the central cylinder (8).

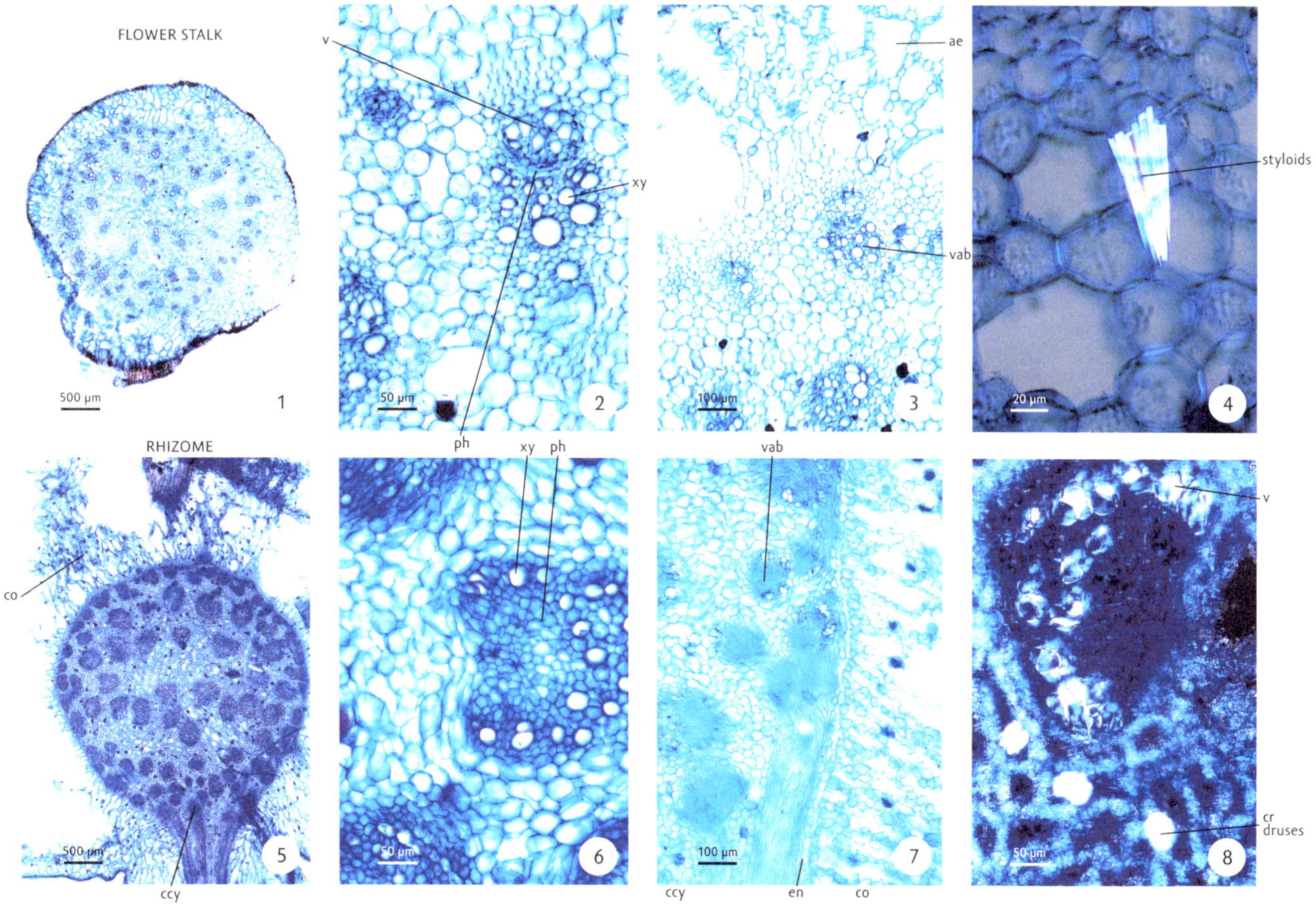

F. H. Schweingruber et al., *Anatomic Atlas of Aquatic and Wetland Plant Stems*, https://doi.org/10.1007/978-3-030-33420-8_72

Sparganium emersum Rehmann **Unbranched Bur-reed** Typhaceae

Perennial floating-leaved, emergent or submerged herb, rhizomatous. Stem erect, sometimes floating, 20–60 cm high, protrudent above water level. Aerial leaves upright, flat and strongly keeled, ribbon-like basal floating leaves. Inflorescence an unbranched raceme, the spherical heads sessile or stalked. Fruit is an elliptical nut, narrowed into beak.

Distribution and ecology

Circumpolar boreo-temperate element. In all CE countries, rooted in mud, fishpond and water ditch margins, slow running, clear and cool waters, prefers alkaline muddy soils, but not calcareous. Also meso- or eutrophic waters, it is tolerant of disturbance and may be frequent along the heavily managed rivers.

Anatomy

Lower part of the circular flower stalk with a diameter of 6 mm (1).

A distinct cortex surrounds a large central cylinder (1, 2). The cortex is divided into an outer part which consists mainly of large parenchyma cells and rudimentary vascular bundles. The inner part consists of very thin-walled, unlignified cells and many laticifer-like ducts (2). The xylem of the closed collateral vascular bundles mainly consists of a few round vessels (3). The phloem is composed of groups of sieve tubes and companion cells (3). Intercellulars and crystals are absent. A few ducts with distinct cell walls occur in the central cylinder (2). Laticifer-like cells occur also in the central cylinder (2).

Rhizome with a diameter of 4–6 mm (5).

A distinct cortex surrounds a large central cylinder (5). The xylem of the concentric vascular bundles mainly consists of an arc of poorly lignified vessels and unlignified parenchyma cells (7). Vessel-wall pits are scalariform. The cortex consists of radially oriented parenchyma cells and large intercellulars (5). Many raphids occur in central cylinder (4).

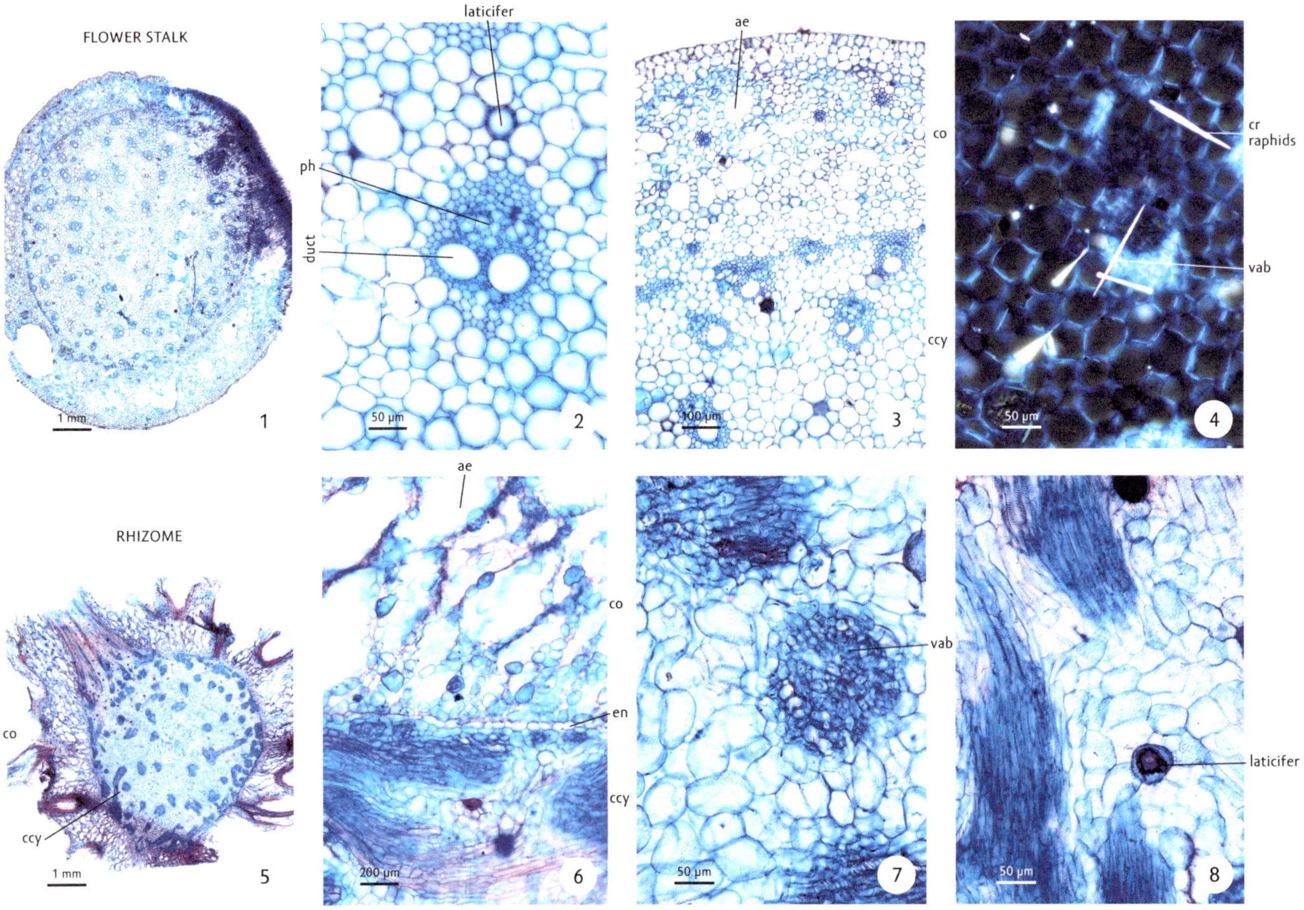

Sparganium erectum L.

Branched Bur-reed

Perennial emergent herb, rhizomatous. Stem stiff, erect, 50–150 cm high, stout, branched above. Leaves all erect, strongly keeled throughout. Inflorescence a branched raceme, each branch with male heads above, female below. Fruit is a nut, narrowed into beak.

Distribution and ecology

Circumpolar temperate element, but absent from eastern N America. In all CE countries, from lowlands to mountains. In stagnant, shallow waters, margins of lakes and fishponds, slow rivers and canals, rare in reed swamps. In mesotrophic or eutrophic habitats, very tolerant of eutrophication. Efficient in fine sediment trapping. Strongly grazed by cattle.

Anatomy

Circular or oval flower stalk with a diameter of 3-6 mm (1).

A distinct cortex surrounds a large central cylinder (1). Central cylinder and cortex are separated by a wavy band of thin- to thick-walled fibers (1-3). The xylem of the closed collateral vascular bundles mainly consists of large air ducts with ring-like thickenings (4) and surrounding unlignified parenchyma cells (2). Vessels are small, angular or absent (3). The phloem is composed of groups of sieve tubes and companion cells (2). Vascular bundles are surrounded by a uniseriate (2, 3) sheath of lignified fibers. Small rudimentary vascular bundles is present in the cortex (1, 2). Intercellulars are very small in the central cylinder (3). The epidermis is thin-walled (1). Bundles of raphids occur in the cortex. Laticifer-like cells occur also in the cortex (2).

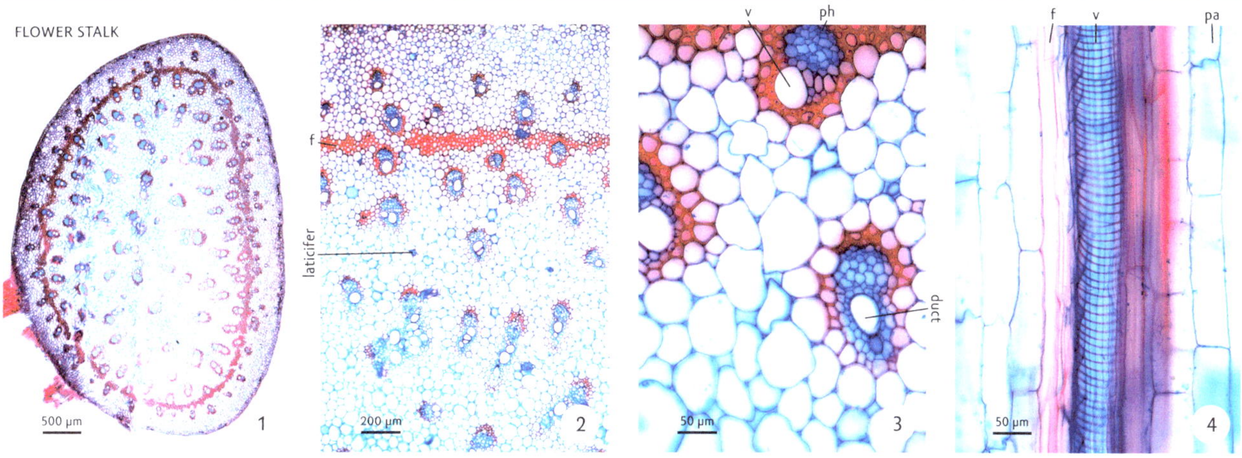

Perennial floating-leaved or emergent herb, rhizomatous. Stems usually floating, rarely erect, up to 50 cm high. Leaves usually floating, barely inflated at base, flat. Inflorescence a simple raceme, the spherical heads sessile. Fruit is a nut, narrowed into beak.

Distribution and ecology

Circumpolar boreo-temperate element. In all CE countries, in shallow mesotrophic waters of acidic lakes, fishponds, pools or ditches, with high organic substrate (peat), often co-occur with different *Utricularia* sp. Rhizomes only short, usually reproduces by seeds.

Anatomy

Circular flower stalk with a diameter of 2–3 mm (1).

A distinct aerenchymatic cortex surrounds a large central cylinder (1, 2). The xylem of the closed collateral vascular bundles consists of an air duct and few small, angular vessels with poorly lignified walls and surrounding parenchyma cells (2). The phloem is composed of groups of sieve tubes and companion cells (2). A few small vascular bundles exist in the cortex (2). A ring of fibers separate the central cylinder and the cortex (1, 2). A net-like aerenchyma forms the cortex (2). The epidermis is thin-walled with lignified external and internal walls (2). Long styloids and bundles of raphids in the cortex.

Rhizome with a diameter of 4–6 mm (3).

A distinct cortex surrounds a large central cylinder (3, 4). The xylem of the almost concentric vascular bundles consists of an arc of lignified vessels and unlignified parenchyma cells (5). Vessel-wall pits are scalariform. The cortex consists of a net-like aerenchyma (3). Long styloids and bundles of raphids and crystal sand occur mainly in the cortex.

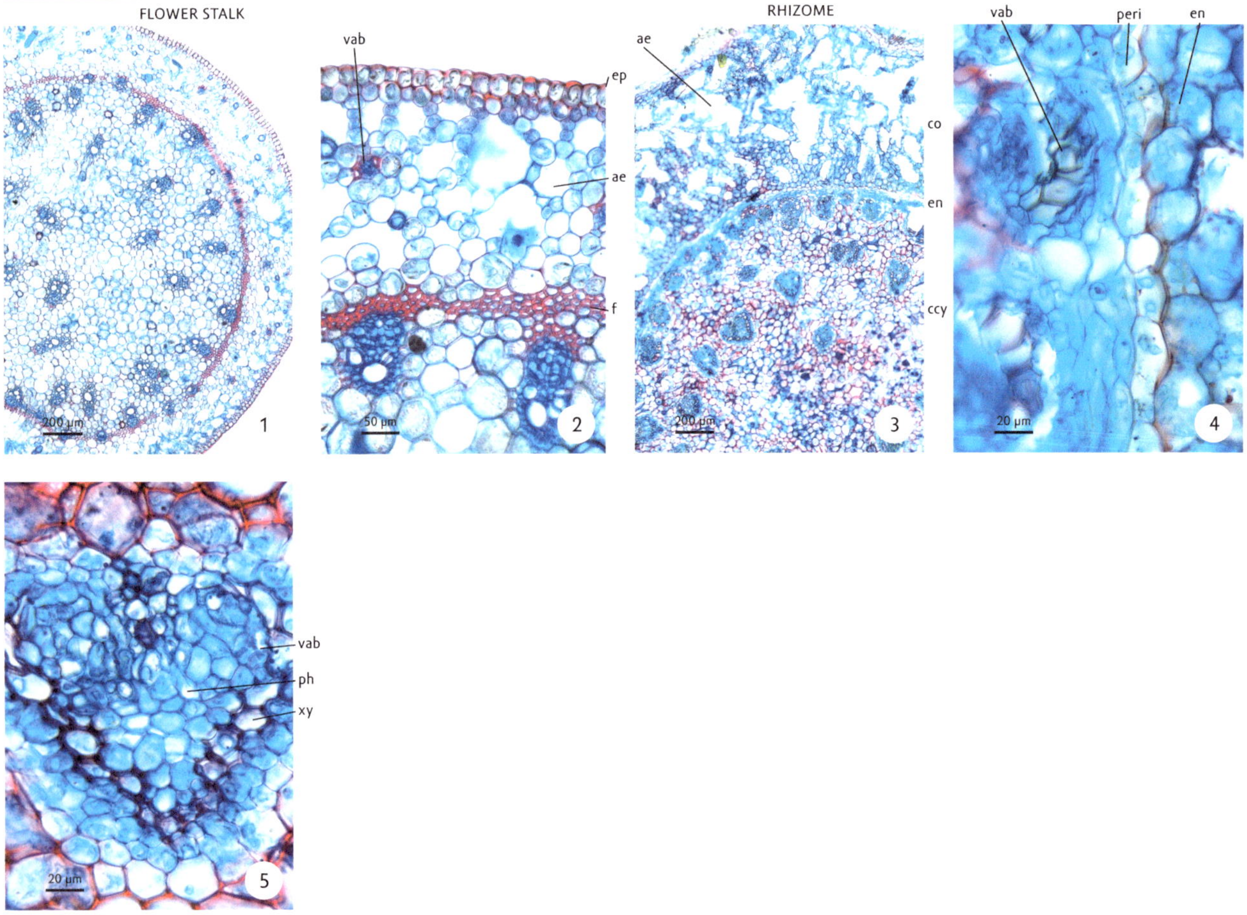

Typha latifolia L. **Bulrush** Typhaceae

Perennial emergent herb, rhizomatous. Stem stiff, up to 3 m high, usually overtopped by some leaves. Stout, roundish leaves, 8–24 mm wide. Inflorescence a cylindric spike, male and female parts of spike rarely <2.5 cm apart. Female part of spike 18–30 mm wide, dark-brown.

Distribution and ecology

Circumpolar southern-temperate element. In all CE countries, in eutrophic up to 1 m deep waters with muddy bottoms, littoral reed swamps, muddy banks of stagnant and slow running waters. It spreads by wind-dispersed fruits, often colonising newly excavated ponds and ditches and subsequently spreading by vegetative growth. Internal gas transport in the shoots via pressurized convective flowthrough of gases.

Anatomy

Circular culm with a diameter of 6–10 mm (1).

A distinct cortex surrounds a large central cylinder (1) but an endodermis is absent. Vascular bundles are surrounded by a sheath of thick-walled, lignified fibers (2). The xylem consists of large round air ducts and small angular lignified vessels and unlignified parenchyma cells (2). The phloem is composed of groups of sieve tubes and companion cells (2). A wavy zone of thick-walled lignified fibers surrounds the peripheral part of the central cylinder (3). A few small vascular bundles exist in the cortex (3). Centripetal cortex consists of large, thin-walled parenchyma cells, small intercellulars (3) and the external part of small, thin-walled cells. The epidermis is thin-walled (3).

Rhizome with a diameter of 4–6 mm (4).

A distinct cortex with a net-like aerenchyma surrounds a large central cylinder (4). The cortex and the central cylinder are separated by a thin-walled endodermis (5). The xylem of the concentric vascular bundles is composed of an arc of lignified angular vessels and unlignified parenchyma cells (5). Vessel-wall pits are scalariform or consist of annular elements (6). The cortex consists of a net-like aerenchyma and many rudimentary vascular bundles (7). Intercellulars in the central part of the stem are small. Raphids occur primarily in the cortex (8). Slimy dark staining substances occur mainly in the cortex.

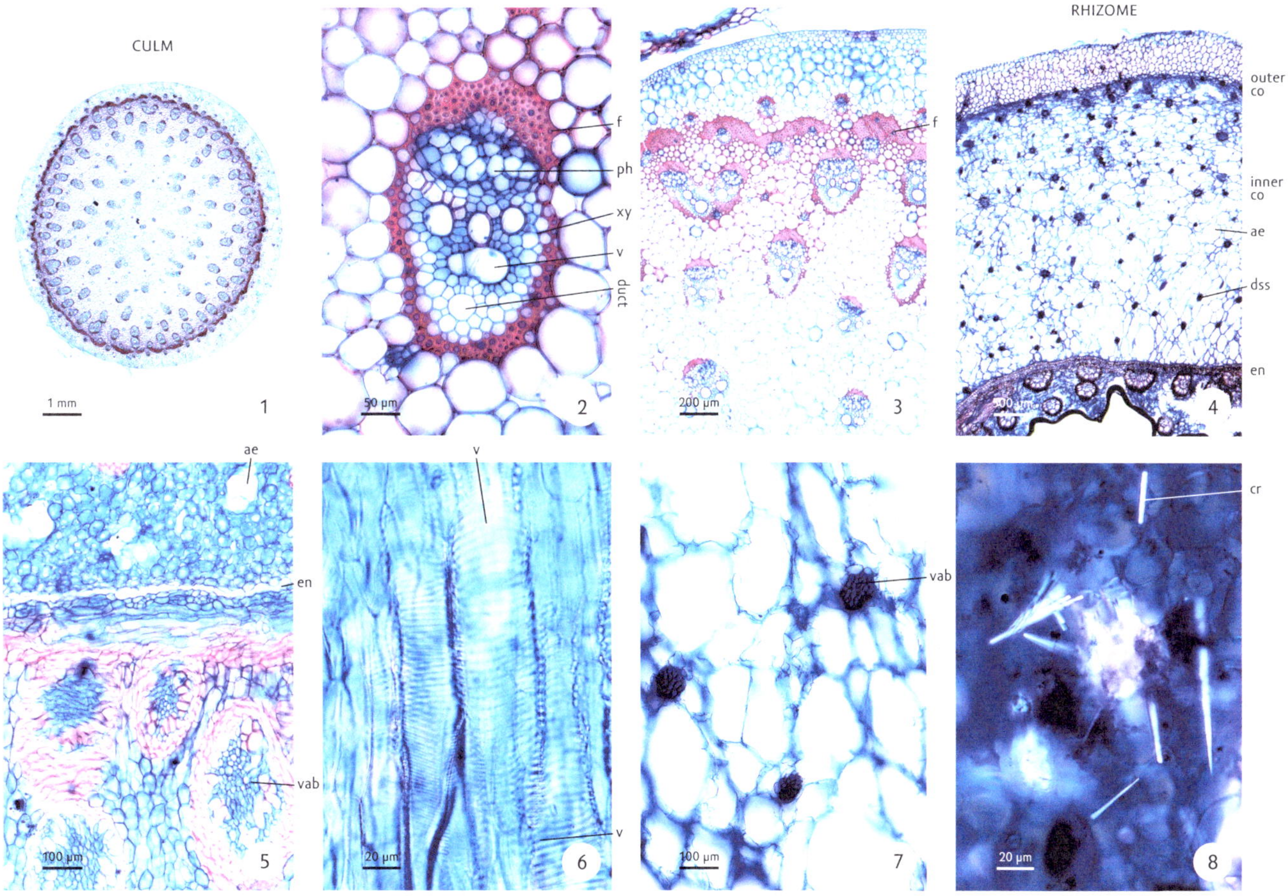

Typha laxmannii Lepech.

Laxman's Bulrush Typhaceae

Perennial emergent herb, rhizomatous. Stem erect, 80–120 cm high, thin. Stout, roundish leaves, 2–4 mm wide. Inflorescence a cylindric spike, the male part is 1.5–4 times longer than the dark brown female part.

Distribution and ecology

Euro-Siberian element, only secondarily in CH, CZ, SK, DE, HU, PL. Banks of stagnant waters. In the south-western limit of its distribution, the species is a synanthropic colonizer of banks of water reservoirs.

Anatomy

Circular culm with a diameter of 4-5 mm (1).

A distinct cortex surrounds a large central cylinder (1). Closed collateral vascular bundles are surrounded by a sheath of thick-walled, lignified fibers (2). The xylem consists of a few, round, large air ducts and large round, lignified vessels and unlignified parenchyma cells (1, 2). The phloem is composed of groups of sieve tubes and companion cells (2). A zone of thick-walled lignified fibers limits the peripheral part from the central cylinder (2). A few vascular bundles exist in the cortex (1). The middle part of the cortex consists primarily of large, thin-walled parenchyma cells, large irregularly formed intercellulars and rudimentary vascular bundles (2) and the external part of small, thin-walled cells (2). The epidermis is thin-walled (1). Raphids occur in the cortex and in the central cylinder.

Rhizome with a diameter of 4–6 mm (3).

A distinct cortex surrounds a large central cylinder (4). The cortex and the central cylinder are separated by a thin-walled endodermis and a few layers of fibers (4). Central vascular bundles are surrounded by a centripetal extended sheath of thick-walled, lignified fibers (5). The xylem mainly consists of an arc of lignified angular vessels and unlignified parenchyma cells. Vessel-wall pits are scalariform or consist of annular elements. The central part of the cortex consists of a net-like aerenchyma and many rudimentary vascular bundles and the external part of thin-walled, round parenchyma cells without intercellulars (4). Intercellulars in the central part of the stem are small. Raphids occur primarily in the cortex (6).

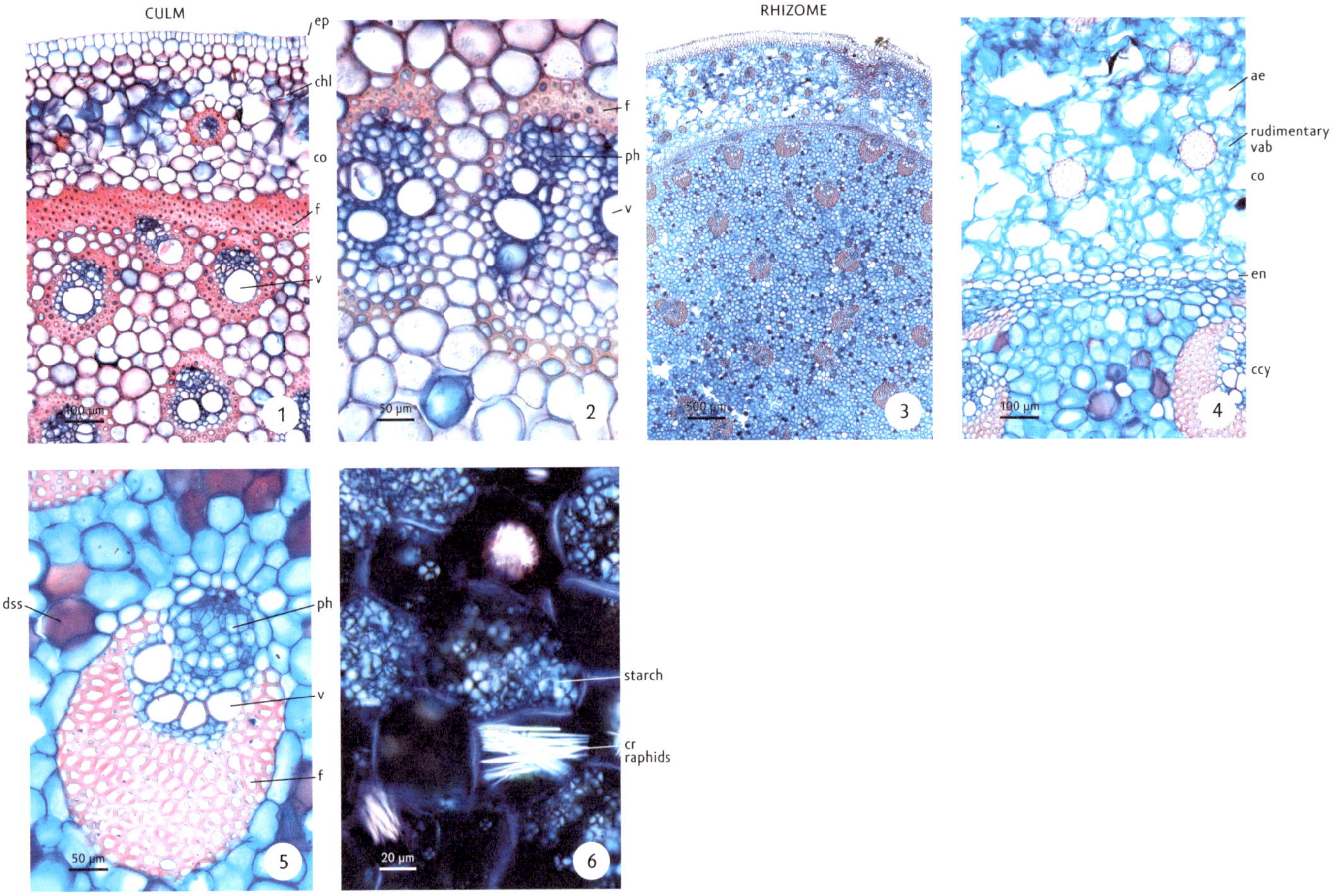

Perennial wetland herb, rhizomatous. Stem 30–75 cm high, thin, erect, below covered with sheaths, leafless above. Leaves only 1–3 mm wide. Inflorescence a cylindric spike; the female part 15–50 mm long.

Distribution and ecology

Euro-Siberian element, only in AT, CH and DE, extinct in CZ. Formerly widespread along fast flowing alpine rivers but is now red-listed as critically endangered, needs regularly disturbed, sparsely vegetated river margins.

Anatomy

Rippled circular culm with a diameter of 2 mm (1).

A small cortex surrounds a large central cylinder (1). Closed collateral vascular bundles are externally limited by a half moon-like sheath of thick-walled, lignified fibers (1, 2). The xylem consists of one or two round air ducts and angular poorly lignified vessels and unlignified parenchyma cells (2, 3). The phloem is composed of groups of sieve tubes and companion cells (3). Vascular bundles in the cortex are connected with the periphery with a beam of thick-walled fibers (1, 2). Assimilatory, unlignified parenchymatic tissues surround the culm (1, 2). Both epidermis and hypodermis are thin-walled (2).

Rhizome with a diameter of 3–5 mm.

A distinct aerenchymatic cortex surrounds a large central cylinder (4). The cortex and the central cylinder are separated by a layer of thin-walled cells (4). Central vascular bundles are surrounded by a centripetal extended sheath of thick-walled, lignified fibers (5). The xylem mainly consists of large, round vessels and unlignified parenchyma cells (5). Vessel-wall pits are scalariform or consist of annular elements. The cortex consists of large, radial intercellulars between small radial parenchymatic stripes (4). Intercellulars in the central part of the central cylinder are small. Slime cells with dark staining substances are located between parenchyma cells (6). Raphids occur in the cortex and the central cylinder.

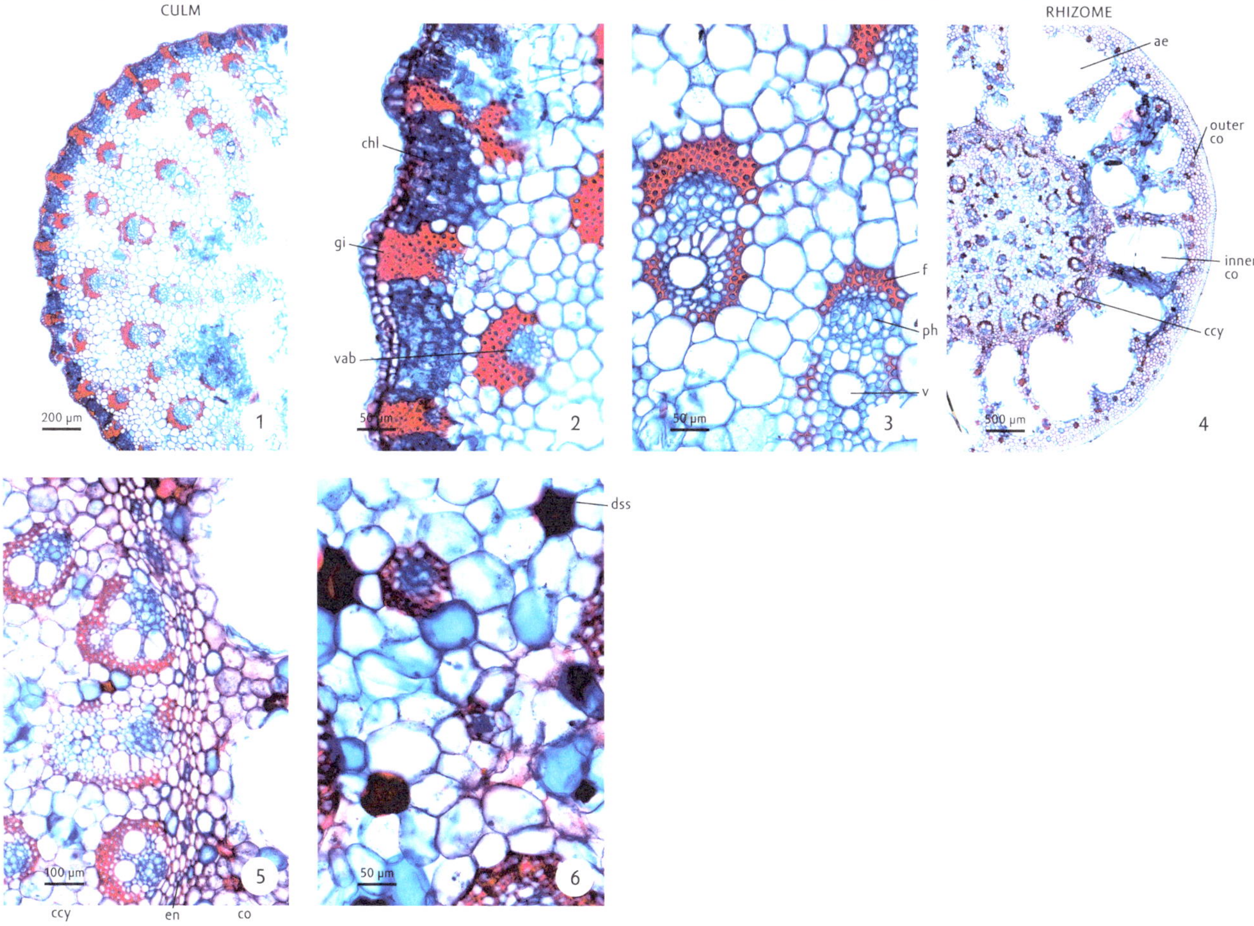

Typha shuttleworthii W.D.J. Koch et Sond.

Perennial wetland herb with stems 0.7–1.5 m tall, rhizomatous. Yellow green to light green leaves are mostly basal, elongate and narrow, with spongy parenchyma. The inflorescence is terminal, cylindrical spike of very dense flowers, with no gap between male and female parts of spike. Female spike has a silvery-grey colour when fruiting. An achene-like fruit with persistent perianth parts, helping in wind dispersal.

Distribution and ecology

Subatlantic-submediterranean distribution in Europe. In all CE countries except HU, only rarely in CZ, PL and SK. It grows usually on banks of oligotrophic waters and in wet meadows from lowlands to uplands.

Anatomy

Circular culm with a diameter of 4–8 mm (1).

A distinct cortex surrounds a large central cylinder (1). Closed collateral vascular bundles are surrounded by a sheath of thick-walled, lignified fibers (2, 3). The xylem consists of a few, round, large air ducts and large, round lignified vessels and unlignified parenchyma cells (2). The phloem is composed of groups of sieve tubes and companion cells (2). A zone of thick-walled lignified fibers limits the peripheral part from the central cylinder (3). A few vascular bundles exist in the cortex (1, 3). The middle part of the cortex consists of primarily large, thin-walled parenchyma cells and large irregularly formed intercellulars and rudimentary vascular bundles and the external part of small, thin-walled cells (3). The epidermis is thin-walled. Raphids occur in the cortex and the central cylinder.

Rhizome with a diameter of 4–6 mm (4).

A distinct cortex surrounds a large central cylinder (4). The cortex and the central cylinder are separated by a thin-walled endodermis and a few layers of fibers (5). The central vascular bundles are surrounded by a centripetal extended sheath of thick-walled, lignified fibers (5). The xylem mainly consists of an arc of lignified angular vessels and unlignified parenchyma cells (6). Vessel-wall pits are scalariform or consist of annular elements. The central part of the cortex consists of a net-like aerenchyma and many rudimentary vascular bundles (5) and the external part of thin-walled, round parenchyma cells without intercellulars (4). Intercellulars in the central part of the rhizome are small (5, 6). Raphids occur primarily in the cortex.

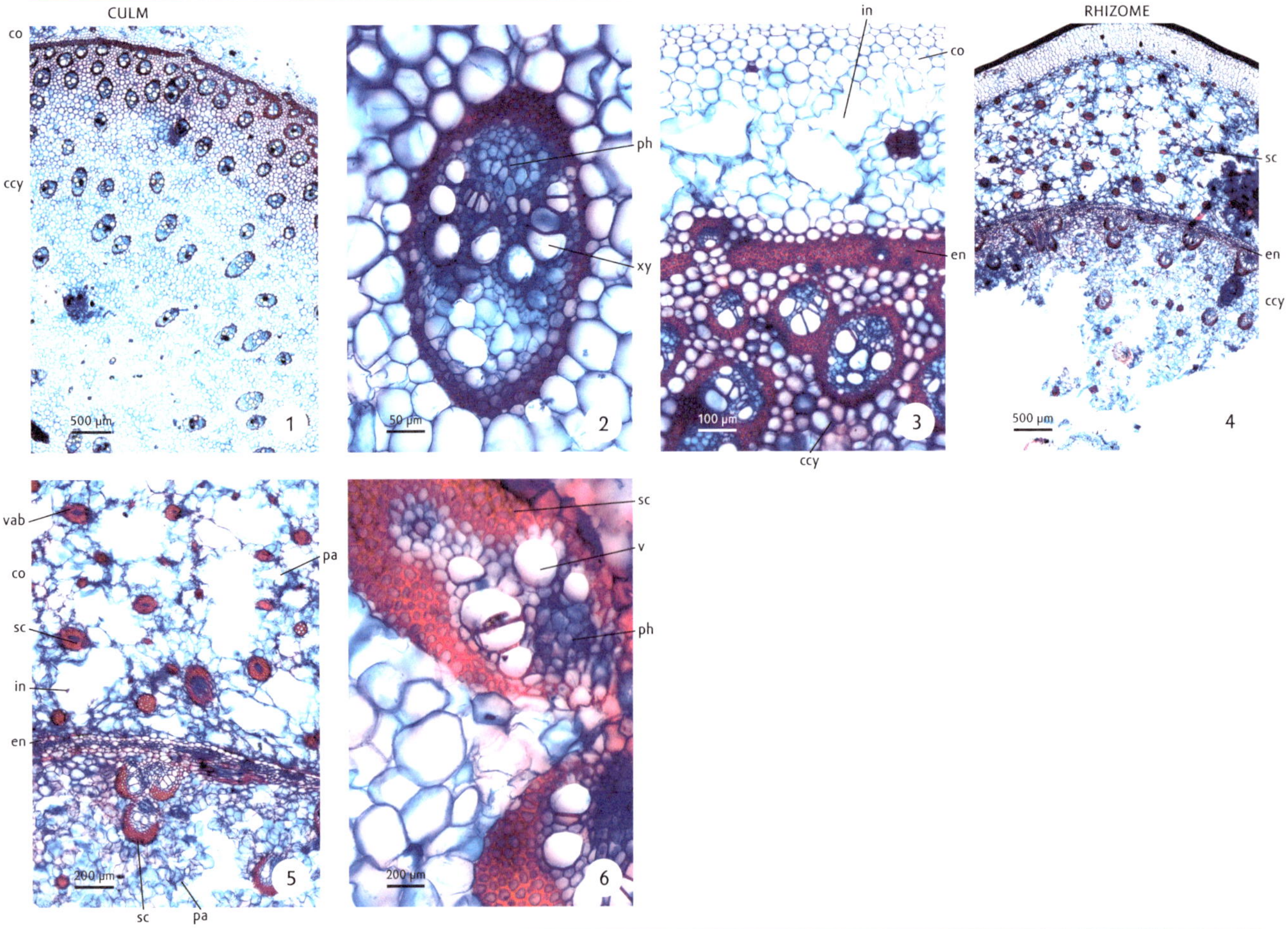

Perennial submerged marine herb, with horizontal rhizomes. Branched flowering stem over 1.0 m long, hairlike green leaves up to 20–50 cm long. The inflorescence is 10 cm long, male and female flowers in separate alternating clusters. The fruit is a nutlet with a transparent coat containing the seed.

Distribution and ecology

Circumpolar wide-temperate distribution. In CE countries only in DE and PL where it grows in the subtidal zone, on gravel, sand or sandy mud. Avoids brackish water or very exposed coasts. It occurs up to depths of 4 m, forming dense sub-tidal meadows. This species declined throughout Europe after a major outbreak of wasting disease in the 1930s, attributed to the microorganism *Labyrinthula macrocystis.*

Anatomy

Round shoot with a diameter of 1–3 mm.

A large cortex and an epidermis surround a small central cylinder (1, 2).

Central cylinder: An area of larger and smaller, but poorly differentiated unlignified cells surrounds a center with an air duct (2). Fibers and vessels are absent (2). Angular cells probably represent sieve tubes (2). Cell walls are not structured, therefore it is impossible to determine the type of cells (2). The endodermis consists of unlignified thin- or thin- to thick-walled cells (2).

Cortex: The cortex consists primarily of parenchyma cells (3, 4). The outer part contains small (1) and the central part net-like intercellulars. The epidermis is thin-walled (4). Crystal sand occurs in the outer part of the cortex.

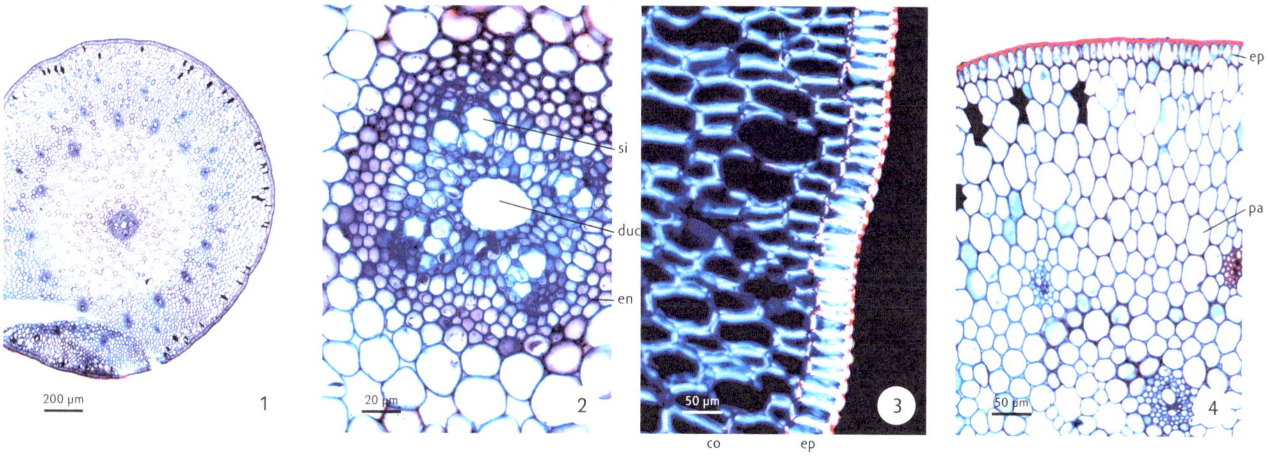

F. H. Schweingruber et al., *Anatomic Atlas of Aquatic and Wetland Plant Stems*, https://doi.org/10.1007/978-3-030-33420-8_73

Synthesis of Anatomical, Ecological and Phylogenetical Data

Dicotyledonous aquatic and wetland plants

This synthesis includes 212 studied species belonging to 124 genera from 51 families. The most species-rich family is Asteraceae (30 taxa), followed by Ranunculaceae (18), Caryophyllaceae (12), Rosaceae (10), Apiaceae (9), Lamiaceae (9), Plantaginaceae (9), Polygonaceae (9), Onagraceae (8), Lentibulariaceae (7) and Primulaceae (6). The most species-rich genera are *Utricularia* (7), *Epilobium* (6), *Ranunculus* (6), *Veronica* (6), *Cirsium* (5), *Mentha* (5), *Persicaria* (5), *Senecio* (5), *Batrachium* (4), *Bidens* (4) and *Myriophyllum* (4). Most of the studied taxa are terrestrial wetland plants (149), followed by emergent plants (30), submerged aquatic plants (23) and rooted floating-leaved aquatic plants (10, hereafter aquatic floating). In terms of growth form, most of the studied species are perennial clonal hemicryptophytes (143), followed by short-lived annual or biennial species (40), perennial nonclonal polycarpic species (11), perennial nonclonal monocarpic species (9 taxa such as *Oenanthe aquatica, Laserpitium pruthenicum, Pedicularis* spp.) and woody species (9).

The analysis of clonal growth organs based on the CLO-PLA classification (Klimešová et al. 2017) revealed that most of the studied aquatic and wetland clonal plant species have short epigeogenous rhizomes (53 taxa, belonging to genera *Cirsium, Nuphar, Nymphaea, Ranunculus, Senecio* or *Viola*), followed by species with longer hypogeogenous rhizomes (42 species mostly from genera *Epilobium, Lysimachia, Mentha* and *Thalictrum*), horizontal stolons (28 taxa from *Batrachium, Callitriche, Myriophyllum, Stellaria* and *Veronica*), and root sprouters (12 taxa including woody species such as *Alnus* spp., *Frangula alnus, Spiraea salicifolia,* and herbs *Inula britannica* and *Sium latifolium*). There were also turion-forming aquatic plants (11 taxa from genera *Aldrovanda, Ceratophyllum* and *Utricularia*). Perennial tap-rooted forbs were less common than clonal species among the studied taxa (15) and were more or less equally represented across all major families, with most representatives in Fabaceae (*Lotus* spp.), Apiaceae (e.g. *Laserpitium pruthenicum, Oenanthe aquatica, Peucedanum palustre*) and Caryophyllaceae (e.g. *Lychnis flos-cuculi*).

The analysis of species habitat preferences – based on 19 bioclimatic CHELSA variables (Karger et al. 2016) calculated for species geographic occurrences in the GBIF (Global Biodiversity Information Facility, www.gbif.org) database – revealed that short-rhizomatous plants prefer colder climates characterized by the lowest mean annual temperatures (Figure 1), the lowest minimum temperature in the coldest month and higher annual precipitation (Figure 2); short-rhizomatous clonal forbs are typically found in higher elevation wetlands (e.g., *Pinguicula alpina, Ranunculus aconitifolius, Ranunculus pygmaeus, Rumex alpinus, Senecio alpinus, Swertia perennis* and *Viola biflora*), wet oligotrophic upland meadows (e.g., *Cirsium palustre, Succisa pratensis, Tephroseris crispa*), and fresh waters (*Nuphar, Nymphaea*) (Figure 3). Most aquatic plants and terrestrial short-lived and tap-rooted perennial species, on the other hand, prefer habitats with a warmer climate (Figure 4).

F. H. Schweingruber et al., *Anatomic Atlas of Aquatic and Wetland Plant Stems*, https://doi.org/10.1007/978-3-030-33420-8_74

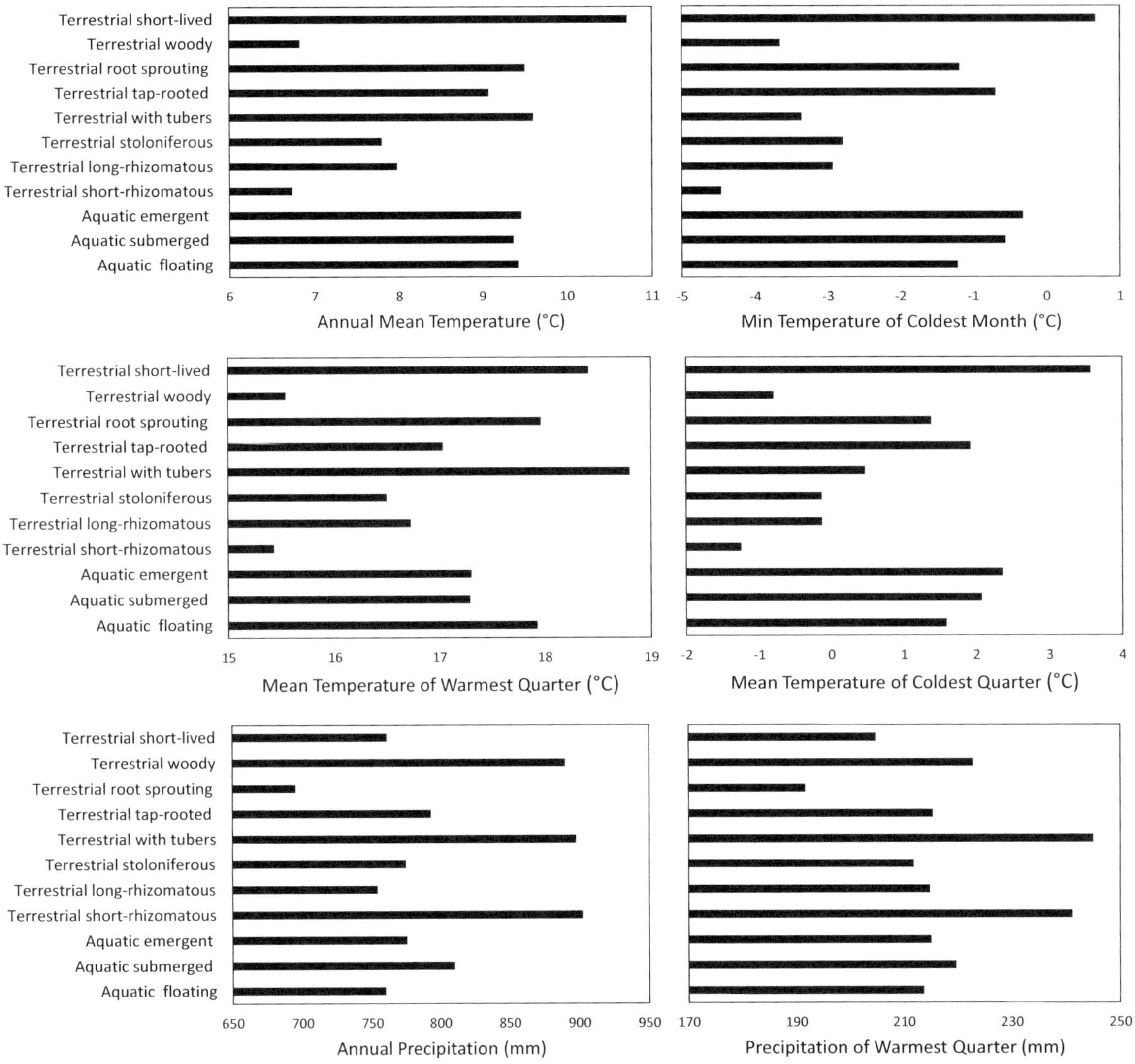

Figure 1. Aquatic and wetland plant life and growth form categories and their relations to selected bioclimatic variables.

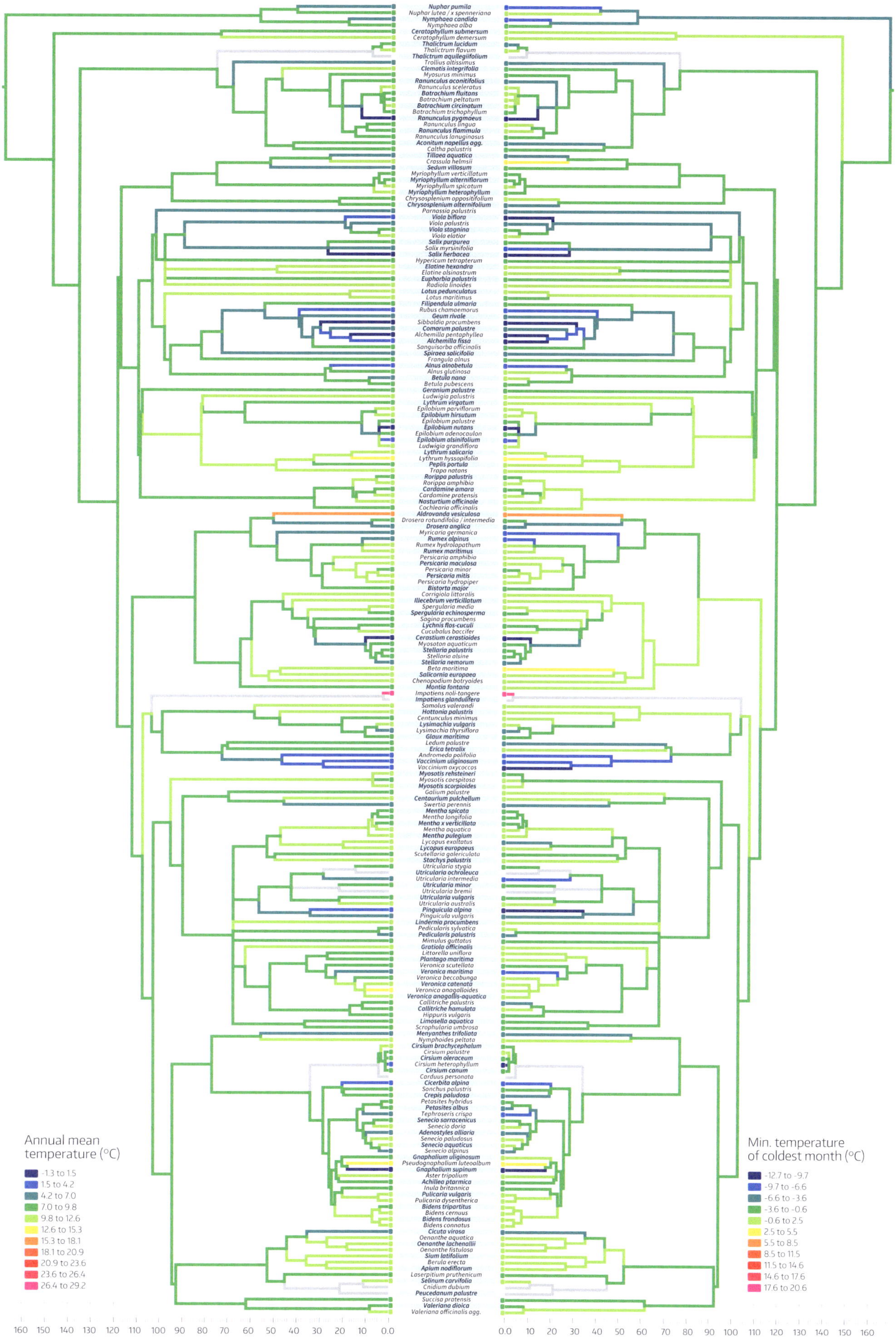

Figure 2. The phylogenetic tree of studied dicot species with a maximum likelihood reconstruction of selected species climatic preferences (Left: Annual mean temperature, Right: Min. temperature of the coldes month).

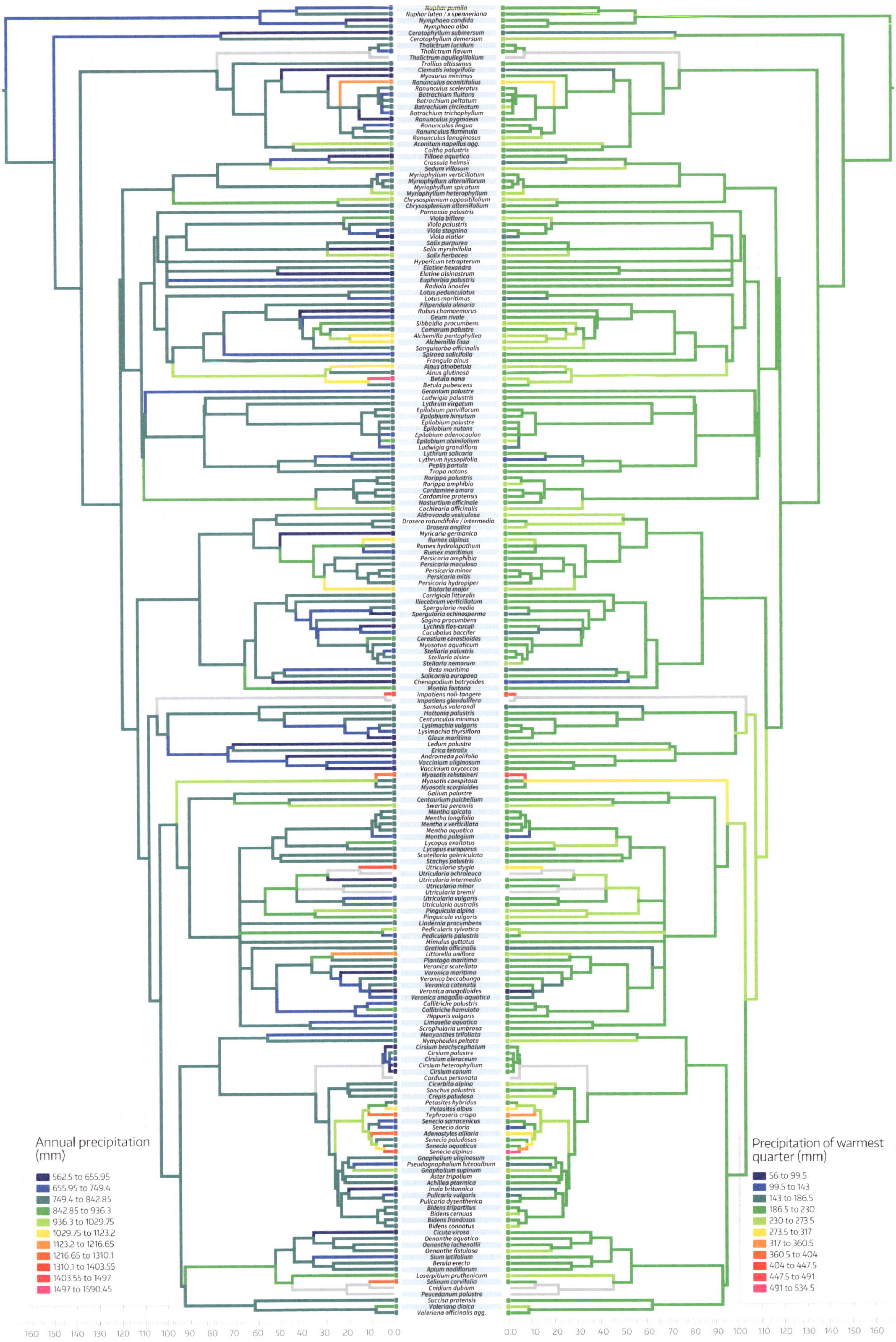

Figure 3. The phylogenetic tree of studied dicot species with a maximum likelihood reconstruction of selected species climatic preferences (Left: Annual precipitation, Right: Precipitation of the warmest quater).

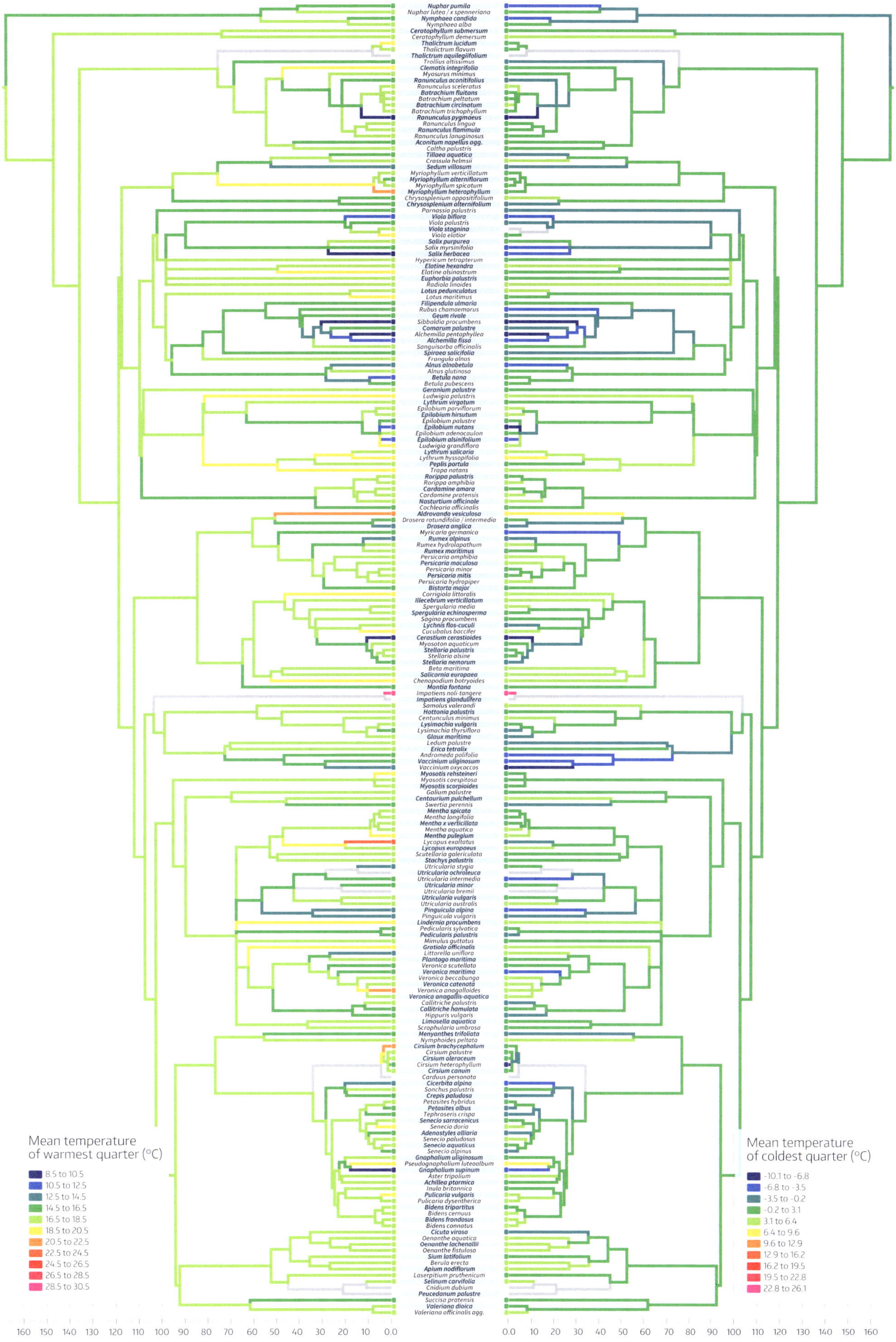

Figure 4. The phylogenetic tree of studied dicot species with a maximum likelihood reconstruction of selected species climatic preferences (Left: Mean temperature of the warmest quater, Right: Mean temperature of the coldest quater).

Analysis of the Ellenberg ecological indicator values (EIV, Chytrý et al. 2018) showed that aquatic floating plants, terrestrial short-lived taxa and the studied trees and shrubs have higher affinities for well-irradiated habitats (Figure 5 and 6), while tap-rooted perennial plants and clonal species with tubers and short rhizomes have higher affinities for lower soil moisture (Figure 7). Short-lived terrestrial plants, root sprouters, tuberous plants and aquatic emergent plants prefer more fertile soils with higher pH values. Tap-rooted perennial species, tuberous plants and short-lived species showed a higher tolerance of saline soils than other plant groups (Figure 8).

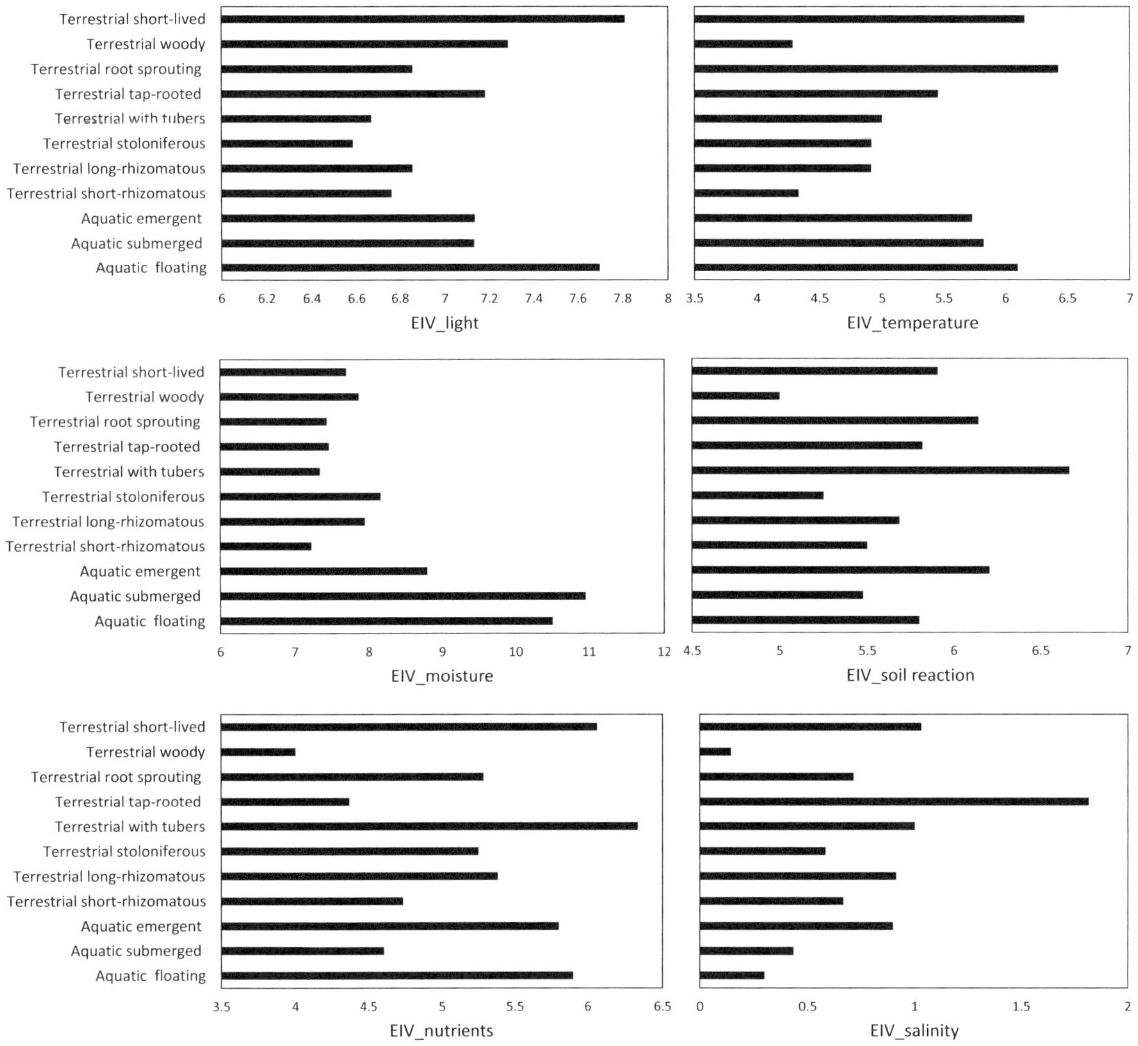

Figure 5. Aquatic and wetland plant life and growth form categories and their relation to ecological indicator values.

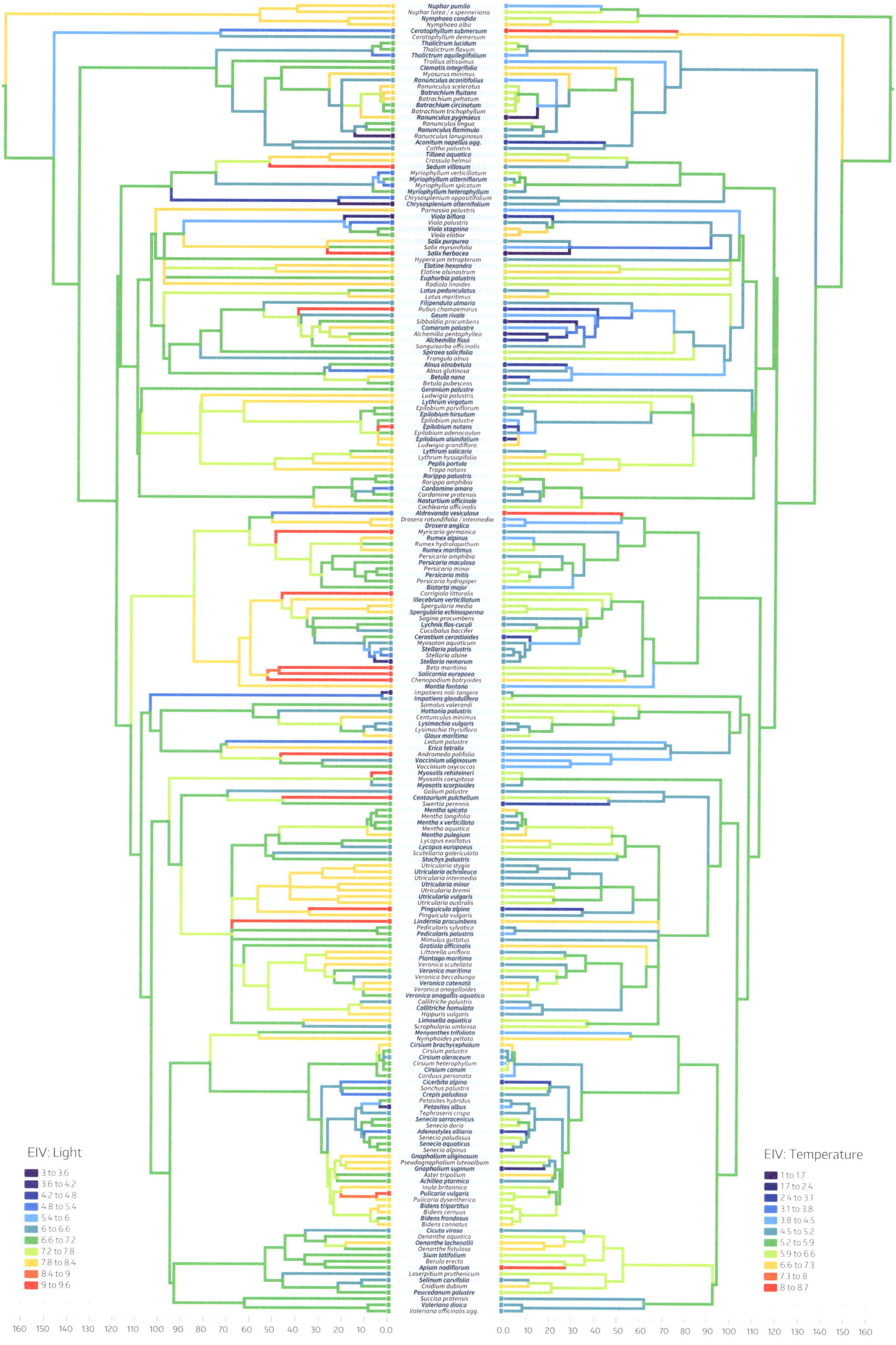

Figure 6. The phylogenetic tree of studied dicot species with a maximum likelihood reconstruction of species indicator values (Left: EIV for light, Right: EIV for temperature).

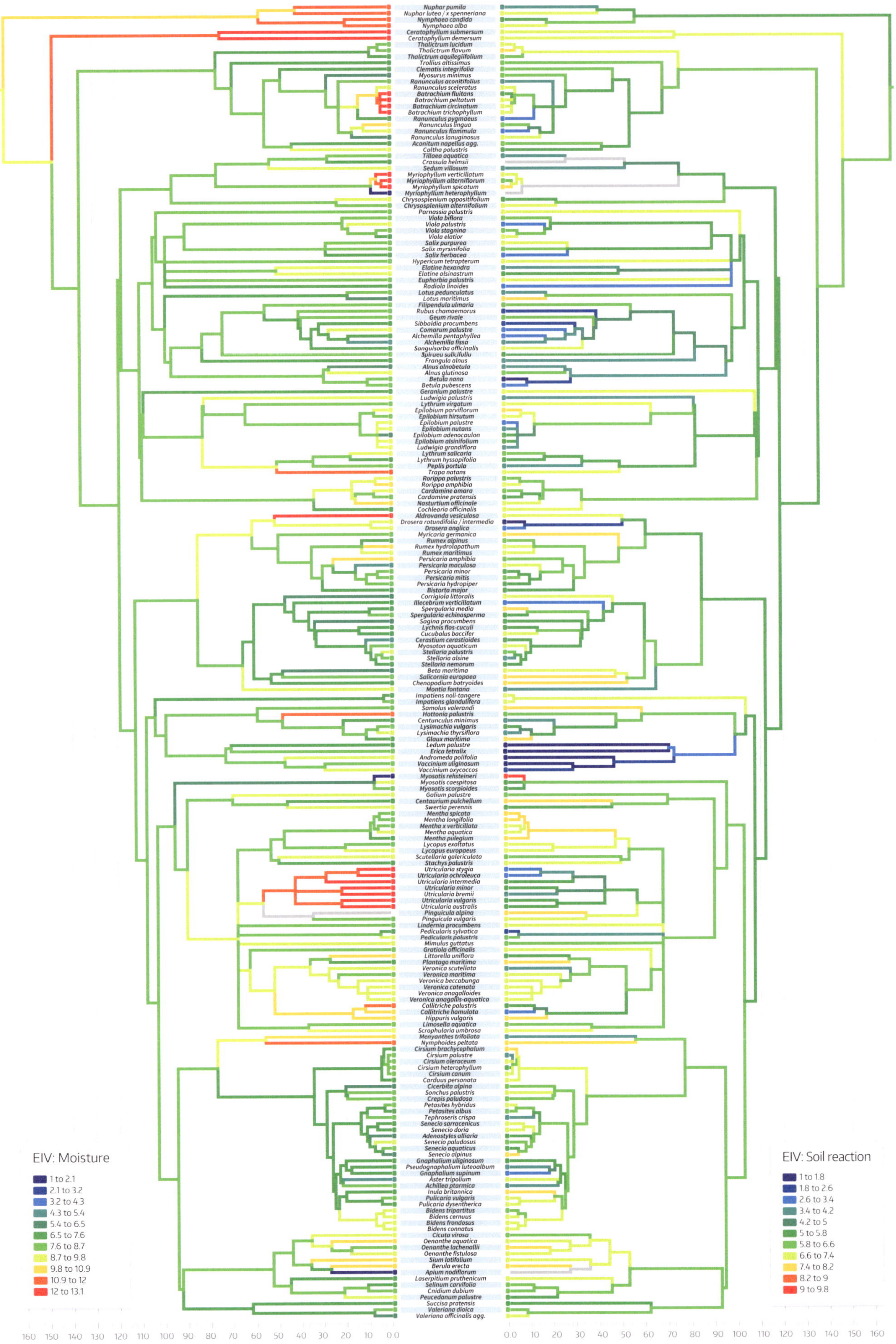

Figure 7. The phylogenetic tree of studied dicot species with a maximum likelihood reconstruction of species indicator values (Left: EIV for moisture, Right: EIV for soil reaction).

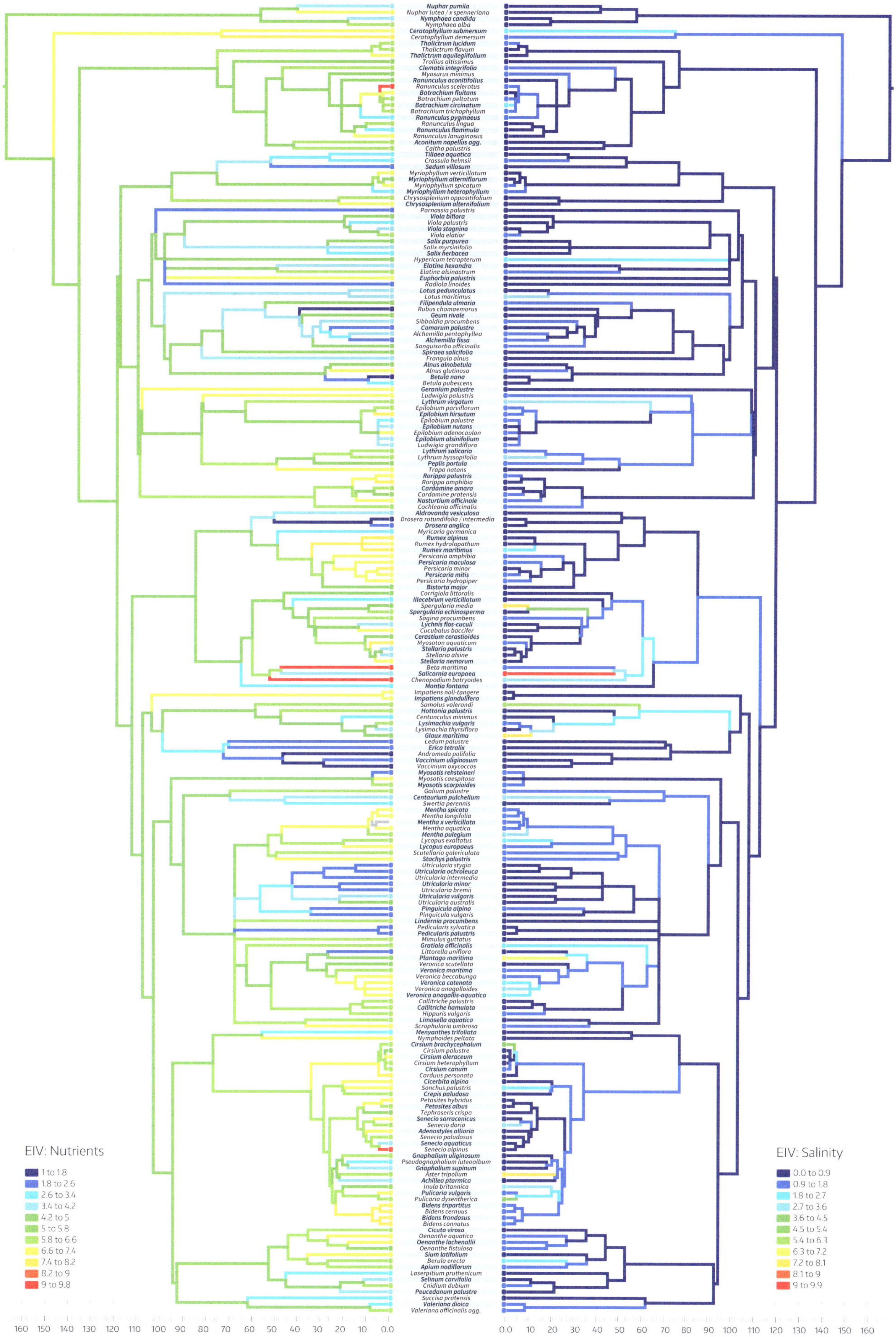

Figure 8. The phylogenetic tree of studied dicot species with a maximum likelihood reconstruction of species indicator values (Left: EIV for nutrients, Right: EIV for salinity).

Analysis of selected functional traits extracted from LEDA (Kleyer et al. 2008) and CLO-PLA databases (Klimešová et al. 2017) showed that the plants achieving the largest stature (plant height) are woody root sprouters and splitters (wetland shrubs, trees and liana such as *Cucubalus baccifer*), followed by tall wetland long-rhizomatous forbs from genera *Epilobium*, *Lysimachia*, *Lythrum*, *Mentha* and *Thalictrum* (Figure 9 and 10). Woody root sprouters and tall long-rhizomatous species have higher leaf dry matter content (LDMC; i.e. lower leaf water content) and lower specific leaf area (SLA) (Figure 11). The opposite is true for aquatic plants with turions and for short-lived plants and wetland plants with stem tubers and stolons, typical of lower LDMC and high SLA. A high ability of clonal propagation characterized by the clonality index and lateral spread is found in aquatic submerged plants and in wetland plants with stolons, long rhizomes, stem tubers and root sprouting.

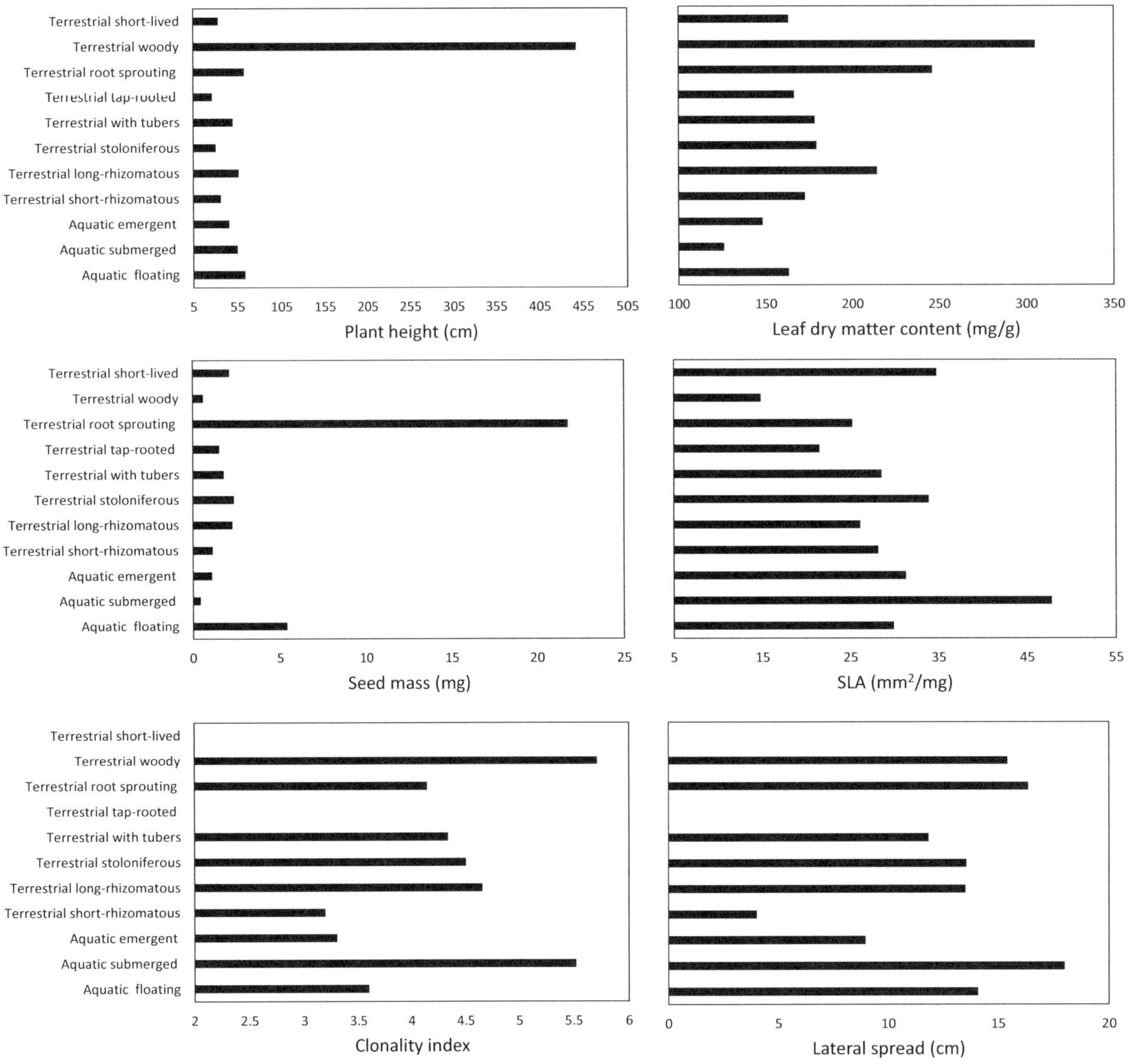

Figure 9. Comparison of selected functional traits among aquatic and wetland plant life and growth form categories containing 212 species.

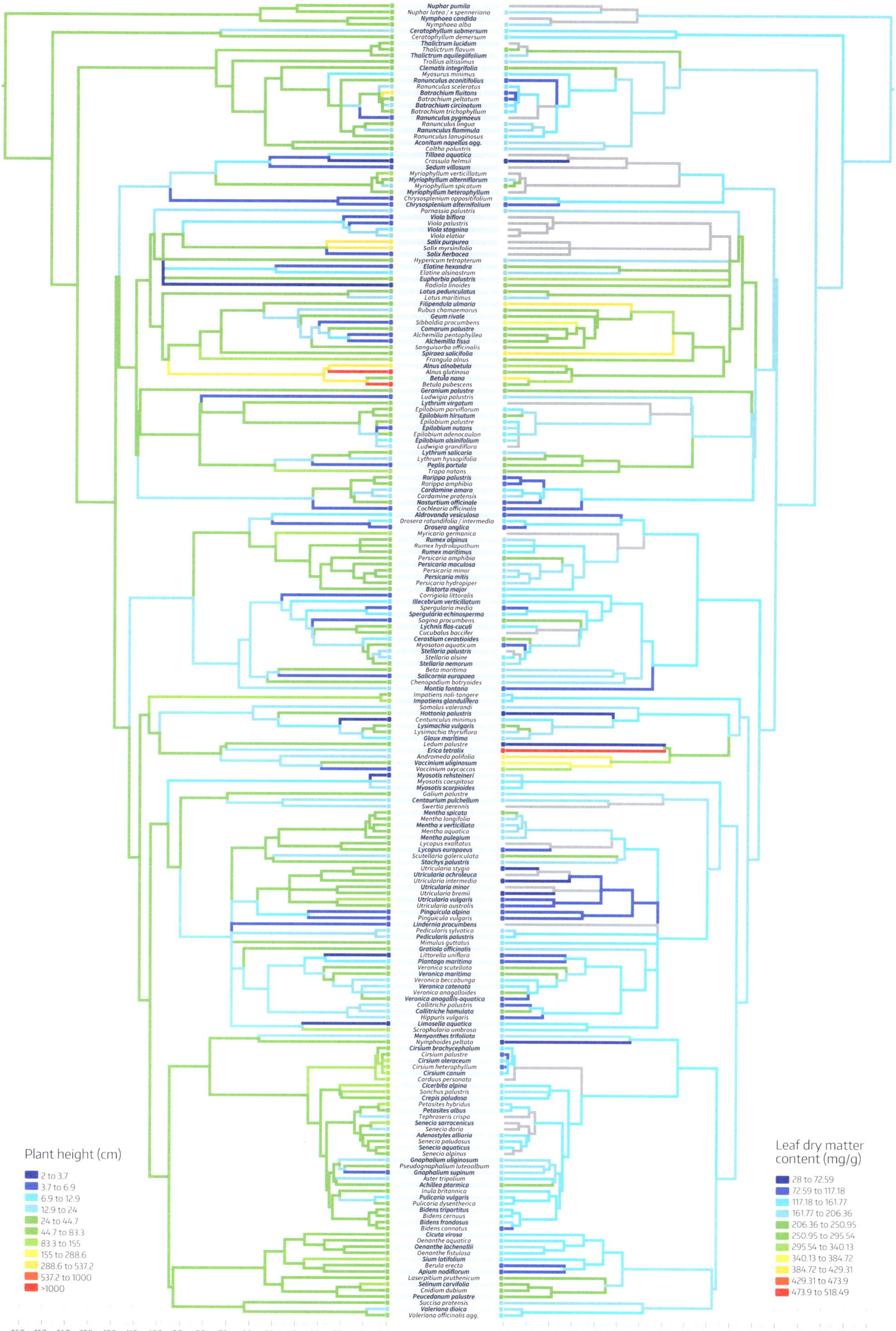

Figure 10. The phylogenetic tree of studied dicot species with a maximum likelihood reconstruction of selected functional traits (Left: Plant height, Right: LDMC).

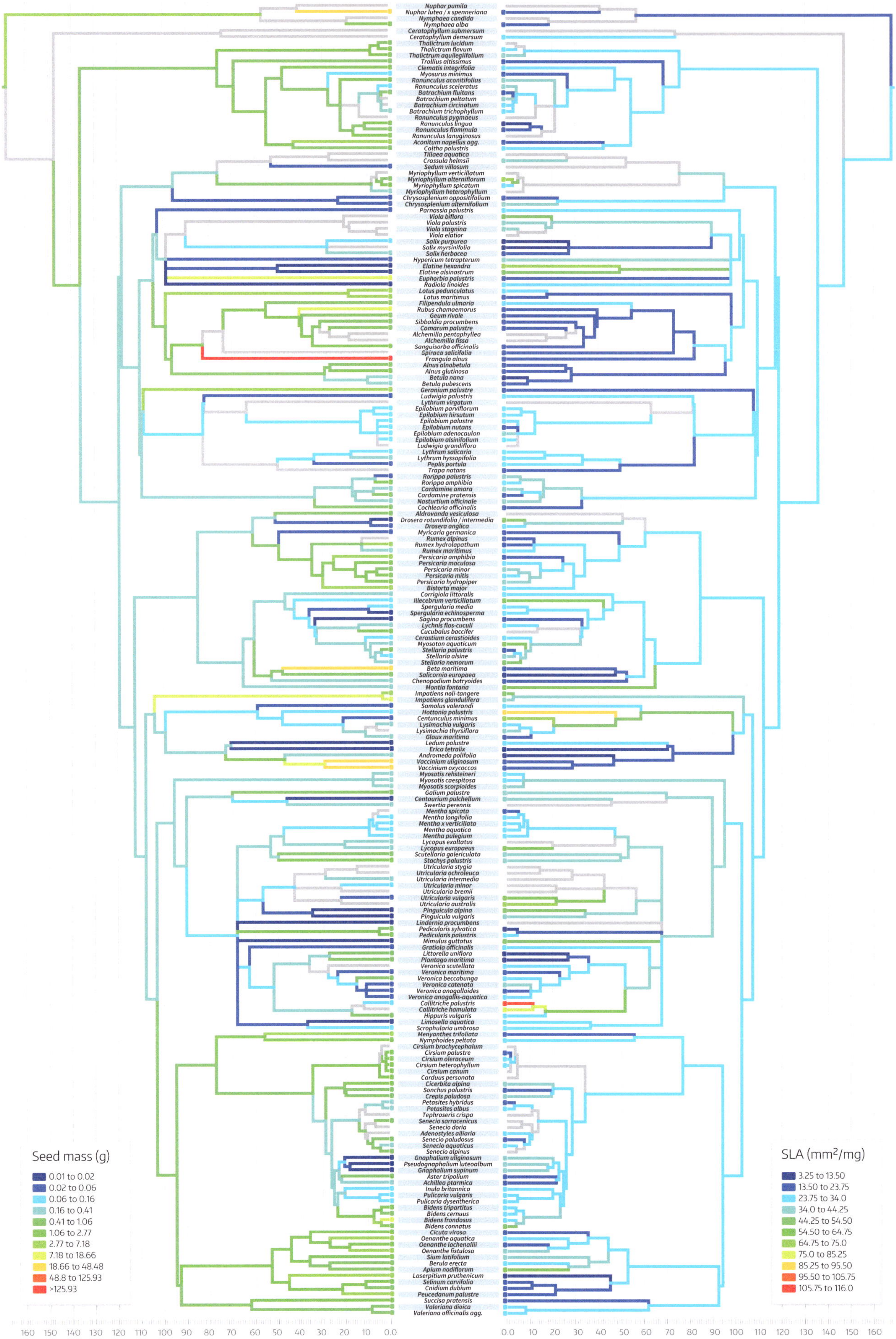

Figure 11. The phylogenetic tree of studied dicot species with a maximum likelihood reconstruction of selected functional traits (Left: Seed mass, Right: SLA).

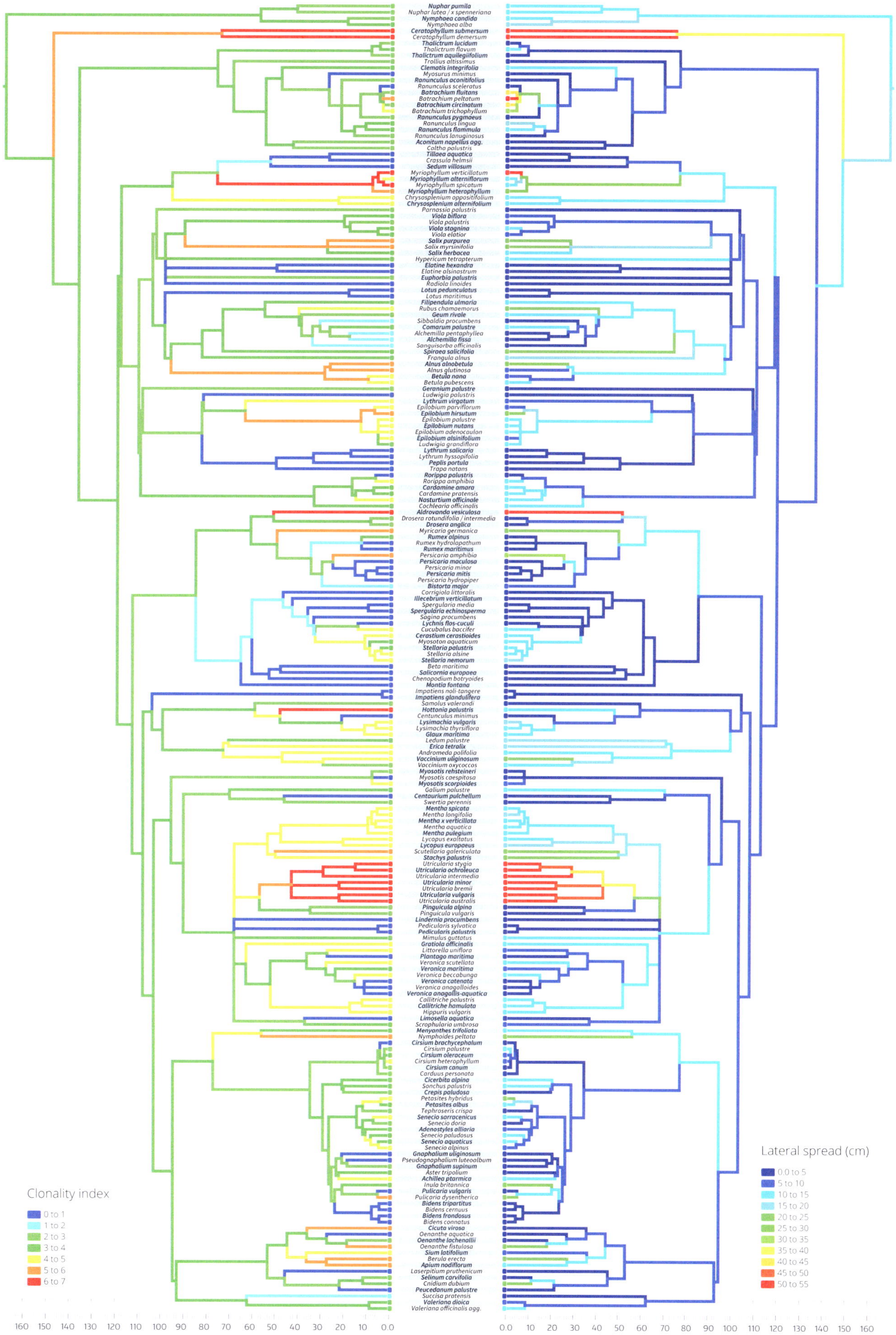

Figure 12. The phylogenetic tree of studied dicot species with a maximum likelihood reconstruction of selected functional traits (Left: Clonality index, Right: Lateral spread).

Anatomy of dicotyledonous species

Principal Component Analysis of all anatomical features of aquatic and wetland dicots has revealed their interrelationships with and affinities to plant growth form and soil and climate drivers. The ecological attributes (climate variables and ecological indicator values, EIV) and growth forms had a significant effect on the divergence of anatomical traits. The main anatomical differences along the first (horizontal) ordination axis are associated with habitat preferences along moisture gradients, from fresh waters to drier banks and floodplain forest floors. This means that the first axis separated aquatic plants, often propagating by turions and belonging to families Ceratophyllaceae, Elatinaceae, Haloragaceae and Lentibulariaceae from perennial species with long hypogeogenous rhizomes or species with partly or entirely woody stems belonging mostly to Asteraceae, Betulaceae, Lamiaceae and Salicaceae. The aquatic plants have anatomical structures characterized by regular round or honeycomb lysigenous (also called expansigenous) aerenchyma, absent or solitary vessels, scalariform intervessel pitting, pervasive parenchyma and mostly solitary sieve elements and missing endodermis, fibers and bark secretory elements. The anatomical structure of wetland tall forbs and woody root sprouters from the opposite end of the moisture gradient is characterized by thick-walled libriform fibers, vessels with a tangential diameter of 50–100 µm arranged in short radial multiples, alternate intervessel pitting, absent or small intercellulars and diffuse-porous rings. The changes along the second axis seem to be associated with decreasing growing season temperature and annual precipitation (Figure 13) from short-lived warm-adapted plants of well-irradiated open habitats (e.g. Crassulaceae, Gentianaceae, Plantaginaceae) to cold-adapted, large wet meadow forbs with short rhizome and subshrubs typically from Ranunculaceae and Rosaceae (Figure 14). Short-lived warm-adapted plants have one ring and the endodermis is largely absent They have simple perforation plates, small vessel diameters, absent fibers or libriform fibers and absent rays. Perennial forbs, though, have distinct rings with the average plant age ranging between 2 and 10 years. They have large multiseriate upright rays and small or absent intercellulars. The endodermis is present, they have small vessels with a tangential diameter of 20–50 µm, they have a vasicentric paratracheal and a pervasive parenchyma.

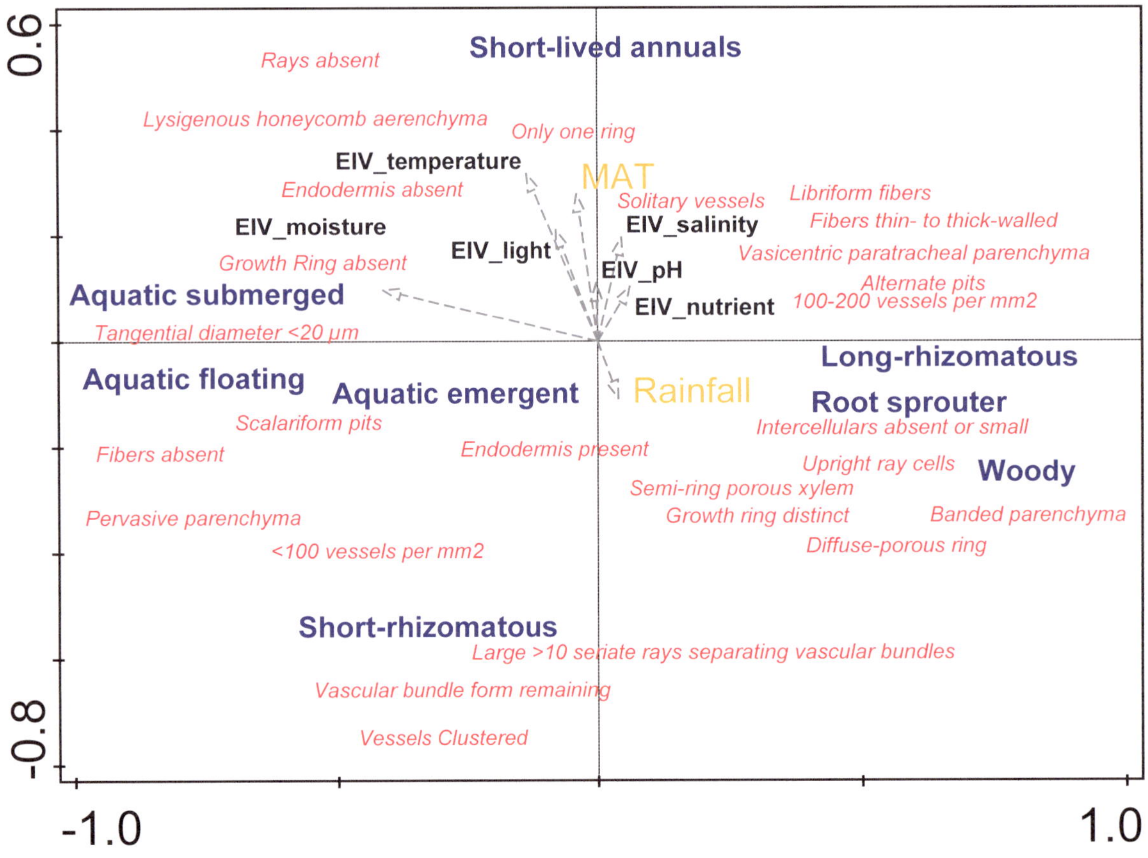

Figure 13. Principal Component Analysis of plant anatomical features in relation to plant growth and life forms, ecological indicator values (EIV), mean annual temperature (MAT) and annual precipitation. Environmental variables explained altogether 21% of the variability in anatomical stem structure.

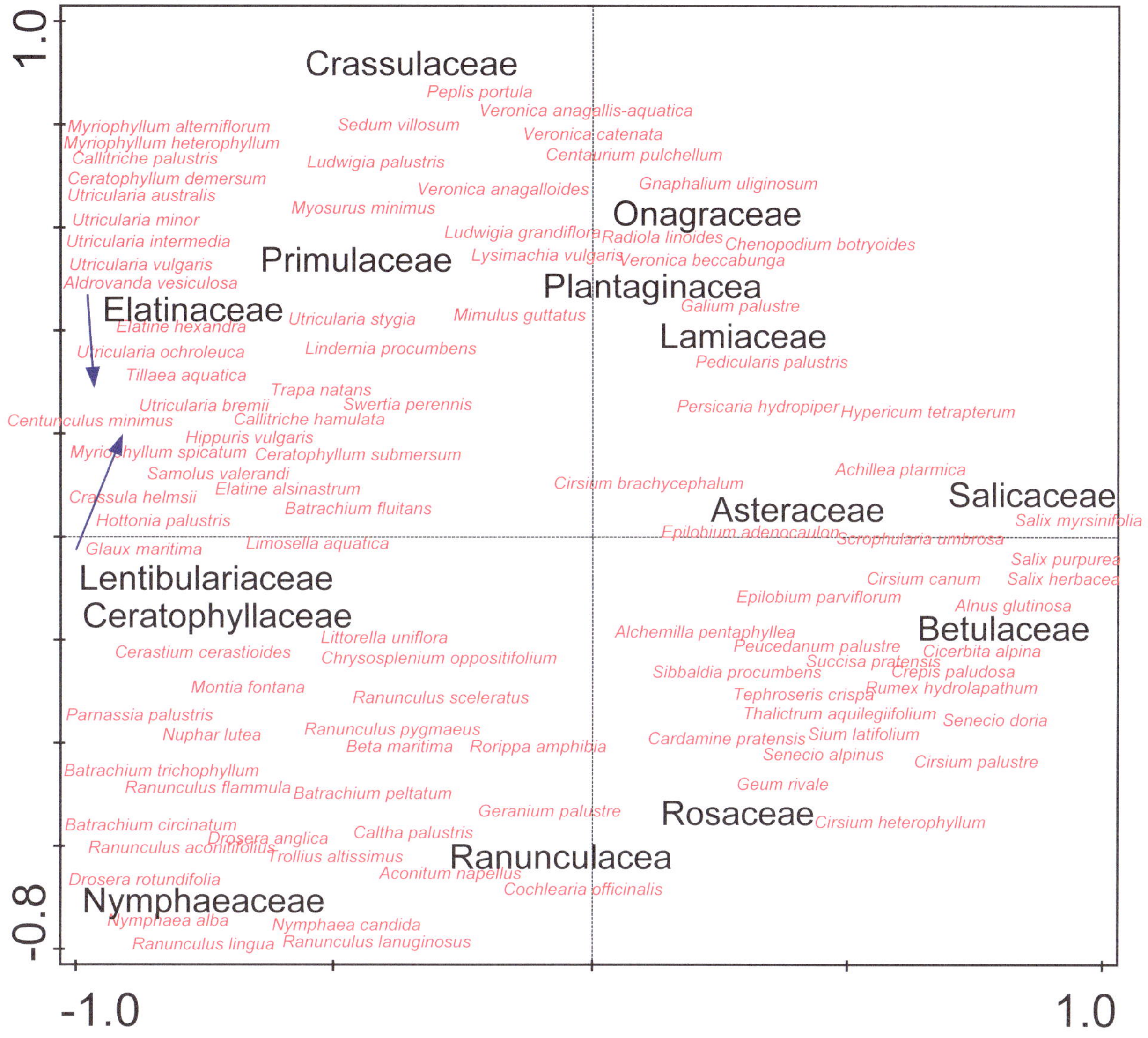

Figure 14. Principal Component Analysis of plant anatomical features in relation to species composition and plant families (for corresponding anatomical features see Figure 13).

Plant age and growth rate

Countable annual rings are present in 42% of the perennial plants studied (Figure 15), however, the degree of distinctness varies, due to taxonomy or environmental conditions. The age can be determined for the majority of tap-rooted species and root sprouters (mostly woody species) and for about half of the rhizomatous forbs. The values vary between families, genera and species. Most aquatic and many wetland plants (e.g., *Amaranthaceae, Haloragaceae, Nymphaeaceae, Ranunculaceae*) have no countable rings. Annual plants have one annual ring. Species 2–10 years-old prevail (91%) among the studied perennial plants. Among the species that were 10–20 years old are forbs (*Lythrum salicaria* and *Sanguisorba officinalis*) and subshrubs (*Andromeda polifolia* and *Sibbaldia procumbens*). A fraction of longer-lived species belongs to wetland subshrubs (e.g. *Erica tetralix, Myricaria germanica, Rubus chamaemorus*), shrubs and trees (*Alnus, Betula* and *Salix* spp.).

The majority of annual rings were between 0.1 to 0.5 mm wide (Figure 15). Shorter vegetation periods and lower summer temperatures limit the growth of high-elevation (montane) wetland short-rhizomatous forbs with radial growth increments <0.5 mm per year. Slow growth and higher longevity is a characteristic of most rhizomatous forbs growing in cold mountainous habitats. Tap-rooted perennial forbs such as *Lythrum salicaria, Spergularia media* and *Lotus maritimus* have increments around 0.5 mm or wider such as in *Laserpitium pruthenicum* (2.5 mm), *Peucedanum palustre* (0.7 mm), *Plantago maritima* (0.6 mm) and *Rumex hydrolapathum* (5 mm). Species with increments >1 mm form only 17% of all studied taxa, with the highest share among the low-elevation shrubs (e.g., *Salix purpurea*), long-rhizomatous forbs (genera *Epilobium, Senecio* and *Sonchus*) and short-lived wetland species such as *Impatiens noli-tangere, Persicaria maculosa, P. mitis* and *Bidens* spp. (Figure 16). The growth rate is clearly positively related to plant height ($P < 0.05$, $R^2 = 0.28$), being smallest in short-rhizomatous plants, and highest in tall herbaceous tap-rooted forbs and long-rhizomatous forbs (e.g. *Filipendula ulmaria, Lysimachia vulgaris, Mentha longifolia, Thalictrum flavum, Sonchus palustris*) and subshrubs (e.g. *Vaccinium uliginosum, Ledum palustre*).

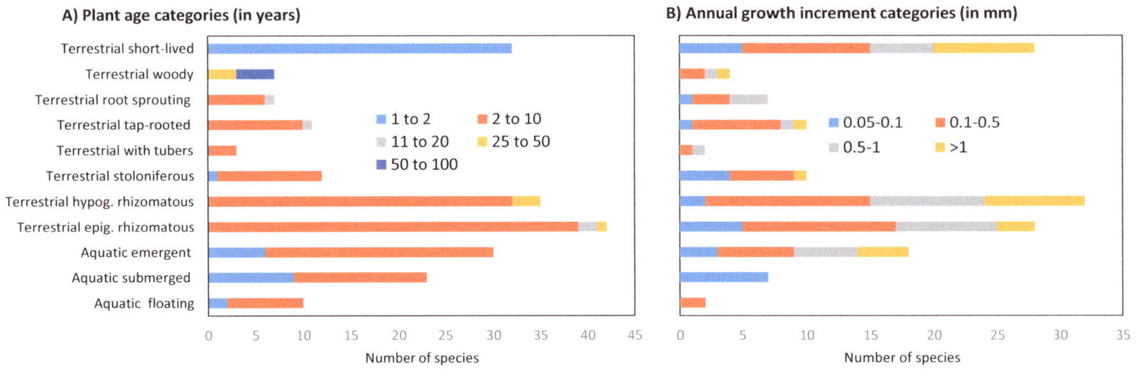

Figure 15. Plant age and annual ring width increments of 212 dicotyledonous plants in relation to their life (aquatic versus terrestrial wetland) and growth (based on clonality) forms.

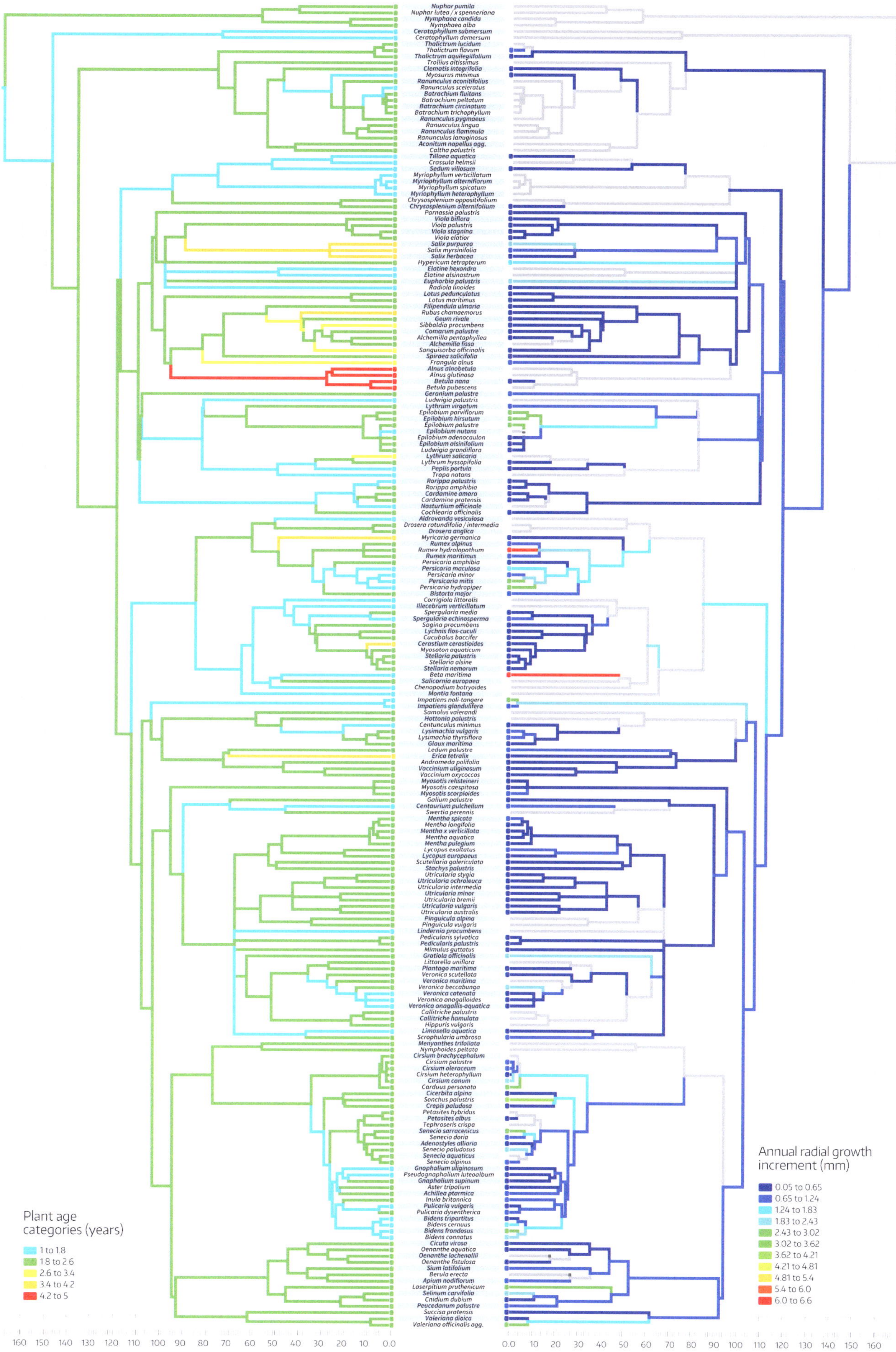

Figure 16. The phylogenetic tree of studied dicot species with a maximum likelihood reconstruction of selected features (Left: Plant age, Right: Annual ring width).

The structure of the conductive system

The structure of the water conductive system varies among the studied species due to taxonomy and environmental conditions. Among species with distinct rings, **diffuse porosity** (56%) and **semi-ring porosity** (42%) dominate. Ring porosity is rare and restricted to woody shrubs such as *Myricaria germanica*. Semi-ring porosity is characteristic of tap-rooted forbs and root-sprouting shrubs such as *Lotus maritimus, Lythrum salicaria, Frangula alnus* and *Spiraea salicifolia* and tall rhizomatous wetland forbs such as *Cirsium oleraceum, Filipendula ulmaria* and *Tephroseris crispa*. Vessels are missing in *Aldrovanda* and *Ceratophyllum* species.

Vessels are primarily solitary (37%) or in short radial multiples (22%), and less frequently in long radial multiples (7%), tangential bends (4%) or clusters (8%) (Figure 17). Solitary vessels are mostly represented in aquatic submerged plants and terrestrial long-rhizomatous forbs, while vessels grouped in multiples are more represented in woody plants and short-rhizomatous forbs, and aquatic emergent forbs such as *Salicornia europaea* having long radial multiples, *Hippuris vulgaris* with vessels in tangential bends and *Elatine alsinastrum* with diagonal/radial vessel pattern. Dendritic vessels are rare among the studied species and restricted to *Epilobium nutans* (Figure 18).

Vessel number varies between 50 and 500 per mm^2 (Figure 17). Species with 100-200 vessels per mm^2 prevail (49%) across all life and growth forms except for aquatic submerged, followed by plants with >200 vessels per mm^2 (38%). The smallest number of vessels <100 per mm^2 is found in aquatic floating and submerged plants, while higher vessel numbers are found in terrestrial rhizomatous perennial forbs, annual species and also in aquatic emergent plants.

Vessel tangential diameter varies between <20 to a maximum of 200 μm. A fraction of species with small diameter vessels <20 μm is high among submerged aquatic plants, while those with larger vessels >50 μm is high among tap-rooted non-clonal forbs and long-rhizomatous plants. Variation in vessel diameter is significantly related to plant height, with small, <10 cm tall annuals (e.g. *Centunculus minimus, Illecebrum verticillatum, Peplis portula* and *Spergularia echinosperma*) and short rhizomatous plants (e.g. *Pinguicula* and *Viola* species) having the smallest diameters (<20 μm) and larger forbs >60 cm tall having the largest diameters (100–200 μm) such as *Impatiens glandulifera* and *Inula britannica*. Most of the woody species from the genera *Alnus, Betula, Frangula* and *Salix* and tall wetland forbs from the genera *Cirsium, Mentha* and *Lythrum* have tangential vessel diameters in the 50–100 μm range.

Simple perforation plates are characteristic for 81% of all species. Scalariform perforation plates are found in only 10% of species across unrelated genera, from short-lived ones (*Elatine alsinastrum, Bidens* spp.) to perennial stoloniferous (*Stellaria* spp.) and rhizomatous herbs (*Menyanthes trifoliata, Senecio paludosus*), to heavily lignified subshrubs (*Erica tetralix, Ledum palustre, Rubus chamaemorus, Vaccinium uliginosum*) and tree species (*Alnus, Betula*). A high proportion of species have alternate intervessel pits (52%), followed by scalariform pits (35%), with a clear habitat- and taxonomy-related variation. Scalariform intervessel pitting is overrepresented in aquatic plants and underrepresented in terrestrial wetland plants. In particular, aquatic submerged plants represent a high fraction of species (*Batrachium*) or exclusive species (*Callitriche, Ceratophyllum* and *Utricularia*) with scalariform intervessel pitting. Vessels with scalariform intervessel pits are frequently embedded in pervasive parenchyma (Figure 18). The absence of tyloses or phenolic plugs in most species (91%) indicates that all vessels are functional during their time of existence. Tannins are present in *Rumex alpinus* and *Sedum villosum*, while tyloses are present in *Petasites hybridus*.

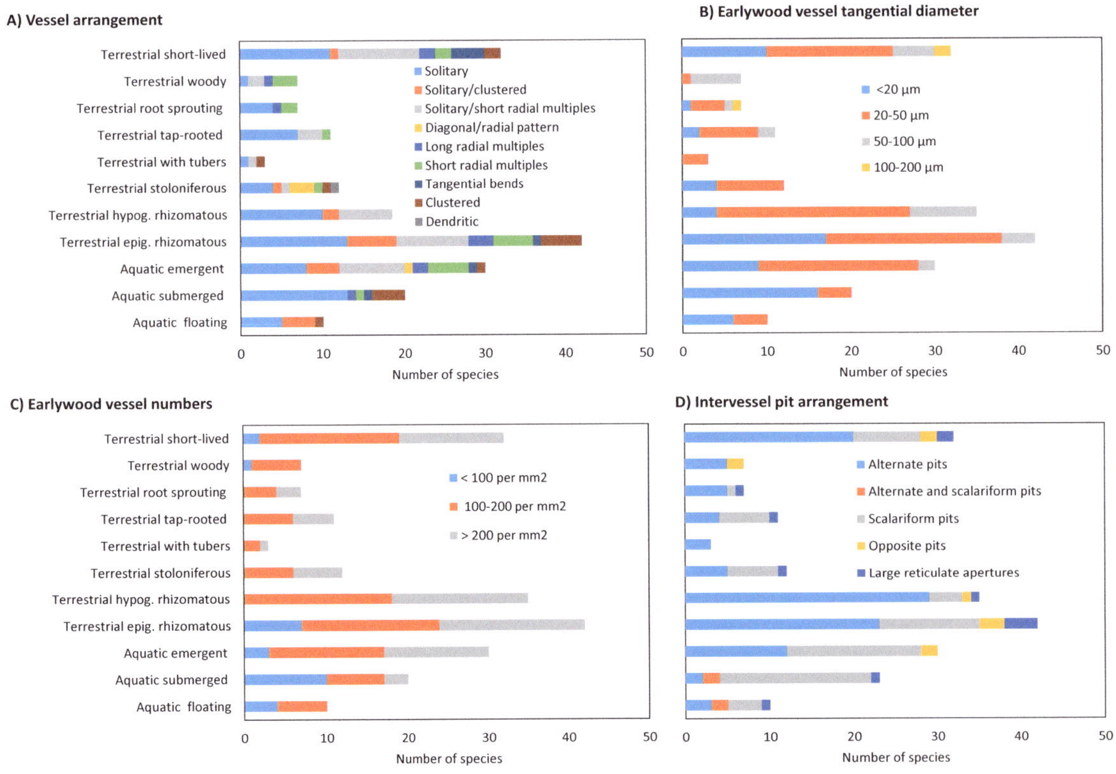

Figure 17. Numbers of species with specific vessel parameters (arrangement, numbers, size, intervessel pitting) in individual life and growth form categories, based on 212 aquatic and wetland plant species.

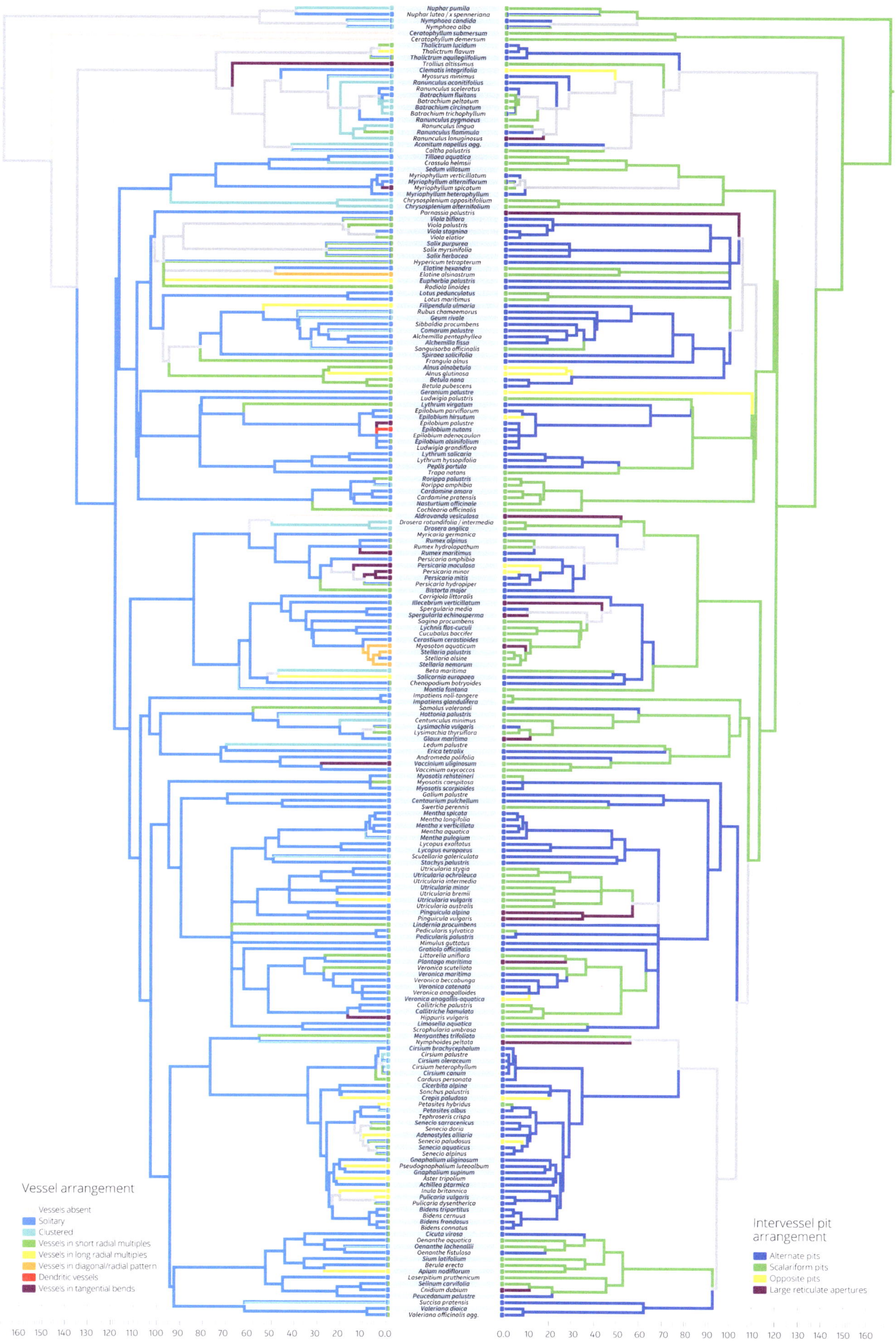

Figure 18. The phylogenetic tree of studied dicot species with a maximum likelihood reconstruction of selected features (Left: Vessel arrangement, Right: Intervessel pit arrangement).

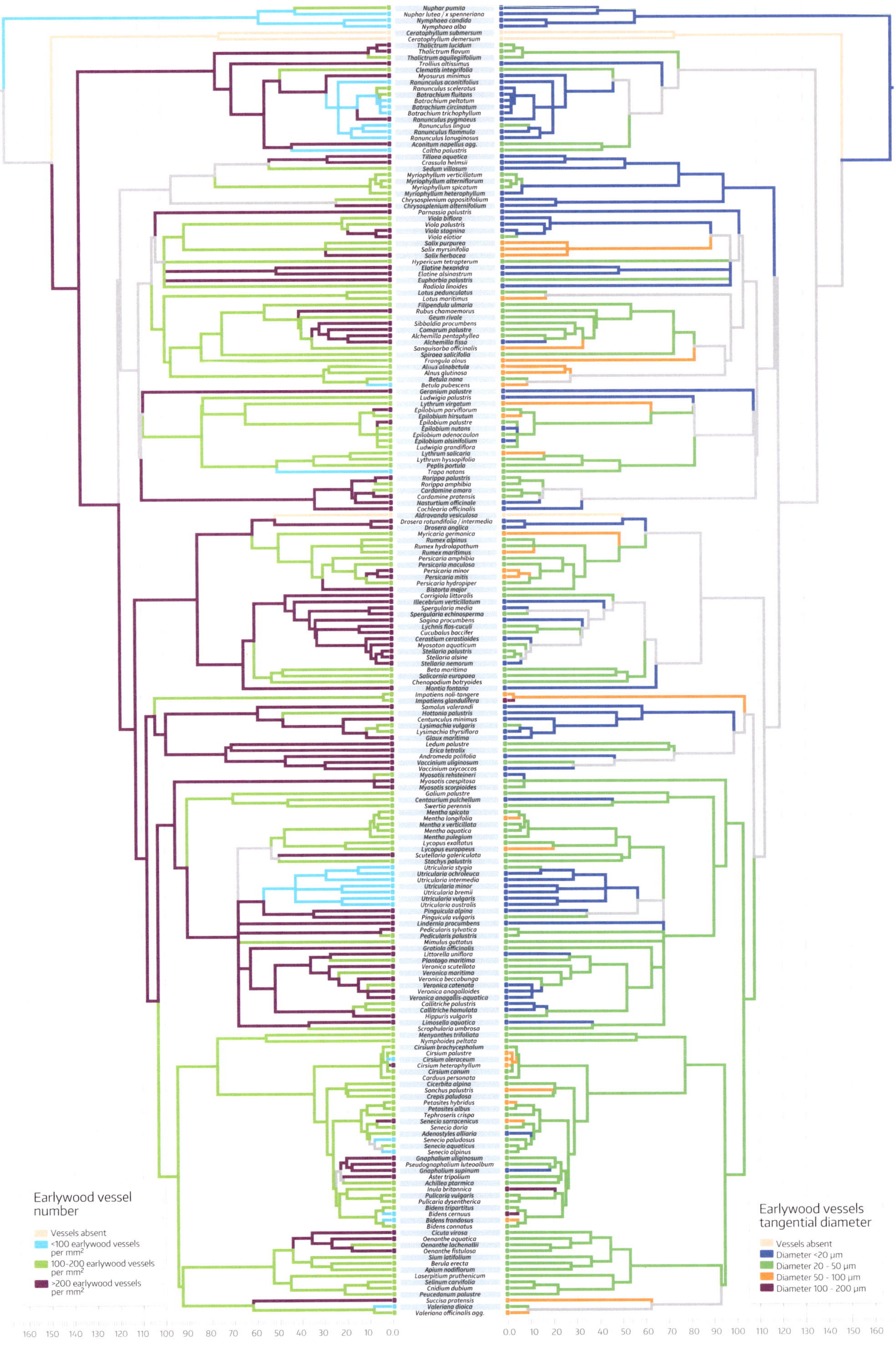

Figure 19. The phylogenetic tree of studied dicot species with a maximum likelihood reconstruction of selected features (Left: Earlywood vessel number, Right: Earlywood vessel tangetial diameter).

The structure of stabilization system and lignification

A high number of the studied plants were without fibers (42%), which can be explained by their small size and aquatic habitat. This means that that the majority of the total biomass is allocated to submerged/below-ground or floating organs and only a small proportion to aboveground/emergent shoots and leaves, resulting in a limited need for mechanical stem support (Figure 20). There is a clear relationship between fiber proportion and elevation and plant size. Of those plants shorter than 10 cm, 66% have no fibers. These are mostly small short-lived wetland plants such as *Plantago maritima, Sagina procumbens, Spergularia* spp., *Tillaea aquatica* and *Chrysosplenium* spp., or alpine plants such as *Pinguicula alpina* and *Ranunculus pygmaeus* (Figure 21). Thin- to thick-walled libriform fibers are found in the majority of plants with fibers (77%), with thick-walled fibers in 10% and thin-walled fibers in 11% of the studied species. Fiber-tracheids occur in unrelated species, all having rather larger stature and sharing the ability of clonal multiplication either through lignifying rhizomes (*Filipendula ulmaria, Geum rivale, Sanguisorba officinalis*) or root-sprouting (*Spiraea salicifolia*). Thick-walled fibers are more represented in tall, rhizomatous forbs (*Oenanthe lachenallii, Petasites hybridus, Veronica maritima*) and root-sprouting forbs and shrubs (*Inula britannica, Spiraea salicifolia*).

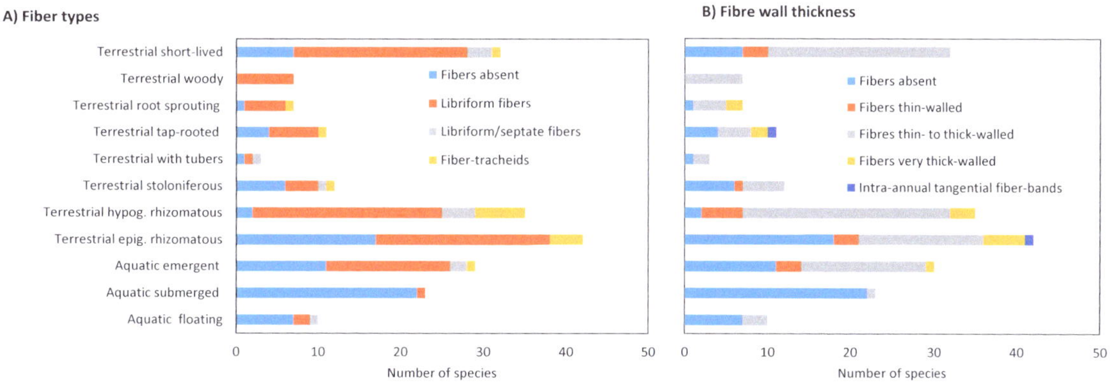

Figure 20. Numbers of species with specific fiber and fiber wall thickness types in individual life and growth form categories, based on 212 aquatic and wetland plant species.

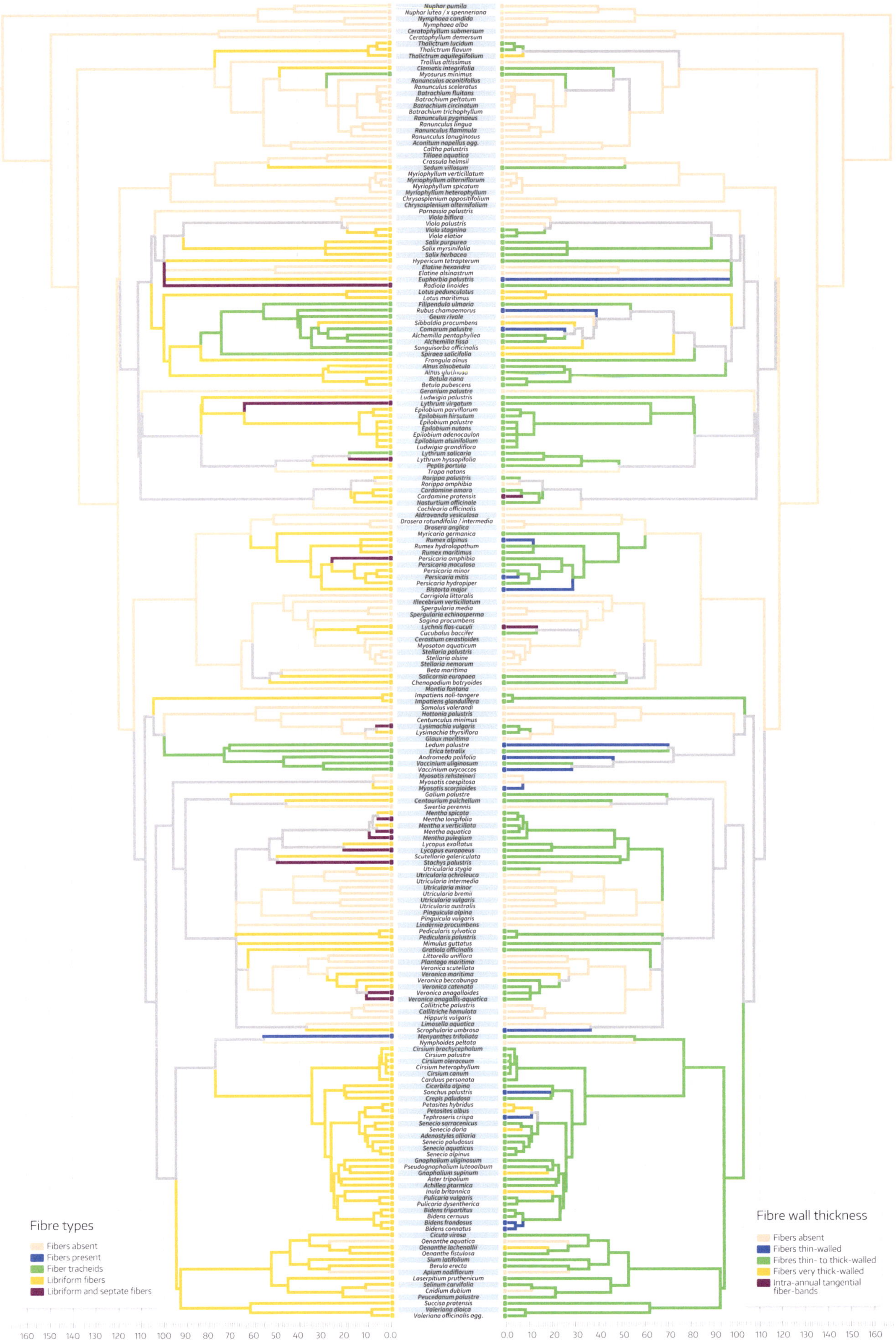

Figure 21. The phylogenetic tree of studied dicot species with a maximum likelihood reconstruction of selected features (Left: Fiber types, Right: Fiber wall thickness).

The structure of storage system

Axial parenchyma cells and rays form the storage system in the xylem (Figure 22). However, the largest part of the storage system can be also in the bark and mainly in the cortex. There was an obvious high proportion of species with much parenchyma in the xylem, mostly pervasive parenchyma (30%, Figure 22) and also large multiseriate ray zones between permanent vascular bundles (41% of all species). Related to aquatic conditions is the dominance of unlignified tissues in the submerged plants and this is indicated by the blue aspect of the Astrablue/Safranin-stained slides. Plants with very large unlignified cortex zones and pervasive parenchyma were dominant (Figure 23). Pervasive parenchyma is an almost exclusive feature in aquatic submerged plants (genera *Aldrovanda, Batrachium, Ceratophyllum, Myriophyllum* and *Utricularia*) and floating-leaved plants (*Hottonia, Nuphar, Nymphaea, Nymphoides* and *Trapa*), and a common feature among emergent aquatic plants (*Elatine, Hippuris, Menyanthes*) and terrestrial short-rhizomatous clonal plants (e.g., *Caltha palustris, Drosera, Parnassia palustris, Valeriana dioica* and *Pinguicula* spp.). A high number of the studied plants have vasicentric paratracheal parenchyma (35%) or combined pervasive and vasicentric parenchyma (17%). The proportion of species with vasicentric paratracheal parenchyma prevails among tall long-rhizomatous forbs such as *Cicerbita alpina, Epilobium hirsutum, Filipendula ulmaria, Scrophularia umbrosa, Mentha* and *Thalictrum* species, and is also common among short-rhizomatous forbs such as *Bistorta major, Cirsium oleraceum, Petasites albus* and *Senecio aquaticus*. Vasicentric paratracheal parenchyma is also the most common parenchyma type in short-lived annual and biennial species such as *Illecebrum verticillatum, Impatiens glandulifera* and *Pulicaria vulgaris*. Apotracheal parenchyma is rare (3%) as it is a feature belonging to trees and shrubs. Banded paratracheal parenchyma is absent in the majority of studied species (89%), being restricted to trees (*Alnus*), shrubs (*Myricaria, Salix*), subshrubs (*Vaccinium uliginosum*) and large forbs (e.g., *Cirsium* spp., *Crepis paludosa, Lysimachia thyrsiflora, Veronica maritima*).

A high proportion of parenchyma in the xylem in any growth form just demonstrates that the survival and regeneration of studied wetland plants is anatomically based on an extended storage system. The plant's need for storing carbohydrates is reflected in the prevalence of multiseriate rays (55%), while exclusively uniseriate rays are less abundant (12%) (Figure 22). Most represented are plants with >10 seriate rays separating vascular bundles (42%), 4–10 seriate (7%) and 1–3 seriate (6%) rays. There is a clear relationship between ray width and temperature, the lower the plant temperature optima, the wider the rays. Multiseriate rays separating vascular bundles are an almost exclusive feature in short-rhizomatous forbs that prevail in alpine wetlands (e.g., *Pinguicula alpina, Ranunculus aconitifolius, Ranunculus pygmaeus, Rumex alpinus, Senecio alpinus* and *Sibbaldia procumbens*) and wet oligotrophic meadows (e.g., *Cirsium palustre, Selinum carvifolia, Succisa pratensis* and *Tephroseris crispa*) (Figure 23). Since parenchyma cells store mobile carbohydrates (starch, fructans) these systems are optimized to survive intra-seasonal frost events or long lasting cold periods. Exclusively uniseriate rays are found in trees, shrubs and large lignified perennial forbs from genera *Epilobium, Mentha, Lythrum* and short-lived herbs such as *Pseudognaphalium luteoalbum, Lythrum hyssopifolia* and *Persicaria maculosa* (Figure 24). Plants without rays form 32%, being the most represented among aquatic floating-leaved and submerged plants (all Ceratophyllaceae, Haloragaceae and Lentibulariaceae), while some submerged rooted plants such as *Batrachium circinatum, B. peltatum, B. trichophyllum* and *Littorella uniflora* have vascular bundles separated by large multi-seriate rays. Most plants have rays with upright cells (67%) and in many cases, they are confluent with the axial parenchyma. Since stem-radial stress in herbaceous plants is minimal there is no need to form prostrate cells. Rays with prostrate (procumbent) ray cells (3%) occur mainly in shrubs and trees (*Alnus, Betula, Frangula*).

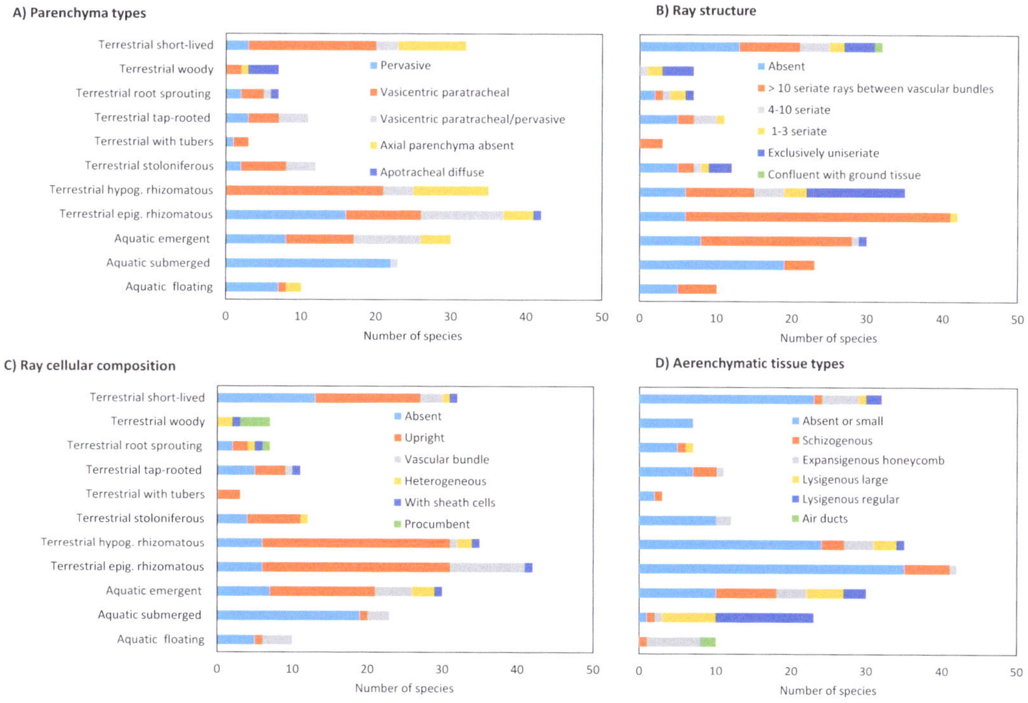

Figure 22. Numbers of species with specific parenchyma, ray and aerenchyma tissue types in individual life and growth form categories, based on 212 aquatic and wetland plant species.

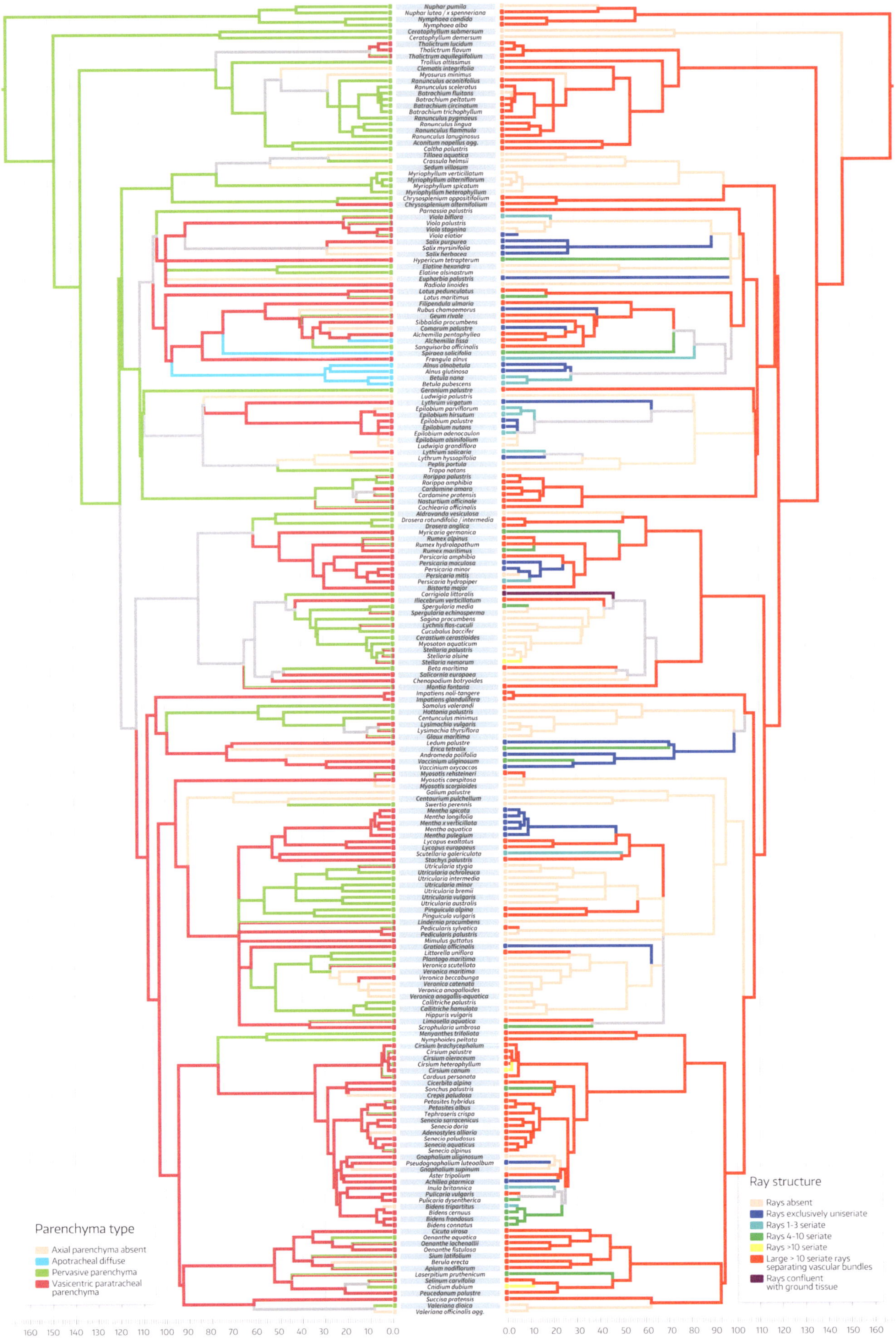

Figure 23. The phylogenetic tree of studied dicot species with a maximum likelihood reconstruction of selected features (Left: Parenchyma type, Right: Ray structure).

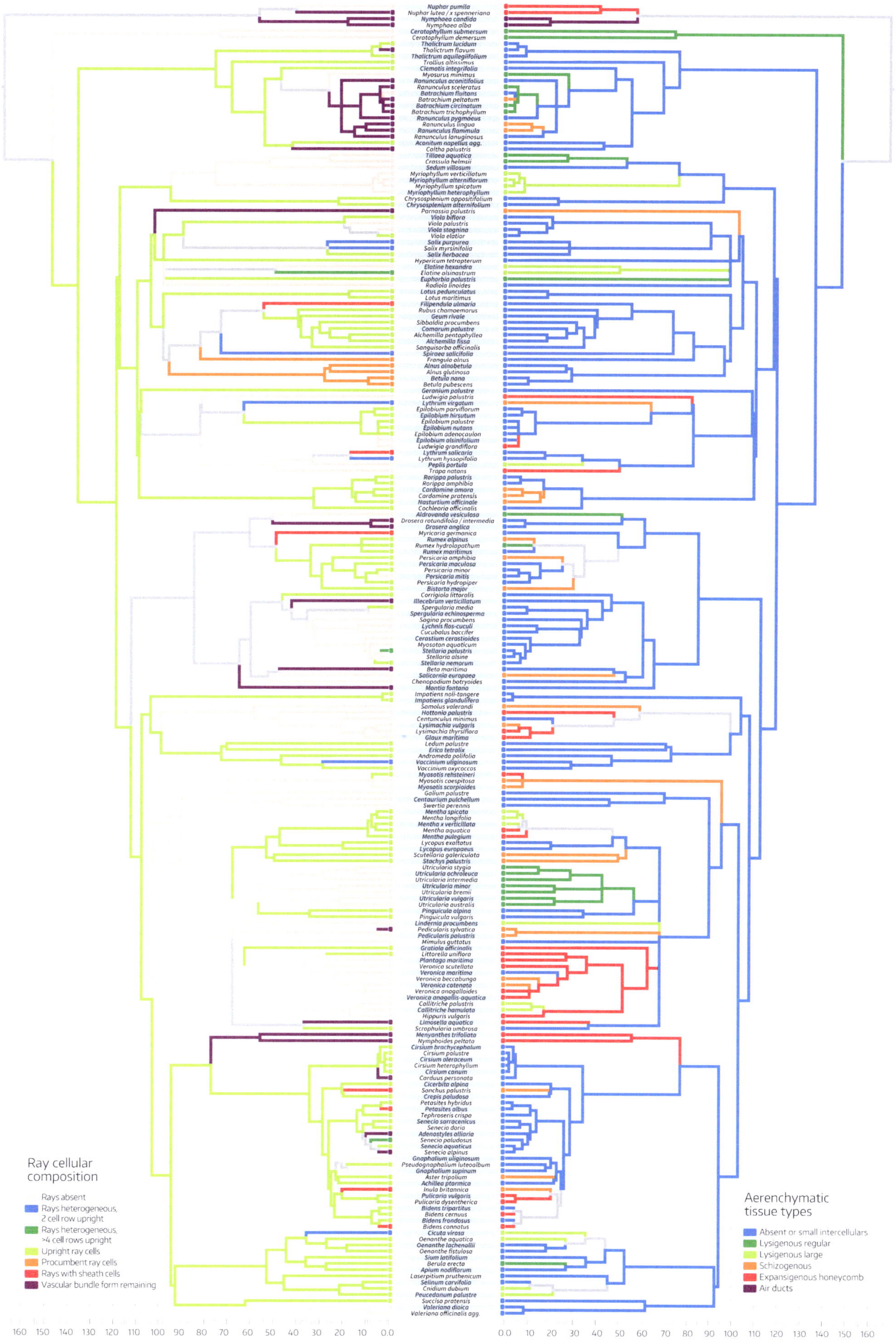

Figure 24. The phylogenetic tree of studied dicot species with a maximum likelihood reconstruction of selected features (Left: Ray cellular composition, Right: Aerenchymatic tissue types).

Aerenchyma (air spaces)

Additional mechanical support and the necessary transport of oxygen towards the roots of most aquatic and many wetland plants is achieved through the formation of conspicuous intercellular spaces (cavities, lacunae, air spaces) in parenchymatic tissue in stems, present in 39% of the studied taxa (Figure 22 and 24). Schizogenous aerenchyma (31%) prevails in aquatic emergent plants (e.g. *Cicuta virosa, Lycopus europaeus, Oenanthe aquatica* and *Veronica beccabunga*), alpine short-rhizomatous (e.g. *Ranunculus aconitifolius, Rumex alpinus*), long-rhizomatous (*Glaux maritima*) and saline forbs (*Samolus valerandi*), but also in short-lived low-stature herbs (*Centunculus minimus, Myosurus minimus, Sedum villosum* and *Veronica catenata*) and root-sprouters (*Cnidium dubium*). Lysigenous types are present in 30% of species with aerenchyma, with honeycomb (23%), regular round cavities (24%) or large irregular or radial (21%) prevailing over lysigenous air ducts (2%).

Aerenchymatic tissues formed by lysigenous canals are found only in aquatic floating (*Nymphaea alba, N. candida*), submerged (*Myriophyllum* spp.) and aquatic emergent (*Cochlearia officinalis*) plants. Lysigenous honeycomb aerenchyma (also called expansigenous aerenchyma, Seago et al. 2005) is the most widespread type among aquatic floating plants, being present across different families such as Nymphaeaceae (*Nuphar lutea, Nuphar pumila*), Menyanthaceae (*Nymphoides peltata*), Onagraceae (*Ludwigia grandiflora, Ludwigia palustris*), Primulaceae (*Hottonia palustris*) and Lythraceae (*Trapa natans*), and less frequently in wetland terrestrial herbs such as rhizomatous forbs (*Gratiola officinalis, Mentha aquatica, Myosotis rehsteineri, Pulicaria dysentherica*) and short-lived annuals (*Bidens cernuus, Limosella aquatica, Veronica anagalloides*). The lysigenous regular round aerenchyma type dominates in many phylogenetically unrelated aquatic submerged plants including all studied *Callitriche, Ceratophyllum, Utricularia* and *Myriophyllum* species (except for *Myriophyllum spicatum* with honeycomb aerenchyma), and also *Aldrovanda vesiculosa, Crassula helmsii, Elatine hexandra* and *Littorella uniflora*.

Secondary intra-xylem periderm and sieve elements

A characteristic of some phylogenetically unrelated perennial aquatic and wetland species is root segmentation as a result of senescence. If the need to support the capacity of water transport and the need to support the stem metabolism is out of balance, living parenchyma cells within the xylem change their storing mode to a meristematic mode to form an inter-xylary periderm towards the pith, separating the metabolic active outer part from a dead central part. This process can occur several times with all segregated parts forming a new phellem around the stem. Tertiary meristematic activities resulting in the formation of a secondary intra-xylem periderm towards the pith were observed in only 4% of species such as *Clematis integrifolia, Gratiola officinalis, Hottonia palustris* and *Succisa pratensis*, while successive cambia simultaneously producing xylem and phloem was observed in 3% of species such as *Senecio doria, Petasites hybridus, Ranunculus lingua* and *Nymphaea candida*; intra-xylary phloem is formed in *Valeriana dioica* (Figure 25).

Sieve elements are mostly solitary (43%), in clusters (33%), radially (15%) or tangentially (8%) arranged. Solitary sieve tubes are slightly more represented than clustered sieve tubes in terrestrial rhizomatous clonal plants and woody species, while the opposite pattern is found in non-clonal tap-rooted forbs and aquatic emergent plants, where clustered sieve tubes are more abundant. Phloem ray dilatations were found in 27% of species, mostly in terrestrial wetland rhizomatous genera *Cicerbita, Cirsium, Myosotis, Senecio* and *Vaccinium* (Figure 26).

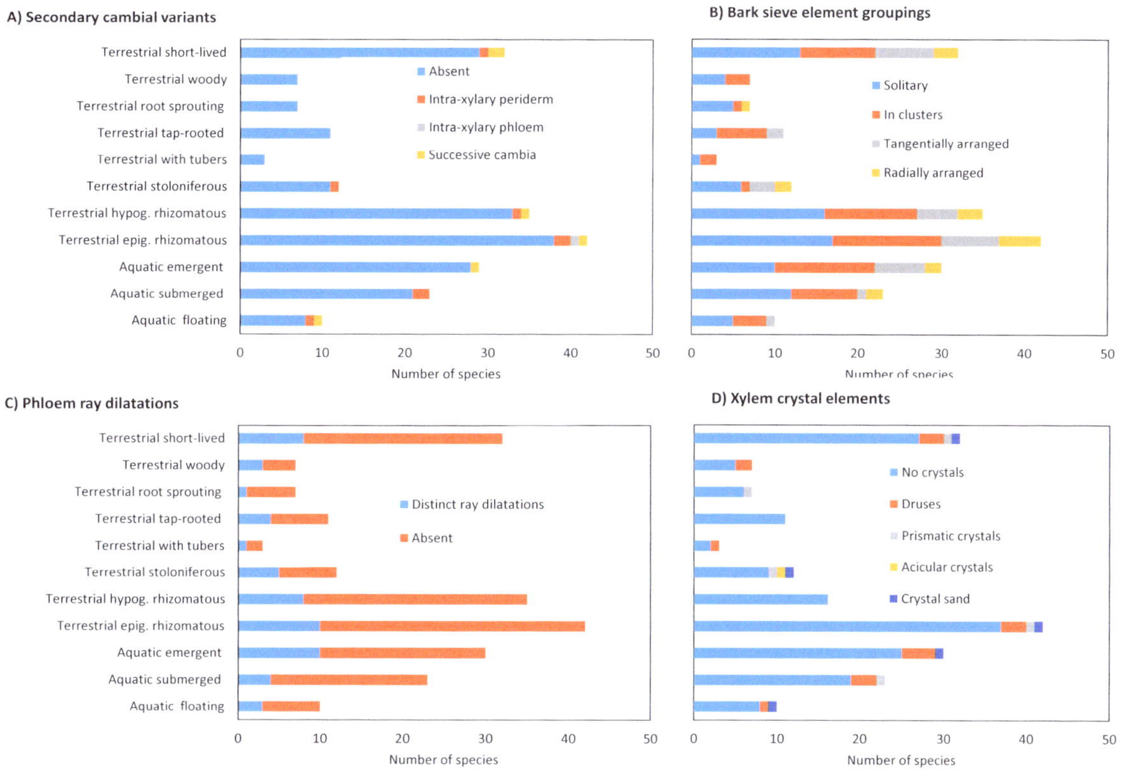

Figure 25. Numbers of species with specific secondary cambial variants, sieve tube groupings and xylem crystal elements in individual life and growth form categories, based on 212 aquatic and wetland plant species.

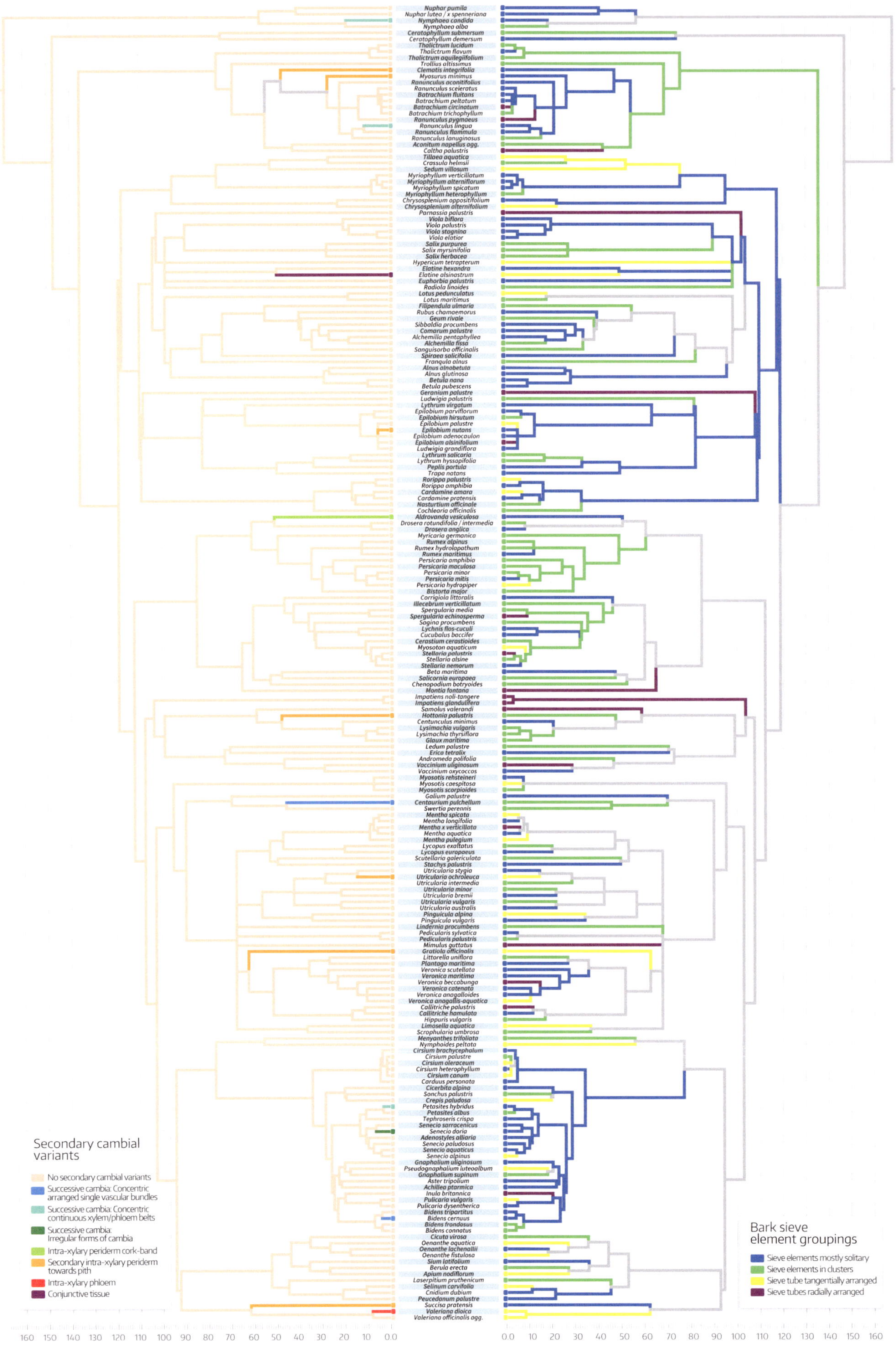

Figure 26. The phylogenetic tree of studied dicot species with a maximum likelihood reconstruction of selected features (Left: Secondary cambial variants, Right: Bark sieve element groupings).

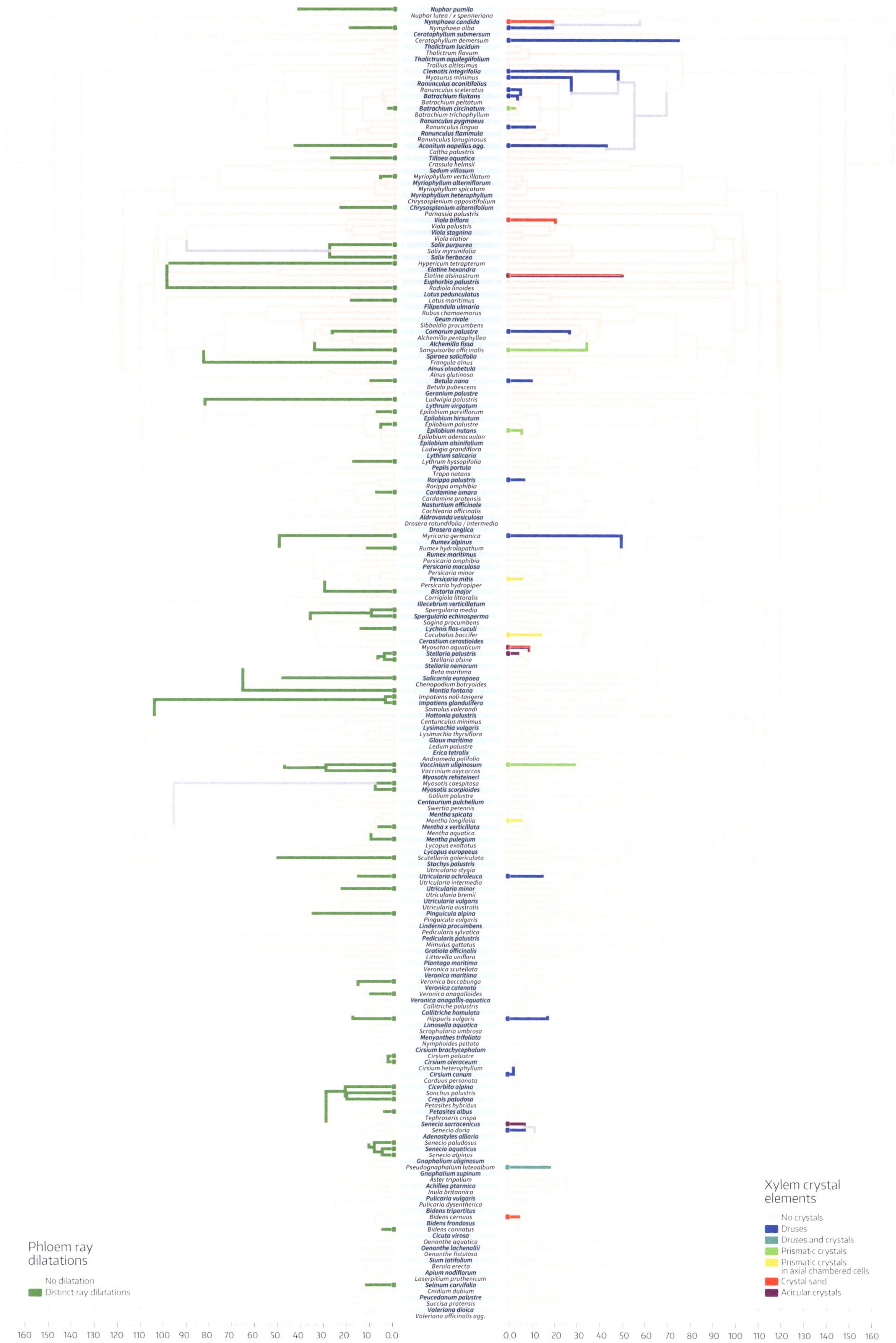

Figure 27. The phylogenetic tree of studied dicot species with a maximum likelihood reconstruction of selected features (Left: Phloem ray dilatations, Right: Xylem crystal elements).

Xylem and phellem secretory and crystal elements

Xylem secretory and crystal elements, mainly in form of druses, are rare among the studied species (15%) (Figure 25). We can only speculate about their significance. They might support the regulation of calcium levels in cells or belong to the defense system (herbivory). Crystal druses are found across unrelated taxa such as *Betula nana*, *Ceratophyllum demersum*, *Cirsium canum*, *Clematis integrifolia*, *Comarum palustre*, *Nymphaea alba*, *Ranunculus lingua* and *Senecio doria*. Acicular crystals occur in *Elatine alsinastrum*, *Senecio sarracenicus* and *Stellaria palustris*, crystal sand in *Nymphaea candida* and *Viola biflora*, while prismatic crystals in *Batrachium circinatum*, *Epilobium nutans*, *Sanguisorba officinalis* and *Vaccinium uliginosum* (Figure 27).

Bark sclereids were found in 30% of the species, mostly in tall rhizomatous terrestrial wetland forbs and aquatic emergent plants (Figure 28), while bark crystal elements occurred in 25% of the species, mostly as druses (13%) and less as prismatic crystals (9%) and sand (3%). Bark secretory elements were found in 13% of the species, mostly as oil ducts, such as in *Betula nana*, *Cirsium heterophyllum*, *Geranium palustre*, *Impatiens glandulifera*, *Radiola linoides*, *Trollius altissimus* and *Vaccinium oxycoccos*. Rhaphides occur in *Viola palustris*, while latex is found in *Erica tetralix*, *Mentha spicata*, *Ranunculus pygmaeus* and *Sonchus palustris*, and tannins in *Betula pubescens* and *Lysimachia thyrsiflora*.

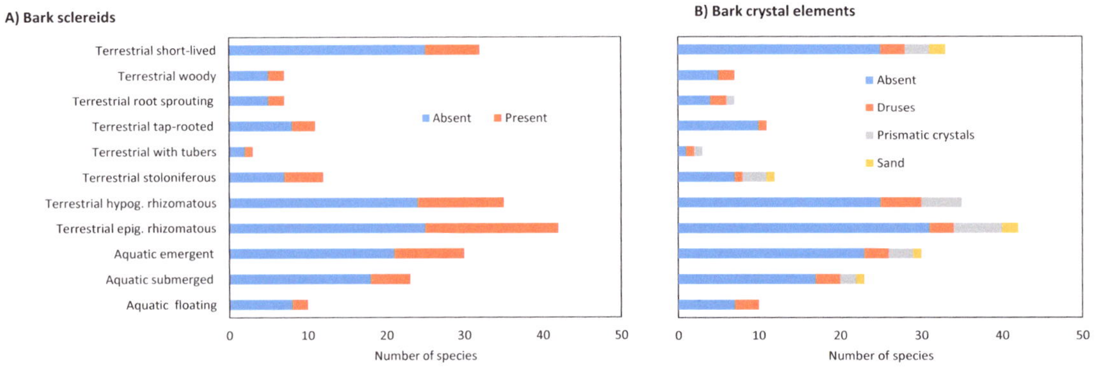

Figure 28. Comparison of phellem secretory and crystal elements among 212 aquatic and wetland plant species categorized into several life and growth forms.

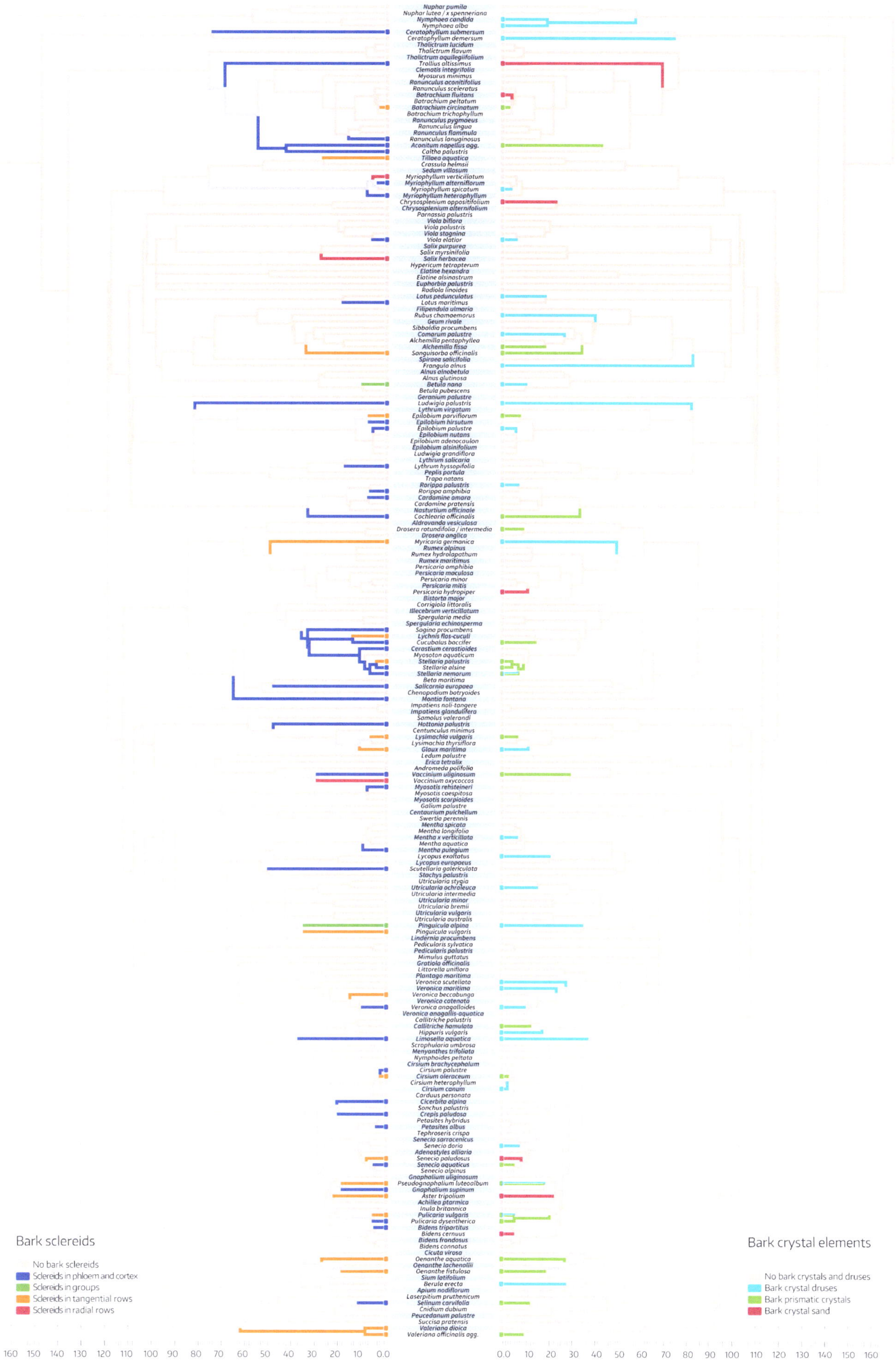

Figure 29. The phylogenetic tree of studied dicot species with a maximum likelihood reconstruction of selected features (Left: Bark sclereids, Right: Bark crystal elements).

Stem anatomy in monocotyledonous aquatic and wetland plants

In this chapter, we present characteristic anatomical structures of a large number of monocotyledonous aquatic and wetland species, 66% of which are graminoids (117 species), and try to relate microscopic features to hydrological conditions and taxonomic origin. The study is based on cross sections of culms. Each section is located at an internode above the leaf sheaths, approximately in the center of the total length of the culms. Since anatomical variation is minimized by the constant internal location of the section within the culm, the whole set of analyzed material is comparable. Double staining gives additional information about the distribution of lignified and unlignified parts. We focus on two anatomical features: the presence of fibers and different aerenchyma tissue types as related to mechanical support and aeration.

The 178 monocot species studied belong to 22 families and 62 genera. The most species-rich family is Cyperaceae (73 taxa, 41%), followed by Poaceae (22 species, 12%), Juncaceae (22, 12%), Potamogetonaceae (18, 10%), Araceae (6, 4%), Alismataceae (4, 2%), Hydrocharitaceae (4, 2%), Sparganiaceae (4, 2%), Typhaceae (4, 2%). The most species-rich genera are *Carex* (44), *Juncus* (22), *Potamogeton* (16), *Eriophorum* (5), *Bolboschoenus* (4), *Cyperus* (4), *Lemna* (4), *Sparganium* (4) and *Typha* (4). Analysis of ecological indicator values (EIV, Chytrý et al. 2018) showed that aquatic rooted leaf-floating and submerged plants have lower affinities for well-irradiated habitats than aquatic surface-floating and emergent plants (Figure 30). Among terrestrial plants, those with short rhizomes have the highest EIV values for light, followed by stoloniferous plants (Figure 31). Terrestrial plants with stem or root tubers (*Dactylorhiza, Gladiolus, Orchis*) and short-rhizomatous plants are associated with warmer climate (highest indicator values for temperature), while terrestrial long-rhizomatous plants with colder climate. In aquatic plants, submerged and surface-floating species are associated with warmer climate, while rooted leaf-floating plants with colder climate. Terrestrial plants with stem or root tubers and aquatic submerged plants (*Elodea, Potamogeton, Najas, Zannichellia* and *Zostera*) are associated with more alkaline soils, while aquatic surface-floating and submerged plants with a more fertile environment (Figure 32). Terrestrial short-rhizomatous plants showed much higher tolerance for saline soils than other plant groups (Figure 33).

The analysis of selected functional traits extracted from LEDA (Kleyer et al. 2008) and CLO-PLA databases (Klimešová et al. 2017) shows that aquatic emergent plants and terrestrial tuberous and long-rhizomatous plants achieve the larger stature (plant height) (Figure 34). High values in aquatic submerged and aquatic rooted leaf floating plants correspond to the total length of their stems supported (partly or fully) by the water column. Conspicuously low leaf dry matter content and high specific leaf area in aquatic surface floating, rooted leaf floating and submerged plants is associated with a different construction of their leaves (higher leaf water and/or air content, very thin leaves, etc.) in comparison to the terrestrial wetland plants (Figure 35). The high ability of clonal dispersal is associated with aquatic surface-floating and submerged plants and terrestrial stoloniferous grasses (Figure 36).

F. H. Schweingruber et al., *Anatomic Atlas of Aquatic and Wetland Plant Stems*, https://doi.org/10.1007/978-3-030-33420-8_75

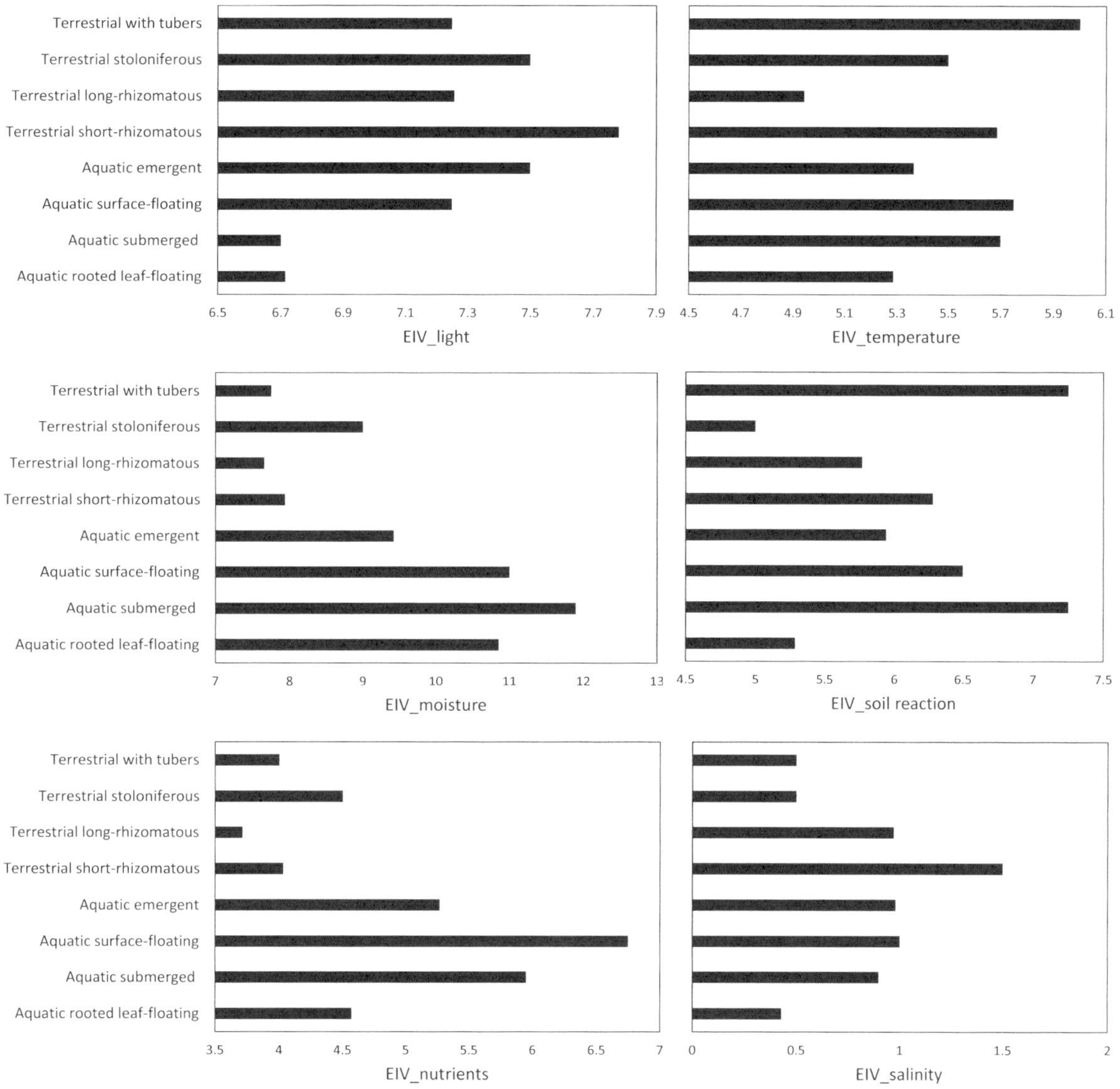

Figure 30. Comparison of ecological indicator values in 178 monocotyledonous aquatic and wetland plant species divided into eight life and growth form categories.

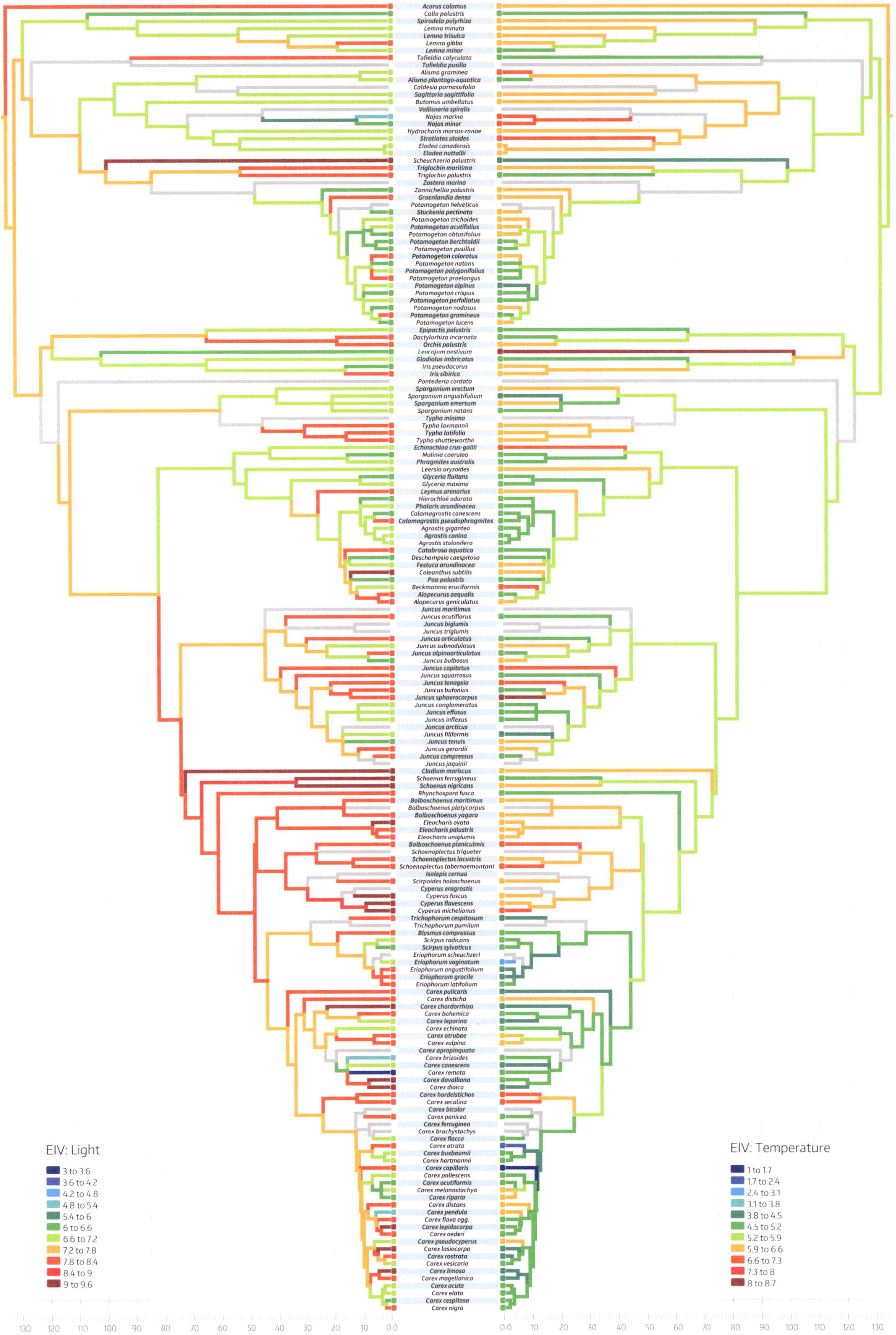

Figure 31. The phylogenetic tree of studied monocot species with a maximum likelihood reconstruction of species indicator values (Left: EIV for light, Right: EIV for temperature).

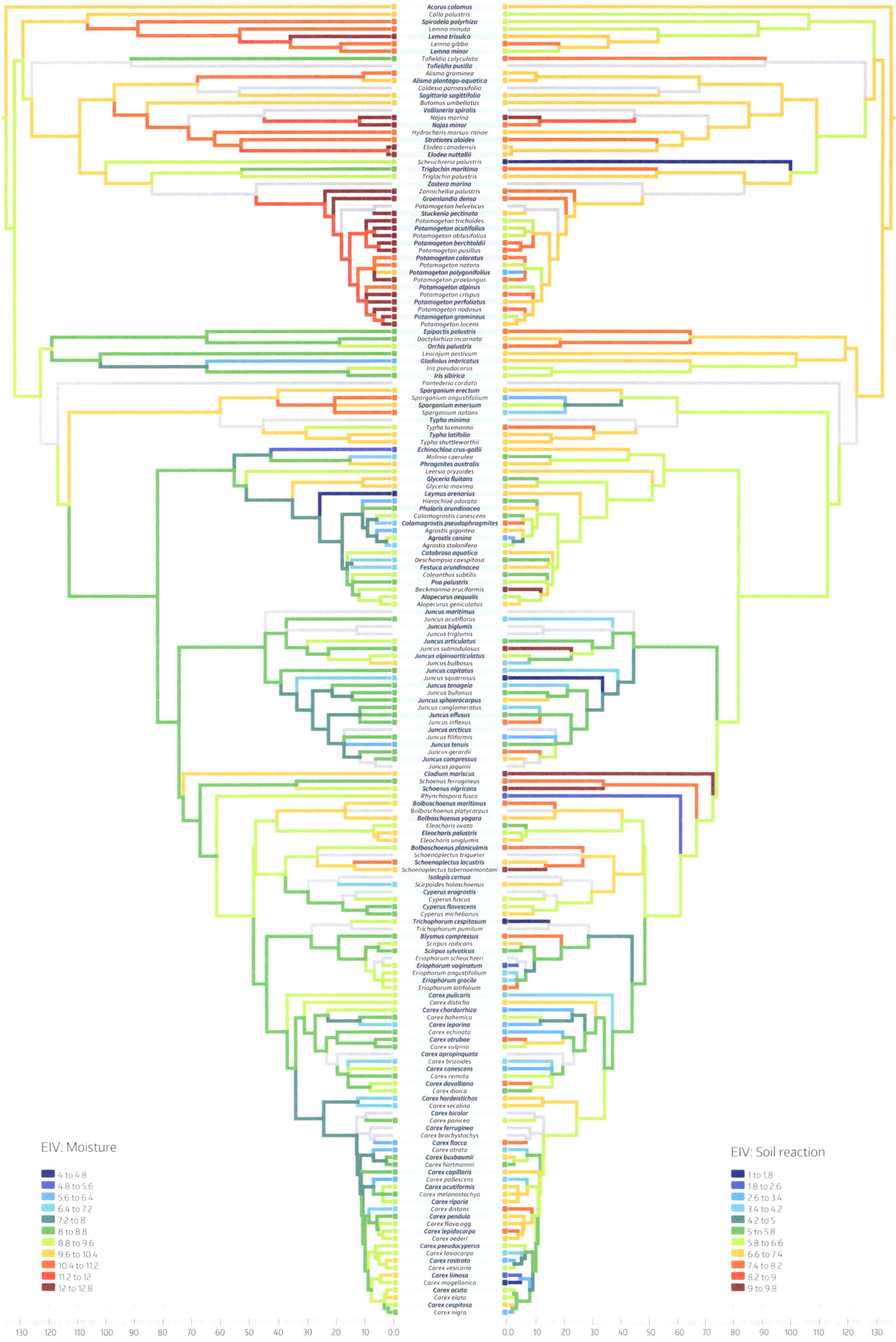

Figure 32. The phylogenetic tree of studied monocot species with a maximum likelihood reconstruction of species indicator values (Left: EIV for moisture, Right: EIV for soil reaction).

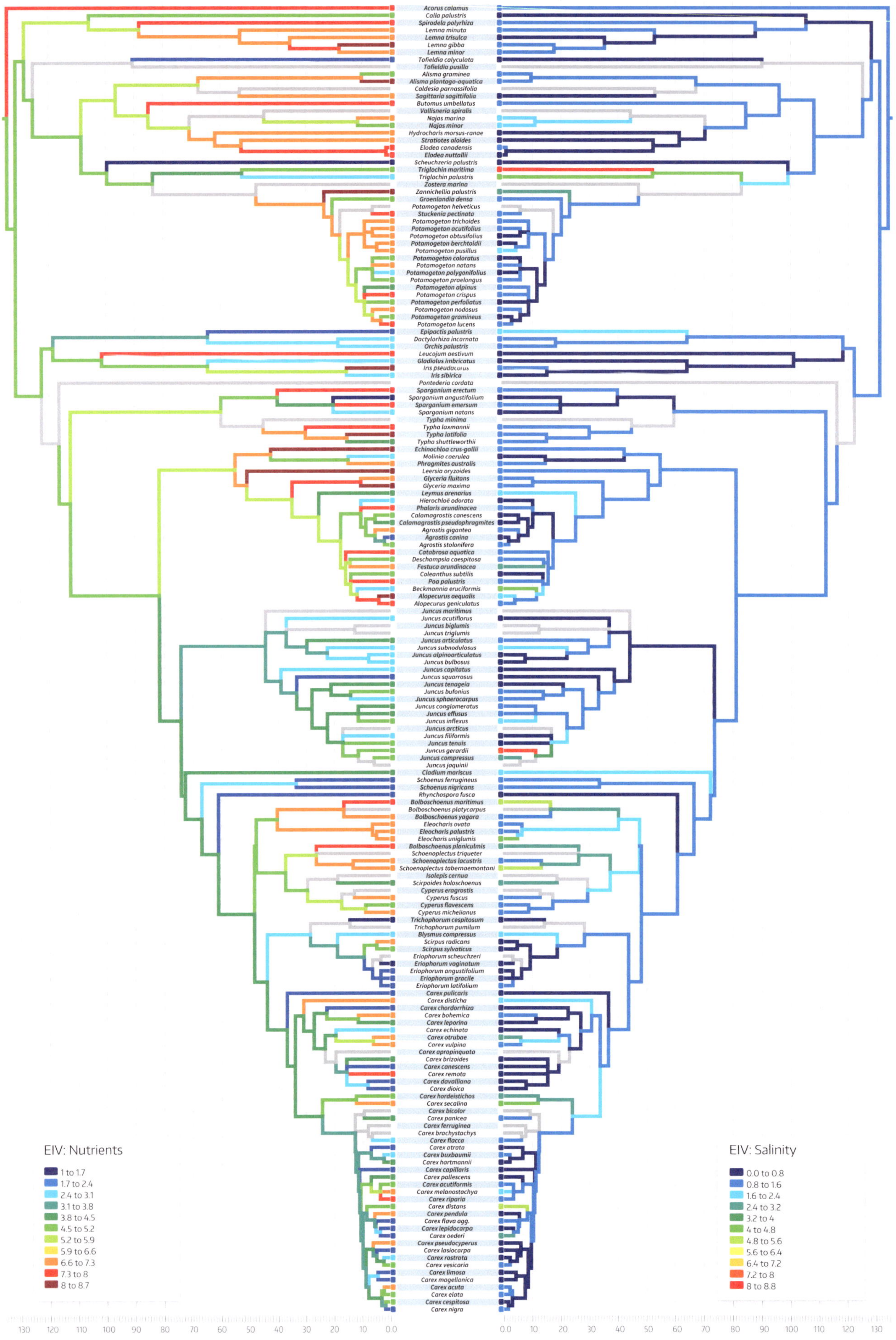

Figure 33. The phylogenetic tree of studied monocot species with a maximum likelihood reconstruction of species indicator values (Left: EIV for nutrients, Right: EIV for salinity).

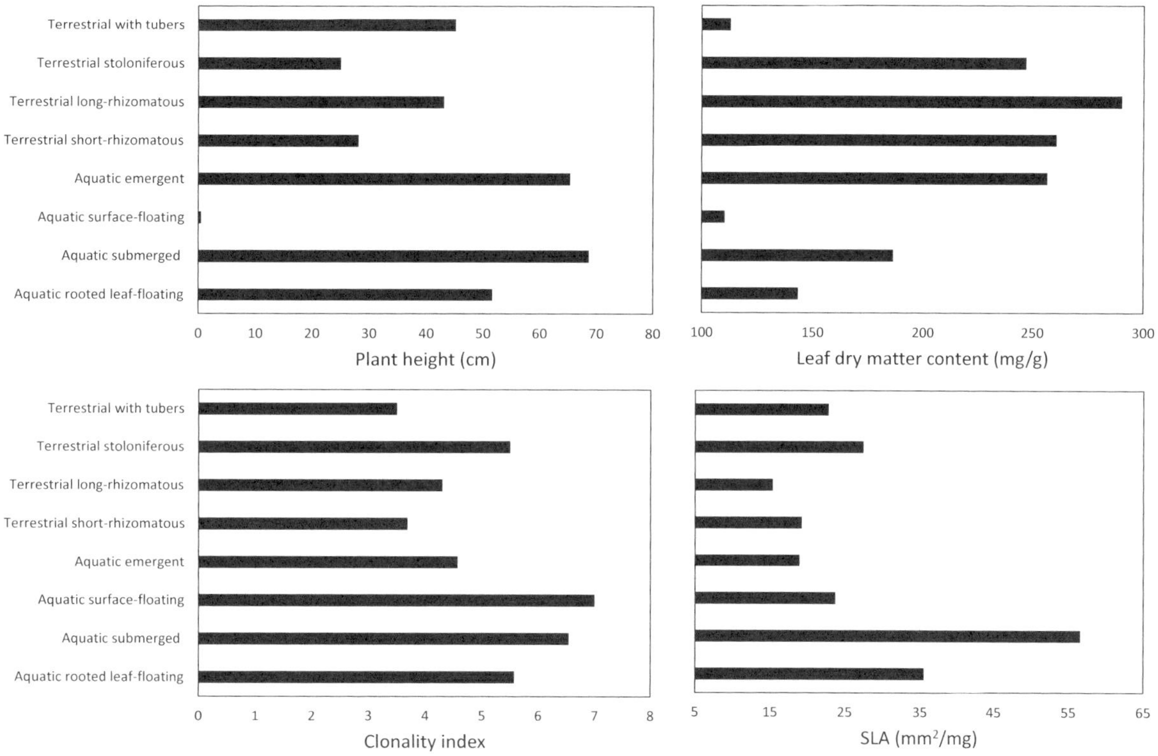

Figure 34. Comparison of selected functional traits among aquatic and wetland plant life and growth form categories containing 178 monocotyledonous species.

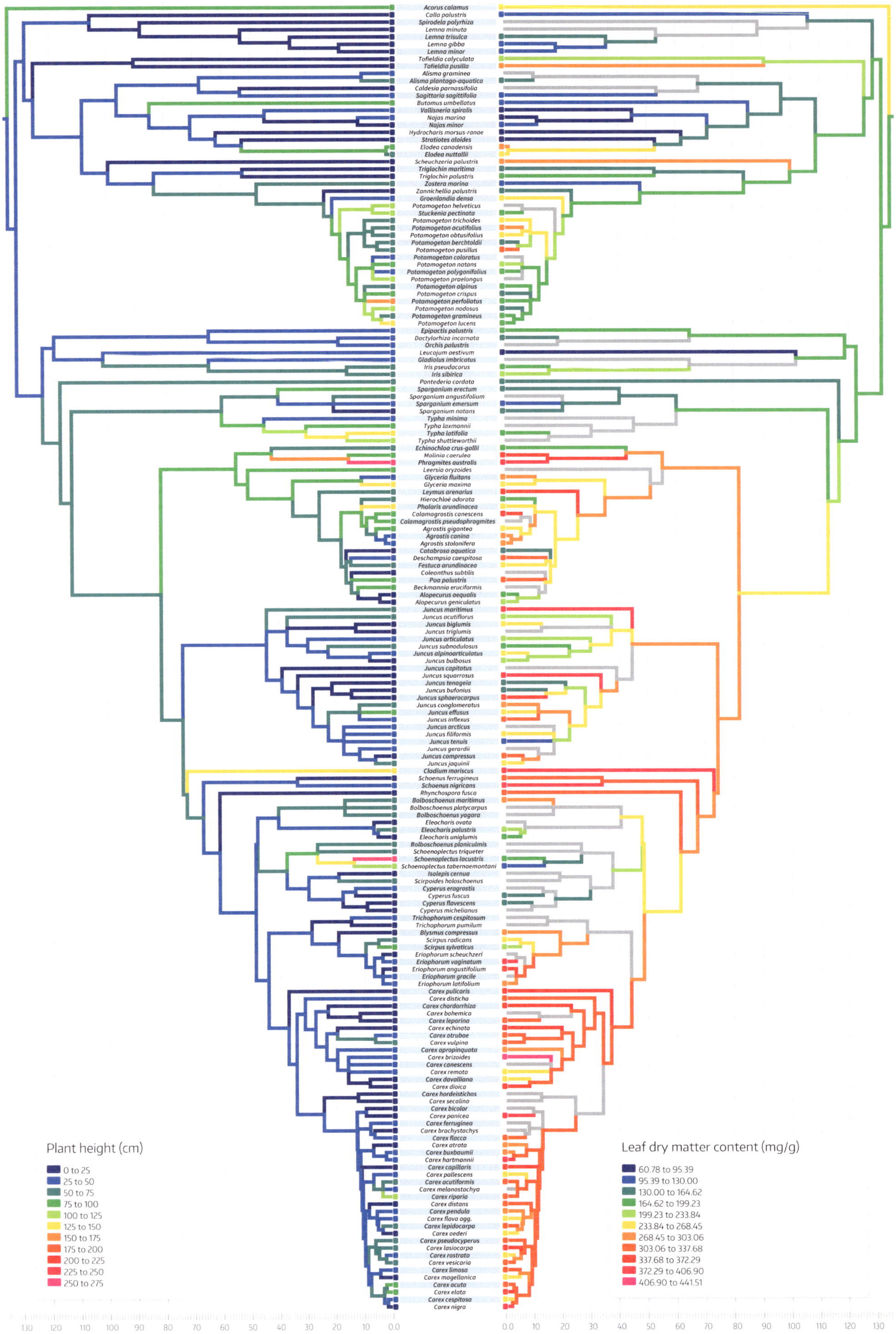

Figure 35. The phylogenetic tree of studied monocot species with a maximum likelihood reconstruction of selected functional traits (Left: Plant height, Right: LDMC).

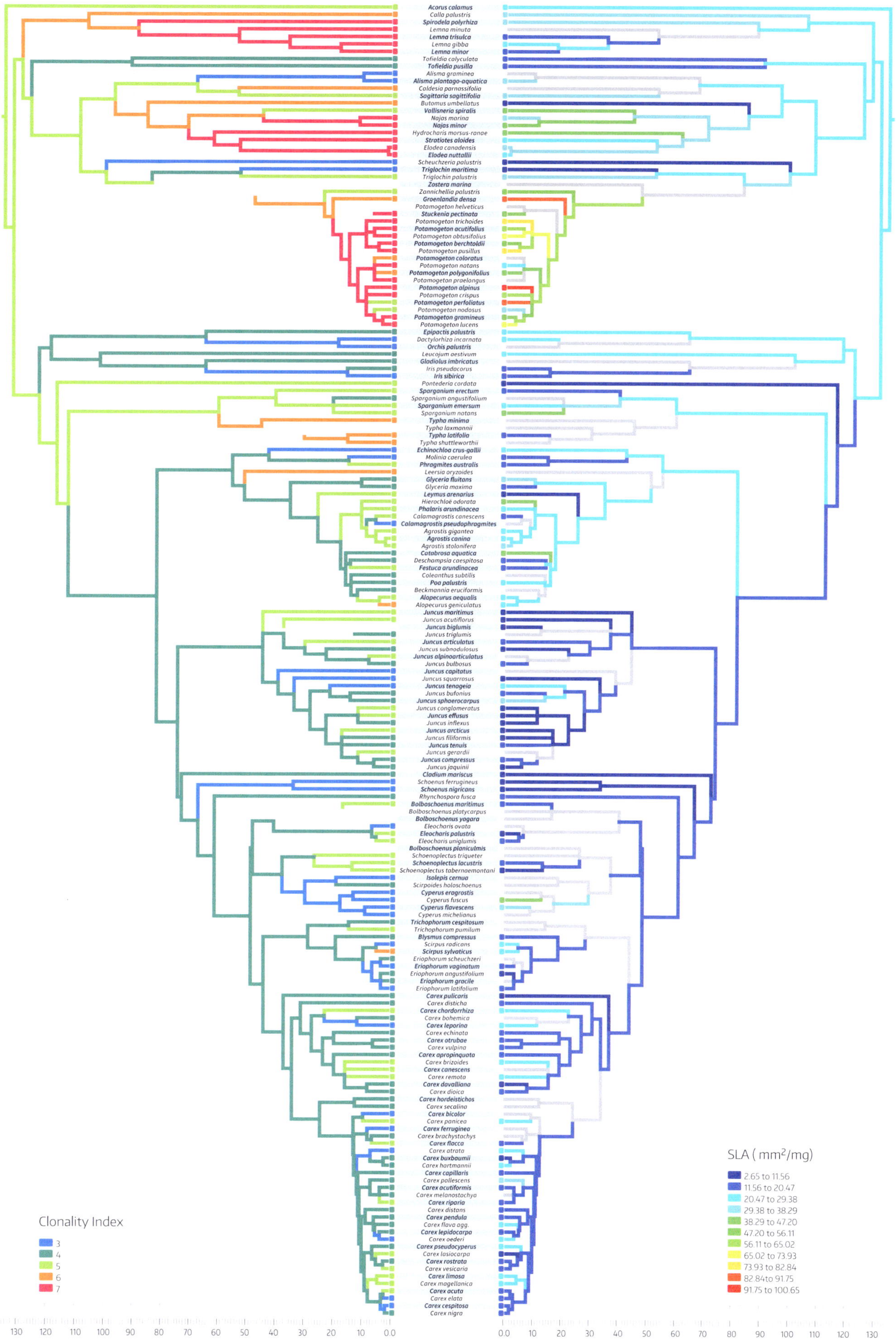

Figure 36. The phylogenetic tree of studied monocot species with a maximum likelihood reconstruction of selected functional traits (Left: Clonality index, Right: SLA).

Mechanical support in the studied monocotyledons is achieved through the formation of lignified fiber sheaths, which are present in 74% of the studied species, which are mostly terrestrial graminoids (Figure 37). Fibers are absent in most aquatic surface-floating, submerged and rooted leaf-floating plants. Additional mechanical support and low construction costs are brought about by a hollow stem and conspicuous intercellular spaces (aerenchyma) in the parenchymatic tissue. Hollow culms surrounded with many large thin-walled unlignified cells are found in 40% of studied taxa, mostly among terrestrial long-rhizomatous plants (75% species in this category), and partly also short-rhizomatous and aquatic emergent species such as tall grasses *Calamagrostis canescens*, *Glyceria maxima*, *Phalaris arundinacea* and *Phragmites australis*, and sedges *Carex otrubae* and *Carex vesicaria* (Figure 38). Full culms are found in aquatic plants with emergent stems such as *Eleocharis*, *Eriophorum* and *Typha* species, while culm centers filled with polygonally arranged parenchyma are present in *Schoenoplectus* species.

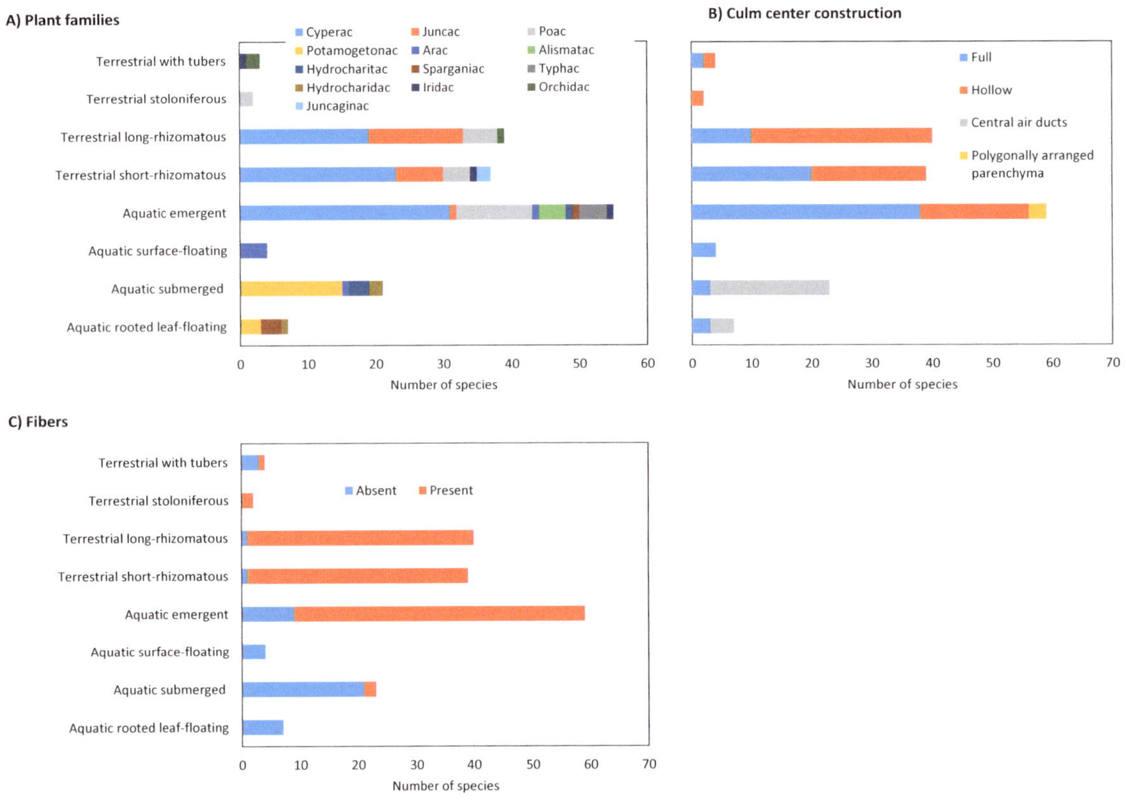

Figure 37. Abundance of plant families, species with fibers and specific culm construction in individual life and growth form categories, based on 178 aquatic and wetland monocot plant species.

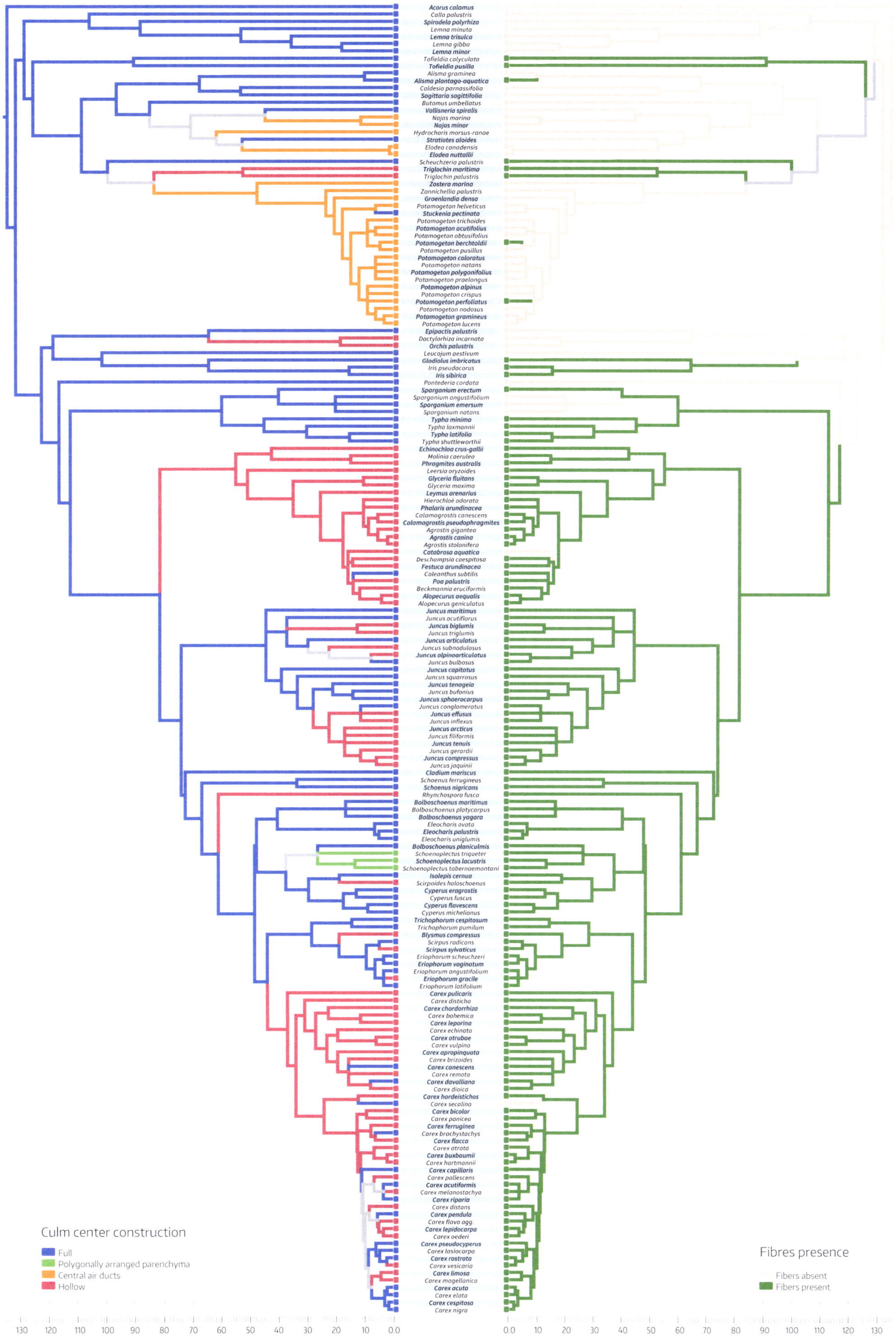

Figure 38. The phylogenetic tree of studied monocot species with a maximum likelihood reconstruction of selected features (Left: Culm center construction, Right: Fibres presence).

Additional mechanical support and transport of oxygen to the roots is achieved through the formation of conspicuous intercellular spaces (aerenchyma) in the central cylinder, vascular bundles, cortex and chlorenchyma (Figure 39). Air ducts in the central cylinder are present in 14% of the studied taxa, and exclusively in aquatic submerged (*Elodea, Najas, Potamogeton*) and rooted floating-leaved plants (*Hydrocharis morsus-ranae*). Air ducts and intercellulars in vascular bundles are present in 40% of species, mainly in aquatic emergent plants (*Acorus, Bolboschoenus, Scirpus, Typha*) and rooted floating-leaved species (*Potamogeton natans, Sparganium*) (Figure 40). Culm intercellulars in cortex and chlorenchyma are present in 77% of the studied monocotyledons, with the lysigenous large radial type of aerechyma prevailing among aquatic emergent taxa and rhizomatous terrestrial plants, while the expansigenous honeycomb type dominates among aquatic submerged plants.

Aerenchyma is also developed in the rhizomes of many terrestrial (26%) and aquatic emergent plants (21% of all the studied species). An expansigenous honeycomb aerenchyma is present in several aquatic emergent plants (*Acorus calamus, Alisma plantago-aquatica, Butomus umbellatus, Cladium mariscus, Scheuchzeria palustris, Schoenoplectus* sp., *Typha latifolia, T. laxmannii*), and occasionally also in aquatic rooted leaf-floating (*Hydrocharis morsus-ranae, Sparganium natans*) and terrestrial rhizomatous species (*Carex disticha, Scirpus sylvaticus*) (Figure 41). Large lysigenous radial aerenchyma is present in many aquatic emergent plants (*Carex elata, C. lasiocarpa, C. limosa, C. riparia, Eleocharis palustris, Rhynchospora fusca, Scirpus radicans, Typha minima*) and terrestrial rhizomatous plants (*Blysmus compressus, Carex echinata, C. nigra, C. panicea, Juncus conglomeratus, J. filiformis, J. gerardii, J. inflexus, J. squarrosus*).

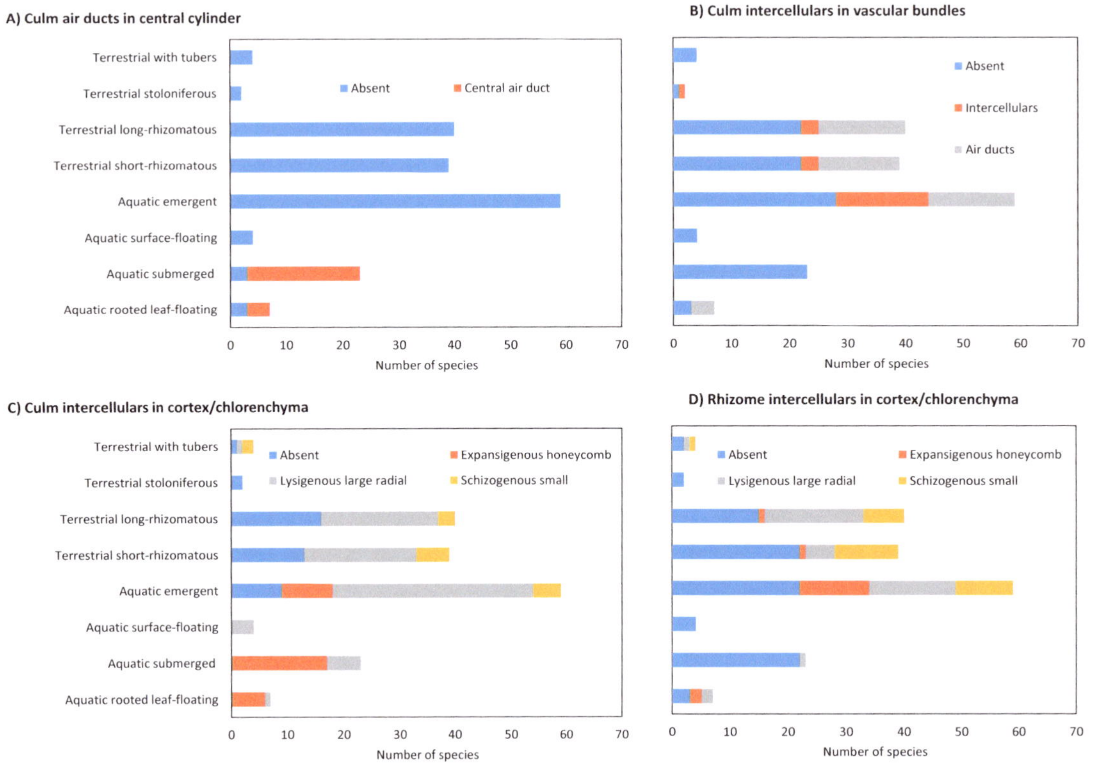

Figure 39. Numbers of species with air ducts and intercellulars in culm central cylinder, cortex and vascular bundles in individual life and growth form categories, based on 178 aquatic and wetland monocot plant species.

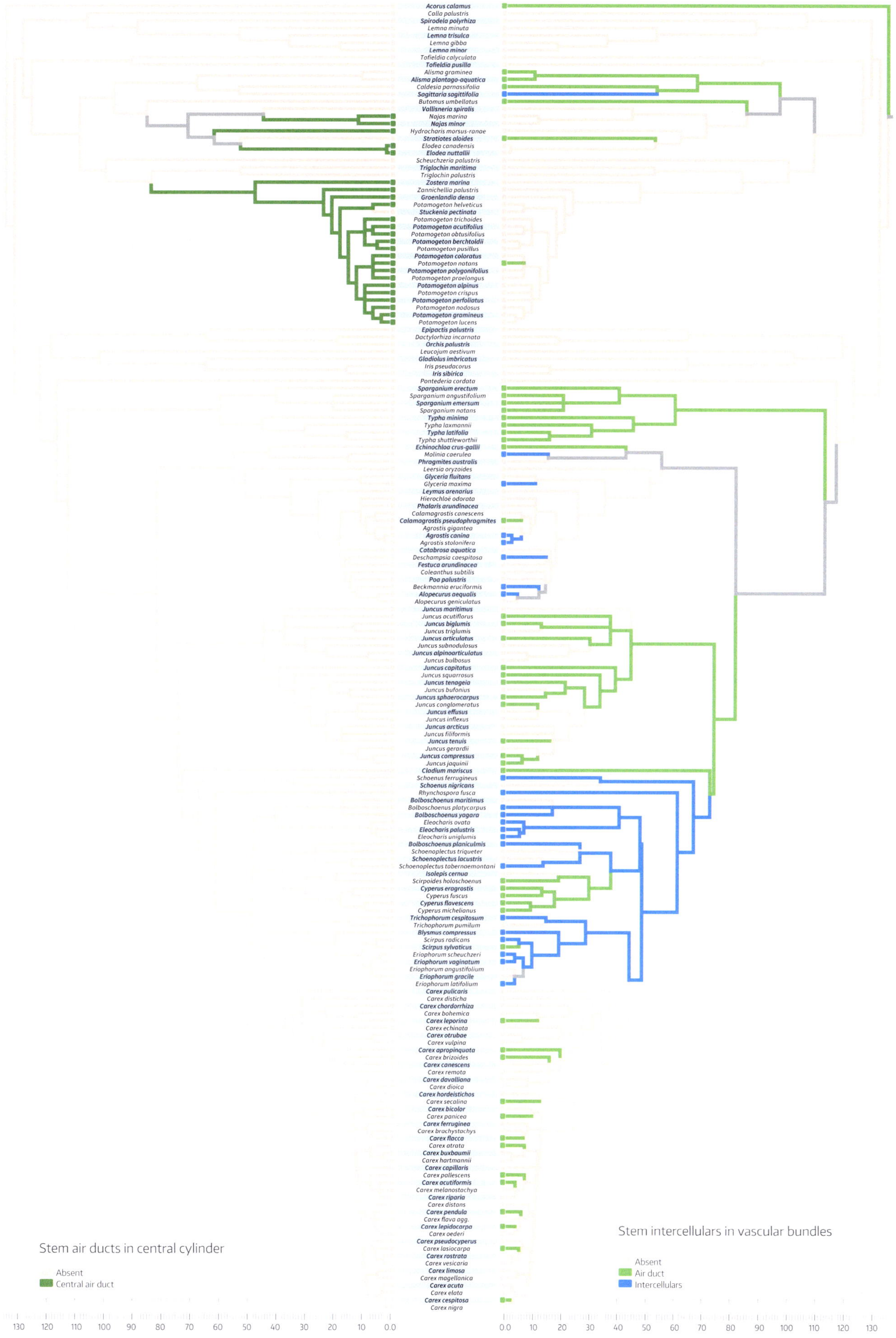

Figure 40. The phylogenetic tree of studied monocot species with a maximum likelihood reconstruction of selected features (Left: Stem air ducts in central cylinder, Right: Stem intercellulars in vascular bundles).

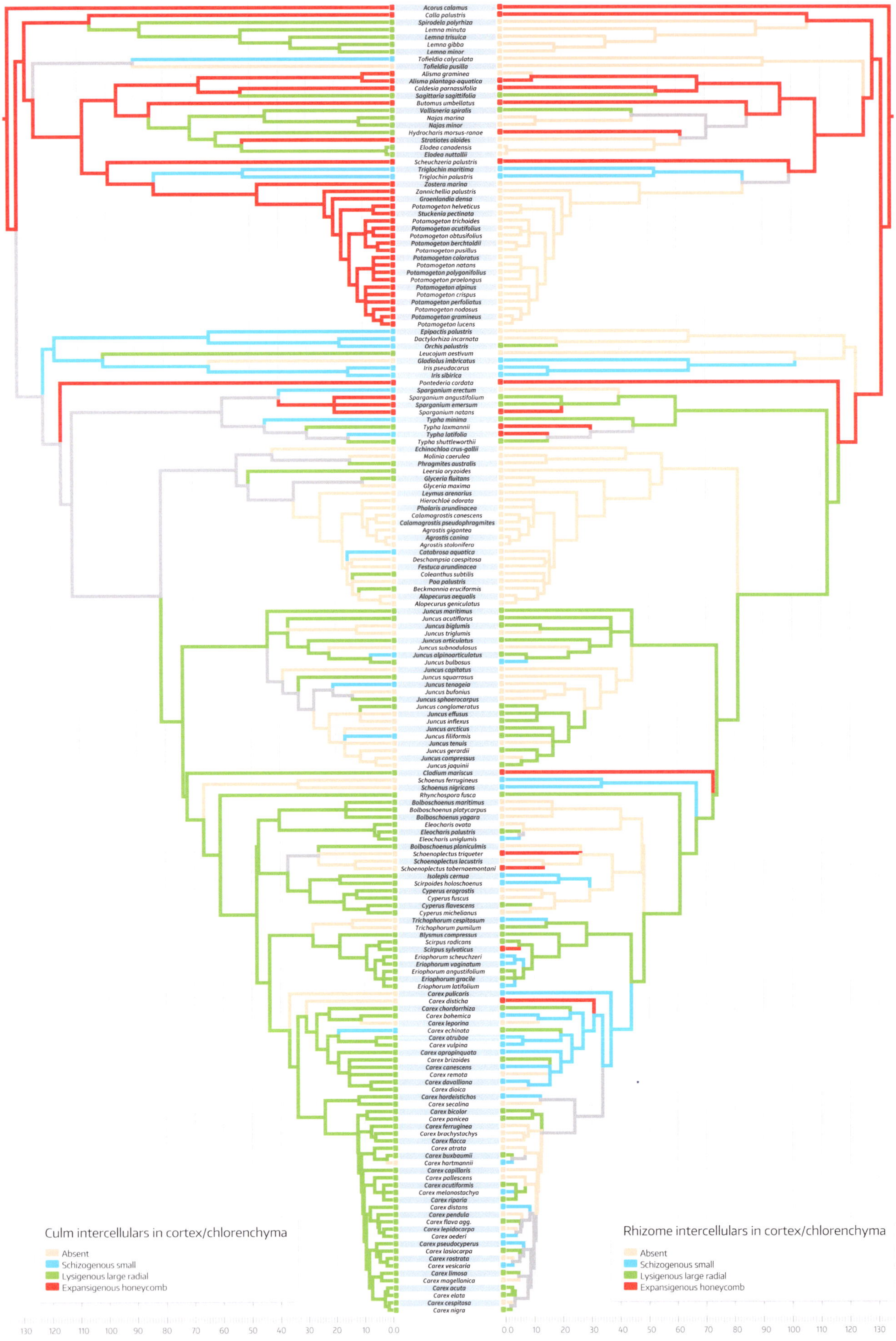

Figure 41. The phylogenetic tree of studied monocot species with a maximum likelihood reconstruction of selected features (Left: Culm intercellulars in cortex/chlorenchyma, Right: Rhizome intercellulars in cortex/chlorenchyma).

Acknowledgements

The authors of the book want to express their gratitude to all who have contributed to preparation of the book in any way. Thanks are due to all who have provided us with plant material or photos of plants included in the book. Sincere thanks are due to Prof. Rolf Rutishauser (Zurich University, Switzerland), Prof. Georg Janauer (University of Vienna, Austria) and Dr. Jan Květ (University of South Bohemia, České Budějovice, Czech Republic) for special correction of the text. Special thanks are due to Pavel Říha for the help with phylogenetic analyses and to Dr. Brian G. McMillan (University of Strathclyde, Glasgow, U.K.) for the correction of language of the introduction and synthesis text. We are indebted to Mr. Milan Krištůfek (České Budějovice, Czech Republic), who has elaborated the book design, made the layout and adjusted all photos. For AK, LA and JD, the work on the book was supported by the Czech Science Foundation (GACR 17-19376S) and the long-term research development project of the Czech Academy of Sciences No. RVO 67985939. We are also indebted to the staff of the Collection of aquatic and wetland plants of the Institute of Botany, CAS, at Třeboň who provided us with many plant species for this study. They include Dr. Jana Navrátilová and Mrs. Kateřina Dvořáková. Peter Sigrist embedded some delicate hydrophytes in paraffin and sectioned them with a rotary microtome.

Image credits

Thanks are due to all who have provided us with photos of plants included in the book. All images donors are listed below in the alphabetical order, with corresponding page numbers indicating figures for which the credit is given. Special thanks also to the photogaleries http://www.botanickafotogalerie.cz, http://flora.upol.cz and the Saxifraga foundation (http://www.freenatureimages.eu) which have enabled the use of their photos. All other images and figures represent the author's own work.

Barendse R.: 55, 119, 202, 332, 350
Bogosavljević S.: 64
Cibulka R.: 243
Čepelová B.: 78, 81, 94, 122, 143, 192, 197, 223, 249, 254, 277, 338, 423
Čtvrtlíková M.: 39, 214
Duchoslav M.:69, 382
Dvorský M.: 393
Dvořák V.: 88
Ekrt L.: 75, 179, 281, 310
Fajmon K.: 194
Flísek J.: 124, 154
Grulich V.: 52, 232, 233, 419
Gutzerová N.: 46, 102
Gygax A.: 98
Hlásek J.: 66, 67, 68, 153, 237, 239, 339, 348, 370, 407, 424, 430
Holubová D.: 85, 114, 196, 205, 278
Hroneš M.: 221
Kameníček J.: 204
Kohoutek M.: 146
Koutecká E.: 80, 163, 175, 209, 321, 322, 373
Koutecký P.: 48, 54, 92, 127, 132, 145, 148, 177, 201, 238, 248, 267, 281, 294, 358, 414
Kučera T.: 93, 58, 399
Meininger P.: 150
Navrátil J.: 290, 381, 383
Navrátilová J.: 57, 65, 72, 107, 111, 124, 135, 149, 154, 182, 183, 186, 189, 199, 209, 217, 226, 236, 255, 259, 261, 286, 293, 298, 306, 313, 320, 324, 344, 349, 351, 374, 376, 377, 379, 386, 429, 431, 432
Nejeschleba V.: 47, 87, 380
Rybka V.: 85, 103, 337, 343, 355, 363, 396
Skočná J.: 170
Stikvoort E.: 56
Štech M.: 251, 253, 260, 305, 372
Šumberová K.: 50, 206, 265
Topic J.: 49
van der Straaten J.: 55
van Kruijsbergen W.: 105
Veselý P.: 59, 204

F. H. Schweingruber et al., *Anatomic Atlas of Aquatic and Wetland Plant Stems*, https://doi.org/10.1007/978-3-030-33420-8

References

Adamec L (1997a) Mineral nutrition of carnivorous plants: A review. Bot Rev 63:273-299

Adamec L (1997b) Relations between K^+ uptake and photosynthetic uptake of inorganic carbon by aquatic plants. Biol Plant 39:599-606

Adamec L (2000) Rootless aquatic plant *Aldrovanda vesiculosa*: physiological polarity, mineral nutrition, and importance of carnivory. Biol Plant 43:113-119

Adamec L (2003) Ecophysiological characterization of dormancy states in turions of the aquatic carnivorous plant *Aldrovanda vesiculosa*. Biol Plant 47:395-402

Adamec L (2008) Respiration of turions and winter apices in aquatic carnivorous plants. Biologia 63:515-520

Adamec L (2009) Photosynthetic CO_2 affinity of the aquatic carnivorous plant *Utricularia australis* (Lentibulariaceae) and its investment in carnivory. Ecol Res 24:327-333

Adamec L (2010) Tissue mineral nutrient content in turions of aquatic plants: does it represent a storage function? Fundam Appl Limnol 176:145-151

Adamec L (2011) Dark respiration and photosynthesis of dormant and sprouting turions of aquatic plants. Fundam Appl Limnol 179:151-158

Adamec L (2014) Different reutilization of mineral nutrients in senescent leaves of aquatic and terrestrial carnivorous *Utricularia* species. Aquat Bot 119:1-6

Adamec L, Kučerová A (2013) Overwintering temperatures affect freezing temperatures of turions of aquatic plants. Flora 208:497-501

Aeschimann D, Lauber K Moser DM, Theurillat J-P (2004) Flora alpina. Paul Haupt Verlag, Bern, Stuttgart, Heidelberg

Agami M, Waisel Y (1986) The ecophysiology of roots of submerged vascular plants. Physiol Vég 24:607-624

ap Rees T, Jenkins LET, Smith AM, Wilson PM (1987) The metabolism of flood-tolerant plants. In: Crawford RMM (ed) Plant life in aquatic and amphibious habitats. Blackwell Scientific Publications, Oxford, U.K., p 227-238

Armstrong W, Armstrong J, Beckett PM (1996) Pressurised ventilation in emergent macrophytes: the mechanism and mathematical modelling of humidity-induced convection. Aquat Bot 54:121-135

Armstrong W, Justin SHFW, Beckett PM et al (1991) Root adaptation to soil waterlogging. Aquat Bot 39:57-73

Barko JW, Gunnison D, Carpenter SR (1991) Sediment interactions with submersed macrophyte growth and community dynamics. Aquat Bot 41:41-65

Bartha D, Király G, Schmidt D, Tiborcz V (eds) (2015) Distribution atlas of vascular plants of Hungary, University of West Hungary Press, Sopron

Bartley MR, Spence DHN (1987) Dormancy and propagation in helophytes and hydrophytes. Arch Hydrobiol (Beih) 27:139-155

Benková VE, Schweingruber FH (2004) Anatomy of Russian woods. Paul Haupt Verlag, Bern

Boston HL, Adams MS, Madsen JD (1989) Photosynthetic strategies and productivity in aquatic systems. Aquat Bot 34:27-57

Botany.cz (online). https://botany.cz/en/. Accessed 10 Jan 2018

Braendle R, Crawford RMM (1999) Plants as amphibians. Perspect Plant Ecol Evol Syst 2:56-78

Chytrý M, Tichý L, Dřevojan P et al (2018) Ellenberg-type indicator values for the Czech flora. Preslia 90: 83-103

Colmer TD, Pedersen O (2008). Underwater photosynthesis and respiration in leaves of submerged wetland plants: gas films improve CO_2 and O_2 exchange. New Phytol 177:918–926

Cook CDK (1990) Aquatic plant book. SPB Acad. Publ., The Hague

Crawford RMM (1989) Studies in plant survival. Studies in ecology, vol 11. Blackwell Scientific Publications, Oxford

Crivellaro A, Schweingruber FH (2015) Stem anatomical features of Dicotyledons. Xylem, phloem, cortex and periderm characteristics for ecological and taxonomical analysis. Forstbuchverlag, Remagen-Oberwinther, Germany

Cutler DF (1969) Juncales. In: Metcalfe CR (ed) Anatomy of the Monocotyledons, vol IV. Clarendon Press, Oxford, U.K.

Dacey JWH (1987) Knudsen-transitional flow and gas pressurization in leaves of *Nelumbo*. Plant Physiol 85:199-203

Danihelka J, Chrtek J Jr, Kaplan Z (2012) Checklist of vascular plants of the Czech Republic. Preslia 84:647-811

Database of the Czech flora and vegetation (online). https://pladias.cz/en/. Accessed 12 Jan 2018

Denny P (1987) Mineral cycling by wetland plants – a review. Arch Hydrobiol, Adv Limnol Beih 27:1-25

Doležal J, Dvorský M, Börner A, Wild J, Schweingruber FH (2018) Anatomy, age and ecology of high mountain plants in Ladakh, the Western Himalaya. Springer International Publishing AG, Cham, Switzerland

Dostál J (1989) Nová květena ČSSR 1, 2 (Flora of Czechoslovakia). Academia, Prague

Evans DE (2003) Aerenchyma formation. New Phytol 161:35-49

Fischer MA, Oswald K, Adler W (2008) Exkusionsflora für Österreich, Liechtenstein und Südtirol. 3rd edit Biologiezentrum der Oberösterreichischen Landesmuseen, Linz

Fitter R, Fitter A (1984) Collins guide to the grasses, sedges, rushes and ferns of Britain and Northern Europe. W. Collins Sons & Co Ltd., London

Floraweb (online). http://www.floraweb.de. Accessed 10 Dec 2017

Gärtner H, Lucchinetti S, Schweingruber FH (2015) A new sledge microtome to combine wood anatomy and tree-ring ecology. Iawa J 36:452-459

Gärtner H, Schweingruber FH (2013) Microscopic preparation techniques for plant stem analysis. Forstbuchverlag, Remagen-Oberwinther, Germany

Gopal B, Květ J, Löffler H, Masing V, Patten BC (1990) Definition and classification. In: Patten BC et al (eds) Wetlands and shallow continental water bodies, vol 1. SPB Academic Publishing, The Hague, p 9-15

Greguss P. (1945) Bestimmung der mitteleuropäischen Laubhölzer und Sträucher auf xylotomischer Grundlage. Verlag des Ungarischen Naturwissenschaftlichen Museums, Budapest

Grigore M-N, Ivanescu L, Toma C (2014) Halophytes: An integrative anatomical study. Springer, Cham, Heidelberg, New York, Dordrecht, London

Grosse W (1996a) The mechanism of thermal transpiration (= thermal osmosis). Aquat Bot 54:101-110

Grosse W (1996b) Pressurised ventilation in floating-leaved aquatic macrophytes. Aquat Bot 54:137-150

F. H. Schweingruber et al., *Anatomic Atlas of Aquatic and Wetland Plant Stems*, https://doi.org/10.1007/978-3-030-33420-8

Grosse W, Büchel HB, Tiebel H (1991) Pressurized ventilation in wetland plants. Aquat Bot 39:89-98

Haeupler H, Muer T (2000) Bildatlas der Farn- und Blütenpflanzen Deutschlands. Ulmer, Stuttgart

Hejný S (1960) Ökologische Charakteristik der Wasser- und Sumpfpflanzen in den slowakischen Tiefebenen (Donau- und Theissgebiet). SAV, Bratislava

Hejný S, Segal S, Raspopov IM (1998) General ecology of wetlands. In: Westlake DF, Květ J, Szczepański A (eds) The production ecology of wetlands. Cambridge University Press, Cambridge, U.K., p 1-77

Hejný S, Slavík B (eds) (1988) Květena České socialistické republiky 1 (Flora of the Czech Socialist Republic). Academia, Prague

Hejný S, Slavík B (eds) (1990–2008) Květena České republiky (2–8) (Flora of the Czech Republic). Academia, Prague

Hutchinson GE (1975) A treatise on limnology, vol 3. Limnological Botany. John Wiley and Sons, New York, London, Sydney, Toronto

Hyldgaard B, Brix H (2011) Plasticity in carbon acquisition of the heterophyllous *Luronium natans*: An endangered freshwater species in Europe. Aquat Bot 94:127–133

Info Flora (online). https://www.infoflora.ch. Accessed 1 Dec 2017

Jackson MB (1990) Hormones and developmental change in plants subjected to submergence or soil waterlogging. Aquat Bot 38:49-72

Jeschke WD (1976) Ionic relations of leaf cells. In: Lüttge U, Pitman MG (eds) Transport in Plants II, Part B, Tissues and Organs. Encyclopedia of Plant Physiology, New Series, vol 2. Springer-Verlag, Berlin, Heidelberg, New York, p 160-194

Jongduk J, Seung CL, Hong-Keun C (2008) Anatomical patterns of aerenchyma in aquatic and wetland plants. J. Plant Biol 51:428-439

Karger DN, Conrad O, Böhner J, Kawohl T, Kreft H, Soria-Auza RW, Kessler M (2016). CHELSA climatologies at high resolution for the earth's land surface areas (Version 1.1). World Data Center for Climate. doi:10.1594/WDCC/CHELSA_v1_1

Kaźmierczakowa R, Bloch-Orłowska J, Celka Z, Cwener A, Dajdok Z, Michalska-Hejduk D, Pawlikowski P, Szczęśniak E, Ziarnek K (2016) Polska czerwona lista paprotników i roślin kwiatowych (Polish Red List of ferns and flowering plants). IOP PAN, Cracow

Kaźmierczakowa R, Zarzycki K, Mirek Z (2014) Polska czerwona księga roślin. Paprotniki i rośliny kwiatowe (Polish Red Book of plants. Ferns and flowering plants). IOP PAN, Cracow

Keating RC (2002) Acoraceae and Araceae. In: Gregory M, Cutler DF (eds) Anatomy of the Monocotyledons, vol IX. Clarendon Press, Oxford, U.K.

Király G (2009) Új magyar füvészkönyv. Magyarország hajtásos növényei (New Hungarian herbal. The vascular plants of Hungary. Identification key). Aggteleki Nemzeti Park Igazgatóság, Hungary

Klavsen SK, Madsen TV, Maberly SC (2011) Crassulacean acid metabolism in the context of other carbon-concentrating mechanisms in freshwater plants: a review. Photosynt. Res. 109:269-279

Kleyer M, Bekker RM, Knevel IC et al (2008) The LEDA Traitbase: a database of life-history traits of the Northwest European flora. J Ecol 96: 1266-1274

Klimešová J, Danihelka J, Chrtek J et al (2017) CLO-PLA: a database of clonal and bud-bank traits of the Central European flora. Ecology 98: 1179

Koizumi Y, Hara Y, Yazaki Y et al (2011) Involvement of plasma membrane H^+-ATPase in anoxic elongation of stems in pondweed (*Potamogeton disctinctus*) turions. New Phytol 190:421-430

Končalová H (1990) Anatomical adaptations to waterlogging in roots of wetland graminoids: limitations and drawbacks. Aquat Bot 38:127-134

Kubát K, Hrouda L, Chrtek J Jr, Kaplan Z, Kirschner J, Štěpánek J (eds) (2002) Klíč ke květeně České republiky (Key to the flora of the Czech Republic). Academia, Prague

Maberly SC, Madsen TV (2002) Freshwater angiosperm carbon concentrating mechanisms: processes and patterns. Funct Plant Biol 29:393-405

Maberly SC, Spence DHN (1983) Photosynthetic inorganic carbon use by freshwater plants. J. Ecol. 71:705-724

Madsen TV, Sand-Jensen K (1991) Photosynthetic carbon assimilation in aquatic macrophytes. Aquat Bot 41:5-40

Metcalfe CR (1971) Cyperaceae. In: Metcalfe CR (ed) Anatomy of the Monocotyledons, vol V. Clarendon Press, Oxford, U.K

Meusel E, Jäger EJ (eds) Vergleichende Chorologie der Zentraleuropäischen Flora, Band I, II and III (online). http://chorologie.biologie.uni-halle.de. Accessed 12 Dec 2017

Mirek Z, Piękoś-Mirkowa H, Zając M, Zając M (2002) Flowering plants and pteridophytes of Poland. A checklist. Biodiversity of Poland. W. Szafer Institute of Botany, PAN, Cracow

Mirek Z, Zarzycki K, Wojewoda W, Szeląg Z (2006) Czerwona lista roślin i grzybów Polski (Red list of plants and fungi in Poland). W. Szafer Institute of Botany, PAN, Cracow

Neumann K, Schoch W, Détienne P, Schweingruber FH (2001) Woods of the Sahara and the Sahel. Paul Haupt Verlag, Bern, Stuttgart, Wien

Nielsen LT, Borum J (2008) Why the free floating macrophyte *Stratiotes aloides* mainly grows in highly CO_2-supersaturated waters. Aquat Bot 89:379-384

Ogden EC (1974) Anatomical patterns of some aquatic vascular plants of New York. Bull New York State Mus Sci Serv 424:1–133

Online Atlas of the British and Irish Flora (online). www.brc.ac.uk/plantatlas. Accessed 10 Dec 2017

Pedersen O, Colmer TD, Sand-Jensen K (2013) Underwater photosynthesis of submerged plants – recent advances and methods. Front Plant Sci 4:140

Pfandenhauer JS, Klötzli FA (2014) Vegetation der Erde. Grundlagen, Ökologie, Verbreitung. Springer-Verlag, Berlin, Heidelberg

Płachno BJ, Adamec L, Kozieradzka-Kiszkurno M et al (2014). Cytochemical and ultrastructural aspects of aquatic carnivorous plant turions. Protoplasma 251:1449-1454

Pokorný J, Květ J (2004) Aquatic plants and lake ecosystems. In: O'Sullivan PE, Reynolds CS (eds) The lakes handbook, vol. 1, Limnology and limnetic ecology. Blackwell Sci. and Blackwell Publ., Malden, Oxford, Carlton, p 309-340

Prins HBA, de Guia MB (1986) Carbon source of the water soldier, *Stratiotes aloides* L. Aquat Bot 26:225-234

Prins HBA, Elzenga JTM (1989). Bicarbonate utilization: function and mechanism. Aquat Bot 34:59-83

Rejmánková E, Rejmánek M, Květ J (1990) Maximizing duckweed (Lemnaceae) production by suitable harvest strategy. In: Whigham DF, Good RE, Květ J (eds) Wetland ecology and management: Case studies. Kluwer Academic Publ., Dordrecht. p 39-43

Ridge I (1987) Ethylene and growth control in amphibious plants. In: Crawford RMM (ed) Plant life in aquatic and amphibious habitats. Blackwell Scientific Publications, Oxford, U.K., p 53-76
Rudall P (1995) Iridaceae. In: Cutler DF, Gregory M (eds) Anatomy of the Monocotyledons, vol VIII. Clarendon Press, Oxford, U.K.
Sand-Jensen K (1989) Environmental variables and their effect on photosynthesis of aquatic plant communities. Aquat Bot 34:5-25
Sand-Jensen K, Krause-Jensen D (1997) Broad-scale comparison of photosynthesis in terrestrial and aquatic plant communities. Oikos 80:203-208
Schenk H (1886) Vergleichende Anatomie der submersen Gewächse. Bibliotheca Botanica. Abhandlungen aus dem Gesamtgebiet der Botanik, Bd. 1. Verlag Theodor Fischer, Cassel
Schuette JL, Klug MJ (1995) Evidence for mass flow in flowering individuals of the submersed vascular plant *Myriophyllum heterophyllum*. Plant Physiol 108:1251-1258
Schuurkes JAAR, Kok CJ, Den Hartog C (1986) Ammonium and nitrate uptake by aquatic plants from poorly buffered and acidified waters. Aquat Bot 24:131-146
Schweingruber FH (1990) Anatomy of European woods. Paul Haupt Verlag, Bern, Stuttgart, Heidelberg
Schweingruber FH, Berger H (2019) Anatomy of culms and flower stalks of monocotyledonous plants, vol 3. Forstbuchverlag, Remagen-Oberwinther, Germany
Schweingruber FH, Börner A, Schulze ED (2011) Atlas of stem anatomy in herbs, shrubs and trees, vol 1. Springer-Verlag, Berlin, Heidelberg
Schweingruber FH, Börner A, Schulze ED (2013) Atlas of stem anatomy in herbs, shrubs and trees, vol 2. Springer-Verlag, Berlin, Heidelberg
Sculthorpe CD (1967) The biology of aquatic vascular plants. Edward Arnold, London
Seago JL Jr, Marsh LC, Stevens KJ et al (2005) A re-examination of the root cortex in wetland flowering plants with respect to aerenchyma. Ann Bot 96:565-579
Signorini G, Di Giulio G, Fiorovanti M (2014) Il legno nei beni culturali. Guida all determinazione delle specie legnose. Universita degli studi Firenze
Smits AJM, Laan P, Thier RH et al (1990) Root aerenchyma, oxygen leakage patterns and alcoholic fermentation ability of the roots of some nymphaeid and isoetic macrophytes in relation to the sediment type of their habitat. Aquat Bot 38:3-27
Sorrell BK (1991) Transient pressure gradients in the lacunar system of the submerged macrophyte *Egeria densa* Planch. Aquat Bot 39:99-108
Sorrell BK, Dromgoole FI (1987) Oxygen transport in the submerged freshwater macropyte *Egeria densa* Planch. I. Oxygen production, storage and release. Aquat Bot 28:63-80
Sorrell BK, Dromgoole FI (1988) Oxygen transport in the submerged freshwater macropyte *Egeria densa* Planch. II. Role of lacunal gas pressures. Aquat Bot 31:93-106
Taylor N, Taylor EL, Krings M (2009) Palaeobotany. The biology and evolution of fossil plants. Academic Press Elsevier, Amsterdam
Tokarska-Guzik B, Dajdok Z, Zając M, Zając A, Urbisz A, Danielewicz W, Hołdyński C (2012) Rośliny obcego pochodzenia w Polsce ze szczególnym uwzględnieniem gatunków inwazyjnych (Plants of total origin in Poland in particular consideration of invasive species). GDOŚ, Warszaw
Tomlinson PB (1982) Helobieae (Alismatidae). In: Metcalfe CR (ed) Anatomy of the Monocotyledons, vol VII. Clarendon Press, Oxford, U.K.
Villanueva VR, Simola LK, Mardon M, (1985). Polyamines in turions and young plants of *Hydrocharis morsus-ranae* and *Utricularia intermedia*. Phytochemistry 24:171-172
Voesenek LACJ, Colmer TD, Pierik R et al (2006) How plants cope with complete submergence. New Phytol 170:213-226
Vollenweider RA, Talling JF, Westlake DF (1969) A manual on methods for measuring primary production in aquatic environments. IBP Handbook 12. Blackwell Scientific, Oxford
Westlake DF (1975) Macrophytes. In: Whitton BA (ed) River ecology. Blackwell, Oxford, p106-128
Wetzel RG, Brammer ES, Forsberg C (1984) Photosynthesis of submerged macrophytes in acidified lakes. I. Carbon fluxes and recycling of CO_2 in *Juncus bulbosus* L. Aquat Bot 19:329-342
Winkel A, Borum J (2009) Use of sediment CO_2 by submersed rooted plants. Ann Bot 103:1015-1023
Winston RD, Gorham PR (1979a) Turions and dormancy states in *Utricularia vulgaris*. Can J Bot 57:2740-2749
Winston RD, Gorham PR (1979b) Roles of endogenous and exogenous growth regulators in dormancy of *Utricularia vulgaris*. Can J Bot 57:2750-2759

Species index

F. H. Schweingruber et al., *Anatomic Atlas of Aquatic and Wetland Plant Stems*, https://doi.org/10.1007/978-3-030-33420-8